Frontiers in Sedimentary Geology

Frontiers in Sedimentary Geology

A.H. Bouma
Editor-in-Chief

Submarine Fans and Related Turbidite Systems
A.H. Bouma, W.R. Normark, and N.E. Barnes (editors)

New Perspectives in Basin Analysis
K.L. Kleinspehn and C. Paola (editors)

Microstructure of Fine-Grained Sediments: From Mud to Shale
R.H. Bennett, W.R. Bryant, and M.H. Hulbert (editors)

R.H. Bennett W.R. Bryant M.H. Hulbert
Editors

Microstructure of Fine-Grained Sediments

From Mud to Shale

W.A. Chiou, R.W. Faas, J. Kasprowicz, H. Li,
T. Lomenick, N.R. O'Brien, S. Pamukcu, P. Smart,
C.E. Weaver, and T. Yamamoto
Associate Editors

With 458 Illustrations in 708 Parts

Springer-Verlag
New York Berlin Heidelberg London
Paris Tokyo Hong Kong Barcelona

Richard H. Bennett
Seafloor Geosciences Division
Naval Ocean Research and Development Activity
Stennis Space Center, MS 39529-5004 USA

William R. Bryant
Department of Oceanography
Texas A&M University
College Station, TX 77843 USA

Matthew H. Hulbert
IMC/Pitman-Moore, Inc.
Terre Haute, IN 47808 USA

Series Editor
Arnold H. Bouma
School of Geoscience
Louisiana State University
Baton Rouge, LA 70803-4101 USA

Library of Congress Cataloging-in-Publication Data
Microstructure of fine-grained sediments, from mud to shale / R.H. Bennett, W.R. Bryant, M.H. Hulbert, editors.
p. cm. — (Frontiers in sedimentary geology)
Includes index.
ISBN 0-387-97339-7 (alk. paper)
1. Clay—Analysis. 2. Sediments (Geology) 3. Rocks, Sedimentary—Analysis. I. Bennett, Richard Harold, 1939– II. Bryant, William R. III. Hulbert, Matthew H. IV. Series.
QE471.3.M53 1990
551.3′04—dc20 90-37343
CIP

Printed on acid-free paper

Typeset by Publishers Service of Montana, Bozeman, Montana.
Printed and bound by Edwards Brothers Inc., Ann Arbor, Michigan.
Printed in the United States of America.

9 8 7 6 5 4 3 2 1

ISBN 0-387-97339-7 Springer-Verlag New York Berlin Heidelberg
ISBN 3-540-97339-7 Springer-Verlag Berlin Heidelberg New York

Series Preface

The *Frontiers in Sedimentary Geology* series was established for the student, the researcher, and the applied scientist to enhance their potential to stay abreast of the most recent ideas and developments and to become familiar with certain topics in the field of sedimentary geology. The series deals with subjects that are in the forefront of both scientific and economic interests. The treatment of a subject in an individual volume, therefore, is a combination of topical, regional, and interdisciplinary approaches. The interdisciplinary aspects are becoming more and more important because most studies dealing with the natural sciences cannot any longer effectively stand alone. Although this thrust may sound simple, in reality it is not, basically because each discipline has developed its own jargon and definitions of terms. Communication among disciplines is a major issue and can be accomplished more constructively when people with different backgrounds join together at the same symposium and can read from the same volume rather than confining themselves only within the world of their own specialty meetings and journals. Books in this series provide this connective link between disciplines.

Each book in this series provides a continuous and connected flow of concepts throughout the volume by the use of introductory chapters that outline a topic sufficiently to grasp its problems and to understand the contributions that follow. The individuality of each paper will be maintained as much as possible by preserving the major aspects of the personal style of the author(s). Such an approach reduces the jumpiness typical of an edited book. A broad integrated presentation, on the other hand, is difficult to be discussed by a single person.

The *Frontiers in Sedimentary Geology* series thus strives to present books in such a manner that the reader can become well informed in a minimum amount of time. This volume, *Microstructure of Fine-Grained Sediments: From Mud to Shale*, fits the criterion of the series very well. The publication is the result of a workshop held in 1988 that was attended by a large group of leading scientists and engineers representing several disciplines and subdisciplines. It is fascinating to realize how many parts of our life deal with aspects of fine-grained sediments, both the consolidated and unconsolidated types. A rather complete list is presented in the last chapter of the book and it suffices to say that it not only includes geology, sedimentology, and the oil and gas industry, but even more seriously the waste disposal, groundwater contamination, preservation, biology, chemistry, ceramics, agriculture, and at least another dozen other fields. Such a broad utilization of the fine-grained sediments and the large number of study directions involved makes communication extremely difficult but nevertheless worthwhile and rewarding. Organization of this book proceeds from very basic concepts and principles to the applied problems. The reader quickly appreciates the importance of a thorough understanding of the basic scientific principles and processes for effective utilization of fine-grained soft sediments and rock in practical applications. The breakdown of all contributions into a number of major categories with each category having its own introduction facilitates understanding of some of the nonfamiliar chapters.

The symposium itself was a very successful event and it helped to bridge some of the communication gaps. The organizers took the task on themselves to play the roles of editors and subeditors. As a series editor I am very pleased that the editors opted for their book to become part of the *Frontiers in Sedimentary Geology* series.

Arnold H. Bouma

Preface

This book presents a comprehensive cross section of recent research and present concepts of basic and applied studies on the clay microstructure of fine-grained sediments and argillaceous rocks. The book is a compilation of peer-reviewed technical papers, most of which were presented at an international symposium-workshop entitled "The Clay Microstructure of Fine-Grained Terrigenous Marine Sediments: From Muds to Shale," convened at the Naval Ocean Research and Development Activity, Stennis Space Center, Mississippi, U.S.A. in October 1988. The theme of the symposium and the thrust of this book focus on (1) the fundamental role of microstructure (microfabric and physicochemistry) as a determinant of fine-grained sediment properties, (2) the significance of microstructure to diagenetic, mass physical and mechanical, static, dynamic, and geoacoustic properties, and to microscale sedimentary features, and (3) the relationship and importance of these factors to basic research and applied problems.

The book is organized into three major sections under the titles Basic Clay Microstructure, Applied Clay Microstructure, and Future Research Directions and Recommendations. The contents of these major comprehensive sections focus on the following topics: microstructure signatures, environmental processes, modeling, measurements and techniques, and applications. Each major section is preceded by an overview that provides the reader with a synopsis of the chapters to follow and ties the book together as a coherent entity. The third major section directs the reader to future directions and recommendations that summarize the consensus of the workshop participants who defined significant future research areas and recommendations focused on clay microstructure. A major element in the task of identifying these areas of future research was the input by outstanding researchers and engineers representing a wide range of specialties and technical fields. This last section also emphasizes the significance of clay microstructure to an enormous variety of technical disciplines, and an important objective of this recommendation section is to stimulate and to guide new thrusts in research and engineering practice involving clayey sediments and argillaceous rocks. The book directs the reader to numerous areas where a more complete knowledge of the microstructure would add significantly to understanding of the fundamental physics, chemistry, and sedimentology of fine-grained sediments.

Review of the book reveals that new breakthroughs in highly specialized disciplines and the advancement of technology will require an understanding of clay microstructure, which will come as a result of effective interdisciplinary studies addressing the very fundamental nature of the interacting organic and inorganic parts that together comprise the complex sedimentological system as a whole. Technical papers confirm, again, the principle that basic research is the backbone of technology. Without such basic research as a foundation to decision making, important controls on our environment will not be effectively realized as man continues to utilize the rocks and sediments for waste disposal, agriculture, industrial and military applications, and construction.

Although the field of clay microstructure is presently in a rudimentary and somewhat primitive posture, the concepts, observations, and principles presented in this book will stimulate new thought and be a "springboard" for exciting novel research. As editors, we expect that this comprehensive book will minimize the convoluted and sometimes static and dead-end research paths that were common during the early history of research on clay microstructure and related sciences. These early years of pioneering research and engineering, dating back to Henry Sorby, Arthur Casagrande, and Karl Terzaghi, focused on the disciplines of colloid chemistry and the physics of fine-particle behavior in relationship to sedimentological processes and the geotechnical properties of soils. These studies, although rudimentary in nature, provided an important underpinning for the advancement of the field of clay microstructure that is now making use of emerging high technology instrumentation and computer applications that are introduced in this book.

The importance of microstructure to sedimentary geology and the underpinning concepts are based on the principle that the microfabric and the physicochemistry of fine-grained geologic materials are the fundamental "building blocks" of clayey sediments and rocks. Thus the microstructure determines the characteristic physical, mechanical, and geoacoustic properties of fine-grained sediments and rocks under static and dynamic stresses. Basic research studies presented in this book that address the fundamentals of clay sediments and argillaceous rocks are prerequisite to achieving a functional understanding of the sediment and rock properties and their response to natural and man-induced stresses, and to the resulting effects on our environment. An important goal will be achieved if this publication stimulates new thought and insight in clay microstructure and provides a new rational basis for understanding the developmental history of fine-grained sediments by revealing important fundamentals of the chemistry, physics, and biology, and the interplay of processes. The success of engineering practice applied to most sediments and rocks depends critically on a thorough quantitative understanding of basic and applied clay microstructure. The symposium-workshop and this book clearly reveal the importance of an integrated multidisciplinary approach. Basic research and engineering studies of clay microstructure benefit enormously from the interaction among different disciplines, as demonstrated by the numerous chapters in this book.

Richard H. Bennett
Naval Ocean Research and Development Activity
Stennis Space Center, Mississippi, USA

Acknowledgments

The efforts of numerous contributors and reviewers of the technical papers and a few key sponsors made this book possible. The sponsors who recognized the importance of advancing fundamental research and engineering studies of the microstructure of fine-grained sediments and rocks include the Naval Ocean Research and Development Activity (NORDA), the Office of Naval Research/Contract Research Division (ONR/CRD), the Department of Energy (DOE)/Chicago Office, and the Oak Ridge National Laboratory (ORNL). Special appreciation is given to Dr. William Moseley (NORDA), Dr. Herbert Eppert (NORDA), Mrs. Halcyon Morris (NORDA), Dr. Joseph Kravitz (ONR), Mr. John Kasprowicz (DOE), Dr. Steven Stow (ORNL), and Dr. Thomas Lomenick (ORNL), each of whom gave encouragement in making the symposium-workshop (October 1988) a success and this book possible. NORDA, ONR, and DOE/ORNL provided financial support for the symposium-workshop, and NORDA supported the first editor during the compilation of the numerous peer-reviewed papers in this book. The efforts of Ms. Margaret Chenault of the staff of the University of Southern Mississippi's Division of Lifelong Learning are gratefully acknowledged for assistance in administering the workshop.

Several NORDA staff members assisted in numerous activities that made the workshop and book possible. These persons are due special recognition as follows: appreciation is extended to Ms. F. Lee Nastav for her faithful assistance in preparing the workshop program booklet with abstracts and for assisting in various and numerous tasks during the assembly of this book; the late Mr. John Burns provided invaluable logistics during the symposium-workshop and, as always, interfaced effectively with visitors in the NORDA tradition; Mr. Samuel Tooma gave encouragement during the planning of the workshop and throughout the preparation of this book; Dr. Huon Li assisted in various aspects of the symposium-workshop logistics and in the interchange with numerous participants. The encouragement and assistance of Dr. Li on various parts of this book are greatly appreciated. Sharon Vreeland arranged the domestic and foreign travel for the workshop participants.

The encouragement and support by the management of Pitman-Moore, Inc., for the third editor, are gratefully acknowledged. The associate editors provided valuable assistance during the technical paper peer-review process, and their efforts were very important in providing high quality manuscripts and well-organized sections of this book. Thanks go to the many critical reviewers for their efforts and work with the session chairmen. The editors appreciate the suggestions and assistance of Dr. Arnold Bouma, Series Editor for Springer-Verlag, during the planning of this book.

This publication has been approved for public release by NORDA with distribution unlimited. NORDA special project #056:360:88.

Contents

Part II Applied Clay Microstructure
B. Measurements/Techniques/Sampling Strategy

Part II Applied Clay Microstructure
C. Applications: Present Requirements

Part III Future Research Directions and Recommendations: Basic and Applied

Contributors

P.D. ALLEY, Department of Geosciences, University of Houston, Houston, TX 77204, USA

A.G. ALTSCHAEFFL, Department of Civil Engineering, Purdue University, West Lafayette, IN 47907, USA

K. ANDO, Geo-Acoustics Laboratory, Rosenstiel School of Marine and Atmospheric Science, University of Miami, Miami, FL 33149, USA

K.W. ASPREY, Geological Survey of Canada, Bedford Institute of Oceanography, Dartmouth, Nova Scotia B2Y 4A2, Canada

N.B. AUGHENBAUGH, Department of Geology and Geological Engineering, University of Mississippi, University, MS 38677, USA

ROY J. BAERWALD, Department of Biological Sciences, University of New Orleans-Lakefront, New Orleans, LA 70148, USA

KEITH W. BEDFORD, Department of Civil Engineering, The Ohio State University, Columbus, OH 43210-1275, USA

RICHARD H. BENNETT, Seafloor Geosciences Division, Naval Ocean Research and Development Activity, Stennis Space Center, MS 39529-5004, USA

SHOBHA K. BHATIA, Department of Civil and Environmental Engineering, Syracuse University, Syracuse, NY 13244-1190, USA

GEORGE D. BRUNTON, Geology and Geological Engineering Department, School of Engineering, University of Mississippi, University, MS 38677, USA

WILLIAM R. BRYANT, Department of Oceanography, Texas A&M University, College Station, TX 77843, USA

PATTI J. BURKETT, Seafloor Geosciences Division, Naval Ocean Research and Development Activity, Stennis Space Center, MS 39529-5004, USA

P.K. CARPENTER, Division of Geological & Planetary Sciences, Caltech, Pasadena, CA 91125, USA

B. CARSON, Department of Geological Sciences, Lehigh University, Bethlehem, PA 18015, USA

JOS CENENS, Laboratorium voor Oppervlaktechemie, K.U. Leuven, K. Mercierlaan, 92, B-3030 Leuven, Belgium

F. TOM CHANG, Department of Civil Engineering, Lehigh University, Bethlehem, PA 18015, USA

WEN-AN CHIOU, Department of Materials Science & Engineering, Northwestern University, Evanston, IL 60208, USA

W.B. DADE, School of Oceanography, University of Washington, Seattle, WA 98195, USA

KATHLEEN A. DADEY, Graduate School of Oceanography, University of Rhode Island, Narragansett, RI 02882, USA

DAVID K. DAVIES, David K. Davies and Associates, Inc., Stonehollow Place, Kingwood, TX 77339, USA

D.J. DEMASTER, Department of Marine, Earth and Atmospheric Sciences, North Carolina State University, Raleigh, NC 27695, USA

FRANS C. DE SCHRYVER, Laboratorium voor Moleculaire Dynamica en Spectroscopie, Departement Scheikunde, K.U. Leuven, Celestijnenlaan, 200F, B-3030 Leuven, Belgium

W. DROST-HANSEN, Laboratory for Water Research, Department of Chemistry, University of Miami, Coral Gables, FL 33124, USA

DEAN A. DUNN, Department of Geology, University of Southern Mississippi, Southern Station, Hattiesburg, MS 39406-5044, USA

ROBERT EHRLICH, Department of Geological Sciences, University of South Carolina, Columbia, SC 29208, USA

RICHARD W. FAAS, Department of Geology, Lafayette College, Easton, PA 18042, USA

FEDERICO FERNANDEZ, Geotechnical Research Centre, Department of Civil Engineering, The University of Western Ontario, London, Ontario N6A 5B9, Canada

R.E. FERRELL, JR., Basin Research Institute and Department of Geology and Geophysics, Louisiana State University, Baton Rouge, LA 70803-4101, USA

KATHLEEN M. FISCHER, Seafloor Geosciences Division, Naval Ocean Research and Development Activity, Stennis Space Center, MS 39529-5004, USA

AKIRA FUKAMI, Department of Physics, College of Humanities and Sciences, Nihon University, Sakurajousui, Setagaya-ku, Tokyo 156, Japan

KURIO FUKUSHIMA, Department of Physics, College of Humanities and Sciences, Nihon University, Sakurajosui, Setagaya-ku, Tokyo 156, Japan

J.E. Gillott, Department of Civil Engineering, The University of Calgary, Calgary, Alberta T2N 1N4, Canada

Donn S. Gorsline, Department of Geological Sciences, University of Southern California, Los Angeles, CA 90089-0740, USA

Eliza Grabowski, Department of Civil Engineering, The University of Calgary, Calgary, Alberta T2N 1N4, Canada

Gao Guorui, Nanjing Architectural and Civil Engineering Institute, Nanjing, Jiangsu, People's Republic of China

T.M. Hariu, Department of Geological Sciences, University of South Carolina, Columbia, SC 29208, USA

D.E. Heffler, Geological Survey of Canada, Bedford Institute of Oceanography, Dartmouth, Nova Scotia B2Y 4A2, Canada

M.W. Hounslow, School of Environmental Sciences, University of East Anglia, Norwich NR4 7TJ, England

Matthew H. Hulbert, IMC/Pitman-Moore, Inc., Terre Haute, IN 47808, USA

L.K. Hyder, Environmental Sciences Division, Oak Ridge National Laboratory, Oak Ridge, TN 37831-6317, USA

P.A. Jumars, School of Oceanography, University of Washington, Seattle, WA 98195, USA

J.D. Kasprowicz, Department of Energy/Chicago Operations, Argonne, IL 60439, USA

R. Kirby, Ravensrodd Consultants, Ltd., Taunton TA1 4XW, Somerset, England

Norihiko Kohyama, National Institute of Industrial Health, Ministry of Labor, 6-21-1, Nagao, Tamaku, Kawasaki 213, Japan

Kate Kranck, Department of Fisheries and Oceans, Bedford Institute of Oceanography, Dartmouth, Nova Scotia B2Y 4A2, Canada

S.A. Kuehl, Department of Geological Sciences, University of South Carolina, Columbia, SC 29208, USA

S.Y. Lee, Environmental Sciences Division, Oak Ridge National Laboratory, Oak Ridge, TN 37831-6317, USA

Margaret Leinen, Graduate School of Oceanography, University of Rhode Island, Narragansett, RI 02882, USA

X. Leng, Civil Engineering Department, Glasgow University, Glasgow G12 8QQ, Scotland

G.P. Lennon, Department of Civil Engineering, Lehigh University, Bethlehem, PA 18015, USA

J.N. Leonard, Department of Oceanography, Texas A&M University, College Station, TX 77843, USA

HUON LI, Seafloor Geosciences Division, Naval Ocean Research and Development Activity, Stennis Space Center, MS 39529-5004, USA

CHARLES LIBICKI, Department of Civil Engineering, The Ohio State University, Columbus, OH 43210-1275, USA

THOMAS F. LOMENICK, Oak Ridge National Laboratory, Oak Ridge, TN 37831, USA

FRED T. MACKENZIE, Department of Oceanography, University of Hawaii, Honolulu, HI 96822, USA

I. McCONNOCHIE, Civil Engineering Department, Glasgow University, Glasgow G12 8QQ, Scotland

ASHISH J. MEHTA, Coastal and Oceanographic Engineering Department, University of Florida, Gainesville, FL 32611, USA

BRUNO MIASCHI, SACLANT Undersea Research Centre, San Bartolemeo 400, I-19026, La Spezia, Italy

ENRICO MUZI, SACLANT Undersea Research Centre, San Bartolemeo 400, I-19026, La Spezia, Italy

MYSORE S. NATARAJ, Department of Civil Engineering, University of New Orleans, New Orleans, LA 70148, USA

C.A. NITTROUER, Marine Sciences Research Center, State University of New York, Stony Brook, NY 11794, USA

A.R.M. NOWELL, School of Oceanography, University of Washington, Seattle, WA 98195, USA

NEAL R. O'BRIEN, Geology Department, S. U. N. Y.–Potsdam College, Potsdam, NY 13676, USA

THOMAS H. ORSI, Planning Systems Incorporated, PSI Engineering Center, Industrial Park, Long Beach, MS 39560, USA

SIBEL PAMUKCU, Department of Civil Engineering, Lehigh University, Bethlehem, PA 18015, USA

EMMANUEL PARTHENIADES, Department of Aerospace Engineering, Mechanics and Engineering Sciences, University of Florida, Gainesville, FL 32611, USA

J.W. PIERCE, Museum of Natural History, Smithsonian Institution, Washington, DC 20560, USA

ROBERT M. QUIGLEY, Geotechnical Research Centre, Department of Civil Engineering, The University of Western Ontario, London, Ontario N6A 5B9, Canada

F.R. RACK, Department of Oceanography, Texas A&M University, College Station, TX 77843, USA

CHAPTER 1

The Signatures of Clay Microstructure—Overview

Richard W. Faas and Neal R. O'Brien

Microstructure preserved in fine-grained sediments and rocks may be used to identify and recognize specific sedimentary environments and processes, thus serving as a distinct imprint or "signature" of that environment. The following important questions are addressed in this series of papers: (1) What microfabrics are produced in various modern sedimentary environments? (2) What are the actual processes and mechanisms that determine clay signatures? (3) What fabric signatures are found in shales and how may they be used to provide clues to ancient environments and processes? (4) Can shales be classified according to fabric types, which, in turn, may be used to interpret shale properties?

The paper by Bennett, O'Brien, and Hulbert describes three fundamental processes that influence clay and shale microfabric signatures: (1) physicochemical, (2) bioorganic, and (3) burial diagenesis. The processes are discussed in terms of specific mechanisms that drive microfabric development. By way of scanning and transmission electron micrographs, they provide a pictorial frame of reference that characterizes the fabrics associated with each process. One significant contribution shows the unique fabric signature of electrochemically flocculated clays compared to biosediment aggregates (e.g., "marine snow") and to that of randomly oriented fabric produced by bioturbation.

Microfabrics produced in modern sedimentary environments have been recorded by using SEM, TEM, and thin section techniques. Fabrics of recent clayey sediment from the Amazon subaqueous delta show in thin section unique features related to sedimentary processes. In their paper, Kuehl, Hariu, Sanford, Nittrouer, and DeMaster show how various primary structures in the mud are useful environmental indicators. A physical model of clay behavior is discussed by Stepkowska in terms of the physics of the clay–water system. The role of clay microstructure in the strength of the "system" is examined at the level of fundamental structural elements and the chemistry of the interacting components. The importance of aggregation processes is discussed in relation to the stability or instability of sedimentary formations and a laboratory method to define the potential rate and extent of aggregation that may occur in the field is proposed.

Wartel, Sethi, and Faas examined clay microfabrics in sediments from modern marine North Sea shelf and estuarine (Schelde) tidal flat environments and the Rijkevorsel Clay Member of the Campine Formation, an ancient (Pleistocene) tidal flat. Fabric signatures occurring in both tidal flat deposits are similar and recognizably different from the marine sedimentary microfabric. Three hypotheses for the genesis and evolution of gas-derived microvoids are proposed and mechanisms are invoked to explain microstructure signatures, defined as "nonstressed" and "stressed" microvoids. The development and preservation of microvoids in the Rijkevorsel Clay Member are interpreted as reflecting a "secondary" postdepositional microfabric that developed under periglacial climates.

The unique fabric signatures of recent interbedded pelagic, bioturbated pelagic, and turbiditic clays from the Southern Nares Abyssal Plain are recorded by Shepard and Rutledge. Their TEM study indicates a turbiditic clay fabric of random clay domains in large floccules mixed with silt grains that contrasts with a pelagic facies fabric showing many domains arranged in chains of an edge-to-face orientation.

The microfabric changes and physical properties of consolidated clay are revealed by detailed studies of two different geological environments reported in this section. Bryant, Bennett, Burkett, and Rack discuss changes in a consolidated clay section from the Weddell Sea, and Tribble, Mackenzie, and Urmos describe the microfabric signatures resulting from processes in a deeply buried, clay-rich sediment of the Barbados convergent margin. These studies show the importance of different stress regimes on microfabric development and the observable microfabric signatures.

Schieber used textural features found in thin sections of Mid-Proterozoic shales from Montana to distinguish shales formed on a shallow, low-energy shelf from deeper water facies. His micrographs show evidence of subaqueous microbial mats, of features due to storm erosion and deposition, of suspension settling and bottom flowing currents, and other sedimentary processes. Appalachian Devonian shales are the focus of the paper by Davies, Bryant, Vessell, and Burkett in which they show the relationships between porosity, permeability, and shale microfabric. Their results show that microfabric controls shale permeability, e.g., shales with parallel microfabric possess lower permeabilities. They also present a classification of shale types based on fabric that is in turn related to sedimentary environments and processes.

It is apparent from the results presented in this section that the fabric of fine-grained sediments and rocks provides useful clues not only in understanding the processes by which they are formed, but also in determining sedimentary environments. Results also show that significant attention should be paid to post-depositional fabric development, particularly that arising from gravitational, biological, and geochemical processes acting through geological time. Clay microstructure is currently being studied from different points of view, using different levels of magnification, and investigating sediment types from various environments. Multiple techniques, applied to the solution of problems, range from gross-scale X-radiography, through optical microscopy, to the scales revealed by SEM and TEM.

What must be done in future microstructure studies? Perhaps instead of attempting to describe fabrics more precisely, we should seek answers to the following questions: What is the environmental significance of a specific fabric? Are there multiple environments within which similar microfabrics originate? How may fabric information prove valuable in predicting properties of argillaceous sediment and rocks? Finally, advances in our understanding and interpretation of sedimentary environments from microfabrics will depend on how well we can recognize and differentiate the many fabric signatures associated with processes and mechanisms of modern and ancient depositional environments.

CHAPTER 2

Determinants of Clay and Shale Microfabric Signatures: Processes and Mechanisms

Richard H. Bennett, Neal R. O'Brien, and Matthew H. Hulbert

Introduction

The energy sources that result in sediment particle associations, reorientation, and disaggregation are presented in terms of processes and mechanisms. Based on electron microscopy observations and theoretical considerations, the observed and modeled microfabric forms and signatures are associated with processes and mechanisms operating in various micro- and macroenvironments. The interplay of geological, chemical, and biological processes and mechanisms during transport, deposition, and burial of particulate material largely controls and ultimately determines the physical nature, properties, and observable micro- and macrocharacteristics of soft sediments and their indurated equivalents. Discrete events such as suspended sediment transport, flocculation, and slumping may be identified and/or observed in the field or laboratory. More often, the sedimentary material is studied to understand and infer processes and mechanisms responsible for its fundamental properties, origin, significance, and stratigraphic position in the geological record. The particle-to-particle development and ultimate nature of a sedimentary deposit and its variability in time and space depend on multiple processes that include some important mechanisms that occur extremely fast and others that progress over eons. As defined in this study, mechanisms are the specific energy sources that drive microfabric development. Two or more related mechanisms constitute that broader classification termed process. In the continuum of microfabric development, the fundamental processes in which the individual mechanisms operate are described as (1) physicochemical, (2) bioorganic, and (3) burial diagenesis (Fig. 2.1).

The properties of a sedimentary deposit at the time of initial deposition are determined by (1) particle size, (2) mineralogy, (3) particle sized distribution, and (4) microstructure (fabric and physicochemistry). The sediment particles are derived not only from detrital but also from biogenic sources. Postdepositional authigenic mineralization is important in some environments. The energy regimes characteristic of the particulate transport pathways and postdepositional environments significantly affect the time-dependent development of the microfabric of suspensions and sedimentary deposits. Discriminating analysis and discernment of the developmental stages of sediment microstructure, as a function of processes and mechanisms, are antecedent to gaining a functional understanding of the relationship and importance of microstructure to the developmental history of a deposit.

The microstructure is a crucial fundamental property that largely determines the sediment's physical and mechanical properties and behavior under static and dynamic stresses. Important properties such as porosity, permeability, and stress–strain behavior are intimately tied to the microfabric and the physicochemical characteristics (Bennett et al., 1977, 1989). Improvement in predictive models for science as well as for the solution of practical problems is achieved by continual investigation into the complex interactive processes that influence the fundamental static and dynamic properties of sediments and rocks. Important scientific and technical problems abound in the disciplines of geoacoustics, physical and historical geology, waste disposal, hydrology, agriculture, and the myriad of other activities that depend on and can ultimately utilize a knowledge of the microfabric of sedimentary deposits.

The first objective of this paper is to formulate a rational theme that describes the *major* interactive processes and mechanisms important in microfabric development temporally and spatially. Various sources of energy are intrinsically coupled to the processes and mechanisms that determine clay sediment and shale microfabric signatures. Although some microfabric models are reasonably well known in terms of the processes that determine microfabric signatures, other models that describe

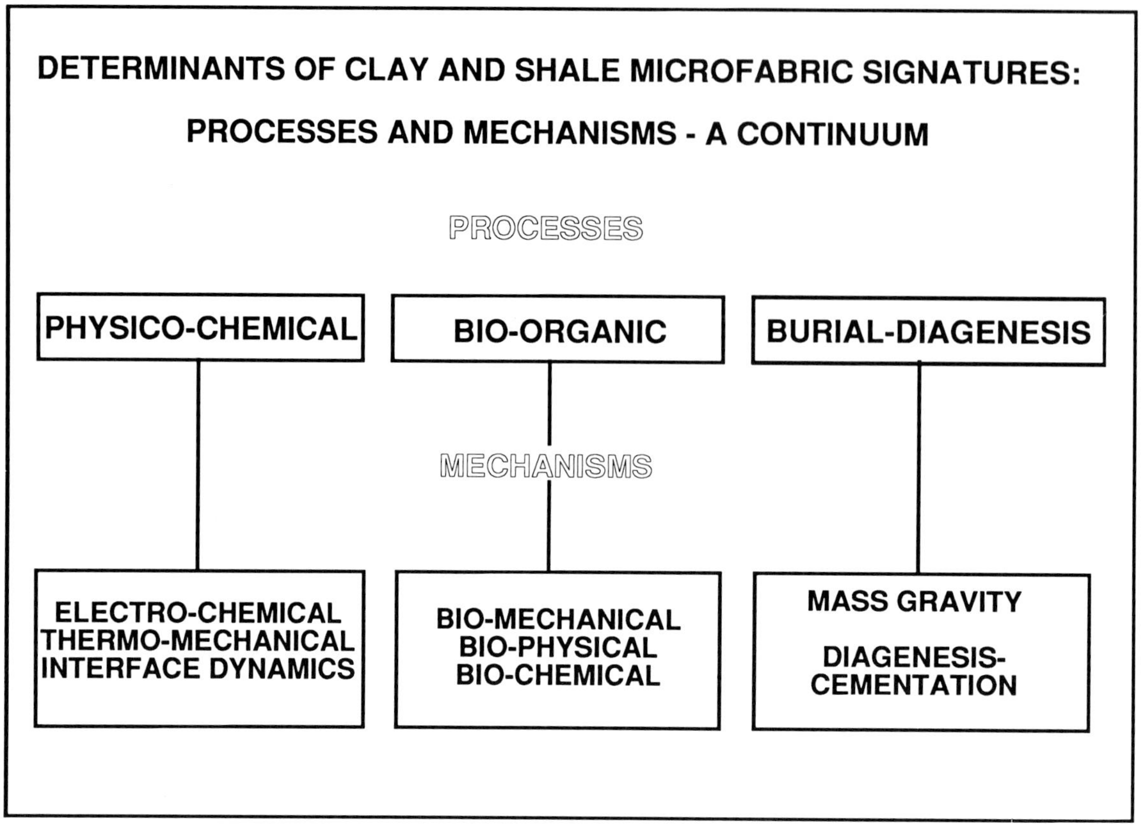

Figure 2.1. Diagram depicting the major processes and mechanisms that determine microfabric signatures in the macro and micro geological environments. Processes and mechanisms represent a continuum during the developmental history of clay sediment and shale microstructure.

the history of microfabric development are only in the formative stages of quantitative understanding. This study presents a conceptual framework of the current state of understanding of the processes and mechanisms that drive microfabric development in order to set the stage for future understanding of the complex microstructure processes and the evolution of advanced microfabric models. The second objective is to describe the macro- and microenvironments in which the processes and mechanisms and the particular energy regimes predominate during microfabric development. The third is to describe the time and physical scales over which the various processes and mechanisms prevail. The physical scales vary from submicron and molecular to meters and in some cases particular processes may extend over distances of kilometers. The fourth is to suggest how these processes and mechanisms can be incorporated into descriptive and predictive models that reveal important intrinsic properties and attributes of direct relevance to geology, geophysics, geoacoustics, geotechnique, and interpretation of the geologic record. Present qualitative and semiquantitative understanding of the microstructure of fine-grained sediment permits only rudimentary models of microfabric development. More complex models that include temporal and spatial scales are required for successful solution of practical problems. The fifth is to reveal the deficiencies in our knowledge of the interacting processes and mechanisms and to suggest future areas of potential research. The main thrust of this study is to present a unified approach to understanding the processes and mechanisms driving microfabric development. An important underpinning of this study is to delimit the macro- and microenvironments where the active energy regimes dominate along the complex and often convoluted sediment transport pathways: from "sediment source-to-sink."

Background

The term clay microstructure refers to two fundamental properties; the fabric and physicochemistry. Early usage and definition of these terms are found in Mitchell (1956), Lambe (1958), and Foster and De (1971). Clay fabric is defined as the orientation and arrangement or spatial distribution of the solid particles and the particle-to-particle relationships. Clay minerals are hydrous-aluminum silicates (classified as phyllosilicates) and are generally less than about 4 μm in size as determined by standard techniques (Lambe, 1951; Krumbein and Pettijohn, 1938). The physicochemistry relates to the interparticle forces of the sediment. These forces result from both the physical interactions arising from gravitational forces and the electrical nature of the particle and the surrounding fluids (Bennett et al., 1977). During geological time when sediment experiences increasing overburden with increasing subbottom depth, the gravitational forces dominate electrical forces between particles and burial diagenesis ensues. Often organic material has a significant influence on the strength of interparticle bonds in clay sediments (Pusch, 1973). Because of the ubiquitous nature of organic materials, clay–organic interactions are believed to be significant mechanisms in the developmental history of microfabric (Bennett et al., 1988). Thus knowledge of the fabric and physicochemistry, the fundamental "building blocks," is essential to our understanding of the nature, properties, and large-scale developmental history of sedimentary deposits.

The importance of microfabric in the development of sedimentary deposits was suggested by Sorby (1908). Terzaghi (1925) and Casagrande (1932) proposed primitive fabric models to help explain the bonding and sensitivity of cohesive sediments. During the following 50 years, geologists, soils engineers, and soil scientists proposed numerous simplified models of sediment fabric to account for factors such as the electrolitic environment surrounding the clay particles (Goldschmidt, 1926; Lambe, 1953, 1958; van Ophlen, 1963, 1977; Von Engelhardt and Gaida, 1963; O'Brien, 1970a, 1971), the process of consolidation and compaction (Quigley and Thompson, 1966; Smart, 1967; Ingles, 1968), and soil dynamics and behavior (Pusch, 1970; Yong, 1972). A "quantum jump" in observational techniques was realized with the advent of the transmission electron microscope (TEM) and advanced fabric models emerged (Rosenqvist, 1959); Pusch, 1966; Bowles, 1968; Bowles et al., 1969; Moon, 1972; Collins and McGown, 1974; Bennett et al., 1977). The more recently developed scanning electron microscope (SEM) has provided a means of observing the microfabric of shales and surfaces of clay minerals (Keller, 1976, 1978; O'Brien, 1968, 1970b; O'Brien and Hisatomi, 1978; Weaver, 1984).

Early models of microfabric were developed on the basis of simplified assumptions of the physical chemistry of the fine-grained minerals (single platelets) and the electrolitic chemistry of the fluid suspension. Recent electron microscopy observations (Moon, 1972; Yong and Sheeran, 1973; Collins and McGown, 1974; Bennett, 1976; Bennett et al., 1977, 1981) have revealed the presence of multiplate particles (domains) as the predominant fundamental particle type rather than the thin, single plate particles proposed in early models (Terzaghi, 1925; Casagrande, 1932; Lambe, 1953, 1958). Thus, high-resolution observational evidence has demonstrated that the single clay particle fabric model of sediment is not wholly tenable and the sediment is more completely represented by domain, aggregate, and linking chain particle arrangements (Bennett, 1976; Bennett and Hulbert, 1986). A domain is defined as a multiplate particle composed of parallel or nearly parallel plates that may be stacked either as sheets in a book or with an offset or stair-step arrangement (Fig. 2.2). Diagrammatic examples of domains were given by Moon (1972), and computer analysis verified that domains are important elements in submarine sediments (Bennett, 1976; Bennett et al., 1977). A domain is considered to have significant structural integrity and to behave in a functional sense as a unit particle for a finite period of time under an applied stress regime. Thus, it is important to note that "a particle" can be defined in terms of its morphology as well as its function. Important microfabric elements of a hypothetical sediment are depicted in Figures 2.2 and 2.3.

Discussion

Throughout the microfabric development of a sediment, beginning with erosion of material from parent sources to the time of deposition and burial, particulate material passes through a wide variety of macro- and microenvironmental conditions. To study and trace the history of microfabric development, the environmental conditions have been organized here into a framework described as a continuum of processes and mechanisms. These sources of energy associated with the processes and mechanisms are the determinants of clay and shale microfabric signatures. The major fundamental processes active in the continuum include (1) physicochemical, (2) bioorganic, and (3) burial diagenesis. Processes include two or more important environmental forcing functions or energy sources defined here as mechanisms (Fig. 2.1). Models depicted in Figure 2.4 reveal simplified graphic examples of their function. The mechanisms are largely responsible for the specific particle-to-particle interactions during a particular time in which a general process is active. Two or more mechanisms (such as electrochemical and interface dynamics) that drive microfabric development may operate contemporaneously but typically only one fundamental process dominates at a specific time.

Physicochemical processes play a major role in microfabric development during fluvial and aeolian transport stages of particulates and on their contact with a depositional interface. The fundamental mechanisms operating on particulate materials in the physicochemical process include thermomechanical, electrochemical, and interface dynamics. Bioorganic processes are important in marine and coastal environments, during transport and sedimentation of particulates in organic-rich waters, in areas of high productivity, at the depositional interface, and in surficial sediments. The mechanisms that are important in the bioorganic regime include biophysical, biomechanical, and biochemical. Processes of burial diagenesis drive microfabric development when overburden or tectonic stresses dominate physicochemical and bioorganic bonding energies. Mass gravity effects (slumping, sliding, creep, and consolidation) and diagenesis–cementation are important postdepositional/mechanogravity mechanisms in the

Figure 2.2. Diagram of the fundamental particle units called domains that comprise the "building blocks" of clay microfabric in sediments and rocks (single particles are rare; stipple indicates individual clay layer). (A) neatly stacked multiplate particle called the domain seen in plane view. (B) Cross-sectional view of a neatly stacked domain as would be observed in an ultrathin section prepared for TEM. (C) A cross-sectional view of a domain but showing the frequent offset arrangement of plates. (D) Plane and cross-sectional views of the possible shingle type arrangements of a domain with offset plates. (E) Linkages of domains in stepped face-to-face arrangement that form long chains (cross-sectional view). (F) Plane view of edge-to-edge contacts of domains that also form long chains. Note the difference in the potential strength of the domain formed chains in E and F. (G) Cross-sectional view of edge-to-face domains that are commonly found in marine environments; the microfabric develops large void space between domains. (H) Plane view of a chain of clay plates formed by stepped face-to-face arrangements.

process of burial diagenesis. Cementation at particle contacts may alter fabric morphology and inhibit particle reorientations within a deposit during burial or when the sediments are subjected to tectonic stresses.

The microfabric signatures resulting from the various processes and mechanisms are often recorded in the sediments and rock and are revealed by direct electron microscopy observation. Some of the signatures are difficult to capture in the undisturbed state and thus they are difficult to evaluate and still remain to be studied in detail. In addition, some mechanisms produce very fragile and delicate microfabrics that stretch technology beyond present observational and measurement limits. Thus simplified models, such as depicted in Figures 2.2, 2.3, and 2.4 are a great asset in developing a conceptual understanding of particle-to-particle interactions and resulting microfabrics. The remainder of this chapter presents examples of the microfabric signatures of sediment and argillaceous rock revealed by electron microscopy observations. Simplified microfabric models are presented for those particulate materials and sediments where direct observations have not yet been made. Verification of these models await future technological developments.

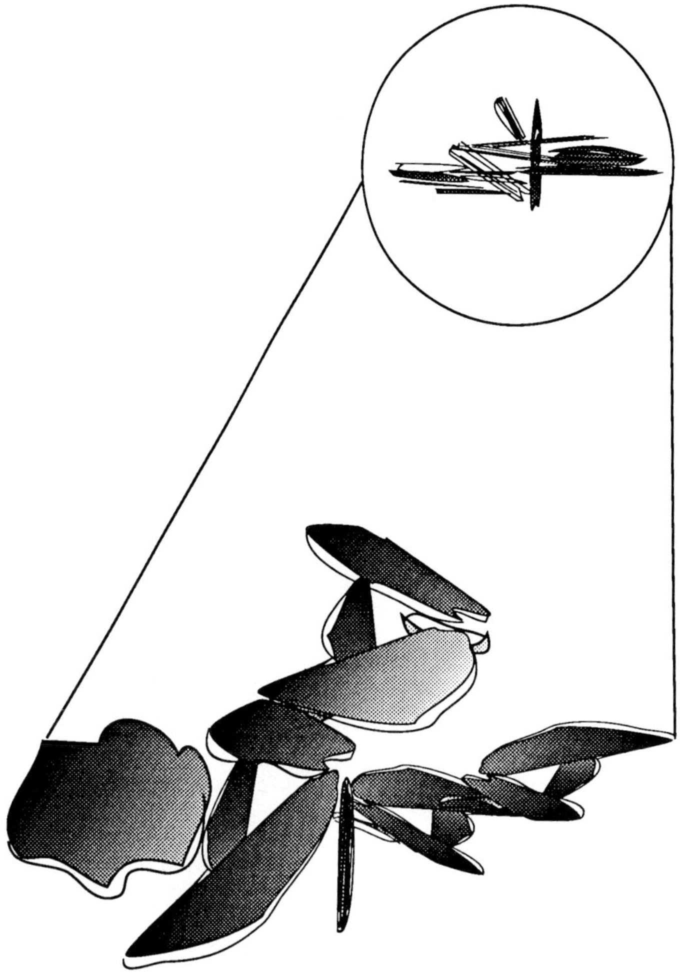

Figure 2.3. Plane and cross-sectional views of the microfabric of a hypothetical sediment constructed of numerous domains in "random" arrangement.

Physicochemical Processes

As particles are transported through water and air and become incorporated into the sediment, physical and chemical processes operate to bring them together into aggregates, to hold them together, to break aggregates apart, and to reorient particles within aggregates and the sediment mass. The resultant fabric of the sediment depends on the balance between the mechanisms operating to bring and hold particles together and those tending to disrupt particle-to-particle contacts. Energy mechanisms included in these processes include electrostatic interaction and chemical bonding (electrochemical mechanisms), thermally driven movement of particles (thermomechanical mechanisms), and interactions occurring at the surfaces of contact between the various materials and phases (interface dynamics mechanisms).

Electrochemical Mechanisms

The same forces responsible for the chemical bonding that holds particles together internally also bind particles to one another. These forces are included in the mechanism termed electrochemical. At a point of contact between two particles of the same material, the bonding is similar to and possibly even indistinguishable from that within the bulk material, i.e., covalent and ionic bonds, London–van der Waal's attraction, and extremely short-range Born repulsion (Bennett and Hulbert, 1986).

For particles in near contact (separation distances ranging from atomic diameters to dimensions of clay particles), the electrochemical forces of greatest importance are electrostatic interactions and van der Waal's attraction. The van der Waal's attraction (force) exists between all particles. The force between two particles increases as the particles are moved closer together and as their area of overlap increases. Electrostatic interactions exist between electrically charged particles; particles of opposite charge attract each other and those of the same charge repel. The greater the electrical charge and the closer the particles approach each other, the stronger the electrostatic interaction.

In an aqueous medium, the electrochemical interactions become complex. Almost all materials develop an electrical charge on their surfaces when immersed in water. The magnitude and even the sign of the electrical charge on a particle in water may be quite sensitive to the pH and E_h of the system and on the types and concentrations of dissolved salts (Stumn and Morgan, 1981). This sensitivity to the composition of the surrounding aqueous medium results from the sorption of ions from water onto the particle surface and the release of ions from the particle to water. Hydrogen ions and hydroxide ions are often of

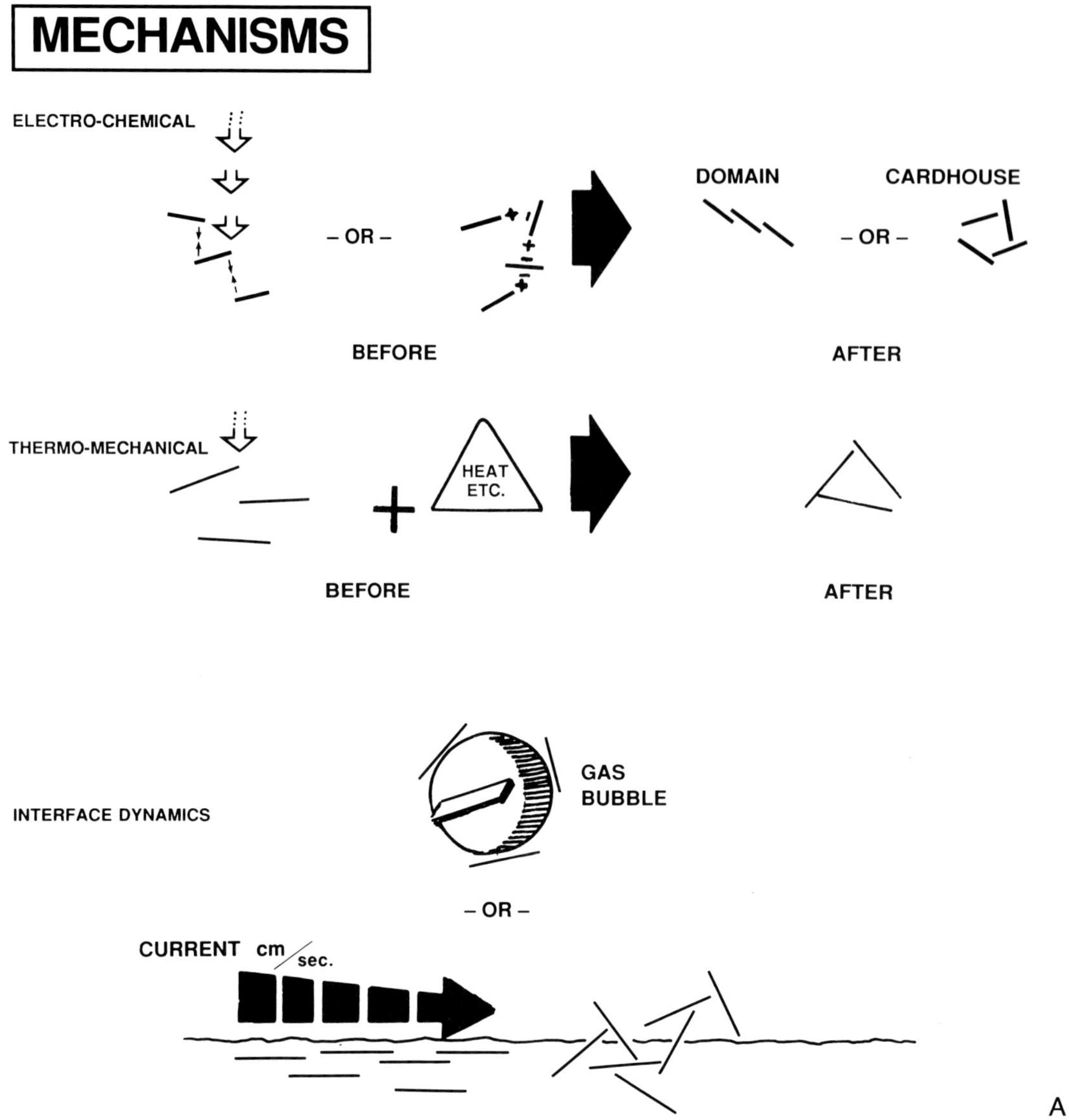

Figure 2.4. Determinants of clay sediment and shale microfabric signatures: processes and mechanisms—a continuum. Diagrams depict the major processes and mechanisms that influence the microfabric signatures in the macro- and microgeological environments. (A) Physicochemical processes. (B) Bioorganic processes. (C) Burial–diagenesis processes.

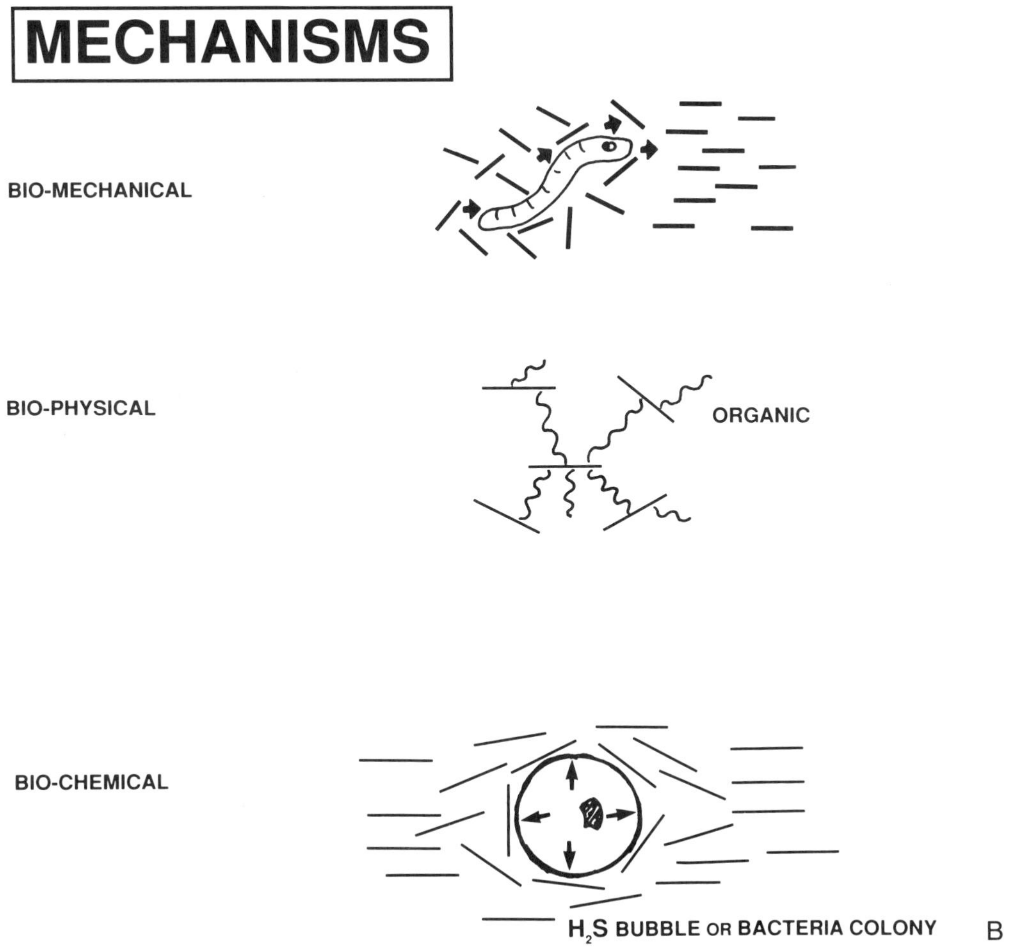
MECHANISMS
BIO-MECHANICAL
BIO-PHYSICAL
ORGANIC
BIO-CHEMICAL
H2S BUBBLE OR BACTERIA COLONY
B

Figure 2.4B

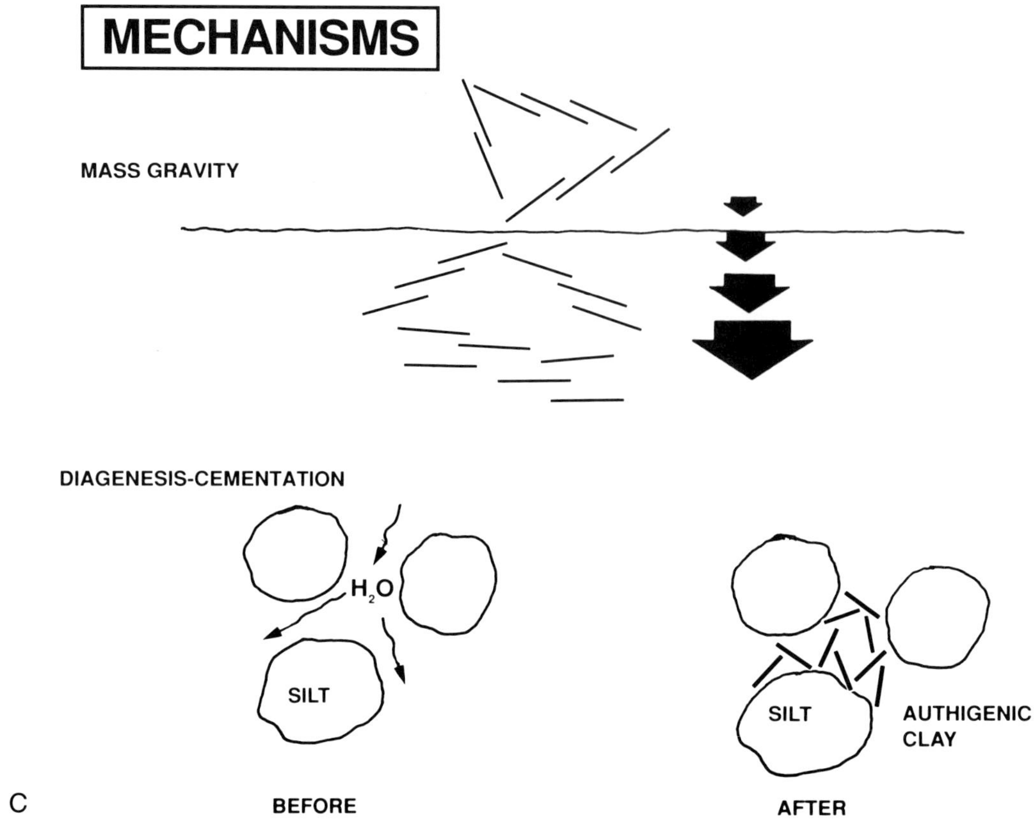

Figure 2.4C

particular importance in establishing a charge. The electrical charge on the surface of a particle in a given aqueous solution will depend not only on the chemical identity of the particle, but may even be of opposite sign on different crystallographic surfaces of a single particle.

Clay minerals in natural waters typically have a net surface charge of negative sign (van Olphen, 1963; Grim, 1968). The distribution of charge on the surface of a clay particle is expected to be patchy rather than homogeneous, and regions of positive charge may develop at edges, at defects, and at sites of sorption of positive ions. The surface charge of the particles is balanced by the net charge of the mobile ions surrounding it in the aqueous medium. In the water nearby there is a net surplus of ions with charge opposite to the particle charge and of total charge equal to the particle charge. These excess positive ions surround the negative particle in a diffuse cloud of charge. As two clay particles in water approach one another, the repulsive interaction between the two clouds of positive charge tends to prevent the particles approaching one another sufficiently closely so that shorter range van der Waal's attraction may pull them together. As the salt content of the water increases, the effective distance of separation of the positive charges from the particle surfaces decreases, and the effective reach of the electrostatic repulsion decreases along with an increase in the potential of van der Waal's attraction. Now the particles have a greater probability of approaching sufficiently close to permit aggregation. Graphic examples of the above clay particle interactions in saline environments are given by Bennett and Hulbert (1986).

Initial aggregation tends to be edge-to-face, probably controlled by differences in charge density on the different portions of the particle exterior (Figs. 2.5–2.7). The resulting aggregates are open with a high water content and the areas of particle contact and of particle overlap are minimal. In the absence of strong bonding at the point of contact, reorientation to a face-to-face configuration with a resulting increase in van der Waal's attraction may be expected (Fig. 2.8). Rotation about the line of contact between particles will result in an offset shingle-type fabric. Once such a fabric with a relatively large area of particle overlap is created, a much larger amount of energy would be required to convert it to a neatly stacked book-type fabric.

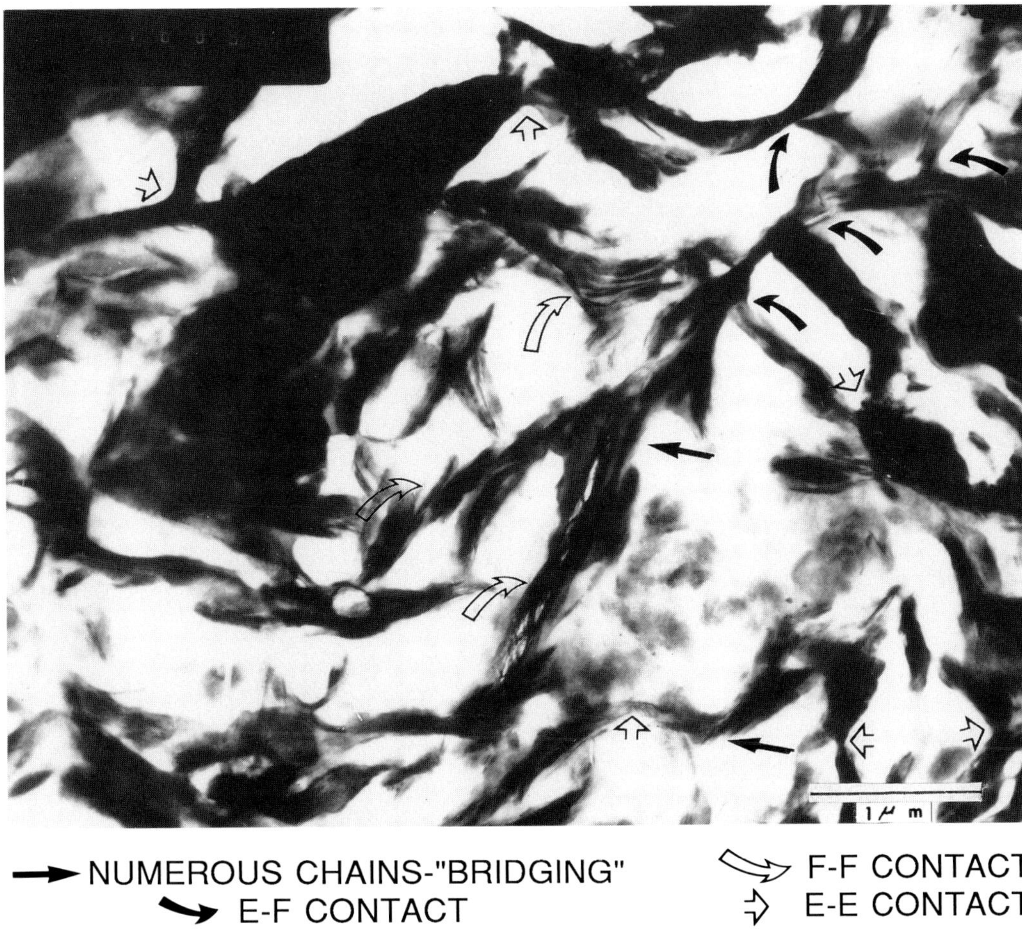

Figure 2.5. Transmission electron microscope (TEM) photomicrograph of clay fabric characteristic of the smectite–illite-rich Mississippi Delta submarine sediments. Note the delicate "bridging" between particles (domains), the chains, and the various modes of particle association, E-E, E-F, and stepped F-F and some E-E contacts.

Thermomechanical Mechanisms

The motion of particles that results in their initial approach and their reorientation after contact is one result of the mechanism termed thermomechanical (Fig. 2.9A). At ordinary environmental temperatures, the thermal energy of water is observable in the Brownian motion it imparts to suspended particles smaller than about 10 μm in diameter (Feynman et al., 1963). The kinetic energy of small particles is significant relative to repulsion forces at typical temperatures of the marine environment and becomes greater in high-temperature regions such as those near hydrothermal vents (Fig. 2.9A).

Another thermomechanical mechanism is the modification of sediment fabric by the freezing of interstitial water. Particles are excluded as water freezes and may be squeezed together if they happen to be trapped between two approaching freezing fronts or between a freezing front and a barrier (Fig. 2.9B). The resulting sediment microfabric may be characterized by relatively dense sediment units surrounding large voids filled with ice (Fig. 2.9A, B). When the ice melts the sediment tends to retain the modified fabric while salts excluded by the freezing process are redissolved (see for example Fig. 4.1 of Bennett and Hulbert, 1986).

Temperature differences between adjacent water masses leads to microscale laminar flow and turbulence at the boundaries. Shear from this microturbulence may disrupt suspended aggregates or reorient particles within aggregates (Gibbs, 1981). Clearly the scale of the turbulence is critical—turbulence on a dimensional scale much greater than the size of the aggregate will merely translate it through space. A much less probable result of microturbulence is to bring suspended particles into effective contact to form larger particles (Koh, 1984).

Interface Dynamics Mechanisms

Microturbulence also may be considered more generally as an aspect of mass and energy transfer at boundaries between contacting water masses or contacting particles and the surrounding

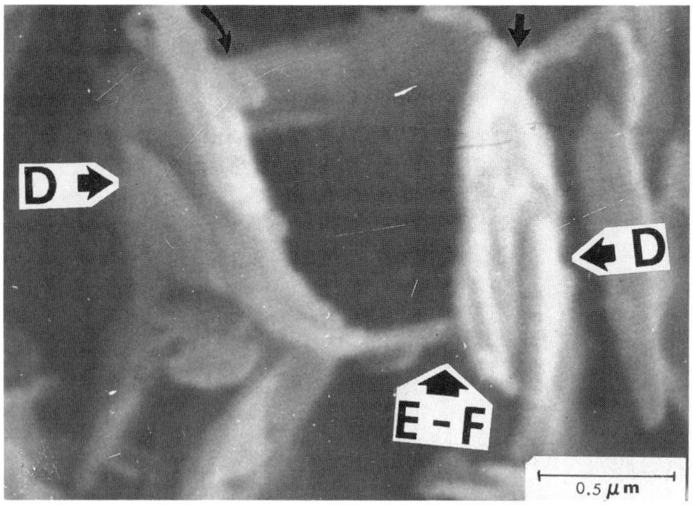

Figure 2.6. Typical microfabric of surficial submarine sediment from the Mississippi Delta. SEM depicts example of a single floccule composed of domains (D) arranged edge-to-face (E-F) in surrounding material of randomly arranged domains.

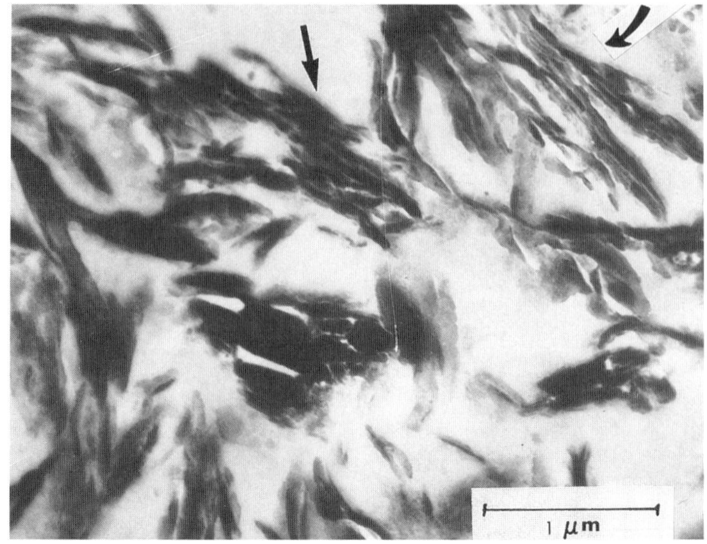

Figure 2.8. TEM showing clear examples of domains typical of the microfabric of Mississippi Delta sediment. Arrows point to face-to-face (F-F) orientation of domains.

fluid. This mechanism is termed interface dynamics. Energy sources include mass fluid flow driven by wave, current, and gravity forces and gravitational settling of particles (Fig. 2.10). In general, mass fluid flow has its greatest impact on microfabric when it impinges on the surface of the sediment. It may serve to reorient particles attached to the surface or it may drive suspended particles onto the surface. In the vicinity of a surface of significant area, the perpendicular component of fluid motion

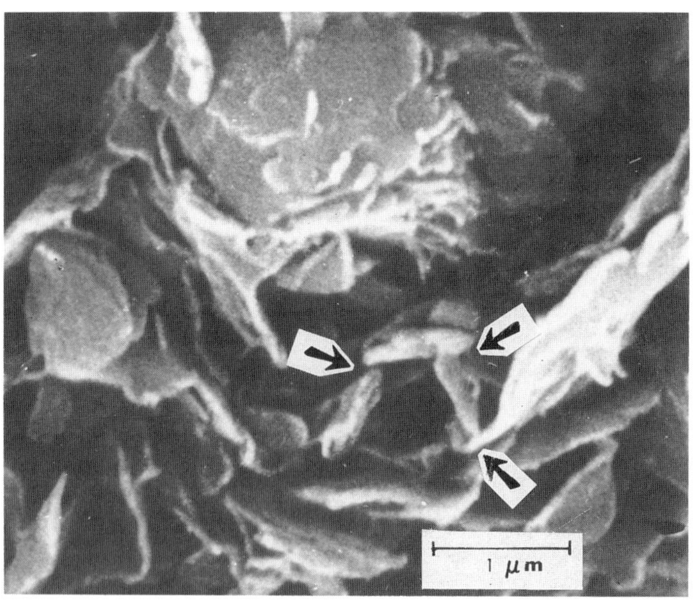

Figure 2.7. SEM of Mississippi Delta Sample depicting some edge-to-edge (E-E) and face-to-face (F-F) particle contacts. Note large domain in upper center of figure and the large and randomly shaped voids.

approaches zero. Therefore shear forces overcome interparticle binding forces and gravity forces only in regions of relatively rapid flow. For unconsolidated clay-size particles the minimum flow for particle entrainment is about 2.0 cm/sec (Boggs, 1987); a range of flow rates (0.6–3.0 cm/sec) was suggested recently by Li and Bennett (this volume). An important factor not considered above and often omitted from discussions of fabric development under dynamic conditions is that, although individual grains may be of clay size, the kinetic unit is often a multigrain aggregate. Thus, consolidated clay sediments may resist erosion under flow rates of 1 m/sec (Boggs, 1987); and, conversely, multigrain particles (chains, Fig. 2.2) may protrude many single grain diameters above the plane of the sediment surface, and be quite subject to disturbance at even minimal flow.

Particles may be brought into contact by any of a number of mechanisms that result in differential rates of particle movement (Montgomery, 1985). Such a differential may arise under gravitational settling between particles of different sizes or densities, for example, and is particularly characteristic of dynamic effects at interfaces. In the vicinity of a static boundary, the velocity of flow of water or air becomes less the nearer the boundary is approached. Particles swept along in the fluid attain a similar velocity gradient and the rate with which they collide is approximately proportional to their concentration and to the velocity gradient (Swift and Friedlander, 1964). Attachment of particles to gas bubbles in water, which may be greatly enhanced by the presence of organic matter, results in the particles being moved in the direction opposite to the general gravitational settling and may be moderately effective in bringing particles into contact (Leja, 1982). These particle-to-particle contacts occur both during bubble movement through the water and on the accumulation of bubbles at the surface. Differential motion of particles attached to a

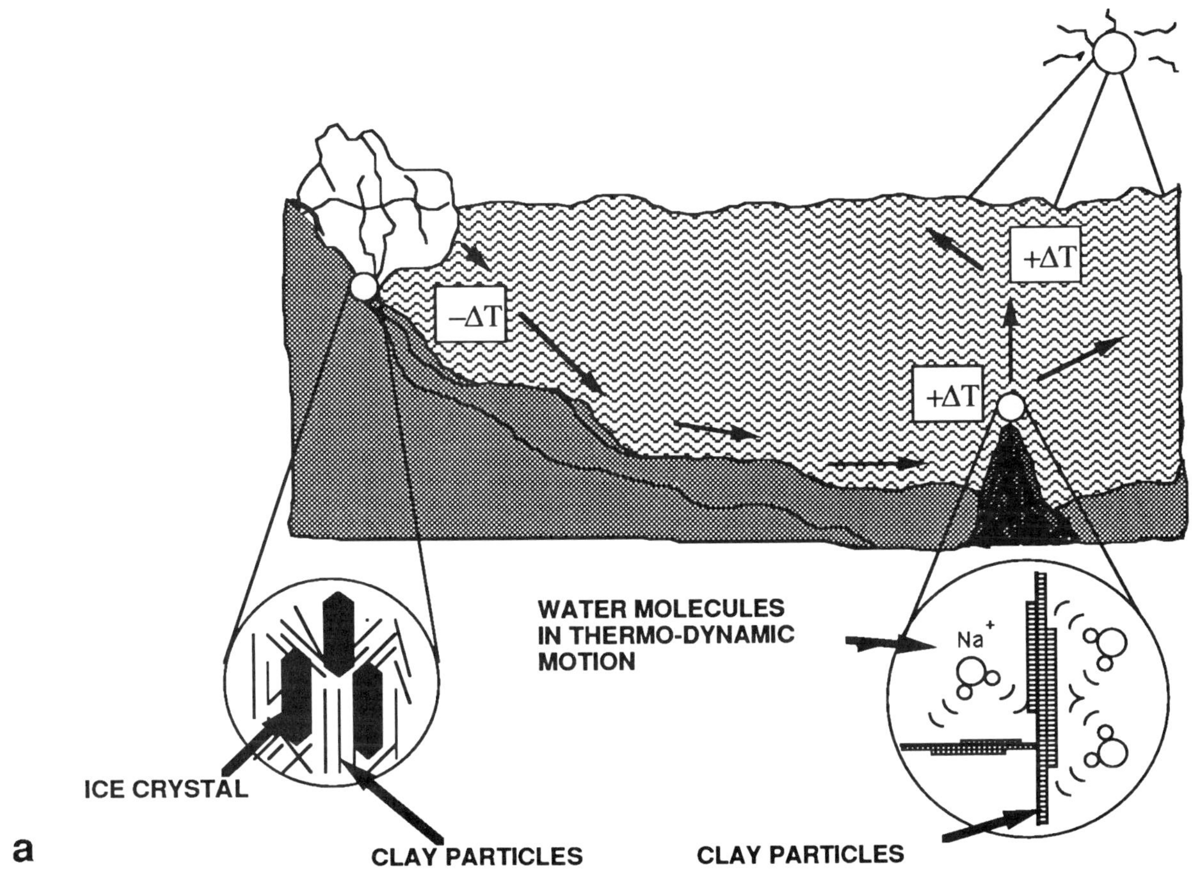

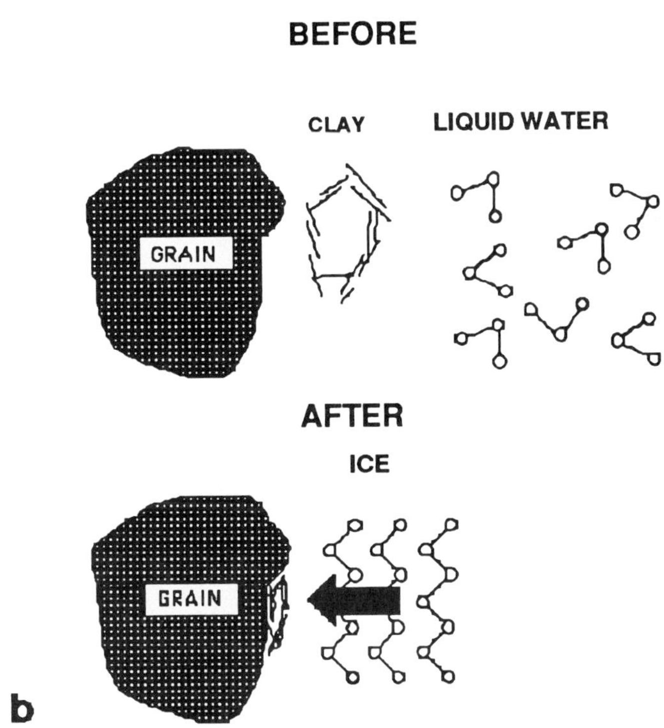

Figure 2.9. (a) Fabric modification by thermomechanical mechanism. Thermally driven mass flow (solar and hydrothermal heating and atmospheric cooling). Arrows depict relative motion of the water mass; crystallization of ice in sediment; Brownian motion of water molecules and suspended particles. (b) The effects of freezing (surrounding water) on the reorientation of clay particles as the ice crystals press the particles against a large solid such as a quartz grain or pebble.

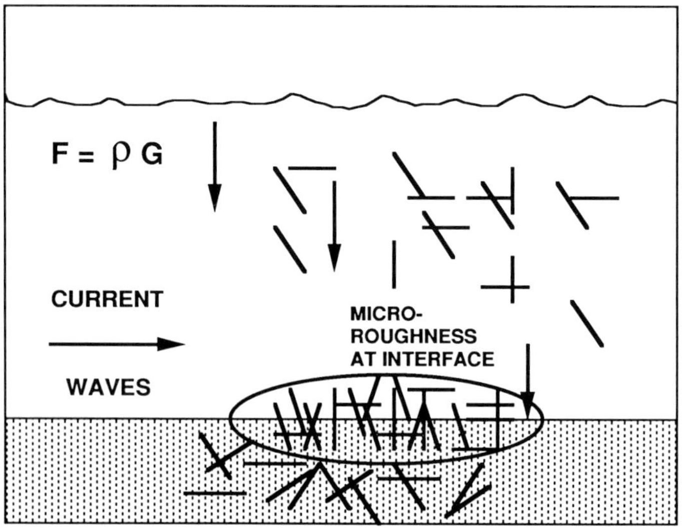

Figure 2.10. Diagram depicting some modes of fabric modification by interface dynamics mechanisms: differential motion of settling particles under the influence of gravity, differential flow of water masses of differing density, impact of particles on sediment interface, and flow at interface (note the importance of microroughness on particle orientation)

single bubble becomes even more pronounced as the water film surrounding the bubble thins when it rises to the water surface or as the gas of the submerged bubble dissolves.

Bioorganic Processes

Bioorganic processes represent direct effects of living organisms on sediment properties and indirect effects mediated by the chemicals (organic materials) they produce. The term bioorganic is constructed from *bio* meaning living and *organic* meaning the chemical products of the life process (Morrison and Boyd, 1983). The mechanical alterations of sediment fabric by activities of organisms are classified as biomechanical. Included are activities that take place in the sediment such as bioturbation and activities that take place in the water column such as ingestion and reorientation of particles and clusters of particles by filter feeders. In the latter case, biophysical mechanisms such as the binding of particles together by organic matter also may be important. Biophysical mechanisms include the adherence of particles to sticky organic mucus and particle binding by polymer bridging. Biochemical influences on sediment microstructure result from changes in the microenvironment brought about by the metabolic activities of organisms. These include the breakdown of materials such as the polymers, which are important in polymer bridging and the production of biogenic materials such as cementing precipitates of pyrite. Biogenic gases, including methane and carbon dioxide, can have a locally important impact on the chemical environment surrounding sediment particles and, in shallow-water settings, can disrupt sediment by bouyant release (see Wartel et al., this volume).

Biomechanical Mechanisms

Bioturbation, "the churning and stirring of a sediment by organisms" (Bates and Jackson, 1987), is a significant mechanism responsible for producing the particle orientation in a sediment and ultimately in a rock. This biomechanical mechanism produces a clay microfabric that possesses a randomness similar to the primary fabric of a flocculated (edge-to-face) clay, however, it is mainly characterized by randomly oriented individual particles (seen in SEM) rather than the random domains common in flocculated clays formed by other processes. Another distinguishing feature observed by the second author, especially apparent in bioturbated shales and mudstones, is the abundance of silt grains mixed in with the bioturbated clayey material.

In bioturbated sediment, the randomly oriented fabric of individual platelets is a result of mixing by organisms. Mixing disturbs the original primary flocculated fabric. Rhodes and Boyer (1982) indicate that "to facilitate burrowing and feeding, some metazoa, especially bivalves, also liquify the sediment by injecting water anteriorly into the bottom . . . this causes an instantaneous local increase in pore water pressure and the liquid limit of the sediment is temporarily exceeded." At this point the primary flocculated fabric is not only disrupted but silt size grains are mobilized and mixed in with the liquified clayey sediment. On burial and lithification the bioturbated sediment often retains its random fabric. Preservation of randomness in extensively bioturbated sediment is attributed to binding of particles by mucus secreted by burrowing organisms (Rhodes and Boyer, 1982; O'Brien, 1987).

Figure 2.11 illustrates typical microfabric produced by the biomechanical mechanism. Hand sample and X-ray radiographic viewing was first done on each of these poorly fissile gray shales or mudstones to demonstrate that they had been extensively bioturbated. In radiographs all samples show burrows or other evidence of sediment mixing. It should be stressed that to be certain of positive identification of a bioturbated fabric one should combine radiography or thin-section analysis with SEM viewing. Body fossils are not common in the samples, however, various types of trace fossils are apparent. The two important features which characterize this microfabric are (1) a randomness of individual clay flakes, and (2) silt size grains (quartz?) mixed in with the clay matrix.

Fecal pellets leave another fabric signature representing a biomechanical mechanism of sediment aggregation. An SEM study by Syvitski and Lewis (1980) revealed the characteristics of marine zooplankton fecal pellets. The ingestion of sediment and its expulsion by organisms as fecal pellets are common. Pryor (1975) studied the feeding activities and excretory products of the marine decapod *Callianassa major* and the

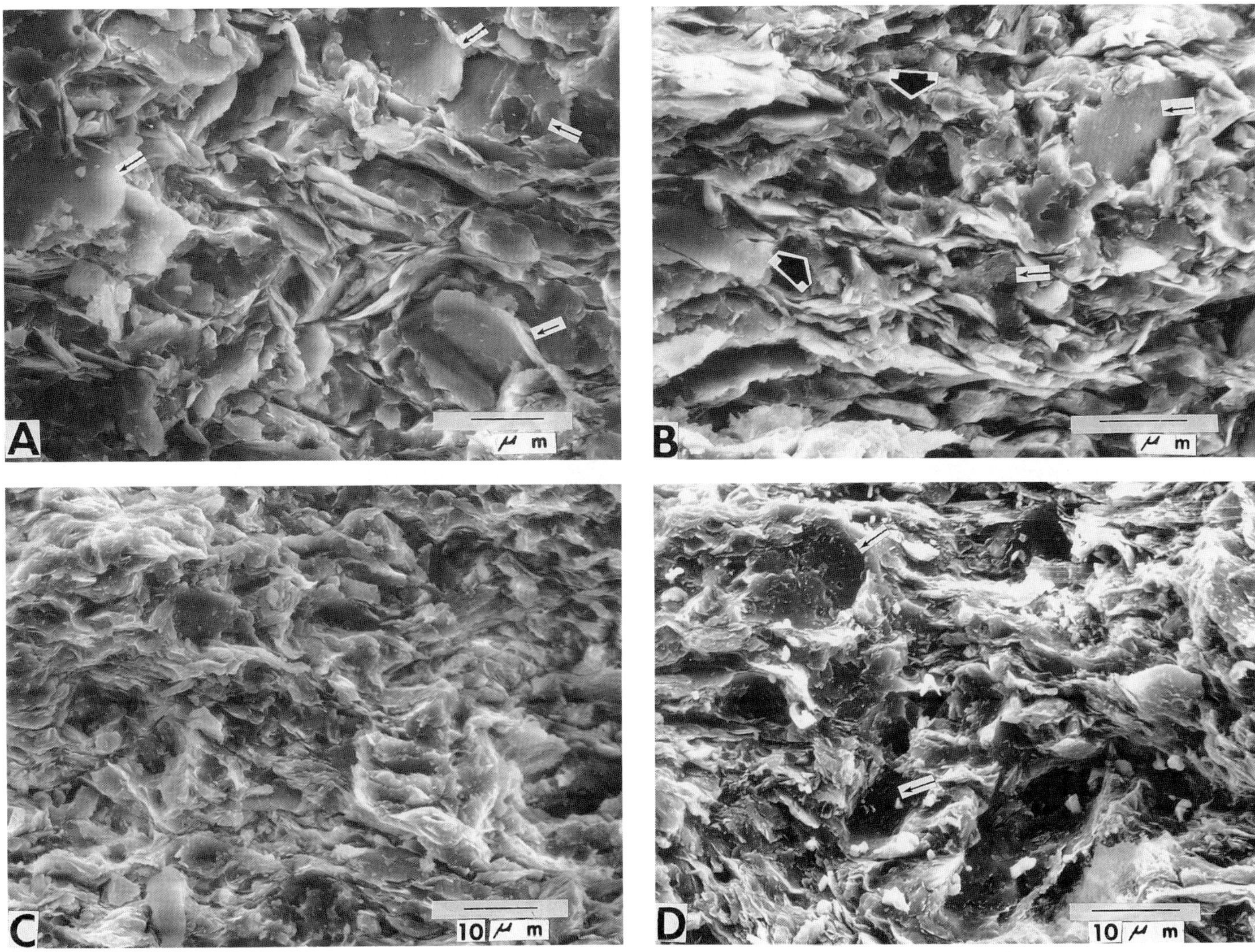

Figure 2.11. SEM photomicrographs of particle association produced by the biomechanical (bioturbation) mechanisms. (A) Silt grains (small arrows) mixed with individual platy clay flakes: Cashaqua shale (Devonian, Wyoming County, NY). Scale = 10 μm. (B) Individual randomly oriented flakes (large arrows) mixed with silt grains (small arrows) in the bioturbated Huron shale (Devonian, Mason County, WV). Scale = 10 μm. (C) Bioturbated mudstone fabric. Notice swirled nature of platy flakes. Penn Yan shale (Devonian, Livingston County, NY). Scale = 10 μm. (D) Typical bioturbated fabric (arrows show silt grains or cavities formerly occupied by grains dislodged during preparation) (Jurassic, gray shale, Yorkshire, England).

marine annelid *Onuphis microcephala* and found that in some areas they removed argillaceous particles from suspension, and deposited layers of fecal mud as thick as 4.5 mm/year, producing as much as "12 metric tons (dry weight) of pelleted mud per square kilometer per year."

The fabric of densely packed randomly oriented fine particles in a well-defined ellipsoidal pellet is easily recognized at low magnification in an SEM. Both recent (Fig. 2.12A, B), and ancient fecal pellets (Figs. 2.13A, B, 2.14A, B) reveal biogenic and lithogenic particles. However, when viewed at magnification greater than ×1000 it is often difficult to distinguish fecal pellet fabric from that produced by physicochemical processes (Fig. 2.13B). At lower magnification the sharp border of the pellet is easily identified and the fabric difference between pellet and enclosing rock becomes obvious (Fig. 2.14A). Figure 2.14A and B-2 show the typical fossil fecal pellet microfabric characterized by dense packing of randomly oriented nonplaty material that is in contrast to the preferred orientation of platy minerals composing the surrounding shale (see arrows in Fig. 2.14A, B-2).

Biophysical Mechanisms

Clusters of randomly oriented particles are preserved in some organic-rich argillaceous rocks. Here the term biosediment aggregate refers to clusters that are believed to have formed

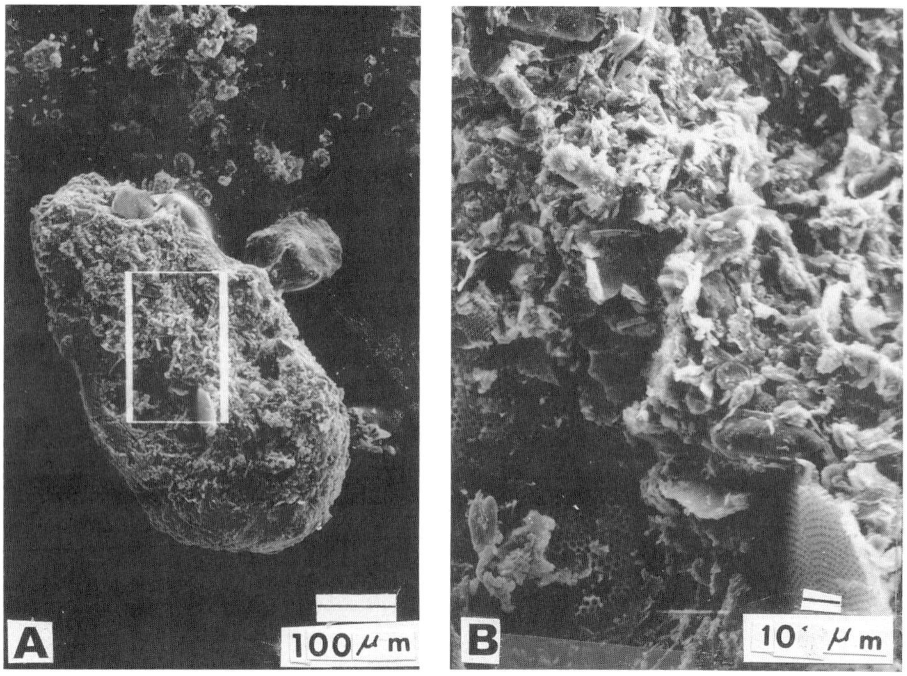

Figure 2.12. Fabric of a recent fecal pellet. (A) Recent fecal pellet (organism unknown) found in marine sediment, Dixon Entrance, Queen Charlotte Is., British Columbia. Scale = 100 μm. (B) Close-up view of pellet area shown in box outlined in A. Notice the random mix of biogenic (diatom fragments) and lithogenic (clay?) matter. Scale = 10 μm.

when lithogenic sedimentary material adhered to sticky organic mucus nets and/or formed by a clay–polymer bridge interaction. Hence, they represent another mechanism by which sediment may aggregate and settle through the water column.

Our investigation of certain organic-rich marine shales reveals the presence of numerous aggregates scattered randomly throughout a matrix of predominantly preferred platy matter. Their presence should not be surprising since McCave (1984) stated that fine-grained marine sediment settles as aggregates produced by biochemical bonding or electrostatic attractions. There is abundant evidence supporting the role of organisms and organic matter in promoting recent sediment aggregation (see for example, Riley, 1963; Kane, 1967; MacLean and Smart, 1978; Silver et al., 1978; Trent et al., 1978; Mullins, 1980; Shanks and Trent, 1980; Syvitski and Murray, 1981). The interaction between organic and inorganic matter is considered responsible for biosediment aggregate formation described here. Examples of their microfabric are shown in Figure 2.15A and B.

Biosediment aggregates found in this study occur only in nonbioturbated rocks that are well laminated, thus eliminating postdepositional biogenic mixing as a mechanism responsible for their formation. They are relics of the original sediment fabric. Because of their association with highly organic sedimentary rock [all shales in which they occur have a total organic content (TOC) of >3%] it is concluded that they record the complex organic–clay interaction which took place during sediment deposition. Modern analogs of this interaction are numerous.

Coccolithophoroid mucus is reported to play a major role in formation of aggregates of particles in recent sediment (Honjo, 1982). Pierce and Siegel (1979) observed that the major portion of suspended solids in estuarine and organic water is composed of aggregates of mineral grains, soft organic matter, biogenic debris, and phytoplankton, all bound by a matrix of organic matter. Alldredge (1986) has shown how the planktonic gelatinous zooplankton *Appendicularia* (Chordata, Tunicata) produces an external mucus structure that becomes clogged with sediment and is eventually discarded. Fungal mycelia and filamentous bacteria in Lake Tahoe, Nevada, were found aggregated with detrital particles (Pearl, 1973).

The work of Avnimelech et al. (1982) is very significant to our study in illustrating the morphology of what we interpret as recent biosediment aggregates (Fig. 2.16). They found that aggregation of clay was promoted by algae that secrete large amounts of polysaccharides and other polymers that produce sticky surfaces, which in turn promote aggregation of clay on algal surfaces. Notice, in Figure 2.16, that the platy clay flakes appear caught on the fine algal filaments much like a fly trapped in a spider's web. The sticky algal surfaces that bind the clay contain polymers that act as bridges via cations or anions to the clay particles. Avnimelech and Menzel (1984) and Reed (personal communication) have used this aggregation mechanism to clarify muddy ponds simply by stimulating algal growth by adding fertilizer to the water, which causes sedimentation of algal–clay clusters. Cyanobacterially induced flocs have been produced in clarifying suspensions of finely divided mineral wastes from phosphate ore beneficiation (Leslie et al., 1984).

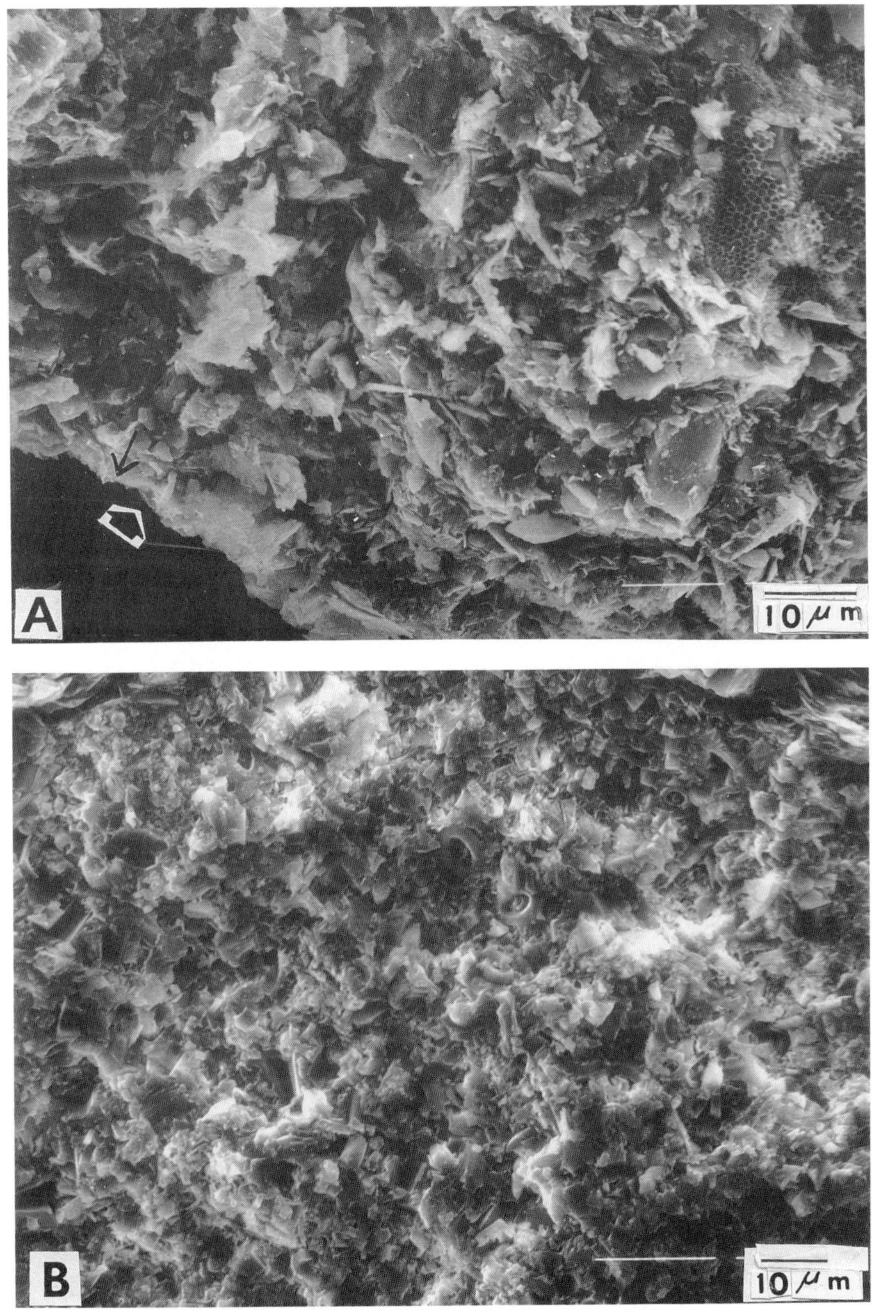

Figure 2.13. Comparison of the microfabric of a recent and Jurassic fecal pellet. (A) Magnified view of the microfabric of recent fecal pellet in marine sediment, Emerald Basin, off Nova Scotia. Note the random particle reorientation. Scale = 10 μm. (B) Magnified view of fossil fecal pellet fabric, Bituminous Shale (Jurassic) Yorkshire, England. Note the random particle orientation and dense packing. Coccolith fragments are scattered throughout. Scale = 10 μm.

Organic polyelectrolytes have also been reported by numerous investigators to destabilize colloidal suspensions by a "polymer bridging mechanism" (Ruehrwein and Ward, 1952; Lamer and Healey, 1963; Black et al., 1965; Stumn and Morgan, 1981). The bridge forms when part of the polymer chain attaches to a solid particle surface (at an adsorption site) whereas other sections of the chain may extend out into the solution and become attached to a free site on a second particle (Fig. 2.17). This mechanism produces an aggregate of particles and polymers (see also Syvitski, this volume). Rashid (1985) described this mechanism as it applies to the sedimentation of floccules of clay and organic complexes in the marine environment and stressed the role of saline water in promoting flocculation. It is significant that our observations show aggregates in marine organic-rich argillaceous rocks.

We propose that large masses of sediment could aggregate and settle out in an organic-rich environment by the biophysical (sediment aggregate) mechanism. Initially large macromolecules

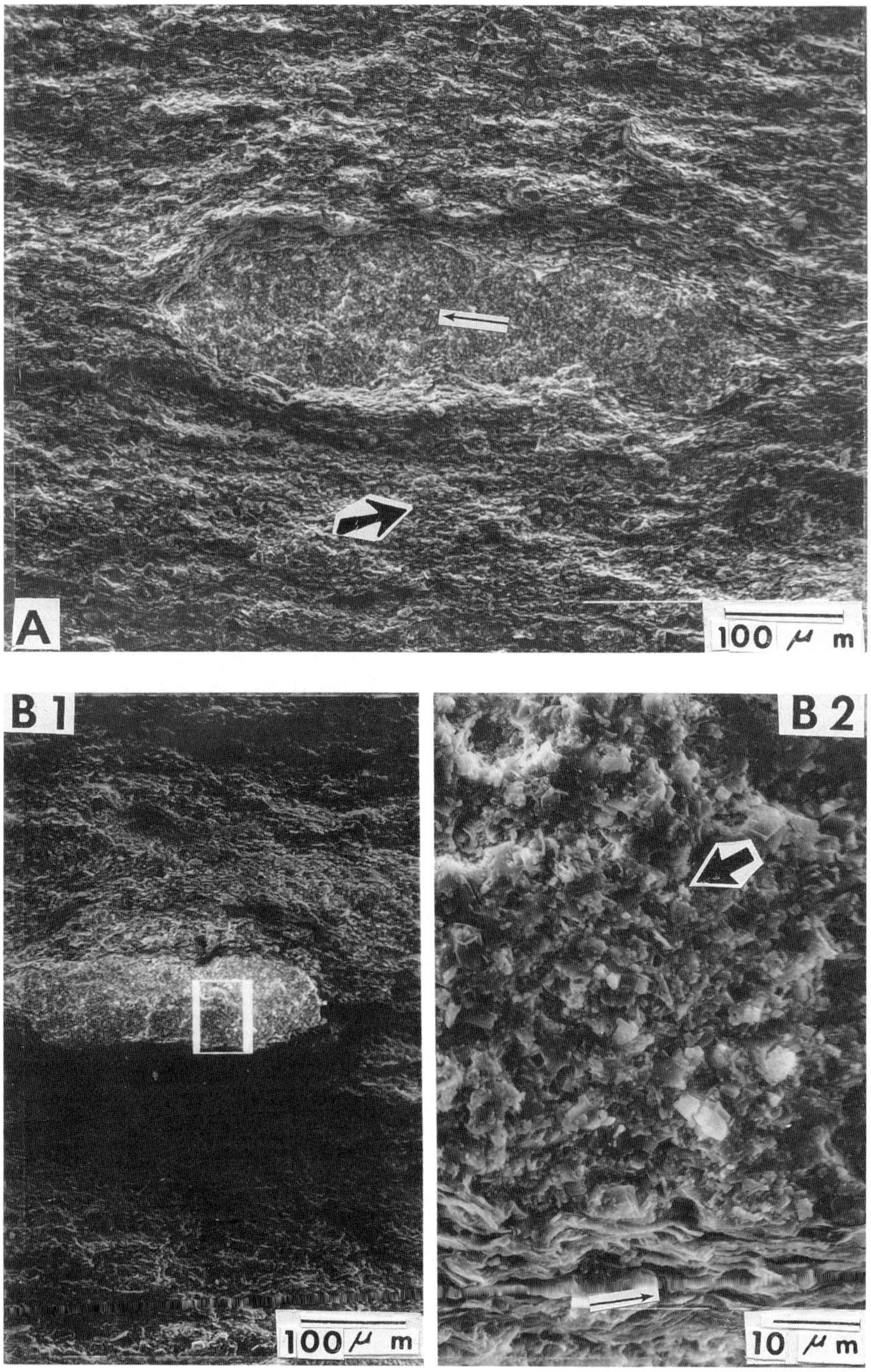

Figure 2.14. Microfabric of fossil fecal pellets. (A) View of fossil fecal pellet at low magnification. Note contrasting fabrics of pellet (small arrow) and surrounding shale matrix (large arrow). Scale = 100 μm. (B-1) Fossil fecal pellet in Bituminous Shale (Jurassic) Yorkshire, England. Scale = 100 μm. (B-2) Close-up view of area shown in outlined box in B-1. Note fabric difference between pellet (large arrow) and preferred orientation in the surrounding shale (small arrow). Scale = 10 μm.

and organic mucus strings suspended in ocean water act as substrates onto which clay, silt, and other detritus aggregates (Fig. 2.17). As this mass settles, other suspended particles are swept up by being caught in the sticky mucus web. The actual aggregation mechanism consists of an organic–sediment interaction (i.e., polymer bridging) as polymer chains link particles into a lacy network of organic and inorganic detritus. Once incorporated within the bottom sediment, the degree to which a biosediment-aggregate resists compaction and particle reorientation relates to the amount of polymer bridges, which in turn relates to the original organic concentration (and organic type?) in the seawater. One has to explain the numerous aggregates scattered

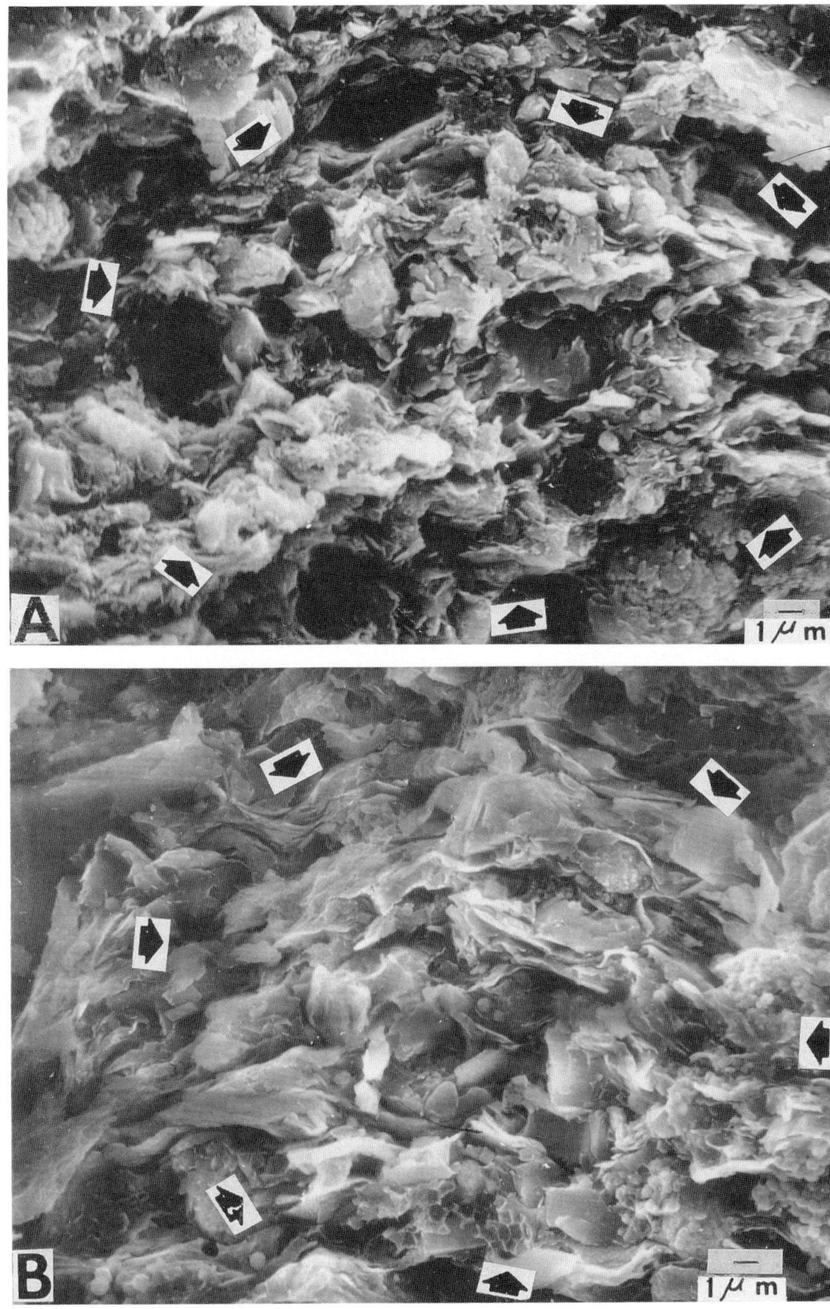

Figure 2.15. Typical examples of biosediment aggregates. (A) Loose packing of randomly oriented platy particles in an organic rich biosediment aggregate shale, Swope Fm. (Pennsylvanian, Adair County, Iowa). Scale = 1 μm. (B) Biosediment aggregate microfabric characterized by an open texture of randomly oriented particles. Note the authigenic framboids (lower right) in the surrounding shale. (Jurassic, Ravenscar, England). Scale = 1 μm.

throughout a shale with dominantly preferred particle orientation. The preferred orientation may represent totally collapsed aggregates whereas the clumps of random particles represent some aggregates whose gel strength allowed them to resist reorientation. The concentration of organics is important. A small concentration of polymers in suspension could produce aggregates composed of only a limited number of weakly formed bridges. Black et al. (1965) found, for example, that if too few sites are occupied, the bridging will be too weak even to withstand shearing forces by agitation. Thus, in a sedimentary environment that is anoxic but contains a low content of organic matter, the stress exerted by bottom flowing currents or sediment overburden could be a factor in disturbing the stability of most weakly bridged aggregates and thus produce a dominantly preferred particle fabric surrounding those few clumps of the original biosediment aggregates which resisted deformation.

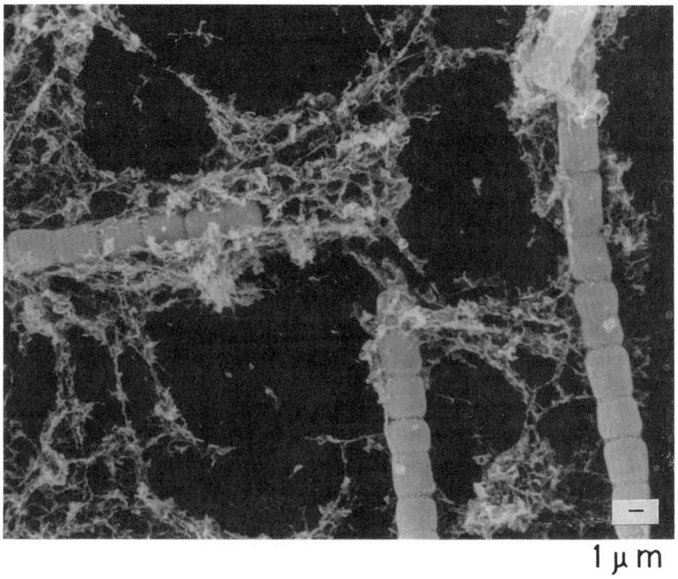

Figure 2.16. A biosediment/algae-clay aggregate from Avnimeleck et al. (1982). Note the fine algal filaments and large (white) particle possibly some clay particle and/or skeletal debris and the overall morphology of the aggregate. Scale = 1 μm. (Reprinted with permission from Y. Avnumelech et al., 1982. Fig. 2a, 216, p. 63–65, 2 April, Science, Mutual flocculation of algae and clay: evidence and implication. Copyright Science, 1982 by the AAAS.)

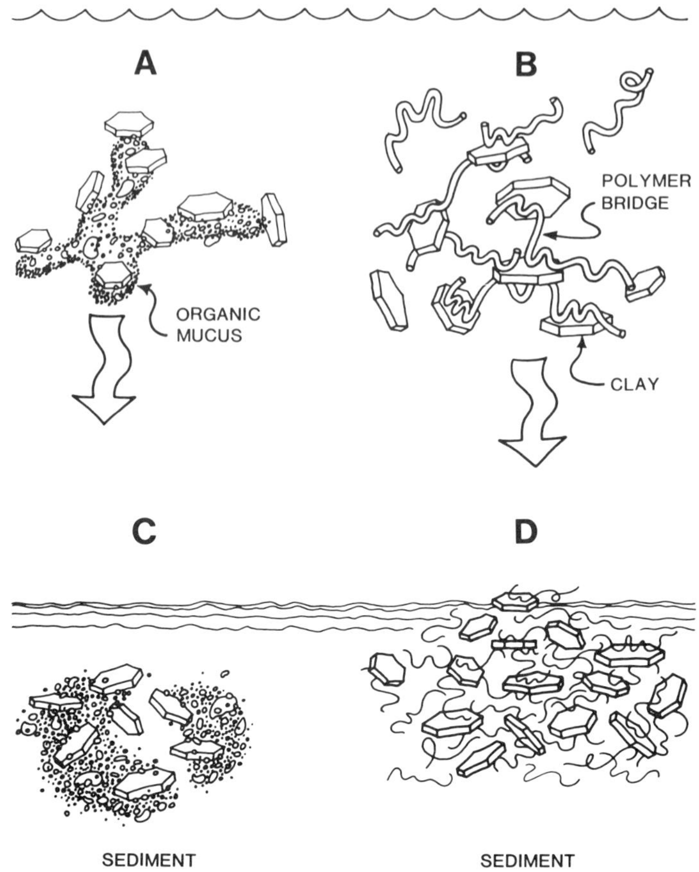

Figure 2.17. Steps in the formation of biosediment aggregates. (A) Large clumps of organic mucus sweep up suspended detritus during settling. (B) The actual contact between organics and sediment is a result of particle linking by polymer bridges. Notice linkage of clay flakes by polymers. (C–D) Orientation of biosediment aggregates in loosely consolidated sediment.

Although the exact understanding of their preservation is not clear, aggregates of randomly oriented particles observed in this study are in organic-rich rocks in the geologic record and presumably the biosediment aggregates occur also in some modern sediments. It is suggested that biosediment aggregates may be a result of a polymer bridging mechanism.

Biochemical Mechanisms

Organisms alter their environments by the production and destruction of many chemical entities. The impacts on sediment microstructure that these chemical alterations cause are included in the term biochemical mechanisms. Some of the simpler chemical conversions of interest are illustrated (Fig. 2.18). Not only are sugars, proteins, oils, and the rest of the complex array of biomolecules made by organisms, but also these compounds are decomposed and result in a variety of changes in inorganic constituents. The decay of polysaccharides and other polymers responsible for polymer bridging discussed above, for example, may allow particles in aggregates to be more easily moved apart or reoriented after sedimentation.

Gases produced by organisms in sediment masses also may alter sediment microstructure as they are released. A spectacular example of macroscale disruption of deep-sea sediment by presumed gas release was reported recently (Prior et al., 1989). In this case, about 2 million cubic meters of sediment was excavated and redeposited about the resulting blowout crater. The measured crater depth was 58 m, and the total amount of sediment excavated corresponded to a sedimentation time of about 200,000 years. Much less spectacular but much more common venting of biologically derived gases such as CO_2 and CH_4 in shallow water churns and resuspends sediments.

An example of a biochemical mechanism reflected through inorganic species is the formation of authigenic pyrite by the combination of ferrous ions and disulfide ions generated metabolically in anaerobic environments (Lynch, 1983). The formation of authigenic pyrite framboids in fine-grained sediment also influences clay fabric. Although framboid growth does not produce sediment aggregates, it does exert a significant influence on microfabric by causing particle reorientation in specific areas of the sediment. The individual framboid (composed of iron sulfide) found in shale is spheroidal (Fig. 2.19). The spheroidicity

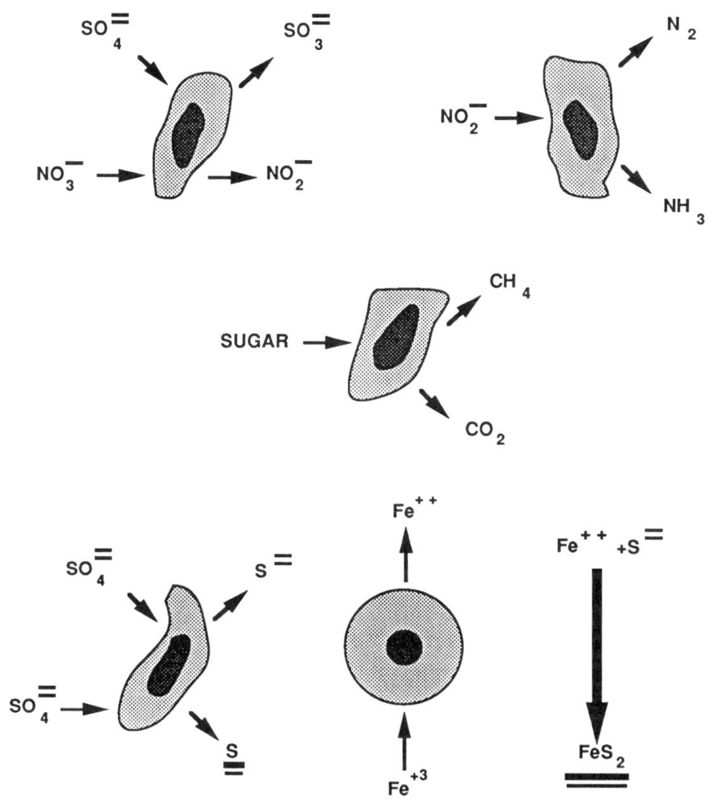

Figure 2.18. Diagram depicting simple biochemical conversions of materials, including representation for the formation of authigenic pyrite. Organisms are depicted by stippled form showing uptake of chemical species such as nitrate, which is reduced to nitrite. Nitrite can be processed by an organism expelling the diatomic molecule (N_2), nitrogen gas, and also producing ammonia NH_3.

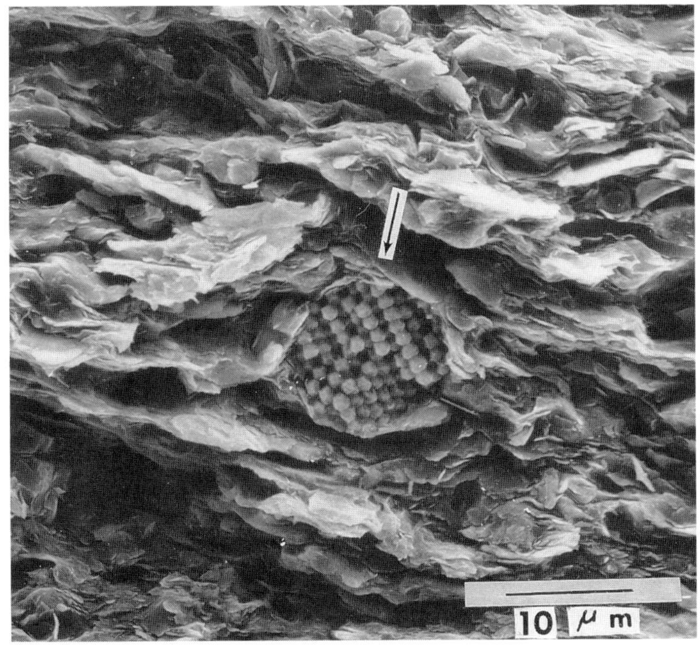

Figure 2.19. Pyrite framboid in Bedford shale formation (Mississippian, Ohio). Notice doming of platy particles (arrow) around framboid and well developed preferred orientation in surrounding shale. Scale = 10 μm.

is attributed to pseudomorphism of a preexisting spherical body—e.g., immiscible organic globules or infilling of gaseous vacuoles (Rickard, 1970). A biochemical origin for framboid sphericity has been proposed (Kalliokoski, 1974; Jabor and Mountjoy, 1976) and is discussed here. Bacteria or microflora colonies may produce the spheroidal morphology of framboids and also may be responsible for the biochemical reactions leading to their formation in sediment.

The typical framboid found commonly in organic-rich shales displays platy flakes wrapped around the sphere and is associated with a doming of adjacent particles. The adjacent particle orientation is interpreted as indicating that the sphere (be it a gas bubble or bacterial colony) grew in soft, easily deformed, flocculated clay sediment. Such a sediment would exist in the early stage of deposition. Framboids are found today at relatively shallow depth (<3 m) forming as a feature of early sediment diagenesis (Love, 1967). It is proposed that some framboids found in organic-rich shales originally grew in soft flocculated clay under the reducing conditions existing in the sediment soon after deposition. These conditions favored bacterial growth. As a bacteria colony enlarged (generating sulfur in the process) the soft flocculated clay sediment was easily deformed adjacent to it causing clay flakes immediately adjacent to the colony to wrap around its surface and reorienting and doming the sediment farther away in response to the growing sphere (Fig. 2.19). Subsequent sediment loading due to deeper burial reoriented the surrounding flocculated clay into the typical preferred fabric of shale. During early diagenesis, iron diffusion from the organic-rich anoxic sediment into the bacterial cavity resulted in iron sulfide precipitation.

This is only one possible explanation for the origin of pyrite framboids in clayey sediment. What is important is that this example illustrates they may form by a biochemical mechanism that in turn exerts an influence on the final clay or shale microfabric.

Burial Diagenesis Processes

Postdepositional alteration of microfabric during the process of burial diagenesis is driven largely by two mechanisms: mass gravity and diagenesis–cementation. Both mechanisms may proceed simultaneously. However, mass gravity stresses within a deposit are ever present regardless of the degree of chemical activity and mineralogical alterations and often mass gravity acts independently without influence of significant chemical activity associated with diagenesis–cementation. Burst (1976)

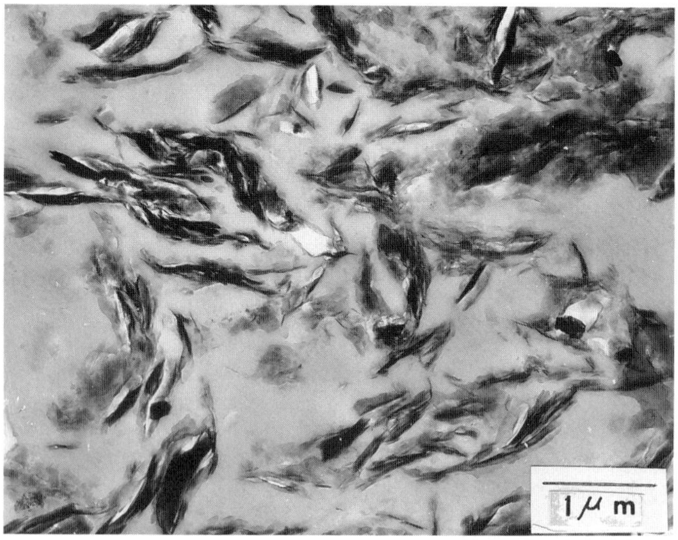

Figure 2.20. Microfabric typical of the surficial submarine sediment (1.3 m subbottom) of the Mississippi Delta observed by TEM. Note random arrangement of domains and large, irregular-shaped, voids. Scale = 1 μm.

Figure 2.21. Microfabric of Mississippi Delta sediments (30 cm subbottom) observed by SEM. Compare with TEM in Figure 20. Note domains, large voids, and short chains. Scale = 1 μm.

suggests that gravity is the dominant consolidation–compaction mechanism during dewatering of argillaceous sediment in the upper 914 m, but he contends that mineral diagenesis of clays, and osmotic and aquathermal pressuring are significant contributing mechanisms at burial depth.

Mass gravity mechanisms are used here to include not only consolidation and compaction, but also the dynamics of slumping, creep, orbital bed motion and deformation, and seismic shock. For this study, diagenesis–cementation mechanisms include mineral alterations, interstitial fluid transport, cementation and leaching, and organic-clay and gas interactions. A very limited number of studies have addressed the significance of these various mechanisms on microfabric development. The most popular topics have included study of fissility versus depth of burial (Hedberg, 1936; Weller, 1959; White, 1961; Gillott, 1969; Rieke and Chilingarian, 1974) and the association of organic material with the development of preferred particle orientation in shales (Gipson, 1965; Odom, 1967; O'Brien, 1968, 1989).

Mass Gravity Mechanisms

Limited but revealing sediment microfabric signatures that developed as a function of post depositional/mechanogravity mechanisms have been observed in electron micrographs. The depositional microfabric of high porosity sediment characterized by randomly oriented domains (Figs. 2.20, 2.21) was observed to have responded to overburden stress by mechanical rearrangement of particles normal to the direction of stress (Figs. 2.22, 2.23). Initial depositional porosities of 70–75% contrasted sharply with consolidated sediment porosities of approximately 50% at burial depths of only 90–150 m (overburden stresses of 6–12 kPa). Reorientation of particles is observed for both field and laboratory consolidation (Bowles et al., 1969; Bennett et al., 1977, 1981). Hedberg (1936) recognized the importance of mechanical rearrangement of particles during the early stages of compaction and his observations appear to have been well documented by later studies (Bowles et al., 1969; Bennett et al., 1977, 1981; Faas and Crocket, 1983).

The consolidation process involves volume reduction as a result of sediment dewatering under an imposed load. The geologist often refers to this process as compaction. The importance of the mechanisms involved in sediment dewatering was recognized as early as 1908 by Sorby. He observed significant reductions in porosity as a function of compaction. As noted above, volume reduction on compaction also is associated with platy particle reorientation. Randomness in the original flocculated clayey sediment changes to more preferred orientation with depth. This volume reduction and particle reorientation are important early burial diagenesis processes that influence the final microfabric of resultant rocks. White (1961), Gipson (1965, 1966), and Gillott (1969) concluded that the high degree of clay mineral preferred particle orientation was associated with fissile shales and that random orientation of particles was characteristic of the more massive argillaceous rocks. Scanning electron microscopy observations of shales by O'Brien (1968, 1970b) revealed strong preferred particle orientation in a fissile Pennsylvanian black shale (Fig. 2.14A) and random particle orientation in nonfissile argillaceous rocks (Fig. 2.11A).

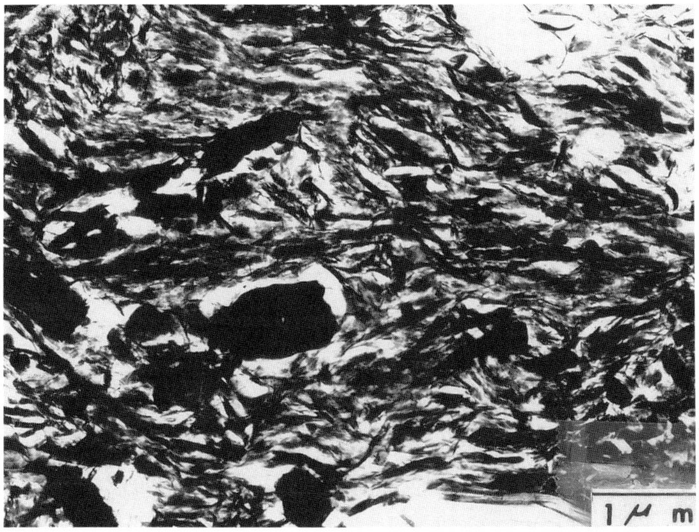

Figure 2.22. Microfabric of consolidated sediments from the Mississippi Deleta recovered from 120 m subbottom observed by TEM. Note the highly oriented clay plates that lie in a direction normal to the direction of the overburden stress. Larger particles prevent perfect alignment of all the platelets. Voids become long and linear as compared to surficial sediments (see Figs. 2.21 and 2.22).

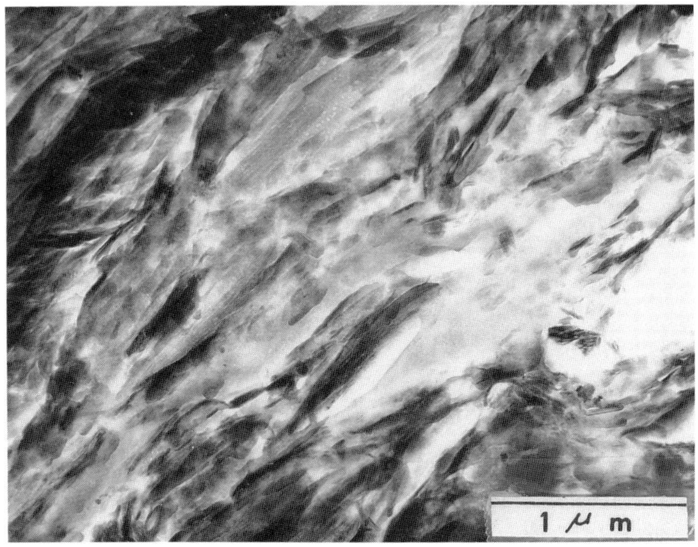

Figure 2.23. High TEM magnification of consolidated sediment recovered from 150 m subbottom, Mississippi Delta. Note the strong degree of preferred particle orientation. Scale = 1 μm.

The most conclusive evidence to date indicates that fissility is related to preferred particle orientation, but the degree of preferred orientation is not directly related to depth of burial in most cases. Thus the decrease in porosity with depth of burial varies considerably among various sediment types. Evidence from numerous studies of the relation between sediment and shale microfabric and compaction has shown the importance of other factors such as geochemical conditions, environments of deposition, biological activity, and related geological and dynamic factors that set the stage and influence the initial formation of microfabric in suspension and at the depositional interface. The initial depositional microfabric and effects of biological activity most certainly influence the response of the particles in the postdepositional development of microfabric with increasing depth of burial and when the deposit is subjected to external environmental stresses (Bennett, 1976, 1977).

Downslope movement and shearing of sedimentary deposits also have a profound influence on the postdepositional development of microfabric. Massive blocks of sedimentary material can move downslope without significant resuspension of particles (McGregor and Bennett, 1981) and, conversely, slumping can produce massive amounts of resuspended particulates that can travel many kilometers subaerially or on the sea floor as turbidity currents (Middleton and Bouma, 1973).

The Mississippi Delta, a "natural laboratory" for the study of coastal sedimentary processes, is an excellent example of a sedimentary environment experiencing massive downslope movement of fine-grained deposits (Coleman and Garrison, 1977; Coleman and Prior, 1978). "Crust" zones, defined on the basis of shear strength profiles with depth of burial, typically show a significant strength reduction or "cut back" in shear strength at depths of about 8–14 m subbottom (Doyle et al., 1971; Bea and Arnold, 1973). These zones are related to submarine slumping of soft, high porosity, deltaic sediments and the movement of massive amounts of sediment seaward on gentle slopes. The microfabric of these smectite–illite-rich high porosity sediments above and below the "crusts" is characterized by randomly arranged clay particles and domains (Figs. 2.24–2.26). This type of microfabric is typical of the unconsolidated surficial submarine sediments of the Mississippi Delta (Bowles, 1968; Bennett et al., 1977, 1981). Detailed microfabric studies of the "crusts" zones (Figs. 2.27, 2.28) have revealed a significant difference in the particle arrangements and orientations compared to the fabric above and below the "crusts." The microfabric in the "crusts" zones is characterized by areas of preferred particle alignment but with an overall appearance of remolding (Bennett et al., 1977; Bohlke and Bennett, 1980). As the sediment shears, remolds, and dewaters, the microfabric undergoes particle rearrangement. These microfabric signatures are characteristic of remolded sediment that has been verified by Bennett (1976) and Bennett et al. (1977, 1981). Pusch (1970) studied the effects of sediment shearing on microfabric and showed the distortion of links and chains and reorientation of particles in the direction of the principle shear stress. In shallow coastal environments, energy from surface waves reach the sea floor and produce orbital motion of the sediment (Yamamoto, 1982). These motions reduce the sediment strength and when shearing stresses exceed the sediment shear

Figure 2.24. SEM of surficial sediments recovered from 30 cm subbottom above the "crust" zone of the Mississippi Delta in relatively undisturbed material relative to the sheared characteristic of the "crusts." Note similar microfabric as observed in other subenvironments of the delta (compare Figs. 2.20 and 2.21). Scale = 1 μm.

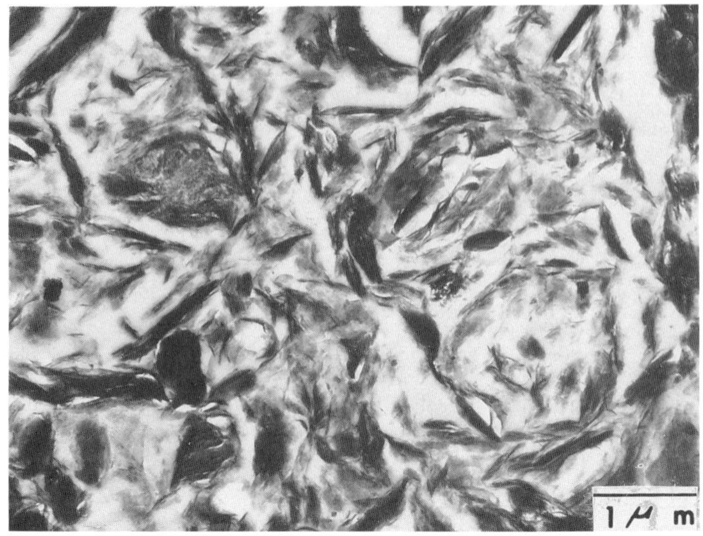

Figure 2.25. TEM of microfabric characteristic of sediments above the "crust" zone (compare with Fig. 2.24). Note high porosity, large voids, delicate particle-to-particle contacts, and short chains. Scale = 1 μm.

strength the deposit fails and submarine sliding often occurs (Dunlap et al., 1978; Henkel, 1970; Bea, 1971).

Diagenesis–Cementation Mechanisms

Diagenesis of mineral phases and cementation operate over a wide range of thermal regimes that can alter the character of the fabric. The literature is replete with studies of mineral transformations, mud to shale diagenesis, compaction (consolidation) versus depth of burial, and so on, however, meager attention has been given to the relationships of mineral alterations and microfabric. A recent study by Howard (1987) pointed out the importance of microfabric in controlling fluid flow properties; differences in permeability were shown to have a significant influence on the reaction rates of mineral transformation in shales. Reaction rates clearly depend on complex interacting processes and mechanisms that include not only fluid flow properties but also the availability and presence of inorganic chemical species in solution, the mineralogy of the solid particles, and the organic compounds present. The presence of organics in sedimentary sequences was shown to be very significant in burial diagenesis (Johns, 1979). The complexity of the geochemistry in cementation and mineral diagenesis was revealed clearly in the work of Curtis: "Chemical changes within clay mineral assemblages cannot be attributed to reactions among clays alone and indeed all the available evidence suggests other minerals are involved. Clastic sediment sequences include shales, silts, and sands. Shales start life as muds with upward of 70% pore water and relatively reactive, unstable minerals derived from soil profiles. As well as the clay minerals themselves, amorphous compounds of iron, aluminum, and silicon are important, as is organic matter. Carbonate, reactive silica, and more organic matter may be added from depositional waters" (Curtis, 1985, pp. 91–92).

Dunoyer De Segonzac (1970) pointed out that mineral transformations, neoformation, and recrystallization are accompa-

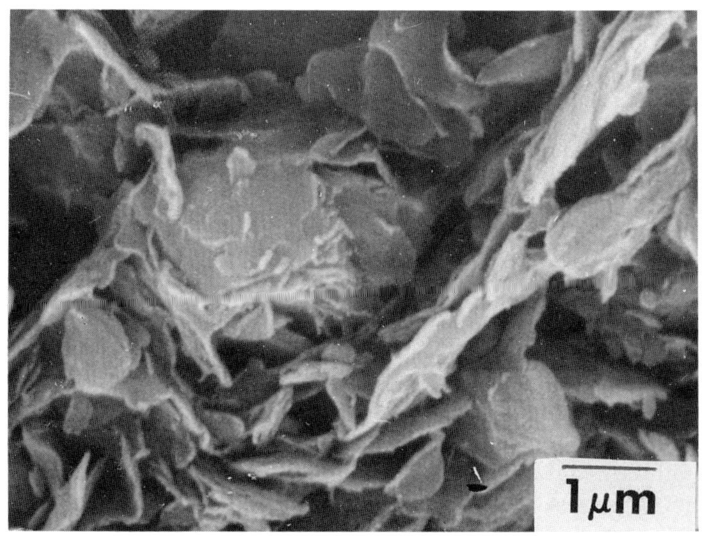

Figure 2.26. SEM of microfabric characteristic of sediments below the sheared sediments of the "crust" zone. Note similarity with the undisturbed surficial sediments. Scale = 1 μm.

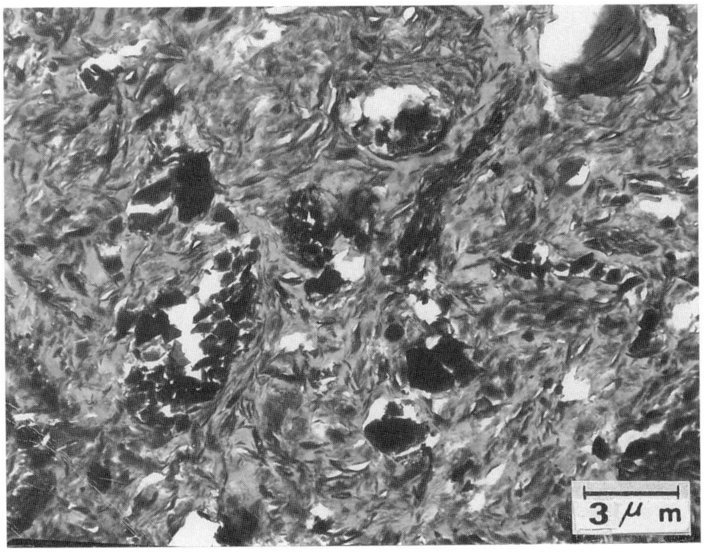

Figure 2.27. Microfabric characteristic of the "crust" zone of the Mississippi Delta as observed by TEM. Note swirled features and localized areas of preferred particle orientation. Clay platelets orient in the form of an "onion" skin around larger particles. The microfabric is typical of remolded clay sediment (Bennett et al., 1977, 1981). Scale = 3 μm.

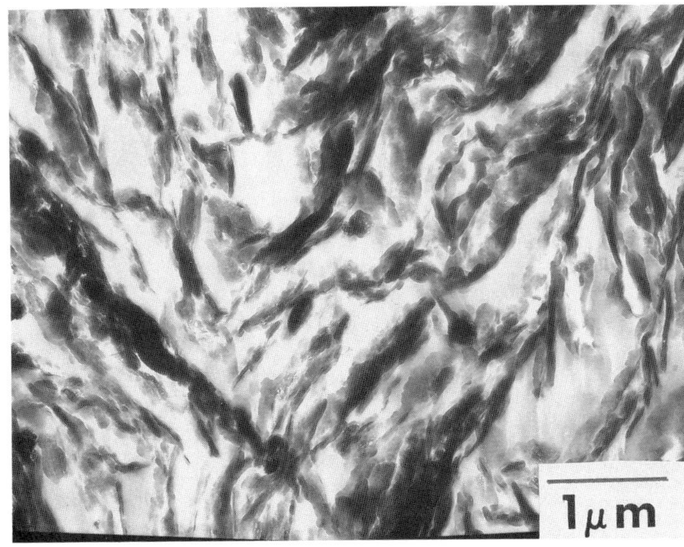

Figure 2.28. High magnification of remolded sediment characteristic of the "crust" zone depicting a localized area of preferred particle orientation (8 m subbottom). 16.5 mm = 1 μm.

nied by crystalline growth and changes in particle morphology, which translates directly to changes in microfabric. The complex microfabric of "red clays" from the Pacific Ocean Basin was revealed in scanning and transmission electron micrographs (Bryant and Bennett, 1988); authigenic mineralization was shown to have a profound influence on not only the clay microfabric features but also on the physical and mechanical properties of the sediment (Fig. 2.29). The overconsolidation characteristics of the soft clays were attributed to the strong bonding of argillaceous shale clasts and quartz grains by X-ray amorphous and well-developed authigenic smectite (Bryant and Bennett, 1988). Considerably more research is required to develop meaningful microfabric models and to discern microfabric signatures that are directly related to diagenetic mechanisms including authigenic mineralization, mineral transformations, solution and deposition of minerals and cementing agents.

Summary

The microfabric signatures resulting from the various processes and mechanisms are often recorded in the sediments and rock and can be revealed by electron microscopy observations. Based on numerous detailed studies of sediment and shale microstructure and the related environmental factors affecting the development of microfabric, three dominant processes have been identified and studied. The important processes that produce microfabric development of clay sediments and shales are physicochemical, bioorganic, and burial diagenesis (Figs. 2.1, 2.4).

Mechanisms associated with physicochemical processes include (1) electrochemical mechanisms from the interaction of particles in response to the electrolytic nature of the surrounding medium, the chemistry of associated organic materials, and the "fixed" electrical character of the specific minerals; (2) thermomechanical mechanisms that arise from thermally driven forces in the water column (Fig. 2.4A and from dynamic interaction of particles driven by wave, current, and gravity forces (Fig. 2.10); and (3) interface dynamics mechanisms in which particles impinging on the sediment–water interface are in dynamic motion and collide with other particles in a variety of configurations depending on the microrelief of the bottom. Electrochemical and thermomechanical mechanisms produce high intravoid (edge-to-face) flocs and face-to-face sheet like multiplate particle associations. Domains are common (Fig. 2.2).

Mechanisms associated with bioorganic processes are (1) biomechanical or bioturbation, which produces a random clay microfabric similar to the primary fabric of flocculated (edge-to-face) clay; however, often it is characterized by randomly oriented individual particles (as resolved by SEM) rather than thick domains and/or stepped face-to-face particles; (2) biophysical mechanisms produce biosediment aggregates, which are clusters of randomly oriented clay–silt particles formed when lithogenic matter adheres to sticky organic mucus or is bound together by polymer bridging; and (3) biochemical mechanisms, which cause microfabric changes in sediment due to chemical transformations mediated by organisms.

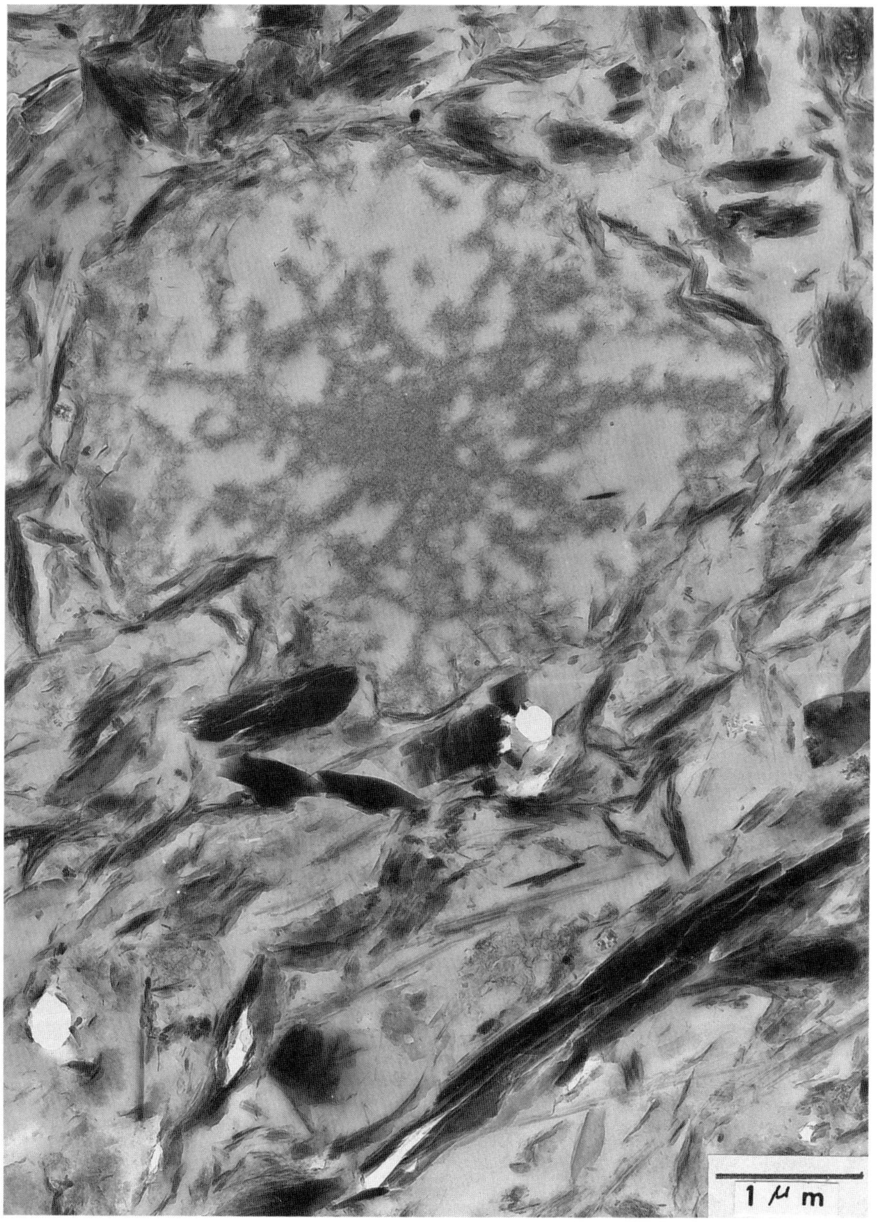

Figure 2.29. The microfabric of authigenic mineralization in a red clay of the northwest Pacific deep-sea basin (22.5 m subbottom) as observed by TEM. Note the lacy, finely divided crystals that form a very porous network. The mineral "smectite" appears to have developed by expansion of the crystal network pressing larger crystals of illite and smectite radially outward while reorienting the larger particles in a direction normal to the direction of stress. Note how the electron-dense particles appear to outline the circumference of the authigenic mineral. Sample recovered by DSDP Leg 86, Hole 576 (water depth 6217 m, 32° 21.38'N, 164° 16.52'E) in the basin east of Shatski Rise. Scale = 1 μm.

Burial diagenesis processes involve (1) postdepositional/mechanogravity mechanisms that arise from overburden stress and gravity-driven vertical and downslope forces that modify microfabric during consolidation and shearing; and (2) diagenesis–cementation mechanisms that alter the original character of the microfabric such as the formation of authigenic minerals.

All of the above-mentioned mechanisms produce microfabric signatures. Some signatures may be revealed in sediments and rocks by detailed electron microscopy observations. Recognition of a unique signature coupled with other observed properties (e.g., percent organics, porosity, and geochemistry) and larger scale features such as observed with X-ray radiography thus enables the investigator to determine or infer the dominant processes and mechanisms responsible for the observed microfabric. Collectively, these macro- and microproperties provide a qualitative and quantitative data base for understanding the geological record, and environments of deposition. These properties also provide crucial input properties for predictive modeling of the static and dynamic behavior of sediments and rocks for a wide variety of basic and applied research studies and for solving practical problems.

The processes and mechanisms that determine the microfabric signatures operate over various physical and time scales that are largely specific to a particular mechanism. For example, the

Table 2.1. Summary of processes and of the fabric signatures and physical and temporal scales associated with various mechanisms.

Processes	Mechanisms	Fabric signatures (predominant)*	Scales physical	Scales time	Remarks
Physicochemical	Electrochemical	E-F	Atomic and molecular to ~ 4 μm	μsec to msec	Two particles may rotate F-F
	Thermomechanical	F-F (some E-F)	Molecular to ≲0.2 mm	msec to min	Initial contacts E-F then rotation to F-F; common in selective environments
	Interface dynamics	F-F and E-F	μm to ~ ≲0.5 mm	sec	Some large compound particles may be possible at high concentrations
Bioorganic	Biomechanical	E-F	~0.5 mm to >2.0 mm	sec to min	Some F-F possible during bioturbation
	Biophysical	E-E and F-F	μm to mm	sec to min	Some very large clay organic complexes possible
	Biochemical	Nonunique (unknown)	μm to mm	hr to yr	New chemicals formed, some altered
Burial diagenesis	Mass gravity	F-F localized swirl	cm to km	≳yr	Can operate over large physical scales
	Diagenesis–cementation	Nonunique (unknown)	molecular	≳yr	New minerals formed, some altered, changes in morphology

*E-F, edge-to-face; E-E, edge-to-edge; F-F, face-to-face.

energy regimes driving microfabric development can be important over physical scales that can range in size from atomic dimensions to as large as kilometers and over time scales that are important in the microsecond range or up to periods of years or greater in the geological sense (Table 2.1). Examples of these extremes are found by comparing the electrochemical mechanism, where atomic and molecular dimensions are important over micro- and milliseconds during microfabric development, with the mass gravity mechanism, that occurs over dimensions of centimeters and kilometers for periods of thousands of years.

Examination of Table 2.1 reveals that although the various mechanisms produce predominant fabric signatures, similar particle-to-particle associations are common to different processes (compare physicochemical processes with bioorganic processes). Obviously more than a single criterion is usually required to interpret the processes of microfabric development. Such information may include percent organic carbon, interstitial water chemistry, porosity and/or permeability of the geological material, position in the stratigraphic column, and association with other particulates.

An important factor in understanding the microfabric of sediments and rocks is the aspect of observational scale. This factor can range from large-scale field observations, through the scale of X-ray radiography, to the scale of angstroms revealed by techniques of electron microscopy. Often many levels of observation are required to develop an adequate understanding of the particular problems(s) being addressed whether it be in disciplines of geology, environmental engineering, geotechnical engineering, petroleum exploration, or the recovery of hydrocarbons. Understanding of the processes and mechanisms responsible for microfabric development is an important factor in gaining a functional understanding of the relationship of microstructure to the developmental history and the bulk physical and mechanical properties of a deposit.

The microstructure is a crucial fundamental property that plays a significant role in determining the sediment and rock physical and mechanical properties and its behavior under static and dynamic loads. Many of the rock and sediment properties, important to a variety of scientific and technical disciplines, are intimately tied to the microfabric characteristics (Bennett et al., 1977, 1989). Delineation of the complex interrelationships among microfabric, processes and mechanisms is significant in terms of (1) understanding the role of specific energy regimes during the stages of microfabric development and sediment diagenesis and (2) development of predictive models (acoustical, mechanical, geological) that depend critically on the nature of the microstructure and basic physical and mechanical properties. Future investigations of the subtle difference in microfabric and its relationship to environmental processes and mechanisms, including detailed geochemical factors (organic and inorganic) and large and small scale features, could add important dimensions to scientific and technical knowledge and the ultimate applications of microstructure to practical problems.

Acknowledgments

The authors appreciate the assistance of Ms. F.L. Nastav in the preparation of the figures and in proofing of the manuscript. The first author was supported by NORDA under Program Element Number 61153N. The authors gratefully acknowledge the critical discussions with Drs. Huon Li and N. Currier regarding many of the ideas and concepts. The second author appreciates the support from donors of the Petroleum Research Fund, administered by the American Chemical Society, and from the National Science Foundation Grant EAR-8611608 for funding. Ms. P.J. Burkett provided the TEM photomicrograph of the clay sample from the Shatski Rise. Field support, samples, and laboratory support from Drs. Michael House, Philip Heckel, John Gray, Richard Young, William Kirchgasser, Egon Matijevic, James Syvitski, Mrs. Flona Cole, Roy Madison, and the Iowa Geological Survey are greatly appreciated.

References

Alldredge, A.L., 1986. Aggregate dynamics Biological processes which form, alter and destroy aggregates in the ocean. In: Alldredge, A.L., and C.O. Hartwig (eds.), Office of Naval Research, Aggregate Dynamics in the Sea Workshop Report, American Institute of Biological Sciences, 211 p.

Avnimelech, Y., and R.G. Menzel, 1984. Coflocculation of algae and clay to clarify turbid impoundments. Journal of Soil and Water Conservation, May–June, p. 200–203.

Avnimelech, Y., B.W. Troeger, and L.W. Reed, 1982. Mutual flocculation of algae and clay: evidence and implications. Science, v. 216, p. 63–65.

Bates, R.L., and J.A. Jackson, 1987. Glossary of Geology, 3rd ed. American Geological Institute, Alexandria, VA, 778 p.

Bea, R.G., 1971. How sea-floor slides affect offshore structures. The Oil and Gas Journal, p. 88–92.

Bea, R.G., and P. Arnold, 1973. Movements and forces developed by wave induced slides in soft clay. Fifth Annual Offshore Technology Conference, Paper 1899, Houston, TX, v. 2, p. 731–742.

Bennett, R.H., 1976. Clay Fabric and Geotechnical Properties of Selected Submarine Sediment Cores from the Mississippi Delta. Ph.D. dissertation, Texas A&M University, 269 p.

Bennett, R.H., and M.H. Hulbert, 1986. Clay Microstructure. International Human Resources Development Corporation Press, Prentice Hall, Boston, MA, 161 p.

Bennett, R.H., W.R. Bryant, and G.H. Keller, 1977. Clay fabric and geotechnical properties of selected submarine sediment cores from the Mississippi Delta. NOAA Professional Paper No. 9, U.S. Department of Commerce/NOAA/ERL, 86 p.

Bennett, R.H., W.R. Bryant, and G.H. Keller, 1981. Clay Fabric of selected submarine sediments: fundamental properties and models. Journal of Sedimentary Petrology, v. 51, no. I, p. 0217–0232.

Bennett, R.H., N.R. O'Brien, and M.H. Hulbert, 1988. Determinants of clay and shale microfabric signatures: processes and mechanisms. Abst. 4 in: Clay Microstructure: The Microstructure of Fine-Grained Terrigenous Marine Sediments-From Muds to Shales. Conference and Workshop, NORDA, Stennis Space Center, MS, Oct. 4–7.

Bennett, R.H., K.M. Fischer, D.L. Lavoie, W.R. Bryant, and R. Rezak, 1989. Porometry and fabric of marine clay and carbonate sediments: determinants of permeability. Journal of Marine Geology, v. 89, p. 127–152.

Black, A.P., F.B. Birkner, and J.J. Morgan, 1965. Destabilization of dilute clay suspensions with labeled polymers. Journal of American Water Works Association, v. 57, p. 1547–1560.

Boggs, S., Jr., 1987. Principles of Sedimentology and Stratigraphy. Merrill Publishing Company, Columbus, OH, p. 45–53.

Bohlke, B.M., and R.H. Bennett, 1980. Mississippi prodelta crusts: a clay fabric and geotechnical analysis. Marine Geotechnology, v. 4, p. 55–82.

Bowles, F.A., 1968. Microstructure of sediments: investigation with ultrathin section, Science. 159, 1236–1237.

Bowles, F.A., W.R. Bryant, and C. Wallin, 1969. Microstructure of unconsolidated and consolidated marine sediments. Journal of Sedimentary Petrology, v. 39, p. 1546–1551.

Bryant, W.R., and R.H. Bennett, 1988. Origin, physical, and mineralogical nature of red clays: the Pacific Ocean basin as a model. Geo-Marine Letters, Special Issue, v. 8, p. 189–249.

Burst, J.F., 1976. Argillaceous sediment dewatering. Annual review of Earth and Planetary Science, v. 4, p. 293–318.

Casagrande, A., 1932. The structure of clay and its importance in foundation engineering. Contributions to Soil Mechanics, Boston Society of Civil Engineers, (1940) p. 72, and Journal of Boston Society of Civil Engineers, v. 15 (April 1932).

Coleman, J.M., and L.E. Garrison, 1977. Geological aspects of marine slope stability northwestern Gulf of Mexico. Marine Geotechnology, v. 2, p. 9–44.

Coleman, J.M., and D.B. Prior, 1978. Submarine landslides in the Mississippi River Delta. Offshore Technology Conference, OTC paper 3170, p. 1067–1071.

Collins, K., and A. McGown, 1974. The form and function of microfabric features in a variety of natural soils. Geotechnique v. 24, p. 223–254.

Curtis, C.D., 1985. Clay mineral precipitation and transformation during burial diagenesis. Philosophical Transactions of the Royal Society of London A, v. 315, p. 91–105.

Doyle, E.H., B. McClelland, and G.H. Ferguson, 1971. Wire-line vane probe for deep penetration measurements of ocean sediment strength. Offshore Technology Conference Preprints, Paper No. 1327, v. 2, p. 21–32.

Dunlap, W.A., W.R. Bryant, R.H. Bennett, and A. Richards, 1978. Pore pressure measurements in underconsolidated sediments. Offshore Technology Conference Proceedings, Paper No. 3168, v. 2, p. 1049–58.

Dunoyer De Segonzac, G., 1970. The transformation of clay minerals during diagenesis and low-grade metamorphism: a review. Sedimentology, v. 15, p. 281–346.

Faas, R.W., and D.S. Crocket, 1983. Clay Fabric Development in a Deep Sea Core: Site 515, Deep Sea Drilling Project Leg 72. Initial Report DSDP, v. LXXII, (Leg 72) U.S. Government Printing Office, Washington, DC, p. 519–535.

Feynman, R.P., R.B. Leighton, and M. Sands, 1963. The Feynman Lectures on Physics, v. 1, Addison-Wesley, Reading, MA, p. 41-1–41-10.

Foster, R.H., and P.K. De, 1971. Optical and electron microscopic investigation of shear induced structures in lightly consolidated (soft) and heavily consolidated (hard) kaolinite. Clays and Clay Minerals, v. 19, p. 31–47.

Gibbs, R.J., 1981. Floc breakage by pumps. Journal of Sedimentary Petrology, v. 51, p. 670–672.

Gillott, J.E., 1969. Study of the fabric of fine-grained sediment with the scanning electron microscope. Journal of Sedimentary Petrology, v. 39, p. 90–105.

Gipson, M., Jr., 1965. Application of the electron microscope to the study of particle orientation and fissility in shale. Journal of Sedimentary Petrology, v. 35, p. 408–414.

Gipson, M., Jr., 1966. A study of the relations of depth, porosity and clay mineral orientation in Pennsylvanian shales. Journal of Sedimentary Petrology, v. 36, p. 888–903.

Goldschmidt, V.M., 1926. Undersokelser over lersedimenter. Nordisk-jord-brugsforskuing, no. 4-7, p. 343–445.

Grim, R.E., 1968. Clay mineralogy, 2nd ed., McGraw-Hill Co., New York, 596 p.

Hedberg, H.D., 1936. Gravitational compaction of clays and shales. American Journal of Science, v. 31, p. 241–287.

Henkel, D.J., 1970. The role of waves in causing submarine landslides. Geotechnique, v. 20, p. 75–80.

Honjo, S., 1982. Seasonality and interaction of biogenic and lithogenic particulate flux at the Panama Basin. Science, v. 218, p. 883–884.

Howard, J.J., 1987. Influence of shale fabric on illie/smectite diagenesis in the Ologocene Frio Formation, South Texas. In: Schultz, L.G., H. van Olphen, and F.A. Mumpton (eds.), Proceedings of the International Clay Conference, Denver 1985, The Clay Minerals Society, Bloomington, IN, p. 144–150.

Jabor, B.J., and E.W. Mountjoy, 1976. Late proterozoic microbiota of the Mietta Group, Southern British Columbia. Geology, v. 4, p. 111–119.

Johns, W.D., 1979. Clay minerals catalysis and petroleum generation. Annual Review of Earth and Planetary Science, v. 7, p. 183–198.

Kalliokoski, J., 1974., Pyrite framboid: animal, vegetable, mineral? Geology, v. 2, p. 26–27.

Kane, J.E., 1967. Organic aggregates in surface waters of the Ligurian Sea. Limnology and Oceanography, v. 12, p. 287–294.

Keller, W.D., 1976. Scan electron micrographs of kaolins collected from diverse environments of origin—I. Clays and Clay Minerals, v. 24, p. 107–113.

Keller, W.D., 1978. Classification of kaolins exemplified by their textures in scan electron micrographs. Clays and Clay Minerals, v. 26, p. 1–20.

Koh, P.T.L., 1984. Compartmental modelling of stirred tank for flocculation requiring a minimum critical shear rate. Chemical Engineering Science, v. 39, p. 1759–1764.

Krumbein, W.C., and F.J. Pettijohn, 1938. Manual of Sedimentary Petrography. Appleton-Century-Crofts, New York, 549 p.

Lambe, T.W., 1951. Soil Testing for Engineers. Wiley, New York, 165 p.

Lambe, T.W., 1953. The structure of inorganic soil. American Society of Civil Engineers Proceedings, v. 79, p. 1-49, Separate 315, October.

Lambe, T.W., 1953. The engineering behavior of compacted clay. Journal of Soil Mechanics and Foundations Division, Proceedings of the American Society of Civil Engineers, v. 84, p. 1–35.

Lamer, V.K., and T.W. Healy, 1963. Absorption-flocculation reactions of macromolecules at the solid-liquid interface. Reviews of Pure and Applied Chemistry, v. 13, p. 112–113.

Leja, J., 1982. Surface Chemistry of Froth Flotation. Plenum Press, New York, p. 1–59.

Leslie, J.F., R.L. Greasham, and M.H. Hulbert, 1984. A cyanobacterial system for consolidation of phosphate slimes. Geomicrobiology Journal, v. 3, p. 343–358.

Love, L.G., 1967. Early diagenetic iron sulphide in recent sediments of the Waoh England, Sedimentology, v. 9, p. 327–352.

Lynch, J.M., 1983. Soil Biotechnology. Blackwell Scientific Publications, Oxford, 191 p.

MacLean, P.A., and P. Smart, 1978. Effect of bacteria on soil structures. Geotechnique, March, p. 122–123.

McCave, I.N., 1984. Erosion, transport and deposition of fine-grained marine sediments. In: Stow, D.A.V., and D.J. Piper (eds.), Fine-grained Sediments: Deep Water Processes and Facies. Geological Society of London, Special Publication no. 15, p. 35–69.

McGregor, B.A., and R.H. Bennett, 1981. Sediment failure and sedimentary framework of the Wilmington Geotechnical Corridor, U.S. Atlantic Continental Margin. Sedimentary Geology, v. 30, p. 213–234.

Middleton, G.V., and A.H. Bouma, 1973. Turbidites and deep water sedimentation. Lecture Notes. Pacific Section, Society of Economic Paleontologists and Mineralogists, 157 p.

Mitchell, J.K., 1956. The fabric of natural clays and its relation to its engineering properties. Proceedings of Highway Research Board, v. 35, p. 693–713.

Montgomery Consulting Engineers, J.M., 1985. Water Treatment Principles and Design. Wiley, New York, p. 116–151.

Moon, C.F., 1972. The microstructure of clay sediments, Earth-Science Reviews, v. 8, p. 303–321.

Morrison, R.T., and R.N. Boyd, 1983. Organic Chemistry, 4th ed., Allyn and Bacon, Boston, MA, p. 1–3.

Mullins, M.M., 1980. Interactions between marine zooplankton and suspended particles. In: Kavanaugh, M.C., and J.O. Leckie (eds.), Particulates in Water, Characterization, Fate, Effects, and Removal. Washington, DC, American Chemical Society, Advances in Chemistry Series, v. 189, p. 233–241.

O'Brien, N.R., 1968. Electron microscope study of black shale fabric. Naturwissenschaften, v. 55, p. 490–491.

O'Brien, N.R., 1970a. Fabric of flocculated clay sediment. Geological Society of America Abstracts with Programs, v. 2, p. 637.

O'Brien, N.R., 1970b. The fabric of shale: an electron microscope study. Sedimentology, v. 15, p. 229–246.

O'Brien, N.R., 1971. Fabric of kaolinite and illite floccules. Clays and Clay Minerals, v. 19, p. 353–359.

O'Brien, N.R., 1987. The effects of bioturbation on the fabric of shale. Journal of Sedimentary Petrology, v. 57, p. 449–455.

O'Brien, N.R., 1989. The origin of lamination in middle and upper Devonian black shales, New York State. Northeastern Geology, v. 11, p. 159–165.

O'Brien, N.R., and K. Hisatomi, 1978. Sedimentological study of a turbidite cycle, Kii Peninsula, Japan. Memoirs, The Faculty of Science, Koyoto University, Series of Geology and Mineralogy, v. XLV, no. 2, Kyoto, p. 177–186.

Odom, I.E., 1967. Clay fabric and its relation to structural properties in mid-continent Pennsylvanian sediments. Journal of Sedimentary Petrology, v. 37, p. 610–623.

Pates, R.L., and J.A. Jackson, 1987. Glossary of Geology, 3rd ed., American Geological Institute, Alexandria, VA, 788 p.

Pearl, H., 1973. Detritus in Lake Tahoe: structural modification by attached microflora. Science, v. 180, p. 496–498.

Pierce, J.W., and F.R. Siegel, 1979. Particulate material suspended in estuarine and oceanic waters. Scanning Electron Microscopy, SEM, Inc., v. I, p. 555–562.

Prior, D.B., E.H. Doyle, and M.J. Kaluza, 1989. Evidence for sediment eruption on deep sea floor, Gulf of Mexico. Science, v. 243, p. 517–519.

Pryor, W.A., 1975. Biogenic sedimentation and alteration of argillaceous sediments in shallow marine environments. Geological Society of America Bulletin, v. 86, p. 1244–1254.

Pusch, R., 1966. Quick clay microstructure. Engineering Geology, v. 1, p. 433–443.

Pusch, R., 1970. Microstructural changes in soft quick clay at failure. Canadian Geotechnical Journal, v. 7, p. 1–7.

Pusch, R., 1973. The influence of organic matter on the geotechnical properties of clays. National Swedish Building Research Summaries. Stockholm, Sweden, 64 p.

Quigley, R.M., and C.D. Thompson, 1966. The fabric of anisotropically consolidated sensitive marine clay. Canadian Geotechnical Journal, v. 3, p. 61–73.

Rashid, M.A., 1985. Geochemistry of Marine Humic Compounds. Springer-Verlag, New York, 300 p.

Rhodes, D.C., and L.F. Boyer, 1982. The effects of marine benthos on physical properties of sediments. A successional perspective. In: McCall, P.L., and M.J.S. Tevesy (eds.), Animal-Sediment Relations, the Biogenic Alteration of Sediments. Plenum Press, New York and London, p. 3–52.

Rickard, D.T., 1970. The origin of framboids, Lithos, v. 3, p. 269–293.

Rieke, H.H., III, and G.V. Chilingarian, 1974. Compaction of Argillaceous Sediments. Elseiver Scientific Publishing Co., New York, 424 p.

Riley, G.A., 1963. Organic aggregates in seawater and the dynamics of their formation and utilization. Limnology and Oceanography, v. 8, p. 372–381.

Rosenqvist, I. Th., 1959. Physico-chemical properties of soils: soil-water systems. Journal of Soil Mechanics and Foundations Division, Proceedings American Society of Civil Engineers, v. 85, p. 31–53.

Ruehrwein, R.A., and D.W. Ward, 1952. Mechanism of clay aggregation by polyelectrolytes. Soil Science, v. 73, p. 485–492.

Shanks, A.L., and J.D. Trent, 1980. Marine snow: sinking rates and potential role in vertical flux. Deep-Sea Research, v. 27A, p. 137–143.

Silver, M.W., A.L. Shanks, and J.D. Trent, 1978. Marine snow: microplankton habitat and source of small-scale patchiness in pelagic populations. Science, v. 201, p. 371–373.

Sorby, H.C., 1908. On the application of quantitative methods to the study of the structure and history of rocks. Quarterly Journal Geological Society of London, v. 64, p. 171–233.

Stumm, W., and J.J. Morgan, 1981. Aquatic Chemistry—An Introduction Emphasizing Chemical Equilibria in Natural Waters. Wiley, New York, 280 p.

Swift, D.L., and S.K. Friedlander, 1964. The coagulation of hydrosols by Brownian motion and laminar shear flow. Journal of Colloid Science, v. 19, p. 621–647.

Syvitski, J.P.M., and A.G. Lewis, 1980. Sediment ingestion by *Tigriopus californius* and other zooplankton: mineral transformation and sedimentological considerations. Journal of Sedimentary Petrology, v. 50, p. 869–880.

Syvitski, J.P.M., and J.W. Murray, 1981. Particle interaction in fjord suspended sediment. Marine Geology, v. 39, p. 215–242.

Terzaghi, K., 1925. Erdbaumechanik auf Bodenphysikalischer Grundlage. Franz Deuticke, Leipzig und Wien, 399 p.

Trent, J.D., A.L. Shanks, and M.W. Silver, 1978. In-situ and laboratory measurements on macroscopic aggregates in Monterey Bay, CA. Limnology and Oceanography, v. 23, p. 626–635.

van Olphen, H., 1963. An introduction to Clay Colloid Chemistry. Wiley-Interscience, New York, 301 p.

van Olphen, H., 1977. An introduction to Clay Colloid Chemistry, 2nd ed. Interscience Publishers, New York, 318 p.

Von Engelhardt, W., and K.H. Gaida, 1963. Concentration changes of pore solution during the compaction of clay sediments. Journal of Sedimentary Petrology, v. 33, p. 919–930.

Weaver, C.E., 1984. Shale-slate metamorphism in southern Appalachians. Elsevier, New York, 239 p.

Weller, J.M., 1959. Compaction of sediments. Bulletin of American Association of Petroleum Geology, v. 43, p. 273–310.

White, W.A., 1961. Colloid phenomena in the sedimentation of argillaceous rock. Journal of Sedimentary Petrology, v. 31, p. 560–570.

Yamamoto, T., 1982. Experiments of wave-driven soil transport in clay beds. Geo-Marine Letters, p. 205–208.

Yong, R.N., 1972. Soil technology and stabilization. In: Moh, F.C. (ed.), Proceedings, Fourth Asian Regional Conference on Soil Mechanics and Foundation Engineering, v. 2, p. 111–124.

Yong, R.N., and D.E. Sheeran, 1973. Fabric unit interaction and soil behavior. In: International Symposium on Soil Structure, Gothenburg, Sweden, p. 176–183.

CHAPTER 3

Millimeter-Scale Sedimentary Structure of Fine-Grained Sediments: Examples from Continental Margin Environments

S.A. Kuehl, T.M. Hariu, M.W. Sanford, C.A. Nittrouer, and D.J. DeMaster

Introduction

Sedimentary structures in fine-grained sediments have been largely ignored as indicators of depositional environment, perhaps because the only common primary macroscopic structure observed in muds is horizontal lamination (Potter et al., 1980). However, microscopic examination of muds reveals a variety of structures (e.g., micro-cross-laminations, graded laminae, plasmic fabric) that could be diagnostic of particular environmental conditions (e.g., Stow, 1979; Broadhead and Potter, 1980; Kuehl et al., 1988). This study examines the millimeter-scale sedimentary structure of sediment cores collected from three continental margin environments: the Amazon and Bengal continental shelves, and the continental slope near Wilmington Canyon, U.S. East Coast. The occurrence of diagnostic sedimentary structures from these contrasting fine-grained sedimentary environments is examined to provide criteria for interpretation of mudrocks.

Amazon Shelf

Setting

The Amazon River discharges about 1.2×10^9 tons/yr of sediment, predominantly silt and clay, to the Atlantic Ocean (Gibbs, 1967; Meade et al., 1985). The shelf seaward of the Amazon is influenced by strong tidal currents (exceeding 100 cm/sec within a few meters of the seabed; Nittrouer et al., 1986), and by moderate surface waves (~ 1.5 m average wave height; Meserve, 1974). A sediment budget constructed for the Amazon shelf indicates that about half of the sediment discharge of the river accumulates on the adjacent Brazilian shelf (Kuehl et al., 1986a). The remainder bypasses the Brazilian shelf to the northwest, and Amazon River sediments are found as far northwestward as the Orinoco River (Reyne, 1961). Shallow seismic reflection profiles indicate that Amazon River sediments accumulate on the Brazilian shelf as a subaqueous delta, with topset, foreset, and bottomset stratigraphy (Figueiredo et al., 1972; Nittrouer et al., 1986). Sediment accumulation rates on the Amazon shelf generally increase seaward from negligible values nearshore to a maximum of about 10 cm/yr near the seaward edge of the topset region and in the foreset region. Accumulation rates decrease further offshore in the bottomset region and are negligible on the outer shelf where relict sands are exposed (Kuehl et al., 1986a).

Sedimentary Structures

Macrostructures

Based on radiographic examination of macroscopic sedimentary structures from the Amazon subaqueous delta, five sedimentary facies are recognized, including (1) physically stratified sand, (2) interbedded mud and sand, (3) proximal-shelf sandy silt, (4) faintly laminated mud, and (5) mottled mud (Fig. 3.1) (Kuehl et al., 1986b). In addition, organic-rich laminae are present in facies near the outer topset and in the foreset regions. In general, physical sedimentary structures are observed where accumulation rates are high (>4 cm/yr) or where physical reworking is intense (<30 m water depth). Biological structures are observed offshore where accumulation rates are reduced and where physical reworking is negligible.

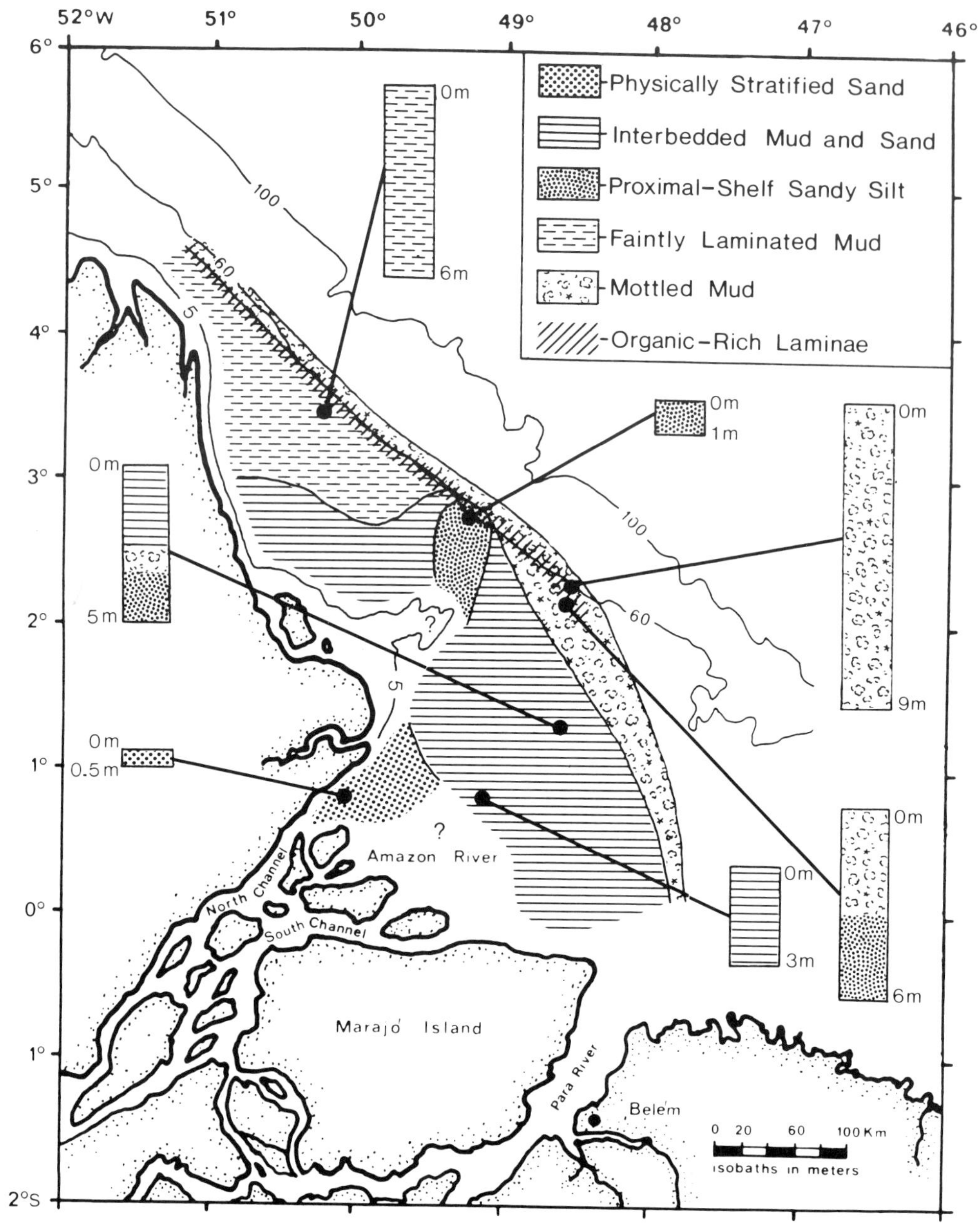

Figure 3.1. Distribution of macroscopic sedimentary structures in the Amazon subaqueous delta based on radiographic examination of sediment cores. (From Kuehl et al., 1986b.)

Microfabric

Thin sections were prepared from undisturbed core samples following the techniques outlined in Kuehl et al. (1988). In the facies of interbedded mud and sand, laminae consist of thin (~1 mm) layers of rounded to subangular quartz silt particles within a matrix of clay (Fig. 3.2A). Basal contacts of the laminae generally are sharp, indicating that lamina formation is usually preceded by increased bottom shear stress and consequent erosion of the seabed. This is in contrast to the classic environmental interpretation of laminae in muds, which suggests that laminae form from deposition from suspension under quiescent conditions (Pettijohn, 1957). Basal contacts sometimes are irregular, and micro-rip-up clasts of mud sometimes occur (Fig. 3.2B), indicating currents of sufficient magnitude to erode semiconsolidated muds. Laminae that consist of equidimensional heavy minerals, elongate woody fragments, and ellipsoidal fecal pellets also are observed (Fig. 3.2C–E). This observation suggests that congregational sorting (bedload sorting by particle weight, density, and shape) (Moss, 1963; Kuenen, 1966; Piper,

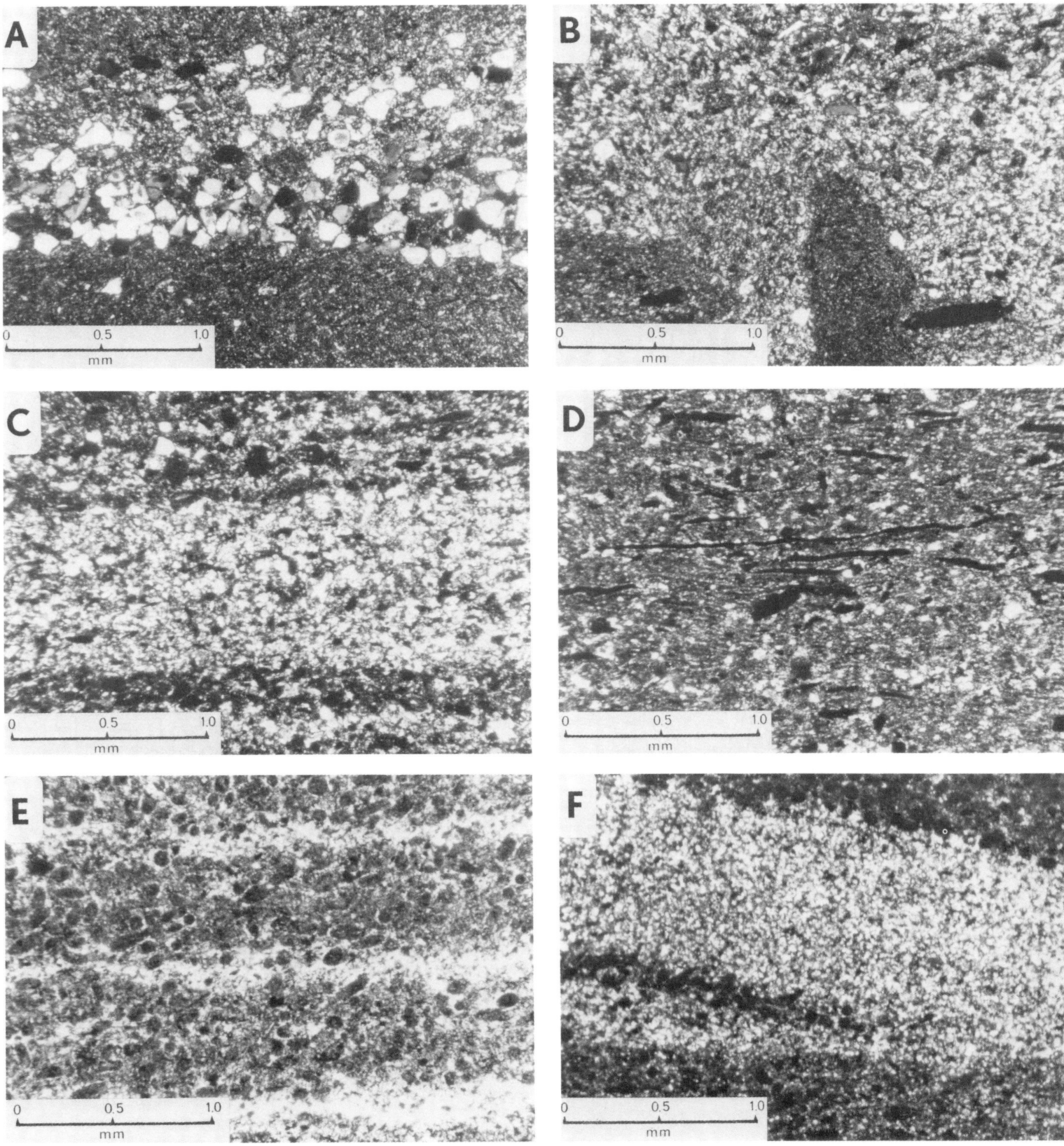

Figure 3.2. Photomicrographs from the facies of interbedded mud and sand. (A) Quartz lamina within a matrix of platy minerals. The sharp basal contact probably indicates scouring of the seabed prior to formation of the lamina. (B) Micro-rip-up clast of mud within a quartz lamina. (C) Laminae composed of heavy minerals (opaque grains) within a matrix of quartz. (D) Lamina containing a high concentration of woody fragments (dark elongate grains). (E) Interlamination of quartz and fecal pellets. (F) Lamina containing micro-cross-lamination. The foreset laminae are composed of alternating layers of quartz and fecal pellets. (From Kuehl et al., 1988.)

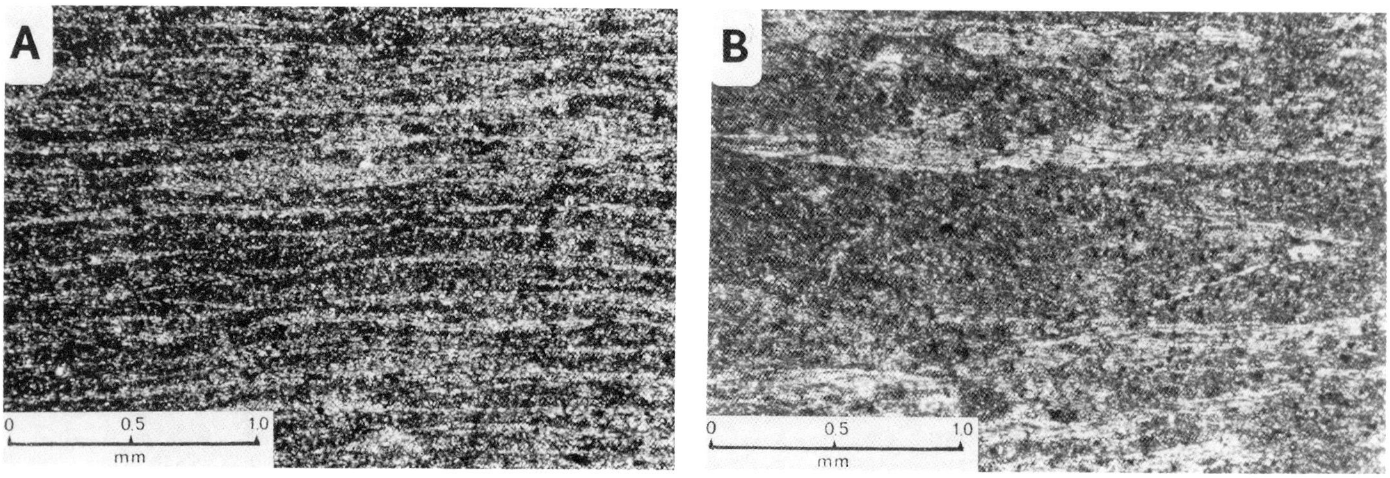

Figure 3.3. Photomicrographs from the facies on faintly laminated mud (crossed nicols oriented 45° with respect to bedding) displaying plasmic fabric. (A) Photomicrograph displaying layered unistrial fabric. Layers of relatively coarse platy minerals oriented parallel to bedding (light layers) are interlayered with finer, unoriented platy minerals (dark layers). (B) Photomicrograph displaying thickly layered unistrial fabric. The individual minerals within oriented (light) layers are about 0.1 mm long.

1972) occurs on the shelf. A lamina containing micro-cross-laminations was found (Fig. 3.2F), indicating currents sufficient to produce bedform migration.

Plasmic Fabric

A separate category of sedimentary structure is identified in Amazon muds based on extinction phenomena of constituent platy minerals. Alignment of platy clay minerals and mica produces extinction patterns when observed using the polarizing microscope that have been referred to as plasmic fabric by soil scientists. Two patterns are observed from Amazon muds and are members of a specific category of plasmic fabric termed unistrial fabric by Brewer (1964). The first displays a layered extinction pattern that consists of thin (~0.01 mm), oriented layers of clays that have lateral continuity of about 1 mm (Fig. 3.3A). These oriented layers are vertically interspersed at ~0.1-mm intervals with layers of finer grained, apparently randomly arranged platy minerals. Unusually thick oriented layers (~0.5 mm) with large platelets (~0.1 mm long) are observed in a thin section from the facies of faintly laminated mud (Fig. 3.3B). Unlike the silt laminae described above (see Microfabric), oriented layers of platy minerals are not observed in radiographs. The second extinction pattern observed in Amazon muds results from "uniform" orientation of platy minerals, which are oriented parallel to bedding and which display no apparent vertical change in orientation or grain size on scales less than 1 mm.

Although the mechanism for formation of layered plasmic fabric is not known with certainty, any proposed mechanism must consider the fine-scale sorting of platy minerals that results in the interspersed coarse- and fine-grained layers. The concept of shear sorting in the boundary layer (Stow and Bowen, 1980) could explain this phenomena. This concept suggests that when clay flocs and quartz silt particles carried by a turbidity current begin to settle from suspension, increased shear in the boundary layer will initially cause clay flocs to disaggregate (Krone, 1962). Because of their reduced settling velocity, individual clay particles will remain in suspension; however, silt particles will settle through the boundary layer, forming a silt lamina. As sediment is supplied to the boundary layer from above, a critical concentration of clay particles eventually is reached such that coagulation occurs, and the clay is rapidly deposited. Continuation of this cycle leads to the formation of a thinly laminated mud deposit. Although the concept of shear sorting was proposed to explain silt laminae in turbidites, the general mechanism also may be responsible for the observed interlayering of platy minerals. Disaggregation of flocs in the boundary layer may allow sorting of constituent platy minerals during deposition. The preferred orientation of platy minerals appears in some cases to be related to the size of the individual platelets. In the layered fabric, particles within relatively coarse layers are oriented parallel to bedding, whereas the particles within fine layers appear unoriented.

Bengal Shelf

Setting

The Ganges, Brahmaputra, and Meghna rivers join together near Dhaka, Bangladesh, and together discharge an estimated 1.7 billion tons/yr to the Bay of Bengal, ranking this system as the world's largest supplier of sediment to the oceans (Fig. 3.4) (Milliman and Meade, 1983). Similar to the Amazon shelf, the

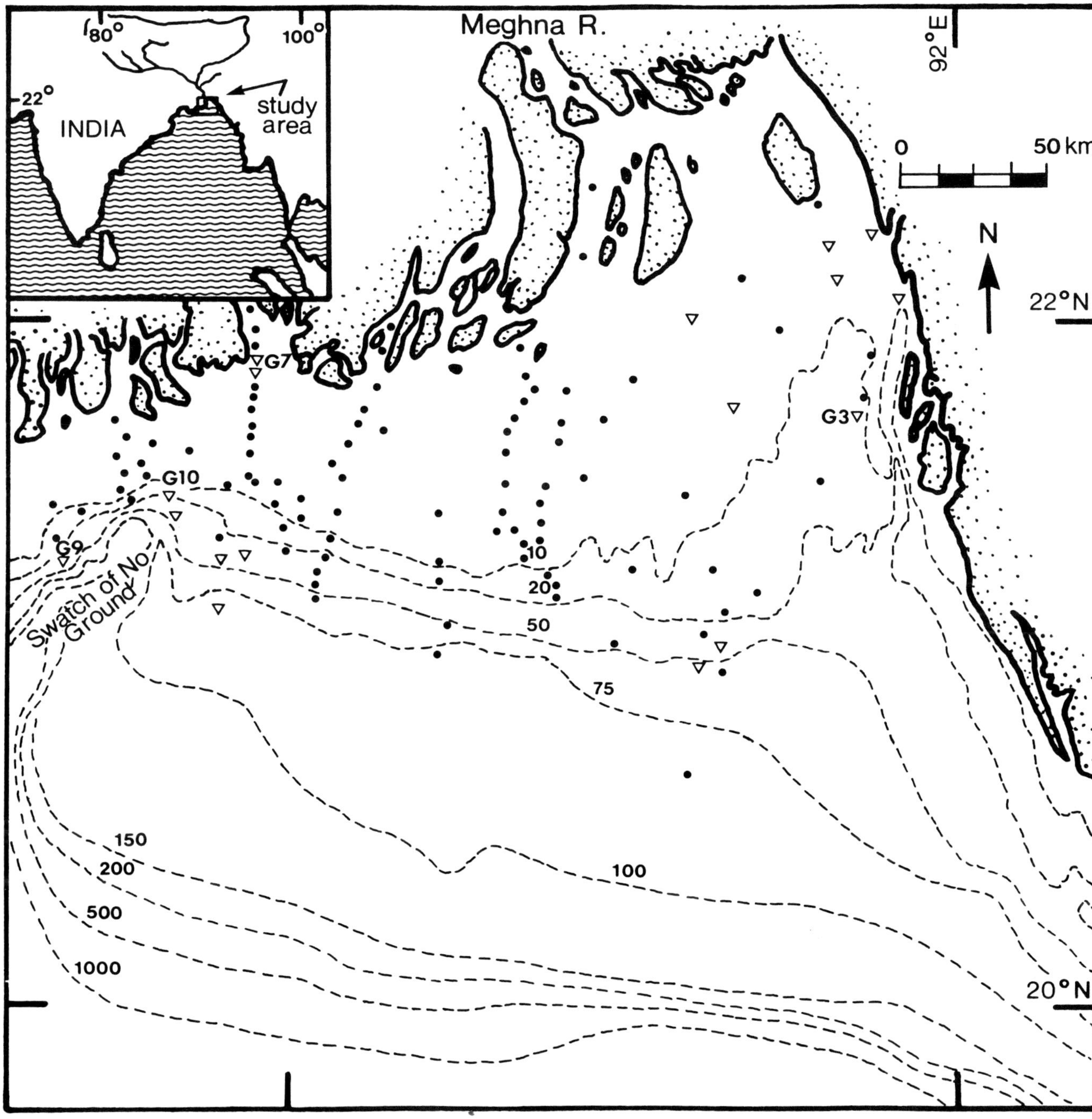

Figure 3.4. Bathymetric chart of the Bengal shelf showing the location of cores (open triangles) and grab samples (solid circles).

Bengal shelf is characterized by a dynamic oceanographic environment affected by strong tides (tidal range as much as 6 m with associated tidal currents as high as 3.8 m/sec) and moderate wave activity (Eysink, 1983). Sediments presently discharged by the Ganges–Brahmaputra river system accumulate on the shelf in water depths less than about 80 m, forming a subaqueous delta similar to that observed off the Amazon river (Kuehl et al., 1989). The inner portion of the shelf (<20 m) is mantled primarily with fine sands and coarse silts, with the exception of thin (<1 m) mud deposits, which are present on the western portion of the shelf and east of the main river channel (Fig. 3.5). The fine sands and coarse silts on the inner shelf grade seaward and westward into a mid-shelf mud deposit that is intercepted by a major submarine canyon called the "Swatch of No Ground" on

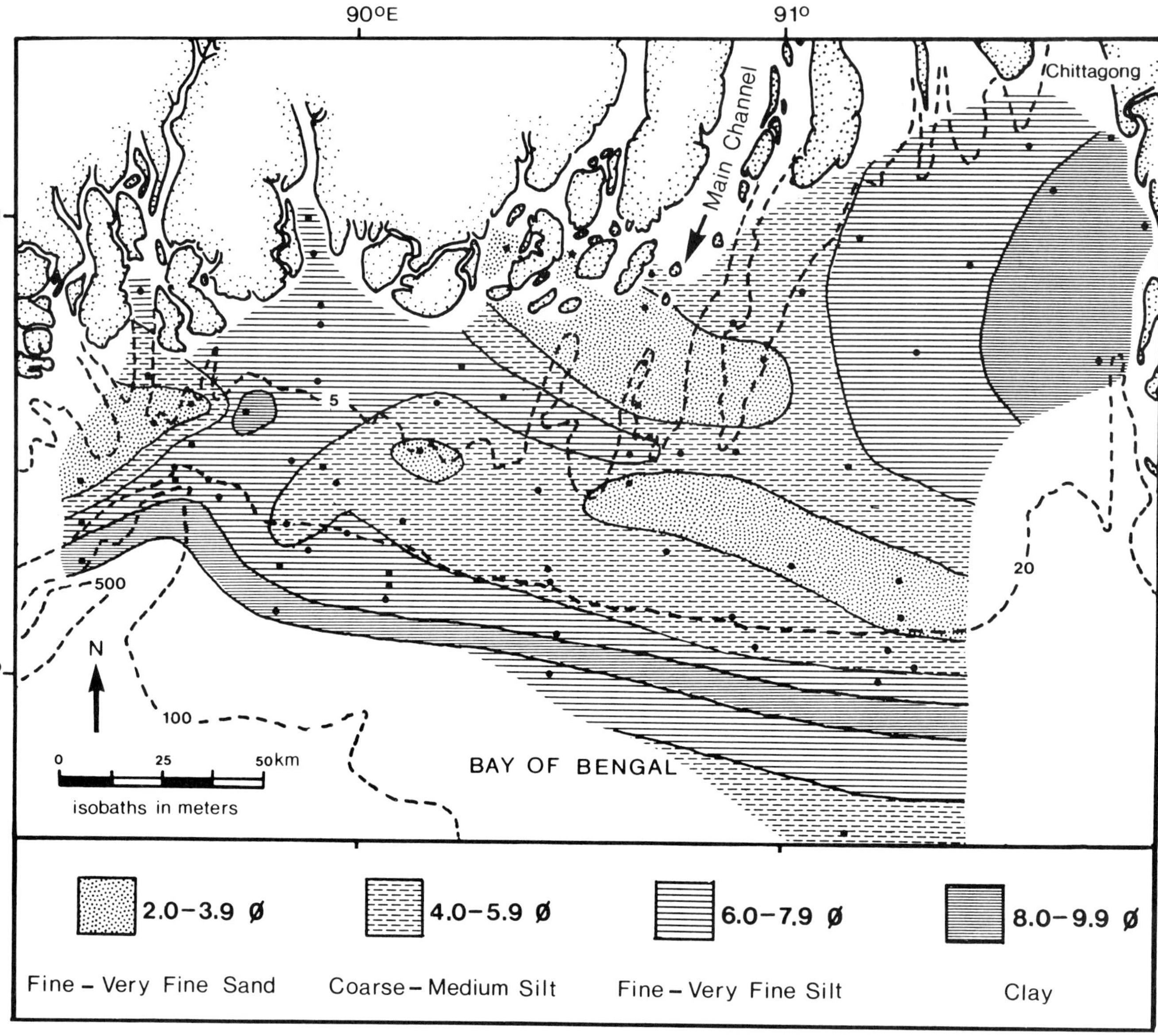

Figure 3.5. Distribution of mean grain size on the Bengal shelf based on analyses of grab samples and core tops.

the western part of the Bengal shelf. Accumulation rates on the Bengal shelf increase from low rates nearshore to >8.6 cm/yr in the foreset region near the "Swatch of No Ground" (Kuehl et al., 1989). Accumulation rates on the shelf decrease further seaward and apparently are negligible on the outer shelf where a carbonate-rich shell hash is present.

Sedimentary Structures

Cores collected from the nearshore mud deposits primarily reveal physical sedimentary structures, with little evidence of bioturbation. Radiographs from the nearshore mud deposits typically contain thin laminations of fine sand and/or coarse silt within a muddy matrix (Fig. 3.6A). Although physical structures are predominant, some evidenced of bioturbation is observed (Fig. 3.6B). Microscopic examination of the cores reveals that laminae are composed of a variety of materials (quartz, wood fragments, heavy minerals) and have sharp basal contacts. The observed laminae are similar to those previously described from the Amazon shelf and shown in Figure 3.2.

Radiographs from the mid-shelf mud deposit also are dominated by physical sedimentary structures. Cores collected near the "Swatch of No Ground" reveal faintly laminated muds (Fig. 3.7A) and laminated muds and silts (Fig. 3.7B). Many of the laminated muds and silts near the canyon display unusual struc-

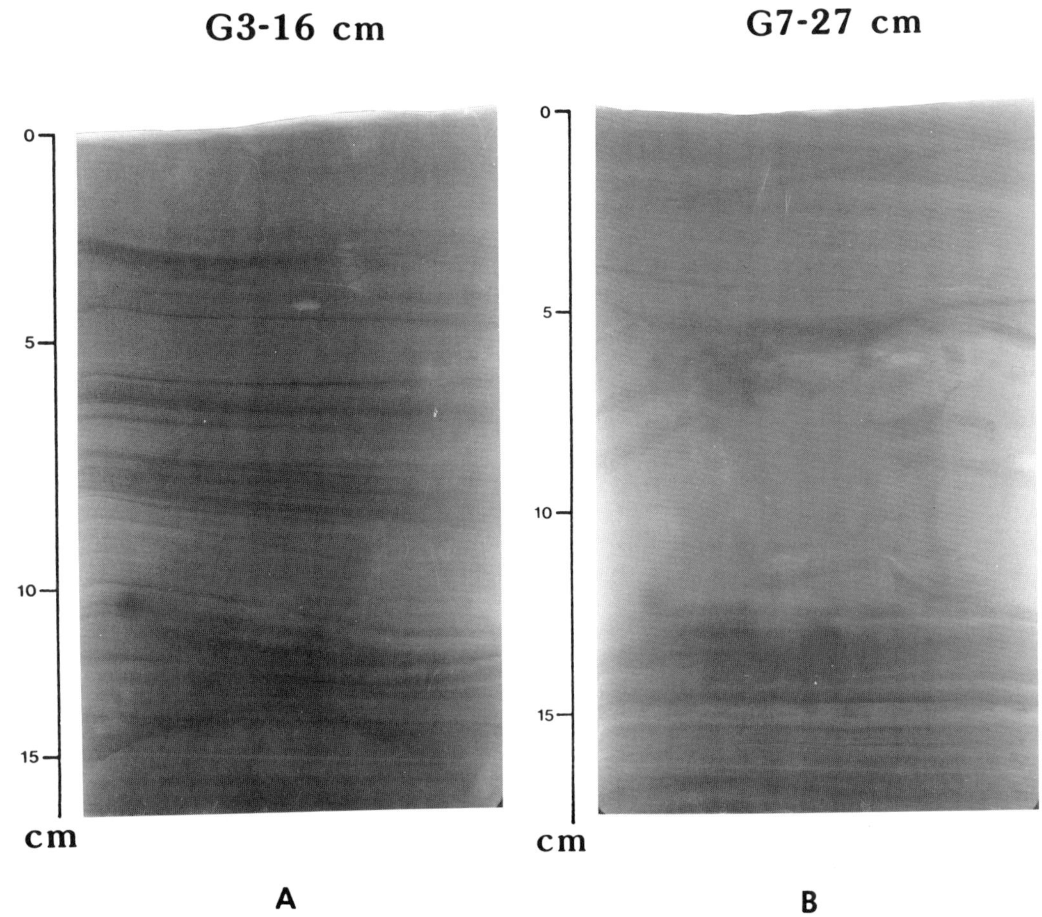

Figure 3.6. Radiographs (positives) from the nearshore mud deposits. (A) Core revealing fine interlaminations of fine sand/coarse silt and mud. (B) Core revealing fine interlaminations of fine sand/coarse silt and mud, and some evidence of bioturbation.

tures that appear similar to mottling caused by bioturbation. However, closer examination of this structure in thin sections reveals penecontemporaneous deformation features (Fig. 3.8) characteristic of rapid deposition of alternating layers of sand and mud (Kuenen and Prentice, 1957). This observation is reasonable because these cores were collected from an area of high accumulation rate on the shelf.

Wilmington Canyon Area

Setting

The continental slope off southern New Jersey is extensively dissected by two distinct types of canyons: meandering canyons that incise the shelf (e.g., Wilmington Canyon, Baltimore Canyon) and straight canyons that begin on the upper slope (e.g., South Wilmington Canyon, North and South Heyes Canyons) (Fig. 3.9). Sea MARC I images from the continental slope in the Wilmington Canyon area have revealed an extensive system of gulleys feeding into the canyons and producing a dendritic drainage pattern (McGregor et al., 1982). The location of gullies on canyon walls with steep seafloor gradients suggests that they result from mass wasting. Previous ALVIN dives and television surveys in Wilmington Canyon have provided evidence for erosion of the canyon walls (Stanley et al., 1972; Stubblefield et al., 1982). Rubble recovered from base of wall outcrops suggests erosion and transport down the walls (Stanley et al., 1986).

Sedimentary Structures

Radiographic examination of 73 cores collected in the study area using DSRV ALVIN and the RV Atlantis II reveals little evidence of primary sedimentary structures. Most of the cores are mottled to nearly homogeneous, with occasional burrows and tubes (Fig. 3.10A, B). A few of the radiographs, however, display some evidence of physical stratification. A core collected from the base of a gulley in North Heyes Canyon shows what appears to be horizontal bedding (Fig. 3.11A). Inspection of the preserved core slab reveals that the bedding results from a patchy distribution of a pale brown mud within typical hemipelagic, light olive-grey mud. Thin sections from this core show sharp contacts between the hemipelagic and pale brown

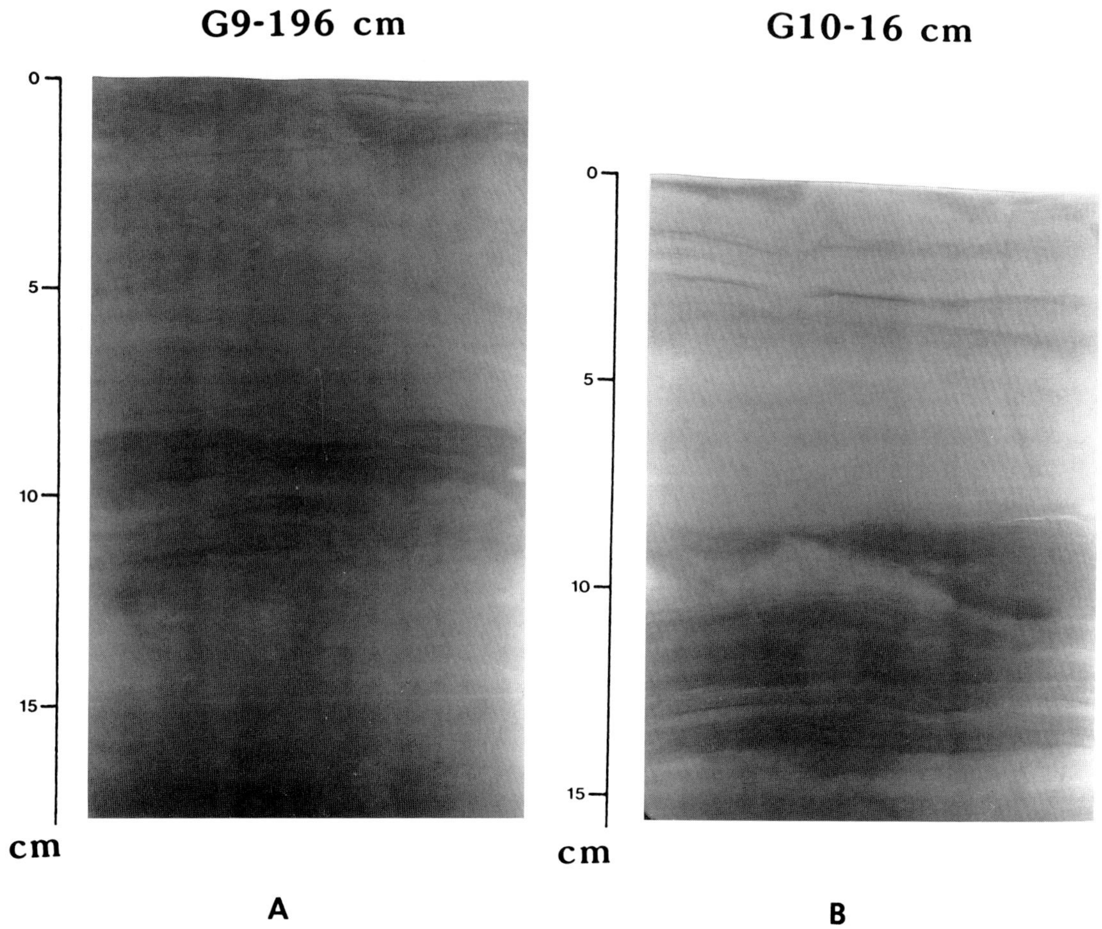

Figure 3.7. Radiographs (positives) from near the "Swatch of No Ground." (A) core revealing faint silt laminations in nearly homogeneous mud. (B) Core revealing fine interlaminations of fine sand/coarse silt and mud, and some apparent bioturbation.

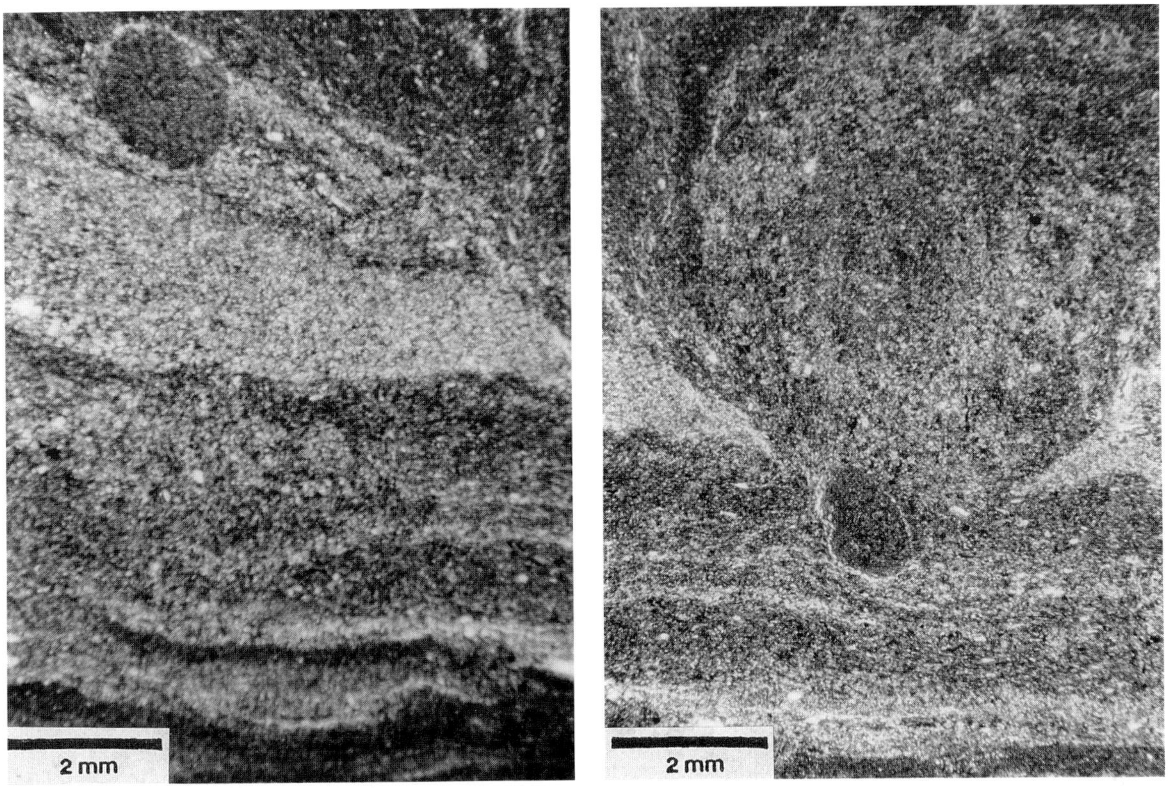

Figure 3.8. Photomicrographs of cores collected near the "Swatch of No Ground" revealing small-scale ball and pillow features resulting from rapid deposition of alternating layers of silt and clay.

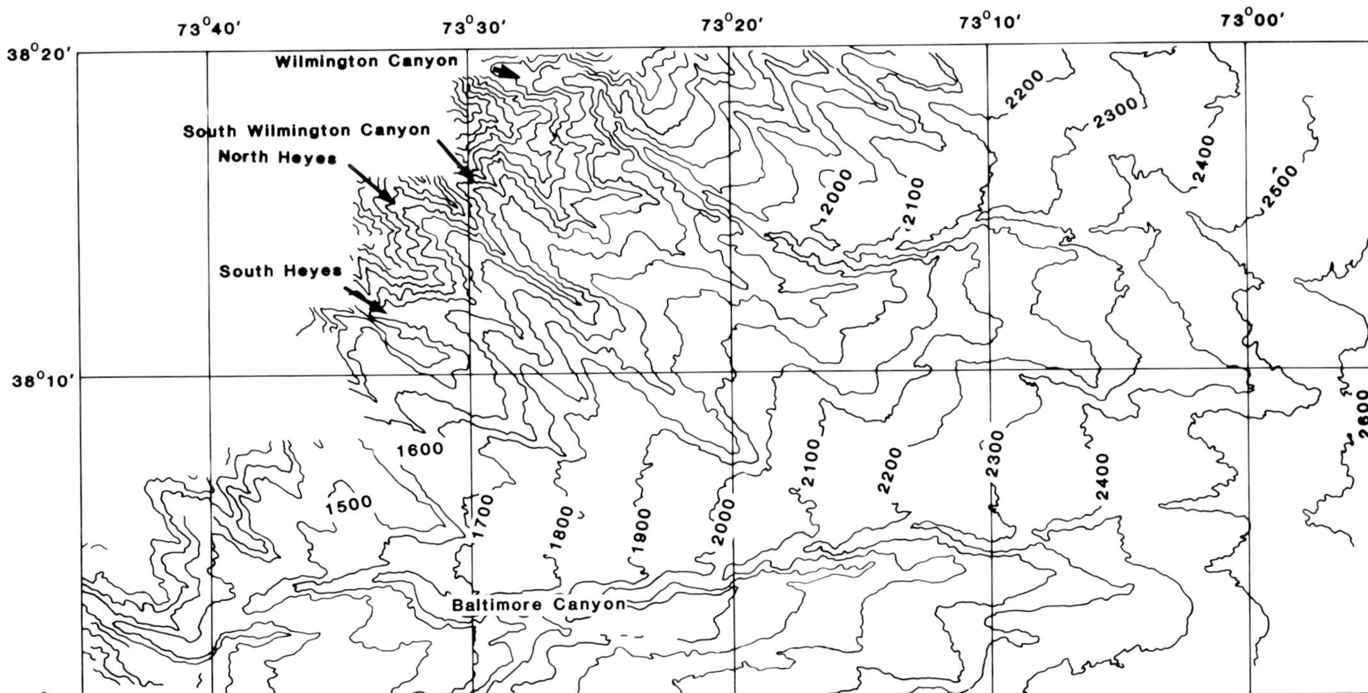

Figure 3.9. Modified from a Seabeam map of the continental slope seaward of southern New Jersey showing meandering (e.g., Wilmington and Baltimore) and straight (South Wilmington, North and South Heyes) canyons in the study area.

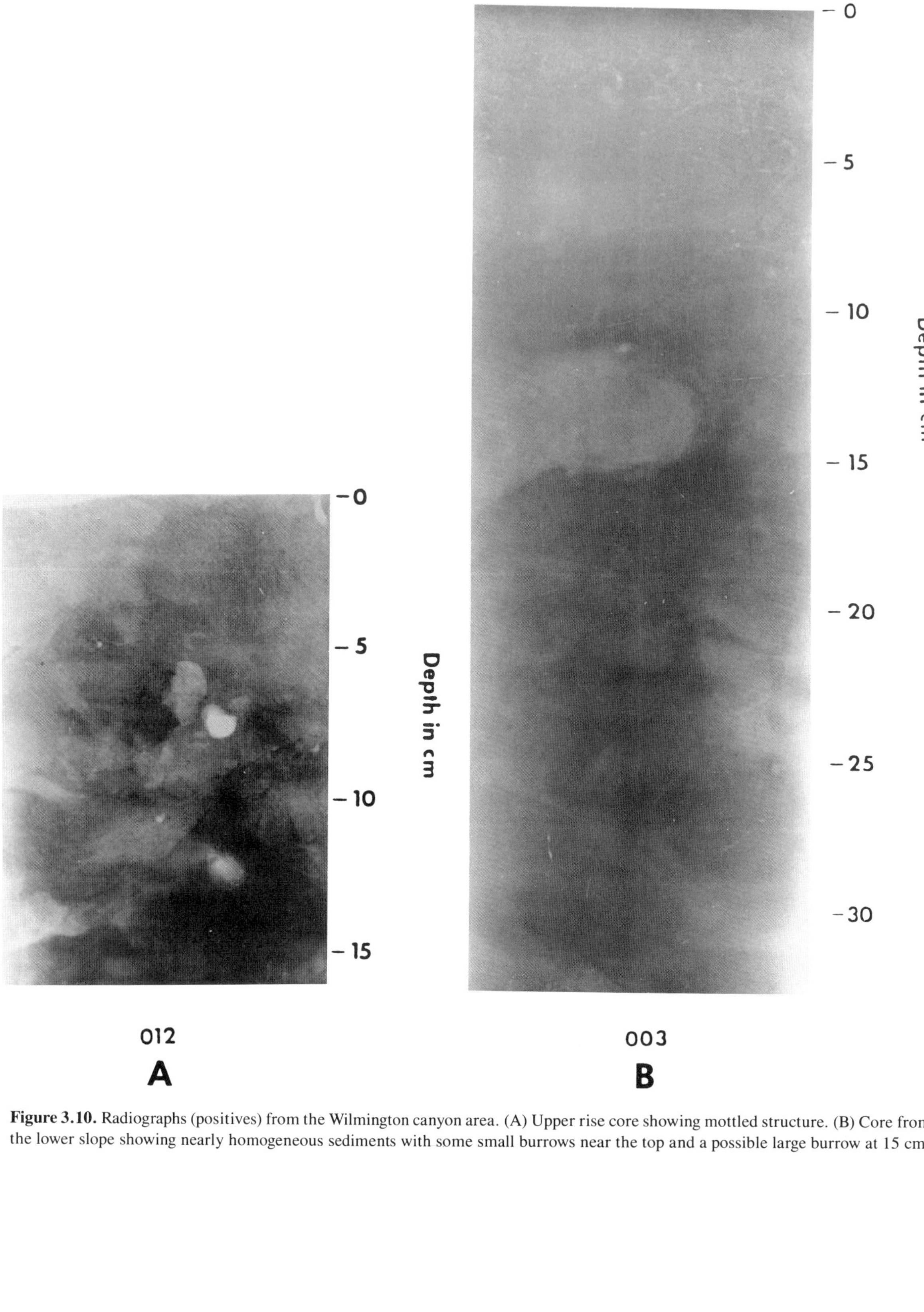

Figure 3.10. Radiographs (positives) from the Wilmington canyon area. (A) Upper rise core showing mottled structure. (B) Core from the lower slope showing nearly homogeneous sediments with some small burrows near the top and a possible large burrow at 15 cm.

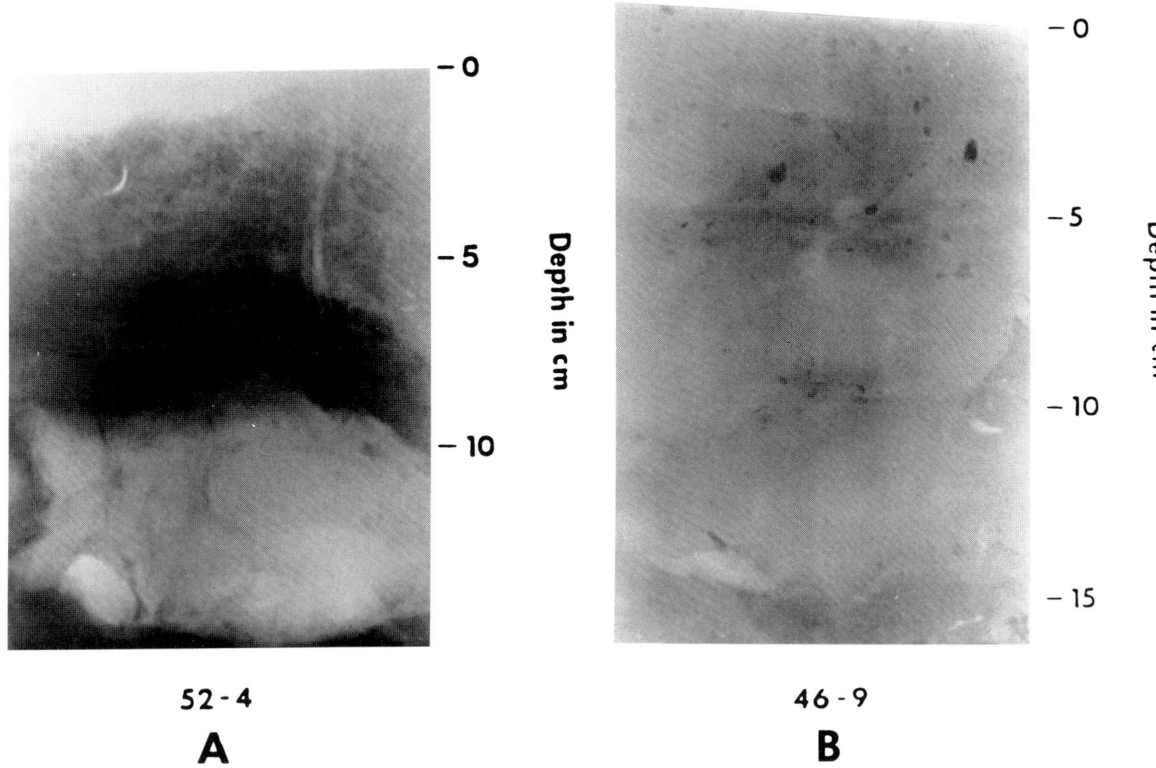

Figure 3.11. Radiographs (positives) from the Wilmington canyon area. (A) Bedding in a gulley core from North Heyes canyon. (B) Core taken near a meander in Wilmington canyon revealing indurated mud clasts (dark fragments) throughout nearly homogeneous sediment.

muds, and rust colored clay balls included within the matrix of both mud types (Fig. 3.12A, B). Granular inclusions of similar color and texture were also found in the Wilmington canyon. Some of the larger inclusions can be seen in several of the radiographs from Wilmington Canyon as small, dark fragments (Fig. 3.11B). Geochronological studies and observations from DSRV ALVIN indicate that these clasts are derived from bioerosion of the canyon walls and have accumulated during the past 100 years (Sanford et al., 1990).

Summary and Conclusions

Examination of millimeter-scale sedimentary structures in fine-grained sediments from the Amazon and Bengal shelves, and from the continental slope off the U.S. East Coast provides important clues concerning depositional conditions. Analyses of thin sections using a petrographic microscope reveals diagnostic structures that cannot be observed using other techniques (e.g., radiography, SEM). A unique category of sedimentary structure, termed plasmic fabric, is observed from fine-grained sediments of the Amazon subaqueous delta. Plasmic fabric results from preferential alignment of anisotropic clay minerals and mica, and in Amazon muds often occurs as horizontal laminae ($\sim$0.1 mm thickness) of oriented minerals that are interlaminated with randomly oriented minerals, producing a "striated" extinction pattern when observed using a polarizing microscope. The fabric is considered a primary (physical) structure; however, the mechanism for its formation is unknown. Characteristics of silt laminae for Amazon and Bengal muds often can be used as environmental indicators. For example, micro-cross-lamination and abrupt basal contacts indicate strong currents, with bedload transport and scouring, respectively. On the Bengal shelf, silt laminae with a convoluted appearance were observed in an area of extremely high sediment accumulation rate ($>$10 cm/yr). Although these laminae appear bioturbated in radiographs, microscopic examination suggests that the structure results from penecontemporaneous deformation of the rapidly accumulating muds. Cores collected from the Wilmington canyon area, U.S. East Coast, generally reveal few physical structures in radiographs; however, features observed in thin sections provide important clues concerning recent evolution of these canyon systems. Some cores from the canyon floors reveal millimeter-scale clasts of pale brown mud in a matrix of modern olive-gray hemipelagic mud. The presence of these clasts, along with radiochemical and observational data, suggests

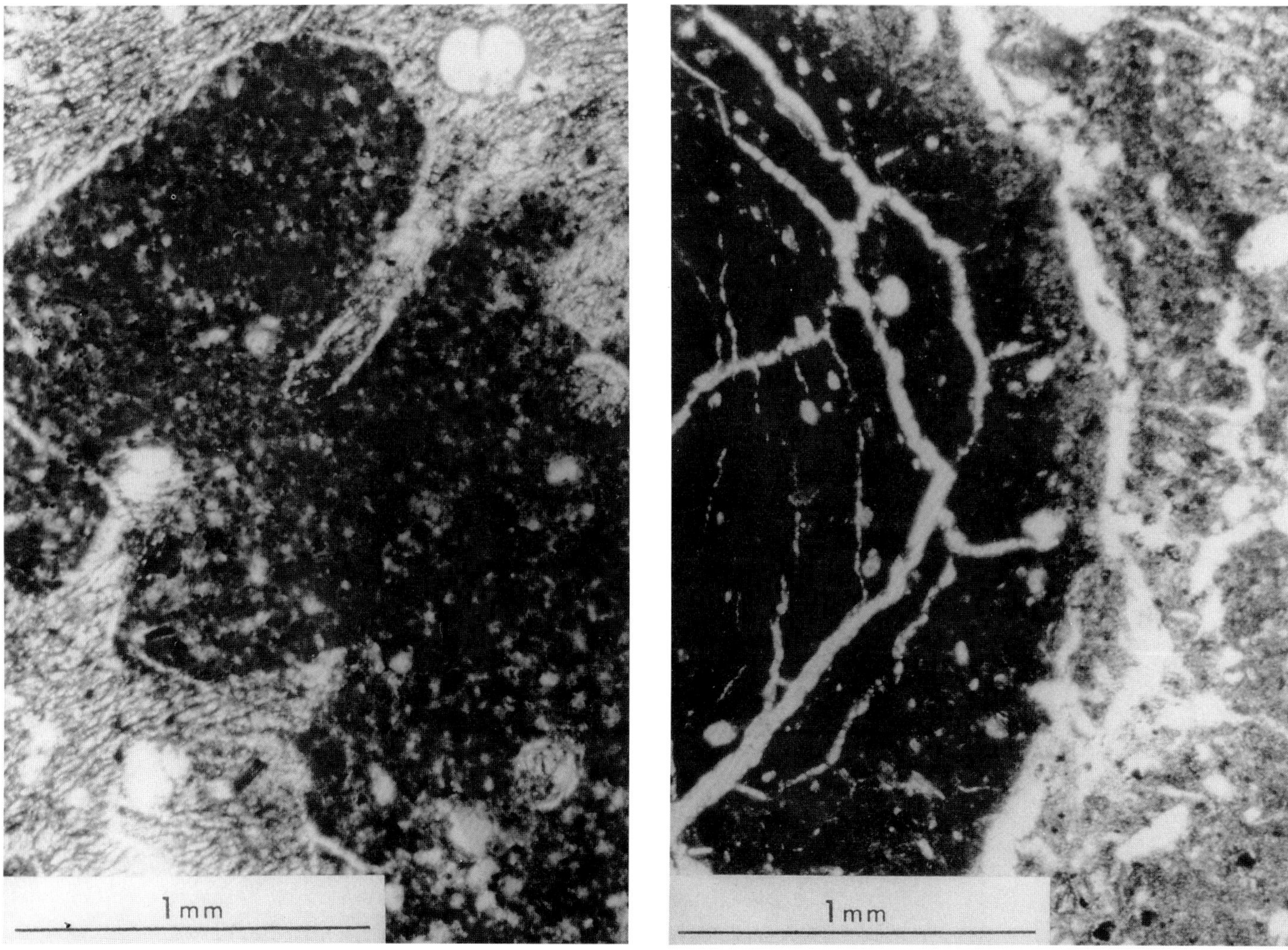

Figure 3.12. Photomicrographs from a core collected from Wilmington canyon. (A) Well indurated mud clast. (B) Mud ball. These mud clasts are derived from erosion of canyon walls and have accumulated during the past 100 years.

that the canyon walls in this area presently are eroding, and are supplying sediment to the canyon floors. The above observations of microscale structures in modern continental margin environments provide important criteria for interpretation of ancient mudrocks.

Acknowledgments

This research was supported by the National Science Foundation (Grant OCE-8117709, OCE-8415413, and OCE-8614159) and the Carolina Venture Fund. ALVIN dives and ship time for the Wilmington Canyon study were supported by NOAA URP. The authors thank Marylin Segall for drafting and correcting some of the figures.

References

Brewer, R., 1964. Fabric And Mineral Analysis Of Soils. Wiley, New York, 470 p.

Broadhead, R.F., and P.E. Potter, 1980. Petrology of the devonian gas-bearing shale along Lake Erie helps explain gas shows. U.S. Department of Energy Technical Report, 49 p.

Eysink, W.D., 1983. Basic considerations on the morphology and land accretion potentials in the estuary of the lower Meghna River. Land Reclamation Project, Bangladesh Water Development Board, Technical Report No. 15, 85 p.

Figueiredo, A.G., L.A. Gamboa, M. Gorini, and E. Costa Alves, 1972. Natureza da sedimentacao atual do Rio Amazonas-testemunhos E geomorfologia submarina, (canyon) Amazonas testemunhof submarinos. Congresso Brasileiro de Geologia, v. 26, p. 51–56.

Gibbs, R.J., 1967. The geochemistry of the Amazon River system: part I. The factors that control the salinity and the composition and concentration of the suspended solids. Geological Society of American Bulletin, v. 78, p. 1203–1232.

Krone, R.B., 1962. Flume studies of the transport of sediment in estuarial shoaling processes: Hydraulic Engineering Laboratory and Sanitary Engineering Research Laboratory, University of California, Berkeley, 91 p.

Kuehl, S.A., D.J. DeMaster, and C.A. Nittrouer, 1986a. Nature of sediment accumulation on the Amazon continental shelf. Continental Shelf Research, v. 6, p. 311–336.

Kuehl, S.A., C.A. Nittrouer, and D.J. DeMaster, 1986b. Distribution of sedimentary structures in the Amazon subaqueous delta. Continental Shelf Research, v. 6, p. 311–336.

Kuehl, S.A., C.A. Nittrouer, and D.J. DeMaster, 1988. Microfabric study of fine-grained sediments: observations from the Amazon subaqueous delta. Journal of Sedimentary Petrology, v. 58, p. 12–23.

Kuehl, S.A., T.M. Hariu, and W.S. Moore, 1989. Shelf sedimentation off the Ganges-Brahmaputra river system: evidence for sediment bypassing to the Bengal Fan. Geology, v. 17, p. 1132–1135.

Kuenen, Ph. H., 1966. Experimental turbidite lamination in a circular flume. Journal of Geology, v. 74, p. 523–545.

Kuenen, Ph. H., and J.E. Prentice, 1957. Flow markings and load-casts. Geological Magazine, v. 94, p. 173–174.

McGregor, B.A., W.L. Stubblefield, W.B.F. Ryan, and D.C. Twichell, 1982. Wilmington submarine canyon: a marine fluvial-like system. Geology, v. 10, p. 27–30.

Meade, R.H., T. Dunne, J.E. Richey, U. DeM. Santos, and E. Saliti, 1985. Storage and remobilization of suspended sediment in the lower Amazon River of Brazil. Science, v. 228, p. 488–490.

Meserve, J.M., 1974. United States Navy Marine Climatic Atlas of the World, Volume I, North Atlantic Ocean. U.S. Printing Office, Washington, D.C., 371 p.

Milliman, J.D., and R.H. Meade, 1983. World-wide delivery of river sediment to the oceans. Journal of Geology, v. 91, p. 1–21.

Moss, A.J., 1963. The physical nature of common sandy and pebbly deposits, part II. American Journal of Science, v. 261, p. 297–232.

Nittrouer, C.A., T.B. Curtin, and D.J. DeMaster, 1986. Concentration and flux of suspended sediment on the Amazon Continental shelf. Continental Shelf Research, v. 6, p. 151–174.

Pettijohn, F.J., 1957. Sedimentary Rocks. Harper & Row, New York, 718 p.

Piper, D.J.W., 1972. Turbidite origin of some laminated mudstones, Geological Magazine, v. 109, p. 115–126.

Potter, P.E., J.B. Maynard, and W.A. Pryor, 1980. Sedimentology of Shale. Springer-Verlag, New York, 303 p.

Reyne, A., 1961. On the contribution of the Amazon River to accretion of the coast of the Guianas. Geologie en Mijnbouw, v. 40, p. 219–226.

Sanford, M.W., S.A. Kuehl, and C.A. Nittrouer, 1990. Modern sedimentary processes in the Wilmington canyon area, U.S. East Coast, Marine Geology, v. 92, p. 205–226.

Stanley, D.J., P. Fenner, and G. Kelling, 1972. Currents and sediment transport at the Wilmington Canyon shelf-break as observed by underwater television. In: Swift, D., D. Duane, and O. Pilkey (eds.), Shelf Sediment Transport: Process and Pattern. Dowden, Hutchinson and Ross, Stroudsburg, PA, p. 621–644.

Stanley, D.J., S.J. Culver, and W.L. Stubblefield, 1986. Petrologic and foraminiferal evidence for active downslope transport in Wilmington Canyon. Marine Geology, v. 69, p. 207–218.

Stow, D.A.V., 1979. Distinguishing between fine-grained turbidites and contourites of the Nova Scotian deep water margin. Sedimentology, v. 26, p. 371–387.

Stow, D.A.V., and A.J. Bowen, 1980. A physical model for the transport and sorting of fine-grained sediment by turbidity currents. Sedimentology, v. 27, p. 31–46.

Stubblefield, W.L., B.A. McGregor, E.B. Forde, D.N. Lambert, and G.F. Merrill, 1982. Reconnaissance in DSRV ALVIN of a "fluvial-like" meander system in Wilmington Canyon and slump features in South Wilmington Canyon. Geology, v. 10, p. 31–36.

CHAPTER 4

Problems of Particle Delamination and of Stepwise Aggregation in Clay Swelling

E.T. Stepkowska

Physical Model of Clay Behavior

Comparison of calculated and measured values indicates that diffuse layer repulsion between parallel clay particles is the main cause of swelling of expansive soils (see Morgenstern and Balasubramonian, 1980; Stepkowska 1980). In certain clays an extensive swelling is observed exceeding prognostic values derived both from experiment and theory (e.g., smectite clays in the vicinity of Denver, CO and tunnel shafts in some Japanese clays). On the other hand earth dams may exhibit excessive filtration due to crack formation. Both these phenomena can be explained by the physics of the clay–water system.

The physical model of clay behavior (Stepkowska, 1982, 1988) explains the excessive swelling in terms of particle delamination, i.e., decrease in particle thickness, δ, which is an activation energy process. Crack formation may be explained by gradual aggregation processes that were observed in stored bentonite slurries (Stepkowska and Jefferis, 1983; see also Popescu, 1980).

This physical model is based on consideration of structural elements of the clay–water system in dynamic equilibrium and of their mutual interactions (Fig. 4.1): platy clay particles of variable thickness (mainly 6–40 nm) are arranged parallel to each other in domains that may be oriented edge-to-face, forming clusters ($\sim$ 0.5 μm in smectite) and these are aggregated in larger units (aggregates, $\sim$ 10 μm), which aggregation process is a stepwise phenomenon proceeding in time and resulting in visible macroscopic crumbs. On every step of aggregation these units form a more or less regular "lattice" similar to a crystal lattice with vacancies or macropores filled with water and/or water vapor (or air) depending on external load conditions.

Between structural elements there exist contact bonds of high strength and long-range interactions of smaller magnitude. The strength of the system as a whole is determined by the resistance available in the weakest planes where deformation may proceed; thus long-range interactions are of highest interest.

These long-range interactions are diffuse layer repulsion (energy, V_R, and pressure P_R) and London–van der Waals attraction (energy, V_A, and pressure, P_A) that may be calculated from DLVO theory (see e.g., Stepkowska, 1980) with the additional assumption that V_A depends on the size of interacting bodies and that in accordance with some literature data it is about six times higher on every step of aggregation. Both interactions are functions of the distance between interacting elements.

The total energy, V_T, i.e., the sum of attraction and repulsion, as the function of distance, $2d$, between structural elements indicates some extrema (maximum and minima) where the total pressure, P_T, i.e., the sum of attractive and repulsive pressure, is zero: by definition pressure is the derivative of energy per unit surface against the distance ($P_T = -dV_T/dd = 0$). This is the condition of equilibrium at zero external load and zero pore water pressure, u. Otherwise the condition of equilibrium is

$$-P' + P_R + P_A = 0 \tag{1}$$

where P' is the external effective pressure expressed in units comparable with P_T and the contact bonds are neglected.

External pressure is considered here as energy density per unit volume in accordance with gas pressure $P = nRT/v$ [compare formula (2) in Snethen, 1980]. For the given clay layer of thickness H and of unit weight γ, the average energy density is half the effective stress:

$$-P' = 1/2\sigma' = 1/2(\gamma - \gamma_w)H \tag{2}$$

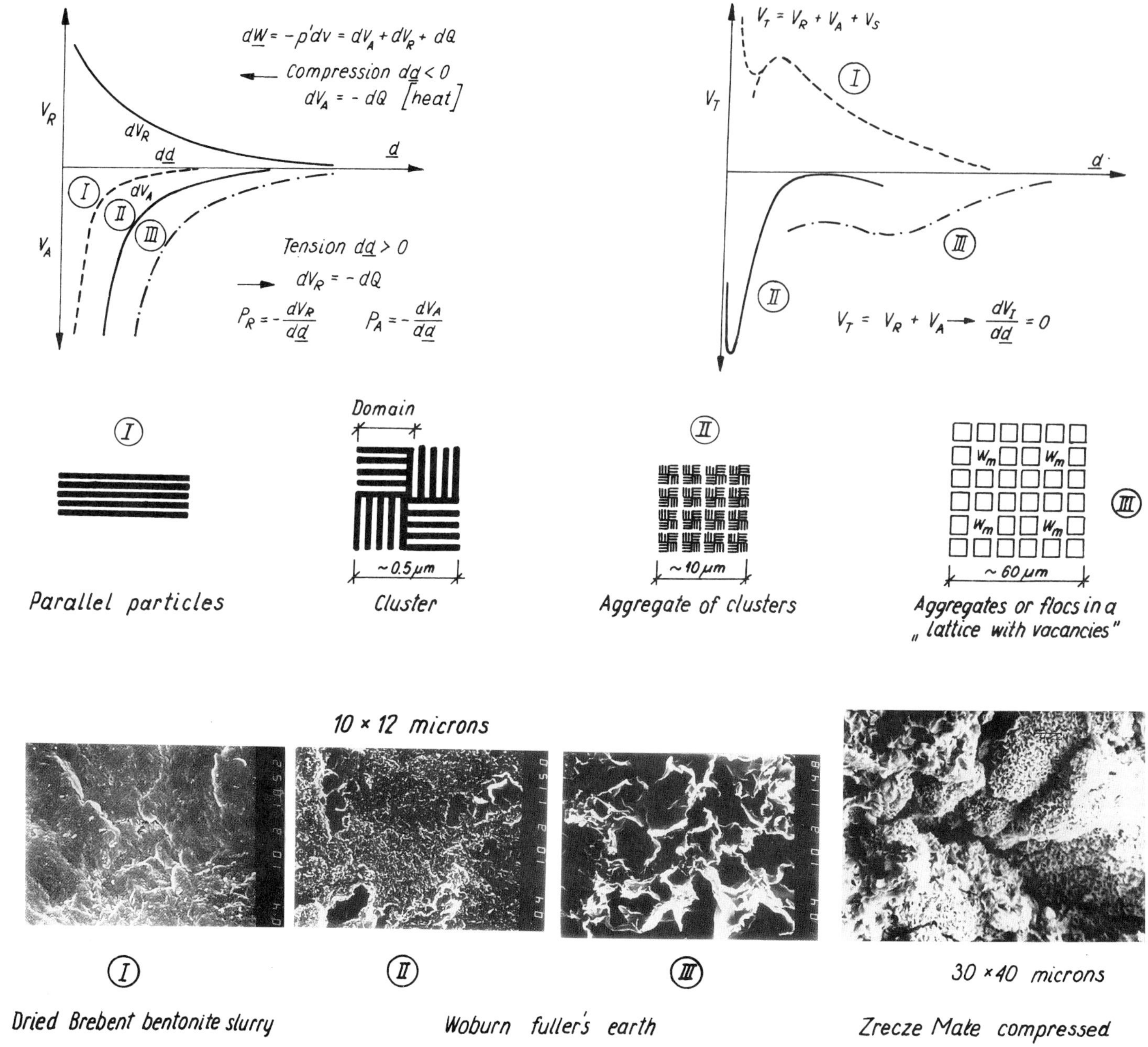

Figure 4.1. Interaction energy V as the function of distance $2d$ and various forms of microstructure.

The condition of equilibrium may be achieved by the following:

1. A change in distance between structural elements, $2d$, i.e.
   - change in water content, which is uniformly distributed on the surfaces of parallel clay particles and determines the interparticle distance, e.g., swelling in unloading, consolidation in loading,
   - change in macropore water content, i.e., rebound in unloading of a closed system, resulting in suction of water from macropores, which is substituted by water vapor with resultant volume increase, or
   - change in particle thickness, δ (see below): delamination or decrease in δ at constant water content results in increase of specific surface, S, and thus in decrease of interparticle distance with increased V_A and V_R. Collapse, i.e., increase in δ at constant water content, results in decrease of specific surface, S, and thus in increase of interparticle distance. This was observed experimentally and this results in a change of long-range interactions (Fig. 4.2).
2. A change in attraction energy and pressure by aggregation or disaggregation processes: e.g., clusters behave as separate units at lower water content indicating increased attraction,

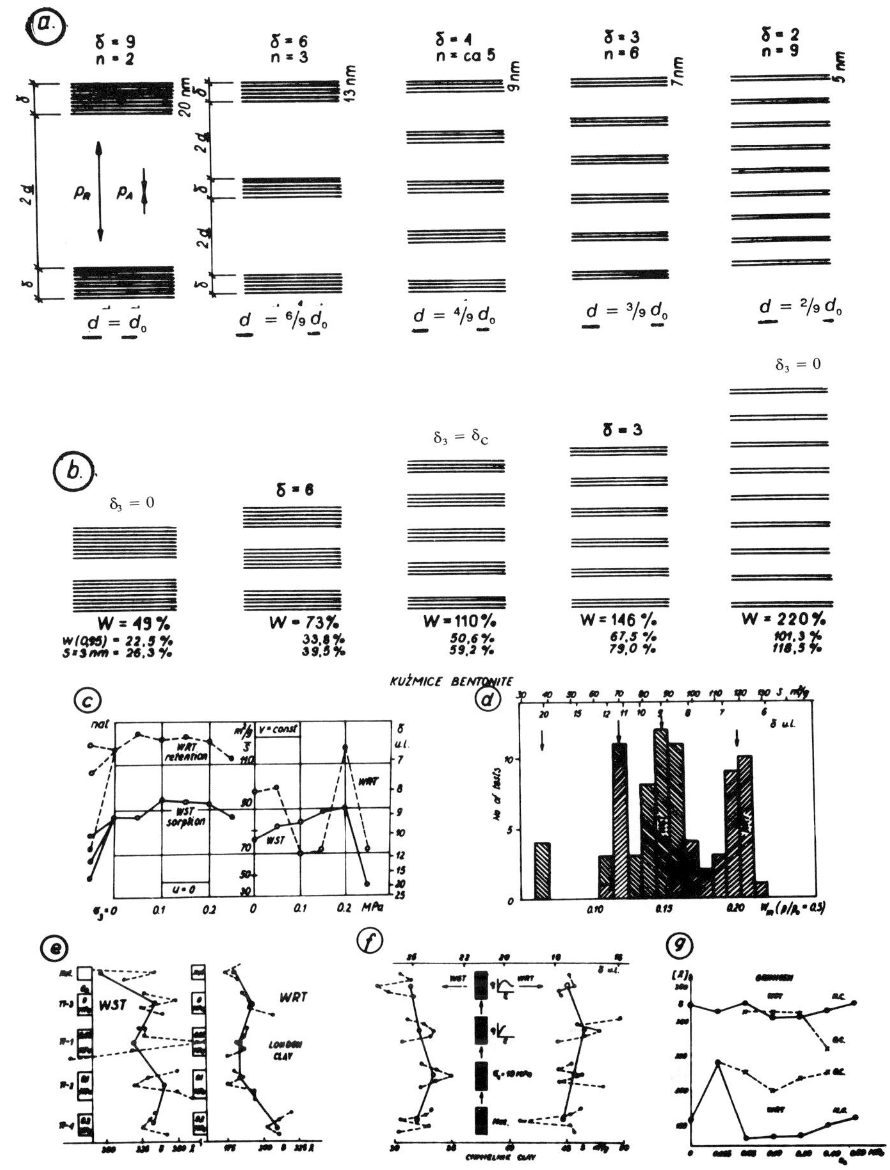

Figure 4.2. Thickness change in clay particle (a) Particle delamination at a constant volume and constant water content (p_R and p_A increase with decrease in δ and $2d$) (b) Swelling due to particle delamination at constant d = 3 nm, u = 0 and variable water content (p_R and p_A are approximately constant). Change in δ as measured after triaxial compression of Kuzmice bentonite (c) and (d); of London clay (e); of Chmielnik clay (f); and of Grimmen clay (g).

whereas at high water content they disintegrate into separate domains indicating interparticle attraction of lower magnitude. Compression strength is usually due to intercluster attraction, the measured tensile strength was equal to interparticle attraction (see below, Fig. 4.6). With time, aggregation processes proceed stepwise resulting in the increase in size, in shape regularity, and in distribution regularity of structural elements, accompanied by the increase in attraction energy. The internal rearrangement of the system may lead to crack formation.

3. Distribution of the system into two or three microstructural phases of either parallel particle arrangement or of cluster structure or of aggregate floc structure (flocs are envelopes of bent particles enclosing water). These three microstructures are observed in natural clays and in bentonite slurries (Fig. 4.1).

Little consideration is given to some of the processes mentioned above, though they are of great importance for soil mechanics problems, e.g., change in particle thickness or aggregation processes.

Change in Particle Thickness

The clay particle or crystallite is composed of several unit layers each of 1 nm thickness in 2:1 layer silicates (Fig. 4.2). Kaolinite, a nonexpansive clay, will not be discussed here. It may be assumed that in the clay–water system there is a three-molecular water layer on the external surface and in expanding smectite a three-molecular water layer in the internal space. The ratio of external sorbed water to the total water sorption is constant in

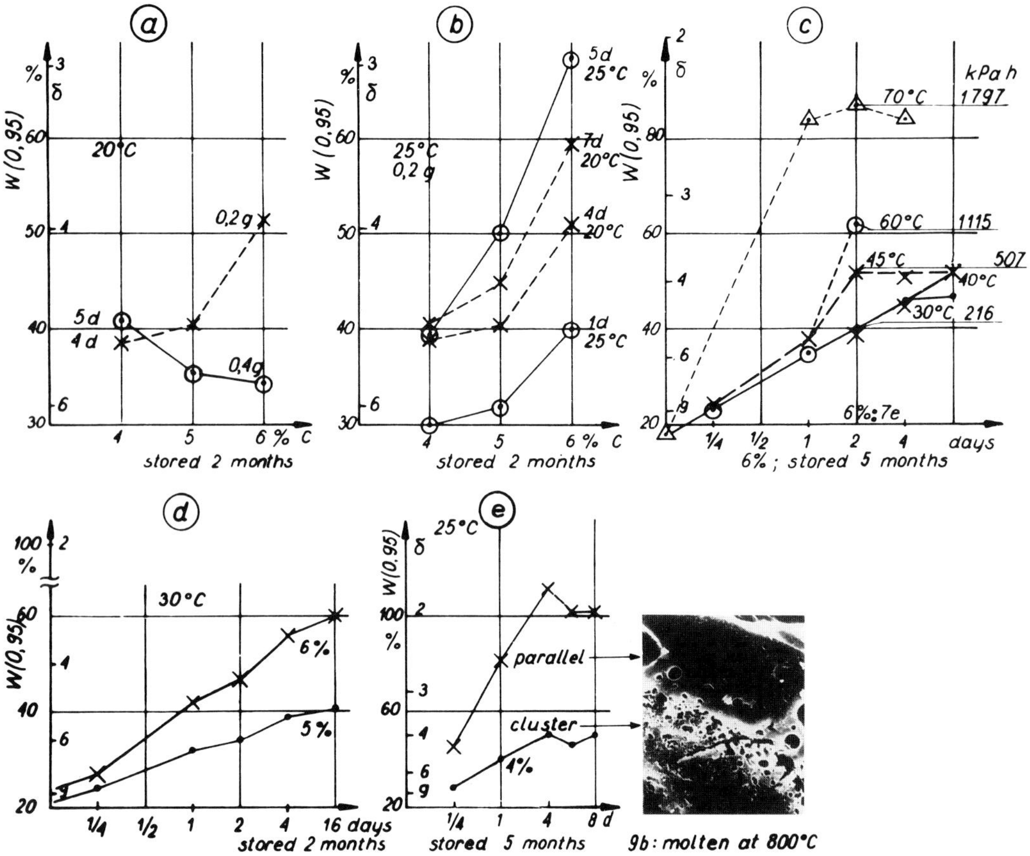

Figure 4.3. Sorption of Berkbent bentonite slurries (2000 RPM, 5′) stored and dried at 45°C; influence of (a) sample size, (b, c) time and temperature, (d) concentration, and (e) microstructure.

most clays, thus the external specific surface, S, and the particle thickness, δ, may be measured from the water sorption at the given relative humidity, e.g. $p/p_0 = 0.95$ as $W_{0.95}$ or at $p/p_0 = 0.5$ as $W_{0.5}$, giving consistent values with other measurement methods (BET, SEM, XRD, glycol retention; see Newman, 1983; Stepkowska, 1982):

$$S = W_{0.5} \times 585 \text{ m}^2 \text{ g}^{-1} \text{ H}_2\text{O} \tag{3a}$$

$$S = W_{0.95} \times 390 \text{ m}^2 \text{ g}^{-1} \text{ H}_2\text{O} \tag{3b}$$

In some clays, S calculated from (3a) may not equal S calculated from (3b), which means delamination in water sorption process.

$$\delta = \frac{S(\text{unit layer})}{S(\text{measured})} \tag{4}$$

where $S(\text{unit layer}) = 790 \text{ m}^2 \text{ g}^{-1} \text{ H}_2\text{O}$.

Cation exchange capacity (CEC) is proportional to water sorption and it may be as well calculated from

$$\text{CEC} = W_{0.5} \times 5.927 \text{ meq g}^{-1} \text{ H}_2\text{O} \tag{5}$$

Here W is expressed in g water per g clay and is determined at 220°C.

The change in particle thickness is an activation energy process or rather that of activation action (energy multiplied by time). Most probably this energy in delamination, i.e., decrease in δ, is the energy of ionic lattice attraction between unit layers (it is equal to activation energy of creep process; see Mitchell, 1976; Stepkowska, 1982) or it is the energy of hydrogen bond. Whereas in collapse, i.e., increase in δ, it is that of repulsion of diffuse layers.

The energy necessary for the decrease or increase in particle thickness may be supplied either as mechanical energy of external load or as energy of water molecules in liquid or gaseous state acting at a given temperature for a given time.

Indeed change in δ was measured after triaxial compression of London clay (Fig. 4.2a), of Kuzmice bentonite (see Stepkowska, 1980), and of Grimmen clay consolidated at δ_c and sheared either at $\delta_3 = \delta_c$ or at $\delta_3 = 0$ (Stepkowska and Störr, 1986) (Fig. 4.2g). It was dependent on lateral pressure σ_3 if compression proceeded at constant volume (v = constant) and at variable pore water pressure, u. It was almost independent of σ_3 if the compression process proceeded at $u = 0$ condition. The histogram of the measured values of δ indicated a stepwise change, i.e., one particle may be composed of ~ 6, 9, 12, or 20 unit layers (see Fig. 4.2c, d).

Delamination was observed also in high-pressure consolidation of Chmielnik bentonite, which could be due to the high pore water pressure at the initial state of the consolidation process. It should be checked whether or not the long-term loading causes the particle collapse, resulting in secondary consolidation. Prin-

Table 4.1. Partial pressure p, of water over the 10% H_2SO_4 solution at various temperatures.

Temperature (°C)	h (0.95) (mm Hg)	p (kPa or 10^4 erg/cm^3)
0	4.38	0.58
20	16.77	2.24
25	22.71	3.03
30	30.42	4.05
40	52.89	7.05
45	68.72	9.16
60	142.83	19.04
70	223.42	29.78

Table 4.2. Estimate of activation action in delamination of Na-bentonite particles at 30°C (Fig. 4.3d, p = 4.058 kPa).

W (0.95)	δ	Time (days)		Activation action	
				c=6%	c=5%
(%)	(u.l.*)	c=6%	c=5%	(kPa h)	
16.9	12		–	–	
22.5	9	~ 1/8	~ 1/4	12.2	24.3
25.3	8	~ 1/4	~ 1/2	24.3	48.6
33.8	6	~ 1/2	~ 2	48.6	194.4
50.6	4	~ 3	–	291.6	–
67.5	3	~ 16	–	1555.2	–
101.3	2	–	–	–	–

*u.l., unit layers.

cipal stress difference may cause the initial decrease in particle thickness followed by its increase, which is accompanied by the drop in the measured strength (residual strength). This was observed in Chmielnik bentonite (Fig. 4.2f, see Stepkowska, 1980).

The gradual delamination process was observed during sorption measurement in dried Na-bentonite slurries: the values of water sorption were dependent on preparation of the slurry (concentration, stirring rate, storing time), on the relative water vapor pressure during the sorption process, on temperature, on time, and on sample size. The measured values were again close to δ = ~ 6–7, 5, 4, and 3 unit layers or rather within the range of δ = 3 to 7 unit layers per particle (Fig. 4.3).

Low δ values were not measured by water sorption unless relative water vapor pressure, temperature, and time were high enough (Fig. 4.3). Slurry prepared at the initial concentration 4% indicated at p/p_0 = 0.95 a limited delamination both at 20° and at 25°C; 5 and 6% slurries delaminated much quicker at 25 than at 20°C and the sorption of a 0.4 g sample was lower than that of a 0.2 g sample (Fig. 4.3a, b). At p/p_0 = 0.5 delamination did not exceed δ = 9 unit layers.

If pressure is considered as volumetric energy density, it may be calculated for the given relative humidity value (e.g., p/p_0 = 0.95 over the 10% H_2SO_4 solution):

$$p = \frac{nRT}{v} = \frac{h}{h_0}\frac{\mathrm{NkT_0}}{v_0} = h \times 0.1333 \qquad (6)$$

expressed in kPa (or in 10^4 erg cm^{-3}) if h is in mm Hg.

Here n is the number of moles, R is the gas constant, T is temperature in K and v is volume, h is the water vapour pressure over the 10% H_2SO_4 solution read from physico-chemical tables for the given temperature (here in mm Hg). h_0 is the normal atmospheric pressure, i.e., 760 mm Hg, N = 6.025×10^{23} molecules per mole is the Avogadro number lc = 1.38×10^{-16} erg. K^{-1} = 1.38×10^{-23} y.K^{-1} is the Boltzmann constant, T_0, is the normal temperature 273.16 K and v_0 = 22 μl mole^{-1} is the molar volume at T_0 and h_0.

The calculated values of energy densities at various temperatures are presented in Table 4.1. These values are proportional to the number of water molecules of the given energy bombarding a unit surface of clay particle in a unit time. If the total time is considered, the total number of molecules striking the surface may be calculated. Whether molecules of higher energy are acting for a shorter time or molecules of lower energy are acting for a longer time, the delamination would be similar, provided a certain minimum energy level is supplied. External effective stress may cause an opposite phenomenon, i.e., particle collapse either under load or at negative pore water pressure, i.e., increase in thickness of all the particles in the system. This also needs a certain energy level to surpass the diffuse layer repulsion and to form the ionic lattice. This phenomenon may be reversible and when the external load is removed, particles may return with time to the initial small thickness, which may be accompanied by swelling exceeding the values measured in short-term consolidometer test.

Pressure levels may be estimated by measuring water sorption proceeding in time at the given p/p_0 and at the given temperature. This was done, for example, for Na-bentonite slurry (Berkbent). Water sorption levels corresponding to the given δ values were attained at p/p_0 = 0.95 and 30°C after the given times (Table 4.2). These values multiplied by the energy density level 4.058 kPa read from the Table 4.1 for 30°C gives the value of activation action and simultaneously indicates the pressure level which would give a similar result after 1 hr.

Activation energy of delamination process should depend on various factors and should

- increase with valency of exchangeable cation,
- decrease with hydration state of the interlayer space (e.g., in Na-bentonite d_{001} = 1.5 nm at 20°C and 1.26 nm at 25°C),

Table 4.3. Calculated water content in Na-smectite depending on particle thickness δ.

δ (u.l.)	S (m^2 g^{-1})	W (0.95) (%)	W (d=3 nm) (%)	Total W (%)
12	65.8	16.9	19.7	36.6
9	87.8	22.5	26.3	48.8
8	98.8	25.3	29.6	54.9
6	131.7	33.8	39.5	73.3
4	197.5	50.6	59.2	109.8
3	263.3	67.5	79.0	146.5
2	395.0	101.3	118.5	219.8

Table 4.4. Properties of clay from Bydgoszcz (depth 3.2 m).

Borehole Nr	Fraction less than 2 μm CEC (meq g^{-1})	S (m^2 g^{-1})	δ (u.l.)	Swellability from air dry W_0 (%)	W_f (%)	Δv (%)	Swelling W_n (%)	W_f (%)	Δv (%)
5	0.71	72.3	10.9	13.9	88.2	47.8	35.1	49.0	7.4
11	0.50	53.8	14.7	5.9	57.8	26.4	17.2	36.9	14.3
13	0.53	56.5	14.0	6.2	56.7	27.6	24.5	35.3	6.1

- increase with presence of bonding factors ("columns") in the interlayer space,
- increase with ordering of stacking of unit layers: this factor might have been disturbed during stirring of 6% bentonite slurry, which decreased the activation energy of the delamination process in comparison with that of the remaining concentrations (see Fig. 4.3b).
- depends on place and magnitude of isomorphic substitution.

Comparing three samples of the clay from Bydgoszcz the highest swelling was detected in the sample, which indicated the highest structural disorder as measured by XRD (see below).

The practical importance of these considerations is great, e.g., in designing excavations in expansive clays. In this case the clay stratum is unloaded and the equilibrium is disturbed. The danger is especially high in case of the presence of water vapor pressure (e.g., under a reservoir) and of high temperature as these factors promote delamination. In Table 4.3 and in Figure 4.2 are presented the expected values of the total water content at the given particle thickness (no macropore water present) and these values manifest dramatically the danger of clay swelling due to particle delamination. This danger is increased by the possible inhomogeneity of the clay stratum, both macroscopic and microscopic, though not observable externally.

Even if a clay sample is prepared to be homogeneous, it may be inhomogeneous due to distribution of the system into various microstructural phases to attain equilibrium state (see above). Various microstructural phases may indicate various swelling, i.e., parallel structure swells more than cluster structure (see Fig. 4.3e). Such nonuniform swelling was measured in three samples of clay from Bydgoszcz (see Table 4.4). The mineral composition of these samples was similar, though sample 5 contained more montorillonite and indicated the most disordered structure in XRD of Mg-glycerol samples. This sample indicated as well the highest swelling and the highest delamination.

In the research program described here the particle thickness was measured by water sorption test (WSTest), i.e., successive storage of oven dried clay samples at $p/p_0 = 0.5$ and 0.95 followed by drying and heating (Stepkowska, 1979) or by water retention test (WRTest), where the sample is stored at decreasing relative humidity conditions. It was observed that clays of nonvariable particle thickness indicate no sorption hysteresis, i.e., similar values in water sorption and in water retention; their

Table 4.5. Sorption properties of various clays.

Sample	Test	W_h (0.5) (%)	W_h (0.95) (%)	Remarks
Z.M. bentonite	WRT	20.05	36.98	Small variability in δ by XRD and other methods
	WST	17.95	29.25	
Kuzmice bentonite	WRT	19.37	29.59	High variability in δ after mechanical treatment
	WST	9.85	24.12	
Dobrzyn Nr 104 (24m) Powder	WST	3.88	20.40	Landslides are observed on slopes in this clay
Undisturbed	WST	4.46	25.28	
Upper lias clay (0.2 g) 6.5 m, Daventry	WRT	4.47 ± 0.34	16.93 ± 1.07	Landslide area (R.J. Chandler, Imp. College London)
	WST	3.62 ± 0.22	6.9–17.3	
London clay	WRT	9.63	20.12	
	WST	5.03	10.07	
Grimmen clay	WRT	7.80	24.64	
	WST	5.34	14.39	
Chmielnik clay	WRT	6.72	19.55	
	WST	5.15	9.78	
Clay from Bydgoszcz (2 μm, Mg) Nr 5	WST	9.95	22.53	High inhomogeneous swelling properties
Nr 11		8.07	17.33	
Nr 13		8.60	16.87	

W_h is related to oven dry weight of clay sample (105°C)

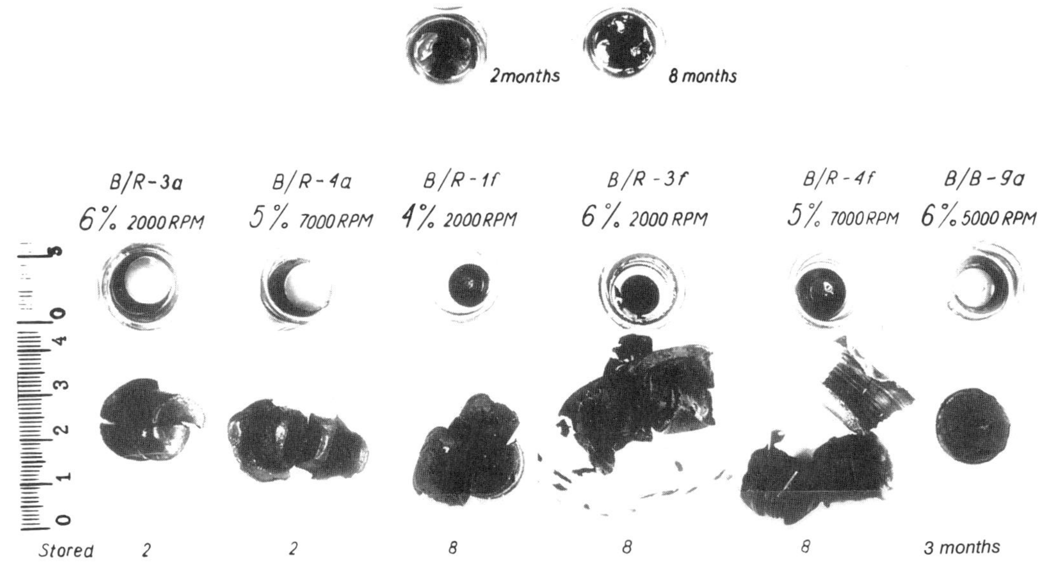

Figure 4.4. Gradual aggregation in Na-bentonite slurries.

ratio of water sorption at $p/p_0 = 0.95$ to that at $p/p_0 = 0.5$ is approximately (compare eqs. 3a and 3b):

$$W_{0.95}:W_{0.5} = \sim 1.5 \tag{7}$$

If these values differ pronouncedly, delamination is possible (see Table 4.5) and such clays may be dangerous in sliding or in swelling (see also Stepkowska, 1980).

Thus the measurement of water sorption at various relative humidities and at various temperatures may be both qualitatively and quantitatively a useful indication of engineering properties of clays that is worth further study.

Possible Danger of Stepwise Aggregation

In the study of microstructure of Na-bentonite slurries the gradual aggregation was observed that started on microscopic scale but after prolonged storage macroscopic aggregates were observed. The 7% Na-bentonite slurry (Berkbent, 5000 RPM, 15') disintegrated after 3 day drying at 45°C into two equal parts if prestored for 2 months or into three equal parts if prestored for 8 months (Fig. 4.4).

Thus this cylindrical sample $\phi = 5$ cm, $h = 5$ cm was composed either of two or of three macroaggregates of suitable size. In other samples the aggregation process manifested itself in dry state in which case samples disintegrated after drying into two or three macroaggregates of suitable size. In other samples the aggregation process manifested itself in dry state in which case samples disintegrated after drying into two or three equal parts depending on slurry preparation and on storage time (Fig. 4.4).

Some of them formed a regular cylinder (B/B-9a). An aggregated system may be very sensitive to varying water content conditions: aggregates shrink during drying and cracks of varying size may open between them. These cracks are sensitive to water flow and they may be disastrous to the whole soil structure.

Aggregation processes do not manifest visibly in the concentrated clay–water system, therefore they attract little attention (crumb structure soils were studied by Popescu, 1980). The strength of the system is determined by resistance of smaller units, e.g., intercluster or interparticle forces may be decisive as the arrangement of these elements must be deformed to deform the whole system (Fig. 4.6).

These aggregation processes may be of highest importance in certain problems as, e.g., landslides: big crumbs of clay may

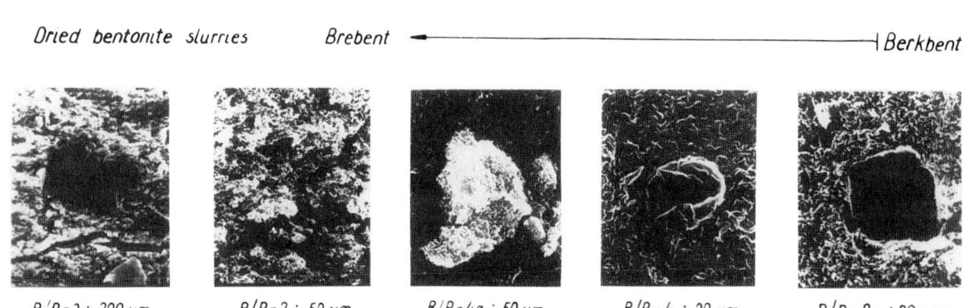

Figure 4.5. Dried bentonite slurries.

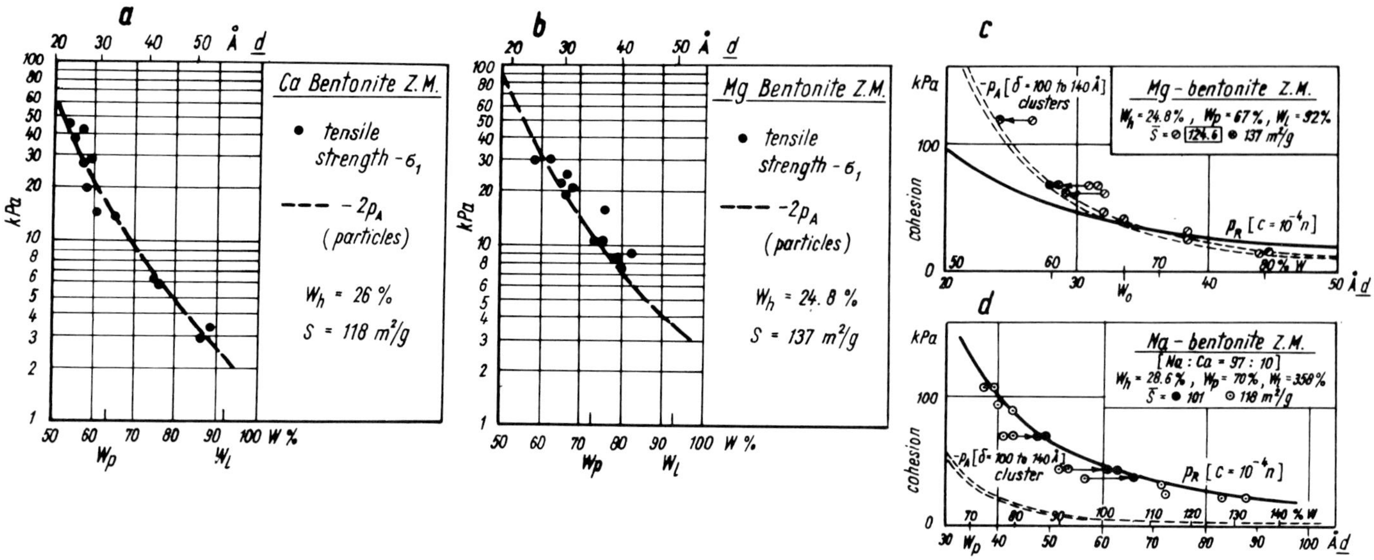

Figure 4.6. Comparison of calculated and measured values. (a) and (b) tensile strength as compared with calculated $2p_A$. (c) cohesion as compared with calculated pA. (d) cohesion as compared with calculated P_R in Na-bentonite.

disintegrate from a steep clay slope. They may be also highly dangerous in earth structures: water will flow primarily between aggregates disturbing their mutual arrangement and their mutual attraction and in unfavorable conditions this may end up in a complete disaster for the whole structure.

Data presented here are preliminary indications only, obtained in a theoretical study, but research should proceed in this direction.

Conclusions

The thickness of the platy clay particle may vary in various processes (e.g., loading or unloading). In the case of a decrease in thickness or delamination, excessive swelling may occur. This is an activation energy process depending on pressure (considered here as volumetric energy density), on temperature, and on time. It may be measured by water sorption at various relative humidities as the function of time and temperature.

From these data the energy and time necessary for delamination may be evaluated and, assuming equivalence of various forms of energy, the external pressure may be estimated that acting for the given time may cause the same result in change of particle thickness. This approach permits a direct transition from physical values and internal forces to engineering parameters.

Variability of clay properties as measured in the laboratory for the same clay (e.g., consistency limits) may be due to variable particle thickness.

Attention should be paid to aggregation processes proceeding in time in the clay–water system as this may lead to crack formation dangerous for the stability of the earth structure.

References

Mitchell, J.K., 1976. Fundamentals of Soil Behavior. Wiley, New York, 422 p.

Morgenstern, N.R., and B. Balasubramonian, 1980. Effects of pore fluid on the swelling of clay-shale. Proc. 4th International Conference on Expansive Soils, v. I, p. 190–205.

Newman, A.C.D., 1983. The specific surface of soils determined by water sorption. Journal of Soil Science, v. 34, p. 23–32.

Popescu, M., 1980. Behavior of expansive soils with a crumb structure. Proc. 4th International Conference on Expansive Soils, v. I, p. 158–171.

Snethen, D.R., 1980. Characterization of expansive soils using soil suction data. Proc. 4th International Conference on Expansive Soils, v. I, p. 54–75

Stepkowska, E.T., 1979. Sorption properties of consolidated and compressed clays. Proc. VIth International Clay Conference, Oxford, 1978, p. 457–464.

Stepkowska, E.T., 1980. Study in physical theory of expansiveness in clays. Proc. 4th International Conference on Expansive Soils, v. I, p. 136–157.

Stepkowska, E.T., 1982. Study in physics of clay behavior. Archiwum Hydrotechniki, v. XXIX, no. 4, p. 469–490.

Stepkowska, E.T., 1988. Structural model of clay behaviour. Rorprawy Hydrotechnierne Nr. 50, p. 227–237.

Stepkowska, E.T., and S.A. Jefferis, 1983. Study in microstructures of clay slurries. Archiwum Hydrotechniki, v. XXX, no. 2, p. 193–211.

Stepkowska, E.T. and M. Störr 1986. Variability in water sorption by Mechanical Treatment in the clay from Grimmen, GDR. 10th Conf. on Clay Min and Petrology, Ostrava, 1988, p. 197–207. Ed. J. Konta.

CHAPTER 5

The Nature and Significance of Gas-Generated Microvoids as "Secondary" Microfabric Features in Modern and Pleistocene Marine and Estuarine Sediments

Stanislas Wartel, Sethi Parvinger Singh, and Richard W. Faas

Introduction

"Clay fabric refers to the spatial distribution, orientations and particle to particle relations of the solid particles (generally those less than 3.9 microns in size) of sediment" (Bennett and Hulbert, 1986). The particular fabric attained is a function of the physical, chemical, biological, and climatic factors that constitute a specific environment of sedimentary deposition, within which the *primary fabric* developed, i.e., that original particle orientation that occurs during or soon after deposition under undisturbed (i.e., nonbioturbated) conditions (O'Brien, 1987; Bennett et al, this volume).

A review of the literature (Lambe, 1958; Grim, 1962; Smart, 1967, 1975; Barden, 1972; Moon, 1972; Collins and McGown, 1974; Chiou, 1980; Bohlke and Bennett, 1980; Faas and Crocket, 1983; Bennett and Hulbert, 1986) shows that until recently, the primary goals of clay fabric research had concentrated on gaining an understanding of the processes of flocculation and dispersion, domain formation, a description of the different configurations between the clay particles, and the relationships between the microfabric and the geotechnical properties of the material. There seemed to be little attempt to utilize microfabric information toward the recognition and understanding of the nature of the depositional environment except for some rather obvious and generalized environments, e.g., black shales accumulating under anoxic conditions. This may reflect an implicit belief that with few exceptions, all fabrics are *secondary*, having formed postdepositionally through tectonic or overburden pressures.

We believe that there may be some types of microfabrics that are retained as *primary* (i.e., formed during or shortly after deposition) or *secondary* (i.e., developed prior to lithification) within a sedimentary unit that may be used to infer and reconstruct original depositional conditions of older environments. This work concentrates on the occurrence and distribution of gas-derived void spaces as recorded in marine and estuarine sediments.

Reference to these void spaces have not been found in the literature although an abundance of references discuss the presence of gas in fine-grained sediments (Reeburgh, 1969; Schubel, 1974; Martens, 1976). Voids in sediments caused by gas bubbles most commonly range between 0.5 and 5 mm (Anderson, 1974), however, there seems to have been little serious attempt to describe and classify these void spaces. In this work we will slightly modify the terminology of Smart (1975) and assign the term "micro" to those voids between 0.2 and 6 μm diameter, i.e., microvoids. "Mesovoids" will refer to voids between 6 and 200 μm (0.2mm), and "macrovoids" range between 0.2 and 6 mm. A new terminology of "nonstressed" and "stressed" gas-derived voids is proposed. Although the "nonstressed" gas-derived voids (with straight sided, undeformed, and nonforced particles) were observed in samples from all three study areas, the "stressed" gas-derived voids (with curled, warped, and deformed particles) were observed only in the Rijkevorsel Clay Member. Further references to "stressed" and "nonstressed" voids will omit the term "gas-derived," while tacitly assuming that evolution and expansion of gas is the original generating mechanism. In addition, the size term "micro" will be used throughout inasmuch as most of the voids observed fall within that size range.

Particle orientation and fabric description will use the following designations: FF, face-to-face; EF, edge-to-face; EE, edge-to-edge. Various combinations of these orientations will be designated as D, domain, e.g., DFF, domain, face-to-face; DSFF, domain, stepped face-to-face; or SP, single particle in either an SPFF, single particle face-to-face or SPEF, single particle edge-to-face configuration.

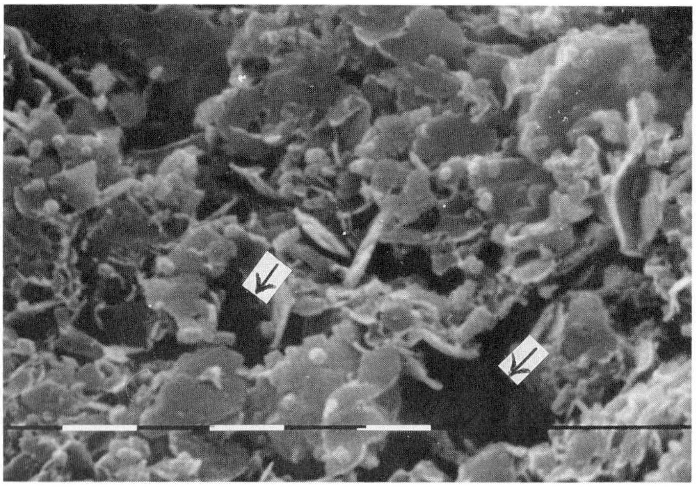

Figure 5.1. North Sea/Box core 88A25/Sample 821. Randomly oriented particles, connected with EF linkages. "Nonstressed" microvoids (arrow). FF domains are also observed. White bar on photograph is 1 μm long.

Figure 5.2. Schelde/Box core 88B23. Randomly oriented particles connected with EF linkages. Large domains with FF linkage occur. Near the center of the plate a large floc with predominantly FF linked particles is seen. White bar on photograph is 10 μm long.

Studied Areas

Seven samples from the North Sea were obtained from an area of near-shore mud deposition off the Belgian continental shelf. These mud deposits are primarily located between the offshore subaqueous sand ridges and the inshore zone. Origins proposed for these muds include (1) erosion of Tertiary clays outcropping on the North Sea floor (Bastin, 1971), (2) erosion of Holocene clay in the swales between the banks (Houbolt, 1968), and (3) a net-seaward transport of suspended matter from the Schelde estuary (Terwindt, 1977). The area is referred to as the "Coastal Wave Shadow Zone" as it lies in the shadow zone of the NE trending Flemish and Hinder banks. Seven samples were also obtained from different mud depositional areas within the well-mixed to partially mixed section of the Schelde estuary (Wartel, 1977).

Eight samples were obtained from the Rijkevorsel Clay Member (Lower Pleistocene, Tiglian) of the Campine Formation (Nova-pit, Beerse in Northern Belgium). This sedimentary unit is considered to have been deposited under estuarine conditions (Paepe and Vanhoorne, 1970). Where sampled, the Rijkevorsel Clay Member consists of an approximately 2-m-thick clay layer overlying a 3-m-thick unit of tidal deposits consisting of alternating sand and clay layers. The clay unit has a bluish-gray color and a compact form with a nodular structure. No other structures are discernable, even through radiographs. The Rijkevorsel Clay Member is overlain by cryoturbated aeolian sands (Lower Pleistocene, Eburonian) and cover sands (Upper Pleistocene), together totaling 3 m in thickness.

Methods

Stainless-steel boxes (20 × 16 × 3.5 cm) were used to obtain oriented and essentially undisturbed samples from the Rijkevorsel Clay Member. A Reineck box corer was used to obtain unconsolidated, essentially undisturbed and oriented samples from the uppermost 60 cm of the sea bed from the North Sea and the Schelde estuary. From these, subsamples using the stainless-steel boxes were obtained.

In the laboratory, subsamples (3 × 4 × 2 mm) were taken from the stainless-steel box core and oriented parallel to the bedding plane of the sediment. The samples were split and trimmed (into cubes of 2–3 mm sides) using a razor blade and transferred into a porous cup, keeping their orientation. The peeling technique for cleaning (Barden and Sides, 1971) was used and salt removal and water replacement was done by washing with ethanol. Samples were critical point dried using liquid carbon dioxide (Bennett et al., 1979), gold-coated, and viewed with the SEM. More than 350 SEM photographs were analyzed.

Observations

The clay microfabric of the sediments from the North Sea is characterized by randomly oriented single particles in an edge-to-face (SPEF) configuration. Some flocs are seen but very few domains are observed (Fig. 5.1).

Samples from the Schelde estuary exhibit a clay microfabric characterized by randomly oriented single particles or face-to-face domains (DFF) appearing as single particles under the SEM in an edge-to-face (SPEF) configuration together with some domains (DFF and DSFF) connected to each other in an edge-to-face (DFF-EF, DSFF-EF, and DFF-DSFF-EF) manner (Fig. 5.2).

Samples from different layers within the Rijkevorsel Clay Member show different microfabrics. One shows single particles with face-to-face (SPFF) and stepped face-to-face (SFF) contacts (Fig. 5.3). The other is dominated by randomly oriented domains

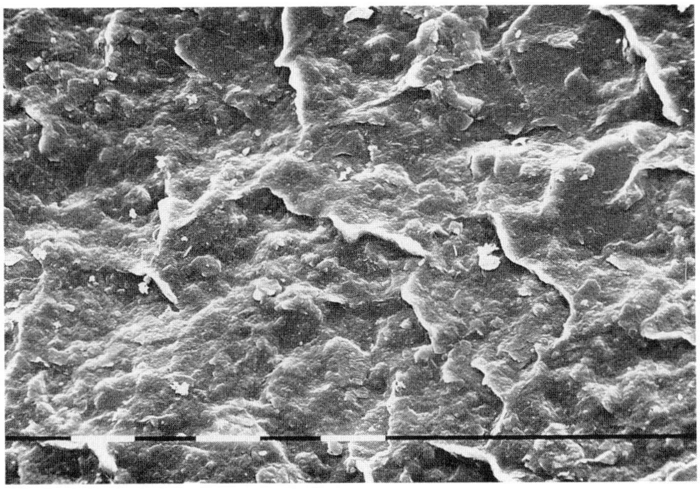

Figure 5.3. Campine Formation/Box core 88B11/Sample S4. Very smooth fabric consisting of single particles with FF orientation. White bar on photograph is 10 μm long.

Figure 5.4. Schelde/Box core 88B23. Randomly oriented particles connected with EF linkages. Large domains with FF-linkage occur. Near the center of the plate a large floc with predominantly FF linked particles is seen. White bar on photograph is 10 μm long.

(DFF and DSFF) of variable sizes, connected to each other with weak chains in an edge to face (EF) orientation (Fig. 5.4).

Microvoids

Gas voids, from a few to several tens of micrometers in diameter, result from the decay of organic matter contemporaneous with sedimentation. The gas can escape through the sediment while the sediment is still in a semifluid state (Martens et al., 1976). The escape of gas is also a function of the effective overburden pressure, which can vary due to the tidal range effect (Wartel et al., 1985). Gas escape will be of greater importance in shallow depositional areas than in deeper areas, since the hydrostatic overburden pressure, due to tidal range effects, is more pronounced in the shoaler environments. For example, for sediments accumulating in water depths between 6 and 10 m, a tidal range of 3–4 m could be significant in allowing gas to escape through differential hydrostatic pressure events.

Microvoids were observed in samples from all three of the areas studied. However, all of these microvoids do not exhibit the same structure but rather show at least two different kinds of voids, depending on the structure and disposition of the surrounding clay particles. In some, the surrounding clay particles are seen to be straight-sided and undeformed (Fig. 5.4), whereas in others the surrounding clay particles were seen to be curled, warped, and bent outwards (Fig. 5.5 and 5.6). This difference in the structure of the surrounding clay particles can be explained by considering the forcing effect caused by an expanding medium, which could cause the curling and flaring outward of the surrounding clay particles. To discuss these microstructural features, we introduce two new terms, "nonstressed" microvoids and "stressed" microvoids.

1. "Nonstressed" microvoids exhibit peripheral margins that are formed by straight-sided, nondeformed, and nonforced particles, and show no evidence of outward expansion (Fig. 5.4).
2. "Stressed" microvoids are periferally rimmed by curled, warped, and deformed clay particles and show evidence of expansion effects, which cause outward flaring of the peripheral margin of the void (Figs. 5.5 and 5.6).

Discussion

Three hypotheses are proposed to explain the origin of "stressed" microvoids and their evolution from the "nonstressed" to "stressed" state, as observed in the samples of the Rijkevorsel Clay member.

1. The transformation of a "nonstressed" microvoid into a "stressed" microvoid.
   a. While sediment is still in a semifluid state, the pressure of the gas within the microvoid will increase due to continued organic decomposition.
   b. Lateral forcing due to increased gas pressure causes domain shifting and microvoid formation, creating spherical passages (maximum volume) for the escape of gas.
   c. Continuing increase in lateral pressure would force the domains into bent shapes until a critical pressure is reached.
   d. Eventually, escape of gas, with a "bursting-effect" (Figs. 5.7 and 5.8) would occur, causing the flaring outward of the edges of the clay particles, thereby producing a "stressed" microvoid.
2. The transformation of a "nonstressed" microvoid into a "stressed" microvoid due to saturation of the voids with water

Figure 5.5. Campine Formation/Box core 88B18/Sample S1. Domains and single particles in an edge to face (DEF–SPEF) configuration. "Stressed" microvoid (arrow) with curled and warped particle edges that flare outward due to forcing effect of the expanding medium. White bar on photograph is 10 μm long.

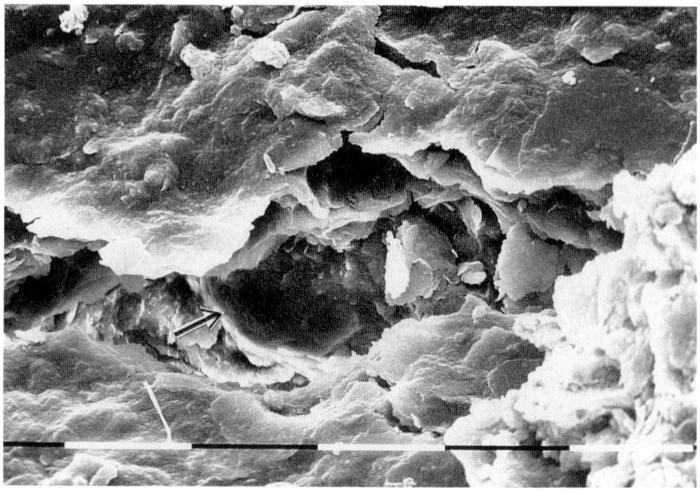

Figure 5.6. Campine Formation/Box core 88B11/Sample S4. "Stressed" microvoid (arrow) perpendicular to the rather flat (DFF) surface. Arrow shows particle edges with very slight outward flaring. White bar on photograph is 10 μm long.

after gas escape. This condition effectively resists deformation due to overburden pressure. The proposed sequence of change is as follows:

a. Development of microvoids as in 1a and b. After gas escape, the microvoids become filled with water, this process occurring just after sedimentation.
b. Continuing sedimentation increases the overburden pressure on the particles, without a significant change in permeability.
c. During the following initial compaction stage permeability is decreased with increasing lateral pressure on the particles surrounding the voids, causing curling, bending, and flaring outward as the pore water resists compression.
d. A "stressed" microvoid with deformed particles and domains causing a flared perimeter surrounding the void is formed.

3. The transformation of a "nonstressed" microvoid into a "stressed" one is due to freezing and consequent volume expansion of water within the void under high pressure. The proposed sequence of events is as follows:
   a. An initial "nonstressed" microvoid in which domain shifting occurs due to increasing gas pressure, cf. 1a and b above.
   b. Escape of gas and filling of the microvoid with water, cf. 2a above.
   c. Finally, freezing and consequent volume expansion of water, causing particle deformation, curling, and flaring outward under high pressure, thereby forming a "stressed" microvoid.

Conclusions

A SEM study of the clay microfabric from Recent marine (North Sea), estuarine (Schelde estuary), and Quaternary (Campine Formation—Lower Pleistocene) deposits reveals that specific differences exist in the types of clay microfabric. We have recorded the occurrence of some previously unrecognized, microstructures and have defined them as "nonstressed" microvoids and "stressed" microvoids. Inasmuch as all samples received identical preparation, we believe the observed differences are real ones and express differences in depositional setting and geological history. Although several mechanisms are invoked to explain the origin of the "stressed" microvoids in the Rijkevorsel Clay Member, we suggest that hypothesis (3),

Figure 5.7. Campine Formation/Box core 88B18/Sample S2. "Stressed" microvoid; particles bending around the void. Particles connected with EF and FF linkages. White bar on photograph is 1 μm long.

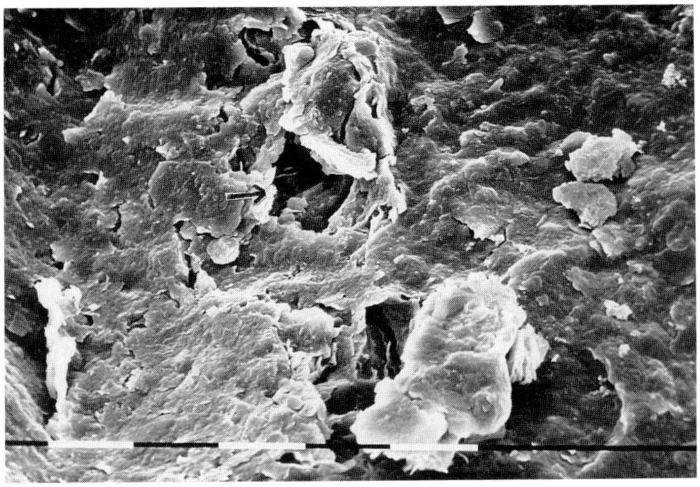

Figure 5.8. Campine Formation/Box core 88B11/Sample S4. "Stressed" microvoid exhibiting the "bursting effect" (arrow) perpendicular to the rather flat (DFF) surface. White bar on photograph is 10 μm long.

involving the freezing of pore water inside the microvoids under periglacial climates during the Pleistocene, is most probable. Our evidence is hypothetical but it seems reasonable that the brackish interstitial water in periodically exposed tidal flat deposits will freshen through desalting during freeze–thaw cycles in periglacial climates. This freshwater saturated sediment will freeze readily with a consequent expansion of the pore water. This expansion will create pressures in excess of the overburden pressure and will result in the existence of "stressed" microvoids within the unconsolidated sediment. We believe that the presence of "stressed" microvoids in Lower Pleistocene estuarine sediments is of considerable interest and importance inasmuch as they record conditions that must have developed soon after deposition and have been retained as early *secondary* features over considerable geologic time. It seems reasonable to suggest that there are other geological situations wherein the present microfabric may still retain some *primary* or early *secondary* characters and be useful in interpreting the original depositional conditions. If so, we may yet be able to utilize microfabrics as indicating certain specific paleoenvironmental conditions and depositional processes.

Acknowledgments

The authors are indebted to the Management unit of the North Sea and Schelde estuary mathematical model for providing ship facilities. The SEM photographs were made by Mr. J. Gillis of the Royal Belgian Institute for Natural Sciences (KBIN). Thanks are expressed to Dr. R. Paepe of the Free University of Brussels and to Dr. L. De Vos of the Université Libre de Bruxelles.

References

Anderson, A.L., 1974. Acoustics of gas-bearing sediments. Applied Research Laboratories Technical Report No. 74-19 (ARL-TR-74-19), Applied Research Laboratories, The University of Texas at Austin, 162 p.

Barden, L., 1972. The influence of structure on deformation and failure in clay soils. Geotechnique, v. 22, p. 159–163.

Barden, L., and G.R. Sides, 1971. Sample disturbance in the investigation of clay structure. Geotechnique, v. 21, p. 211–222.

Bastin, A., 1971. Carte lithologique de fond de la Mer du Nord devant la cote belge et de l'estuaire de l'Escaut au moyen de mesures de la radio-activite naturelle. Panel Meeting International Atomic Energy Agency, Paris.

Bennett, R.H., 1979. Clay fabric and related pore geometry of selected submarine sediment: Mississippi Delta. Scanning Electron Microscopy, v. 1, p. 519–524 and 424.

Bennett, R.H., and M.H. Hulbert, 1986. Clay Microstructure. International Human Resources Development Corporation Press. Prentice-Hall, Boston, 161 p.

Bohlke, B.M., and R.H. Bennett, 1980. Mississippi prodelta crusts: a clay fabric and geotechnical analysis. Marine Geotechnology, v. 4, p. 55–82.

Collins, K., and A.M. McGowan, 1974. The form and function of micro-fabric features in a variety of natural soils. Geotechnique, v. 24, p. 223–254.

Chiou, W., 1980. A new technique in preparing in situ marine sediments for clay fabric study. In: Bailey, G.W. (ed.), 38th Annual Proceedings Electron Microscopy Society of America, San Francisco: Oak Ridge, Tennessee, Electron Microscopy Society of America, p. 204–205.

Faas, R.W., and D.S. Crocket, 1983. Clay fabric development in a deep-sea core: site 515, Deep Sea Drilling Project, Leg 72. In: Barker, P.F., R.L. Carlson, D.A. Johnson, et al. (eds.), Initial Reports of the Deep Sea Drilling Project, v. 72. U.S. Government Printing Office, Washington, D.C., p. 510–535.

Grim, R.E., 1962. Clay mineralogy. Science, v. 135, p. 890–898.

Houbolt, J.J.H., 1968. Recent sediments in the southern bight of the North Sea. Geologie en Mijnbouw, v. 47, p. 245–273.

Lambe, T.W., 1958. The structure of compacted clay. Journal of the Soil Mechanics and Foundation Division, Proceedings of the American Society of Civil Engineers, v. 84, p. 1654-1–1654-34.

Martens, C.S., 1976. Control of methane sediment-water bubble transport by macroinfaunal irrigation in Cape Lookout Bight, North Carolina. Science, v. 192, p. 998–1000.

Moon, C.F., 1972. The microstructure of clay sediments. Earth-Science Reviews, v. 8, p. 303–321.

O'Brien, N.R., 1987. The effects of bioturbation on the fabric of shale. Journal of Sedimentary Petrology, v. 57, p. 449–455.

Paepe, R., and R. Vanhoorne, 1970. Stratigraphical position of periglacial phenomenon in the Campine Clay of Belgium, based on palaeobotancial analysis and palaeomagnetic dating. Bulletin Société Belge de Géologie, v. 79, p. 201–211.

Reeburgh, W.S., 1969. Observations of gases in Chesapeake Bay sediments. Limnology and Oceanography, v. 14, p. 368–375.

Schubel, J.R., 1974. Gas bubbles and the acoustically impenetrable or turbid character of some estuarine sediments. In: Kaplan, I.R. (ed.), Natural Gases in Marine Sediments. Plenum Press, New York, p. 275–298.

Smart, P., 1967. Soil structure, mechanical properties and electron microscopy. Ph.D Thesis, Cambridge.

Smart, P., 1975. Soil microstructure. Soil Science, v. 119, p. 385–393.

Terwindt, J.H.J., 1977. Mud in the Dutch Delta Area. Geologie en Mijnbouw, v. 56(3), p. 203–210.

Wartel, S., 1977. Composition, transport and origin of sediments in the Schelde estuary. Geologie en Mijnbouw, v. 56, p. 219–233.

Wartel, S., G. Verreet, and W. Viaene, 1985. Internal structures in mud sediments. (The occurrence of gas and its implications.) Proceedings Progress in Belgian Oceanographic Research, Brussels, March 1985, p. 220–222.

CHAPTER 6

Clay Fabric of Fine-Grained Turbidite Sequences from the Southern Nares Abyssal Plain

L.E. Shephard and A.K. Rutledge

Introduction

Deep-water fine-grained sediments are transported and deposited by a continuum of processes including pelagic settling, reworking by bottom currents, and resedimentation by mass transport (Gorsline, 1984; McCave, 1984; Stow and Piper, 1984). All of these processes are significant in deep-water environments but are particularly relevant in the abyssal plain environment that represents the ultimate sink for sediment deposition. Sediment facies models have been developed that incorporate the primary structural, textural, and compositional characteristics resulting from deep-water sedimentation (Piper, 1978; Stow and Bowen, 1980; Stow and Piper, 1984; Stow, 1985). These characteristics are reflected in the geotechnical properties of a sediment and may be evident in the clay fabric. The fine-grained turbidite sequences within the Nares Abyssal Plain, consisting of interbedded pelagic clays and turbidites, provide an opportunity to evaluate the clay fabric of these deep-water facies and to attempt to delineate interrelationships between the facies types and associated fabric characteristics.

Geological Setting

The Nares Abyssal Plain is an east–west elongated feature centered about 450 km northeast of Puerto Rico (Fig. 6.1). Water depths in the central abyssal plain range between 5800 and 5900m with a regional northeastward trending seafloor gradient of less than 1:2000 (Shephard et al., 1988). The Nares is bounded on the north by the Bermuda Rise and on the south by the Greater Antilles Outer Ridge. Vema Gap forms the western boundary, connecting the Nares with the southern distal Hatteras Abyssal Plain. Sediments consist of fine-grained turbidites interbedded with pelagics and hemipelagics. The turbidites are derived largely from the North American continental margin where they are transported across the Hatteras Abyssal Plain and through Vema Gap prior to deposition in a complex sediment dispersal pattern that precludes correlation of major lithologic units over distances of more than a few kilometers (Kuijpers and Duin, 1986).

A turbidite sequence within the Nares Abyssal Plain typically consists of a series of fine-grained silt beds or laminae (mean grain size ~6 μm) often interbedded with turbiditic clays (mean grain size ~3 μm) that grade upward into homogeneous gray or brown clays (mean grain size ~3 μm) that are commonly burrowed and bioturbated (Fig. 6.2; Shephard et al., 1988). The transition between the turbiditic and nonturbiditic clays (hemipelagics, pelagics, resuspended sediments, etc.) is extremely difficult to delineate on the basis of grain size, structure, or composition. The silt layers consist largely of micas, with lesser quantities of chlorites, quartz, and feldspars, and range in thickness from a few millimeters to a maximum of 30 cm (average thickness ~1–3 cm). The nonturbiditic sediments compose approximately 60% of the sedimentary section sampled in the central Nares (Shephard et al., 1988).

Geotechnical property analyses performed on 21 standard piston cores (up to 13 m long; Shephard et al., 1987) and on 4 Stacores (up to 26 m long; Shephard et al., 1989) suggest gradational index property changes (i.e., water content, porosity, and bulk density) between the sediment facies composing a turbidite sequence. Differences in index properties between turbidite sequences can be significant, however, with changes in water content exceeding 70% over depth intervals of a few centimeters. These results reflect the interrelationship between the dominant

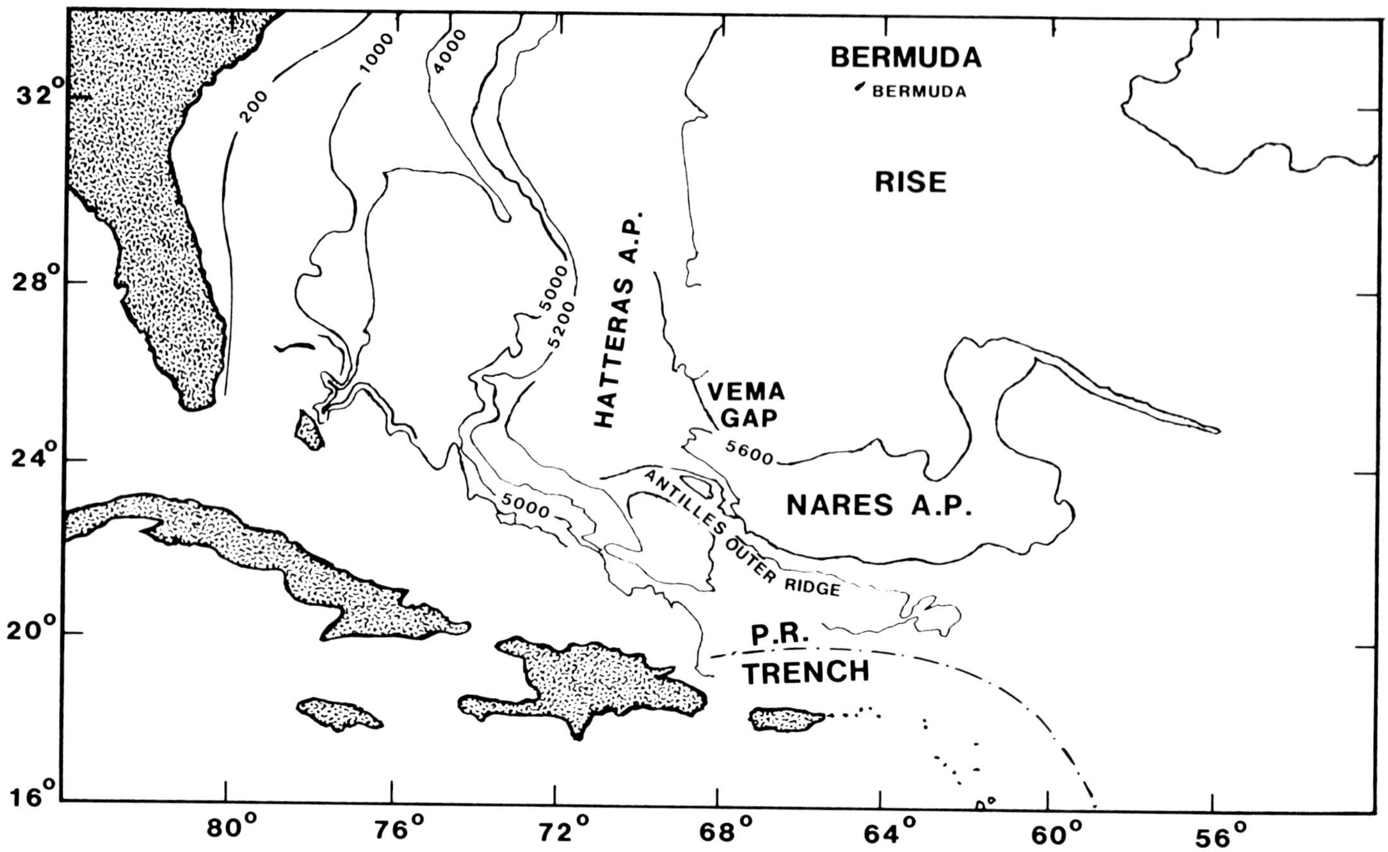

Figure 6.1. Location of the Nares Abyssal Plain in the northwestern Atlantic Basin.

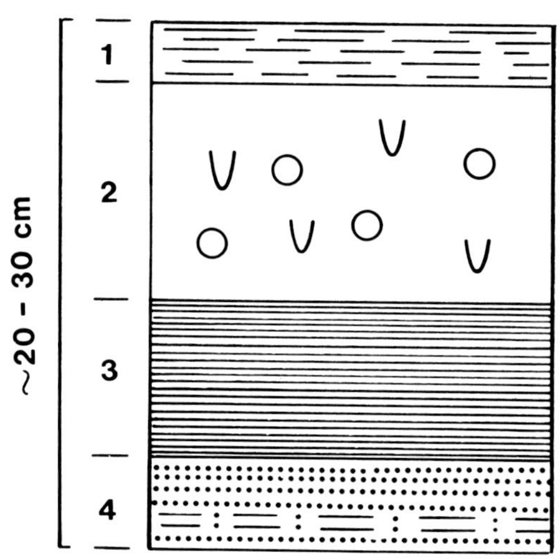

Figure 6.2. Generalized turbidite sequence typical of sediment recovered in the southern Nares Abyssal Plain (unit 1 represents pelagic clay; unit 2 represents bioturbated pelagic clay; unit 3 represents turbiditic clay; and unit 4 represents the basal turbidite silt bed).

depositional processes controlling sediment accumulation and the resulting geotechnical properties.

Techniques for Fabric Sample Preparation and Analyses

The clay fabric sample preparation techniques used followed those described in Bennett et al. (1981) and Chiou et al. (1983). Each fabric sample was obtained from a split core section using a 1.5-cm-diameter piston subsampler and trimmed into cubes with nominal dimensions of 0.5 cm x 0.5 cm x 1 cm. The initial step in the sample preparation was the sequential replacement of the interstitial water with ethyl alcohol, then with amyl acetate, and finally, with liquid carbon dioxide (CO_2) during the critical point drying process. After critical point drying, the samples were placed in BEEM capsules and impregnated with SPURR's resin, a low viscosity ($\sim$60 cps) epoxy resin. To ensure thorough infiltration, each BEEM capsule was filled half-way with the SPURR's, put under vacuum for a minimum of 2 hr, then filled

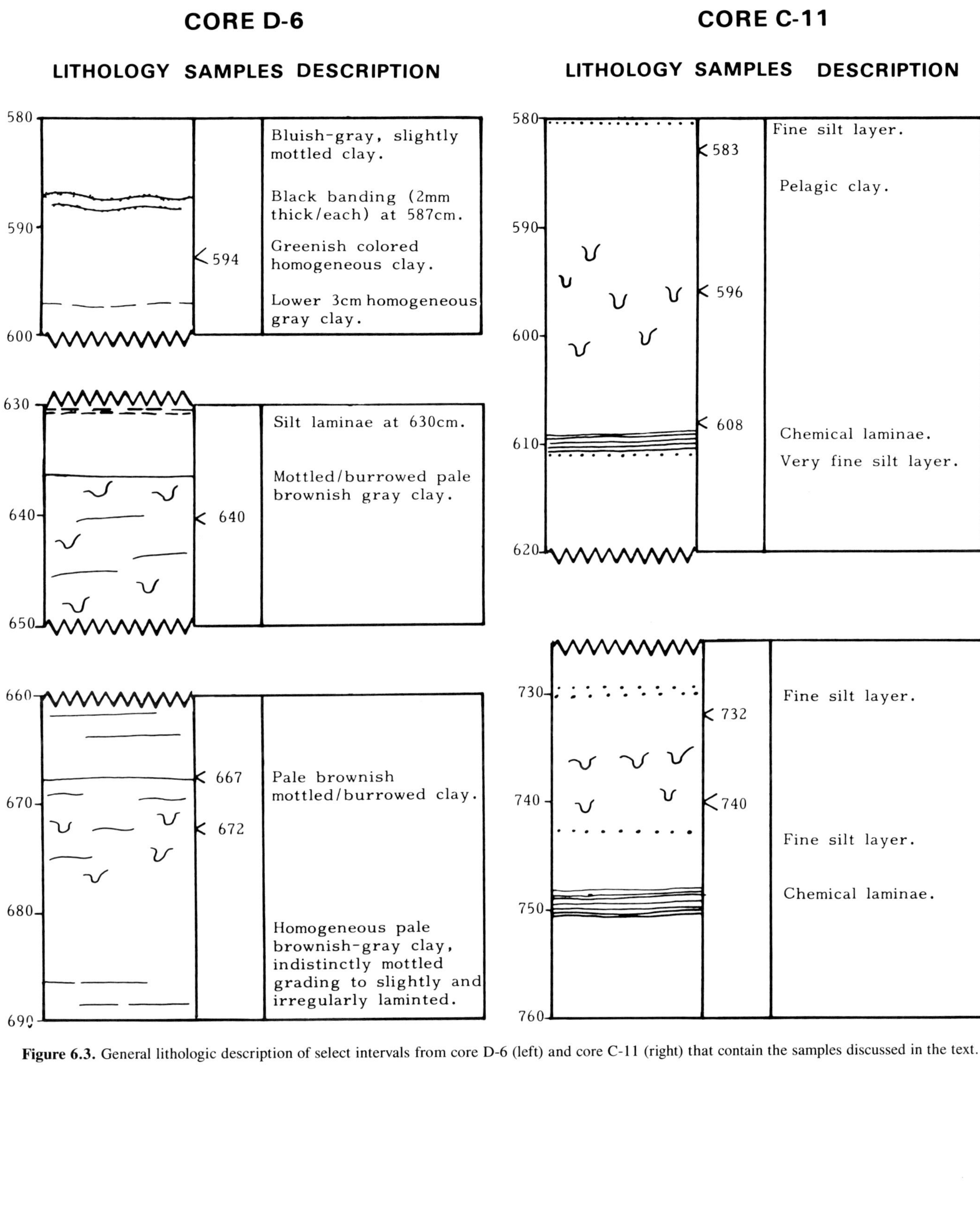

Figure 6.3. General lithologic description of select intervals from core D-6 (left) and core C-11 (right) that contain the samples discussed in the text.

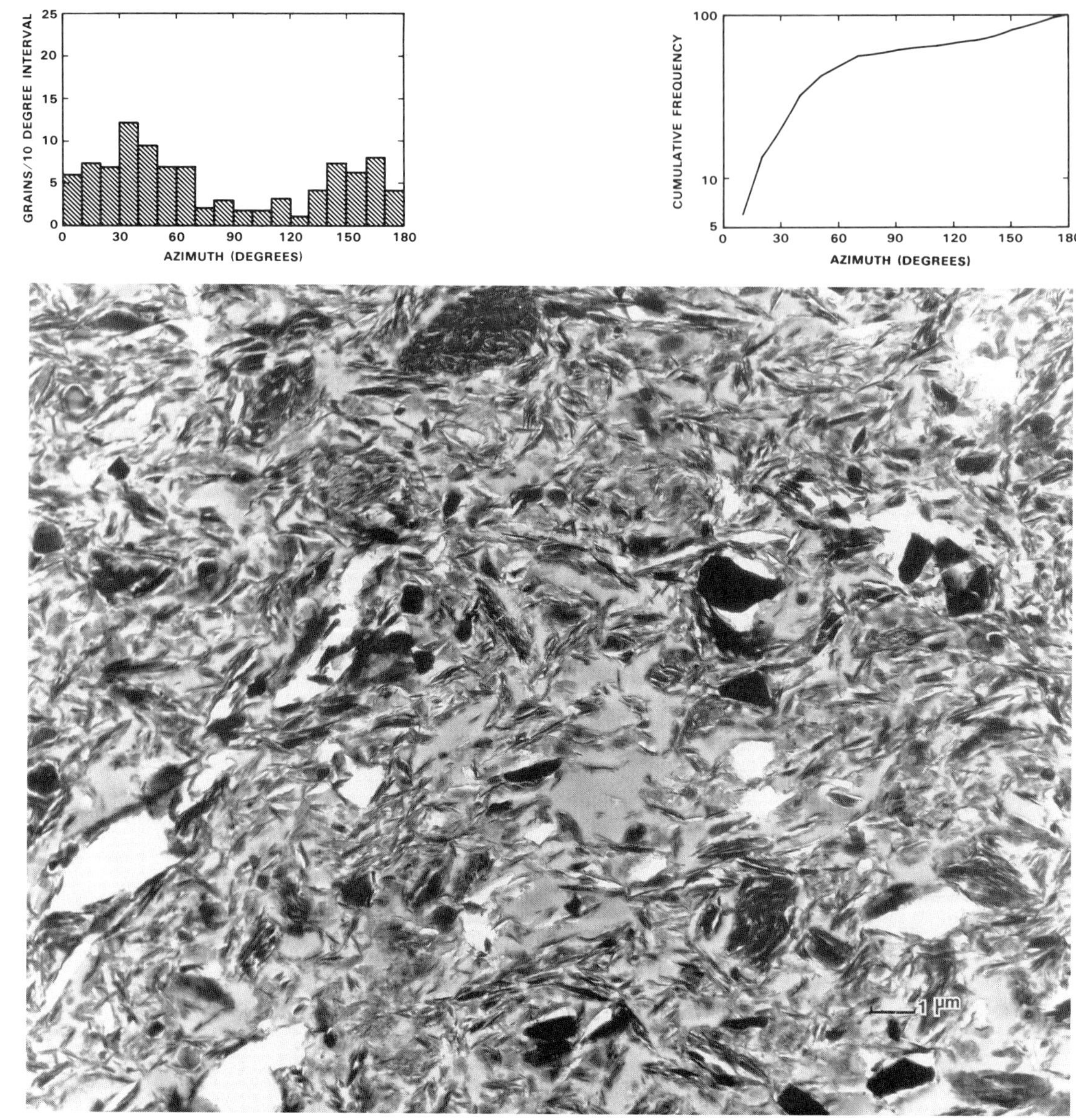

Figure 6.4. Transmission electron micrograph of pelagic clay facies located in core D-6 at 594 cm. Histogram and cumulative distribution plots for the micrographs reflect relative orientation of particles.

completely with the SPURR's and returned to vacuum for another 24 hr. After curing at 23 °C for 24 hr, the samples were removed from the BEEM capsules for ultrathin sectioning. The ultrathin sections (~800–1000 Å) were cut using a diamond knife on a Porter-Blum MT-2B ultramicrotome. The samples were then placed on copper grids, and lightly sputtered with carbon to stabilize them during transmission electron microscopy(TEM) examination. The micrographs were taken at an accelerating voltage of 200 kV on a JEOL 200 cx transmission electron microscope.

The methods used to qualitatively assess the degree of orientation are consistent with those described in Chiou (1981). The only modifications made to Chiou's technique were to utilize an IBM PC/AT personal computer and a Houston Instruments graphics tablet to make the orientation measurements and store the data. The measurements were made along scan lines to avoid biasing the data. One hundred to three hundred orientation measurements were made for each micrograph depending on its magnification; silt grains were not measured.

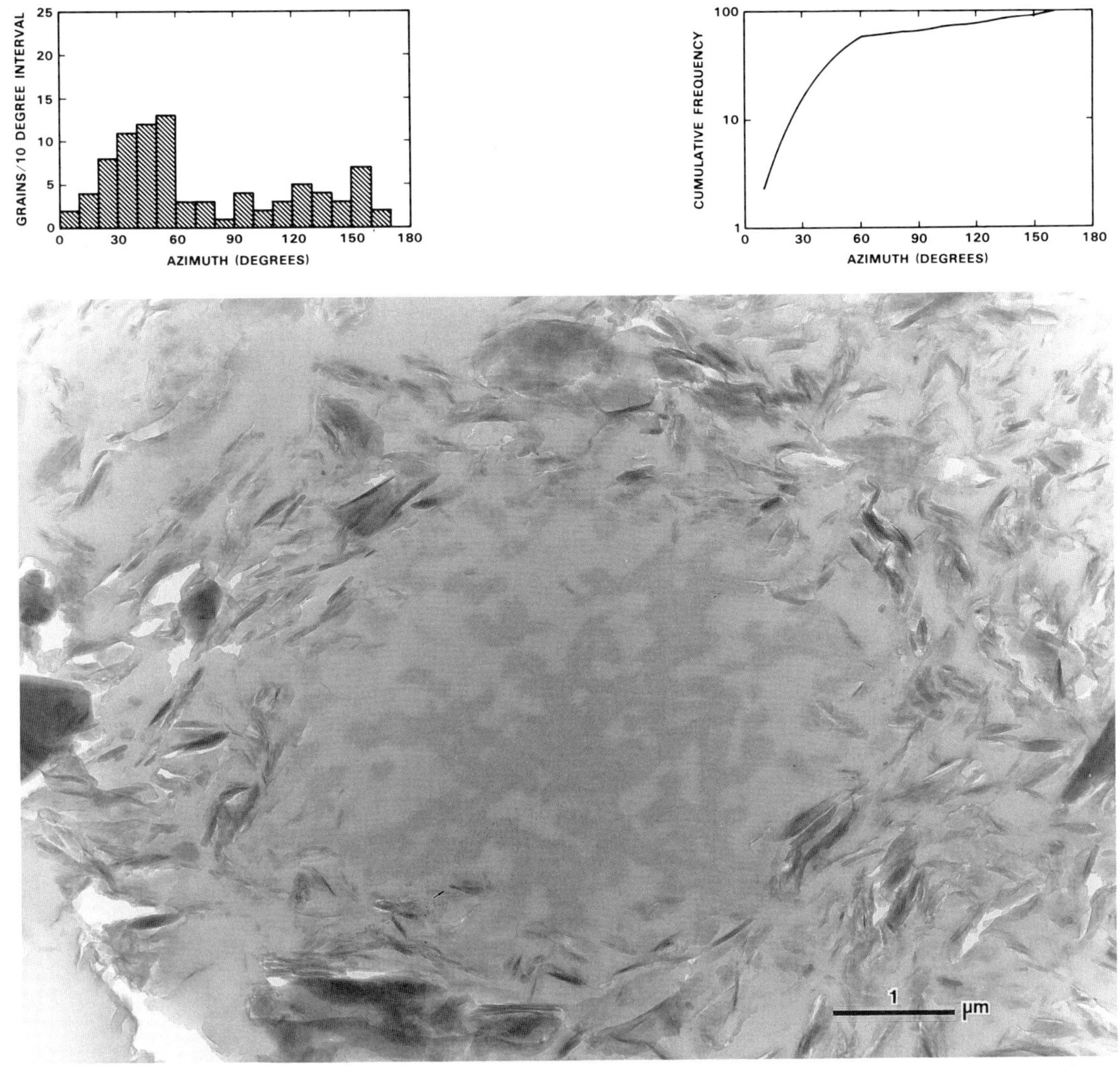

Figure 6.5. Transmission electron micrograph of pelagic clay facies located in core C-11 at 583 cm. Histogram and cumulative distribution plots for the micrographs reflect relative orientation of particles.

Using the graphics tablet, a micrograph was overlain by a sheet of clear plastic marked with the scan lines and the reference axes. Since the x and y axes are user defined for each micrograph, the orientation measurements reflect only the *degree* of orientation for a particular micrograph and cannot be used to relate the particle orientations between two or more micrographs. After measuring the direction of elongation of each clay particle on the scan lines, the data were grouped into eighteen 10° intervals and plotted on histograms that were subsequently used to construct cumulative frequency curves. The histograms and cumulative distribution curves are illustrated for several samples that will be combined as part of an ongoing evaluation to assess the degree of orientation associated with different facies types.

Results of Clay Fabric Analyses

Micrographs of fabric samples obtained from turbidite sequences recovered in piston cores D-6 and C-11 will be used to illustrate the general fabric characteristics associated with each of the dominant clay facies types present in the Nares Abyssal Plain (Figs. 6.4 to 6.11). Samples from core D-6 were obtained from two different turbidite sequences and represent both the pelagic and turbiditic facies (Fig. 6.3). Micrographs presented for samples obtained from core C-11 were obtained from two turbidite sequences ranging in depth from 575 to 612 cm and from 730 to 760 cm (Fig. 6.3). Samples recovered from the homogeneous pelagic clay, bioturbated pelagic clay, and

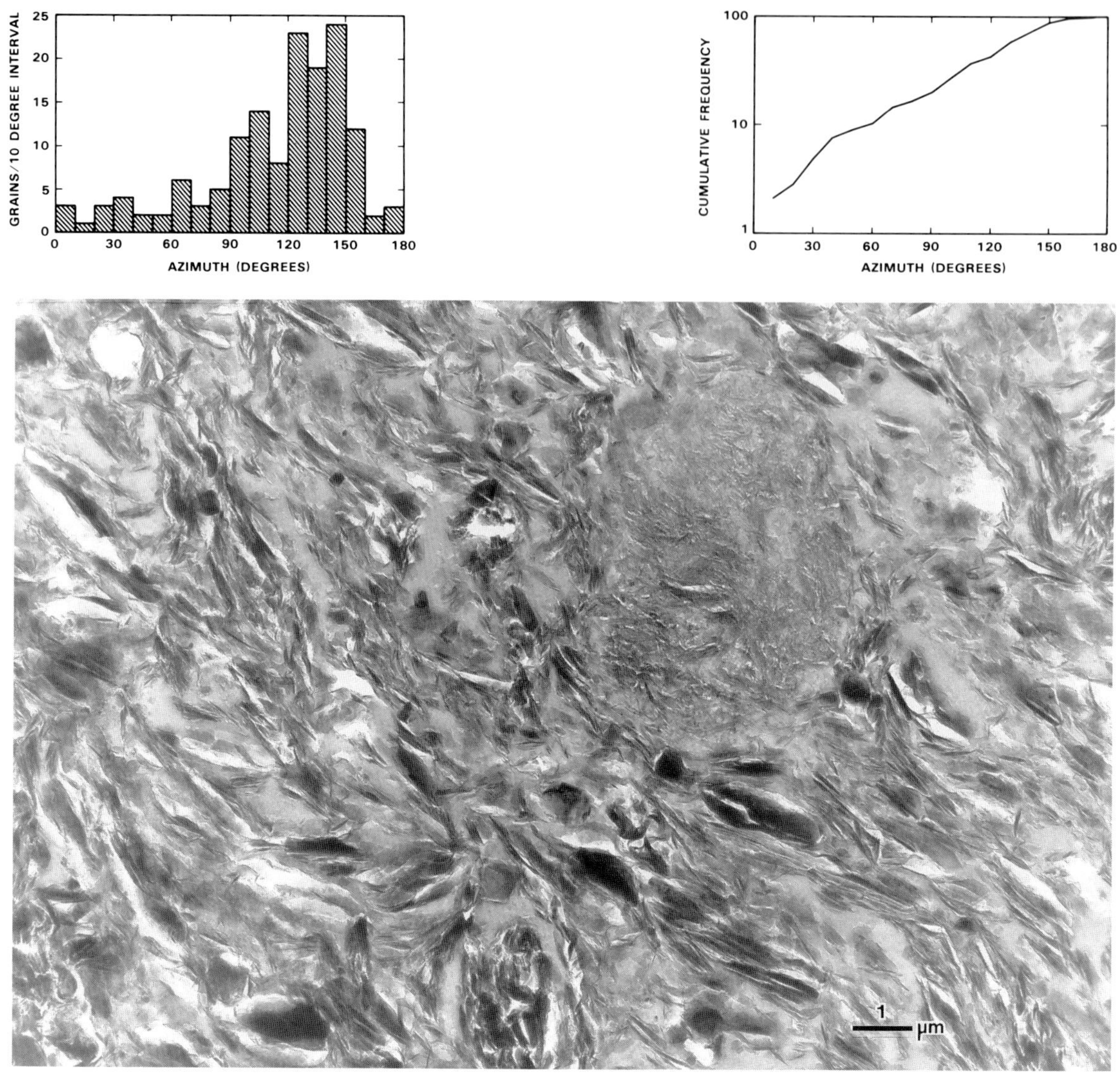

Figure 6.6. Transmission electron micrograph of pelagic clay facies located in core C-11 at 732 cm. Histogram and cumulative distribution plots for the micrographs reflect relative orientation of particles.

turbiditic clay facies from both cores will be described to illustrate similarities and differences.

Pelagic Clay Facies

Samples D-6 594 cm, C-11 583 cm, and C-11 732 cm were obtained within a homogeneous pelagic clay facies at the top of a turbidite sequence (Figs. 6.4, 6.5, and 6.6, respectively). Micrographs from D-6 594 cm are characterized by randomly oriented, face-to-face clay domains interspersed between larger clay floccules and fine silt-size grains. Micrographs from sample C-11 583 cm show well-defined clay particle domains oriented face-to-face. The clay domains appear preferentially oriented around a large pore space that contains a lacey, microcrystalline material, possibly smectite. Voids of this type have been observed in pelagic clay samples recovered in the north-central Pacific abyssal hill province and may possibly be an indicator of very slow sediment accumulation rates. Micrographs from C-11 732 cm reveal clay domains consisting of face-to-face particle domains often arranged in chains interspersed between opaque grains nominally 1–2 μm in size. The large (~ 4–6 μm) aggregate may be a fecal pellet. The degree of orientation determined from these micrographs appears quite

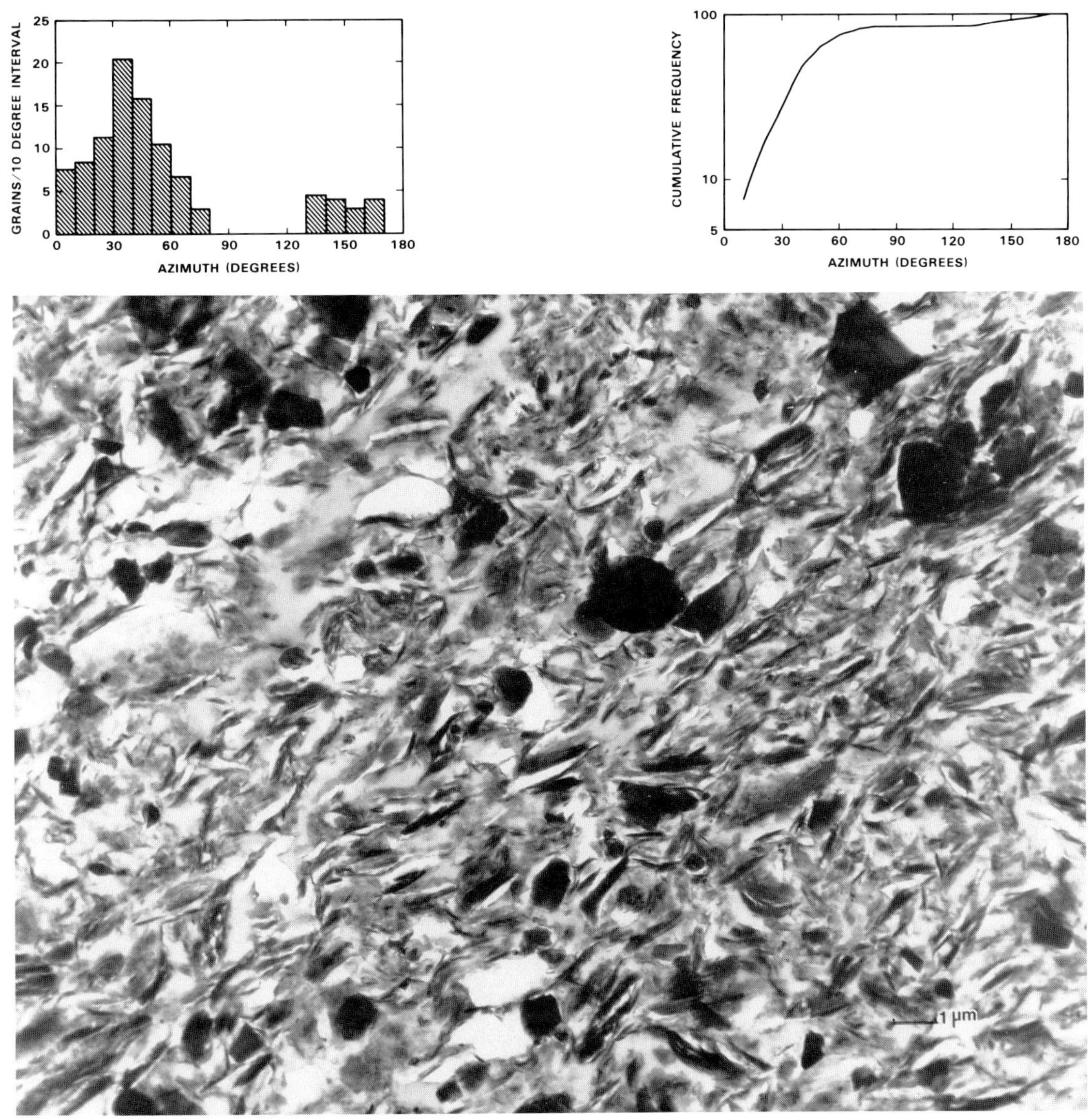

Figure 6.7. Transmission electron micrograph of bioturbated pelagic clay facies located in core D-6 at 640 cm. Histogram and cumulative distribution plots for the micrographs reflect relative orientation of particles.

variable, ranging from random for sample D-6 594 cm to moderate for sample C-11 732 cm.

Bioturbated Pelagic Clay Facies

Samples obtained from D-6 640 cm and C-11 596 cm and 740 cm were recovered from the bioturbated pelagic clay interval of the turbidite sequence (Figs. 6.7, 6.8, and 6.9, respectively). Micrographs from sample D-6 640 cm show numerous fine silt grains interspersed within a matrix consisting of clay particle domains and floccules. Micrographs from sample C-11 596 cm show a series of clay domains characterized largely by face-to-face particle contacts, often aligned in chains with edge-to-face contacts. Small particle aggregates are often present between the larger domains. Striations on the large void are knife marks produced during sectioning. Sample C-11 740 cm is characterized by a series of clay floccules and edge-to-face platelets within a finer grained matrix that also contains opaque grains nominally 1–2 μm in diameter.

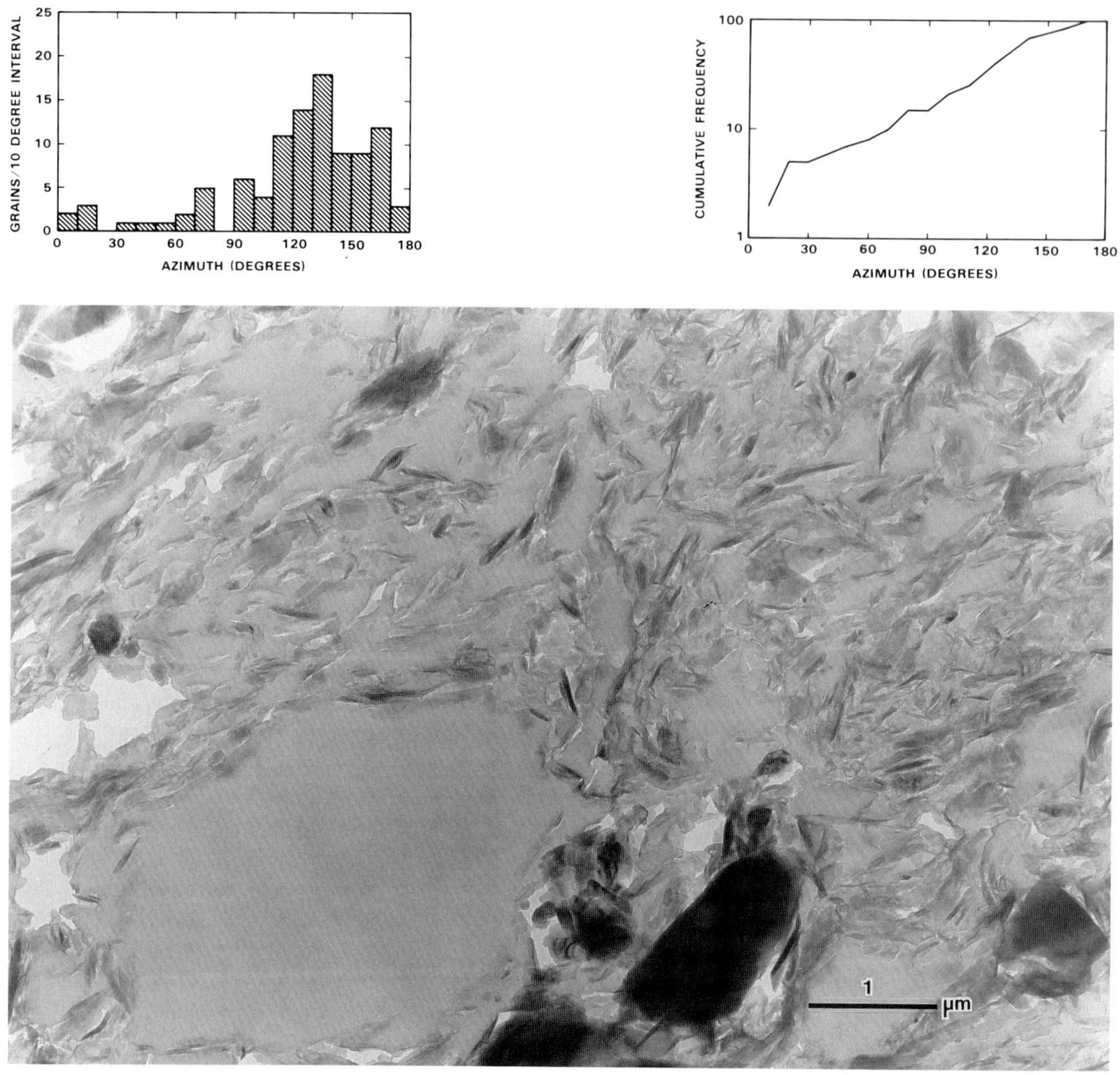

Figure 6.8. Transmission electron micrograph of bioturbated pelagic clay facies located in core C-11 at 596 cm. Histogram and cumulative distribution plots for the micrographs reflect relative orientation of particles.

Turbiditic Clay Facies

Samples D-6 667 cm and C-11 608 cm were obtained from the turbiditic clay facies (Figs. 6.10 and 6.11, respectively). Micrographs from core D-6 667 cm show randomly distributed clay floccules and silt grains within a randomly distributed clay particle matrix. Micrographs from sample C-11 608 cm located at the base of the turbidite sequence shows several features that may be representative of the turbidite facies. The micrographs show a random distribution of clay particle domains interspersed within opaque silt particles and clay floccules. The degree of orientation determined for these micrographs appears random, although the limited number of individual clay particles limits confidence in this assessment.

Observations and Discussion

Preliminary observations based on these and other micrographs suggest that the fabric of the turbidite sequences can be characterized on the basis of the dominant constituents observed. The pelagic facies is dominated by a fairly uniform grain size with a very limited number of silt grains larger than 1–2 μm. Clay floccules and fecal pellets are frequently observed. The clay particles are often arranged in chains except where disrupted by clay floccules, silt grains or other constituents. The bioturbated pelagic clay facies has characteristics observed in both the pelagic and turbiditic facies types. The clay particles are generally arranged in well-defined domains. A limited number of

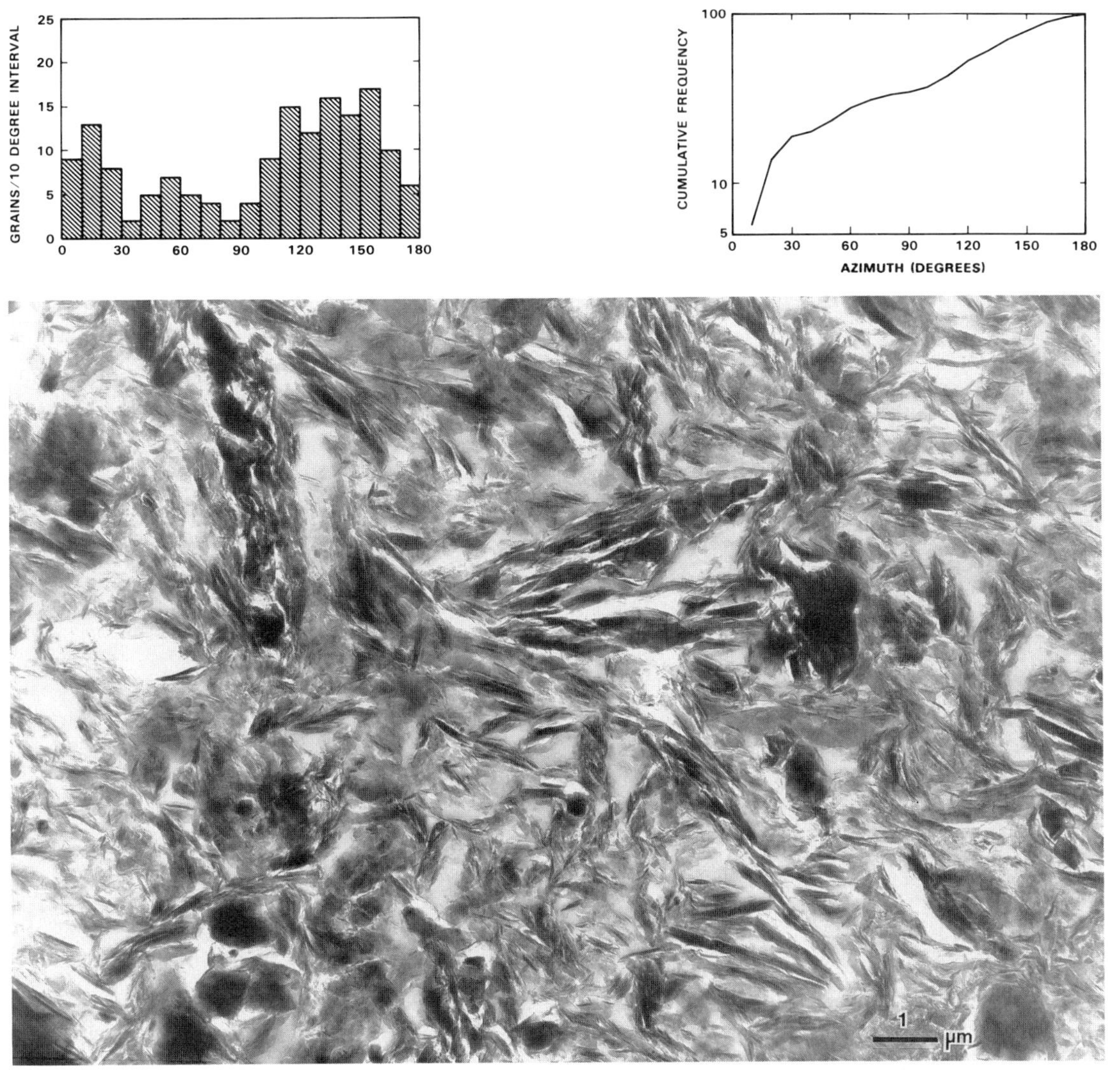

Figure 6.9. Transmission electron micrograph of bioturbated pelagic clay facies located in core C-11 at 740 cm. Histogram and cumulative distribution plots for the micrographs reflect relative orientation of particles.

chains are apparent, possibly because of numerous other non-clay constituents disrupting the chains. The turbidite facies contain numerous clay floccules and silt grains interspersed within a clay matrix. A limited number of chains are apparent, although the domains appear interspersed between the larger particles.

Results from O'Brien et al. (1980) indicate a pronounced change in particle orientation between hemipelagic and turbiditic silts and siltstones (generally $> 60\%$ silt) using scanning electron microscopy. The hemipelagic clays exhibited preferred orientation, whereas the turbidites appeared randomly oriented. They attributed this variation in the degree of orientation to changes in the conditions controlling sediment transport and deposition.

Results from our study also suggest differences in the clay fabric between the pelagic and turbiditic clays, but these differences are less pronounced and reflected more by changes in particle type than by particle orientation. This may suggest that the accumulation rates and possibly the energy regime controlling deposition of the clay facies associated with the turbidites were lower than those associated with the turbidites discussed by O'Brien et al. (1980). Changes resulting from postdepositional processes (i.e., bioturbation) may also be more significant in the Nares turbidites. Additional studies on these and other fine-grained turbidite sequences are planned. Samples from several turbidite sequences recovered from long cores (up to 26 m) are being studied to further define these relationships and to

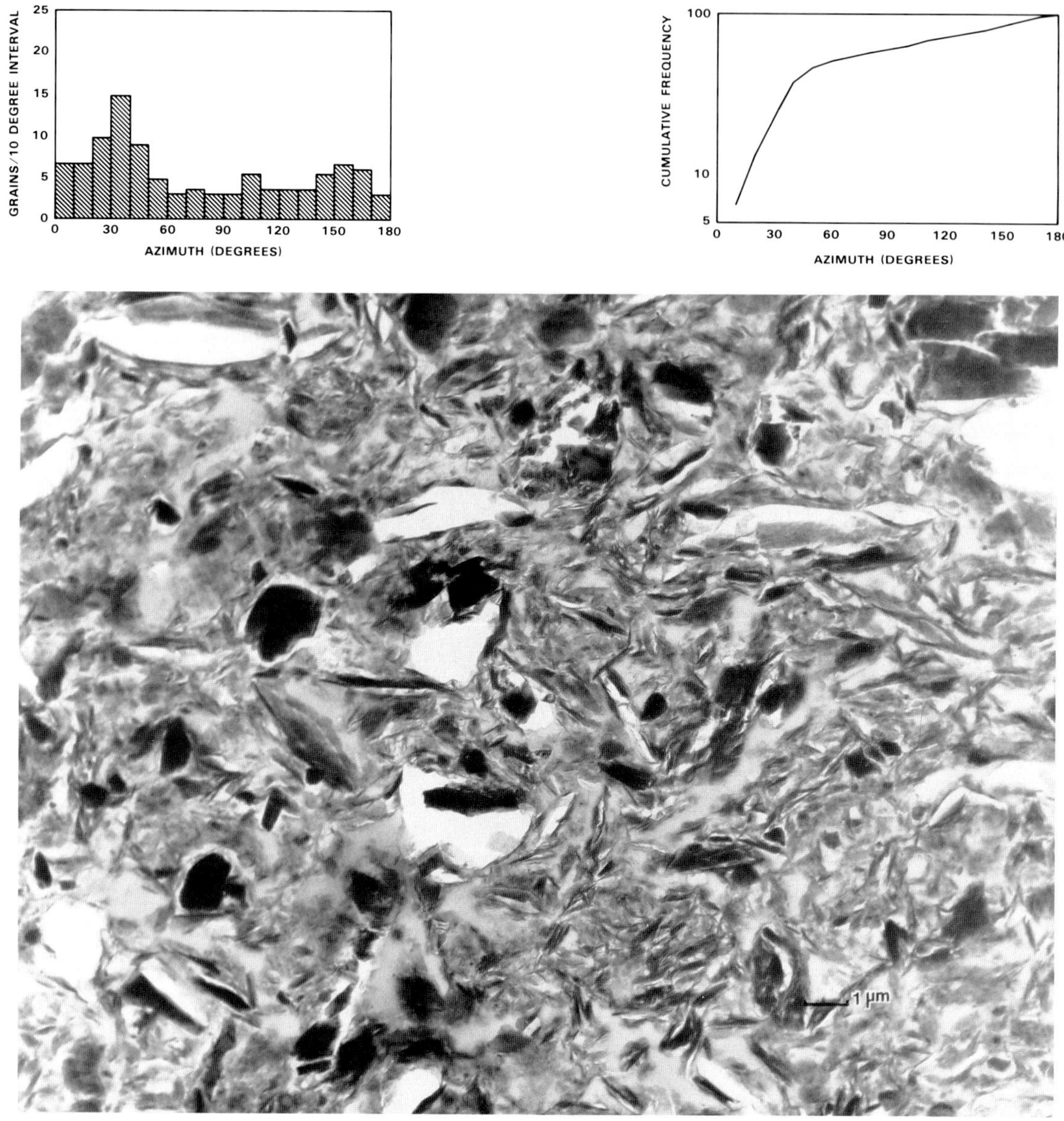

Figure 6.10. Transmission electron micrograph of turbiditic clay facies located in core D-6 at 667 cm. Histogram and cumulative distribution plots for the micrographs reflect relative orientation of particles.

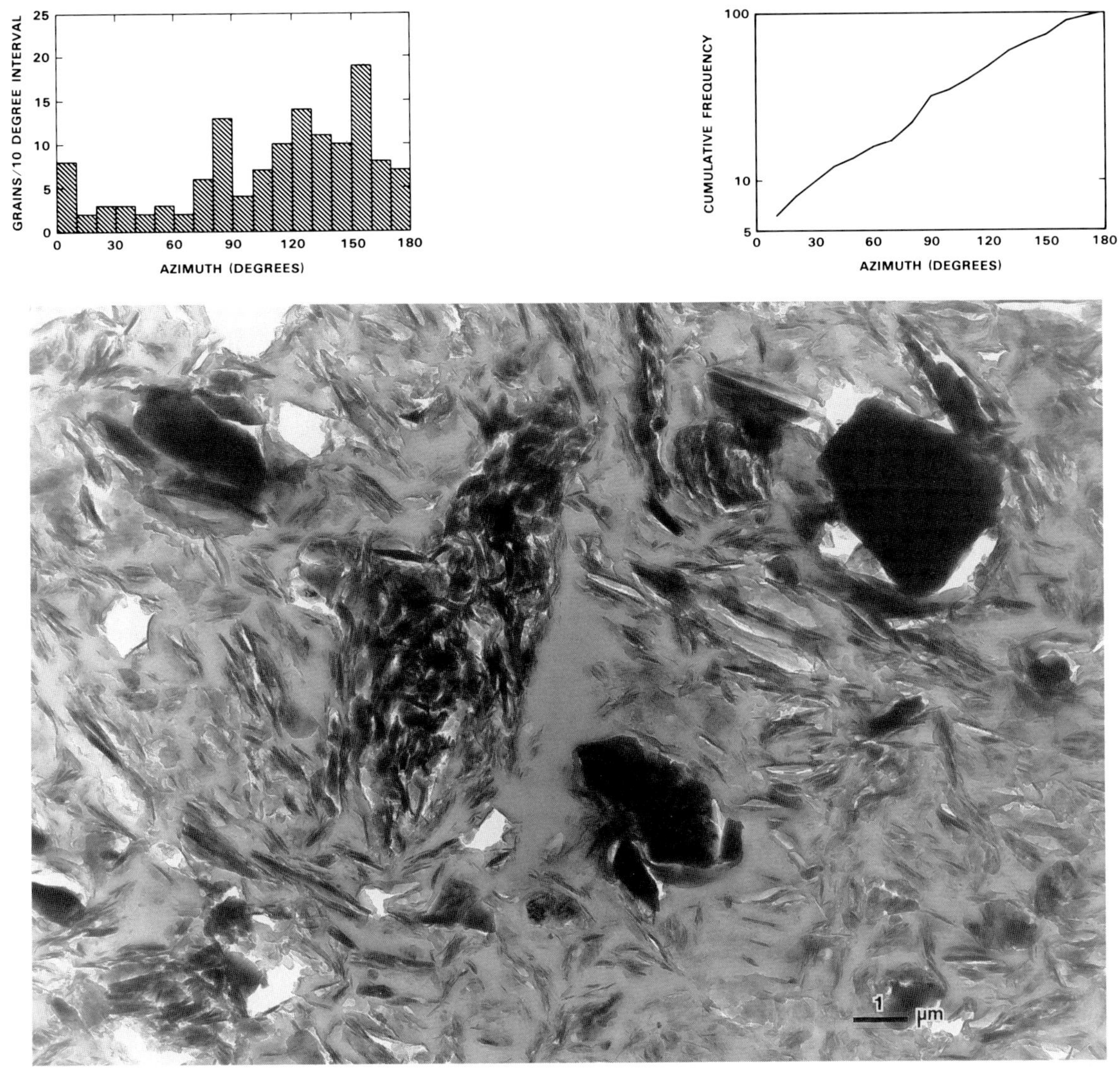

Figure 6.11. Transmission electron micrograph of turbiditic clay facies located in core C-11 at 608 cm. Histogram and cumulative distribution plots for the micrographs reflect relative orientation of particles.

evaluate the effects of consolidation on the particle-to-particle relationships observed within each facies types.

Acknowledgments

This work was supported by the U.S. Department of Energy under contract DE-AC04-DP-00789.

References

Bennett, R.H., W.R. Bryant, and G.H. Keller, 1981. Clay fabric of selected submarine sediments: fundamental properties and models. Journal of Sedimentary Petrology, v. 51, no. 1, p. 217–232.

Chiou, W.A., 1981. Clay fabric of gassy submarine sediments. Ph.D. Dissertation, Texas A&M University, College Station, Texas, 248 p.

Chiou, W.A., L.E. Shephard, W.R. Bryant, and M.P. Looney III, 1983. A technique for preparing high water content clayey sediments for thin and ultrathin section study. Sedimentology, v. 30, no. 2, p. 295–299.

Gorsline, D.S., 1984. A review of fine-grained sediment origins, characteristics, transport and deposition. In: Stow, D.A.V., and D.J.W. Piper (eds.), Fine-Grained Sediments: Deep-Water Processes and Facies. Blackwell, Oxford, p. 17–34.

Kuijpers, A., and E.J. Th. Duin, 1986. Boundary current controlled sediment distribution patterns in the Nares Abyssal Plain. Geo-Marine Letters, v. 6, no. 1, p. 21–18.

McCave, I.N., 1984. Erosion, transport and deposition of fine-grained marine sediments. In: Stow, D.A.V., and D.J.W. Piper (eds.), Fine-Grained Sediments: Deep-Water Processes and Facies. Blackwell, Oxford, p. 35–70.

O'Brien, N.R., K. Nakazawa, and S. Tokuhashi, 1980. Use of clay fabric to distinguish turbiditic and hemipelagic siltstones and silts. Sedimentology, v. 27, no. 1, p. 47–61.

Piper, D.J.W., 1978. turbiditic muds and silts on deep-sea fans and abyssal plains. In: Stanley, D.J., and G. Kelling (eds.), Sedimentation in Submarine Canyons, Fans, and Trenches. Dowden, Hutchinson and Ross, Stroudsburg, PA, p. 163–176.

Shephard, L.E., A.K. Rutledge, W.R. Bryant, and K.M. Moran, 1987. Geotechnical characteristics of fine-grained turbidite sequences from the Nares Abyssal Plain. In: Weaver, P.P.E., and J. Thomson (eds.), Geology and Geochemistry of Abyssal Plains. Geological Society Special Publication 31. Blackwell, Oxford, p. 131–146.

Shephard, L.E., G.A. Auffret, D.A. Buckley, R.C. Searle, and R.T.E. Schuttenhelm, 1988. Feasibility of disposal of High-Level Radioactive Wastes into the Seabed: Volume 3, Geoscience Characterization Studies. Nuclear Energy Agency/Organization for Economic Co-operation and Development, Paris, France. 318 p.

Shephard, L.E., P.J. Schultheiss, and H. Brandes, 1989. Geotechnical properties of Nares Abyssal Plain sediments: shipboard results from the ESOPE expedition. In: Schuttenhelm, R.T.E., G.A. Auffret, D.E. Buckley, R.E. Cranston, C.N. Murray, L.E. Shephard, and A.E. Spijkstra (eds.), Geoscience Investigations of Two North Atlantic Abyssal Plains—the ESOPE International Expedition, Volume II. CEC-Joint Research Center, JRC Report, in press.

Stow, D.A.V., 1985. Fine-grained sediments in deepwater: an overview of processes and facies models. Geo-Marine Letters, v. 5, no. 1, p. 17–23.

Stow, D.A.V., and A.J. Bowen, 1980. A physical model for transport and sorting of fine-grained sediments by turbidity currents. Sedimentology, v. 27, no. 1, p. 31–46.

Stow, D.A.V., and D.J.W. Piper, 1984. Deep-water fine-grained sediments: facies models. In: Stow, D.A.V., and D.J.W. Piper (eds.), Fine-Grained Sediments: Deep-Water Processes and Facies. Blackwell, Oxford, p. 611–640.

CHAPTER 7

Microfabric and Physical Properties Characteristics of a Consolidated Clay Section: ODP Site 697, Weddell Sea

William R. Bryant, Richard H. Bennett, Patti J. Burkett, and F.R. Rack

Introduction

The consolidation of sediment is the main cause of porosity reduction with depth in the upper 1000 m of the marine sediment column and the process is mostly mechanical: the weight of the overlying sediment drives the rearrangement of the individual particles and groups of sedimentary particles and domains. The mechanics of particle reorientation may be understood best through an examination and study of the sediment microfabric. A clay-rich sediment section 318 m thick, recovered during Ocean Drilling Program Leg 113 from the South Orkney Continental Margin, Site 697 in the Weddell Sea (Fig. 7.1) was examined by scanning and transmission electron microscopy of sediment stubs and ultrathin sections. Reorientation of randomly arranged particles of this fine-grained, high porosity (70–75%) sediment occurs very gradually: porosity decreases to only about 50% at a depth of 318 m, because of the very fine-grained nature of the sediment and the presence of extremely fine-grained smectite, which imparts a very low permeability even at porosities of 50%.

The purpose of this study is to characterize the microstructure of a slowly consolidating clayey sediment mass as it compacts from a mud to a mudstone and evaluate the relationship of sediment pore size, microfabric, and mineralogy to the porosity and the permeability of the material.

Microstructure

Two fundamental properties of a clayey sediment, the fabric and physicochemistry, termed the microstructure, strongly influence and largely control the physical and mechanical properties of a sediment, including its consolidation behavior.

An insight into the nature and behavior under stress of the silt-size particles and clay-size aggregates or domains (integral elements of the microstructure) that exist in the sediments at Site 697 can be obtained using techniques of electron microscopy (EM). The clay minerals and larger sized particles in most sedimentary deposits are not of the same size, shape, and mineralogy; these characteristics, which are intimately related to the microstructure, play a significant role in establishing the ultimate fabric and sediment properties at the time of deposition and throughout the history of the deposit (Bennett et al., 1977). An understanding of clay microstructure and the methods of analysis are important because the results and interpretations have far-reaching effects on the fundamental conclusions and ideas of the basic origin, nature, and sources of sediment components. Understanding of clay microstructure is an important prerequisite to understanding the physical behavior of sedimentary deposits in response to static and dynamic loads and to variations in environmental conditions (Bennett et al, 1977; Lambe, 1958a).

At burial depths greater than 1000 m, the processes of mechanical loading, cementation, and recrystallization appear inadequate as an explanation for porosity loss and fluid expulsion in these clayey sediments. Studies by Hinch (1980) suggest that the process controlling porosity reduction in shales may be thermophysical as well as mechanical. This involves thermodesorption of the structured water as temperature increases with depth of burial.

Since reduction in porosity in the upper 1000 m of sediment is predominantly mechanical, a knowledge of the microstructure of a consolidating sediment mass will shed light on the mechanics of the consolidation process (Bennett et al., 1977; Bryant et al., 1981; Bryant and Bennett, 1988).

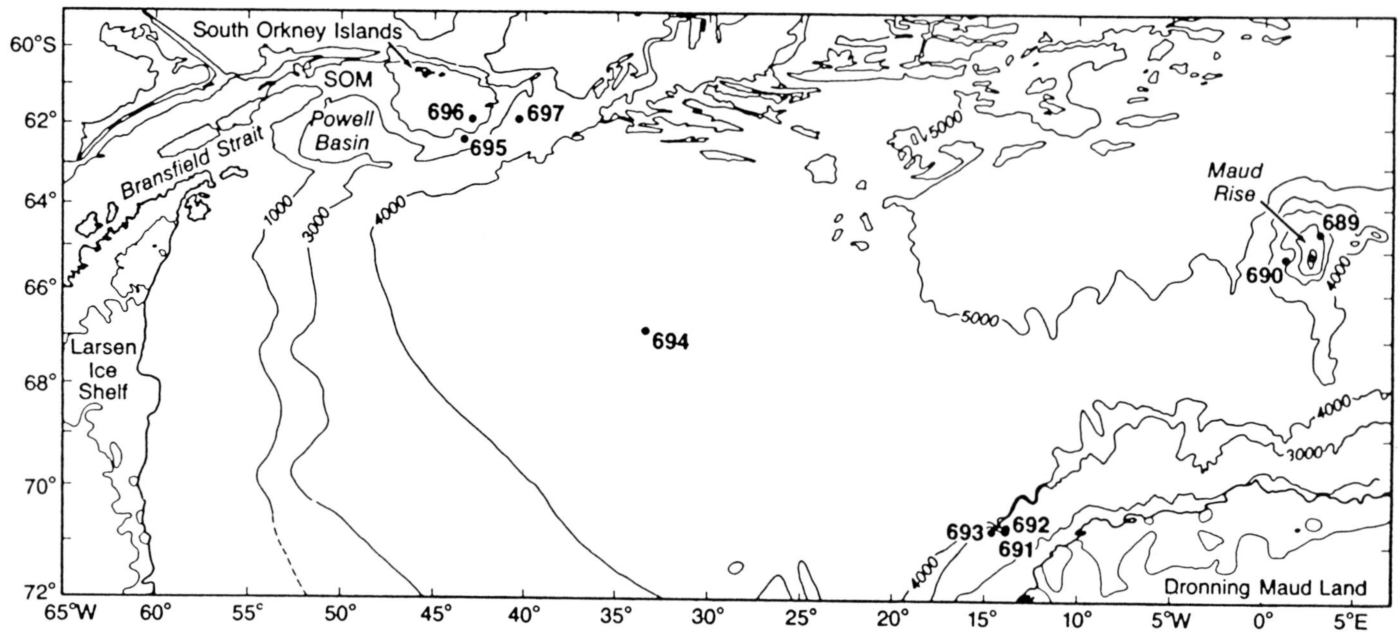

Figure 7.1. Location of Site 697, Jane Basin, Weddell Sea.

Physicochemistry of Clays

The physicochemistry of clays is critical in determining the clay fabric of a marine deposit, especially in the early stage of formation (particles in suspension and at the depositional interface). Detailed studies of clay microstructure and its influence on the physical properties of sediment have been presented by Winterkorn (1948), Bolt (1956), Lambe (1958a), Rosenquist (1959, 1962), Buchanan (1964), and Ingles (1968). Discussions and studies concerning the physicochemistry of clays and colloids are found in Kruyt (1952) and van Olphen (1963).

The physicochemical forces controlling clay mineral structure and the initial framework of clay fabric are essentially electrical in character. Suspended clay particles interact in response to the electrical nature of the particle surface and to the electrochemical characteristics of the surrounding aqueous medium. The phyllosilicate clay minerals are composed of sheets (tetrahedral and octahedral coordination polyhedra) generally in 1-to-1 or 1-to-2 layers, such as two tetrahedral sheets plus one octahedral sheet (smectite, illite, and chlorite). This imparts a platy or sheet-like characteristic to many clay minerals and, for practical purposes, they can be considered as being two-dimensional. In some cases, van der Waals forces and hydrogen bonds hold the two-dimensional clay mineral layers together to form relatively thin particles. During flocculation of suspended clay particles, deposition, and low-pressure diagenesis, clay crystals are not generally broken, nor are the clay layers separated (Lambe, 1958a). The breaking of interlayer bonds appears to indicate the onset of high-pressure diagenesis as revealed through studies of consolidated kaolinite (Cabrera and Smalley, 1971).

Electrostatic interactions are critical in determining clay microstructure, particularly clay fabric, because clay particles carry a net negative charge that is large compared to the particle mass and the net charge varies with the clay mineral type (Thiessen, 1942; van Olphen, 1963; Bennett et al., 1977; Yariv and Cross, 1979; Bennett and Hulbert, 1986; Bryant and Bennett, 1988).

Microfabric

Early studies of clay fabric (Terzaghi, 1925; Goldschmidt, 1926; Casagrande, 1932; Lambe, 1953, 1958b) were followed by clay fabric studies employing X-ray and high-resolution electron microscopy techniques that depicted the actual arrangements and orientation of clay particles. The importance of fabric in determining the physical properties of sediments has been established (Houston and Mitchell, 1969; Mitchell and Houston, 1969; Torrance, 1970). Bonding and effective stress and strength of soil have been intensely studied (Mitchell et al., 1969, 1971; Anderson and Douglas, 1970; Singh, 1970). Olson and Mesri (1970) discussed the influences of both the mechanical and physicochemical mechanisms important in the compressibility of clays. Bryant and Bennett (1988) examined the fabric of Pacific Ocean red clays and revealed that one of the basic components of red clays is aeolian shale clasts. Their paper demonstrates the use of fabric to solve geological problems.

A few typical fabric types have been classified as honeycomb (Terzaghi, 1925; Cassagrande, 1932), cardhouse (Goldschmidt, 1926; Lambe, 1953), turbostratic (Aylmore and Quirk, 1960), bookhouse (Sloane and Kell, 1966), and stairstep (O'Brien,

1971). Recently studies have revealed that the single plate concepts of fabrics are not wholly tenable and that the multiple unit, domain-type fabric is the rule for most sediments (Bennett, 1976; Bennett et al., 1981, 1989; Bennett and Hulbert, 1986). The variation in particle size, shape, and composition, coupled with the depositional environment, physicochemistry, transport mechanisms and energies, and changing environmental conditions, all increase the complexity of the sediment fabric for a particular sedimentary deposit. Numerous studies (Lambe, 1958a,b; Rosenqvist, 1959) have dealt with the fabric and engineering behavior of laboratory-prepared material. Qualitative and ultimately quantitative studies of sediment fabric and microstructure of naturally occurring sediments lead to predictive capabilities of the physical behavior of clay sediments, and provide a clearer understanding of the geotechnical properties, and a better understanding of complex sedimentological processes.

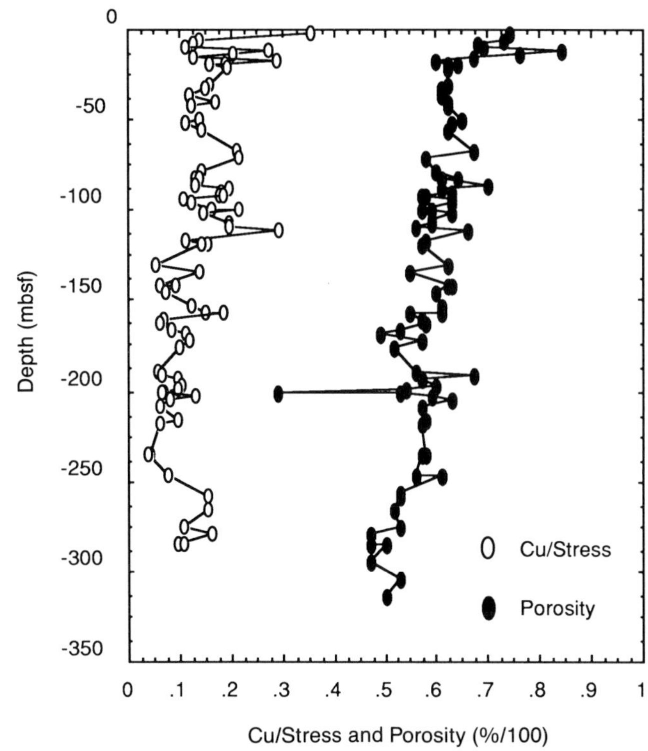

Figure 7.2. Ratio of undrained shear strength to vertical effective stress and decimal percent porosity for Site 697 plotted against depth. Ratios between 0.2 and 0.5 are considered normally consolidated.

Sediment Characteristics

The water depth at Site 697 (Fig. 7.1) is 3484 m. The 322.9 m drilled section has been divided into two units (Barker et al., 1988). Unit I (0–293 mbsf; Pliocene to Pleistocene) contains silty and clayey mud, diatom-bearing silty and clayey mud, clay, and diatom clayey mud. Unit II (293.0–322.9 mbsf; lower Pliocene) consists of silty and clayey mud, and is more coarse grained than Unit I with ice-rafted detritus; diatoms are very rare.

Figure 7.2 illustrates the porosity and ratio of undrained shear strength to vertical stress plotted against depth for sediments at Site 697. The porosity profile at this site is typical of a marine silty clay. The porosity of sediment below 15 mbsf decreases with depth in a more or less steady fashion at 0.04%/m. The low abundance of diatomaceous material, usually less than 15%, is reflected in the steady decrease in porosity with depth. The strength/stress ratio illustrated in Figure 7.2 suggests that sediments at Site 697 are underconsolidated except for several samples above 15 mbsf and between the 70 and 110 mbsf levels. The degree of underconsolidation increases with depth up to 240 mbsf, at which point the sediments become less underconsolidated.

Figure 7.3 illustrates the curves of void ratio versus log vertical stress generated from consolidation test data. The silty clays of Site 697 are highly compressible. It is interesting to note that the final void ratio at the 3200 kPa stress level is similar regardless of the initial void ratio of the sediments. A summary of the consolidation test results is given in Table 7.1. The four samples tested are underconsolidated to highly underconsolidated at greater than approximately 100 m depth. The underconsolidation of these sediments reflects their permeability. The sediments in the deeper portions of the section have permeabilities that range from 9×10^{-9} to 5×10^{-10} cm/sec.

The field consolidation curve for Site 697 is shown in Figure 7.4. The fairly flat slope of this curve and the closer grouping of the data, compared with curves for sediments with large variations in diatom content from other Weddell Sea sites, are typical of marine clay investigated from other areas of the world (Bryant and Rack, 1990). Curves of void ratio and wet bulk density versus depth are depicted in Figures 7.5 and 7.6. Although void ratio and wet bulk density vary considerably, the sedimentary section clearly shows a gradual consolidation and dewatering process as a function of overburden stress (Fig. 7.6B).

The dewatering (consolidation or compaction) of fine-grained sediment is a result of overburden stress and the mechanical rearrangement of the constituent particles. Consolidation is a complex, nonlinear, time-dependent process and the degree of consolidation is a function of not only burial depth (overburden stress) but also numerous complex properties of the particular sedimentary deposit. The "anomalous," poorly understood, consolidation process of deep-sea red clays was studied in detail in terms of the mechanical properties, compressibility, and microfabric response (Bryant and Bennett, 1988), however, a complete understanding of the dewatering of marine sediments has not been realized. A complicating factor in the consolidation process is the variable sedimentation rates that occur throughout the geologic history of a deposit. The sedimentation rates translate directly into rates of loading, or the amount of overburden stress per unit area as a function of time. A plot of the rates of

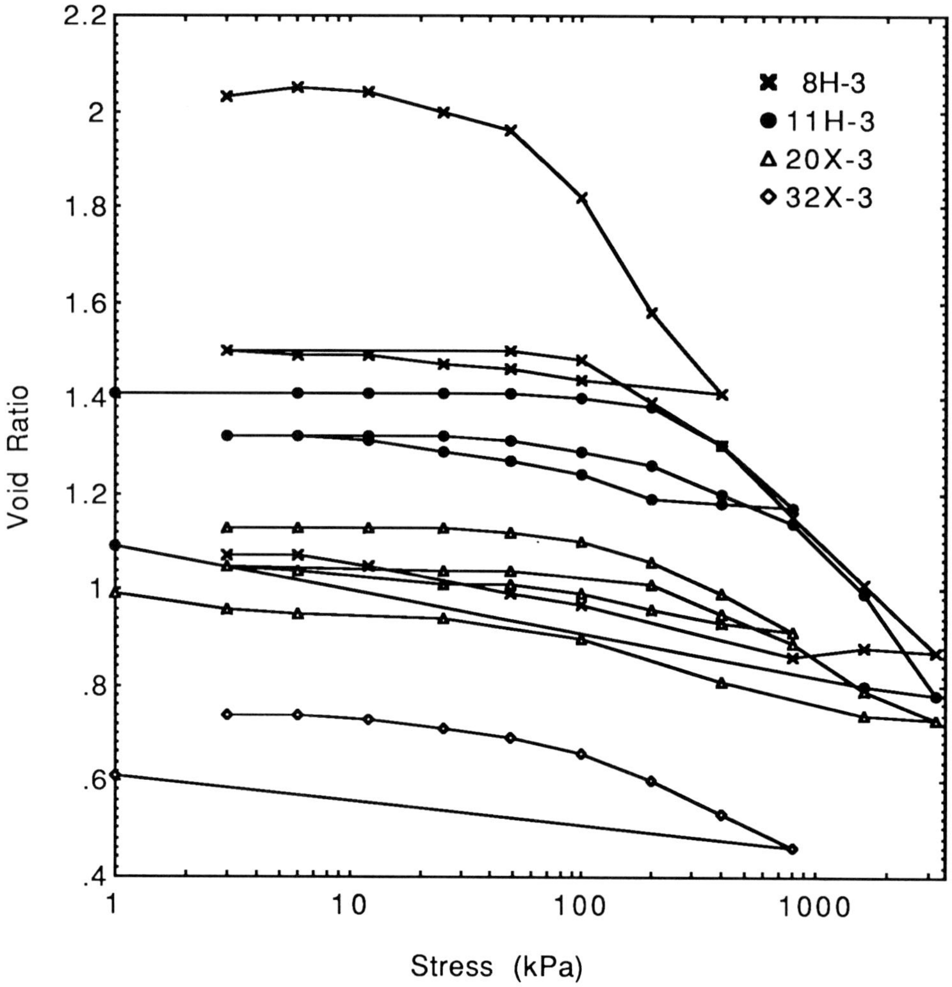

Figure 7.3. Curves of void ratio versus log effective stress for sediments from 90.1, 118.72, 200.6, and 316.66 mbsf at Hole 697B.

Table 7.1. Results of consolidation tests.

	Core ID (hole-core-section)			
	697B-8H-3	697B-11H-3	697B-20X-3	697B-32X-3
Depth (mbsf)	90.1	118.7	200.6	318.7
Diatom content (%)	15	2	15	0
Vertical stress (kPa)	628	851	1483	2530
Porosity (%)	66	58	52	50
Preconsolidation stress (kPa) σ'_c	60	480	270-	90
Overconsolidation ratio	0.09	0.56	0.18	0.04
Overconsolidation difference (kPa)	−568	−251	−1213	−2440
Consolidation characteristic	HUC*	HUC	HUC	HUC
Permeability (cm/sec) at preconsolidation stress	4×10^{-6}	3×10^{-8}	9×10^{-9}	1×10^{-10}

*HUC, highly underconsolidated.

overburden stress versus time and depth depicts the variable stress regimes from the early Pliocene and throughout the Pleistocene (Fig. 7.6C).

During the time of 4.5–4.0 Ma the loading rates averaged approximately 32.66×10^2 kPa/Ma, which diminished to 3.00×10^2 kPa/Ma between the later part of the early Pliocene to the very early Pleistocene (Fig. 7.6C). This reduction in sedimentation amounted to about an order of magnitude decrease in the rates of loading. During the Pleistocene the loading rates average only 2.75×10^2 kPa/Ma, which was about 8% lower than during the late Pliocene. The low rates of sedimentation during the Pleistocene in the Antarctic region is clearly opposite that of nonpolar regions that have been found to experience very high sedimentation rates during glacial times. The massive ice coverage in polar regions during the Pleistocene contrasts sharply with expansive continental land mass exposure of continental shelves as a result of sea level lowering. Thus low latitude basins

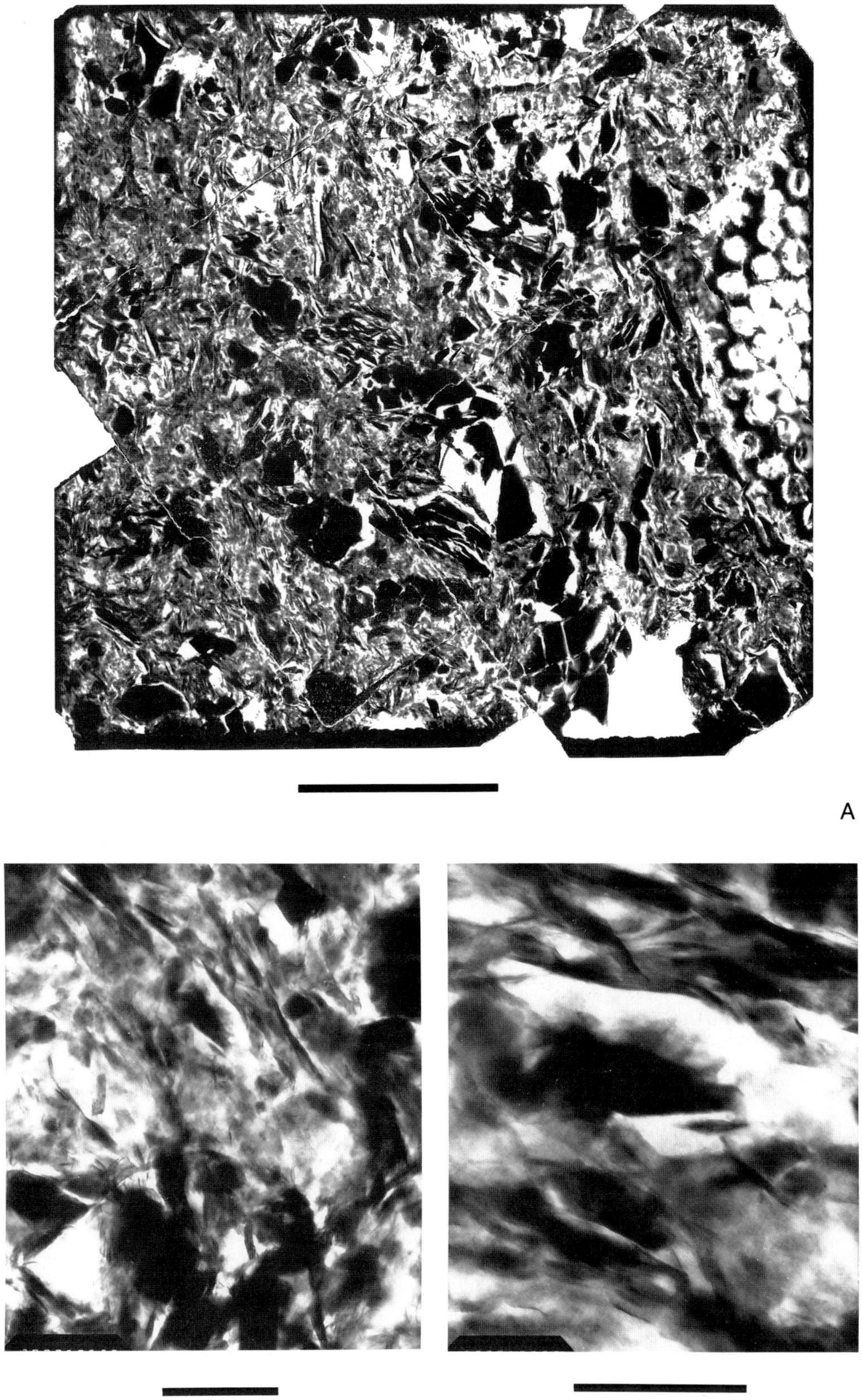

Figure 7.8. (A) Mosaic of TEM photomicrographs of sample 113–697A-2H-02, 55-57 cm, 7.35 mbsf. Note the very fine-grained smectite matrix and portion of diatom test at the right side. Scale bar = 10 μm. (B) TEM photomicrograph of enlarged section of A. Note the local areas of preferred particle orientation of domains. Scale bar = 1 μm. (C) TEM photomicrograph of enlarged section of B. Note the high porosity. Scale bar = 0.5 μm.

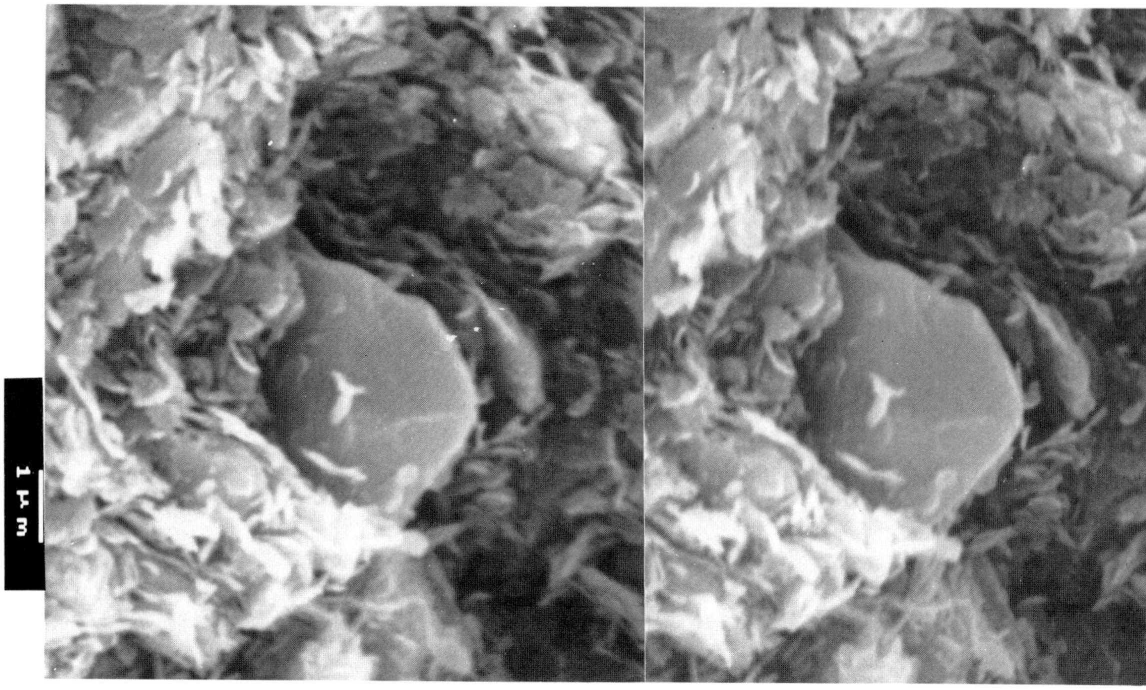

Figure 7.9. Scanning electron photomicrographs (stereo pair) of sample 2H-02, illustrating the nature of the clay particles and other large sediment grains (quartz or shale particles).

portion of Figure 7.10A at the boundary of the diatom test and the aligned particles adjacent to the test. Smectite appears to be a minor constituent and may be X-ray amorphous. This sediment, with a porosity of 62%, is underconsolidated (determined by consolidation tests) and the clay particles do not have a preferred orientation. The vertical effective stress at this stage of fabric development is 200 kPa based on wet bulk density measurements. Figure 7.11 is a scanning electron photomicrograph of the sediment at this level and displays no fine grained or fleecy smectite as is depicted in Figure 7.10. Sediment sample 7H-01 (Fig. 7.12) shows a random arrangement of clay platelets surrounding larger particles, and was taken at a depth of 76.57 mbsf. The sediment at this level consists of an illite-rich clay with 22% quartz and 4% volcanic glass. Chlorite, smectite, and kaolinite are common. The porosity of this sediments is approximately 60% and the vertical effective stress 430 kPa. At low magnification the photomicrograph in Figure 7.12 illustrates a poorly developed orientation of the clay particles. The pores of these sediments measure in the range of 0.1 μm or less. This sediment is underconsolidated and has a permeability of 2×10^{-7} cm/sec.

Porosities are 63% a meter above and a meter below sample 9H-04 taken at a depth of 94.20 mbsf. The bulk densities are 1.67 and 1.74 g/cm^3. These sediments are upper Pliocene in age and consist of diatom ooze; the diatom content was estimated at 60% by smear slide analysis. The stereo pair of scanning electron micrographs (Fig. 7.13) clearly illustrates the diatomaceous nature of this sediment. No transmission electron photomicrographs were obtained of this material due to the difficulty of obtaining usable ultrathin sections.

Sample 10H-02 (Fig. 7.14) was taken at a depth of 106.72 mbsf and shows well-developed domains of parallel layers of large sized clay particles. The sediment at this level is a silty clay composed of illite, chlorite, kaolinite, and smectite and about 14% quartz. The clay is slightly more compacted than in the overlying sample. The supporting clay matrix consists of the amorphous material thought to be authigenic smectite. The porosity of this sediment is 59%, similar to the porosity at the 76.57 mbsf level. The lack of change in porosity with depth suggests that the sediment is underconsolidated, a suggestion that is supported by the ratio of undrained shear strength to effective stress and by consolidation tests (Bryant and Rack, 1990). Figure 7.14 illustrates the relatively dense appearance of the smectite in this section. The pores in this material are very small and the permeability is very low at 5×10^{-8} cm/sec.

Sample 17X-05 was taken at a depth of 173.75 mbsf. Lower Pliocene silty clays are the dominant sediment type at this location. The diatom content of these sediments was estimated to be 15% by smear slide analysis. Figure 7.15 of scanning electron photomicrographs (stereo pair) displays a predominantly siliceous

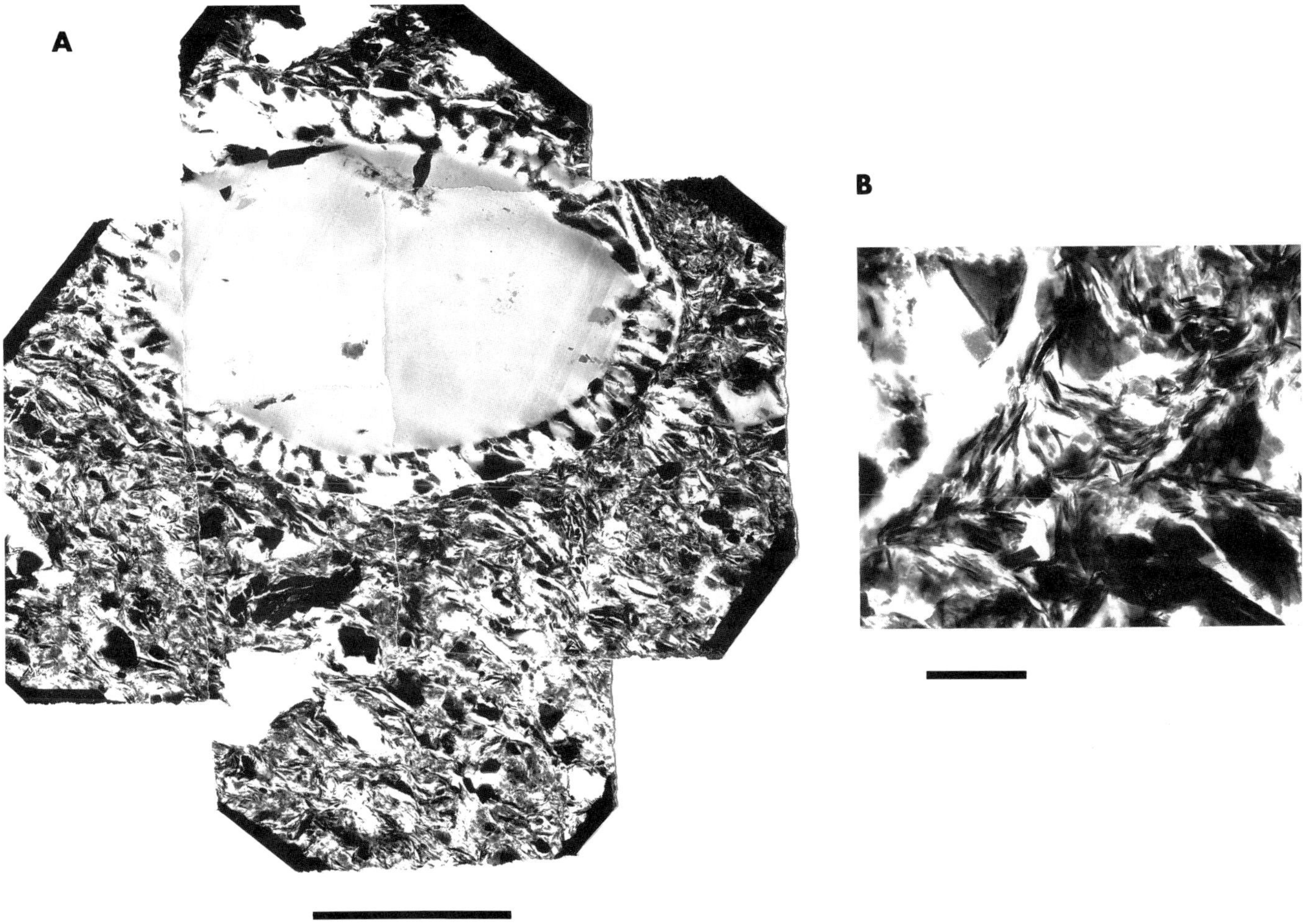

Figure 7.10. (A,B) Mosaic of TEM photomicrographs of sample 113–697B-3H-06, 55–57 cm, taken at a depth of 45.15 mbsf. Note large void created by the diatom test (A) and arrangement of clay particles oriented parallel to surface of test (A and B). Scale bars = 10 and 2 μm.

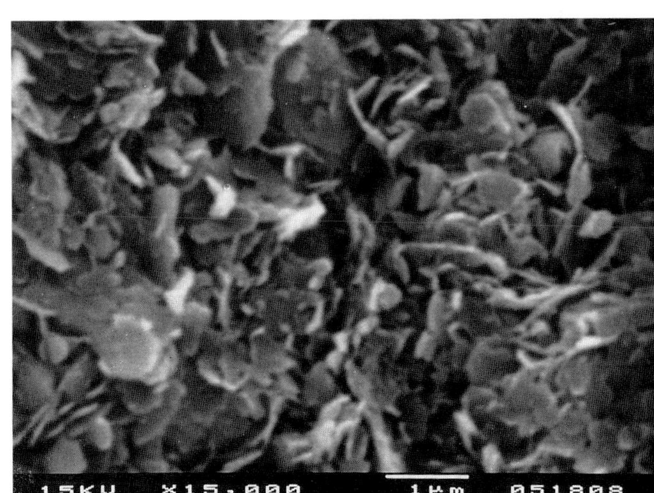

Figure 7.11. Scanning electron photomicrograph of sample 3H-06. Note the lack of fine-grained matrix material (smectite) so prevalent in the TEM illustrations.

Figure 7.12. Mosaic of TEM photomicrographs of sample 113–697B-7H-01, 57–58 cm, taken at a depth of 76.57 mbsf. Note the bimodal distribution of sediment particles and their random arrangement. The larger particles are supported by a "smectite" matrix. Scale bar = 10 μm.

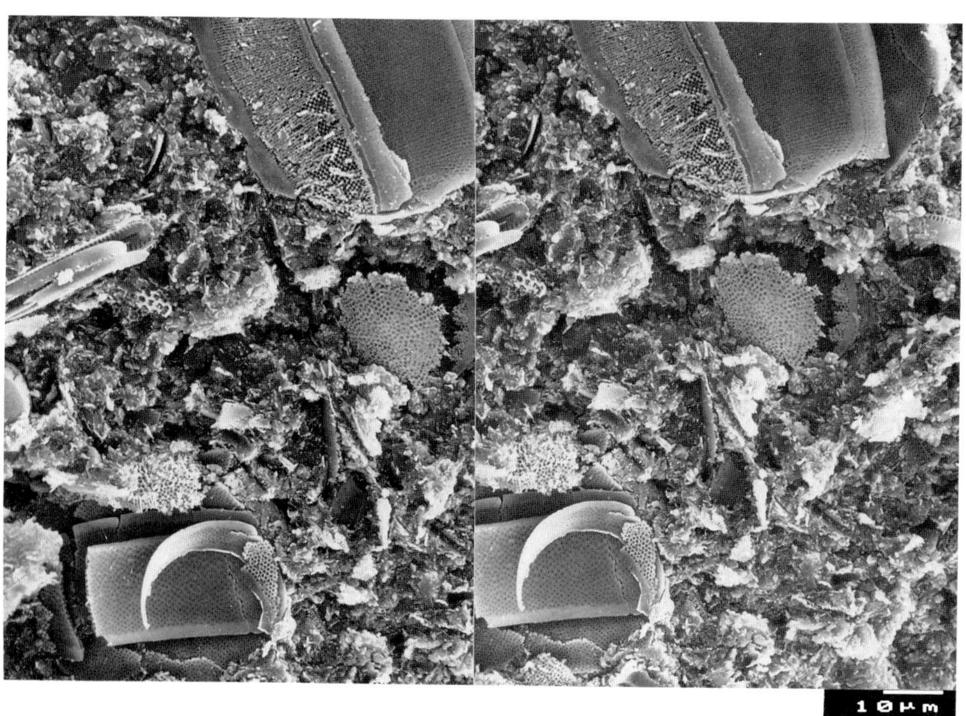

Figure 7.13. Scanning electron photomicrographs (stereo pair) of sample 113-697B-9H-04, 80–84 cm, taken at a depth of 94.20 mbsf. Diatom parts in a clay matrix are the main constituent of this sediment.

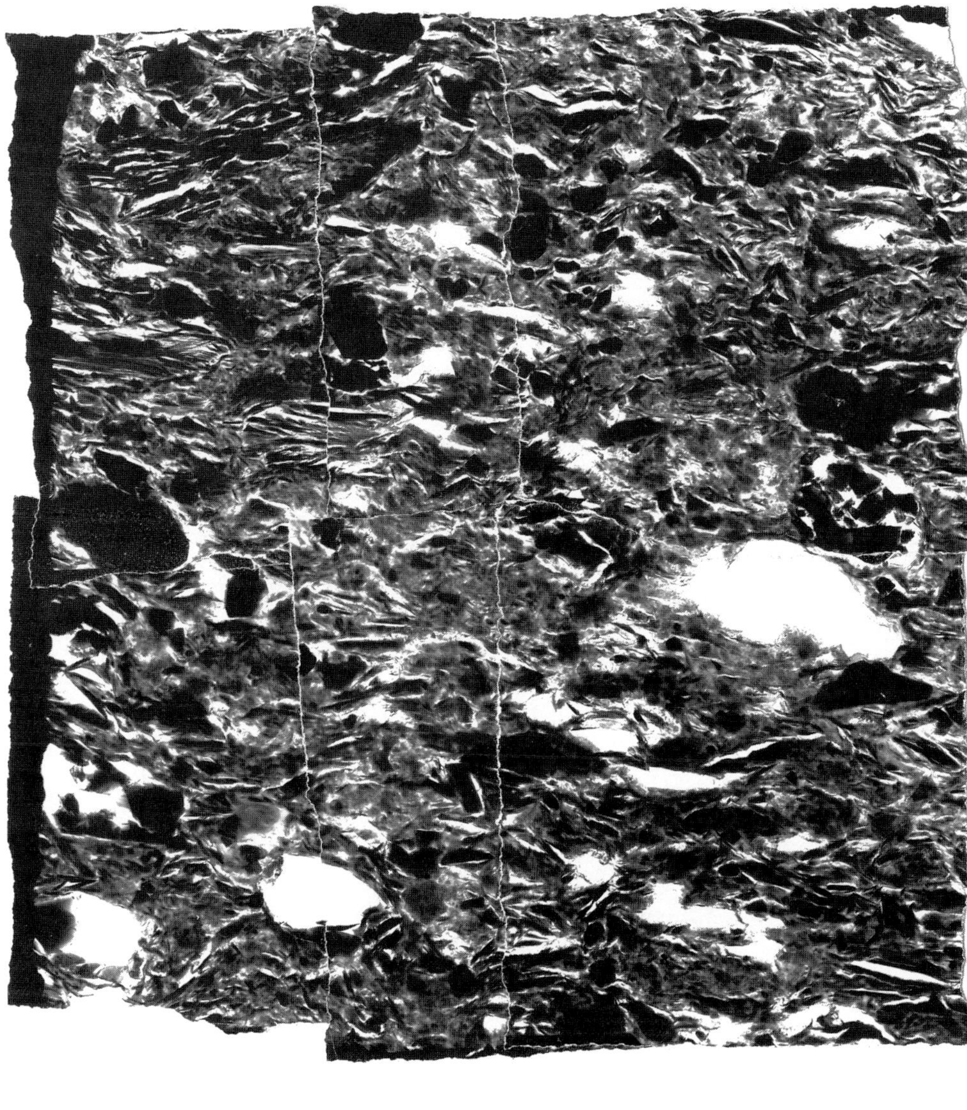

Figure 7.14. Mosaic of TEM photomicrographs of sample 113–697B-10H-02, 52–54 cm, taken at a depth of 106.72 mbsf showing a compressed fabric of a silty clay containing chlorite, kaolinite, illite, and smectite. Particles are randomly oriented but note the apparent particle orientation caused by the microtoming process: linear cuts (white) gives a fictitious orientation. Scale bar = 10 μm.

material mostly consisting of diatoms and diatom parts. An increase in porosity (76%) at this level (Fig. 7.2) expresses the diatomaceous nature of the sediment. No transmission electron micrographs were made of this material due to the difficulty in obtaining usable ultrathin sections.

Sample 21X-02 was taken at a depth of 207.94 mbsf. The sediment in this section consists of lower Pliocene silty clay, with illite, chlorite, kaolinite, and quartz comprising the main components. Smectite was not identified by X-ray diffraction. This sediment (Fig. 7.16) has random particles and domains of 1–2 μm consisting of several clay particles. A supporting matrix of amorphous "smectite" occurs in significant amounts. The TEM photomicrographs in Figures 7.16 and 7.17 are from the same core section and illustrate in detail the highly porous nature of the sediment and random orientation of the major clay minerals in edge-to-face contact and short chains. The porosity of this material was 56%. The effective stress level at this sample depth was calculated to be 1300 kPa and the sediment was judged to be underconsolidated. Slightly more compaction has taken place within the sediment than within sediment at the 106.72 mbsf level.

The lower Pliocene sediments (sample 27X-02) taken at a depth of 265.94 mbsf contain silty clay with 5% quartz. Illite is the most abundant clay; chlorite, kaolinite, and smectite are commonly in association. The sediment porosity of 52% is slightly lower than at the 207.94 mbsf level. The vertical effective stress at this level was calculated to be 1940 kPa. Figure 7.18 shows a slight increase in orientation of the clay fabric; however, Figure 7.19 shows the preponderance of highly random clay

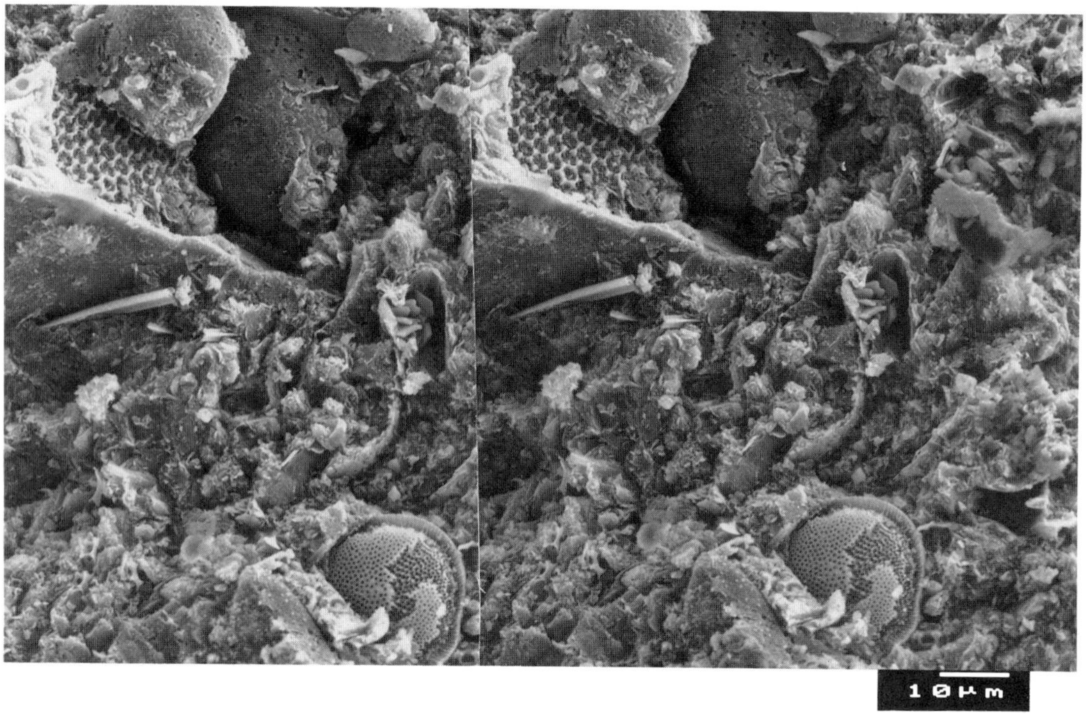

Figure 7.15. Scanning electron photomicrographs (stereo pair) of sample 113–697B-17X-05, 55–57 cm, taken at a depth of 173.5 mbsf. This sample contains mostly diatoms and diatom parts in a clay matrix.

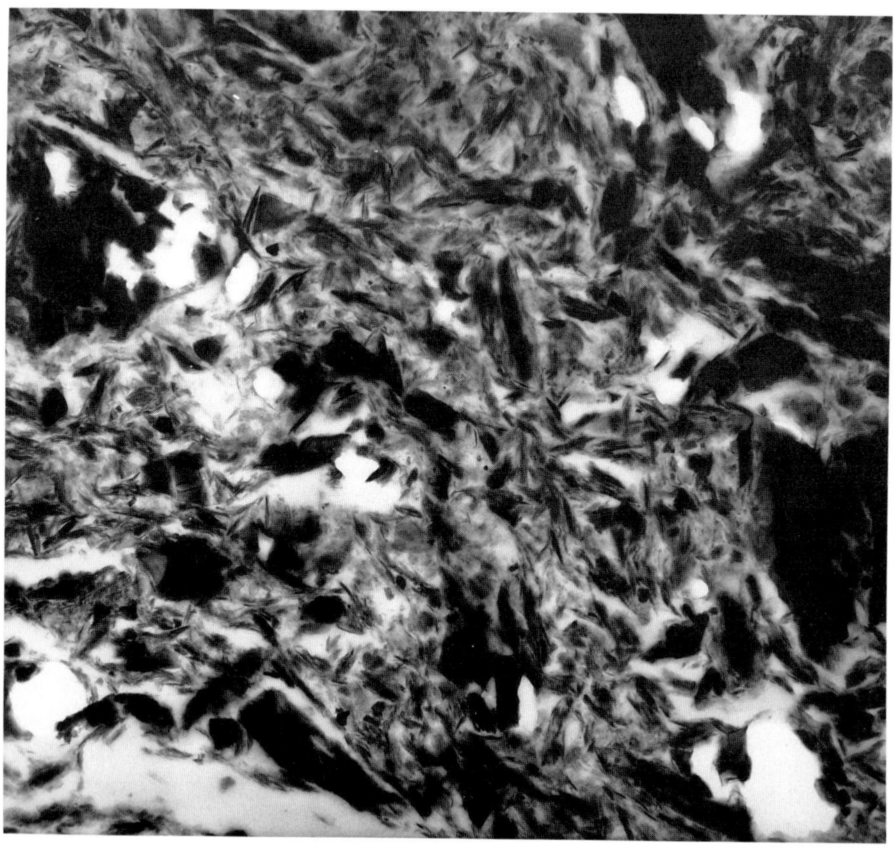

Figure 7.16. Mosaic of TEM photomicrographs of sample 113–697B-21X-02, 54–56 cm, taken at a depth of 207.94 mbsf showing the high degree of randomness of the clay particles and domains. Porosity remains relatively high at 56%. Scale bar = 1 μm.

Figure 7.17. High magnification of sample 113–697B-21X-02, 54–56 cm, depicting the relatively high porosity and randomness of domains in short chains and in edge-to-face contact. Note the fleecy amorphous "smectite" that clusters around the electron-dense domains. Scale bar = 0.5 μm.

domains of illite and smectite. The pores of this material appear quite large relative to the size of the particles. Note the very fine-grained nature of the material.

The sediment at a depth of 318.74 mbsf (sample 32X-05) is lower Pliocene silty clay, and is predominantly illite and chlorite. Kaolinite and smectite are present in rare amounts, with 10% quartz and 10% accessory minerals. The porosity of the material is 50%, making this the most compacted mudstone observed in this study. The calculated vertical effective stress at this depth is 2100 kPa. The calculated preconsolidation pressure was determined to be 100 kPa characterizing the sediment as underconsolidated. Figure 7.20 shows the fabric to be a highly chaotic clay mass with a slight degree of preferred orientation and the predominance of clay aggregates many in a semicircular configuration. The fabric at high magnification appears to consist of packets of very fine-grained clay particles in the 0.1 μm range surrounded by extremely small smectite. Figure 7.21 illustrates the nature of the smectite. It is interesting that even at this degree of compaction, the large clay domains are supported in a matrix of finer clay particles, consisting mostly of smectite, some of which are small enough to be X-ray amorphous. The permeability of this material is extremely low, 1×10^{-10} cm/sec. An SEM photomicrograph of similar clayey sediment (Fig. 7.22) significantly lacks any of the very small smectite. This may be due to the collapse of the smectite in a dehydrated state and the poorer resolution of the scanning microscope compared with

Figure 7.18. TEM photomicrograph of sample 113–697B-27H-02, 54–56 cm, from a depth of 266.4 mbsf. Note the relatively high porosity and random orientation of particles. There is a slight preferred particle orientation in the upper right-hand portion of the figure. Scale bar = 2 μm.

Figure 7.19. TEM photomicrograph of sample 113–697B-27X-02, 54–56 cm, showing the highly random orientation of clay domains of illite and smectite. Note the "smectite" clusters around the electron-dense domains. Scale bar = 0.5 μm.

TEM. The comparison between SEM and TEM fabric analysis of a clayey sediment is an excellent example of the much more complex and detailed nature displayed by TEM photomicrographs than observed by SEM.

Summary and Conclusions

TEM and SEM examination of the microfabric of the silty clayey sediments at Site 697 illustrates the subtle changes that occur in the fabric as a clay mass consolidates from a mud to a mudstone. These observations confirm the suggested classic descriptions of clay fabric change and porosity reduction during the consolidation process. Significant fabric response to overburden stress was shown also in other marine sediment (Faas and Crockett, 1983). Bennett et al. (1977) showed a stronger degree of preferred particle alignment for a consolidated sediment section of Mississippi Delta clays at depths of 150 mbsf than is observed for Site 697 sediments. Significant consolidation and clay particle reorientation occurred with a strong preferred particle orientation normal to the overburden stress. This reorientation occurred at effective stresses of 4 to 7 $\times$ 10^2 kPa because of the presence of an interbedded 25-m-thick drainage path (fine sand) overlying the compacted Pleistocene clays (Bennett et al., 1977, 1981).

Under low overburden stresses the fabric of a silty clay sediment appears as a random arrangement of clay particles, aeolian shale clasts, diatoms, quartz particles, and clay domains. At this stage of compaction the material has a high porosity (74%) and a moderately high permeability (1×10^{-4} to 1×10^{-5} cm/sec). With increasing depth of burial, increased mechanical forces in the form of overburden stress slowly reorient the clay particles. Reorientation in turn decreases porosity, expels intergranular pore fluids, decreases the permeability, and lengthens the time of the compaction process. The drastic decrease in permeability (from 3 to 10 orders of magnitude faster than porosity: Bryant et al., 1986) imparts an underconsolidation characteristic to the sediment. This underconsolidation is common to many clayey marine sediments found in the deep ocean basins. We suggest that the onset of preferred fabric orientation of these sediments below a depth of 200 mbsf at this site produces an anisotropic material with respect to permeability and acoustic velocity.

The identification of fractured illite shale clasts ("fractillites") and associated quartz grains in the upper Pleistocene portions of

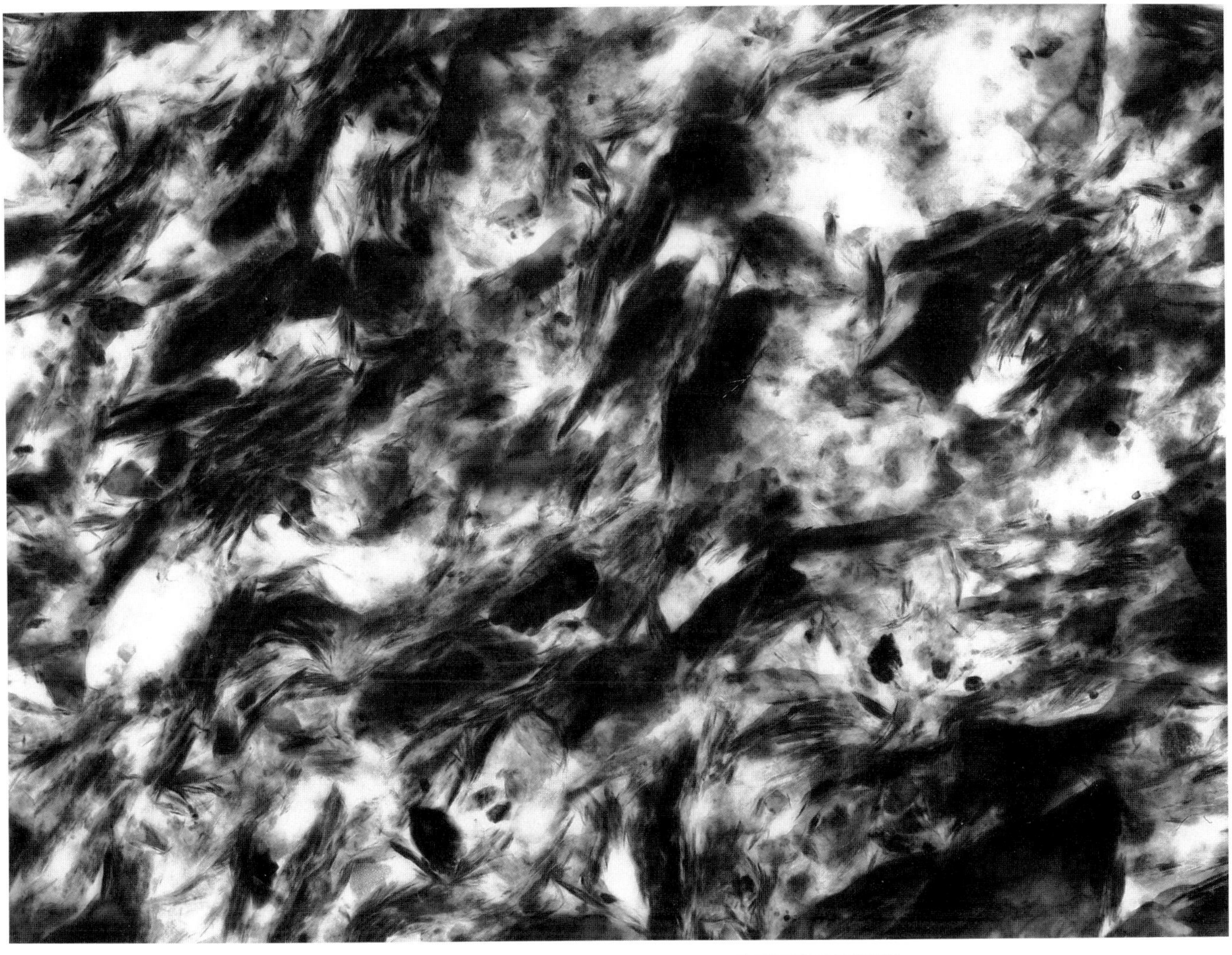

Figure 7.20. High magnification TEM photomicrograph of Core 697B-32X-04, 54–56 cm, from a depth of 317.24 mbsf. Local areas depict random orientation of numerous small particles but overall there is the beginning of preferred orientation consisting of semicircular domains of very fine clay particles and cluster of particles supported by a matrix of smectite. Scale bar = 1 μm.

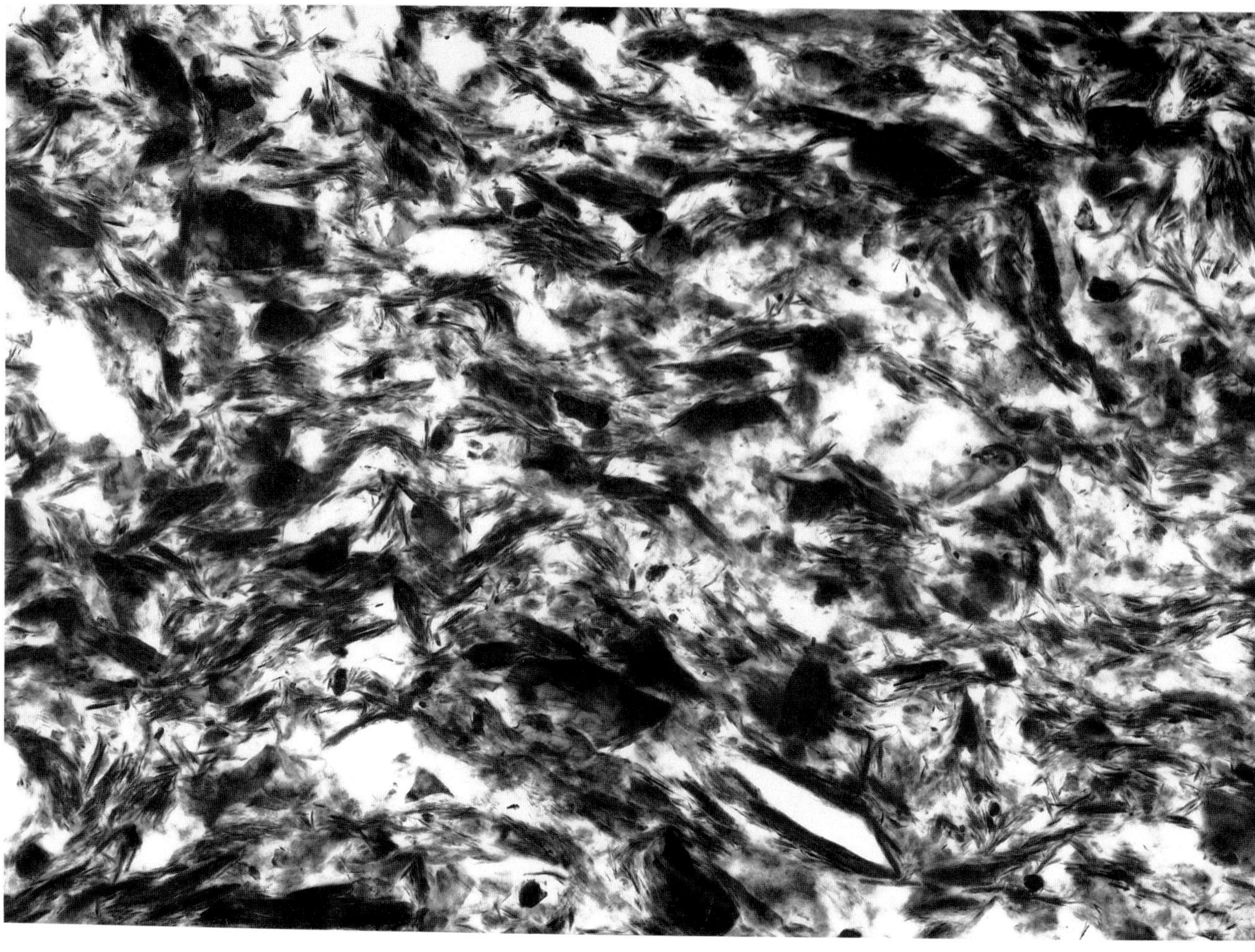

Figure 7.21. Same sample as Figure 7.20. Notice at lower magnification, numerous particles show some preferred orientation normal to the principal stress direction. Scale bar = 1 μm.

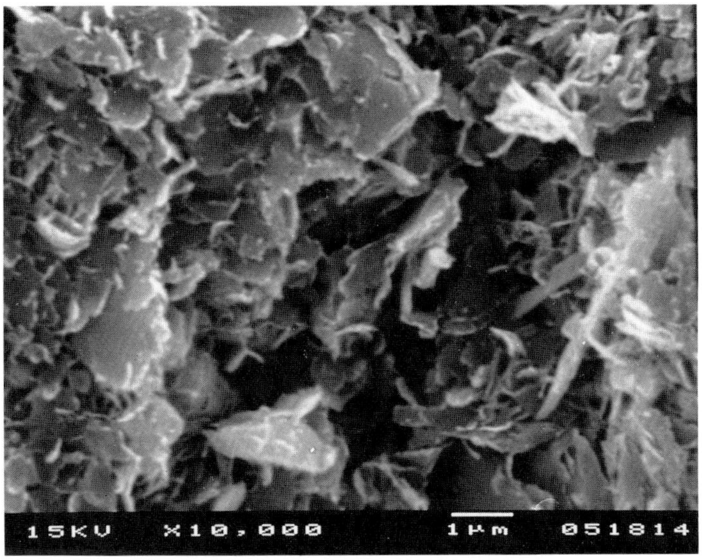

Figure 7.22. Scanning electron photomicrograph of sample 32-04, same location as Figures 7.20 and 7.21, illustrating a random arrangement of clay particles. Note the absence of the supporting matrix of smectic, which is a common constituent in the TEM photomicrographs.

Hole 697 indicates that these fractions of the sediment could be of an aeolian origin. TEM and SEM observations revealed important features of the microfabric. Noteworthy, however, is that SEM does not depict fine smectite particles because of dehydration effects and poor resolution as compared with TEM. SEM provides a somewhat three-dimensional view of the microfabric and pore geometries. TEM provides excellent resolution of particle-particle contacts.

Acknowledgments

Partial support for this research was provided by USSAC of JOI. The second author of this paper was supported by the Naval Oceanographic and Atmospheric Research Laboratory, Stennis Space Center, MS. Thanks are extended to those earlier reviewers of this paper and the Electron Microscope Center, Texas A&M University. Special appreciation is extended to Ms. Lee Nastav for assistance in preparing the final version of this paper.

References

Aylmore, L.A.G., and J.P. Quirk, 1960. Domain or turbostratic structure of clay. Nature, v. 187, p. 1046–1048.

Anderson, O.B., and A.G. Douglas, 1970. Bonding, effective stresses, and strength of soils. Journal Soil Mechanics and Foundation Division, Proceedings American Society Civil Engineers, v. 96, p. 1073–1077.

Barker, P.F., J.P. Kennett, and Scientific Party, 1988. Weddell Sea paleoceanography: Preliminary results of ODP Leg 113. Palaeogeography, Palaeoclimatology, Palaeoecology, v. 67, p. 75–102.

Bennett, R.H., 1976. Clay fabric and geotechnical properties of selected submarine sediment cores from the Mississippi Delta. Ph.D. dissertation, Texas A&M University, College Station, TX, 269 p.

Bennett, R.H., and M.H. Hulbert, 1986. Clay Microstructure. International Human Resources Development Corporation, Boston, 161 p.

Bennett, R.H., W.R. Bryant, and G.H. Keller, 1977. Clay fabric and geotechnical properties of selected submarine sediment cores from the Mississippi Delta. NOAA Professional Paper 9, U.S. Department of Commerce, 86 p.

Bennett, R.H., W.R. Bryant, and G.H. Keller, 1981. Clay fabric of selected submarine sediments: fundamental properties and models. Journal of Sedimentary Petrology, v. 51, no. 1, p. 217–232.

Bennett, R.H., K.M. Fischer, D.Lavoie, W.R. Bryant, and R. Rezak, 1989. Porometry and fabric of marine clay and carbonate sediments: determinants of permeability. Marine Geology, v. 89, p. 127–152.

Bolt, G.H., 1956. Physico-chemical analysis of the compressibility of pure clays. Geotechnique, v. 6, p. 86–93.

Bryant, W.R., and R.H. Bennett, 1988. Origin, physical, and mineralogical nature of red clays: the Pacific basin as a model. Geo-Marine Letters, v. 8, p. 189–249.

Bryant, W.R., and F.R. Rack, 1990. Consolidation characteristics of Weddell Sea sediments: results of ODP Leg 113. In: Barker, P.F., et al. (eds.), Final Report Ocean Drilling Project Volume 113, U.S. Government Printing Office, Washington, D.C. In press.

Bryant, W.R., R. Bennett, and C. Katherman, 1981. Shear strength, consolidation, porosity, and permeability of oceanic sediments. In: Emiliani, C. (ed.), The Sea, 7: The Oceanic Lithosphere. Wiley, New York, p. 1555–1616.

Bryant, W.R., A. Wetzel, E. Taylor, and W. Sweet, 1986. Consolidation characteristics and permeability of Mississippi Fan sediments. In: Bouma, A.H., et al. (eds.), Initial Reports Deep Sea Drilling Project Volume 96, U.S. Government Printing Office, Washington, D.C., p. 797–809.

Buchanan, P.N., 1964. Effect of temperature and absorbed water on permeability and consolidation characteristics of sodium and calcium montmorillonite. Ph.D. dissertation, Texas A&M University, College Station, 104 p.

Cabrera, J.G., and I.J. Smalley, 1971. Engineering behavior and structure of compacted clay, discussion and reply. Journal Soil Mechanics Foundation Division, Proceedings American Society Civil Engineers, v. 19, p. 802–803.

Casagrande, A., 1932. The structure of clay and its importance in foundation engineering. Journal Boston Society Civil Engineers, v. 19.

Faas, R.W., and D.S. Crockett, 1983. Clay fabric development in a deep sea core, Site 515. In: Barker, P.F., R.L. Carlson, D.A. Johnson, et al. (eds.) Initial Reports Deep Sea Drilling Project Volume 72, U.S. Government Printing Office, Washington, D.C., p. 519–535.

Goldschmidt, V.M., 1926. Undersokelser over lersedimenter Nordiskjordbrugsforskuing, v. 4–7, p. 434–445.

Hinch, H.H., 1980. Nature of shales in expulsion, problems of petroleum migration. In: Roberts, W., and R. Cordell (eds.), American Association Petroleum Geologists Studies in Geology, No. 10, p. 1–18.

Houston, W.N., and J.K. Mitchell, 1969. Property interrelationships in sensitive clays. Journal Soil Mechanics Foundation Division, Proceedings American Society Civil Engineers, v. 95, p. 1037–1062.

Ingles, O.G., 1968. Soil chemistry relevant to the engineering behavior of soils. In: Lee, I.K. (ed.), Soil Mechanics, Selected Topics. London, Butterworths, p. 1–57.

Kruyt, H.O. (ed.), 1952. Colloid Science, Irreversible Systems. Amsterdam, Elsevier, v. 1, 554 p.

Lambe, T.W., 1953. The structure of inorganic soil. Proceedings American Society Civil Engineers, v. 79, p. 1–49.

Lambe, T.W., 1958a. The structure of compacted clay. Journal Soil Mechanics Foundation Division, Proceedings American Society Civil Engineers, v. 84, p. 1–35.

Lambe, T.W., 1958b. The engineering behavior of compacted clay. Journal Soil Mechanics Foundation Division, Proceedings American Society Civil Engineers, v. 84, p. 1–35.

Mitchell, J.K., and W.N. Houston, 1969. Causes of clay sensitivity. Journal Soil Mechanics Foundation Division, Proceedings, American Society Civil Engineers, v. 95, p. 845–871.

Mitchell, J.K., A. Singh, and R.G. Campanella, 1969. Bonding, effective stresses, and strength of soils. Journal Soil Mechanics Foundation Division, Proceedings American Society Civil Engineers, v. 95, p. 1219–1246.

Mitchell, J.K., A. Singh, and R.G. Campanella, 1971. Bonding, effective stresses, and strength of soils. Journal Soil Mechanics Foundation Division, Proceedings American Society Civil Engineers, v. 97, p. 478.

O'Brien, N.R., 1971. Fabric of kaolinite and illite floccules. Clays and Clay Minerals, v. 19, p. 353–359.

Olson, R.E., and G. Mesri, 1970. Mechanisms controlling compressibility of clays. Journal Soil Mechanics Foundation Division, Proceedings American Society Civil Engineers, v. 96, p. 1863–1878.

Rosenqvist, I. Th., 1959. Physico-chemical properties of soils: soil-water systems. Journal Soil Mechanics Foundation Division, Proceedings American Society Civil Engineers, v. 85, p. 31–53.

Rosenqvist, I. Th., 1962. The influence of physico-chemical factors upon the mechanical properties of clays. Ninth National Conference on Clays and Clay Minerals, v. 9, p. 12–27.

Singh, S., 1970. Bonding, effective stresses, and strengths of soils, Journal Soil Mechanics Foundation Division, Proceedings American Society Civil Engineers, v. 96, p. 1469–1473.

Sloane, R.L., and T.R. Kell, 1966. The fabric of mechanically compacted kaolin. Fourteenth National Conference on Clay and Clay Minerals, v. 14, p. 289–296.

Terzaghi, K., 1925. Principles of soil mechanics settlement and consolidation of clay. England New-Record, p. 874–878.

Theissen, D.A., 1942. Wechselsetige adsorption von kolloiden. Zeitschrift fur Electrochemie, v. 48, p. 675.

Torrance, K.J., 1970. Causes of clay sensitivity. Journal Soil Mechanics Foundation Division, Proceedings American Society Civil Engineers, v. 96, p. 360–361.

van Olphen, H., 1963. An introduction to clay colloid chemistry. Wilely-Interscience, New York, 301 p.

Winterkorn, H.F., 1948. Physico-chemical properties of soils. Proceedings 2nd International Conference Soil Mechanics Foundation Engineering, Rotterdam, p. 23–29.

Yariv, S., and H. Cross, 1979. Geochemistry of Colloid Systems. Springer-Verlag, Berlin, 450 p.

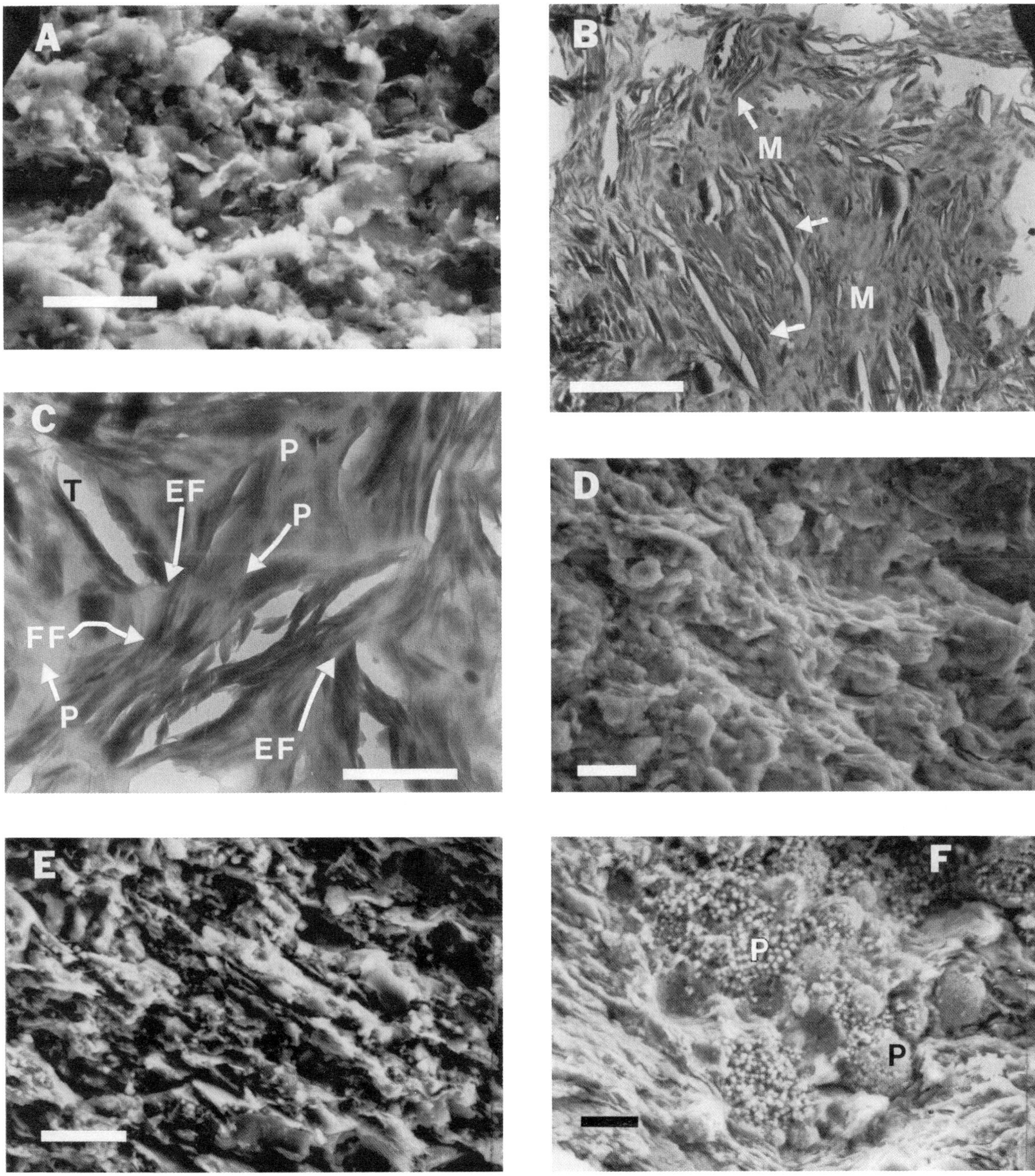

Figure 8.5. TEM and SEM photomicrographs of representative samples. (A) Barbados well sample from 614 m in which no alignment of grains or domains is evident. Scale bar is 10 μm. (B, C) ODP Leg 110 clay from a depth of 110 m at the oceanic reference site. In B, arrows point to domains of clay particles in an unoriented matrix (M). Scale bar is 4 μm. C is a close-up of face-to-face (FF) contact of clay particles within domains and of edge-to-face contact (EF) between domains. Pore spaces (P) and tears (T) in the section are indicated. Scale bar is 1 μm. (D) Well sample from 1162 m showing initial development of a planar fabric. Scale bar is 10 μm. (E) Pervasive planar fabric in a well sample from 2659 m. Scale bar is 10 μm. (F) Clays wrap around early formed framboids of pyrite (P) in a well sample from 2033 m. Scale bar is 10 μm.

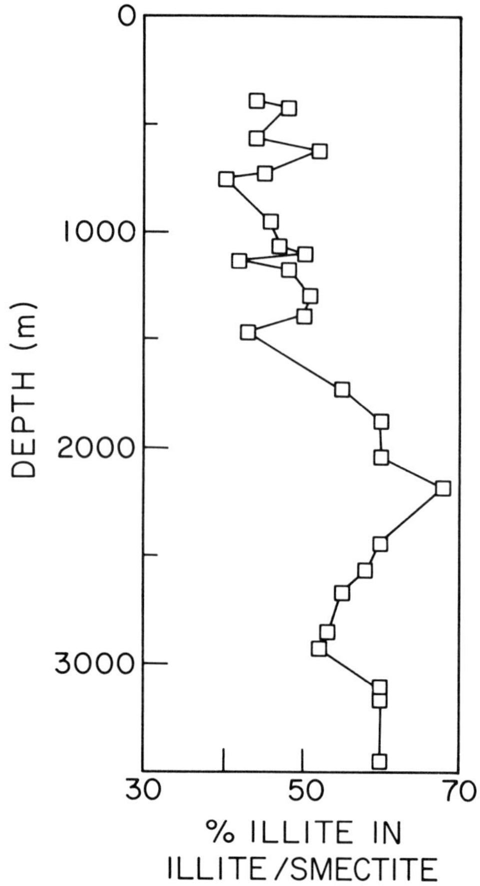

Figure 8.6. Composition of the illite/smectite, in terms of percentage illite interlayers, versus depth for the Barbados well samples. The depth range of most rapid reaction is approximately 1500–2200 m. (Modified from Schoonmaker et al., 1986.)

of domains. Groups of clay platelets aligned face-to-face are scattered throughout a randomly oriented clay matrix (Fig. 8.5B). Several domains in a small region may exhibit similar orientations, but there appears to be no overall direction of alignment. At higher magnification (Fig. 8.5C) the parallel alignment of clay platelets within domains is obvious, whereas adjacent domains may be in edge-to-face contact. The open zones inside the domains are apparently tears in the section formed during slicing. The parallel alignment of clay particles provides zones of weakness along which slip occurs preferentially.

SEM examination of samples from greater burial depths reveals the development of an overall planar fabric. At burial depths of about 1000 m, porosity is reduced to roughly 10% and the sediment is well lithified. A rough lineation of clay domains is apparent from SEM observation (Fig. 8.5D), although zones of random orientation still remain. In deeper samples, a planar fabric is pervasive and cleavage faces of clays are commonly prominent (Fig. 8.5E), leading to the characteristic fissility of these shales.

As mentioned above, the sediments of this study represent both hemipelagic deposits and turbidites. Whereas it is likely that original depositional fabric varied somewhat between samples, it is felt that compaction and perhaps shear are the dominant processes responsible for fabric alteration with depth. In addition, diagenesis may play an important role. The diagenetic conversion of smectite to illite was observed in the well samples from Barbados (Fig. 8.6). The conversion reaction occurs primarily between 1500 and 2000 m, the depth range over which planar fabric becomes evident. Diagenesis may also be important in local disruption of fabric. Pyrite framboids are a common early diagenetic product in the Barbados mudstones (Fig. 8.5F). Groups of framboids frequently disrupt the trend of the planar fabric and fill cracks and small pore spaces. In some instances they are found in association with amorphous iron coatings on clays.

Conclusions

Shallow burial (several 100 m) of clay-rich sediments in the Barbados accretionary prism results in only a slight increase in V_p with bulk density. Muds and mudstones still respond like seawater-dominated systems in which grain–grain interactions are relatively unimportant at burial depths slightly greater than 500 m. The V_p–bulk density trend of these sediments is very similar to that observed for shelf and slope sediments by Hamilton and Bachman (1982) (Fig. 8.3). The fabric of the ODP Leg 110 sediments is characterized by randomly oriented domains of clay particles. Smectite in the ODP samples is derived largely from alteration of volcanic ash (Schoonmaker, 1986; Tribble, 1990). It is possible that domains of smectite grow *in situ* during diagenesis of the sediment. The lack of a net orientation of particles is reflected in the absence of velocity anisotropy in these samples (Fig. 8.4).

Upon deeper burial, the V_p–density gradient increases dramatically, indicating development of rigidity and a framework dominated velocity. Data for samples with bulk densities greater than 2.3 g/cm³ depart from the Hamilton–Bachman curve and are better described by a curve developed for average shales (Hamilton, 1978) (Fig. 8.3), although the V_p–density gradient appears steeper for the Barbados samples than for the average shales. Departure from the Hamilton–Bachman curve is first exhibited by outcrop samples that were probably once buried to depths of several kilometers. It is surprising that sediments buried so deeply retain characteristics similar to those of the shallowly buried sediments described by the Hamilton–Bachman (1982) curve. Several features of sedimentation in the Barbados accretionary prism may explain this behavior. The Barbados samples, combinations of turbidites and hemipelagic sediments, probably had higher depositional rates, and consequently slower compaction rates, than the shelf and slope sediments used by Hamilton and Bachman. In addition, burial in the accretionary prism occurs largely through thrust faulting and therefore is much more rapid than normal burial owing to sedimentation. These factors would combine to yield undercompacted sediments that might have velocity–density characteristics typical of more shallowly buried sediments.

Sediment fabric changes dramatically with increasing depth as clay domains are aligned parallel to one another and a planar fabric develops. Fabric alteration appears to be an important factor influencing velocity increase of the well samples, which have nearly uniform bulk densities. The increased degree of alignment of clay domains in deeply buried samples is also reflected in the development of velocity anisotropy with horizontal velocity significantly greater than vertical velocity. The major mineralogical changes accompanying lithification of these mudstones are the early formation of smectite via alteration of volcanic ash, and the subsequent transformation of the smectite to mixed-layer illite/smectite, and eventually to illite, during deep burial. It is likely that these diagenetic reactions, combined with compaction, are responsible for the changes in overall clay fabric with depth. As noted previously, the Barbados samples have undergone significant tectonism. Initially incorporated into the prism along thrust faults, the sediments were buried deeply and subsequently uplifted. Shear may also play a significant, and little understood, role in the development of planar fabrics and fissility in these rocks (Moon and Hurst, 1984).

Acknowledgments

Samples for this study were provided by the Ocean Drilling Project, Gulf Oil Company, and D. K. Larue. Physical properties measurements performed by Marathon Oil Company and the Illinois Geological Survey are gratefully acknowledged. M. Manghnani provided the laboratory and R. C. Schneider the technical assistance for physical properties and velocity measurements. Comments from R. Wilkens and two anonymous reviewers helped to improve the manuscript. Financial support for this work from the Ocean Drilling Program and the Office of Naval Research is greatly appreciated. Hawaii Institute of Geophysics Contribution No. 2291.

References

Bennett, R.H., W.R. Bryant, and G.H. Keller, 1977. Clay fabric and geotechnical properties of selected submarine sediment cores from the Mississippi Delta. NOAA Professional Paper 9, 86 p.

Birch, F., 1960. The velocity of compressional waves in rocks to 10 kilobars, Part 1. Journal of Geophysical Research, v. 65, p. 1083–1102.

Bohlke, B.M., and R.H. Bennett, 1980. Mississippi Prodelta crusts: a clay fabric and geotechnical analysis. Marine Geotechnology, v. 4, p. 55–82.

Boles, J.R., and S.G. Franks, 1979. Clay diagenesis in Wilcox sandstones of southwest Texas: implications of smectite diagenesis on sandstone cementation. Journal of Sedimentary Petrology, v. 49, p. 55–70.

Bowles, F.A., W.R. Bryant, and C. Wallin, 1969. Microstructure of unconsolidated and consolidated marine sediments. Journal of Sedimentary Petrology, v. 39, p. 1546–1551.

Bryant, W.R., and R.H. Bennett, 1988. Origin, physical, and mineralogical nature of red clays: the Pacific Ocean as a model. Geo-Marine Letters, v. 8, p. 189–249.

Hamilton, E.L., 1976. Variations of density and porosity with depth in deep-sea sediments. Journal of Sedimentary Petrology, v. 46, p. 280–300.

Hamilton, E.L., 1978. Sound velocity-density relations in sea-floor sediments and rocks. Journal of the Acoustical Society of America, v. 63, p. 366–377.

Hamilton, E.L., and R.T. Bachman, 1982. Sound velocity and related properties of marine sediments. Journal of the Acoustical Society of America, v. 72, p. 1891–1904.

Jim, C.Y., 1985. Impregnation of moist and dry unconsolidated clay samples using Spurr resin for microstructural studies. Journal of Sedimentary Petrology, v. 55, p. 597–599.

Larue, D.K., J. Schoonmaker, R. Torrini, J. Clark, M. Clark, and R.C. Schneider, 1985. Barbados: maturation, source rock potential and burial history within a Cenozoic accretionary complex. Marine and Petroleum Geology, v. 2, p. 96–110.

Mascle, A., J.C. Moore, et al., 1988. Proceedings of the Ocean Drilling Program, Initial Reports, 110: College Station, TX (Ocean Drilling Program).

Moon, C.F., and C.W. Hurst, 1984. Fabric of muds and shales: an overview. In: Stow, D.A.V., and D.J.W. Piper (eds.), Fine-Grained Sediments: Deep-Water Processes and Facies. Blackwell Scientific, Palo Alto, CA, p. 579–593.

Schoonmaker, J., 1986. Clay mineralogy and diagenesis of sediments from deformation zones in the Barbados accretionary prism (DSDP Leg 78A). In: Moore, J.C. (ed.), Synthesis of Structural Fabrics of Deep Sea Drilling Project Cores from Forearcs. Geological Society of America Memoir 166, p. 105–116.

Schoonmaker, J., F.T. Mackenzie, and R.C. Speed, 1986. Tectonic implications of illite/smectite diagenesis, Barbados accretionary prism. Clays and Clay Minerals, v. 34, p. 465–472.

Shipley, T.H., 1983. Physical properties, synthetic seismograms, and seismic reflections: correlations at Deep Sea Drilling Project Site 534, Blake-Bahama Basin. In: Sheridan, R.E., F.M. Gradstein, et al. (eds.), Initial Reports of the Deep Sea Drilling Project, v. 76. U.S. Government Printing Office, Washington, D.C., p. 653–666.

Speed, R.C., and D.K. Larue, 1982. Barbados: architecture and implications for accretion. Journal of Geophysical Research, v. 87, p. 3633–3643.

Toksoz, M.N., C.H. Cheng, and A. Timur, 1976. Velocities of seismic waves in porous rocks. Geophysics, v. 41, p. 621–645.

Tribble, J., 1990. Clay diagenesis in the Barbados accretionary complex: potential impact on hydrology and subduction dynamics. In: Moore, J.C., A. Mascle, et al., Proceedings of the Ocean Drilling Program, Scientific Results, 110: College Station, TX (Ocean Drilling Program), 97–110.

Wilkens, R., P. McClellan, K. Moran, J.S. Tribble, E. Taylor, and E. Verduzco, 1990. Diagenesis and dewatering of clay-rich sediments, Barbados accretionary prism. In: Moore, J.C., A. Mascle, et al., Proceedings of the Ocean Drilling Program, Scientific Results, 110: College Station, TX (Ocean Drilling Program), 309–320.

CHAPTER 9

Sedimentary Structures: Textures and Depositional Settings of Shales from the Lower Belt Supergroup, Mid-Proterozoic, Montana, U.S.A.

Jürgen Schieber

Introduction

The Belt Supergroup of the northwestern United States is a thick (20 km) shale-dominated sequence that accumulated in an epicontinental basin between 1450 and 850 Ma (Harrison, 1972; Stewart, 1976). In this chapter shales from the lower portion of the Belt Supergroup are examined and interpreted with respect to potential depositional environment. Shale samples were collected in the eastern portion of the basin from the Newland Formation, and in the central and western portion of the basin from its lateral equivalent, the Prichard Formation (Fig. 9.1).

Shales of the Newland Formation have been studied in considerable detail (Schieber, 1985, 1986, 1989), and six major shale facies types can be distinguished. The overall depositional setting for these shales was probably a shallow, low-energy shelf. Shales of the Prichard Formation were deposited in a deeper portion of the basin, and differ in a number of sedimentary features from those of the Newland Formation. These differences may be used to distinguish between shallow and deep water shales in general. Shales of the Newland Formation are unmetamorphosed (Maxwell and Hower, 1967) except in contact aureolas around plutons. In contrast, shales of the Prichard and Aldridge Formation have commonly undergone greenschist facies metamorphism (Cressman, 1984).

Shales of the Newland Formation

Striped Shale Facies (NS-1)

The characteristic striped appearance of this shale facies is caused by the interlayering of carbonaceous silty shale, dolomitic clayey shale, siltstone, and lithoclast beds (Schieber, 1986). Textural features are shown in Figure 9.2 and summarized in Figure 9.3. Siltstone beds commonly form graded silt/mud couplets with overlying beds of dolomitic clayey shale (Fig. 9.2). These couplets show parallel lamination, cross-lamination, and graded rhythmites in the lower silty portion.

Interpretation

Beds of hummocky cross-stratified sandstones (Harms et al., 1982) occur interbedded with the striped shales, indicating occasional sediment deposition by storms. Interstratified carbonate units contain various indicators of relatively shallow water, such as wave and current ripples, flat pebble conglomerates, and cryptalgal laminites. Irregular wavy-crinkly laminae (Fig. 9.2) and mechanical strength during soft sediment deformation suggest a microbial mat origin for carbonaceous silty shale beds (discussed in depth by Schieber, 1986). The silt/mud couplets strongly resemble modern storm deposits (Aigner and Reineck, 1982) of muddy shelf seas. Their abundance and their association with less abundant thicker storm sand deposits suggests that they were deposited by more frequent but relatively weak storms. The intimate interlayering of silt/mud couplets and beds of carbonaceous silty shale indicates episodic interruption of microbial mat accumulation by storm sedimentation. Because of the absence of any signs of emergence in these shales, and because of the abundant indicators of storm sedimentation, these shales probably accumulated between fairweather and (average) storm wave base.

Carbonaceous Swirl Shale Facies (NS-2)

These shales consist primarily of a clay–dolomite matrix with scattered quartz silt, micas, and carbonaceous flakes (Fig. 9.4).

Figure 9.1. Location map. Stipple pattern shows present day outline of the Belt basin. Newland shale samples were collected from the eastern extension of the basin, Prichard and Aldridge shale samples were collected from the central, western, and northwestern portion of the basin. The pair of solid lines indicates the approximate position of the east–west slice shown in Figure 9.3.

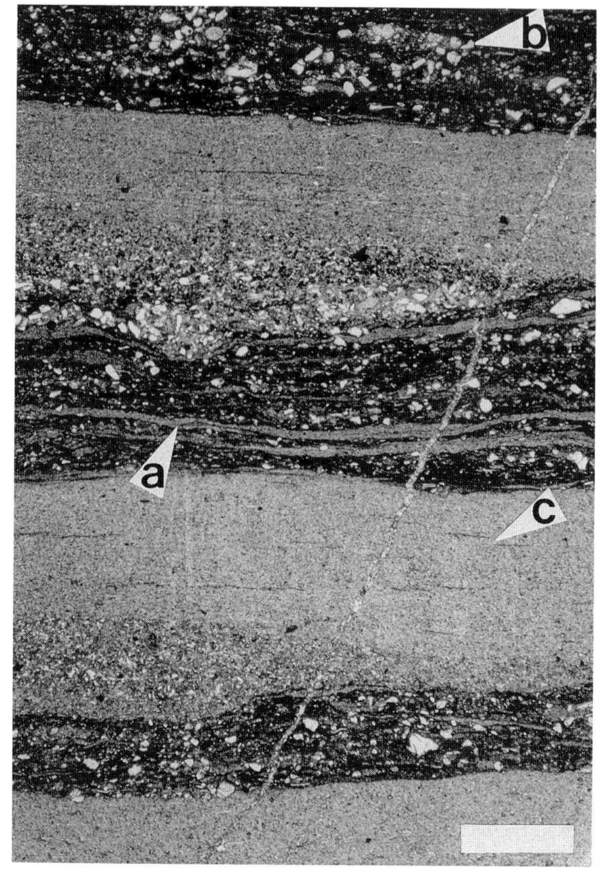

Figure 9.2. Photomicrograph of striped shale. Several beds of carbonaceous silty shale alternate with graded silt/mud couplets. Carbonaceous shale beds consist of wavy-crinkly carbonaceous laminae, alternating with drapes of dolomitic clayey shale (arrow a), and tiny silt-rich lenses (arrow b). These beds have been interpreted as microbial mat deposits. Note carbonaceous flakes in dolomitic clayey shale beds (arrow c). Scale bar is 500 μm.

Bedding planes are irregular and nonparallel, and poor alignment of clays and mica flakes (Fig. 9.4) and crumbled and rolled up carbonaceous flakes are conspicuous (Fig. 9.3). Continuous carbonaceous films and thin laminae of carbonaceous silty shale (identical to those described from facies NS-1) are also found.

Interpretation

These shales are interbedded with fine crystalline dolostones that show a variety of features, such as wavy-crinkly lamination, shrinkage cracks, erosion surfaces, scour and fill structures, ripples, and edgewise conglomerates, that are commonly thought to indicate deposition in restricted shallow shelf lagoons (Wilson, 1975). In analogy to facies NS-1, the carbonaceous films and laminae are interpreted as fossil microbial mats. Crumbled and rolled up carbonaceous flakes are probably due to erosion and redeposition of these thin microbial mat deposits. Because of the absence of signs of emergence, these shales were probably deposited in fairly shallow water, between fairweather and weak storm wave base. Microbial mats colonized the surface in calm periods and were eroded during storms. After storms, with decreasing wave agitation, rapid flocculation from moving water led to irregularly bedded muds with homogeneously mixed and random texture, and with variably deformed microbial mat fragments (carbonaceous flakes).

Carbonaceous Streak Shale Facies (NS-3)

These shales are characterized by silt and carbonaceous flakes (parallel to bedding) that are randomly distributed in a clay–dolomite matrix (Fig. 9.5). Individual beds of this shale type are thick (up to 10 cm), parallel, and lack sedimentary structures. Beds of hummocky cross-stratified coarse sandstone are interbedded with these shales. No signs of erosion and sediment reworking are found except at the base of interbedded hummocky cross-stratified sandstone beds.

Interpretation

Lack of sorting in shale beds suggests rapid sedimentation from suspension (Potter et al., 1980) and absence of reworking. Massive bedding implies deposition from successive prolonged pulses of continuous sedimentation. The presence of hummocky

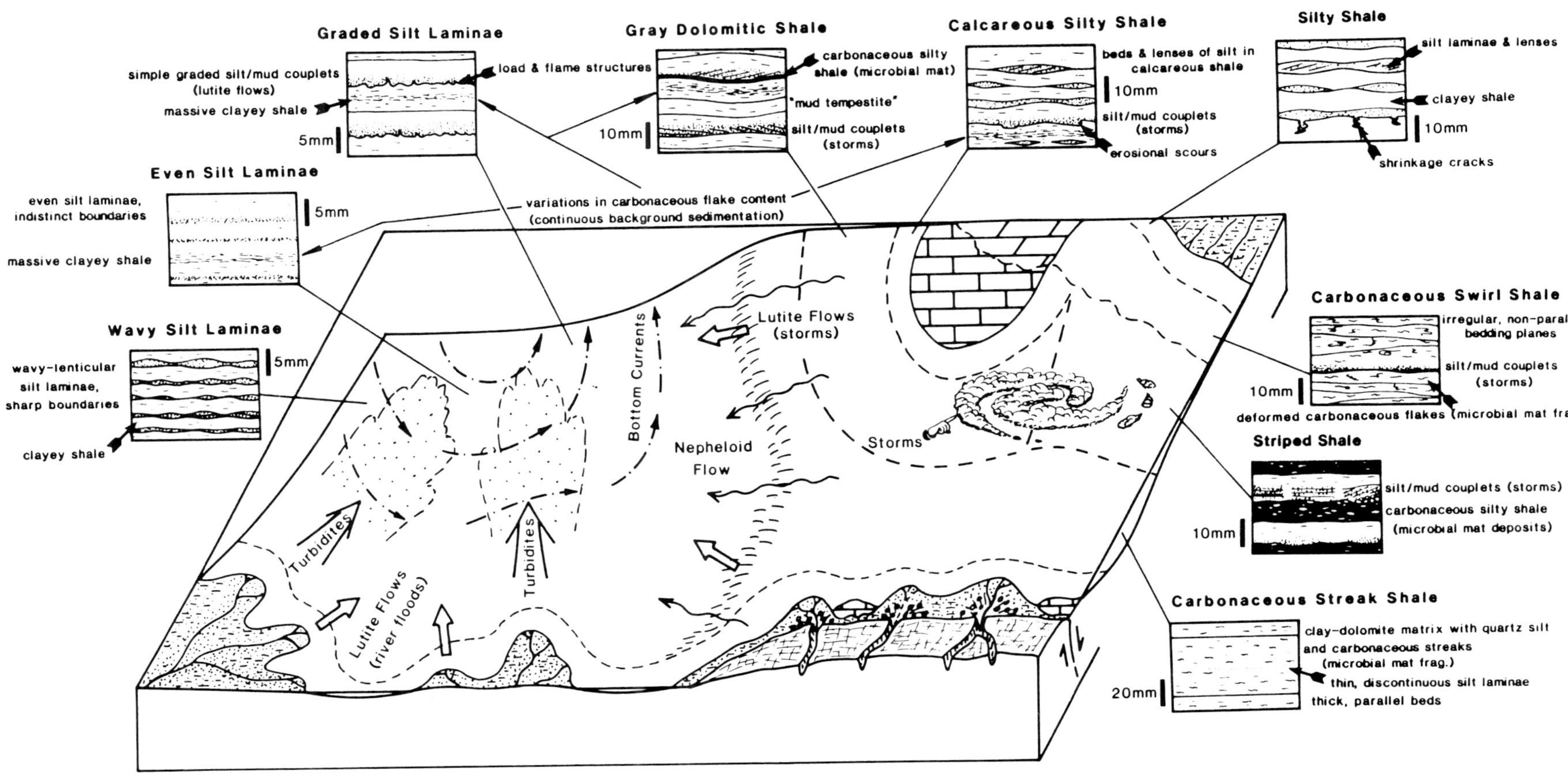

Figure 9.3. Hypothetical east–west slice (viewing north) through the southern portion of the Belt Basin during deposition of Newland (east) and Prichard Formation (central and west). The diagram summarizes major features of shale types, and the author's view of shale facies associations, sediment sources, and sedimentary processes.

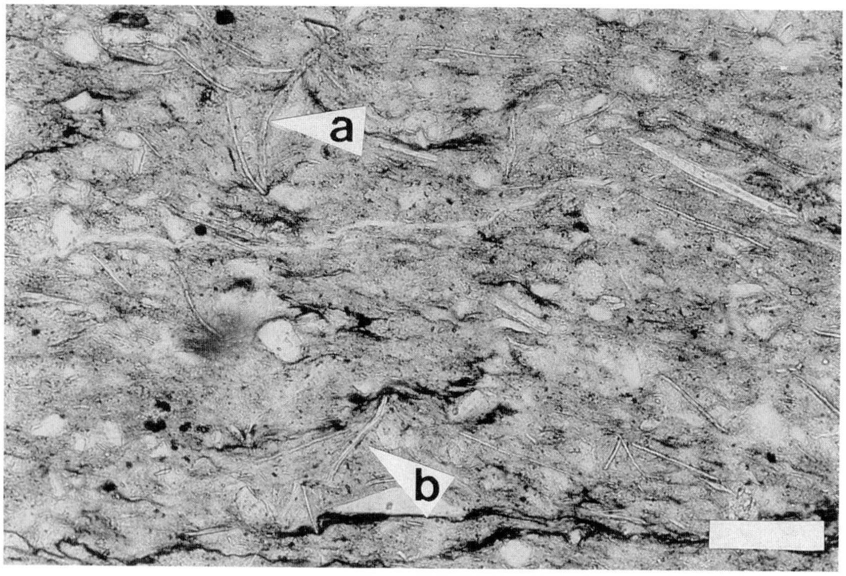

Figure 9.4. Photomicrograph of carbonaceous swirl shale with randomly oriented mica flakes. A fair number of mica flakes are in near vertical position and may show deformation due to compaction (arrows a and b). Scale bar is 100 μm.

cross-stratified sandstone beds and the absence of silt/mud couplets indicates that only very strong storms were able to touch the sea bed (see interpretation of facies NS-1). Thus deposition was probably below average storm wave base. Scattered silt grains in this facies suggest an eolian component to sedimentation. Carbonaceous flakes in this facies are of the same appearance as in NS-1 and NS-2 and are considered microbial mat fragments that were swept in from shallower areas.

Gray Dolomitic Shale Facies (NS-4)

This facies consists of interbedded dolomitic clayey shale (dominant), siltstone, and carbonaceous silty shale. These lithologies are of the same appearance as in facies NS-1 with respect to textures, sedimentary features, and composition. They differ only in their relative abundances. Silt/mud couplets as observed in facies NS-1 are common, and are analogously interpreted as storm deposits. Carbonaceous silty shale beds on the other hand are much less abundant and very thin (1 mm or less). In analogy to NS-1 a microbial mat interpretation is applied to the carbonaceous beds. The paucity and thinness of these deposits may indicate inhibition of mat growth, possibly because of a continuous background sedimentation component that diminished light penetration of the water column. The similarity of this facies to NS-1 with respect to the silt/mud couplets suggests that deposition occurred offshore between fairweather and (average) storm wave base.

Figure 9.5. Photomicrograph of carbonaceous streak shale. Carbonaceous flakes (arrows) float in a matrix of clay, fine crystalline dolomite, and silt. Alignment of carbonaceous fragments indicates bedding plane orientation. Scale bar is 100 μm.

Figure 9.6. Photomicrograph of ML-1 (Prichard Formation) showing simple graded silt/mud couplet with load and flame structures at base (arrow). Scale bar is 200 μm.

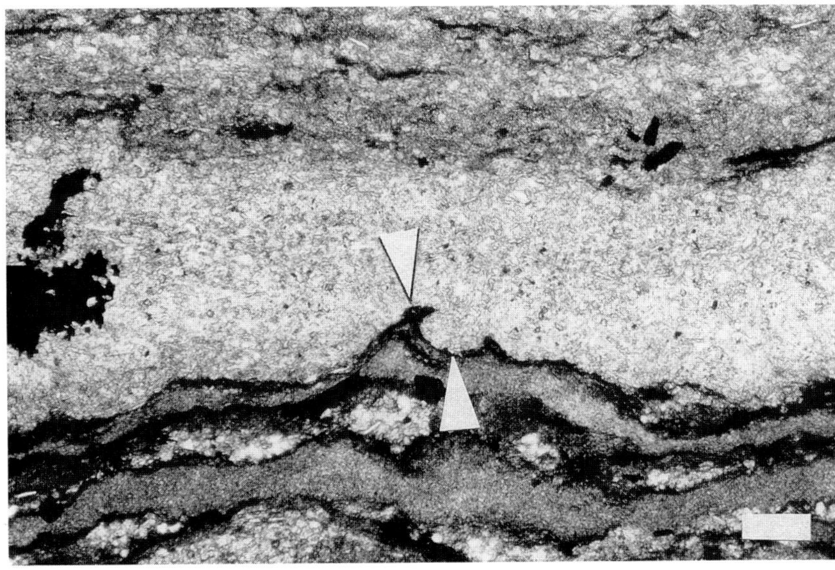

Calcareous Silty Shale Facies (NS-5)

These shales consist of clayey shale with variable amounts of dolomite, calcite, quartz silt, and carbonaceous flakes. Silt occurs as thin laminae, lensy-wavy beds, and trains and lenses of cross-laminated ripples. Silt layers commonly have sharp upper and lower boundaries, and may show erosional scours at the base (Fig. 9.3). Silt/mud couplets, as described from facies NS-1, also occur, but are not prominent.

Interpretation

Carbonates and sandstones that are interbedded with these shales show evidence of wave reworking and episodic strong currents, such as ripples with form-discordant internal structure, cross-stratal offshoots, symmetrical ripples with spillover aprons, imbricated flat pebble conglomerates, planar cross-bedding, and erosive channels. The common occurrence of erosional features in these shales indicates that their deposition took place in shallower water than that of NS-1 and NS-4.

Silty Shale Facies (NS-6)

These shales consist mainly of alternating laminae of siltstone and clayey shale, with interbedded packages of siltstone and sandstone. Silt laminae are usually wavy or consist of individual silt lenses, and have sharp boundaries with laminae of clayey shale (Fig. 9.3). Sedimentary features of siltstone intervals are cross-lamination, wavy and lenticular bedding, mudcracks, clay galls, load-casts, and erosional rills. Sandstone packages contain conglomerate beds, wave ripples, and erosive channels.

Interpretation

Sedimentary features in associated sandstones and siltstones indicate deposition in fairly shallow water. Intermittent exposure is indicated by mudcracks. These shales probably accumulated on nearshore mudflats and in shallow offshore environments.

Depositional Setting

Storm deposits, growth and erosion of benthic microbial mats, wave reworking, erosional channels, mudcracks, and mudclasts all indicate that shales of the Newland Formation accumulated in all likelihood in a shallow, low-energy shelf to shoreline setting. Examination of sedimentary features in interbedded carbonates and sandstones leads to the same conclusion (Schieber, 1985).

Comparison with Shales from the Prichard Formation

The Prichard Formation is characterized in many places by interbeds of thin to medium fine sandstone beds. On the basis of sedimentary features (flute casts, grading, climbing ripples, rip-up clasts etc.), these sandstone beds are commonly interpreted as turbidites (Cressman, 1984). No evidence of wave reworking and exposure is found in the bulk of the Prichard Formation. Even though because of metamorphic overprint the shales of the Prichard Formation are more correctly characterized as argillites, the terms shale, clay, and silt are used here to describe these rocks, primarily for the sake of more direct comparison with shales of the Newland Formation. Distinction of shale facies types in the Newland Formation is relatively easy and is

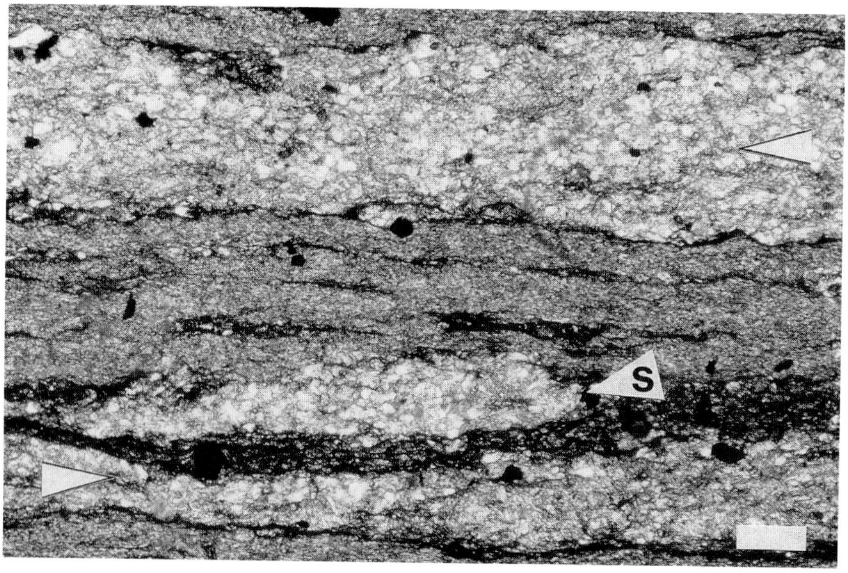

Figure 9.7. Photomicrograph of ML-2 (Prichard Formation) showing wavy-lenticular silt laminae with sharp boundaries (arrows), and single silt lens (arrow labeled S). Scale bar is 200 μm.

accomplished by using macroscopic features visible in hand specimens. In contrast, sedimentary features and lithological differences in shales of the Prichard Formation are much more subtle. Microscopic examination of shale samples from the Prichard Formation shows that samples can be grouped in terms of a small number of "microlithologies" (ML). Four common types are graded silt/mud couplets (ML-1), alternating wavy-lenticular silt and clay laminae (ML-2), massive clay beds (ML-3), and continuous silt laminae (ML-4).

The graded silt/mud couplets (1–5 mm thick) of ML-1 have a layer of fine to medium siltstone at the base (internally graded), and are in gradational contact with an overlying layer of clayey shale (Fig. 9.6). The silt layers commonly show load and flame structures at their base (Fig. 9.6). Silt/mud couplets in the Prichard Formation (ML-1) differ from those of the Newland Formation by the absence of cross-lamination, parallel lamination, and graded rhythmites, an indication that wave action and currents did not influence silt/mud couplet deposition in the Prichard Formation. Thus it appears that the silt/mud couplets of the Prichard Formation were deposited in deeper water than the silt/mud couplets of the Newland Formation. The complete absence of features that have been observed in muddy turbidites (Stow and Shanmugam, 1980) and storm layers (Aigner and Reineck, 1982), may indicate that these silt/mud couplets are the

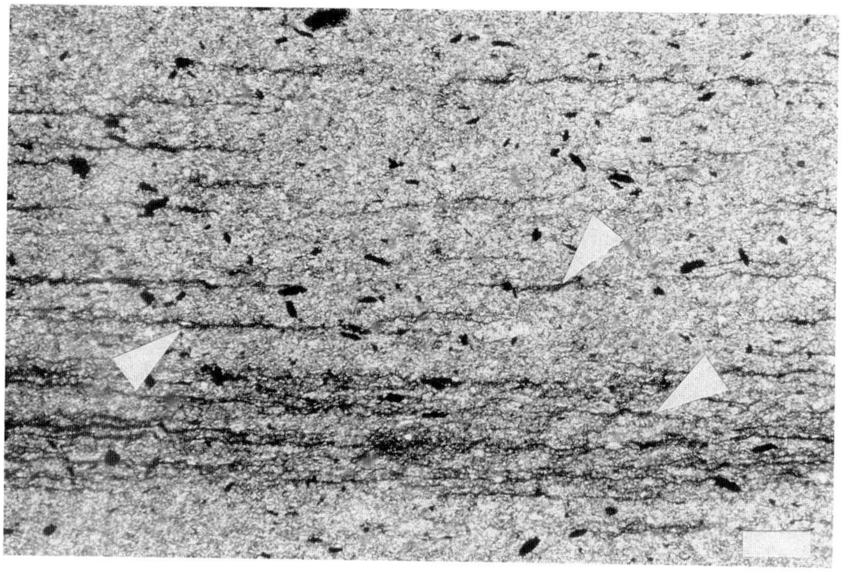

Figure 9.8. Photomicrograph of ML-3 (Prichard Formation) showing massive clayey shale bed with gradual variation in the amount of carbonaceous flakes (arrows). These gradual variations are interpreted to indicate fluctuations in sediment supply during background sedimentation. Scale bar is 200 μm.

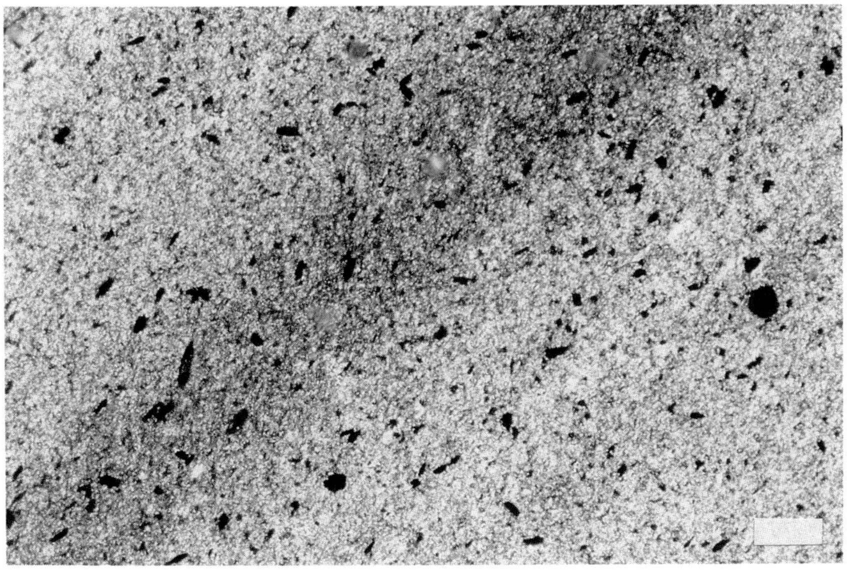

Figure 9.9. Photomicrograph of ML-3 (Prichard Formation) showing gradual variations in clay content within massive clayey shale beds. Clay-rich laminae are darker in color. Variations are interpreted as fluctuations in background sedimentation. Scale bar is 200 μm.

result of short lived lutite flows, possibly due to storms on the shelf or to river floods (McCave, 1972).

ML-2 consists of wavy-lenticular laminae of coarse silt to fine sand with sharp upper and lower boundaries, alternating with laminae of clayey shale (laminae 0.3–2 mm thick). The latter contain streaks of carbonaceous material. The silt laminae contain small amounts of detrital mica, and in places it is apparent that they consist of tiny silt lenses "overriding" each other (Fig. 9.7). These silt lenses may represent tiny starved ripples migrating over a mud surface. The sharp upper boundaries and the wavy appearance of the silt laminae suggest that these silt laminae did not simply settle out of suspension in stagnant water, but rather that currents transported silt over the seabed in bedload. Because silt requires current velocities only on the order of a few ten centimeters per second for bedload transport (Harms et al. 1982), and because silt laminae alternate with laminae of clayey shale, ML-2 is interpreted to be the product of episodic sediment reworking by weak bottom currents.

ML-3 consists of massive layers (up to 10 mm thick) of more or less homogeneous clayey shale and may represent pelagic background sedimentation. Subtle and gradual variations in the content of carbonaceous flakes and clay in these beds (Figs. 9.8 and 9.9) indicate continuous background sedimentation with slight variations in the supply of sedimentary components

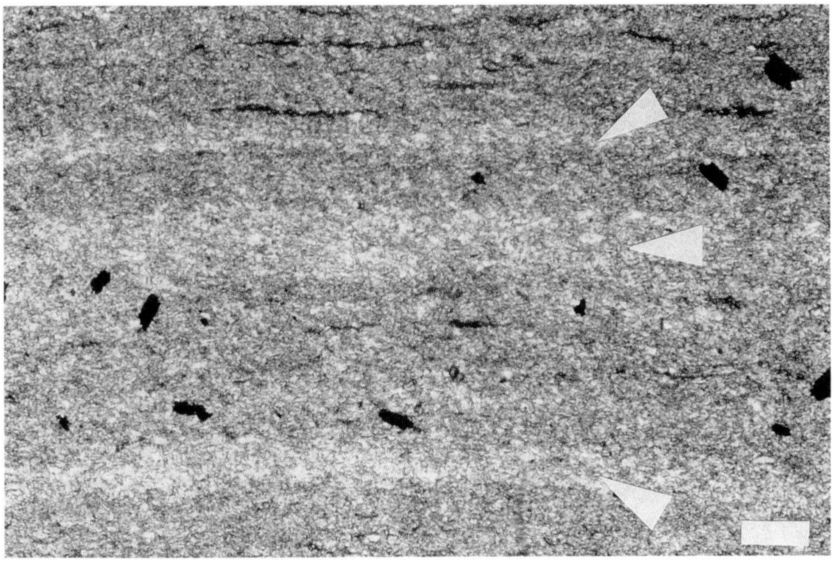

Figure 9.10. Photomicrograph of ML-4 (Prichard Formation) showing even silt laminae (arrows) with indistinct upper and lower boundaries in ML-3. Scale bar is 200 μm.

(Schieber, 1989). Within massive beds of ML-3, continuous even laminae of fine to medium silt (0.25–1 mm thick) with indistinct upper and lower boundaries (ML-4) are found (Fig. 9.10). These occur singly or as bundles of closely (0.2–1 mm) spaced laminae, and are perhaps due to gradually shifting sediment supply by nepheloid flows (Moore, 1969). This kind of silt laminae was not observed in shales of the Newland Formation.

Even silt laminae that bear similarity to ML-4 were described by Huebschman (1973) from the upper portion of the Prichard and Aldridge Formations (the Aldridge Formation is the correlative of the Prichard Formation in Canada). These silt laminae were described as being 1–20 mm thick, and individual laminae were correlated from outcrop to outcrop over a distance of more than 200 km. The wide areal distribution as well as the even and continuous appearance of these silt laminae led Huebschman (1973) to the conclusion that these silt bands were deposited in quiet water (below storm wave base), and that the silt was transported across the basin by dust storms.

Conclusion

Sedimentary features in Newland and Prichard Formations show that shale facies can be interpreted in terms of depositional setting, and that shales of the Newland Formation were deposited in shallow water. The absence of certain features in shales of the Prichard Formation, such as microbial mat deposits and cross-lamination in silt/mud couplets, points toward deposition in a deep basin, and sediment supply by nepheloid flows and reworking by bottom currents are suggested by indistinctly bounded and sharply bounded silt laminae, respectively. Possible eolian sediment input is indicated by scattered silt in "deep" water shale facies of the Newland Formation, and by even silt laminae of large areal extent in the Prichard Formation. Small scale sedimentary features are still readily recognizable in shales that have undergone low-grade metamorphism. Thus, an approach to shale sedimentology as shown here can be applied also to metamorphosed shale sequences.

References

Aigner, T., and H.-E. Reineck, 1982. Proximality trends in modern storm sands from the Helgoland Bight (North Sea) and their implications for basin analysis. Senckenbergiana Maritima, v. 14, p. 183–215.

Cressman, E.R., 1984. Paleogeography and paleotectonic setting of the Prichard Formation—a preliminary interpretation. Montana Bureau of Mines and Geology Special Publication 90, p. 8–9.

Harms, J.C., J.B. Southard, and R.G. Walker, 1982. Structures and sequences in clastic rocks. SEPM Short Course No. 9, Society of Economic Paleontologists and Sedimentologists, Tulsa, 249 p.

Harrison, J.E., 1972. Precambrian Belt basin of northwestern United States: its geometry, sedimentation, and copper occurrences. Bulletin of the Geological Society of America, v. 83, p. 1215–1240.

Huebschman, R.P., 1973. Correlation of fine carbonaceous bands across a Precambrian stagnant basin. Journal of Sedimentary Petrology, v. 43, p. 688–699.

Maxwell, D.T., and J. Hower, 1967. High-grade diagenesis and low-grade metamorphism of illite in the Precambrian Belt Series. American Mineralogist, v. 52, p. 843–856.

McCave, I.N., 1972. Transport and escape of fine-grained sediment from shelf areas. In: Swift, D.J.P., D.B. Duane, and O.H. Pilkey (eds.), Shelf Sediment Transport: Process and Pattern. Dowden, Hutchinson and Ross, Stroudsbourg, PA, p. 225–248.

Moore, D.G., 1969. Reflection profiling studies of the California continental borderland: structure and Quaternary turbidite basins. Geological Society of America Special Paper 107, 142 p.

Potter, P.E., J.B. Maynard, and W.A. Pryor, 1980. Sedimentology of Shale. Springer-Verlag, New York, 306 p.

Schieber, J., 1985. The relationship between basin evolution and genesis of stratiform sulfide horizons in Mid-Proterozoic sediments of Central Montana (Belt Supergroup). Ph.D. Dissertation, University of Oregon, 811 p.

Schieber, J., 1986. The possible role of benthic microbial mats during the formation of carbonaceous shales in shallow Mid-Proterozoic basins. Sedimentology, v. 33, p. 521–536.

Schieber, J., 1989. Facies and origin of shales from the Mid-Proterozoic Newland Formation, Belt basin, Montana, U.S.A. Sedimentology, v. 36, p. 203–219.

Stewart, J.H., 1976. Late Precambrian evolution of North America: plate tectonics implication. Geology, v. 4, p. 11–15.

Stow, D.A.V., and G. Shanmugam, 1980. Sequence of structures in fine-grained turbidites; comparison of recent deep-sea and ancient flysch sediments. Sedimentary Geology, v. 25, p. 23–42.

Wilson, J.L., 1975. Carbonate Facies in Geologic History. Springer-Verlag, New York, 471 p.

CHAPTER 10

Porosities, Permeabilities, and Microfabrics of Devonian Shales

David K. Davies, William R. Bryant, Richard K. Vessell, and Patti J. Burkett

Introduction

Shales generally are regarded as source rocks and seals. However, in the Illinois, Michigan, and Appalachian Basins, shales of Devonian age commonly are regarded as reservoir rocks. According to Broadhead et al. (1982), economic gas production from nonfractured shales has been an established fact in these basins for several decades. In some areas, natural fractures are an important control on production. In other areas, gas productive shale intervals lack natural fractures.

There is a general dearth of geological information and understanding regarding shales, particularly compared to that available on more conventional reservoir rocks (sandstones and carbonates). Shales are difficult to analyze because of their fine grain size, small pore size, and low values of porosity and permeability which approach the resolution levels of existing equipment. The pore structure and reservoir properties of shales are not well understood, despite their potential economic significance and importance in the geologic record.

This chapter reports on the structure of shale porosity, and documents relationships among porosity, permeability, and shale microfabric through combined analysis of geological and engineering data. It is part of an ongoing effort by the Gas Research Institute to understand and maximize Devonian shale production in the southern Appalachian Basin.

Data Base

The data base for this study consists of 118 drilled sidewall core samples from 17 wells in southeastern Ohio, eastern Kentucky, West Virginia, and Virginia (Fig. 10.1). Each sidewall core was drilled horizontally from the well bore, and is a small ($\approx 1 \times 3$ in.) conventional core. Core samples were analyzed to determine (1) petrographic characteristics (composition, texture, fabric, pore structure), (2) physical properties (porosity, permeability, saturation, and grain density), and (3) organic geochemistry (total organic carbon). Sample locations were not selected by the authors, but by engineers involved with log analysis and well stimulation. From a geological viewpoint, sample locations can be considered essentially random.

Prior to detailed analysis, each core was examined with a binocular microscope. Naturally fractured cores were not included in this study. Each core was cut into two portions, perpendicular to bedding. The first portion was used for geologic analysis and the second for physical properties analysis (Fig. 10.2). A complete set of analytical results is available in an open file report (Truman and Campbell, 1986).

Petrographic analyses involved routine thin section, X-ray diffraction (bulk and fine fraction) and scanning electron microscope (SEM) analysis of 118 core samples. Transmission electron microscope (TEM) analysis of ultrathin sections ($\approx$ 700 Å in thickness) was undertaken on 13 samples, using preparation techniques described by Bennett et al. (1977). These analyses were all carried out by the authors. Total organic carbon content (TOC) was determined by a commercial laboratory on 109 of the core samples.

Reservoir properties measurements used in this study include porosity (vol%), permeability (microdarcies, μd), and grain density (g/cm^3). The measurement of accurate values for these properties is difficult in low porosity rocks. A complete suite of results was unobtainable for some of the core samples due to small sample size, microfractures, and permeabilities below the minimum limits of detection ($<$0.001 μd). Results were obtained for the following numbers of cores: porosity, 113 samples; permeability, 68 samples; grain density, 113 samples.

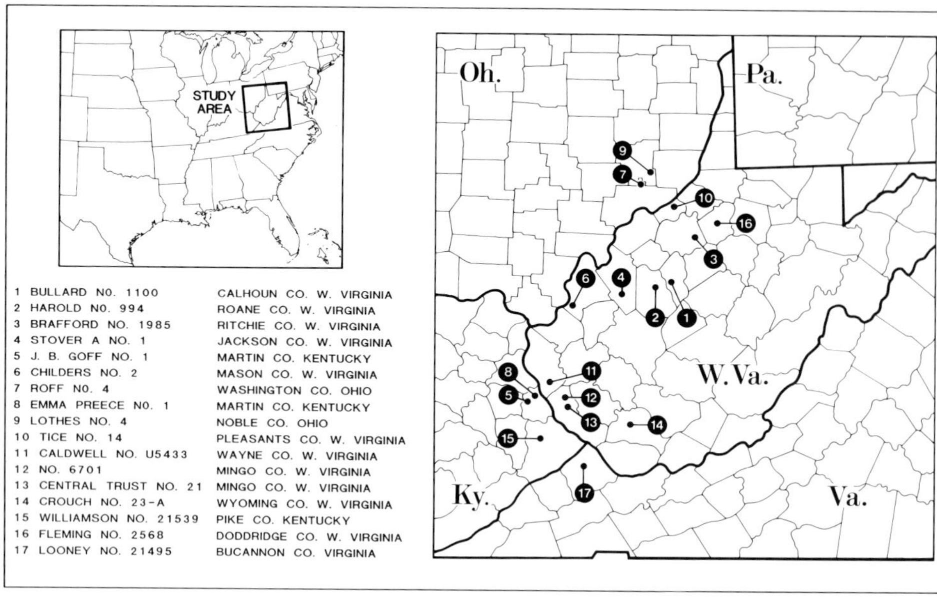

Figure 10.1. Location of study area and sampled wells.

Porosity was determined by subtraction of bulk volume from grain volume. Water was extracted from the cores using a Dean Stark extractor, after which samples were soxhlet cleaned with a chloroform–methanol mixture and dried in an oven at 40°C. Grain volume was measured in a Boyle's law apparatus with helium. No confining pressure was applied to the sample. Bulk volume was measured with calipers on cylindrical samples and by mercury immersion on irregular shapes.

Gas permeability was measured in a pulse-decay apparatus with nitrogen at a confining pressure of 2500 psi and a pore pressure of 1000 psi. This technique for the measurement of permeability is a transient system that minimizes the Klinkenberg gas-slippage effect. A small (25–50 psi) differential gas pressure is exerted across each core, and the change of gas pressure with time is recorded. Permeability is calculated from the rate of pressure decay with time (Walls, 1982; Walls et al., 1982). Bulk

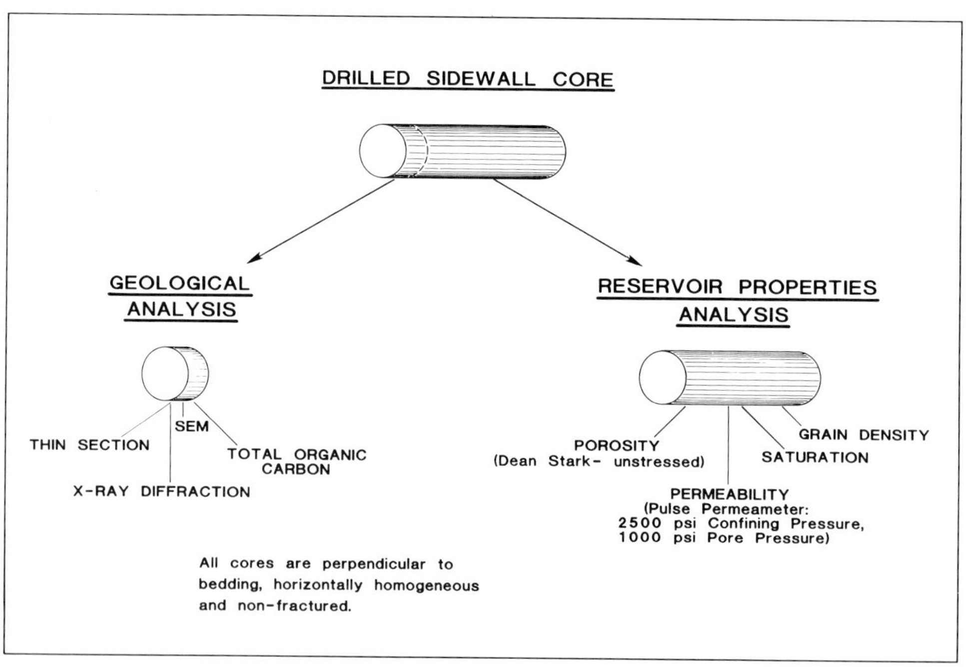

Figure 10.2. Analytical procedures attempted on all core samples.

Table 10.1. Microfabric and average mineralogy of each of the three types of Devonian Shale in the southern Appalachian Basin.*

Shale type	Shale fabric	Mineralogy (wt %)						TOC (wt %)
		Quartz + feldspar	Clay†	Pyrite	Siderite	Calcite	Dolomite†	
A	Chaotic–burrowed	47	47	3	1	1	1	0.5
B	Chaotic–subparallel	39	50	5	4	2	1	0.6
C	Parallel	48	45	6	Tr	Tr	Tr	4.7

*Based on SEM, X-ray diffraction, thin section, and organic geochemical analysis.
†Clay includes mica; dolomite includes ankerite.

density was calculated as preserved weight divided by bulk volume. All reservoir property measurements were performed by a commercial laboratory that specializes in the analysis of core samples with low values for porosity and permeability.

Devonian Shale Lithologies

The Devonian Shale sequence in the southern Appalachians averages some 3000 ft (1000 m) in thickness. This sequence consists dominantly of shales with occasional, thin, interbedded sandstones and siltstones. In this chapter the term "shale" is used for all rocks finer than sand grade that contain more than 35% clay and mica (Krynine, 1948, p. 154–155). Devonian shales are indurated and may or may not have fissility.

The average Devonian shale sample analyzed in this study contains approximately 45% detrital clay (including fine grained mica). Highly crystalline illite is virtually the only clay mineral recorded in X-ray diffraction analysis. The groundmass of the shale consists of an intermixture of finely crystalline illite and muscovite. Angular grains of detrital quartz ($\approx$ 40%) and feldspar (<5%) are scattered irregularly throughout the groundmass of illite/muscovite. Quartz and feldspar grains are usually of silt or very fine sand-size grade.

Three types of shales may be recognized in the sequence. These we refer to as shale Types A, B, and C. Discrimination among shale types is based on a combination of microfabric, mineralogy, and color distinctions (Table 10.1). The three varieties of shale grade into one another. Shale Types A and C may be regarded as end-members in a continuum that grades through shale Type B.

Shale Type A (Fig. 10.3)

This shale type is characterized by chaotic microfabric, commonly the result of intense burrowing activity. Burrowing activity has resulted in disruption of the original depositional fabric, and the plate-like illite/muscovite components have random orientation. Scattered silt grains supported the disrupted fabric during burial compaction. Traces of the original, laminated, depositional fabric may still be seen where burrowing is not intense. These shales are gray in color, contain less than 1% organic carbon, and have a low abundance of reducing minerals, e.g., pyrite (Table 10.1). They are interpreted as being deposited under aerobic bottom conditions.

Shale Type B (Fig. 10.4)

These shales contain less quartz and more authigenic minerals characteristic of reducing conditions than shale Type A (Table 10.1). Authigenic pyrite, siderite, and calcite are ubiquitous, and can be abundant locally. For example, the siderite content of individual cores ranges up to 35% and the pyrite content ranges up to 14%. High concentrations of these minerals in the small cores probably reflect the presence of nodules rather than continuous beds. Shale Type B has the highest average grain density of all rocks analyzed from the Devonian Shale sequence (2.77 g/cm^3, Table 10.2). This is the result of the abundance of "heavy" authigenic sulfide and carbonate minerals. Minor quantities of kerogen occur in the shale.

The microfabric is generally indistinct to subparallel, owing to a tendency for a combination of chaotic and subparallel grain alignment. The general absence of burrows in these rocks suggests deposition in dysaerobic bottom conditions, that shales of Type B may be of deeper water origin than the silty shales of Type A.

Shale Type C (Fig. 10.5)

This shale variety is generally referred to as "black shale" in the Appalachian Basin. Shales of Type C are characterized by high

Table 10.2. Mean values of reservoir properties for individual shale types, Devonian shales, southern Appalachian Basin.

Shale type	Porosity (%)	Permeability (μd)	Grain density (g/cm^3)
A	3.7	3.7	2.75
B	3.6	1.7	2.77
C	4.1	0.9	2.63

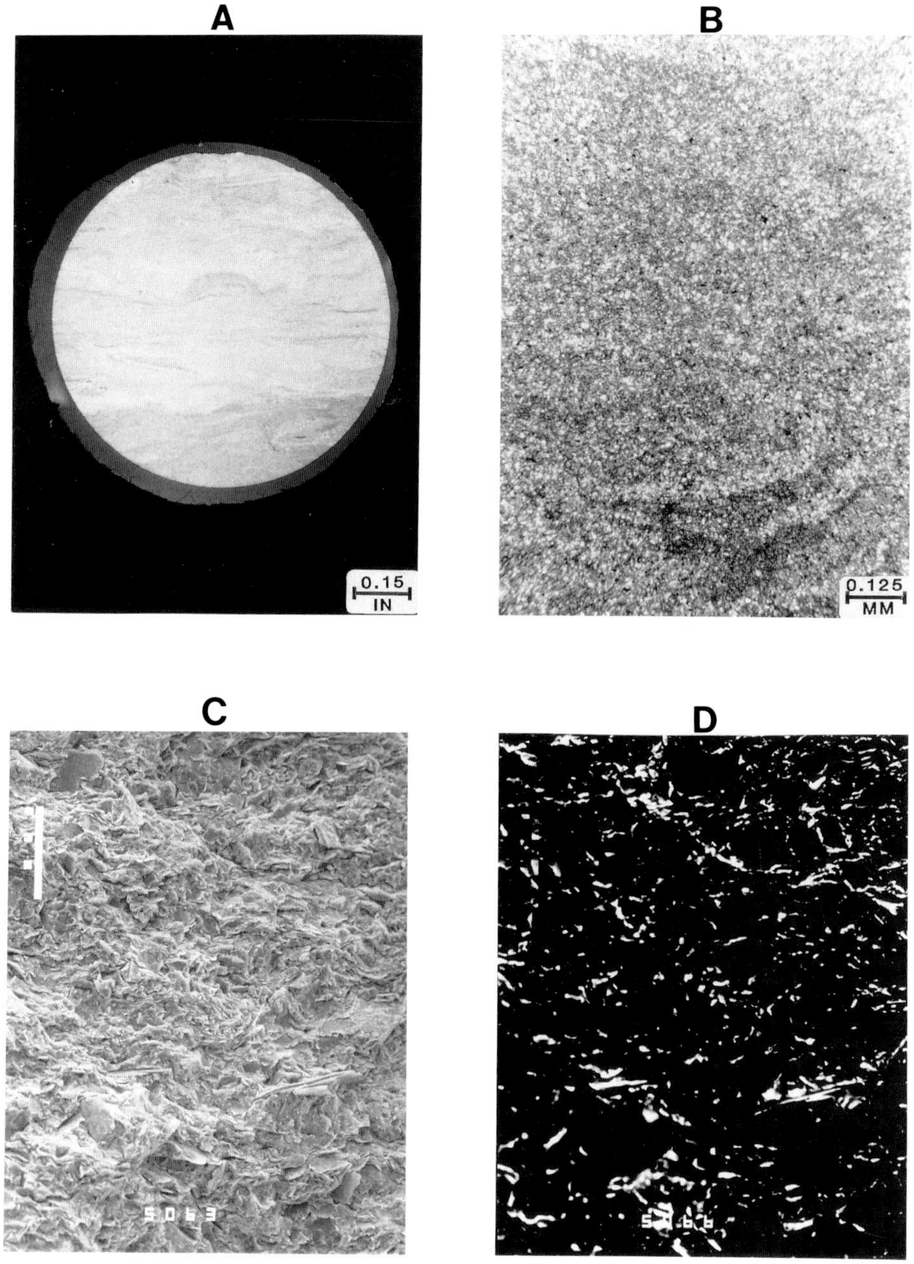

Figure 10.3. Photographs of shale Type A, illustrating lithology and microfabric. (A) View of the end of the core plug, cut perpendicular to bedding. Note the disruption of fabric due to burrowing (transmitted light). (B) Thin section view, at low magnification (plane polarized light). (C) Low magnification, SEM view shows the chaotic arrangement of detrital grains in this shale type. Scale = 100 μm. (D) Inverse polarity SEM view of the same area seen in C. Note the disorganization of the fabric elements (white).

values of TOC (Table 10.1). The total organic carbon content is always high, ranging from 1.24 to 21.35%. The organic material occurs as discrete kerogen fragments that can be large enough to be seen with the naked eye. Pyrite is also abundant and can compose up to 17% of the core samples in those intervals with pyrite nodules. Despite the pyrite content (grain density ≈ 4.99 g/cm^3), this rock type has the lowest grain density of all rocks analyzed. This reflects the presence of significant quantities of the low-density kerogen component (grain density ≈ 1.25 g/cm^3).

The shale fabric is dominated by abundant, well-defined, small-scale laminations. Small-scale parallel laminae are the result of grain segregation: silt-clay-rich laminae alternate with clay–organic-rich laminae. Laminations are frequently enhanced by plates of muscovite and elongate fragments of kerogen.

Black shales of the Devonian section are interpreted as the deepest water shales in the section. It is hypothesized that bottom conditions were anaerobic. This conclusion is supported by the absence of bioturbation and abundance of kerogen in these rocks. The anoxic bottom conditions inhibited life but were con-

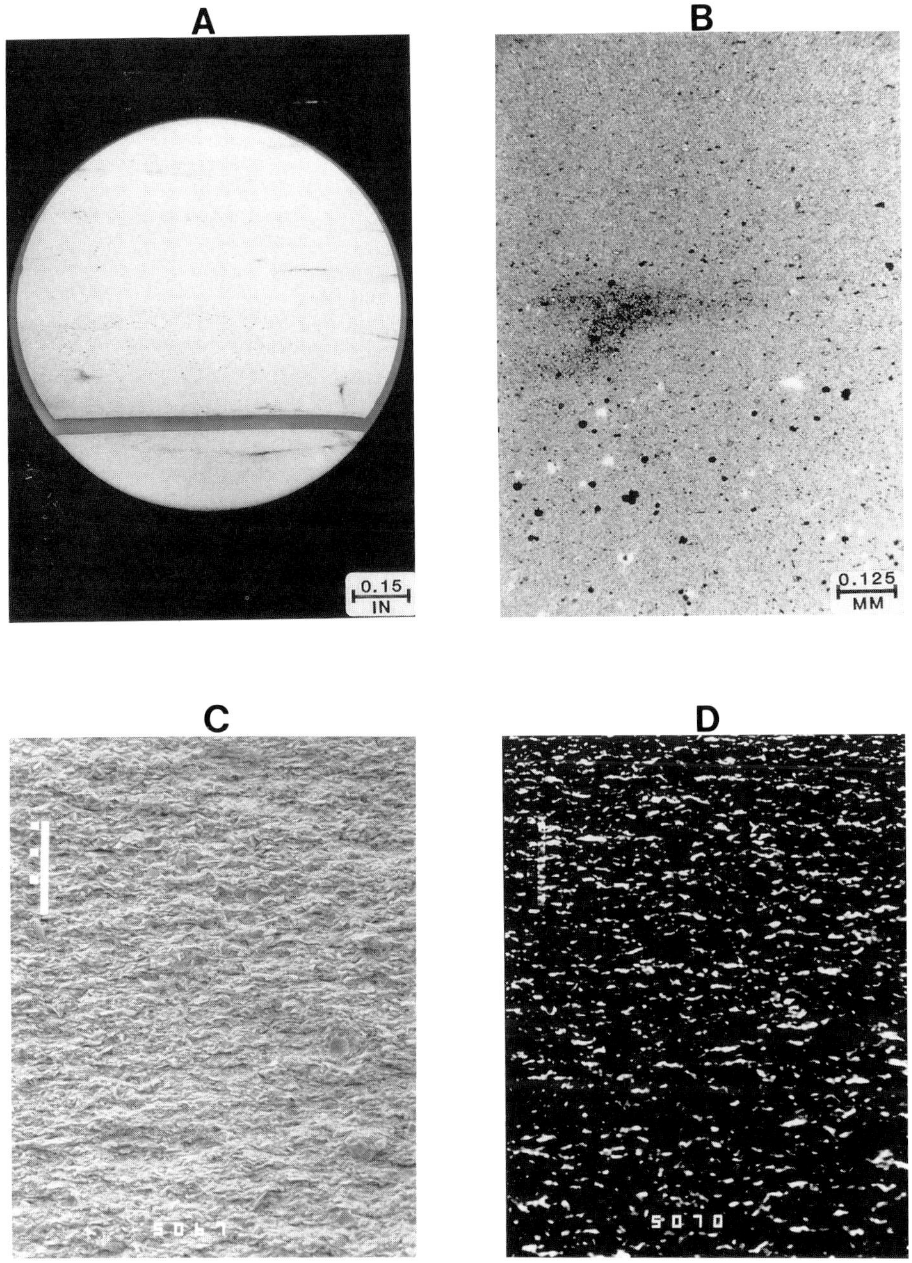

Figure 10.4. Photographs of shale Type B, illustrating lithology and microfabric. (A) View of the end of the core plug, cut perpendicular to bedding. Note the ill-defined microfabric. Plug was broken during thin sectioning, resulting in the horizontal fracture seen towards the base of the core end (transmitted light). (B) Thin section view, at low magnification. Black grains are authigenic pyrite (plane polarized light). (C) Low magnification, SEM view reveals that fabric elements have a greater degree of parallelism than in shale Type A. Scale = 100 μm. (D) Inverse polarity SEM view of the same area seen in C. Fabric elements (white) are generally short and have subparallel to parallel arrangement over much of the area in this view.

ducive to the preservation of lipid-rich, amorphous, and structured marine organic matter that would be destroyed under normal oxidizing conditions. The abundant pyrite resulted from the action of sulfide-reducing bacteria that flourished under these reducing conditions.

Porosity, Permeability, and Microfabric

The results of laboratory analysis of porosity and permeability are summarized in Table 10.2. Mean porosity values for each shale type range from 3.6 to 4.1% (Table 10.2). The highest porosity value and greatest range of porosity values (1.2–7.6%) occur in shale Type C. Evaluation of log-derived values of porosity also suggests that shale Type C contains streaks of relatively high porosity. For this reason, several operators in the study area preferentially perforate the black shales. The slightly higher values of porosity may indicate streaks of better sorting in these distal, deep water, Type C shales.

Mean values for permeability are strongly controlled by shale type (Table 10.2). These differences are not considered to be the result of mineralogical or lithological variations. There are no

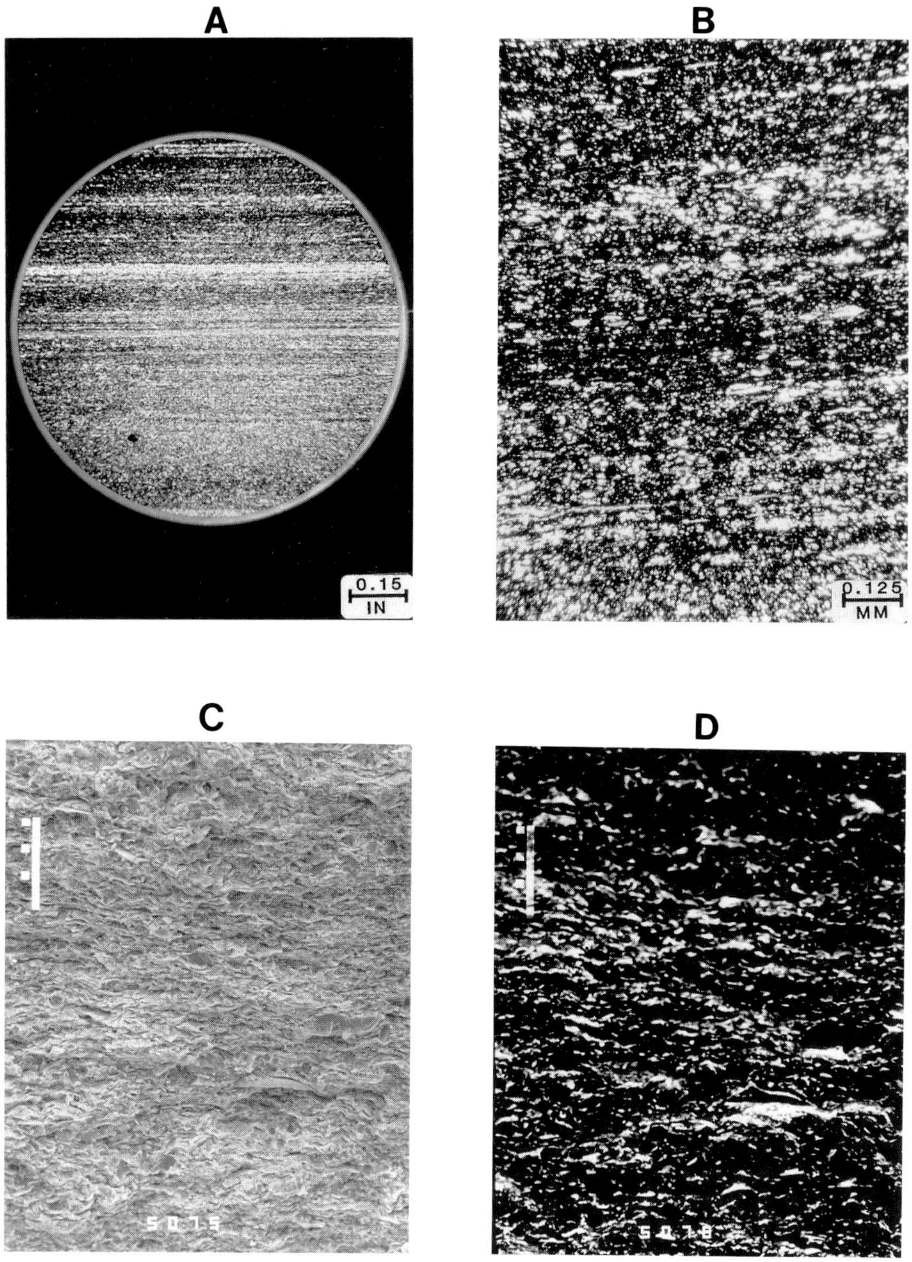

Figure 10.5. Photographs of shale Type C, illustrating lithology and microfabric. (A) View of the end of the core plug, cut perpendicular to bedding. Note the well laminated microfabric (transmitted light). (B) Thin section view, at low magnification. Lamination is well defined through segregation of quartz- and shale-rich laminae (plane polarized light). (C) Low magnification, SEM view reveals that fabric elements are well ordered and parallel. Scale = 100 μm. (D) Inverse polarity SEM view of the same area seen in C. Fabric elements (white) are longer more closely spaced and better defined than in shale Type B.

differences in the clay mineralogy or in the ratio of grains to matrix that would explain the observed permeability differences. In our opinion, microfabric is the fundamental control on shale permeability. Shale Type A, with random microfabric, has the highest permeability. Shale Type C, with laminated microfabric, has the lowest permeability. Shale Type B, with a fabric intermediate between shale Types A and C, has intermediate permeability. These results indicate that, for a given porosity, the permeability will be lowest in shales with laminated microfabric (black shales of Type C) and highest in shales with chaotic fabric (gray shales of Type A).

The fundamental controls on the fabric of Devonian shales appear to be factors present in the depositional environment. Shallow water shales, which have chaotic fabric, have been subjected to burrowing. Deep water shales, which are strongly laminated, show the absence of penecontemporaneous burrowing and retain their presumed, original depositional fabric. Similar environmental controls on shale fabric have been reported by O'Brien et al. (1980) and O'Brien (1987).

Relationships documented here between fabric and measured permeability have previously been recognized only qualitatively. In his work on Devonian shales, Soedder (1988, p. 122) noted

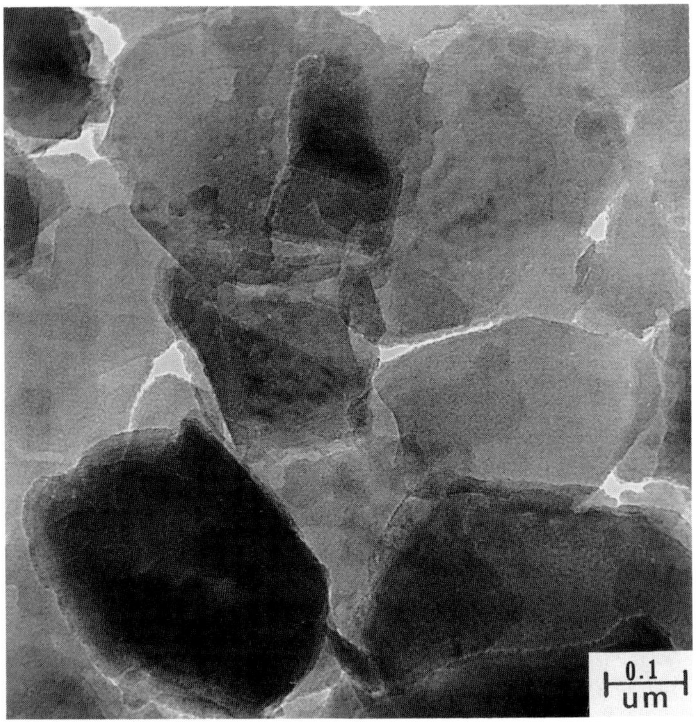

Figure 10.6. TEM photomicrograph of well-sorted, very fine-grained shale. Note the overlapping arrangement of detrital particles (gray/black) and intergranular pores (white). In all TEM views, black, opaque grains are ≈ 700 Å thick (the thickness of the ultrathin section). The thinner the particle, the lighter the gray shade. Random particle stacking produces alternating grains and pores within the 700 Å A plane of the section. White areas represent pores that extend completely through the section.

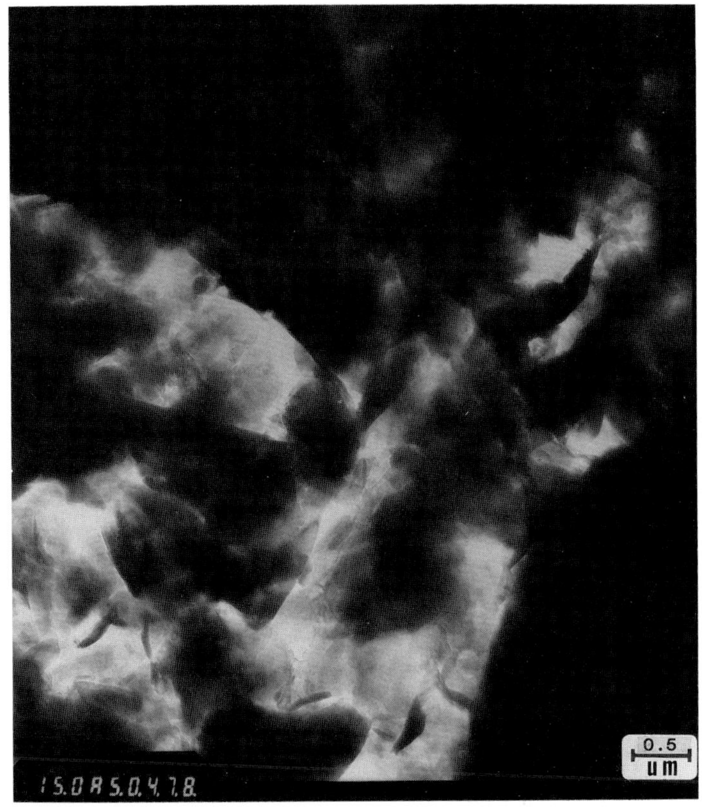

Figure 10.7. TEM photomicrograph of poorly sorted shale. Individual grains are illite and have disparate sizes and shapes. Intergranular pores (white) are common.

that "Differences between the depositional environment. . . had an influence on the pore geometry and microstructure of the resultant lithified shale. . .The presence or absence of oriented clay flakes would appear to have consequence with respect to pore size and connectivity in shales and requires further investigation."

Shale Texture and Pore Structure

Visual analysis of shale grain size, sorting, and pore structure was undertaken using ultrathin sections (thickness ≈ 700 Å) in a transmission electron microscope. Preparation difficulties restricted our views of shale pore systems to views that are generally parallel to the basal plane of the clay and mica components.

Detrital components in Devonian shales of all types range from well to poorly sorted (Figs. 10.6 and 10.7). Most are poorly sorted. Thin, plate-like flakes of detrital illite and mica overlap in random fashion. Several different flakes can overlap in the 700 Å thickness of the ultrathin section, indicating that the thinnest individual detrital clay particles have a thickness of 200–300 Å. Grain size is variable from sample to sample, the finest measured grain size of any sample in the data set being 0.2 μm. Visual porosity is of intergranular origin.

The poorly sorted shales are considerably more abundant than the well-sorted shales. Grain size is highly variable and difficult to measure in TEM because of the small size of the sample used in analysis. In the poorly sorted shales, large (sand and silt-sized) mica flakes, kerogens, and detrital quartz grains are surrounded by smaller flakes of illite that have widely varying sizes. It is not yet possible to section the quartz grains, thus results of TEM analysis in this chapter are selective for the illite-rich, finer grained portions of the rock. Figure 10.7 displays the general textural characteristics of the clay-rich portions of a poorly sorted shale. Most grains have irregular shape. Intergranular porosity is common: modal pore diameters range between 1 and 2 μm.

Image analysis of TEM negatives demonstrates the widespread distribution of porosity around the edges of detrital particles (Fig. 10.8). The pore system in the shales appears to be almost exclusively intergranular. The shape and size of the pores are variable, a function of the irregular size and shape of the surrounding grains. There is a relationship between the size of the detrital particles in the shales and the size of the associated

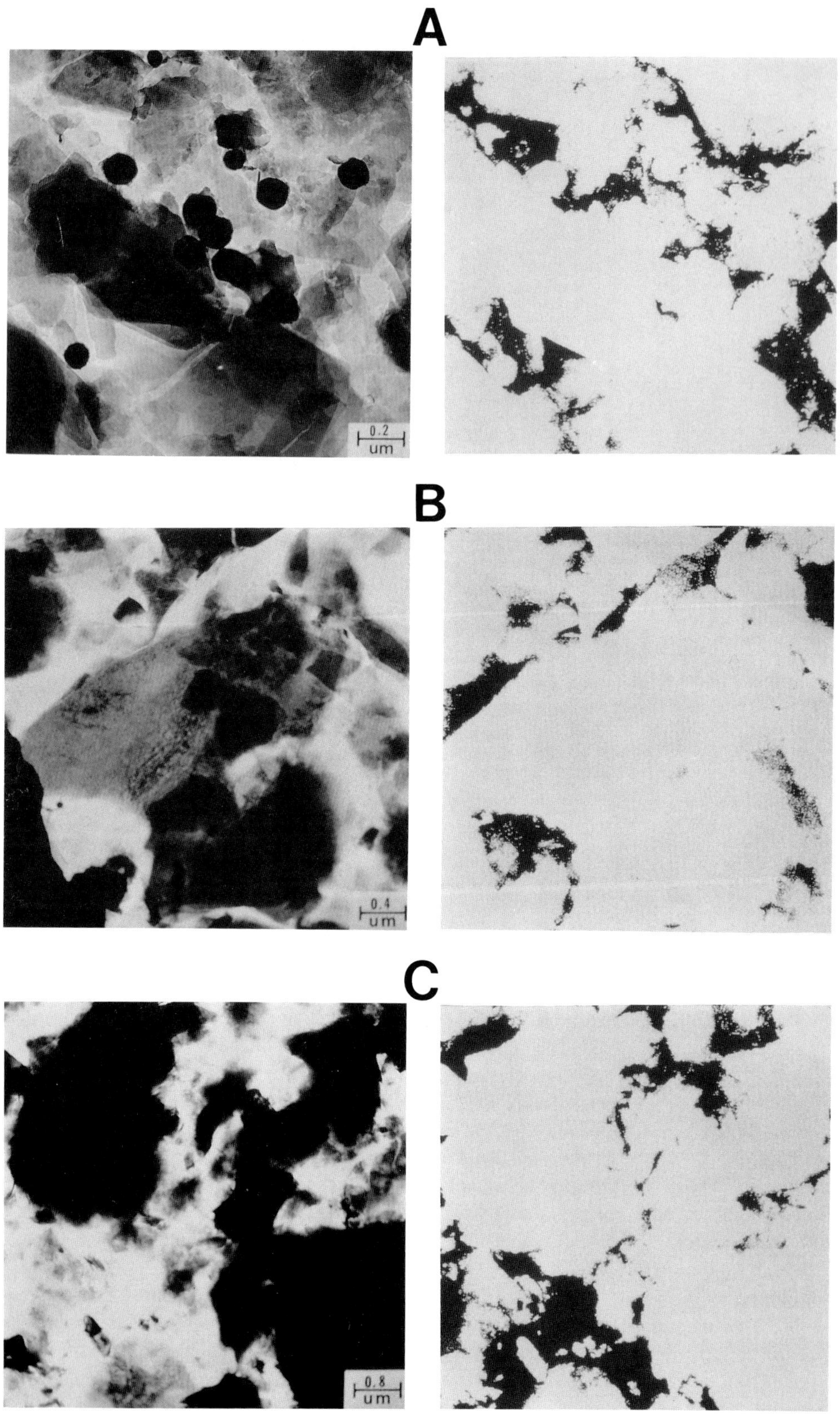

Figure 10.8. TEM photomicrographs (left) and matching, computer-generated images showing grains and pores (right). Porosity is white in the TEM photomicrographs and black in the matching images. (A) This sample has well-developed intergranular porosity. Octohedrons of authigenic pyrite (black) occur within the intergranular pore system. (B, C) These samples have porosity but lack authigenic pyrite. (Computer images are generated using quantitative values for gray shades taken from TEM negatives. These values are used to define grain and pore outlines.)

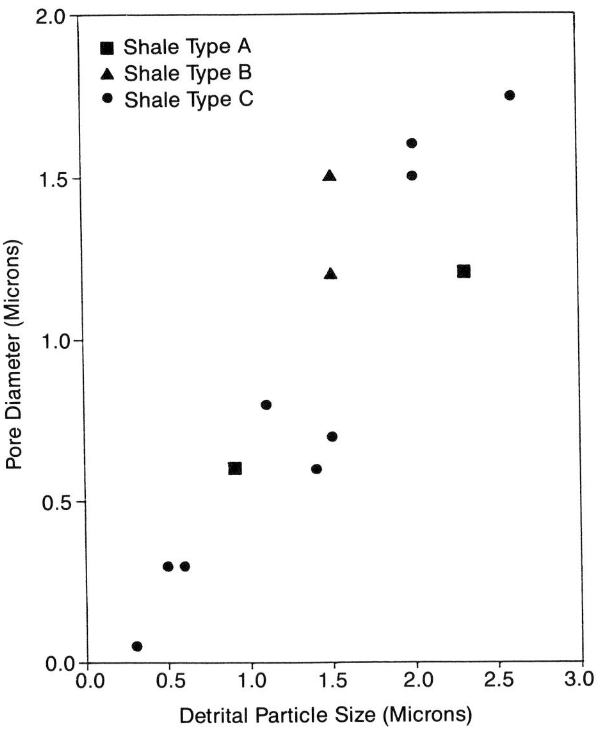

Figure 10.9. Cross-plot illustrating the relationship between particle size and pore diameter. Measurements were made using TEM images. Each data point represents a different image.

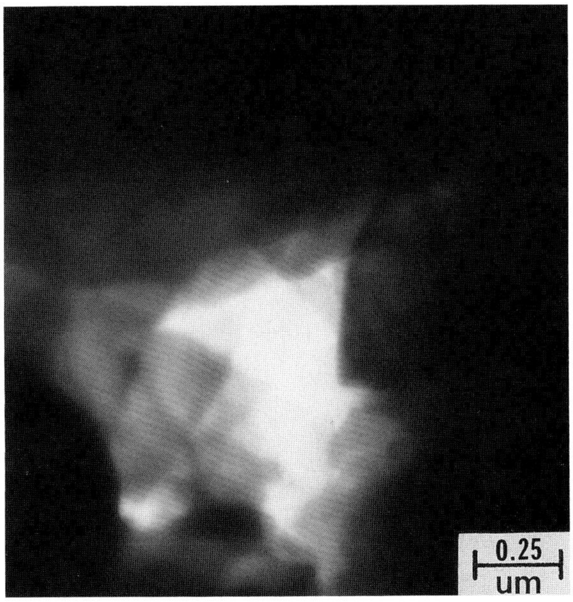

Figure 10.10. TEM photomicrograph of an intergranular pore (white) that is surrounded by illite flakes. Note that the illites are characterized by straight edges, interpreted as diagenetically developed crystal faces. (Photograph is slightly fuzzy due to high magnification.)

pores: the largest pores occur with the largest particles (Fig. 10.9). (This is the standard relationship that exists in sandstone pore systems of intergranular origin.) The relationship becomes increasingly diffuse as the sizes of grains and pores increase (Fig. 10.9).

High magnification views of pores reveal straight edges on illite flakes that may represent crystal faces (Fig. 10.10). The highly crystalline state of the illite indicates considerable diagenetic alteration. Diagenesis has also resulted in the precipitation of authigenic mineral phases in the intergranular shale pores (Figs. 10.8A and 10.11A). Pyrite octohedra are the most common of the authigenic, pore lining minerals. Unidentified, tabular minerals also occur. These may be apatite or a variety of zeolite. The authigenic minerals commonly occur along the pore walls (Fig. 10.11A). Three-dimensional reconstruction of a typical 2 μm pore (Fig. 10.11B) demonstrates that (1) the pores have irregular surfaces, (2) pore height increases away from the grain edges, and (3) 2-μm-long shale pores have a vertical dimension greater than the 700 Å thickness of the thin section.

Permeability and Porosity Considerations

Permeability values for Devonian shales in this study range from 10 to 10^{-2} μd (mean ≈ 2 μd). Permeability values are controlled strongly by microfabric. Shales with parallel microfabric consistently have lower permeabilities (<1 μd) than shales with chaotic microfabric (>1 μd). The commercial measurement techniques used in these analyses are considered state of the art (Walls, 1982; Walls et al., 1982). The absolute values reported here are at variance, however, with theoretical considerations and with evidence derived from direct observation of the shale pore system.

Laboratory experiments in consolidated muds and clays with porosity values of 10–20% yield permeabilities on the order of 10^{-5} μd. These values for permeability are several orders of magnitude lower than the values derived for this study using pulse permeametry.

Even if the absolute values for permeability reported here for Devonian shales are too high, there exists a consistent relationship between permeability and microfabric. Shales with chaotic fabric have higher permeability than shales with laminated microfabric. This relationship between permeability and fabric is supported by independent evidence. Soedder (1988, p. 123) pointed out that "Most gas shows occur in gray shales (Shale Types A and B)... not in the black units (Shale Type C)... At least some of the gray shale is fairly permeable to gas and may have been overlooked as a productive formation."

The porosity of Devonian shale samples, as measured by Dean Stark analysis, ranges from 1 to 8% (mean ≈ 4%). Values for porosity are less affected by changes in shale microfabric than are values for permeability. TEM observations of Devonian shale pore systems reveal that the pores are larger and more

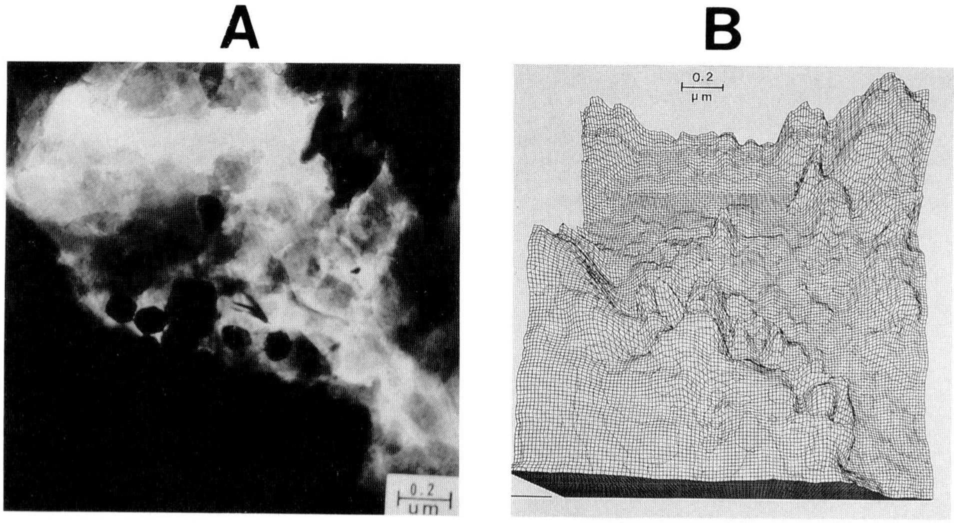

Figure 10.11. (A) TEM view of a single pore (maximum dimensions ≈ 2 µm × 1 µm × 700 Å. (A) The lower edge of this pore is lined with pyrite octohedra. The pore increases in depth away from the edges of the detrital grains. (B) Three-dimensional reconstruction of the pore seen in A, based on image processing of gray levels in the TEM negative.

abundant than would be anticipated from the results of commercial porosity analyses. Image analysis of TEM micrographs indicates that Devonian shales have values for TEM-derived porosity (optical porosity) that are higher than values derived from Dean Stark analysis. Most values for optical porosity average 7–8%. However, the TEM sample size is small (13 samples) and these results should be treated with caution.

Additional, indirect evidence for "true" shale porosity is available using production data derived from Devonian Shale gas wells. These wells almost always overproduce their estimated gas reserves, which are calculated on the basis of core- or log-derived porosity values. This is sometimes ascribed to adsorption of gas in kerogens. Alternately, or perhaps additionally, overproduction could be the result of the use of overly pessimistic values for porosity in the calculation of reserves.

Porosity values derived from TEM and Dean Stark analysis are considerably lower than would be anticipated from compaction experiments and theory. However, both sets of values for Devonian shales generally agree with values derived from ancient (Tertiary and Paleozoic) basins in Japan and Oklahoma (Athy, 1930; Hosoi, 1963). This suggests that geological age is an important control on the ultimate porosity of shales. Magara (1978, p. 31) stated, "This idea . . . contradicts Terzaghi's basic concept on stress balance among total stress, effective stress and fluid pressure." Magara (1978) further pointed out that Terzaghi's concept does not take into account possible effects on porosity that result from shale diagenesis.

TEM observations of shale pore systems in this study revealed that diagenetic modification is common. Pore space reduction has occurred as a result of (1) the precipitation of new mineral phases (Figs. 10.8A and 10.11A) and (2) possible increase in the size of detrital grains due to clay recrystallization (Fig. 10.10). Such changes require time and will reduce porosity.

What of the effects of diagenesis on basic particle orientation? Results presented here suggest that the Devonian shale retained much of the general characteristics of the original depositional fabric during burial diagenesis. Compaction obviously has reduced grain-to-grain spacing but does not appear to have influenced significantly the relative orientation of the individual particles. Shales deposited with random fabric during the Devonian still retain random fabric despite the influence of burial compaction. The growth of siderite, pyrite, and calcite concretions can be expected to alter original grain-to-grain relationships when such growth is displacive. This, however, is a relatively small scale, localized phenomenon, of minor importance when compared with the overall thickness of the Devonian Shale interval.

Comparison of the results of commercial laboratory analysis with the results of direct observation and theoretical considerations suggests that either (1) a large number of shale pores are not interconnected or (2) problems exist in current state-of-the-art commercial measurement techniques, particularly the measurement of permeability. Lack of connectivity of pores would result in Dean Stark porosity values that underestimate true porosity. This would explain the observed differences in porosity values. However, this explanation is inconsistent when applied to the results of the permeability analyses. Low values for measured porosity should be accompanied by low values for permeability. The fact that they are not indicates that significant problems exist with regard to currently used measurement techniques.

Conclusions

1. Microfabric is the fundamental control on permeability in non-fractured shales. Chaotic microfabric has the highest permeability; laminated microfabric has the lowest permeability.
2. At the range of magnification provided by the TEM, pores are dominantly intergranular in the Devonian shales of the study area.

3. Shale pores contain authigenic minerals, commonly pyrite.
4. Shale permeability may be influenced by the composition of the dominant clay mineral that controls grain shape and grain-to-grain relationships.
5. Experimental data suggest that permeability values derived from pulse permeametry may be too high and porosity values from Dean Stark analysis may be too low. This apparent discrepancy is a strong argument in favor of the necessity for changing or improving significantly on existing procedures for the measurement of porosity and permeability in tight, fine-grained rocks.

Acknowledgments

This project was funded principally by the Gas Research Institute, Contract Number 5086-213-1390, Dr. Richard Schepper, project director. Additional support was received from Texas A&M University, Department of Oceanography, and the Office of Naval Research.

References

Athy, L.F., 1930. Density, porosity and compaction of sedimentary rocks. Bulletin of the American Association of Petroleum Geologists, v. 14, p. 1–24.

Bennett, R.H., W.R. Bryant, and G.H. Keller, 1977. Clay fabric and geotechnical properties of selected submarine sediment cores from the Mississippi Delta. U.S. Dept. of Commerce, NOAA Professional Paper No. 9, 86 p.

Broadhead, R.F., R.C. Kepferle, and P.E. Potter, 1982. Stratigraphic and sedimentologic controls of gas in shale—example from upper Devonian of northern Ohio. Bulletin of the American Association of Petroleum Geologists, v. 66, p. 10–27.

Hosoi, H., 1963. First migration of petroleum in Akita and Yamagata Prefectures. Journal of Japanese Association of Mineralogists, Petrologists, and Economic Geologists, v. 49, p. 43–55, 101–114.

Krynine, P.D., 1948. The megascopic study and field classification of sedimentary rocks. Journal of Geology, v. 56, p. 130–165.

Magara, K., 1978. Compaction and fluid migration: practical petroleum geology. Developments in Petroleum Science. Elsevier, New York, 319 p.

O'Brien, N.R., 1987. The effects of bioturbation on the fabric of shale. Journal of Sedimentary Petrology, v. 57, p. 449–455.

O'Brien, N.R., K. Nakazawa, and S. Tokuhashi, 1980. Use of clay fabric to distinguish turbiditic and hemipelagic siltstones and silts. Sedimentology, v. 27, p. 47–61.

Soedder, D.J., 1988. Porosity and permeability of eastern Devonian gas shale. SPE Formation Evaluation, v. 3, p. 116–138.

Truman, R.B., and R.L. Campbell, 1986. Devonian shale well log interpretation. Quarterly Report of the Gas Research Institute, Contract No. 5085-213-1148 (unpubl.), 29 p.

Walls, J.D., 1982. Measurement of fluid salinity effects on tight gas sands with a computer controlled permeameter. SPE Paper 11092, 57th Annual Technical Conference Proceedings, New Orleans, Louisiana, p. 1–7.

Walls, J.D., A.M. Nur, and T. Bourbie, 1982. Effects of pressure and partial water saturation on gas permeability in tight sands: experimental results. SPE Journal Petroleum Technology, v. 34, p. 930–936.

I

Basic Clay Microstructure

B. Environmental Processes: A Continuum

CHAPTER 11

Environmental Processes: A Continuum—Overview

William R. Bryant

The initiation of microfabric begins in the water column and on the seafloor with the formation of floccules. Although this process has been widely discussed and its results measured in great detail, the direct observation of the flocculation of clay particles has never been viewed, nor has the microstructure of the end product of the diagenetic process, shales and argillites, been observed in detail due to the extremely small size and density of the particles involved. Regardless of these limitations, the response of the fine-grained particles to the physical, chemical, and mechanical environment and biological activity encountered in the continuum associated with particle transport, deposition, and burial is recorded by the microfabric. This record can be observed and studied at different scales to establish the major processes controlling microfabric variations.

In the early stages of formation (particles in suspension and at the depositional interface) the physicochemistry of clay is critical in determining the clay fabric of a marine deposit. The physicochemical forces controlling clay mineral structure and the initial framework of clay fabric are essentially electrical in character. Suspended clay particles interact in response to the electrical nature of the particle surface and to the electrochemical characteristics of the surrounding aqueous medium. The important electrical forces are classified as primary valence bonds, hydrogen bonds, van der Waals forces, and simple electrostatic interactions. The phyllosilicate clay minerals are composed of sheets (tetrahedral and octahedral coordination polyhedra) generally in 1-to-1 or 2-to-1 layers, such as two tetrahedral sheets plus one octahedral sheet (smectite, illite, and chlorite). This imparts a platy or sheet-like characteristic to many clay minerals and for practical purposes they can be considered as being two-dimensional. Most clay particles are colloids, which range in size from about 1 mm to as small as 0.001 mm; however, some clay minerals are larger. In some cases, van der Waals forces and hydrogen bonds hold the two-dimensional clay mineral layers together to form relatively thin particles. During flocculation of suspended clay particles and deposition and low pressure diagenesis, clay crystals are not generally broken, nor are the clay layers separated. The breaking of interlayer bonds appears to indicate the onset of high pressure diagenesis.

Electrostatic forces involve attraction between oppositely charged entities and repulsion between similarly charged ones. Electrostatic interactions are critical in determining clay microstructure, particularly clay fabric, because clay particles carry a net negative charge that is large compared to the particle mass and varies with the clay mineral type. The small positive charge carried by particle edges under certain conditions complicates physicochemical interaction. Recent studies have revealed that the single plate concepts of fabrics are not wholly tenable and that the multiple unit, domain-type fabric is the rule for most sediments. The variation in particle size, shape, and composition coupled with the depositional environment, physicochemistry, transport mechanisms and energies, and changing environmental conditions all increase the complexity of the sediment fabric for a particular sedimentary deposit. Only a few studies of the fabric of natural sedimentary material have been made, however, numerous studies have dealt with the fabric and engineering behavior of laboratory prepared material. Qualitative and ultimately quantitative studies of sediment fabric and microstructure of naturally occurring sediment are leading to reliable predictive capabilities of the physical behavior of clay sediments, a clearer understanding of the geotechnical properties, and a much better understanding of complex sedimentological processes.

In this section, the record is examined to delineate the dominant processes responsible for spatial and temporal variations in the microfabric that occurred during the gradual transformation from individual particles to aggregates (multiplate particles and domains) and floccules and ultimately to clay sediments and to shales. The totality of atmospheric, fluvial, biological, chemical, and geotechnical processes is included. The first part of the section consists of three studies: the relationship of grain size and flocculation by Kranck; the association of microfabric and flocculation, agglomeration, and pelletization by Syvitski; and the microstructure of suspensates in their transition from stream to shelf environments by Pierce.

Fischer discusses the importance of organic flux on the deposition of clays in the marine environment and the implication to microstructure and microfabric development. Dade, Nowell, and Jumars demonstrate the importance of mass arrival mechanisms as they relate to clay deposition at the seafloor. The same theme is examined in Kirby's presentation on distinguishing features of layered muds deposited from high concentration suspensions and Partheniades discusses the dynamics of deposition and resuspension of cohesive sediments in a turbulent flow field and its effects on the microstructure of muds. Ross and Mehta extend the observation of clay deposits by studying the fluidization of soft mud in estuarine and lacustrine environments by waves. The interactions occurring in the vicinity of the seafloor are detailed by Li and Bennett to explain the significance of sediment flow dynamics on the microfabric of riverine and continental shelf sediments.

Reynolds and Gorsline include samples of continental borderland deep sea mudstones in the study of silt microfabric by backscattered electron microscopy. The physical properties and microstructural response to deeply buried sediment to accretion–subduction in the Barbados Forearc are examined by Taylor, Burkett, Wackler, and Leonard. The role of microstructure in the characterization of red clays is discussed by Dadey, Leinen, and Silva. They report on the anomalous stress history of older, very deep water, red clay in the northwest Pacific. The chapter by Dadey et al. is the result of the activity of the Deep Sea Drilling Project and the Ocean Drilling Program.

In summary, this section shows that the record has been examined, the complexities have been assessed, and the value of microfabric studies related to the continuum of clay depositional characteristics has been documented.

CHAPTER 12

Interparticle Grain Size Relationships Resulting from Flocculation

Kate Kranck

Microfabric Diagrams

Past studies of microfabrics resulted in a considerable body of information on the internal arrangement of individual grains in a sediment. A number of workers have published their conclusions with diagrams illustrating the conceptual models they have formulated. These illustrations are a useful synthesis of observations not easily summarized by other methods of presentation due to the qualitative nature of the observations. Casagrande (1932), based partly on an earlier illustration by Terzaghi (1925), showed a honeycomb structure of smaller grains connecting larger silt particles (Fig. 12.1). A noteworthy feature is that similar sized grains everywhere appear to be separated from one another by linkages formed by grains of a different size. Pusch's well known sketches (Fig. 12.2) also show similar sized grains within a floc never touching each other (Pusch, 1962, 1970). When several flocs come together they are joined by the smallest grains, implying that these form the strongest bonds. These drawings also bring out the fact that each relict aggregate unit or floc is composed of a large range of grain sizes. The intergrain arrangement in these diagrams is such that the effect of dividing one or a series of flocs along a random line would be to produce two parts with constituent size distributions similar to that of the whole.

Flocculation of Suspended Matter

As a part of research on recent marine sediments Kranck (1975, 1980, 1986) and Kranck and Milligan (1985) studied the role of constituent grain size in flocculation processes. This work has included studies of suspended and bottom sediment from different natural environments, and laboratory experiments on particle settling carried out both in still water and in turbulent suspensions in agitated containers. Principal analytical tools have been precision size analysis using a Coulter counter supplemented by microscopic studies using an inverted plankton microscope. From these observations, information has emerged on the behavior of different sized grains during transport in fluid environments, and on the grain size characteristics of the sediments resulting from settling suspensions.

The two principal controlling factors in fine-grained sediment deposition are gravitational settling and flocculation, with parent source material constituting a less significant third. The latter's relative lack of importance is related to the fact that the rate of settling varies largely as the square of the diameter and results in a range of almost four orders of magnitude difference in settling rates of fine sediment grains. This is in contrast to the less than one order of magnitude variation in concentration of different grain sizes in the unsorted diamicts derived from terrestrial erosion and weathering that are the source of most sediments. The effect of gravitational settling of unsorted sediment suspensions is to deposit bottom sediment for which the concentration of material in progressively larger logarithmic size classes increases as the square of the class mid-diameters. The frequency distribution or spectra of this distribution (plotted on log–log axis) is a straight line with a slope of approximately two. This slope shapes the fine-grained end of the distribution along with a relatively rapid exponential coarse end fall off resulting from a time- or turbulence-dependent limit to the maximum size of particles in the suspension. These two factors combine to form constant, well-sorted slightly skewed distributions. This distribution spectra

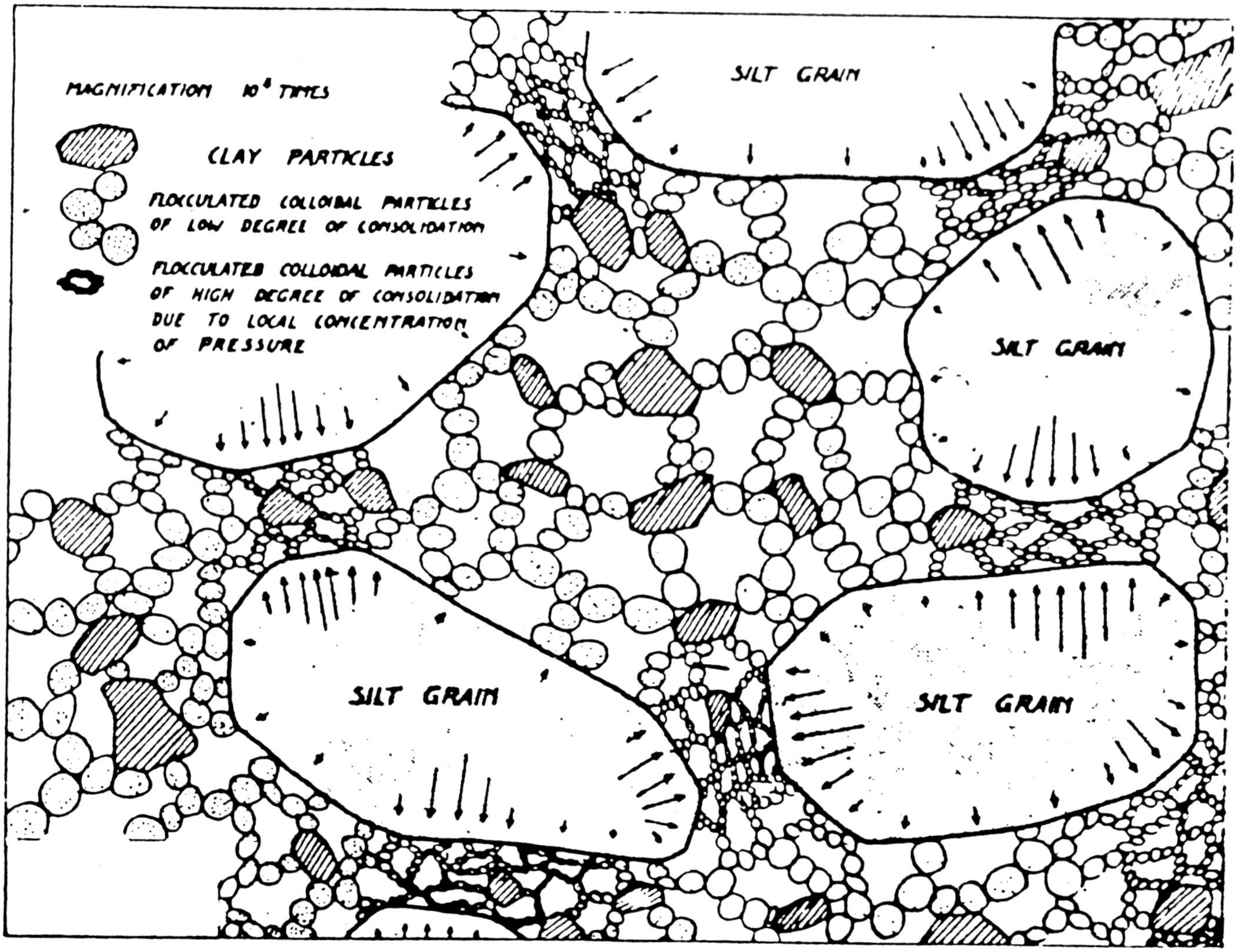

Figure 12.1. The structure of undisturbed clay as suggested by Casagrande. (Modified from Casagrande, 1932.)

would be the universal size composition of fine-grained sediment except for the complicating effect of flocculation. Whereas single grain settling sorts the sediment, flocculation serves to preserve the unsorted character of the parent source suspension. This is because each floc contains similar relative proportions of all grain sizes, smaller than a given size, as found in the source suspension. Thus when the flocs deposit, the fine sizes of the resulting sediment duplicate the source distribution from which it deposited.

The diverse effects of single grain and floc settling are obvious from the shapes of bottom sediment size spectra. In Figure 12.3 equations defining sediment grain size distributions have been fitted to size analysis data (squares) using standard linear regression techniques (Kranck, 1986). The grain and floc settled material is shown by dotted lines and the total of the two as a solid line. Each fitted curve is defined by three constants the values of which are related to source material and depositional conditions. The agreement between the analytical and model data shows the exact response sediments exhibit with respect to the physical environment. Depositional environments where the suspended sediment is in a highly flocculated state such as Rotterdam Harbour, characterized by near bottom "fluid mud" layers (Parker and Kirby, 1982), result in a bottom sediment with size distributions very similar to the poorly sorted suspended sediment distributions. Areas where flocculation is limited, such as Saint John Harbour, where high turbulence inhibits flocculation, produce well-sorted bottom sediment with only a minor tail of poorly sorted fine sediment resulting from settling of flocs. Most marine environments fall somewhere between these two extremes (Kranck and Milligan, 1985).

Mechanics of Flocculation

Experimental evidence and field observations indicate that some portion of all particle sizes up to some maximum size is flocculated in a given suspension. The settling rate of the largest flocs never exceeds the settling rate of the largest grains in the suspension (Kranck et al., 1988). This is the case even when the turbulence is such that much faster settling flocs if present could remain suspended. To explain this observation Kranck (1975) suggested that flocculation occurs only between particles with dissimilar dynamic settling rates. This explanation can apply only when the collision mechanism is due to differential settling. Van der Veen (1981) has shown that flocculation due to shear is inhibited between very different size particles. These opposing theories and the experimental evidence are explained if very similar particles seldom collide with each other, whereas very dissimilar size particles on collision have too large a difference in momentum for their individual movements to be countered by attractive forces. It is likely that coarse silt or sand grains will not flocculate with each other, but remain unflocculated until the floc size has grown to some critical size resulting in bonding between the large relatively inert silicate grain and the fluffy, organic-rich low-density floc. From this scenario it follows that the constituent grains of each floc will have a very heterogeneous size distribution and each grain will have a specific size relationship to its nearest neighbors.

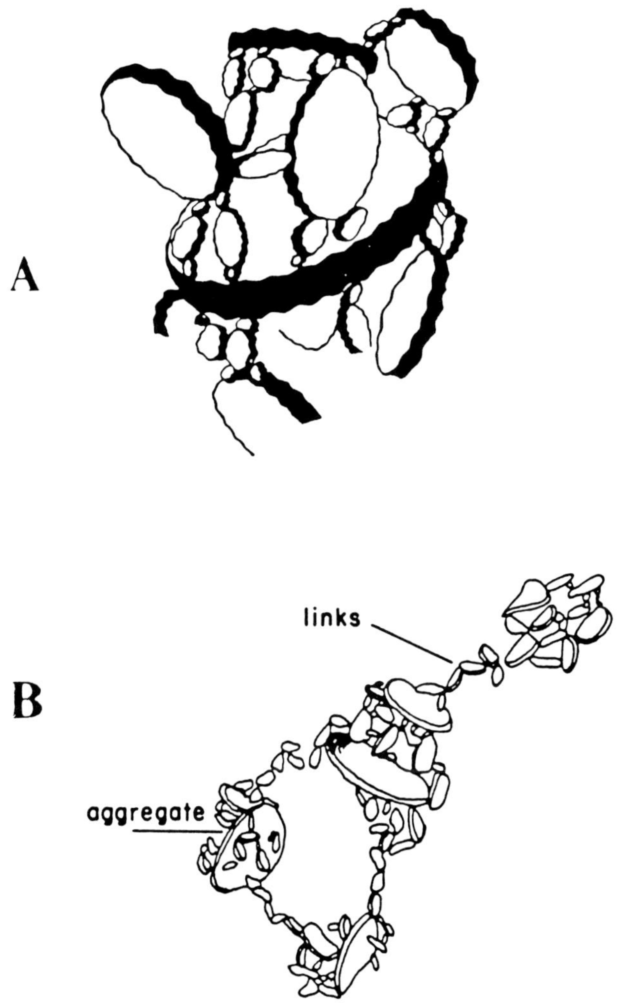

Figure 12.2. Internal grain arrangement of flocs. (A, Redrawn from Pusch, 1962; B, Redrawn from Pusch, 1970.)

Floc Composition

Flocculation encompasses all particulate components in the suspension including organic detritus, anthropogenic contaminants, and living cells. Discussion of the mechanisms of adhesion is outside the scope of this chapter but it is worth noting that, as pointed out by Terzaghi in 1925, flocculation is not limited to physicochemical attraction between clay minerals and other layered silicates. This process does occur but appears to be limited to the clay-sized inorganic grains less than a few microns in diameter, suspended in media with low concentrations of dissolved organic matter (Mushenheim et al., 1990). Flocculation of grains up to the size of sand has been observed in both fresh and seawater. Most flocculation in the natural environment is apparently related to the presence of organic matter and mediated by bacterial activity, producing dissolved organic compounds that adsorb to particle surfaces.

A typical riverine or estuarine suspension consists of large low-density flocs and smaller higher density unflocculated single grains. It is reasonable to suppose that the heaviest (largest) constituent grains of a floc will take the lead in a settling floc followed by the other components resulting in the string or comet-like shapes commonly observed for flocculated aggregates (Silver et al., 1978; Eisma, 1986; Kranck and Milligan, 1980). When this conglomeration reaches the bottom, the coarse constituent grains will land beneath finer material, isolating them from other coarse grains and thereby building up a "floating-grains" texture (Mitchell, 1976).

Is Flocculation Reversible?

Considerable discussion has revolved around the question of whether flocculation is reversible. Bennett and Hulburt (1986) suggest that flocs in a bottom sediment do not deflocculate as a result of changes in the chemical environment. Breakup of flocs at the change from fresh to brackish water has been inferred by Eisma (1986) but not confirmed by other studies. Turbulence appears to place an upper limit to floc size, indicating that at some point the flocs are broken down as fast as they form.

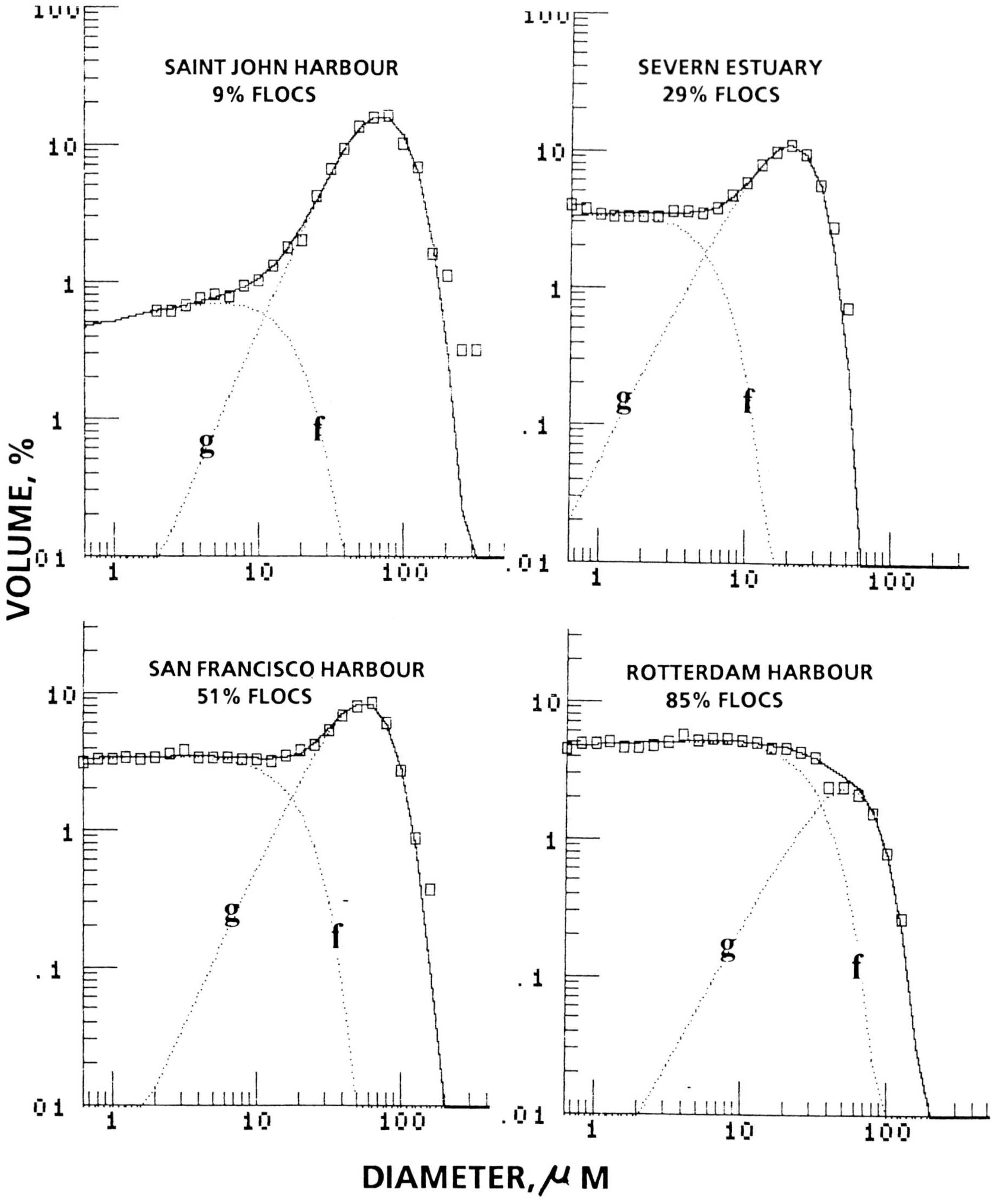

Figure 12.3. Examples of suspended sediment grain size spectra from four different harbors. Squares: Data from Coulter counter analysis showing volume concentration in logarithmic size classes. Dotted line: floc (f) and grain (g) settled portions of distribution. Solid line: total.

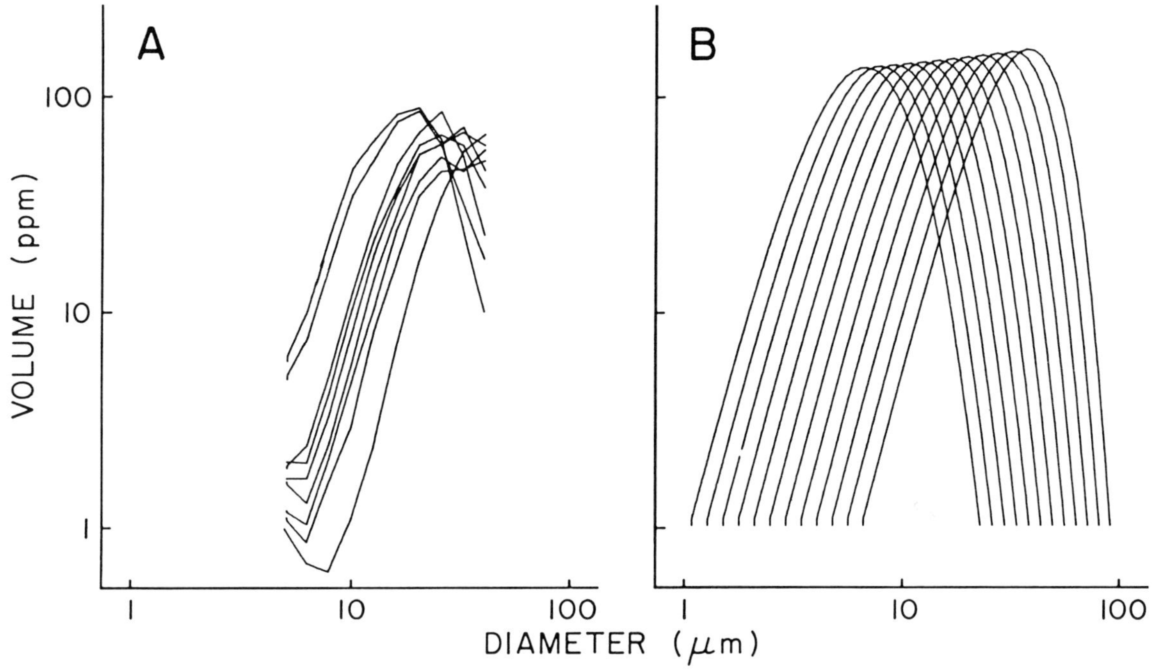

Figure 12.4. Break-up of flocculated suspensions. (A) Coulter counter analysis of a flocculated suspension during continuous stirring on the Coulter counter stand showing a progressive decrease in the modal size due to floc breakup, but preservation of the fine grained slope. (B) Change in particle distribution calculated by assuming that one-half of all particles in each size class break in half each consecutive time interval.

The effect of shear and its resulting floc breakup on floc size distributions was tested in the laboratory by subjecting a flocculated suspension to shear by stirring it on the sample stand of a Coulter counter. When first measured the distribution had a modal size of greater than 60 μm; with continued stirring the size distribution shifted progressively toward smaller sizes, but the finding occurred without any change in the slope of the size spectra (Fig. 12.4A). This pattern of change can be simulated by a model in which, during a given time, one-half of all the particles in each size class breaks into two equal halves due to bulgy deformation and rupture (Thomas, 1964). The resulting shift of particles into smaller size categories causes a gradual shift of the whole distribution toward smaller sizes so that the modal sizes decrease with time while the slope of the fine particle limb of the spectra remains constant (Fig. 12.4B). This suggests that floc break-up preserves the constituent grain size distribution in each floc by forming a series of self-similar particle size distributions with decreasing modal sizes but with the same slope and relative curve shape. Work is still needed on the absolute effect of turbulence on the shear strength of flocs, as this is of significance in the behavior of flocs when part of a sediment bed.

Conclusions

In the above discussion it has been demonstrated that many features of clay-rich microfabrics can be explained by the mechanics of flocculation. These features arise because flocculation is a size-selective process that results in a very specific pattern of grain arrangement analogous to the ordering of atoms in crystals. This interparticle arrangement is retained even when the floc is disrupted by turbulence giving each floc a self-similar, self-replicating distribution relative to the suspension as a whole. It is not surprising that disruption of this network markedly changes the structural behavior of sediments, as illustrated by the difference in the bearing strength between undisturbed and remolded clays and by the difference in shear strength of the highly flocculated estuarine "fluid muds" and less rapidly deposited normal mud bottoms (Faas, 1986).

Acknowledgments

I am grateful to T. G. Milligan for computer and analytical assistance, to R. Kirby and D. Eisma for the Severn Estuary and Rotterdam Harbour samples, respectively, and to P. D. Keizer and D. H. Loring for critical reading of the manuscript.

References

Bennett, R.H., and M.H. Hulbert, 1986. Clay Microstructure, International Human Resources Development Corporation, Boston, 161 p.

Casagrande, A., 1932. The structure of clay and its importance in foundation engineering, contributions to soil mechanics, Boston Society of Civil Engineers, 1925–1940, p. 72–112.

Eisma, D., 1986. Flocculation and de-flocculation of suspended matter in estuaries. Netherland Journal Sea Research, v. 20, p. 183–199.

Faas, R.W., 1986. Mass-physical and geotechnical properties of surficial sediments and dense near bed suspensions in the Amazon continental shelf. Continental Shelf Research, v. 6, p. 189–208.

Kranck, K., 1975. Sediment deposition from flocculated suspensions. Sedimentology, v. 22, p. 111–123.

Kranck, K., 1980. Experiments on the significance of flocculation in the settling of fine grained sediment in still-water. Canadian Journal of Earth Science, v. 17, p. 1517–1526.

Kranck, K., and T. Milligan, 1980. Macroflocs: production of marine snow in the laboratory. Marine Ecology Progress Series, v. 3, p. 19–24.

Kranck, K., and T.G. Milligan, 1985. Origin of grain size spectra of suspension deposited sediment. Geo-Marine Letters, v. 5, p. 61–66.

Kranck, K., T.G. Milligan, and D.K. Muschenheim, 1988. Flocculation processes in sediment transport. Proceedings of International Conference of Fluvial Hydraulics '88. Research Centre for Resources Development, Budapest, Hungary, p. 37–42.

Mitchell, J.K., 1976. Fundamentals of Soil Behaviour. Wiley, New York, 422 p.

Muschenheim, D.K., P.E. Kepkay, and K. Kranck, 1989. Microbial growth in turbulent suspension and its relation to marin aggregate formation. Netherlands Journal of Sea Research, v. 23, p. 283–292.

Parker, W.R., and R. Kirby, 1982. Time dependent properties of cohesive sediments relevant to sedimentation management—European experience. In: Kennedy, V.S. (ed.), Estuarine Comparisons. Academic Press, New York, p. 573–589.

Pusch, R., 1962. Clay particles their size, shape and arrangement in relation to some important physical properties of clays. Statens rad for byggnadsforskning, Handlingar Nr 40, 150 p.

Pusch, R., 1970. Microstructural changes in soft quick clay at failure. Canadian Geotechnical Journal, v. 7, p. 1–7.

Silver, M.W., A.L. Shanks, and J.D. Trent, 1978. Marine snow: microplankton habitat and source of small scale patchiness in pelagic populations. Science, v. 201, p. 371–373.

Terzagi, K., 1925. Erbaumechanik auf Bodephysikalischer Grundlage. Deuticke, Vienna, 339 p.

Thomas, D.G., 1964. Turbulent disruption of flocs in small particle size suspensions. Journal of the American Institute of Chemical Engineering, v. 10, p. 514–523.

van der Veen, T.G.M., 1981. Effects of polymer bridging on selective shear flocculation. Journal of Colloidal and Interface Science, v. 81, p. 290–291.

CHAPTER 13

The Changing Microfabric of Suspended Particulate Matter—The Fluvial to Marine Transition: Flocculation, Agglomeration, and Pelletization

J.P.M. Syvitski

Introduction

The behavior of suspended particles settling in oceans is of great importance to the understanding of the physical properties of pelagic deposits that cover much of earth's seafloor. Suspended particles largely enter the sea by river plumes that ride buoyantly on the denser marine water. To lesser extents they also enter through the action of coastal erosion, aeolian transport, and submarine volcanic eruptions. A large proportion of marine suspensates is also generated *in situ*, through primary productivity and through authigenic biogeochemical reactions.

It is widely recognized that most particles suspended in the ocean settle in units or packages containing many individual particles (McCave, 1975; Lal, 1977). These units tend to occur in one of four types (Syvitski and Murray, 1981): (1) aggregates—inorganic particles strongly bonded by intermolecular or intramolecular, or atomic cohesive forces (e.g., Ernissee and Abbott, 1975); (2) agglomerates—organic and inorganic matter weakly held together by surface tension, organic cohesion, and adhesion (e.g., Johnson, 1974); (3) planktonic fecal pellets where individual particles are mechanically compacted and packaged, sometimes within a peritrophic membrane (e.g., Syvitski and Lewis, 1980); and (4) floccules—inorganic particles held together by van der Waals force, after the electrostatic repelling forces associated with natural mineral grains are neutralized within a saline solution > 3 ppt (e.g., Sakamoto, 1972). Together these four types of marine suspensates have been referred to as "marine snow." Not all marine particles settle as marine snow (see below), and many particles introduced to the ocean settle in more than one form of marine snow, say initially as a floccule and later as an agglomerate or fecal pellet.

This chapter reviews the particle–particle interactions that are likely to occur during the transit of sediment from the time they enter the ocean at a river mouth to the time they reach the seafloor. It is these complex interactions that affect the sedimentation of particles under river plumes and explain the documented exponential decrease in sedimentation rates seaward of a river mouth (Syvitski et al., 1985, 1988).

Methods

A number of methods have been used to ascertain size of the unit to which an individual particle is attached or of which it is a part. The most common method was to collect water samples at depth using water bottle samplers (e.g., Niskin, van Dorn, GoFlow) either triggered individually using messenger weights, or as part of rosette packages whereby discharged capacitors are used to electrically trigger the sampling bottles. The particles within a water sample may be filtered (e.g., using Nucleopore, Millipore, Selas Flotronics filters) and examined microscopically, or analyzed directly for particle size using an electrical resistance pulse counter (e.g., Coulter Counter, Celloscope). Both methods erroneously assumed that water samplers collected marine particulates in an undisturbed manner and were returned to the sea surface without further changes during their transit (see discussion by Syvitski et al., this volume). In fact, the normal sampling approach breaks up the larger marine snow particulates, especially those of great length (> a few millimeters). Even *in situ* pumps that are designed to filter water at depth may break up these larger particles (Syvitski et al., this volume).

It also was erroneously assumed that instruments like the Coulter Counter did not affect the size of "marine snow" during the analysis. Researchers were initially impressed that marine particles analyzed directly after collection were much larger than if those same particles were disaggregated using dispersants

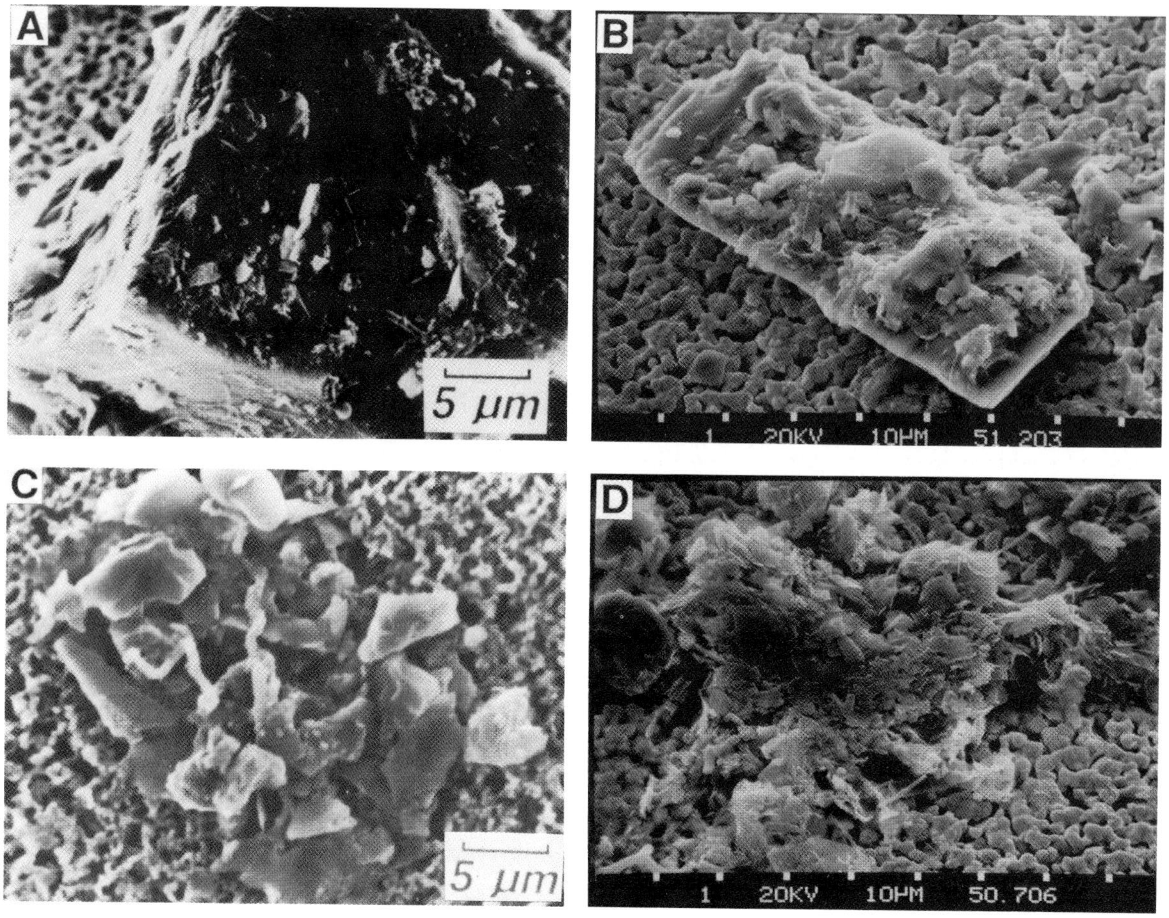

Figure 13.1. Scanning electron micrographs of filtered suspended particles near the discharge outlet of a river. (A) Feldspar grain coated with microflora collected within the Squamish River, British Columbia. Scale = 5 μm. (B) Mud chip resuspended from the seafloor of McBeth Fiord, Baffin Island: sample collected 2 m above the sea floor and at a water depth of 520 m. Scale = 10 μm. (C) Two-dimensional (i.e., very thin) floccule collected in the halocline of the Squamish River plume: at a water depth of 10 m, salinity of 20 ppt, and 12 km from the river mouth, flowing into Howe Sound, British Columbia. Scale = 5 μm. (D) Three-dimensional (i.e., thick) floccule collected 30 m below the surface of the River plume and 20 km from the McBeth river mouth, flowing into McBeth Fiord, Baffin Island. Scale = 10 μm.

and sonification (e.g., Kranck 1975; McCave, 1984). Unfortunately, the stirring mechanism that keeps the particles in suspension and, more importantly, the high flow velocity through a tiny orifice in the detection tube can generate sufficient shear to partly disaggregate the particles and decrease their unit size (Syvitski et al., this volume).

The filtration method also has problems. If the suspensate concentration is too high or the water volume too large, then particles will be filtered on top of one another and the original unit size of "marine snow" impossible to determine. Zabawa (1978) and Syvitski and Murray (1981) both questioned the normal filtration methodology and instituted the practice of immediate and gentle filtration of microsamples so that the probability of one particle landing on one another was less than 1%. Both research teams were then able to classify the individual units of suspended matter, although the above-mentioned sampling problems would still limit the upper size of the "marine snow" detected microscopically. This limitation is substantiated through *in situ* observations by divers and manned submersibles (e.g., Shanks and Trent, 1980; Farrow et al., 1983) and *in situ* stereo photography (Syvitski et al., this volume).

The filtered suspensates may be examined with a combined scanning electron microscope (SEM) and energy-dispersive X-ray analyzer that allows for the determination of composition and particle–particle configuration of the marine particles (e.g., Syvitski et al., 1985). If the filtration is not done gently, there is a danger that some of the particles will be flattened out. This is particularly true for very low density organic-rich particles. This chapter incorporates data generated from the low-concentration and gentle filtration method, followed with energy-dispersive SEM analysis. The samples come principally from prodelta environments within deep coastal basins: from the Pacific

(Syvitski and Murray, 1981; Farrow et al., 1983; Syvitski et al., 1985), the Atlantic (Syvitski et al., 1983), and the Arctic (Syvitski, 1988; LeBlanc et al., 1988).

Particle Interactions in Deep Coastal Basins

Sediment being carried via streams and rivers typically occur as individual particles. This suggests that turbulent shear within the fluvial environment is sufficient to break up most soil particles into their individual mineral components. Rock flour carried via glacial streams is also mostly disaggregated. However, fluvial particles may contain attached hydrous oxides, organic coatings, and freshwater microflora (e.g. Fig. 13.1A). Fluvial erosion of raised marine terraces can also shed balls or clumps of mud during high discharge events (e.g., Schafer et al., 1983).

At the river mouth, there is a radical change in hydraulic conditions between the river and the fluvial plume spreading into the ocean. Normally the river plume will enter the sea as a hypopycnal flow (fresh or brackish water flowing into an ambient basin of saline water). The type of hypopycnal flow, i.e., as dominated by buoyancy, inertia, or friction, will strongly influence the behavior and character of the suspended particulate matter (SPM). In the case of the inertial and friction-dominated flows, mixing between the freshwater and the seawater can be very strong. The plume salinity will therefore be high close to the river mouth and allow the early flocculation of mineral grains (Zabawa, 1978). Frictionally influenced plumes also can be associated with resuspension and thus the input of mud chips off the seafloor. In deep coastal fjords, gravity flows associated with slide-generated turbidity currents or water mass exchanges (deep-water renewal) can similarly resuspend mud chips from the sea floor (Fig. 13.1B).

Most large rivers, however, discharge into the sea via buoyant hypopycnal flows (Syvitski et al., 1985). As the surface layer mixes with the marine water, the salinity of the turbid river plume increases. Inorganic flocculation of silts and clays begins to occur at salinities between 3 and 5 ppt. Phyllosilicate minerals of equal size (below 20 μm) flocculate initially with edge-to-edge orientation (Fig. 13.1C). Nonplaty minerals in the coarse silt to sand range may settle as single entities, especially near the river mouth, although some of these grains can also become attached to the phyllosilicate flocs. Submersible observations through the low salinity waters of a river plume indicate the SPM as a finely divided haze (individual particles are very difficult to identify), rapidly being replaced with more identifiable (> 100 μm) particles through and below the halocline. Salinity within the surface waters will increase with distance out from the river mouth, and eventually almost all suspended particles become flocculated.

In the flocculated state, particles can settle faster than if they were to settle individually and therefore as predicted by Stokes law. For instance, particles in the fine silt and clay size range settle at velocities usually associated with much larger size particles, i.e., 30–100 m/day (Syvitski et al., 1985). The velocity enhancement due to flocculation can exceed four orders of magnitude for clay-size particles (Syvitski et al., 1985). In deep basins, floc size will continue to increase, developing face-to-face and face-to-edge particle orientations until stable three-dimensional flocs are formed (Fig. 13.1D). Obviously at these particle settling rates, the water depth is a limiting factor on a floc size. Eventually growth through particle attachment stops as further attachment would create hydrodynamic instability (Kranck, 1975).

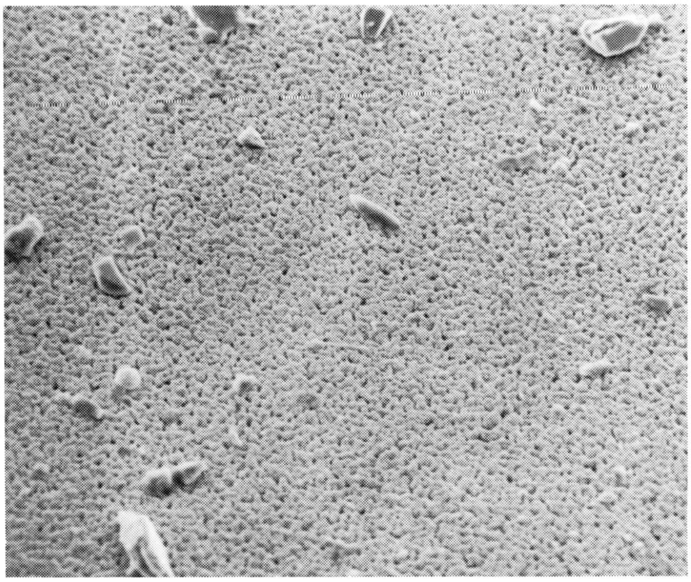

Figure 13.2. Example of suspended particulate matter dominated by wind-borne particles as observed under a scanning electron microscope. Sample collected from the surface waters (20 m) of McBeth Fiord, 30 km from the river mouth during the fall (negligible river discharge) season of 1982 and during a dust storm. Scale = 100 μm.

It is useful here to contrast the initial input of fluvial particles into the marine environment with aeolian input. The primary difference is one of initial concentration, which also happens to be an important requisite for flocculation (Krone, 1972). In the coastal seas of the Canadian Arctic Archipelago, a significant proportion of the suspended load is derived from aeolian input. There, the fluvial season extends only a few weeks to a couple of months (Syvitski et al., 1987) while the "aeolian season" is almost year-round (McKenna-Neuman and Gilbert, 1986). In areas not directly influenced by fluvial discharge, wind-borne particles enter surface waters at a nonorganic particle concentration of ⪡1 mg/liter, and many of the suspended particles settle individually (Fig. 13.2), although the salinity is fully marine.

Except during the initial conditions of flocculation, the influence of organic matter dominates the particle dynamics of mineral matter in the ocean. Much of the terrestrially derived organic particulate matter mixed with the mineral grains occurs as semiinert forms, such as pieces of bark, twigs, leaves, needles,

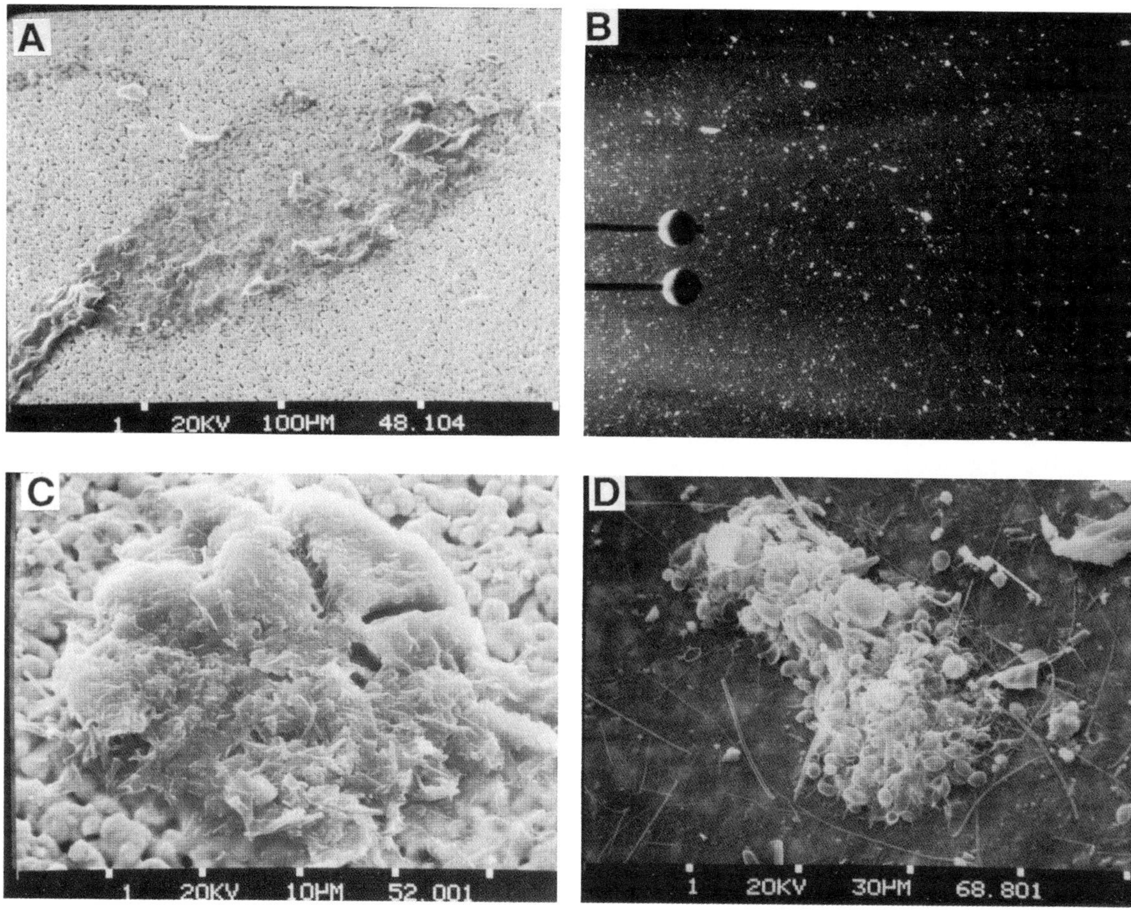

Figure 13.3. Scanning electron micrographs of filtered suspended particles of various sizes and types of agglomerates. (A) Large mucoid agglomerate collected at 30 m water depth and 80 km from the McBeth river mouth, flowing into McBeth Fiord, Baffin Island. Scale = 100 μm. (B) *In situ* Floc Camera photograph of agglomerates within the Bedford Basin, Halifax harbour showing large particles of marine snow. The size of each calibration ball is 15 mm. (C) Agglomerate dominated by mineral matter collected at 5 m water depth and 30 km from the McBeth river mouth, flowing into McBeth Fiord, Baffin Island. Scale = 10 μm. (D) Agglomerate dominated by biogenic detritus, collected at 100 m water depth and 130 km from the Clark river mouth, flowing into Scott Inlet, Baffin Island. Scale = 30 μm.

and a variety of grasses. Through the influence of man, more reactive organic matter from agriculture, sewage discharge, and other industrial by-products can overwhelm those from natural processes. In the more pristine arctic, lichens and grasses are very resistant to biodegradation. In addition, the geological nature of a drainage basin, or seasonality in hydrologic and biologic events, can affect the inorganic–organic agglomeration of marine particulate matter. A river draining a primarily lowland area would, for example, be expected to contain much higher levels of humic materials than a glacier-fed river (Lewis and Syvitski, 1983). Therefore particulates discharged from lowland rivers would be associated with humic material and would be subject to organoclay complexing (Rashid et al., 1972), while those in the glacier-fed river would be associated with the glacial flour and subject simple inorganic flocculation (Buckley, 1972). In temperate regions, the peak runoff of a lowland river affected by rainstorm-related discharge would normally occur during the winter, while peak runoff of a glacier or snow fed river would occur during the spring–summer period. Since the maximum biological productivity occurs during the latter period, organisms would be most affected by the suspended load discharged from glacier- and snow-fed rivers (Lewis and Syvitski, 1983).

During the initial transit of suspended particles within a river plume, mixing processes and entrainment cause the quality of the supporting medium to change from freshwater containing carbon-rich organic matter to seawater containing nitrogen-rich organic matter (Pocklington and Leonard, 1979). Suspended organic matter having an atomic C:N ratio <8.5 is considered to be of a marine origin. Below the halocline, organic cohesive/adhesive forces allow the attachment of organic detritus and biogenic debris (mostly phytoplankton material in various stages of mechanical or chemical destruction) onto the flocs (Syvitski and Murray, 1981). Thus floccules composed wholly of mineral grains are transformed into agglomerates of both organic and inorganic matter. The size that an agglomerate can attain depends on a number of factors. Factors that contribute to large agglomerates include (1) lengthy residence time—the longer the particle remains in suspension the greater the chance for bacterial attachment and interaction with dissolved materials (Lewis and Syvitski, 1983), (2) low populations of zooplankton that would otherwise graze on and thus repackage the agglomerates as fecal pellets (see below), (3) waters rich in dissolved and particulate organic matter, and (4) high primary production and thus production of biogenic detritus. Factors that work to limit the size of an agglomerate are principally (1) turbulence and/or shear within the water column, and (2) grazing by zooplankton. Agglomerates as large as 400 μm (Fig. 13.3A) have been sampled with the filtration/SEM method (LeBlanc et al., 1988). However, *in situ* observations suggest agglomerates may reach 10 to 15 mm (Fig. 13.3B). The composition of agglomerates may range from those dominated by mineral matter (Fig. 13.3C), biogenic detritus (Fig. 13.3D), or mucoid matter (Fig. 13.3A).

Both floccules and agglomerates offer sites for bacterial attachment. Additionally, sorption processes associated with phyllosilicates (e.g., Burba and McAtee, 1981) can transform dissolved nutrients, organic material, and trace metals into a particulate form that can be used as food by suspension feeding planktonic organisms (Lewis and Syvitski, 1983). Zooplankton therefore graze on these flocs and agglomerates, subsequently producing mineral-bearing fecal pellets. Using a manned submersible, zooplankton were observed to be engaged in selective and purposeful feeding on large particles of marine snow—copepods (1–3 mm in length) were found darting from one agglomerate to another, like "bees in a flower field" (Syvitski et al., 1983). Unlike the less dense agglomerates and floccules, fecal pellets are quite compact and hydrodynamically shaped so that they settle rapidly. The settling rate of a fecal pellet depends on its volume, shape, composition, and compaction (e.g., Honjo and Roman, 1978). The composition is important, for if the pellet contains dense minerals they will settle faster (>250 m/day: Syvitski and Lewis, 1980) than if the pellet contained only low-density organic matter.

The rate of fecal pellet production depends on both the distribution of zooplankton in the water column and the zooplankton egestion rate. The spatial distribution and density of zooplankton are highly seasonal, reaching maximum populations in the late spring and early fall. Directly under the halocline (i.e., of a river plume), zooplankton populations are very large, and although they can be found at all depths, they reach a penultimate population maximum near the seafloor. The egestion rate of zooplankton may vary by one order of magnitude depending on the mineral matter "food" made available (very large for montmorillonite, very low for quartz: Syvitski and Lewis, 1980). One copepod is capable of producing up to 19 pellets per day. Proximal to the river mouth, but in deep water, the pellets contain mostly mineral grains, a reflection of the composition of the ambient suspended matter (Fig. 13.4A). In more distal locations, the fecal pellets contain little mineral matter, dominated more by biogenic (phytoplankton) detritus (Fig. 13.4B).

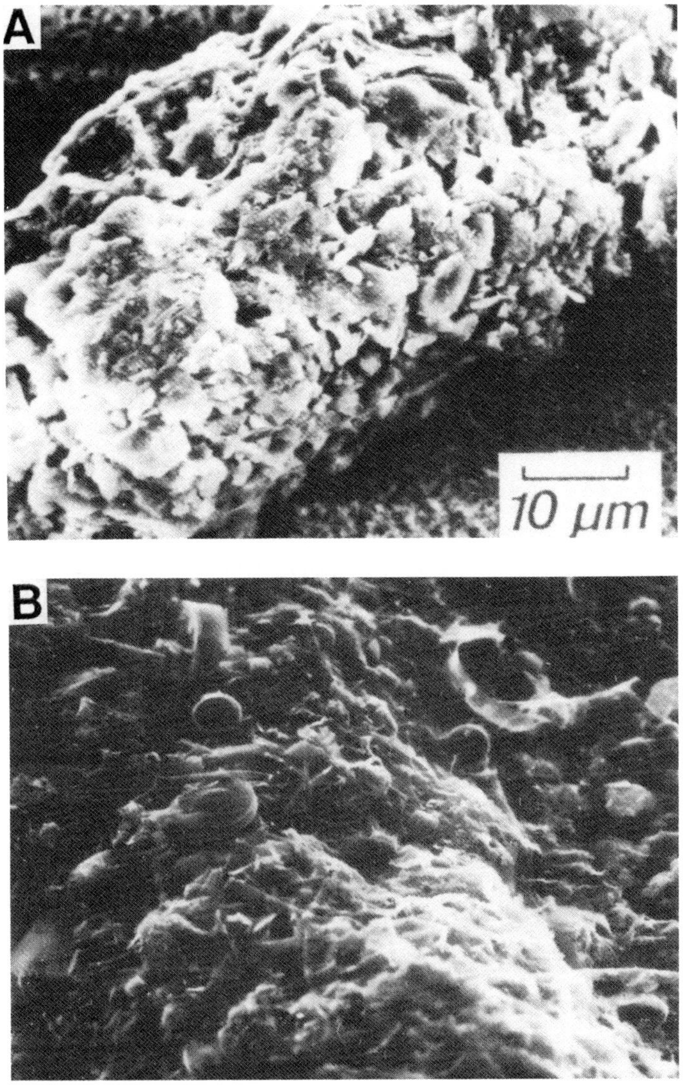

Figure 13.4. Scanning electron micrographs of fecal pellets. (A) Mineral-bearing planktonic fecal pellet collected at 50 m water depth and 15 km from the Squamish river mouth, flowing into Howe Sound, British Columbia. (B) Fecal pellet dominated by biogenic detritus collected at 50 m water depth and 75 km from the Homathko river mouth, flowing into Bute Inlet, British Columbia. Scale = 10 μm.

With an appropriately long residence time, the flocs and agglomerates may eventually become completely encased in organic coatings (Fig. 13.5A). In part this reflects the increasing residence time of these particles. Diatomaceous chains, or mats of intertwined thecate stages of dinoflagellates, are also capable of incorporating mineral matter during their descent. Similarly, the discarded gelatinous housings of *Oikopleura* are capable of

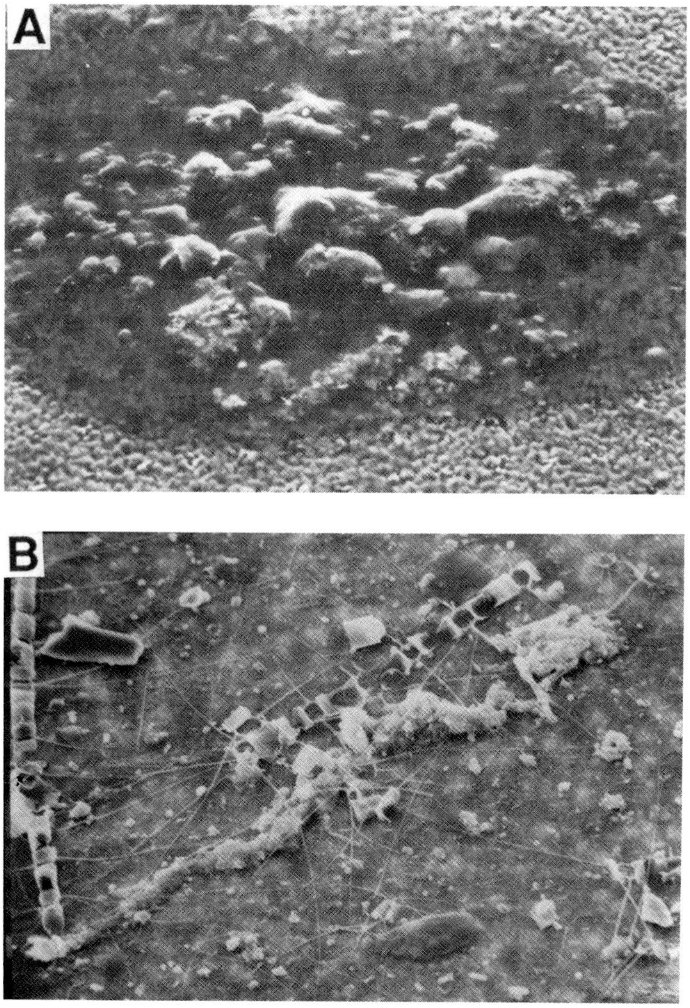

Figure 13.5. Scanning electron micrographs of suspended particles. (A) Organic coating surrounding marine snow particle collected at 20 m water depth and 75 km from the Homathko river mouth, flowing into Bute Inlet, British Columbia. Scale = 100 μm. (B) Stringer collected at 1 m water depth and 40 km from the Clyde river mouth, flowing into Clyde Inlet, Baffin Island. Scale = 30 μm.

picking up particulate matter during their descent (e.g., Syvitski et al., 1983). Mucoid filaments appear to form from the outward growth of bacteria on decaying planktonic fecal pellets and/or decaying planktonic matter. These filaments (also known as streamers or stringers) can vary in size from a few centimeters to over 1 m (Syvitski et al., 1983, 1985). They are long, thin, and delicate, and are found suspended vertically within stable water masses (i.e., those with negligible shear). Filament size and concentration increase with depth and distance seaward, and eventually they become coated with suspended debris (Fig. 13.5B). Many have branched upper and lower ends that may be the result of differential settling and collision of fast settling agglomerates, i.e., a faster settling floc may collide into the middle portion of the filament and pull the two ends together.

Summary

Mineral matter in the marine environment may occur as single particles (especially near river mouths and in more distal regions after delivery by wind), as floccules (in the halocline regions surrounding river plumes), as agglomerates (the most ubiquitous form) and seasonally as planktonic fecal pellets. Although single particles tend to settle very slowly, marine snow tends to settle at tens to hundreds of meters per day. The interrelationships between these different forms of marine snow are complex and poorly understood but play an important role in sedimentary environments dominated by hemipelagic sedimentation.

The final character of the seafloor, in terms of particle–particle bonding, is dependent on this complex scenario of marine sedimentation and on further biogeochemical reactions that await these settling particles after they make contact with the seafloor. Particles of marine snow may be subjected to further modifications after ingestion by benthic organisms, biodegradation by bacteria, and chemical reactions with trace metals diffusing or advecting through the sediment–water interface.

Acknowledgments

This research is supported by the Geological Survey of Canada's Sedimentary and Marine Geology Branch under project 420-8141 "The behaviour of suspended particulate matter (SPM) in natural aqueous environments." I would thank Dr. R. H. Bennett (NORDA) and Dr. Joe Kravitz (ONR) for support during the workshop of Clay Microstructure. I thank C. L. Amos and G. V. Winters for their critical comments.

References

Buckley, D.E., 1972. Geochemical interaction of suspended silicates with river and estuarine water. The 24th International Geological Congress, Sec. 10, p. 282–290.

Burba, J.L., and J.L. McAtee, 1981. Adsorption of ethylene glycol on amine-substituted montmorillonites. Clays and Clay Minerals, v. 29, p. 60–66.

Ernissee, J.J., and W.H. Abbott, 1975. Binding of mineral grains by a species of *Thalassiosira*. In: Simonson, R. (ed.), Third Symposium on Recent and Fossil Marine Diatoms, Kiel, Sept. 9–13, 1974, Royal Octavo, v. 8, p. 241–248.

Farrow, G.E., J.P.M. Syvitski, and V. Tunnecliffe, 1983. Suspended particulate loading on the macrobenthos in a highly turbid fjord; Knight Inlet, British Columbia. Canadian Journal of Fisheries and Aquatic Sciences, v. 40, p. 100–116.

Honjo, S., and M.R. Roman, 1978. Marine copepod fecal pellets: production, preservation and sedimentation. Journal of Marine Research 36, p. 45–57.

Johnson, R.G., 1974. Particulate matter at the sediment-water interface in coastal environments. Journal of Marine Research, v. 32, p. 313–330.

Kranck, K., 1975. Sediment deposition from flocculated suspension. Sedimentology, v. 22, p. 111–123.

Krone, R.B., 1972. A field study of flocculation as a factor in estuarial shoaling processes. Committee on Tidal Hydraulics, Corps of Engineers, U.S. Army, Technical Bulletin No. 19, 62 p.

Lal, D., 1977. The oceanic microcosm of particles. Science, v. 198, p. 997–1009.

LeBlanc, K.W.G., J.P.M. Syvitski, and L. Maillet, 1988. Examination of the suspended particulate matter within arctic fjords. Geological Survey of Canada Open File Report 1773, 301 p.

Lewis, A.G., and J.P.M. Syvitski, 1983. Interaction of plankton and suspended sediment in fjords. Sedimentary Geology, v. 36, p. 81–92.

McCave, I.N., 1975. Vertical flux of particles in the ocean. Deep-Sea Research, v. 22, p. 491–502.

McCave, I.N., 1984. Size spectra and aggregation of suspended particles in the deep ocean. Deep-Sea Research, v. 31, p. 329–352.

McKenna-Neuman, C., and R. Gilbert, 1986. Aeolian processes and landforms in glaciofluvial environments of southeastern Baffin Island, N.W.T., Canada. In: Nickling, W.G. (ed.), Aeolian Geomorphology; Proceedings of the 17th Annual Binghampton Geomorphology Symposium. Geo Books, Norwich, Chapter 12, p. 213–235.

Pocklington, R., and J.D. Leonard, 1979. Terrigenous organic matter in sediments of the St. Lawrence Estuary and the Saguenay Fjord. Journal of the Fisheries Research Board of Canada, v. 36, p. 1250–1255.

Rashid, M.A., D.E. Buckley, and K.R. Robertson, 1972. Interactions of a marine humic acid with clay minerals and a natural sediment. Geoderma, v. 8, p. 11–27.

Sakamoto, W., 1972. Study on the process of river suspension from flocculation to accumulation in estuary. Tokyo Daigaku Jaiyo Kenkyojo Bulletin, v. 5, p. 1–49.

Schafer, C.T., J.N. Smith, and G. Seibert, 1983. Significance of natural and anthropogenic sediment inputs to the Saguenay Fjord, Quebec. Sedimentary Geology, v. 36, p. 177–195.

Shanks, A.L., and J.D. Trent, 1980. Marine snow: sinking rates and potential role in vertical flux. Deep-Sea Research, v. 27, p. 137–144.

Syvitski, J.P.M., 1988. On the deposition of sediment within glacier-influenced fjords: oceanographic controls. Marine Geology, v. 85:301–329.

Syvitski, J.P.M., and A.G. Lewis, 1980. Sediment ingestion by *Tigriopus californicus* and other zooplankton: mineral transformation and sedimentological considerations. Journal of Sedimentary Petrology, v. 50, p. 869–880.

Syvitski, J.P.M., and J.W. Murray, 1981. Particle interaction in fjord suspended sediment. Marine Geology, v. 39, p. 215–242.

Syvitski, J.P.M., N. Silverberg, G. Ouellet, and K.W. Asprey, 1983. First observations of benthos and seston from a submersible in the lower St. Lawrence estuary. Géographie physique et Quaternaire, v. 37, p. 227–240.

Syvitski, J.P.M., K.W. Asprey, D.A. Clattenburg, and G.D. Hodge, 1985. The prodelta environment of a fjord: suspended particle dynamics. Sedimentology, v. 32, p. 83–107.

Syvitski, J.P.M., D.C. Burrell, and J.M. Skei, 1987. Fjords: Processes and Products. Springer-Verlag, New York, 379 p.

Syvitski, J.P.M., J.N. Smith, E.A. Calabrese, and B.P. Boudreau, 1988. Basin sedimentation and the growth of prograding deltas. Journal of Geophysical Research, v. 93, p. 6895–6908.

Zabawa, C.F. 1978. Microstructure of agglomerated suspended sediments in northern Chesapeak Bay estuary. Science, v. 202, p. 49–51.

CHAPTER 14

Microstructure of Suspensates: From Stream to Shelf

J.W. Pierce

Introduction

Models, based on theoretical and speculative concepts, have been proposed for processes leading to the microstructure observed in clays and shales (reviewed by Bennett and Hulbert, 1986). These processes, which occur in the water column, include flocculation, aggregation, disperse settling, and particle-to-particle interactions.

A major portion of the total volume of suspensates is in multicomponent particles even though single grains often outnumber the multicomponent particles in the water column. These particles consist of a spectrum of different silt- and clay-sized mineral grains as well as soft organic matter, plankton debris, and whole plankton tests. These multicomponent particles can attain diameters of centimeters and have been called marine snow for the very large particles found in the marine environment (Silver and Alldredge, 1981; Honjo et al., 1984). Estuarine equivalents, termed macroaggregates by Eisma et al. (1983), are known to exist. Obtaining samples of these macroaggregates is nearly impossible because of the ease with which they break up into individual grains and smaller aggregates (Hamner et al., 1975) and hence most likely would not survive contact with the bottom during deposition. Smaller aggregates and individual grains are the building blocks of the macroaggregates (Eisma et al., 1983; Eisma, 1986).

I have used the term aggregate for most of the multicomponent particles rather than agglomerate or floccule (floc). Aggregate is more widely used in sedimentologic and oceanographic literature than the term agglomerate, which was proposed by Schubel (1982) for multicomponent particles that can be broken up by normal laboratory procedures. Schubel proposed that aggregates be restricted to those complex particles that resist breakage by normal laboratory procedures, although such procedures were left undefined. An argument can be made for use of term floc for a multicomponent particle. Flocculation is a process whereby aggregation of fine-suspended particles, primarily colloidal, occurs in a liquid or semiliquid matrix; this process is generally reversible by agitation and a chemical often is necessary to induce flocculation. Eisma (1986) states that true salt flocculation plays a very minor role in formation of multicomponent particles in estuaries. Thus, I have restricted "floc" to those rather loose, friable-appearing suspensates found in the zone of the first appearance of any salinity in an attempt to differentiate these particles from aggregates. The term aggregate is used for those particles that consist of silt- and clay-sized mineral grains, plankton, and biogenic debris in a matrix of mucus or in which the components appear to be bound together by such material and which are quite cohesive.

Many of these aggregates, which are the components or building blocks of the macroaggregates, survive sampling and filtration and, most likely, would survive deposition and some resuspension. The microstructure of these aggregates can be preserved in sediments and sedimentary rocks. Such microstructures can be altered or destroyed, after deposition, by bioturbation and reorientation of individual grains during compaction.

This chapter examines the microstructure, as revealed by scanning electron micrography (SEM), of aggregates from a variety of environments: fluvial, estuarine, and marine. All samples are from temperate regions and may not be representative of other climatic zones. Fluvial samples were obtained from the streams draining into the Rhode River, Maryland, and from the Rappahannock River, Virginia, both tributaries to Chesapeake Bay. The latter was sampled during the waning stages of a flood. Samples from brackish waters were obtained from the Rhode River, the Rappahannock River, and the mainstem of Chesapeake Bay. Continental shelf waters were sampled off the southeastern coast of the United States and off southern Argentina.

Some of the micrographs used here for illustrative purposes may give the impression that all of the filter is heavily coated with particulate material but uncoated filter surfaces are present outside of the area covered by the micrographs. I have not used any micrographs where particle overlap is evident, i.e., where the entire filter is covered by material or where there is what appears to be overlap created during filtration. This overlap may not be evident in the case of a few grains.

Filters were scanned either along traverses or randomly. Micrographs were taken of single grains or aggregates as they appeared on the screen. This prevented bias of the operator for or against any single type of particle. Magnifications ranged from 500× to 20k×, the higher magnification used to examine microstructure of specific aggregates rather than for scanning the filters. Most of the scanning was done at either 1k× or 2k×.

Microstructure of Suspensates

Freshwater

Data for fluvial environments are limited geographically and in the number of samples. Only one set of samples was acquired in the Rappahannock River. Several limited sets have been obtained during average or low flow conditions from the small streams draining into the Rhode River. It is questionable whether these data sets are representative of all conditions, e.g., rising flood and larger fluvial systems.

Single grains outnumber multicomponent particles over most of the reaches of the streams, although a large portion of the suspensate volume is in multicomponent particles. These particles are simple associations of a few grains (< 10 or so) (Fig. 14.1A). The multicomponent particles make up 74% of the total volume in the Rhode River and 83% in the Rappahannock (Pierce and Nichols, 1986).

Many of the grains in the fluvial environment have well-defined cleavage faces and sharp broken edges of clay flakes (Fig. 14.1B). The individual grains in the multicomponent particles are primarily in edge–edge or face–face contact. This arrangement would tend to substantiate Lambe's (1958) concept that clays do not flocculate in freshwater and that the resulting sediment would have a parallel orientation of clay flakes.

During periods of low stream discharge and low concentrations of suspensates, aggregates are present with the individual grains randomly distributed in a structureless matrix. Because of the low discharge and low concentrations, very little material is in transport at this time. Any sedimentary unit resulting from deposition during these conditions would be of limited thickness and extent and would appear as thin layers in fresh water deposits.

In the lowermost reaches of the streams, very large flocs, associations of many silt- and clay-sized mineral grains, are present (Fig. 14.1C). They constitute over 99% of the particle volume in this segment of the Rappahannock River (Pierce and Nichols, 1986). Similar suspended flocs are present in the Rhode River. The flocs are large, up to 150 μm, they appear to be extremely friable, and have little or no matrix binding the particles together. Other flocs appear to have a skeletal structure with large void spaces between grains (Fig. 14.1D). The grains within the flocs are mostly in face–face or edge–edge contact.

No true edge–face contacts between grains were observed in the micrographs. Particles with such contacts might be present in the lower reaches of fluvial systems but the weak binding forces involved in edge–face contacts could have been disrupted during sampling and filtration. Thus it is highly unlikely that such particles could survive bottom contact if they cannot survive rather gentle sampling and filtration procedures. Microstructure, involving edge–face contacts inherited from the suspensates, should be rare in bottom sediments.

The friable appearance of these flocs on the micrographs of freshwater samples are such that they do not appear to be artifacts of filtration. Also, the lack of a matrix or significant binding agent would seem to preclude resuspension or transport for any great distance during the turbulent and relatively high flow conditions present in the Rappahannock River during sampling. These flocs may be more common than my data suggest because many undoubtedly ruptured during sampling and filtration.

Salinity at the stations where the flocs were sampled was below detection limits of the equipment used during the Rappahannock sampling. Gibbs et al. (1983) reported the presence of similar particles in the Delaware River–upper Delaware Bay where salinities were around $0.3^{0}/_{00}$. If true salt-induced flocculation occurs as a result of the suppression of the double layer, it must be at salinities of $1\ ^{0}/_{00}$ or less, lower than that at which most laboratory experiments on flocculation are conducted.

Brackish water

A major change in the dominant particle type occurs in passing from fresh to brackish water. Very few grains appear to have well-defined cleavage and fracture surfaces and it is assumed that these surfaces have a coating of some material, one possibility being organic matter; those grains that do have well-defined surfaces are of anomalous size (larger than medium silt) compared to the general population of individual grains. These anomalous grains may have been introduced from the nearby shore, be wind blown, or resuspended. The number of aggregates increases over the number in freshwater (55% of the total number of particles in the Rhode River, 50% in the Rappahannock). The volume of the suspensate population is dominated by aggregates (90% of total projected area of all particles in the Rhode River, 94% in the Rappahannock River). Flocs similar to those found in the lower reaches of the freshwater environment are extremely rare.

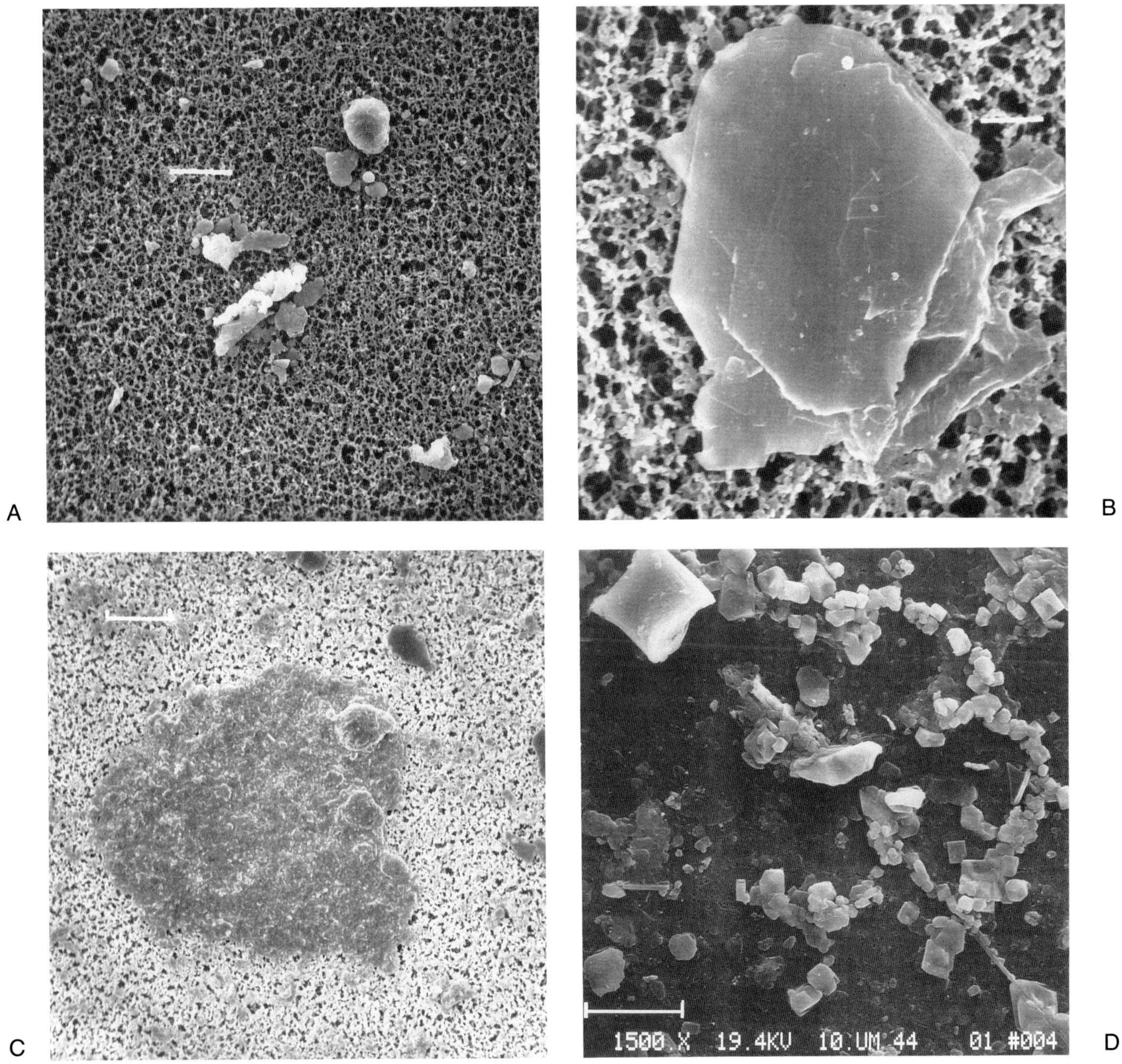

Figure 14.1. Background on micrographs is filter matrix. (A) Simple particles from stream draining into Rhode River. Bar = 10 μm. (B) Single particle with well-defined cleavage and fracture surfaces. Bar = 2 μm. (C) Floc in form of "raft" from Rhode River. Bar = 20 μm. (D) Skeletal floc from Rappahannock River. Bar = 10 μm.

Most of the aggregates have a random association of mineral grains in an apparently structureless matrix, which exhibits no granularity, fracture surfaces, nor cleavage faces (Fig. 14.2A). Associated with the mineral grains in the aggregates are phytoplankton and biotic fragments (Fig. 14.2B). Other aggregates (Fig. 14.2C) may be what have been termed psuedofeces (Schubel, 1982) or pelleted particles (Zabawa, 1978). These pellets are elliptical or oval in outline and presumably were cylindrical or spheroidal in the water column. Aggregates linked in chains, similar to that reported by Pusch (1970), are also present in the water column (Fig. 14.2D). In addition to the mineral particles present in the chains, biogenic debris and phytoplankton tests are often incorporated into these chains. Linkages are primarily by organic matter or by a few grain-to-grain contacts. The major portion of the aggregates in the brackish water regions have irregular shapes although some are elliptical to ovate in outline. Gradations exist from regular forms to no regularly defined geometric shape.

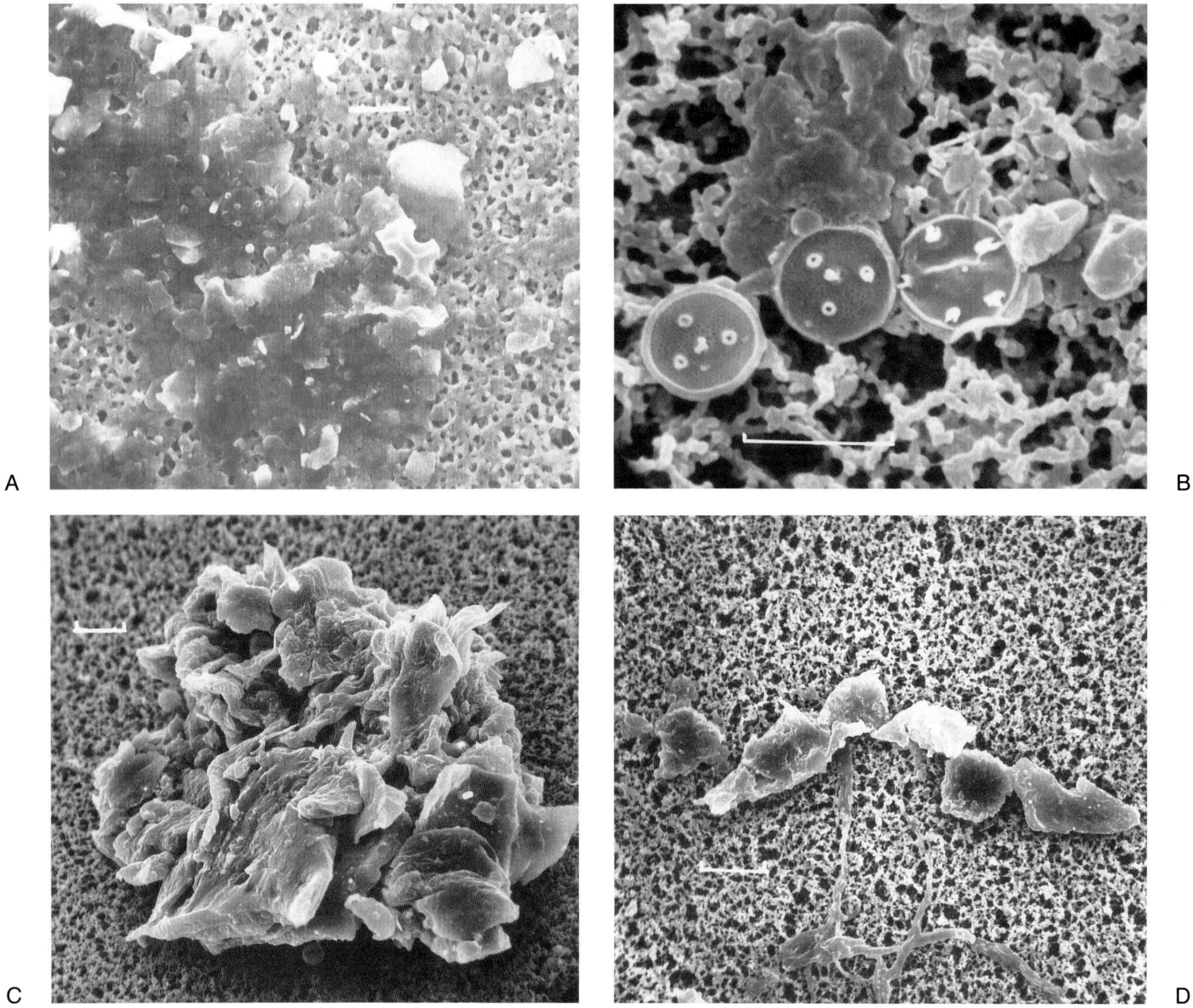

Figure 14.2. Background on micrographs is filter matrix. (A) Aggregate with organic matrix, Rhode River. Bar = 5 μm. (B) Aggregate with attached diatoms, Rhode River. Bar = 5 μm. (C) Pelleted particle, Rhode River. Bar = 10 μm. (D) Aggregates and mineral grains linked in a chain. Bar = 10 μm.

The association of the grains, biotic fragments, and phytoplankton seems to be dictated more by the mucoid matrix rather than by grain-to-grain effects. Insofar as can be determined by the two-dimensional view available on the micrographs, individual grains often are not in contact with other grains but seem to "float" in the matrix. No salt-flocculated structures with edge–face contact (Lambe, 1958) and only one of the aggregated forms of van Olphen (1963) were present in estuarine waters. In the Rhode River, the grains within the aggregates have a smaller average diameter than that of the solitary grains in the suspensate population, 0.9 and 3.5 μm, respectively (Pierce and Siegel, 1979).

The matrix is structureless, presumed to be mucus or other organic matter. I have found no correlation between the amount of matrix in the aggregates and the amount of organic matter in the sample although Plummer et al. (1987) report such a correlation. Organic matter is present in the water column as separate and distinct strands or globules. It is most prevalent in the summer and fall seasons when it dominates over mineral grains and biogenic hard parts in the aggregates. The population of individual mineral grains is lowest at this time of the year because of low sediment input to the estuary and because most of the mineral grains have been incorporated into aggregates.

The idea of an organic matrix has been suggested before (Pierce and Siegel, 1979; Eisma et al., 1983). Meade (1972) noted that organic aggregation and biologic processing are important in estuarine sediment transport processes. Dickson and McCave (1986) found patches of mucus on filters after filtration of samples from the eastern North Atlantic.

The existence of macroaggregates cannot be precluded from the data obtained from the micrographs. Hamner et al. (1975) found that it is nearly impossible to sample or, once sampled, to maintain the macroaggregates as units. Bale and Morris (1987) report an *in situ* mode of 54.9 µm. Later analyses in the laboratory determined a modal size of about 10 µm, suggesting a breakup of macroaggregates into smaller particles. Some of the aggregates and solitary grains appearing in the micrographs probably were part of macroaggregates. Because of the fragile nature of the macroaggregates, it is doubtful that they would survive impact with the bottom and be incorporated into the sediment as a unit. Syvitski et al. (Chapter 30) notes the absence of macroaggregates in the bottom boundary layer, although they were prevalent throughout much of the water column.

True fecal pellets, locally common in the bottom sediments of some estuaries (Syvitski and MacDonald, 1982), are rare in the suspensate population. Fecal pellets, resulting from plankton grazing, are denser than the aggregates and rapidly sink to the bottom. Other fecal pellets, produced by benthic organisms (Haven and Morales-Alamo, 1966), most likely would not been found in the suspensates.

Deposition of the suspensates would result in a poorly mixed sediment, consisting of the units incorporated in the aggregates plus the single grains present in the water column. Decay of the organic matrix could release the grains from the aggregates, resulting in a sediment without any apparent overall organized structure. The result might be similar to the turbostratic structure of Aylmore and Quick (1960) where there are domains with parallelism of platelets but with an overall disorganized structure between domains. Some suggestion of aggregate form could also be preserved in the sediments. Linked chains of small aggregates, present in the water column, might also be preserved. Thus, the fabric would be domains, aggregates, and linking chains as suggested by Bennett et al. (1977). Small grains included within the aggregates would be mixed with those of the population of larger individual grains, giving rise to an even more disorganized structure.

Marine

There appears to be little difference in the microstructure and form of suspensate populations in the marine and estuarine environments. Aggregates make up 51 and 47% of the number of particulates for the shelf waters off the southeastern United States and southern Argentina, respectively. The estimated percentage of total volume in the aggregates is 85 and 93%.

In all likelihood, some of the single grains in the micrographs were part of macroaggregates or marine snow. The breakup of macroaggregates into smaller aggregates and single particles would increase the estimates of the number and volume of single particles in the water column. At a minimum, about 90% of the suspensate volume is in aggregate form in those shelf waters which were sampled. A few individual grains from the shelf waters had fresh cleavage and fracture surfaces.

There is no regular organization of the grains within the aggregates from marine waters, similar to the situation in estuarine samples. There is a close association of phytoplankton with the aggregates. Coccospheres apparently can be bound to aggregates (Fig. 14.3A). A similar situation has been reported by Ernissee and Abbot (1975) in a South Carolina estuary where individuals of *Thallassiosira* were bound to mineral grains in an alternating pattern of grain–diatom–grain, in a linked chain arrangement. For the most part, the phytoplankton shown on the micrographs are part of or in very close proximity to an aggregate, suggesting that those not actually incorporated within the aggregate shown on the micrograph may have become detached during filtration (Fig. 14.3B). Few phytoplankton are found in close proximity to single grains or isolated on the filter surface.

The sedimentary deposit resulting from deposition of these aggregates would be bimodal, similar to that postulated for the estuarine environment, with domains and chains of linked aggregates. The particles within the aggregates appear smaller than the single, loose grains, although no statistical counts were made. Deposition of the suspensates off southern Argentina would have a significant component of large, clean grains.

Discussion

Simple associations of grains were the most prevalent particles present during the waning stages of a flood and during average flow conditions in small streams. Settling of the particles would be essentially on a grain-by-grain basis. At times of extremely low flow, organic matter and grains were incorporated together in aggregates. At times of rising flood, one would expect a large part of the suspended load to be made up of resuspended channel sediments and bank cavings. This would produce a more chaotic sediment with little organization of the clay particles except that inherited from the resuspended material.

In the lower reaches of the streams and at salinities $< 1‰$, most of the volume of the suspensates is in very large particles, which appear to be the result of flocculation, i.e., suppression of the double layer and subsequent grain-to-grain contact. Some resemble "rafts" of sediment while others have a skeletal structure with large pores between the skeletal parts. The sediment resulting from deposition would be similar to that postulated for

Figure 14.3. Background on micrographs is filter matrix. (A) Marine aggregate from off the southeastern United States. Bar = 5 μm. (B) Marine aggregate, possibly pelleted particle, incorporating biotic fragments and phytoplankton. Bar = 5 μm.

settling of individual grains through freshwater and would be quite porous but easily compacted. The flocs are unlike those postulated by Lambe (1958) for salt-flocculated structure. Most of the grain–grain contacts are edge–edge or face–face, which would be similar to one of the flocculated and aggregated forms of van Olphen (1963).

Aggregates contain the major portion of the suspensate volume in both estuarine and marine environments. Most of these aggregates have a mucoid matrix with included silt- and clay-sized mineral grains, biotic debris, and phytoplankton. Many were probably part of macroaggregates that disassociated during sampling and filtration. Others, such as pellets or psuedofeces, are the result of biologic processing.

A sediment resulting from deposition of this suspensate population would appear, after decay of much of the organic material, as consisting of domains, aggregates, and linking chains. Included within the sediment would be individual grains, most of which would be larger than those of the aggregates. These large grains would act as bulwarks against and around which the material of the aggregates would distort.

Recent research on particles suspended in the water column has shown that grain-by-grain deposition is not tenable and that much of the sediment is deposited as compound particles. Some of the microstructure of the bottom sediments would be inherited from that of the suspensates and subsequently modified by overburden pressure and bioturbation.

Acknowledgments

Support for this work over a period of time came from several Smithsonian Scholarly Studies Grants and grants from the Smithsonian Environmental Sciences Program. Thanks are also extended to those many assistants who spent long hours in front of the Scanning Electron Microscope.

References

Aylmore, L.A.G., and J.P. Quick, 1960. Domain or turbostratic structure of clays. Nature, v. 187, p. 1046–1048.

Bale, A.J., and A.W. Morris, 1987. *In situ* measurement of particle size in estuarine waters. Estuarine, Coastal and Shelf Science, v. 24, p. 253–263.

Bennett, R.H., and M.H. Hulbert, 1986. Clay Microstructure. International Human Resources Development Corporation, Boston, MA, 161 p.

Bennett, R.H., W.R. Bryant, and G.R. Keller, 1977. Clay fabric and geotechnical properties of selected submarine sediment cores from the Mississippi Delta. NOAA Prof. Paper 9, 86 p.

Dickson, R.R., and I.N. McCave, 1986. Nepheloid layers on the continental slope west of Porcupine Bank. Deep-Sea Research, v. 33, p. 791–818.

Eisma, D., 1986. Flocculation and deflocculation of suspended matter in estuaries. Netherlands Journal of Sea Research, v. 20, p. 183–199.

Eisma, D., J. Boon, R. Groenewegenn, V. Ittekkot, J. Kalf, and W.G. Mood, 1983. Observations on macro-aggregates, particle size, and organic composition of suspended matter in the Ems Estuary. In: Degens, E.T. (ed.), Transport of Carbon and Minerals in the Major World Rivers. Metteillung aus dem geologisch-Palontologischen Institut der Universitat, Hamburg, v. 55, p. 295–314.

Ernissee, J.J., and Abbot, W.H., 1975. Binding of mineral grains by a species of *Thallassiosira*. In: Simonsen, R. (ed.), Third International Symposium on Recent and Fossil Diatoms, Nova Hedwigia, v. 3, p. 241–252.

Gibbs, R., L. Konwar, and A. Terchurian, 1983. Size of flocs suspended in Delaware Bay. Canadian Journal Fisheries and Aquatic Sciences, v. 40, p. 102–104.

Hamner, W.M., L.P. Madin, A.L. Alldredge, R.W. Gilmer, and P.P. Hamner,

1975. Underwater observations of gelatinous zooplankton: sampling problems, feeding biology and behavior. Limnology and Oceanogography, v. 20, p. 907–917.

Haven, D.S., and R. Morales-Alamo, 1966. Aspects of biodeposition by oysters and other invertebrate filter feeders. Limnology and Oceanography, v. 11, p. 487–498.

Honjo, S., K.W. Doherty, Y.C. Agrawal, and V.L. Asper, 1984. Direct optical assessment of large amorphous aggregates (marine snow) in the deep ocean. Deep-Sea Research, v. 31, p. 221–243.

Lambe, T.W., 1958. The structure of compacted clay. Proceedings American Society of Civil Engineers, v. S91, p. 4–85.

Meade, R.H., 1972. Transport and deposition of sediments in estuaries. In: Nelson, B.W. (ed.), Environmental Framework of Coastal Plain Estuaries, Geological Society America Memoir 133, p. 91–120.

Pierce, J.W., and M.M. Nichols, 1986. Change of particle composition from fluvial into an estuarine environment. Journal of Coastal Research, v. 2, p. 419–425.

Pierce, J.W., and F.R. Siegel, 1979. Particulate material suspended in estuarine and oceanic waters. In: Johari, O.M. (ed.), Scanning Electron Microscopy/1979/I, p. 555–562.

Plummer, D.H., N.J.P. Owens, and R.A. Herbert, 1987. Bacteria particle interactions in turbid estuarine environments. Continental Shelf Research, v. 7, p. 1429–1433.

Pusch, R., 1970. Microstructural changes in soft quick clay. Canadian Geotechnical Journal, v. 7, p. 1–7

Schubel, J., 1982. An eclectic look at fine particles in the coastal ocean. In: Kimrey, L.W., and R.E. Burns (eds.), Proceedings of Workshop, Pollutant Transfer by Particulates, NOAA, Washington, D.C., p. 51–141.

Silver, M.W., and A.L. Alldredge, 1981. Bathypelagic marine snow: deep-sea algal and detrital community. Journal Marine Research, v. 39, p. 510–530.

Syvitski, J.P.M., and R.D. MacDonald, 1982. Sediment character and provenance in a complex fjord; Howe Sound, British Columbia. Canadian Journal of Earth Science, v. 19, p. 1025–1044.

Syvitski, J.P.M., K.W. Asprey, and D.E. Hefflen, 1990. The floc camera: A 3-D imaging system of suspended particulate matter. In: R.H. Bennett, W.R. Bryant, and M.H. Hulbert (eds.), The Microstructure of Fine-Grained Sediments: From Mud to Shale, Springer-Verlag, New York, Chap. 30.

van Olphen, H., 1963. An Introduction to Clay Colloid Chemistry. Wiley Interscience, New York, 301 p.

Zabawa, C.F., 1978. Microstructure of agglomerated suspended sediments in northern Chesapeake Bay Estuary. Science, v. 202, p. 49–51.

CHAPTER 15

The Influence of Organic Carbon Flux on the Deposition of Clays in the Marine Environment: Implications with Respect to Microstructure

Kathleen M. Fischer

Introduction

Over parts of the ocean, primary production varies with the season. Increases in insolation and nutrient concentrations in certain areas combine to produce seasonal blooms of phytoplankton (Fogg, 1975). Upwelling of nutrient-rich waters occurs along some coasts, resulting in periodic bursts of productivity.

Corresponding increases in particle flux have been reported in the waters underlying regions that display periodic increases in surface productivity, and these flux increases are observed in both the organic and inorganic fractions of the settling material. In a study conducted over several years of seasonal variations in particle flux arriving at the 3200 m level in the Sargasso Sea (Deuser and Ross, 1980; Deuser et al., 1981, 1983; Deuser, 1984), increases in the flux of all major components, including elements bound in clay minerals, were concurrent with increases in productivity in the surface waters. Upwelling-related increases in the flux of biogenic components, with simultaneous increases in the flux of lithogenic materials, have been measured in the Panama Basin (Honjo, 1982). The pulses of increased flux at this location were recorded within the same 2-month sampling period in traps spanning over 2600 m of the water column, implying rapid settling. At a site approximately 300 km from the typical seaward extent of coastal upwelling along the Cape Mendocino section of the northern California coast, periods of high flux coincided with times when plumes of cold water transported from upwelling areas extended over the mooring location (Fischer et al., 1983). Although river runoff principally occurred during December through March, maximum detrital flux was measured during the time of peak biogenic flux, which occurred later in the year at this site.

The pathways by which productivity-generated pulses in particle flux travel to the bottom have been elucidated by several studies in recent years. A growing body of data from sediment trap collections throughout the world ocean indicates that the transport of material to bottom sediments occurs via mechanisms that move material rapidly through the water column. The two important mechanisms that have been documented are transport via the fecal pellets of organisms such as zooplankton and via marine snow. Marine snow refers to large ($>$ 500 µm in diameter) amorphous structures formed by the action of physical and biological processes on organic matter and inorganic particles. Marine snow often contains fecal pellets as part of the aggregate. Large, rapidly settling particles and aggregates such as fecal pellets and marine snow appear to be the primary transport mechanism for material in the water column (Fowler and Knauer, 1986) and the main source of material to bottom sediments at most deep ocean sites.

Fecal pellets appear to be an important mode of material transport to the bottom at certain locations. Fecal pellets vary in size (from a few microns to several millimeters), in shape (ovate, ellipsoidal, cylindrical, spherical, and amorphous), in structural integrity (from loose aggregates to densely packed structures), and in sinking speed (from approximately 20 to over 1000 m/day) (Fowler and Knauer, 1986, for review). Inorganic material is frequently a major constituent of fecal pellets. In sediments underlying the Peru upwelling region, dispersal patterns for fine silts and clays derived from adjacent continental fluvial sources were controlled by zooplankton and anchoveta fecal pellet production (Scheidegger and Krissek, 1983; Staresinic et al., 1983). Mineral-rich fecal pellets have been found in other environments as well, such as estuaries and fjords (Syvitski et al., 1983, 1985). Fecal pellets with high clay contents accounted for more than 60% by weight of the material collected in traps in the Santa Barbara Basin (Dunbar and Berger, 1981). High clay content increases the settling rates of pellets, accelerating the transport of material to the bottom (Lewis and Syvitski, 1983). Once formed, fecal pellets may be consumed by other organisms, resulting in the repackaging of settling material. Evidence exists for the repeated repackaging of particles into fecal pellets by zooplankton inhabiting different depth ranges of the water column; pellets of differing size and shape were found at different water depths using sediment traps (Karl and Knauer, 1984). Pellets predominantly composed of material similar in composition to sediments have been found in near-bottom traps along with pellets obviously produced higher in the water column (Honjo, 1980), indicating that meiofauna create pellets near the bottom.

Through the work of several investigators, the importance of marine snow as a transport mechanism to the bottom has been recognized (Syvitski et al., 1983; Honjo et al., 1984; Alldredge and Hartwig, 1986; Fowler and Knauer, 1986; Silver, 1986; Asper, 1987; Syvitski, this volume). These large aggregates have been observed and sampled in shallow waters by divers (Trent et al., 1978), in deep waters by cameras, camera-monitored sediment traps (Honjo et al., 1984; Asper, 1987; Syvitski, this volume), and submersibles (Syvitski et al., 1983; M. W. Silver, personal communication, 1988), and on the surface of sediments in the deep ocean by camera and coring (Billett et al., 1983). Classes of organisms referred to as decomposers are associated with the marine snow throughout deeper sections of the water column, living on the aggregate and consuming the organic matter present during descent of the snow aggregate to the bottom; other organisms in the water column may feed off the aggregate but have a life cycle that is independent of it (Silver, 1986). Studies of fecal pellets and marine snow provide strong evidence for the dynamic processing of settling material by organisms throughout the water column.

In this chapter, flux data from the deployment of multiple sampling sediment traps are presented which relate to the transport and reactivity of organic matter associated with clay sediments. Clay surfaces adsorb many organic compounds (Mortland, 1970; Theng, 1974; Rashid, 1985), and these adsorbed organic compounds, in turn, may react to form bonds between clay particles. The resistance of the adsorbed organic matter to decomposition or alteration and the strength of the bonds between organics adsorbed on clay surfaces may play significant roles in the microstructure of clay-rich sediments. In light of the linked fluxes of organic matter and clay minerals in large particles, biological and physical processes in the water column associated with fecal pellet and marine snow dynamics impact on the early stages of microstructure development of clay minerals in transit to the sea floor.

Description of Sediment Trap Experiment

Six sediment traps were deployed at Site H of MANOP (Manganese Nodule Project). Site H is located in the Guatemala Basin at 6°32′N, 92°50′W in 3575 m of water (Fig. 15.1). The clays at this site are 50–70% montmorillonite with less than 20% illite and chlorite and minor amounts of kaolinite (Griffin et al., 1968). The sedimentation rate at Site H is 0.66 m/10^5 a, and the sediments consist mostly of clays and opal (Fischer, 1984). Although a substantial portion of the flux to the sea floor at this site is biogenic calcium carbonate, little carbonate carbon is preserved in the surface sediments (Fischer et al., 1986) due to the position of the calcium carbonate compensation depth and the slow sediment burial rate. Only a fraction of the organic carbon input to the sediments at this site is preserved (Fischer et al., 1986) due to the consumption of organic carbon by benthic organisms (predominantly by bacteria, but by larger organisms as well) in the surface sediments. Oxygen in the pore waters is consumed within a few centimeters of the sediment–water interface, and bioturbation rapidly decreases in the top 14 cm of the sediment (Kadko and Heath, 1984). Bottom currents at this site average 3 cm/sec, with maximum speeds less than 10 cm/sec (Gardner et al., 1984). Although current speed and direction near the bottom vary over the course of the year, no substantial difference in the variability of speed or direction was apparent between the three sediment trap sampling periods shown in Table 15.1.

OSU (Oregon State University) sediment traps (Fig. 15.2a) with multiple-sampling capability were positioned at water depths of 505, 3075, and 3545 m for a deployment period of 13 months. Each trap was equipped with an electronically controlled valve that turned at preset times sequentially to three cups, dividing the deployment period into three sampling intervals (Table 15.1). Performance of the timers and motors that activate the sample distribution valves was checked in a cold room after trap recovery. A fourth OSU trap, which collected a single sample for the deployment period, was positioned at 3225 m in the water column. Two double-cone Soutar traps (Fig. 15.2b; Soutar et al., 1977), which also collected average samples, were located at water depths of 1465 and 3415 m.

Sodium azide (Alfa Chemical) of 99% purity was added in solid form to the OSU traps. Sodium azide is highly soluble in seawater and is used to reduced bacterial decomposition of organic material (Powell and Fischer, 1982) and to kill larger organisms such as zooplankton, commonly referred to as "swimmers," which enter the trap and feed on collected material. Reagent grade formalin buffered with sodium borate to a pH of approximately 9 was added to the Soutar traps. Formalin reduces bacterial activity, kills swimmers, and chemically binds to organic matter to help preserve the morphology of organic particles. Swimmers larger than 1 mm were removed by sieving after recovery of the trap and were not included in the flux determinations. Organic carbon was measured by hot chromic acid oxidation of the organic material followed by thermal conductivity detection of the evolved CO_2 (Weliky et al., 1983). The analytical precision was $\pm 3\%$. Aluminum was measured by instrument neutron activation analysis (adapted from Conard, 1976) with a precision of $\pm 5\%$. Total nitrogen was measured by determination of ammonia following Kjeldahl digestion (Bremner, 1960). Precision for the nitrogen determinations was $\pm 3\%$. A complete description of the sediment trap experiment can be found in Fischer (1984) and Fischer et al. (1986).

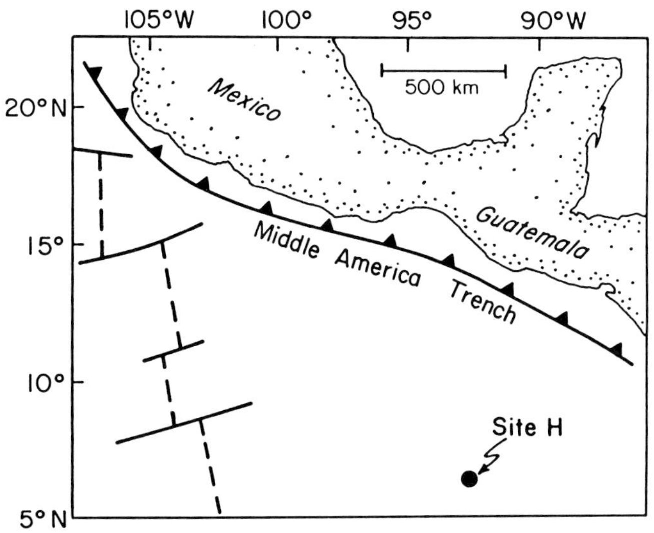

Figure 15.1. Map of the eastern equatorial North Pacific Ocean showing the location of Site H. The East Pacific Rise is shown as the series of dashed lines, with associated fracture zones depicted as solid lines. (Redrawn from Lonsdale and Spiess, 1980.)

A question that always arises with the use of sediment traps is that of collection efficiency, including the absolute efficiency, i.e., the correct flux is measured, and the relative efficiency, i.e., how a given trap compares to other traps currently in use. One approach to addressing the question of absolute efficiency is by measuring the excess ^{230}Th flux collected by the trap. The average particulate flux of excess ^{230}Th at any water depth is controlled by the decay of the parent ^{234}U. (For further discussion of this concept, see Anderson et al., 1983). As shown in Figure 15.3, the average flux of excess ^{230}Th collected in traps above 3415 m is approximately 77–149% of the value expected from ^{230}Th production by ^{234}U (Mahannah, 1984). This suggests that both the OSU and Soutar traps have a reasonable collection efficiency for less reactive elements, such as thorium or aluminum. The excess ^{230}Th fluxes of the two lowest traps (3415 and 3545 m) significantly exceed the theoretical ^{230}Th fluxes, confirming a substantial sedimentary signal from resuspension, as discussed below.

The collection efficiencies of these traps (and other traps currently in use) with respect to more labile components, such as organic material, are more questionable. Accurate organic flux measurement using sediment traps is limited by the efficacy of the bactericide/poison and the ability of the investigator to identify and remove swimmers, as well as by the hydrodynamic collection efficiency of the traps with respect to organic particles (U.S. GOFS Working Group, 1989). The OSU and Soutar traps appear to collect similar organic carbon flux data: the organic carbon flux for the three traps in the lower 350 m of the water column (Fig. 15.3) ranged from 1.41 to 1.57 g/m²/a. The primary objective of this chapter is to report seasonal changes in organic carbon flux, which in all cases were obtained using OSU

Table 15.1. Sampling intervals.

Interval	Open	Closed	Days
1	9-21-80	1-29-81	131
2	1-29-81	7-12-81	164
3	7-12-81	10-14/16-81*	95–97

*Exact date varied with the trap.

traps poisoned with sodium azide. The data from the Soutar traps are presented to provide a more complete picture of organic carbon flux throughout the water column. The Soutar trap has been used in several comparison studies (Spencer, 1981a; Dymond et al., 1981; Fischer, 1984) and, therefore, provides some indication of the performance of OSU traps relative to other traps commonly used in the field.

Results

Organic Carbon Flux

The organic carbon fluxes averaged over the entire deployment for all six traps are shown by the solid line in Figure 15.4. The increases in the average organic carbon flux between 505 and 1465 m and between 3075 m and the three trap levels nearest the bottom are discussed elsewhere (Fischer, 1984; Walsh et al., 1988a,b). Average surface productivity at this site is 110 g C_{org}/m²/a (Owen and Zeitzschel, 1970; Fischer, 1984), of which approximately 18% can be assumed to exit the euphotic zone as new production (Knauer, 1988). The average organic carbon fluxes measured by the traps for the entire deployment range from 1.12 to 1.89 g/m²/a. This indicates that only 6–10% of the new production reaches the water column below 500 m, suggesting rapid recycling of organic carbon between the base of the euphotic zone and the surface 500 m. This result is slightly lower than, but generally consistent with, predicted flux from the modeling of literature trap data from several ocean sites (Berger et al., 1988). The accumulation rate of organic carbon in the surface 0–5 cm of the sediments at Site H is 0.0087 g/m²/a, which is less than 1% of the input of organic carbon estimated from the sediment traps in the lower water column, i.e., 1.12–1.57 g/m²/a. Extensive bioturbation (oxygenation) of the surface sediments and consumption of organic carbon by bacteria and larger benthic organisms produce the low accumulation rate. Thus, only a very small fraction of the organic carbon produced in the euphotic zone is preserved in the sediments.

Organic carbon flux for the three sampling intervals are plotted (dashed lines) versus depth in Figure 15.4. The time of greatest organic carbon flux occurs during interval 1 (October through January) at the three water column depths. The flux of organic carbon at all three depths is greater by at least a factor of 2 during interval 1 compared to interval 2 or 3 (Table 15.2). The period of maximum organic carbon flux occurs during the same

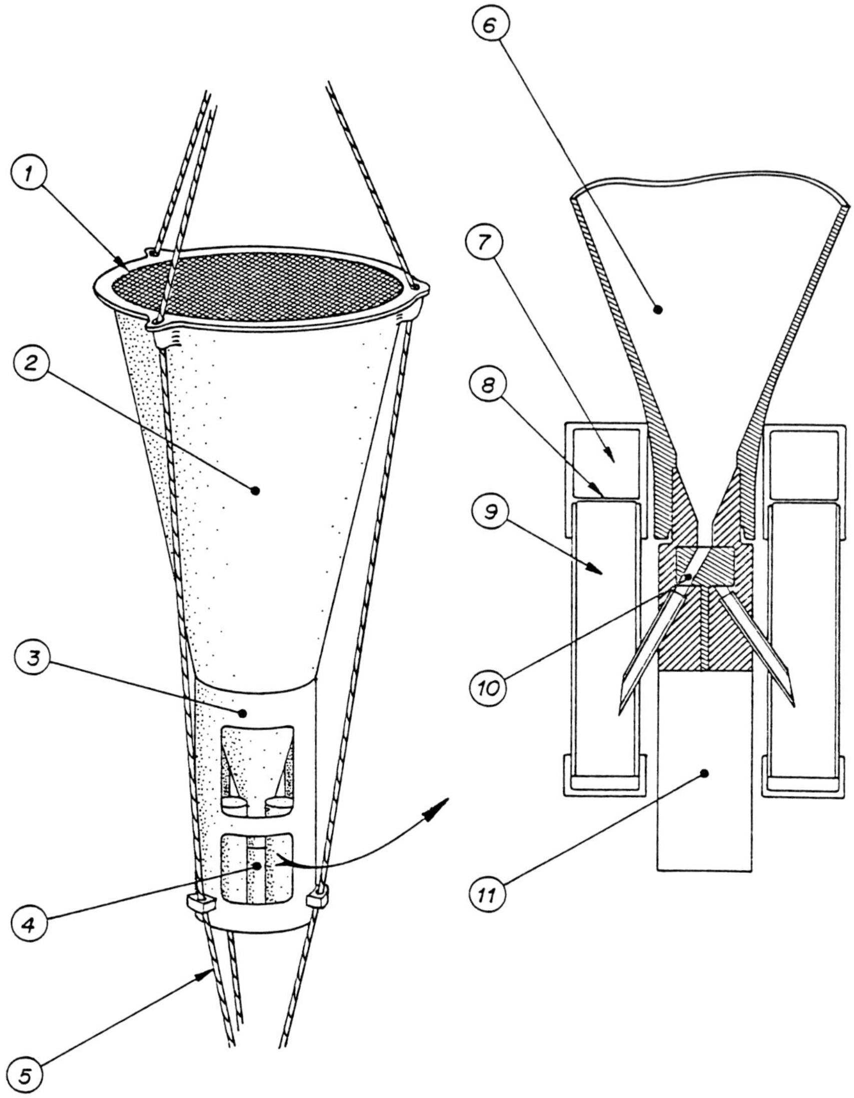

Figure 15.2. (a) The OSU multiple-sampling sediment trap. The mouth area is 0.5 m^2, and the height of the cone is 1.6 m. The labeled parts of the trap are 1, epoxy-coated nylon baffles composed of 1 cm^2 openings which are 5 cm deep; 2, Fiberglas cone; 3, Fiberglas protective sleeve; 4, sample cup rotation apparatus; 5, nylon mooring harness; 6, apex of cone; 7, sodium azide chamber; 8, nylon screen and 0.4 μm Nucleopore filters; 9, PVC sample cup; 10, sample distribution valve; 11, electronics and motor.

131-day interval at all three water depths, suggesting that the settling material transits a 3000-m water column in less than 131 days. An estimate of minimum settling rate can be calculated by dividing the length of the water column (3000 m) by the time (131 days), resulting in a minimum settling velocity of 23 m/day.

The seasonal increase in organic carbon flux arises because of an increase in productivity in the upper region of the water column. Increase in phytoplankton production and the resulting increase in zooplankton and other large animal feeding result in the enhanced formation of large particles. These large particles provide the greater organic carbon flux to the bottom as measured by the traps during interval 1. Seasonal variation in the surface productivity has been recorded during a 1-year study at this site (Owen and Zeitzschel, 1970) that was conducted several years prior to the sediment trap deployments; productivity averaged over 2-month periods varied by a factor of 7 during the course of the year. The period of maximum productivity did not coincide with the interval of maximum flux collected in sediments traps by this study. However, productivity and flux were measured several years apart, and, considering the complex and

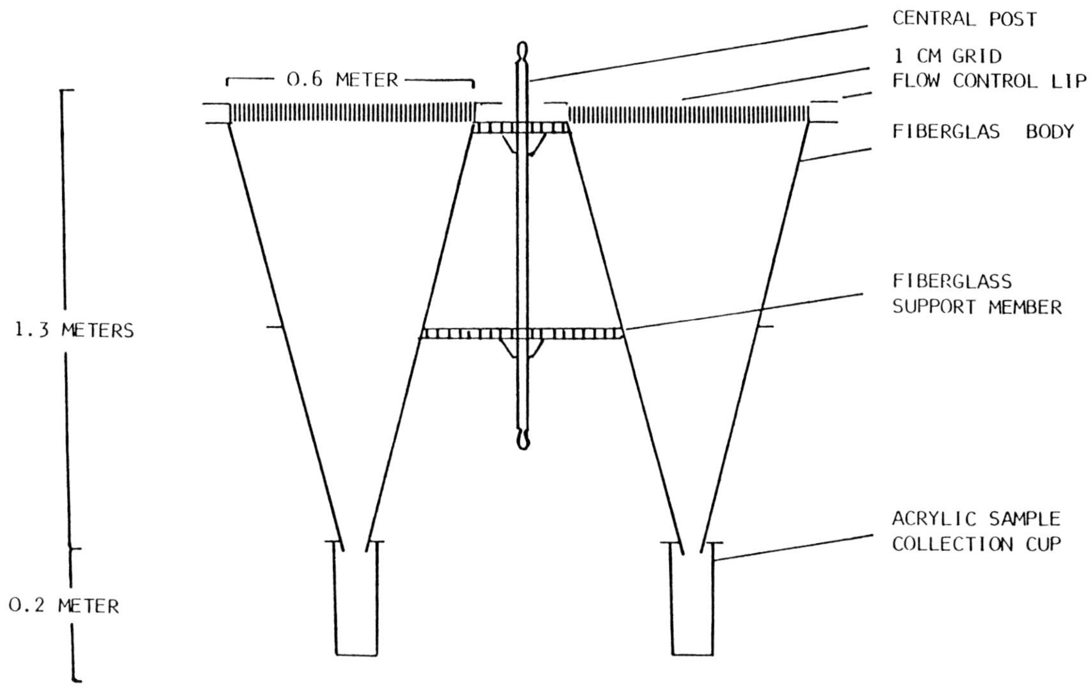

Figure 15.2. (b) The Soutar trap (courtesy of Andrew Soutar, Scripps Institution of Oceanography).

variable physical oceanography of the surface waters at this location, variation in times of maximum productivity from one year to the next is not unexpected. Interannual variability in the magnitude and timing of maximum flux has been recorded in the Sargasso Sea as well (Deuser, 1986).

Aluminum (Clay) Flux

Aluminum flux is used in this study as a measure of the clay flux (Fig. 15.5). Similar to the organic carbon flux, the aluminum flux is greater during interval 1 by a factor of at least 2 compared to interval 2 or 3 at all three water depths (Table 15.2). The simultaneous seasonal increases in aluminum and organic carbon fluxes throughout the water column provide evidence for the association of clays and organic matter during the deposition of material in the ocean. The source of the aluminum flux to Site H and, in particular, the variability of this source over the course of the year are not known. Since Site H is approximately 750 km from the nearest land and no major rivers drain the Central American coast in this region, seasonal input of clays from river runoff is not expected to be a significant source of aluminum input at this location. Potential hydrothermal sources of aluminum are more than 300 km distant. The contributions of hydrothermal material to the site are small, based on a study using a multicomponent mixing model (Fischer, 1984; Fischer et al., 1986). Another possible source of aluminum flux may be terrigenous clays transported to Site H via eolian processes. Shipboard collections of dust in the region of Site H indicate that airborne material is composed predominantly of plagioclase with minor amounts of quartz and trace amounts of kaolinite, chlorite, and smectite (Prospero and Bonatti, 1969). The composition and prevailing wind patterns suggest that this material originates in the arid regions of western and southern Mexico. One factor that argues against a predominantly eolian source of aluminum to Site H is that the mineralogical composition of the airborne particulates does not match the sediments at Site H. Part of the discrepancy may lie in the collection efficiency of the methods used to sample airborne dust. An alternate possibility for the source of aluminum flux at site H is via the transport of suspended aluminum, predominantly in the form of clay minerals, from ocean-sediment boundaries (Spencer, 1981b). The determination of the source of aluminum flux at Site H and the seasonal variability of this source require further detailed study.

Significant increases in average aluminum flux can be seen in the lower 500 m of the water column (Fig. 15.5). These increases in aluminum flux are most likely due to resuspension of bottom clays which are rich in aluminum (Fischer, 1984; Fischer et al., 1986; Walsh et al., 1988b). If a comparison is made between the aluminum fluxes of the 3075-m trap and the 3545-m trap for each sampling interval, it can be seen that maximum resuspension also occurs during interval 1. This suggests that resuspension of recently deposited material is enhanced during periods of high flux and that this resuspended material resettles rapidly (Walsh

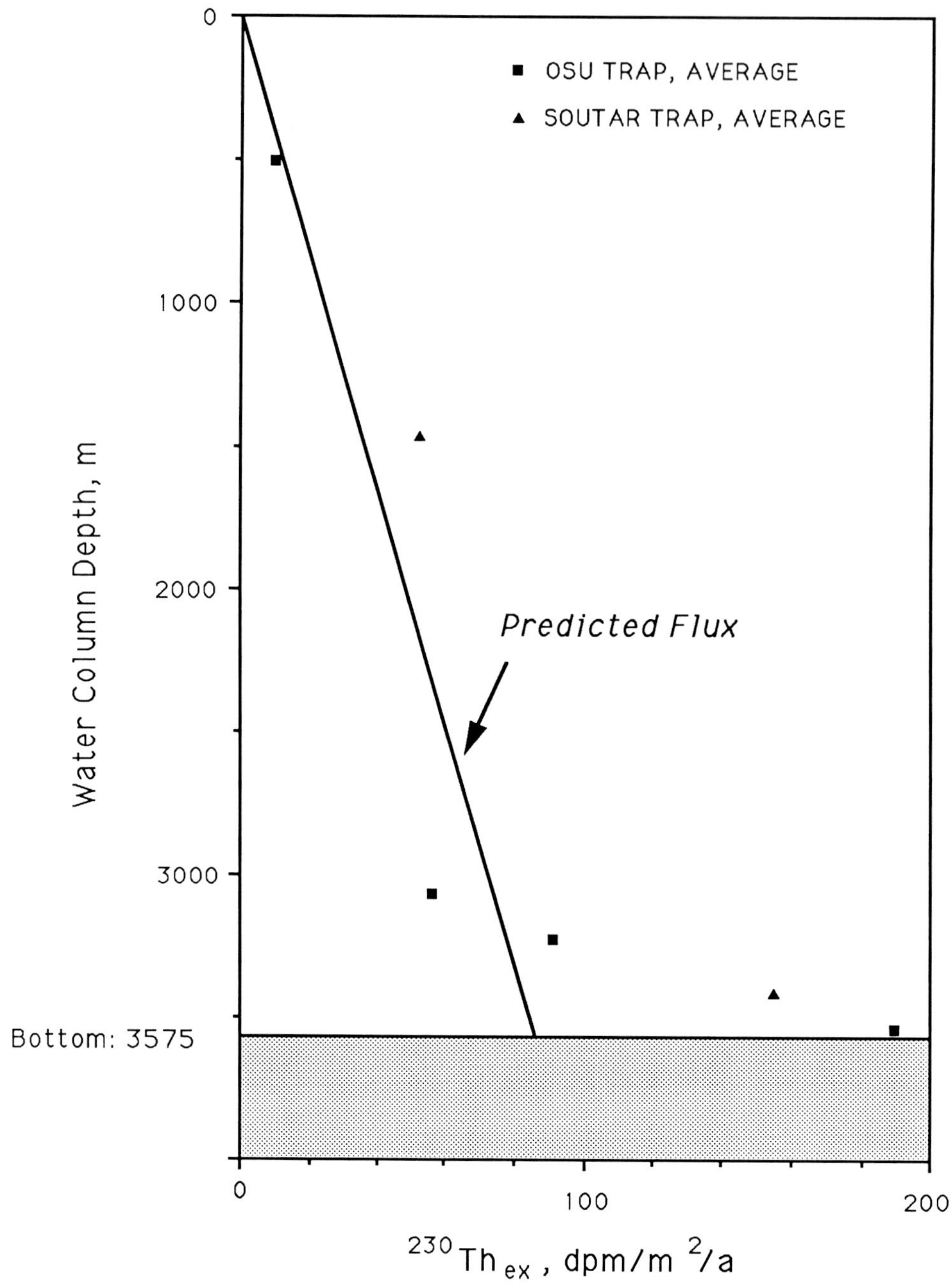

Figure 15.3. Flux of excess ^{230}Th in dpm/m^2/a versus water depth. The straight line represents the flux of excess ^{230}Th, which would be predicted from the decay of dissolved ^{234}U in seawater. (Redrawn from Mahannah, 1984.)

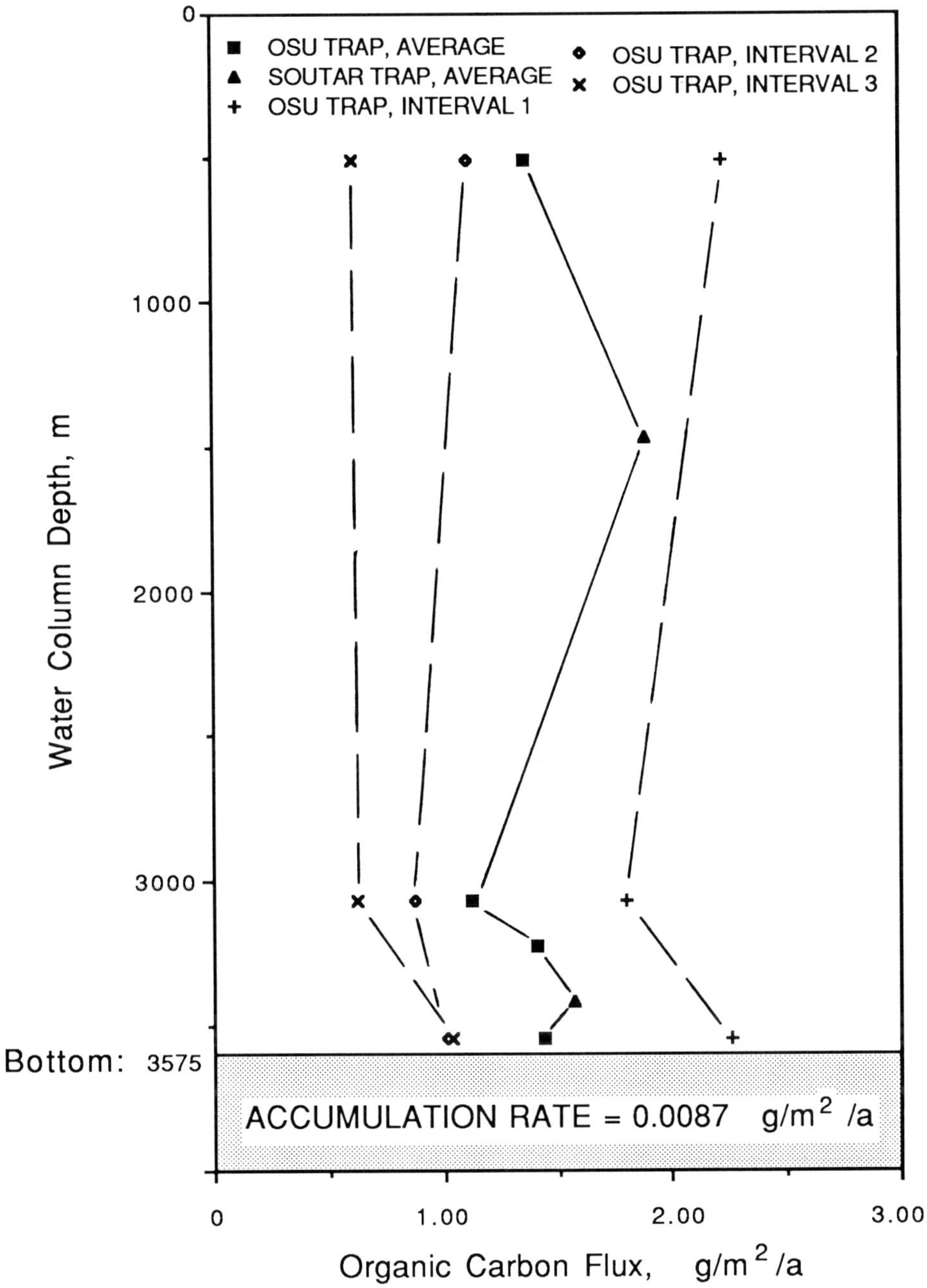

Figure 15.4. Organic carbon flux in g/m²/a versus water depth for the three sampling intervals. The average fluxes for all six traps are also shown. The organic carbon accumulation rate in the surface (0–5 cm) sediments is 0.0087 g/m²/a.

Table 15.2. Organic carbon flux and aluminum flux in $g/m^2/a$ for the three sampling intervals and the average flux for the entire deployment.

Trap depth (m)	Interval	Organic carbon flux	Al flux
505 (OSU)	1	2.23	0.0336
	2	1.11	0.0143
	3	0.607	0.0093
	Average	1.36	0.0195
1465 (Soutar)	Average	1.89	0.0633
3075 (OSU)	1	1.80	0.125
	2	0.878	0.0454
	3	0.627	0.0214
	Average	1.12	0.0662
3225 (OSU)	Average	1.41	0.112
3415 (Soutar)	Average	1.57	0.178
3545 (OSU)	1	2.26	0.435
	2	1.01	0.207
	3	1.04	0.216
	Average	1.44	0.286
Surface sediment, 0–5 cm		0.0087	0.06

Table 15.3. Organic carbon (C_{org}) to total nitrogen (N_t) molar ratios for the three sampling intervals and the average values for the entire deployment.

Trap depth (m)	Interval	C_{org}/N_t*
505 (OSU)	1	7.3
	2	6.5
	3	6.4
	Average	6.9
1465 (Soutar)	Average	8.1
3075 (OSU)	1	7.0
	2	6.6
	3	6.0
	Average	6.6
3225 (OSU)	Average	7.3
3415 (Soutar)	Average	8.3
3545 (OSU)	1	8.7
	2	7.1
	3	6.4
	Average	7.7
Surface sediment, 0–5 cm		5.4
Proteins		~3.
Copepods†		4.3
Fish and fish larvae†		3.9
Polychaetes (benthic organisms)†		3.4

*For sediment trap material, $C_{org}/N_t \approx C_{org}/N_{org}$. For sediments at Site H, $(C_{org}/N_t)/(0.96) \approx C_{org}/N_{org}$ (see text).
†Beers (1966).

et al., 1988b). The aluminum accumulation rate of 0.06 $g/m^2/a$ in the surface (0–5 cm) sediments is in close agreement with the average aluminum fluxes measured by the 1465- and 3075-m traps, 0.0633 and 0.0662 $g/m^2/a$, respectively (Fig. 15.5). The value from the 3075-m trap (0.0662 $g/m^2/a$) is assumed to be the best estimate of aluminum input to the sediments, i.e., this trap is nearest the bottom and appears to contain little or no aluminum input from resuspension. Lateral input of aluminum between the 1465 m level and the 3075 m level does not appear to occur to any appreciable extent based on the flux data for the traps at these depths. Only in the near bottom region, estimated to be less than 500 m above bottom and predominantly within the region 100 m above bottom, does resuspension provide an added source of clay (aluminum) flux to the sediment traps.

C/N Ratios

Organic carbon to total nitrogen molar ratios, C_{org}/N_t, are shown in Table 15.3. Most of the nitrogen in fast-settling particulate matter is in organic form; consequently, total nitrogen flux can be used to approximate organic nitrogen flux in most sediment trap samples, i.e., C_{org}/N_t is approximately equal to C_{org}/N_{org} for the sediment trap material. Table 15.3 also gives the value of this ratio in some marine animals that inhabit the water column and surface sediments. The organic matter in living marine phytoplankton can be approximated with the formula (Riley and Chester, 1971): $(CH_2O)_{106}(NH_3)_{16}H_3PO_4$. This formula gives a molar ratio of organic carbon to organic nitrogen of 6.6. The value of this ratio for particulate organic matter collected by hydrocast in surface waters is close to the value for phytoplankton and increases with depth in the water column (Holm-Hansen et al., 1966; Gordon, 1971), i.e., decomposition results in a preferential loss of nitrogen relative to carbon. Consequently, the ratio is sometimes used as a measure of the state of decomposition of organic material. The ratios from the sediment traps for all sampling periods range between 6.0 and 8.7, and there is no apparent trend with depth. This evidence suggests that the material is relatively fresh, well preserved by the azide, and sinking rapidly. Investigators measuring the magnitudes and chemical composition of selected fractions of organic compounds, such as lipids and amino acids, have found that these fractions change in character throughout the water column, indicating that the organic matter has been altered by the action of various organisms. The types of compounds present indicate that dynamic biological processing of settling organic material by zooplankton, bacteria, and other animals in the water column occurs (Wakeham et al., 1980; Lee and Cronin, 1982, 1984; Wakeham and Canuel, 1988). This type of alteration may not be revealed by C_{org}/N_t.

The value of C_{org}/N_t in the sediments at Site H is 5.4, significantly lower than the ratio measured for the sediment trap samples. In sediments, a significant fraction of the nitrogen measured may be in inorganic form as either exchangeable or fixed ammonium ions associated with clays. Exchangeable ammonium is generally less than 5% of the nitrogen measured in marine sediments; however, fixed ammonium may represent a significant portion of the total nitrogen measured in marine sediments (Müller, 1977). Fixed ammonium is found in the interlayer regions of illites, and the more illite present, the larger the amount of fixed ammonium present (Müller, 1977). The K_2O/Al_2O_3 ratio of the sediment can be used as a measure of the

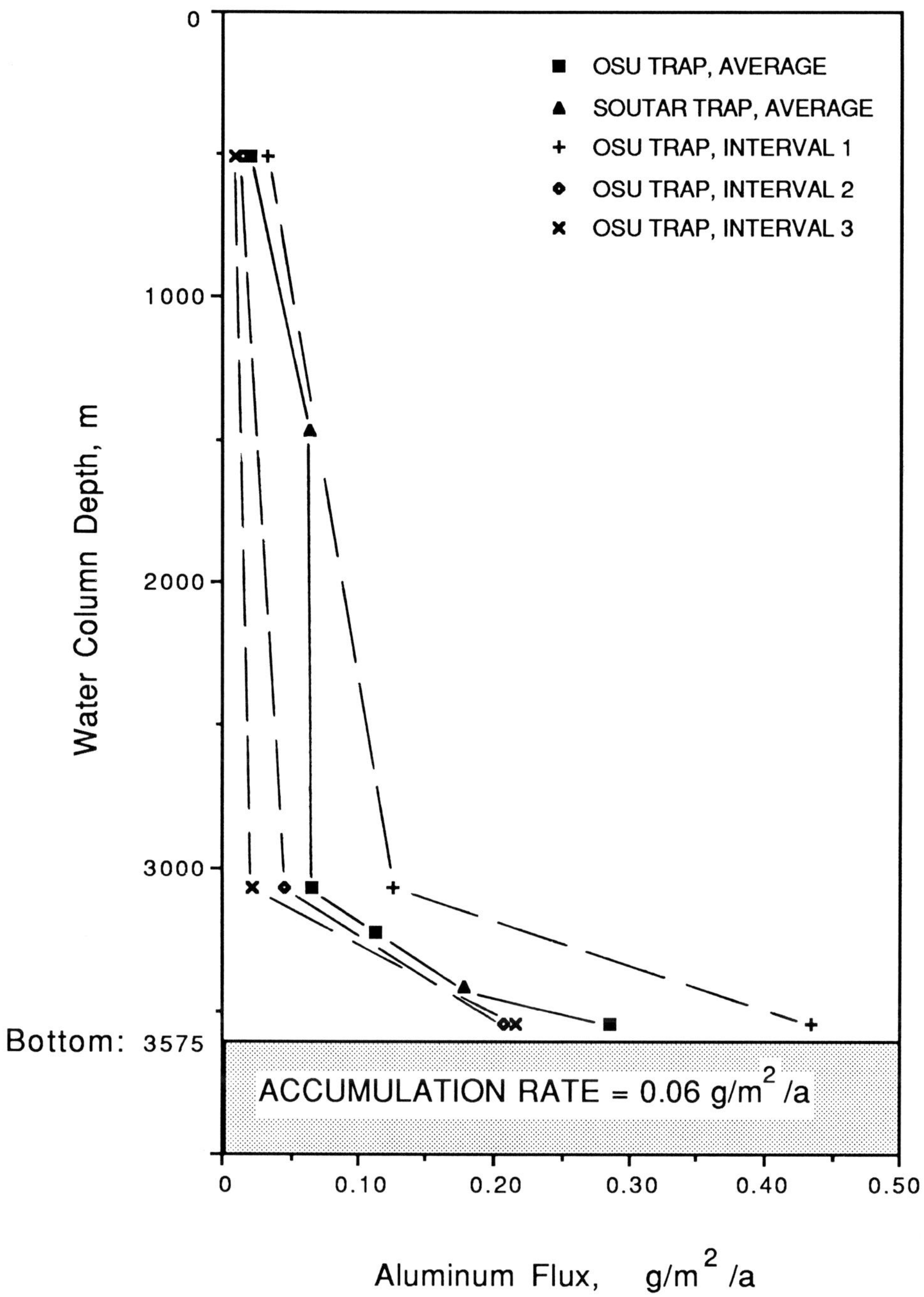

Figure 15.5. Aluminum flux in g/m²/a versus water depth for the three sampling intervals. The average fluxes for all six traps are also shown. The aluminum accumulation rate in the surface (0–5 cm) of the sediments is 0.06 g/m²/a.

amount of illite present in the absence of feldspar. The value of this ratio in the surface sediments at Site H is 0.15. From the graph given by Müller (1977), the nitrogen fixed as ammonium in illites in Site H sediments is approximately 4% of the total nitrogen present; the remaining 96% of the nitrogen in these sediments is organic nitrogen. Accordingly, the molar ratio of organic carbon to *organic* nitrogen (C_{org}/N_{org}) in the surface sediments at Site H is 5.4/0.96 or 5.6. This would mean that the value of this ratio decreases significantly between the water column and surface sediments. This suggests a preferential preservation of organic nitrogen in the sediments relative to organic carbon.

Evidence from deep-sea sediments indicates that a certain fraction of organic matter appears to be highly resistant to breakdown and may be associated with clay surfaces (Müller, 1977). Adsorption of organic matter on clay surfaces may lower C_{org}/N_{org} by preferentially preserving nitrogen relative to carbon in sediments with a low C_{org}/clay ratio (Müller, 1977). A graph by Müller (1977) estimates that C_{org}/N_{org} for Site H is approximately 5. The value of 5.6 estimated for Site H is consistent with the perservation of nitrogen in the sediments, possibly by the adsorption of the organic material onto clay surfaces via bonds involving organic nitrogen (Rashid, 1985). The reactivity of organics adsorbed on clays may be diminished due to the bonds formed in adsorption reactions. In a study of the completeness of removal of organic matter from clays, van Langeveld et al. (1978) found that no chemical treatment completely removed the organic matter present from sediments collected from North Sea, and that the surfaces of the clays appeared to be blocked by the organic matter remaining. This supports the idea that organic matter adsorbed to clay surfaces is strongly bound and is resistant to chemical breakdown. If organic nitrogen is important in the adsorption bonding, then organic nitrogen associated with clay surfaces would be preserved compared to marine organic matter unassociated with clays.

Discussion

Microenvironments

During the formation of fecal pellets, the ingested particles, whether clay particles or the remains of marine plants and animals, must pass through the digestive tract of the organism. This constitutes passing through a microenvironment which may exhibit conditions of pH, E_h, and soluble component concentration that may be significantly different from the conditions found in the open water column. In some cases, the organisms package, and to some extent may consolidate, the ingested particles into a dense structure encased in a peritrophic membrane that slows decomposition and disaggregation of the pellet. Although the residence time of the particles within the organism may be small, the residence time within the environment of the pellet may be significantly greater. The length of the residence time depends on the type of aggregate involved; however, laboratory studies have shown that fecal pellets enclosed in a peritrophic membrane have remained intact for one or more weeks (Small and Fowler, 1973; Honjo and Roman, 1978). On the sea floor, benthic organisms such as polychaetes consume the surface sediments, passing ingested material through their digestive tracts. This also constitutes a microenvironment that may have chemical conditions that differ from the surrounding pore waters. For example, clay diagenesis has been noted in recently produced fecal pellets collected in nature and in pellets produced during feeding experiments conducted in the laboratory (Syvitski and Lewis, 1980; Bayliss and Syvitski, 1982). These results support the concept of microenvironmental conditions within the digestive tracts of organisms and/or within pellets.

Marine snow aggregates also are believed to be microenvironments due to the concentration gradients that may exist in the waters internal to the aggregate relative to bulk seawater (Silver, 1986). For example, Shanks and Trent (1979) found elevated levels of dissolved nutrients NH_4^+, NO_2^-, NO_3^-, and PO_4^{3-} in aggregates, and gradients in oxygen and pH have been found within 1 mm of aggregates 1–4 mm in size (Silver, 1986). The development of microelectrodes for use in the deep sea (Reimers et al., 1986) may improve the ability to measure pH and oxygen concentrations and gradients of microenvironments such as marine snow and sediment burrows. The potential impact of microenvironments on the adsorption of various organics by clay surfaces will need to be considered as more is learned about the chemical conditions of these microenvironments.

Impact on Microstructure

Microstructure refers to the arrangement of solid particles in a sediment and to the strength of the bonding between particles (Bennett and Hulbert, 1986). Both of these factors may influence the physical and geotechnical properties of sediments (Bennett et al., 1977). The most important implications from the sediment trap data presented above are (1) with respect to the effects of biological processing on the orientation of clay particles settling to the sea floor, and (2) with respect to the potential for adsorption of organic matter on clay surfaces and the effect of adsorbed organic matter on the bonding strength between clay particles.

Fecal pellets formed by some organisms, such as copepods, result in the random arrangement of clay particles (Syvitski and Lewis, 1980; Bayliss and Syvitski, 1982; Bennett et al., this volume; Syvitski and Lewis, 1980; Bayliss and Syvitski, 1982). Similarly, the processing of bottom sediments through the digestive tracts of infaunal organisms also results in the random arrangement of clay particles. These are examples of two features of biological processing that affect the fabric of clays prior to burial below the zone of bioturbation in the sediment column.

The potential for chemical conditions in microenvironments such as fecal pellets and marine snow aggregates to differ from bulk seawater is very important to the consideration of the adsorption of organics on clay surfaces. For example, Rashid et al. (1972) showed that the adsorption of humic acids (the major form of organic matter in most marine sediments) increased dramatically when the pH was decreased from 8.1 (approximately the value in the water column at most locations) to 5.5. Within a fecal pellet and internal to a marine snow aggregate, restricted water flow may result in concentration increases or decreases in certain soluble components relative to bulk seawater. For example, as organic matter is decomposed within the pellet or the aggregate, the CO_2 concentration increases, and the acidity of the interstitial water increases (pH decreases). This change in pH could result in the enhanced adsorption of organic material. Similarly, high concentrations in microenvironments of dissolved organic matter, produced by the incomplete decompo-

z height above the bed (L)

z_+ nondimensional height above the bed $z_+ = \left(\frac{u_* z}{\upsilon}\right)$ [Eq. 2a)]

δ_D diffusive sublayer thickness (L) [Eqs. 2d, (6)]

δ_s particle inertial stopping distance $\delta_s = u_*\, 2\, (w_s/g)$ (L) [Eqs. (2c), (6)]

δ_* displacement height in logarithmic velocity profile (L) [Eq. (6b)]

Δ near-bed capture distance (L) [Eq. (6b)]

η_i efficiency of mechanism i, i = G, gravitational settling; R, interception; I; inertial impaction; D, molecular or Brownian diffusion [Eqs. (2), (5)]

μ dynamic viscosity of fluid (M L^{-1} T^{-1})

υ kinematic viscosity of fluid (L^2 T^{-1})

ρ density of fluid (M L^{-3})

ρ_s density of sediment particle (M L^{-3})

Acknowledgments

T. Gross of Skidaway Institute, P. Hill, and other members of our sediments group at U. W. offered valuable comments on this and earlier texts. This study was funded by ONR Grant N00014-87-C-0160 and is University of Washington, School of Oceanography Contribution No. 1812.

References

Bandyopadhyay, P.R., 1987. Rough-wall turbulent boundary layers in the transition regime. Journal of Fluid Mechanics, v. 180, p. 231–266.

Brodkey, R.S., J.M. Wallace, and H. Eckelmann, 1974. Some properties of truncated turbulence signals in bounded shear flows. Journal of Fluid Mechanics, v. 63, p. 209–224.

Browne, L.W.B., 1974. Deposition of particles on rough surfaces during turbulent gas-flow in a pipe. Atmosphere and Environment, v. 8, p. 801–816.

Chamberlain, A.C., 1967. Transport of *Lycopodium* spores and other small particles to rough surfaces. Proceedings of the Royal Society of London Series A, v. 296, p. 45–65.

Cacchione, D.A., D.E. Drake, W.D. Grant, A.J. Williams III, and G.B. Tate, 1983. Variability of seafloor roughness within the Coastal Ocean Dynamics Experiment (CODE) region. Woods Hole Oceanographic Institute Technical Report, WHOI-83-25.

Cantwell, B.J., 1981. Organized motion in turbulent flow. Annual Review of Fluid Mechanics, v. 13, p. 457–515.

Dawson, D.A., and O. Trass, 1972. Mass transfer at rough surfaces. International Journal of Heat and Mass Transfer, v. 15, p. 1317–1336.

Eckman, J.E., 1983. Hydrodynamic processes affecting benthic recruitment. Limnology and Oceanography, v. 28, p. 241–257.

Grass, A.J., 1971. Structural feature of turbulent flow over smooth and rough boundaries. Journal of Fluid Mechanics, v. 50, p. 233–255.

Hinds, W.C., 1982. Aerosol Technology. Wiley, New York, 424 p.

Hunt, J.R., 1986. Particle aggregate breakup by fluid shear. *In*: Mehta, A.J. (ed.), Estuarine Cohesive Sediment Dynamics. Springer-Verlag, New York. p. 85–109.

Jackson, P.S., 1981. On the displacement height in the logarithmic velocity profile. Journal of Fluid Mechanics, v. 111, p. 15–25.

McCave, I.N., 1984. Mechanisms of fine-grained sediment deposition from nepheloid layers. Geo-Marine Letters, v. 4, p. 243–245.

Offen, G.R., and S.J. Kline, 1975. A proposed model of the bursting process in turbulent boundary layers. Journal of Fluid Mechanics, v. 70, p. 209–228.

Owen, P.R., 1960. Dust deposition from a turbulent airstream. International Journal of Air Pollution, v. 3, p. 3–25.

Rubenstein, D.I., and M.A.R. Koehl, 1977. The mechanisms of filter-feeding: some theoretical considerations. American Naturalist, v. 111, p. 981–994.

Sternberg, R.W., 1970. Field measurements of the hydrodynamic roughness of the deep-sea boundary. Deep-Sea Research, v. 17, p. 413–420.

Sumer, M., and R. Deigaard, 1981. Particle motions near the bottom in an open channel, Part 2. Journal of Fluid Mechanics, v. 109, p. 331–337.

Swift, S.A., C.D. Hollister, and R.S. Chandler, 1984. Close-up stereo photographs of abyssal bedforms on the Nova Scotia rise. Marine Geology, v. 66, p. 303–322.

Wheatcroft, R.A., C.R. Smith, and P.A. Jumars, 1989. Dynamics of surficial trace assemblages in the deep sea. Deep-Sea Research, v. 36, p. 71–91.

CHAPTER 17

Distinguishing Features of Layered Muds Deposited from Shallow Water, High Concentration Suspensions

R. Kirby

Introduction

Fine sediment occurs in lakes, rivers, and estuaries, on continental shelves, and in the deep ocean. The transport mechanism by which the sediment is emplaced exerts a major influence on the clay microstructure that develops. Low concentration suspensions settle vertically to form undifferentiated beds, while in contrast high concentration suspensions are advected across the bed prior to deposition leading to a different range of bed structures. Recent research has shown that a breakpoint separating low from high concentration modes of behavior possibly occurs close to 500 mg/liter (Kirby, 1986).

In addition to the differences that these contrasted mechanisms of emplacement impose on the primary fabric, the suspension concentration exerts a major control on the organisms that can inhabit the bed, with the result that secondary or biogenic fabrics that overprint the primary fabric are of a different order of intensity. In low concentration regimes bioturbation may occur on such a scale that the primary fabric is unrecognizable other than by highly sophisticated magnetic remanence measurement techniques. The fabrics developed from high concentration regimes are more likely to remain largely or entirely unbioturbated.

High concentration phenomena and the sedimentary fabrics they give rise to may occur in all environments. In most they occur only episodically, leading to problems of forecasting and hence study. In certain estuaries high concentration phenomena occur on a regular basis arising from some seasonal or tidal control. The shallow water and predictable fluctuations between low and high concentration phenomena make certain tidal estuaries well suited to study.

In the Severn Estuary, UK, a large data base on the three-dimensional structure of the turbidity maximum and the movement of suspensions is complemented by a number of high quality cores from the bed deposits they give rise to.

Suspension Structures

Data has been collected with continuous profiling optical turbidity meters (100–20,000 mg/liter) and high-resolution nuclear densimeters (10,000–400,000 mg/liter). At high tidal energy levels suspended sediment is mixed to the water surface forming homogeneous suspensions. As energy levels decline the mobile suspensions begin to settle leading to the development of one or more layers, or steps, in the concentration profile (Fig. 17.1). The layers are entirely due to concentration effects and variations in primary or in aggregate particle size or mineralogy are not known to occur (Smith, 1982). Such layers were first recognized and described by Kirby and Parker (1983) who coined the term "lutoclines" to describe the discontinuities. Recently the physical mechanism by which the lutoclines are generated and become stabilized was described by Smith and Kirby (1989).

As energy levels continue to diminish, the suspended material settles to form dense near-bed mobile layers, which may be advected to and fro along the estuary axis for great distances over periods of days.

Eventually on low range tides these dense layers come to rest in low-energy areas on the bed of the estuary. Such low-energy areas are frequently coincident with bed deposits of fine sediment. By the time these suspensions stagnate, their concentration has frequently risen beyond the range of optical turbidity meters and their internal concentration structure can be measured only by nuclear densimeters (Fig. 17.2). These stationary suspensions are also commonly layered. Such layering could be

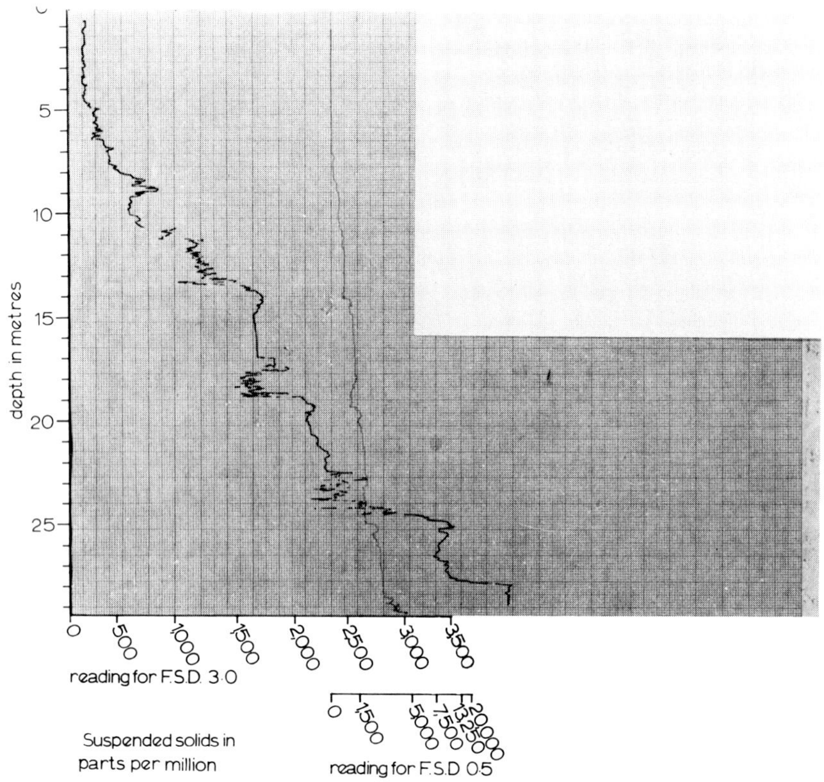

Figure 17.1. Continuous, vertical, electro-optical turbidity profile through mobile suspension showing layered structure. Two meters on different sensitivity settings are used to span the complete turbidity range.

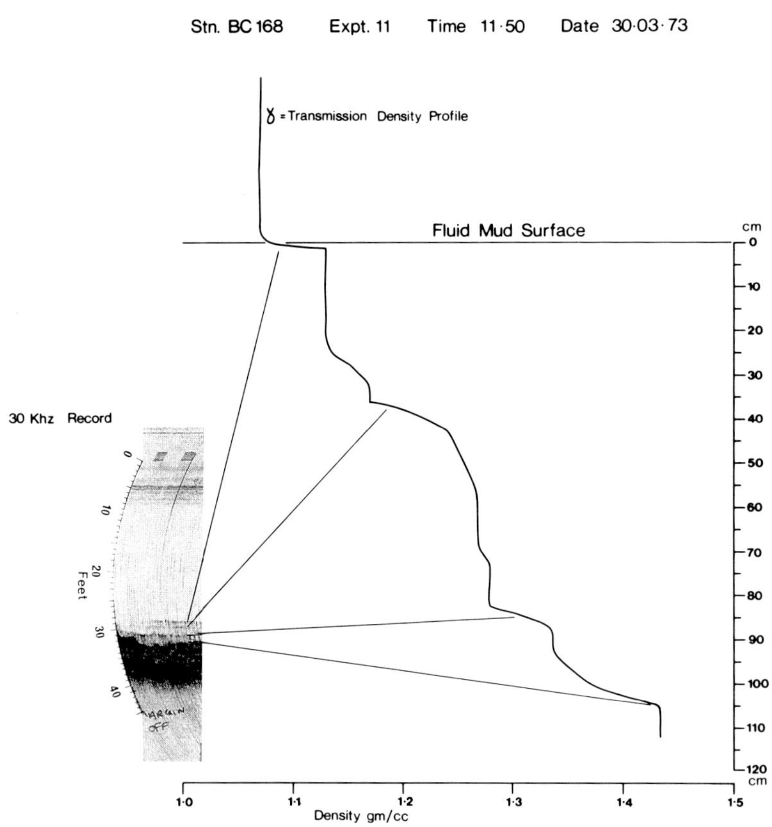

Figure 17.2. Continuous, vertical gamma-transmission profile through stationary suspension showing layered structure. Discontinuities in density correspond to reflectors on fathometer record.

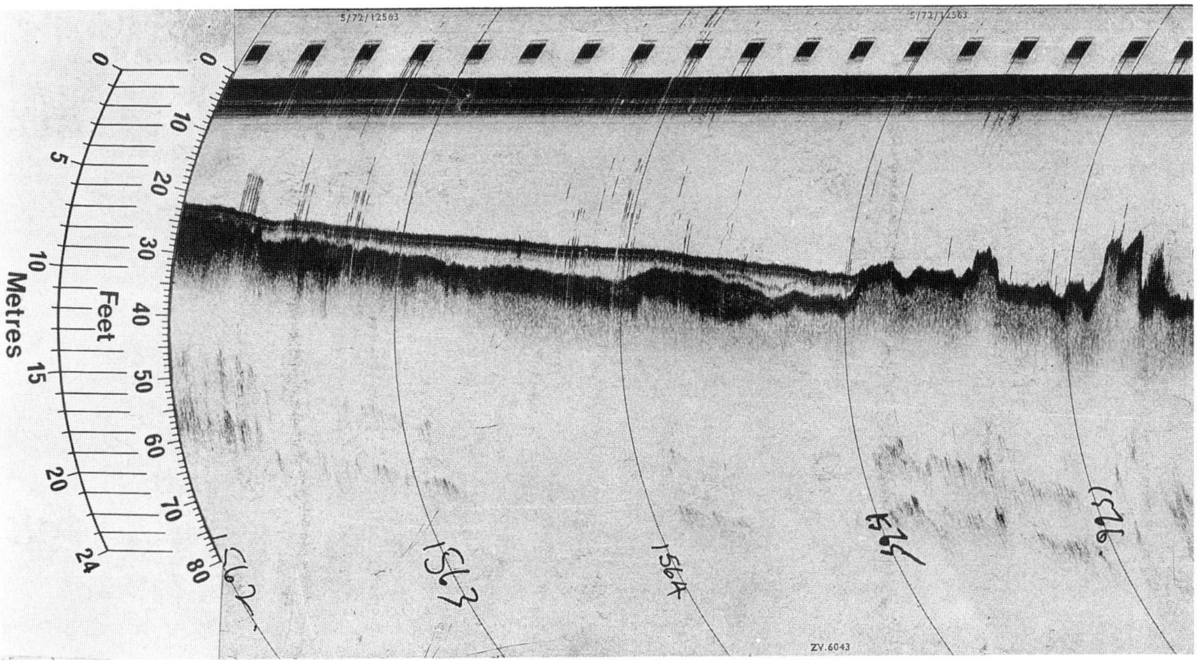

Figure 17.3. Thirty kilohertz echo sounder record showing multilayered stationary suspension overlying settled mud. Between fixes 1564 and 1565 a layer with transitional properties between the suspension and the underlying bed is present.

inherited from the mobile phase or could represent sequential stagnation of dense layers on more than one slack water.

As a general rule these dense layers are reentrained and dispersed on subsequent spring tides, but occasionally and in exceptional circumstances they may consolidate to form bed deposits. Evidence of a genetic link between dense stationary suspensions and settled bed deposits comes from two indirect sources. First, the suspensions frequently overlay mud beds known from radiochemical evidence to be experiencing deposition (Kirby and Parker, 1980). Second, the multilayered stationary suspensions at times show features that are transitional with the underlying bed deposits. Such features include a "draping" of the reflector marking the upper surface of the lowermost suspended layer immediately above the irregular settled mud topography (Fig. 17.3). This is believed to arise from consolidation of an originally more horizontal layer. In addition, this lower layer of the suspension may display the onset of "acoustic turbidity," which is more strongly developed in the underlying bed deposits. In this regime deposition is envisaged as a short-term process separated by longer periods (months or years) in which nondeposition or erosion occurs.

Thus, in this estuary the anatomy of the suspensions is well established. In no case has the precise link between a particular suspension and the bed been unequivocally established. Nevertheless the structure of the bed deposits is well known.

Settled Mud Deposits

The bed deposits have been investigated by remote geophysical techniques and by coring (Kirby and Parker, 1980). Some 40 cores ranging up to 3.5 m in length and with a diameter of 10 cm have been collected and studied. Both destructive and nondestructive tests have been performed on various suites of samples.

Cores for radiochemical analysis were subsampled and frozen on board ship with the result that sedimentological analyses could not be carried out on this suite of cores. Another suite of cores was returned to the laboratory sealed in transparent core liner tubes. The complete cores were first X-rayed in two planes at right angles to permit the true dipping plane of any inclined beds to be recognized. The cores were then halved along this direction.

The internal fabric is evident on two scales: first, the overall stratigraphy (scale of meters), and second, the small scale internal microfabric of the silty clay layers. (Scale of few centimeters to fractions of a millimeter.) Taking these in order, photographs of the halved cores (Fig. 17.4) show a distinctive interbedded thick silty clay and thin sand sequence. A characteristic feature is that there is a well-developed lithological change with depth. Whereas the deeper layers contain regular intercalations of relatively thick sand layers and lenses (0.5–4.0 cm), the thickness and frequency of these layers decrease toward the seabed, until

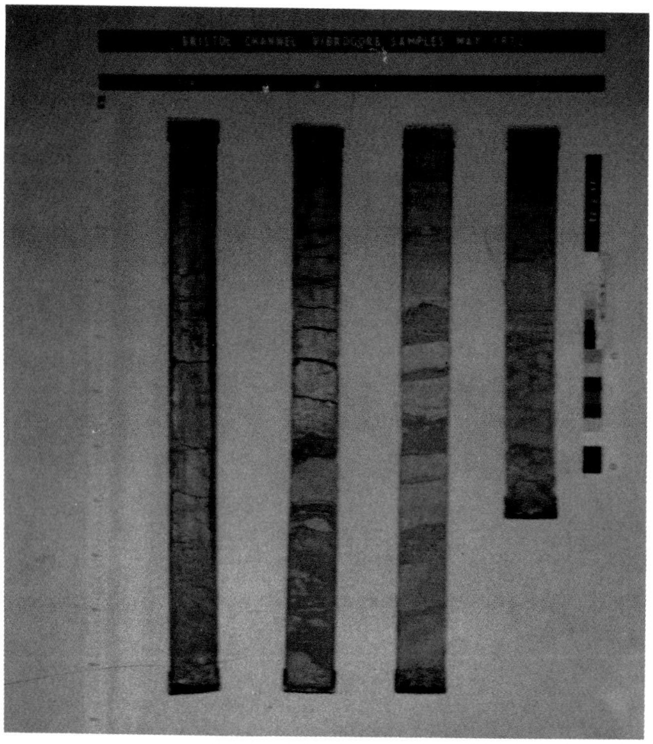

Figure 17.4. Halved section of core from settled mud area showing progressive increase in clay fraction toward seabed (at 93A, 0 cm). Thick clay zones appear structureless until X-rayed.

the top 1.0 m is commonly entirely silty clay, which has an undifferentiated appearance in the cut sections.

Apparently the amount of sand in the system has declined, or the amount of silty clay has increased, or both, over the last few hundred years. It has recently been established (Kirby, 1988) that long-term losses of sand from adjacent beaches and offshore have occurred. Moreover, the tidal mud flats of the estuary are also experiencing long-term erosion. Over the last 600 years at least, a sea wall built to protect the surrounding lowlands has excluded the derived fine sediment from its natural sink inshore. Fine sediment is therefore accumulating preferentially in the subtidal zone fronting the tidal flats.

The variation in thickness of the halved cores across their diameter presents practical problems to X-radiographic analysis. To investigate the internal structure and microfabric of the silty clay zones it has been necessary to cut thin slabs (3–5 mm) from the central axis of the cores. Such thin and constant-thickness slabs can be irradiated by X-radiography using a single uniform exposure setting to produce very high resolution photographs of the clay microstructure. This preparation technique, combined with the initial cutting along the true dip to avoid parallax effects, results in very high resolution X-radiographs of the primary clay microfabric. From X-radiographs it is immediately evident that the apparently structureless silty clays are in reality composed of thousands of layers of varying origin.

Figure 17.5. X-Radiograph of 3-mm-thick slab cut from maximum diameter of BC283 showing several cycles in which "massive" silty clays are overlain by sub-millimeter interbedded silt and clay layers. Lower zone shows reversing cross-bedding directions in silt layers. Diameter = 55 mm.

Figure 17.6. Magnified section of X-radiograph from basal zone of a "massive" silty clay showing "felty" texture and creep-cracks in underlying bed. Creep-crack depth = 0.1 mm.

The meter-long subsections show the primary laminae with only occasional and short-lived evidence of biogenic disturbance. The laminae range from horizontal to angles reaching 30°, often in one core. Magnetic remanence and magnetic property analysis techniques applied to the remaining half-core reveal that the inclined bedding is in all cases an original depositional feature and does not arise from later postdepositional slumping (Imienwanrin, 1989).

The thousands of laminae can be resolved in most cases into two types of silty clay microfabric:

1. massive silty clay units (1–12 cm thick) that are interbedded with
2. thinner (0.1–3 cm) interbedded submillimeter silt and clay units.

A prolonged study of all the 1-m-long X-ray films shows that where a complete sedimentary unit is preserved the submillimeter beds directly overlie the massive units and are apparently genetically linked to them. The features of these distinctive beds are as follows.

Massive Silty Clay Units (Fig. 17.5)

These horizons range from 1 cm to more than 10 cm in thickness. They may have sharply defined bases and tops but otherwise show gradational changes within the bed itself. The major textural features are first, a grading from a lighter toned, silty base to a darker, clay-rich top. A second distinctive feature is that the basal silty sections of such units frequently show a pronounced "felty" texture when examined at high magnification (Fig. 17.6). Such a texture could arise from crushed or sheared macroaggregates. This textural feature could be similar to the "plasma-fabric" described by Kuehl (Chapter 3).

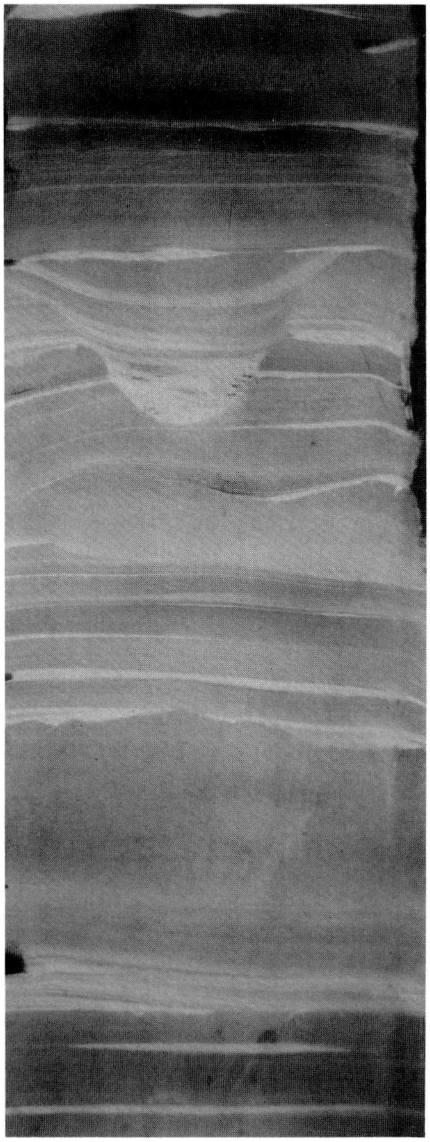

Figure 17.7. X-Radiograph of graded "massive" silty clay layer of BC74GC282 showing eroded top. Diameter = 58 mm.

Another feature occasionally present at the felty textured base of the massive beds is a zone of "pipe-like" features only 0.1 mm deep in the underlying bed (Fig. 17.6). When one of these zones was dissected out of a core the features proved to be a series of short, parallel, creep-cracks. This testifies to the fact that the underlying bed had already reached a consolidated, plastic condition when the overlying bed was deposited on top.

In many cases these massive beds have experienced partial reworking such that their upper surface is marked by a sharp erosion surface (Fig. 17.7).

No sequential sampling to compare the radiochemical composition down such a layer has yet been undertaken and attempts at assessing their remanent magnetic properties have been foiled

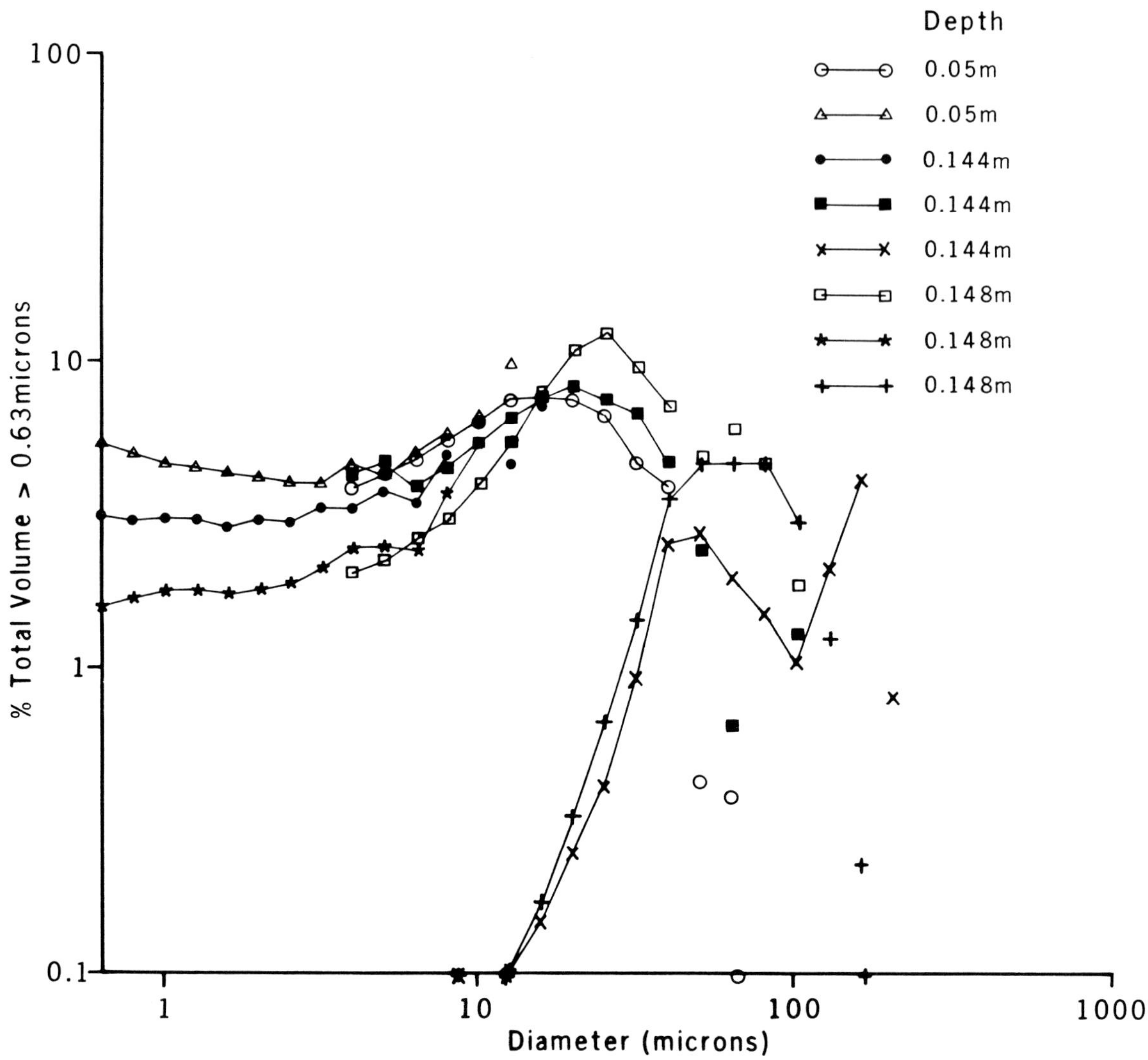

Figure 17.8. Examples of grain-size analyses of subsamples from "massive" silty clay layer (depth = 5.0 cm), submillimeter silt (depth = 14.8 cm), and submillimeter clay (depth = 14.4 cm) from core BC283.

for the moment by the lack of sensitivity of the instrument. However, some evidence that is believed to reflect on the origin of the layers is available from grain-size analysis of subsamples of a massive horizon guided by a true scale X-radiograph. Twelve subsamples (0–12 cm) from one massive layer in BC 283 (Fig. 17.5) were analyzed. The grain size was invariant with depth other than in the basal layer. The result of one analysis is shown in Figure 17.8. The mixed silty clay is believed to contain components of the whole spectrum of grain sizes present in the suspensions. Grain size data for this sample compared to samples from submillimeter layers are shown in Table 17.1.

Submillimeter Laminations (Fig. 17.5)

These horizons generally reach only 1–2 cm in thickness. It has not been possible to discover whether in their most complete form they contain a specific number of dark and light layers. The horizons consist of regular and parallel varve-like alternations of silt and clay laminae each of a similar thickness, commonly 0.1–1.00 mm. In the X-radiographs the light colored laminae are silt and the dark laminae are clay. In some cases the silt horizons may be cross-bedded with alternating horizons having foreset slopes at 180° to each other (Fig. 17.5). This suggests deposition from a reversing tidal current in a low concentration (< 500 mg/liter) regime.

The grain-size analysis of two individual layers dissected from the cores under the guidance of the X-radiographs is shown in Figure 17.8 and Table 17.1. It is apparent that the silt represents the same grain-size population as the coarser fraction of the massive layers, whereas the clay fraction may indicate the finer fraction of the massive layers. The same grain-size population as makes up the massive layers appears to have been fractionated into the submillimeter laminations.

Origin

The geographical proximity to accreting bed deposits and transitional properties of some dense stationary layers provide circumstantial evidence that they may evolve by consolidation into bed deposits of some kind. The massive graded silty clay units have a constant grain size and are interpreted here as being emplaced as a single unit. Their thickness and gradational properties suggest that they represent dewatered dense stationary suspensions that stagnated on a neap tide. If this is the case they represent instantaneous deposition of a mobile layer that has stagnated and become "frozen" or "set" at the bed. Dense layers initially 1.0 m deep will dewater to silty clay units only a few centimeters thick.

There is no evidence for "bleeding" of the coarser silt fraction due to shearing in the laminar sublayer as has been postulated for deep sea turbidites by Stow and Bowen (1978, 1980) and others. Indeed, the graded massive silty clay beds are more silty toward the base. Similarly the fractionated pure silt and pure clay submillimeter beds overlie, rather than underlie, the deposit. Had they developed by being fractionated from the sole of the turbid layer they would underlie the massive beds. The massive beds were deposited in a high concentration regime (> 500 mg/liter).

The submillimeter beds overlying the neap tide massive beds are presumed to arise from fractionation and reworking of the earlier deposits during phases of the tide leading up to springs. The cross-bedded submillimeter silts may represent deposits of a reversing tidal current capped on each occasion by a slack water clay deposit that settled vertically from a suspension of possibly < 500 mg/liter. The submillimeter beds are consequently envisaged as low concentration regime deposits.

The fabrics are not unlike the massive and laminated muds laid down by turbidites in the deep ocean (Chough and Hesse, 1980) Chough et al., 1984; Kranck, 1984).

Conclusion

High-quality data on suspension structures in an estuary that regularly cycles between high and low suspended solids concentrations have been obtained. A close link between high concentration stationary suspensions emplaced by advective transport and the muddy bed deposits developing from them was recognized. Photographs of cut cores show a similar stratigraphy of increasing clay fraction toward the seabed at most coring sites. Very high resolution X-radiographs of apparently structureless mud zones in the cores show two very distinctive clay microfabrics. In the most complete examples submillimeter silt and clay alternations overlay more massive, thicker, graded silty clay units. A tentative link between the advected dense, stationary suspensions, and the massive units is suggested. These are high concentration regime deposits formed at neap tides. In contrast, the overlying submillimeter silt and clay alternations are more likely to arise by conventional bottom traction of the silt layers and vertical slack water settling of the clay layers in a lower concentration regime either side of spring tides.

Table 17.1. Grain-size data from sample BC 283A compared with submillimeter layers.

	BC 283A			% < 63 µm		
	Depth (cm)	1 µm	5 µm	10 µm	20 µm	Peak µm
Massive layer	5.0	4.7	4.2	6.0	7.0	20.0
Submillimeter (fine)	14.4	3.0	3.7	5.1	8.0	20.0
Submillimeter (coarse)	14.8	1.8	2.2	4.0	11.0	25.0

Acknowledgments

The core sampling, photography, and X-radiography were undertaken by Mr. J. O. Malcolm. Grain-size analysis of the samples was undertaken by Dr. K. Kranck, Bedford Institute, Nova Scotia, Canada. The interpretations are my own.

References

Chough, S.K., and R. Hesse, 1980. The Northwest Atlantic Mid-Ocean Channel of the Labrador Sea: III. Head spill vs body spill deposits from turbidity currents on natural levees. Journal of Sedimentary Petrology, v. 50, p. 227–234.

Chough, S.K., G.H. Lee, B.K. Park, and S.W. Kim, 1984. Fine structures of turbidite and associated muds in the Ulleung (Tsushima) Basin, East Sea (Sea of Japan). Journal of Sedimentary Petrology, v. 54, p. 1212–1220.

Imienwanrin, A., 1989. Palaeomagnetism and magnetic fabric of recent sediments from the Severn Estuary system. Ph.D. Thesis, Southampton University, 406 p.

Kirby, R., 1986. Suspended fine cohesive sediment in the Severn Estuary and Inner Bristol Channel, UK. Report to Energy Technology Support Unit of the Department of Energy No ETSU-STP-4042, 243 p.

Kirby, R., 1988. The stability of intertidal mud flats and their relationship to long-term sources and sinks for fine sediment: with particular reference to the Severn Estuary and Bristol Channel. Unpublished report to Severn Tidal Power Group, 103 p.

Kirby, R., and W.R. Parker, 1980. Settled mud deposits in Bridgwater Bay, Bristol Channel, Institute of Oceanographic Sciences Report No. 107, 65 p.

Kirby, R., and W.R. Parker, 1983. The distribution and behavior of fine sediment in the Severn Estuary and inner Bristol Channel. Canadian Journal of Fisheries and Aquatic Science, v. 40, p. 83–95.

Kranck, K., 1984. Grain-size characteristics of turbidites. In: Stow, D.A.V., and D.J.W. Piper (eds.), Fine-Grained Sediments: Deep Water Processes and Facies. Geological Society, p. 83–92.

Kuehl, S.A., T.M. Harlu, M.W. Sanford, C.A. Nittrouer, and D.J. DeMaster, 1990. Millimeter-scale sedimentary structure of fine-grained sediments: Examples from continental margin environments. Microstructure of fine-grained sediments. Springer-Verlag, New York, Chap. 3.

Smith, T.J., 1982. The response of electro-optical turbidity meters to cohesive sediments. Institute of Oceanographic Sciences Report No. 137, 22 p.

Smith, T.J., and R. Kirby, 1989. Generation, stabilization and dissipation of layered fine sediment suspensions. Journal of Coastal Research, v. 5, p. 63–73.

Stow, D.A.V., and A.J. Bowen, 1978. Origin of lamination in deep-sea, fine-grained sediments. Nature, v. 274, p. 324–328.

Stow, D.A.V., and A.J. Bowen, 1980. A physical model for the transport and sorting of fine-grained sediment by turbidity currents. Sedimentology, v. 27, p. 31–46.

CHAPTER 18

Effect of Bed Shear Stresses on the Deposition and Strength of Deposited Cohesive Muds

Emmanuel Partheniades

Introduction

Fine-grain marine sediments consist predominantly of silt and clay of various mineralogical compositions. In almost all natural water environments these sediments flocculate as a result of their net surface physicochemical forces. In a flow field the basic settling and eroding unit is the floc rather than the individual particle. Flocs constitute the primary or first-order agglomeration of fine particles formed by random collision and attachment. The same process may generate second- and third-order agglomerations, or floc aggregates, ultimately forming a continuous aggregate network (Krone, 1963; Partheniades, 1965, 1986a, b, c).

Settling flocs and aggregates are subjected to relatively high stresses when they reach the near-bed zone of high fluid-flow velocity gradients. Flocs with sufficient physicochemical bonds to resist these stresses get attached to the bed permanently while the remaining flocs are broken up and reentrained back into the main flow. It is, therefore, expected that the structure and the fundamental mechanical properties of deposited cohesive beds will be affected by the range of bed shear stresses during deposition. This dependence and the fundamental processes of deposition and resuspension of cohesive sediments are discussed in this chapter.

Structure and Some Basic Properties of Flocs

Fine particles within the flocs can have an edge-to-edge, edge-to-face, and face-to-face attachment. Until recently the single particle was considered as the fundamental unit within the floc (Casagrande, 1932, 1940). In freshwater deposits the dominant fabric was believed to be more open and of the cardhouse type (edge-to-edge and edge-to-face), while in saltwater the clay structure was assumed to be more random with edge-to-edge, face-to-face, and edge-to-face attachements (Partheniades, 1986a, 1986b; Quinn, 1980). Recent electron microscopic studies, however (Quinn, 1980), revealed that in any clay fabric, the basic unit is not the single particle, but the "clay packet," that is a group of particles in a parallel array in the form of a book known as "turbostatic structure" or "multiple aggregate particle fabric." Flocs of kaolinite particles of 1–2 μm of settling diameter formed by settling in quiescent water were examined by a scanning electron microscope after being subjected to a freeze-dried process (Quinn, 1980). Regardless of clay concentration or electrolyte content, no "single particle behavior" was observed; instead, even in extremely dilute suspensions, the fundamental units were clay packets with only a small number of single clay particles. Flocs formed in freshwater displayed a somewhat more open structure than flocs formed in saltwater; however and contrary to early hypotheses (Casagrande, 1932, 1940; Quinn, 1980), edge-to-face and face-to-face flocculation took place in both fresh and salt water. Figures 18.1 and 18.2 show examples of saltwater and freshwater flocs formed at the lowest clay concentration (100 ppm) with both single particles and clay packets clearly discernible in both cases. Figure 18.3 and 18.4 show similar flocs formed at higher particulate concentrations. A higher density of packets and of number of individual platelets per packet is observed suggesting an increase of the overall floc density with particulate concentration. In general, flocs formed at higher particulate concentration suspensions were found to contain packets with higher number of single particle platelets and vice versa. Figures 18.5 and 18.6 show examples of floc aggregates formed in freshwater and saltwater, respectively with an interfloc joint magnified. A greater degree of randomness in the saltwater floc can be observed.

Because of the diversity of particles and packets and the wide variation of the interparticle physicochemical bonds, the density,

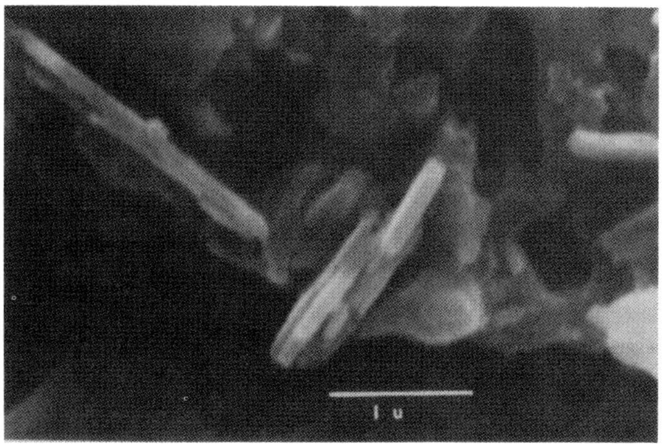

Figure 18.1. Salt water kaolinite floc at $C = 100$ ppm (Quinn, 1980).

Figure 18.2. Fresh water kaolinite floc at $C = 100$ ppm (Quinn, 1980).

size, and average strength of flocs formed by a random flocculation process in a turbulent flow field are expected to vary within wide limits. Krone (1963) estimated the floc density for seven clay muds between 1.44 and 1.64 g/cm^3, which corresponds to a volume fraction of solid particles between 0.084 and 0.253 and to a water volume between 75 and 92% of the total floc volume. Within a given flow regime the floc size is limited by the shear stresses exerted by the fluid on the floc surface. The floc shear strength depends on the magnitude of the interparticle physicochemical bonds, which, in turn, are functions of the clay mineralogy and the water chemistry. Krone derived the following relationship for montmorillonitic clays sampled from the San Francisco bay and tested in a laminar-flow shear field (Krone 1963):

$$\tau r = \frac{16\Delta r}{3\pi} \tau_{\max} \tag{1}$$

where τ is the uniform shear stress in the laminar flow field, r is the aggregate radius, Δr is the radius of the contact area during collision, and τ_{max} is the shear strength of the floc or aggregate. The laminar flow shear field was generated by rotating a cylinder inside another fixed cylinder of a slightly larger diameter while τ was evaluated from the torque necessary to sustain a particular angular velocity. From magnified photographs of settling flocs and from measured shear stress values, τ_{max} was estimated to be about 0.27 N/m^2. Thus, for constant τ_{max} and given τ the floc radius r can be obtained from Eq. (1). The latter has been experimentally verified down to $\tau = 0.006$ N/m^2 (Krone, 1963). Below that shear stress a rapid and erratic behavior was observed, apparently due to the formation of higher order floc aggregates.

The interparticle cohesive bonds are expected to vary statistically (Partheniades, 1965); therefore, the strength of the flocs and aggregates is also expected to vary randomly. If the controlling near-bed shear stress is constant, the maximum value of r will also vary randomly, according to Eq. (1). In an open channel fluid flow, flocs and floc aggregates will eventually approach the near-bed zone, where the shear stress, τ, attains its maximum value. If

$$\tau_{\max} > \frac{3\pi\tau r}{16\Delta r} \tag{2}$$

the floc or aggregate will stick to the bed developing permanent physicochemical bonds with it; otherwise it will be broken up and reentrained into the main flow by turbulence.

A turbulent flow field is characterized by two zones: an upper zone of low shear stresses and near homogeneous turbulence, and a lower near-bed zone of high shear stresses and nonhomogeneous turbulence. The random motion of the upper zone leads to interparticle and interfloc collision and to the formation of flocs and floc aggregates. The maximum instantaneous local shear stresses in the second zone determine, according to Eq. (2), the percentage of those aggregates that will eventually reach the bed and will become part of it. This hydrodynamic interaction between flow, flocs, and bed is described elsewhere (Partheniades, 1965, 1977, 1986a, b, c). Experiments in an open flume with a montmorillonitic mud bed similar to that used by Krone (1963) and with a macroscopic shear strength of about 960 N/m^2 showed that erosion started under a bed shear stress of about 0.10 N/m^2 and became substantial under a shear stress of about 1.4 N/m^2. Thus, the range of eroding shear stresses is of the

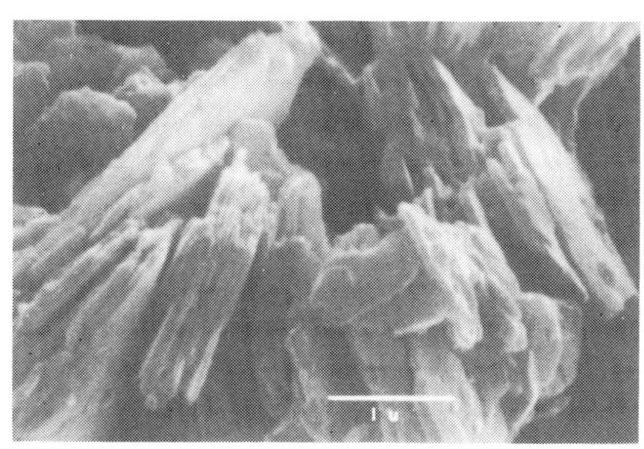

Figure 18.3. Salt water kaolinite floc at $C = 400$ ppm (Quinn, 1980).

Figure 18.4. Fresh water kaolinite floc at C = 400 ppm (Quinn, 1980).

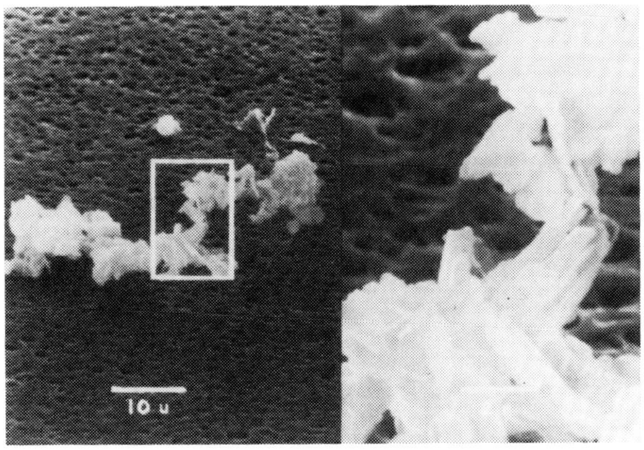

Figure 18.5. Freshwater floc aggregate at C = 500 ppm (Quinn, 1980).

order of the floc strength of 0.27 N/m² (Krone, 1962) rather than of the macroscopic shear strength of the bed. The explanation for this together with a mathematical model of erosion and deposition of cohesive beds is given in Partheniades (1965, 1977) and in a more abbreviated form in Partheniades (1986a,b). This mathematical model and the derived relationships hold for surface erosion only, that is, for erosion taking place by removal of individual particles and/or flocs. This kind of erosion takes place as long as the flow-induced mass shear stresses do not exceed the macroscopic shear strength of the bed. Otherwise, mass erosion occurs by removal of finite soil masses. Under surface erosion conditions, therefore, the interparticle or interfloc forces of the surficial layer control erodibility regardless of the density of the space below that layer.

The Process of Deposition

The depositional properties of suspended cohesive sediments were investigated by Partheniades and co-workers in a special apparatus pictured in Figure 18.7 and outlined schematically in Figure 18.8 (Mehta and Partheniades, 1973, 1975, 1979, 1982; Partheniades, 1986a, b, c; Partheniades and Kennedy, 1966; Partheniades et al., 1968). Its main components are an annular channel containing the water–sediment mixture and an annular ring of slightly smaller width positioned within the channel and in contact with the water surface. The latest version of the annular channel has an average diameter of 60 in. (152.5 cm), a width of 4 in. (10.2 cm), and a depth of 18 in. (45.7 cm). The ring can move vertically within the channel so that the depth of flow can vary to a maximum of 45.0 cm. The system was constructed in 1969 in the Coastal and Oceanographic Engineering Laboratory of the University of Florida. A simultaneous rotation of the two components in opposite directions generates a uniform turbulent flow field free of floc disrupting elements, such as pump blades, return pipes, and diffusers, which have to be present in conventional laboratory flumes. The effect of the rotation-induced secondary currents on the deposition has been eliminated by a proper selection of the speeds of the channel and of the ring so that the sediment deposits uniformly across the width of the channel. These speeds are known as "operational speeds" and can be obtained graphically from a family of operational curves (Mehta and Partheniades, 1973, 1975). At these operational speeds the bed shear stress across the channel measured by a Preston tube was found to be uniform (Partheniades et al., 1968). The average bed shear stress can be directly evaluated either from the total torque on the ring or from an instrumented false bottom. The details of the apparatus are described in Mehta and Partheniades (1973). The general procedure was to have the sediment first completely suspended under the highest possible relative speed of the two components to a uniform initial concentration C_0. The speeds were then adjusted according to the operational curves, so that a predetermined bed shear stress could be obtained. The suspended sediment concentrations were subsequently recorded at specific times.

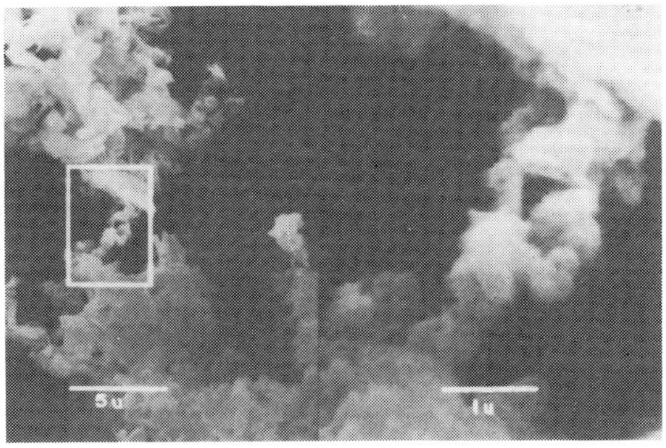

Figure 18.6. Salt water floc aggregate at C = 500 ppm (Quinn, 1980).

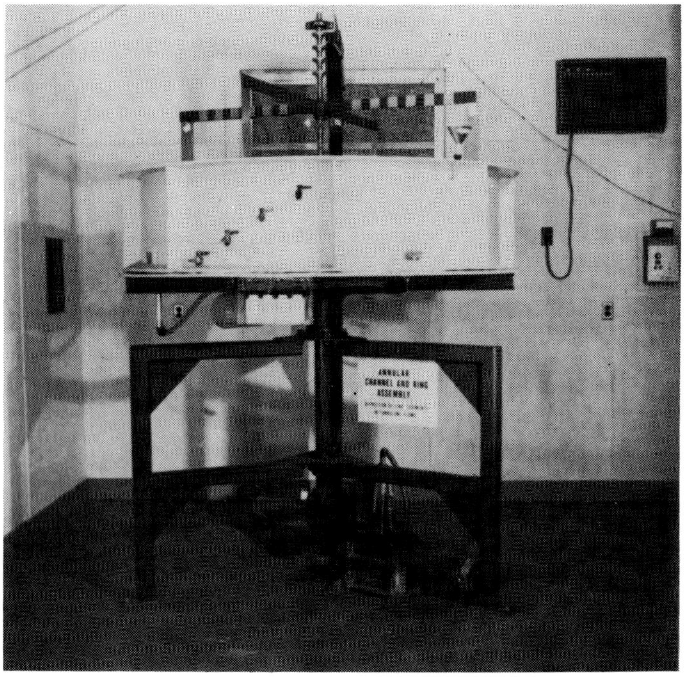

Figure 18.7. Annular channel and ring assembly (Mehta and Partheniades, 1973, 1975).

It was first observed that after a period of transition, the concentration reached a steady-state value, C_{eq}, hereby defined as "equilibrium concentration." The ratio $C^*_{eq} = C_{eq}/C_0$ was found to be independent of C_0 for values of the latter to about 25,000 ppm and to depend on the bed shear stress, τ_b, only. C^*_{eq} represents the degree of retention, which is that portion of the total sediment originally in suspension that forms flocs too weak [$\tau_{max} < 3\pi\tau_b r/16\Delta r$ according to Eq. (2)] to resist the disruptive bed shear stresses. As a result, that part of sediment cannot possibly deposit permanently on the bed; instead, it is broken up and reentrained into the main flow. Likewise, the variable $C^{**}_{eq} = 1 - C^*_{eq}$ represents that part of the total sediment that deposits permanently on the bed and it is termed as the "degree of deposition" being also a function of τ_b.

Figure 18.9 shows an arithmetic plot of C^*_{eq} *versus* τ_b for kaolinite clay in distilled water for different depths. The scatter of the data appears to be random thus defining an average curve intersecting the abscissa at a value indicated as τ_{bmin}. The latter represents that value of τ_b below which no sediment can be maintained in suspension. It constitutes a very important parameter representing the depositional properties of cohesive sediments. In Figure 18.9 the value of τ_{bmin} is about 2 dyn/cm² = 0.2 N/m². Partial deposition can, however, occur to a measurable degree for values of τ_b between 0.2 and 1.10 N/m².

To formulate an analytical relationship for C^*_{eq} in terms of the relevant flow variables and physicochemical properties of the sediment–water system, after several trials, C^*_{eq} was plotted versus the dimensionless parameter $\tau^*_b - 1$, where $\tau^*_b = \tau_b/\tau_{bmin}$. The parameter $\tau^*_b - 1$ can be interpreted as a dimensionless excess bed shear stress. A plot of the data of Figure 18.9 against $\tau^*_b - 1$ on a logarithmic-normal paper in Figure 18.10 gave a straight line suggesting a logarithmic-normal relationship (Mehta and Partheniades, 1973, 1975; Partheniades, 1986a). It was subsequently found that various sediment–water systems when plotted in the form of Figure 18.10 gave parallel straight lines.

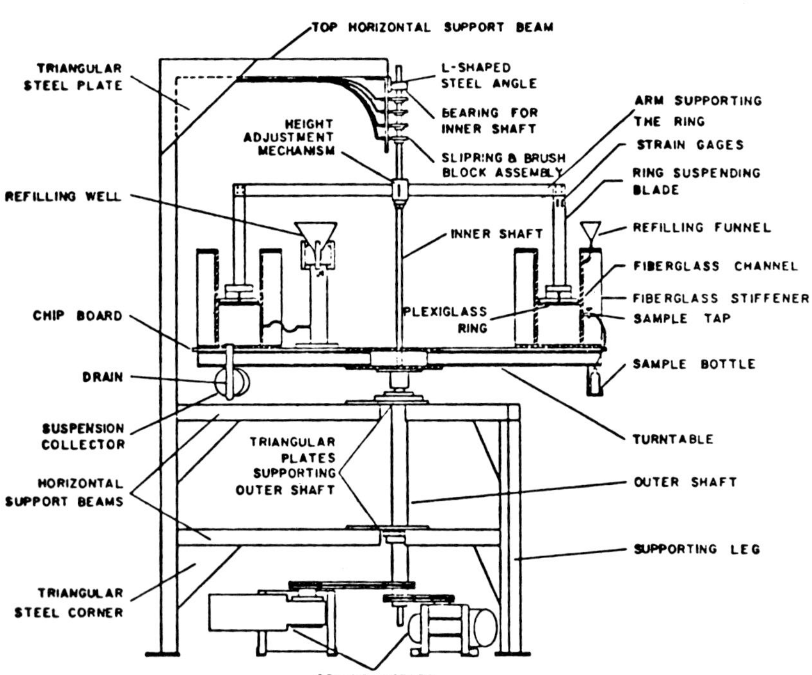

Figure 18.8. Schematic outline of annular channel and ring assembly (Mehta and Partheniades, 1973, 1975).

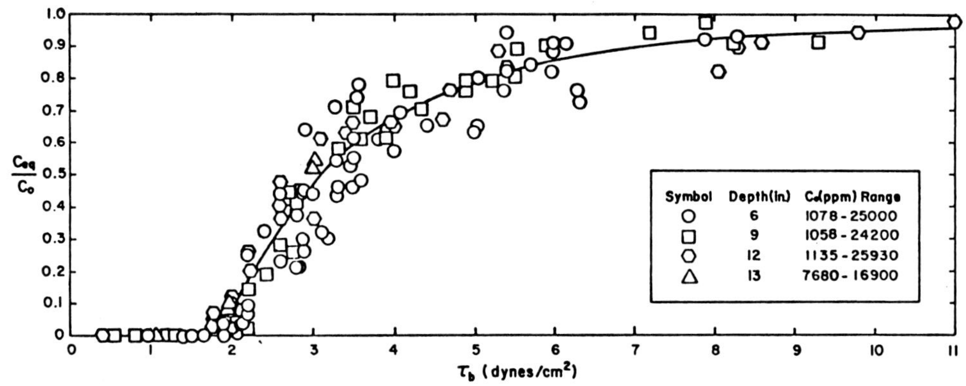

Figure 18.9. Variation of relative equilibrium concentration with bed shear stress (Mehta and Partheniades, 1973, 1975; Partheniades, 1986a, b).

A generalized law can be formulated by plotting C^*_{eq} versus the parameter $(\tau^*_b - 1)/(\tau^*_b - 1)_{50}$ where $(\tau^*_b - 1)_{50}$ is the value of $\tau^*_b - 1$ at which 50% of the total sediment deposits. The results are shown in Figure 18.11. Series A and B are for kaolinite in water at ocean salinity and for depths of 6 in. (15.3 cm) and 9 in. (22.9 cm), respectively. Series C is for a mixture of equal parts of kaolinite and montmorillonite mud from the San Francisco Bay in water at ocean salinity and series D is for San Francisco Bay mud only in water at ocean salinity. The plot leads to the equation

$$C^*_{eq} = \frac{1}{\sqrt{2\pi}} \int_0^Y \exp\left(-\frac{\omega^2}{2}\right) d\omega \tag{3}$$

where

$$Y = \frac{\log[(\tau^*_b - 1)/(\tau^*_b - 1)_{50}]}{\sigma_y} \tag{4}$$

and σ_y is the geometric standard deviation of the distribution. The law, moreover, agrees very closely with earlier results of Partheniades and Kennedy (1966) and Partheniades et al. (1968), with data from open flume studies using San Francisco Bay mud in water at ocean salinity (Partheniades, 1965, 1986a, b) and with open flume data with mud from the Maracaibo estuary in Venezuela (Rosillon and Volkenborn, 1964).

The parameter $(\tau^*_b - 1)_{50}$ was found to be closely related to τ_{bmin} (Mehta and Partheniades, 1973, 1975; Partheniades, 1986a, b, c) so that τ_{bmin} becomes a variable that reflects the effect of the physicochemical properties of the sediment–water system on the average floc strength and on the degree of deposition as represented by C^*_{eq}.

The constancy of C^*_{eq} for constant bed shear stress τ_b leads to the conclusion that a given bed shear stress divides the entire floc population into those with sufficiently high strength to resist the near bed shear stresses and those with insufficient strength. The first floc group reaches the bed, develops physicochemical bonds with it, and remains there; the second group of flocs is broken up and reentrained into the main flow. Therefore, the near-bed shear field determines the degree of deposi-

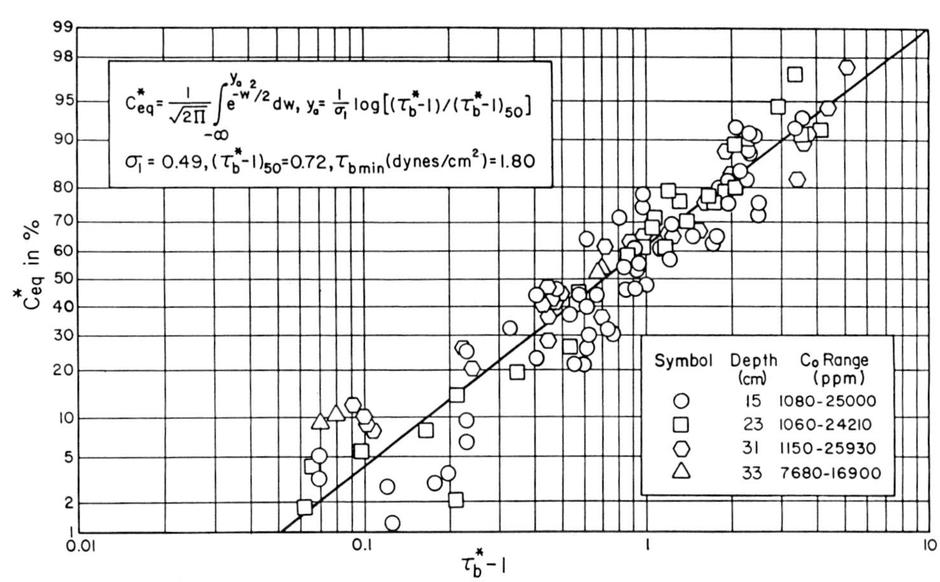

Figure 18.10. Relative equilibrium concentration versus bed shear stress parameter, $\tau^*_b - 1$ (Mehta and Partheniades, 1973, 1975; Partheniades, 1986a, b).

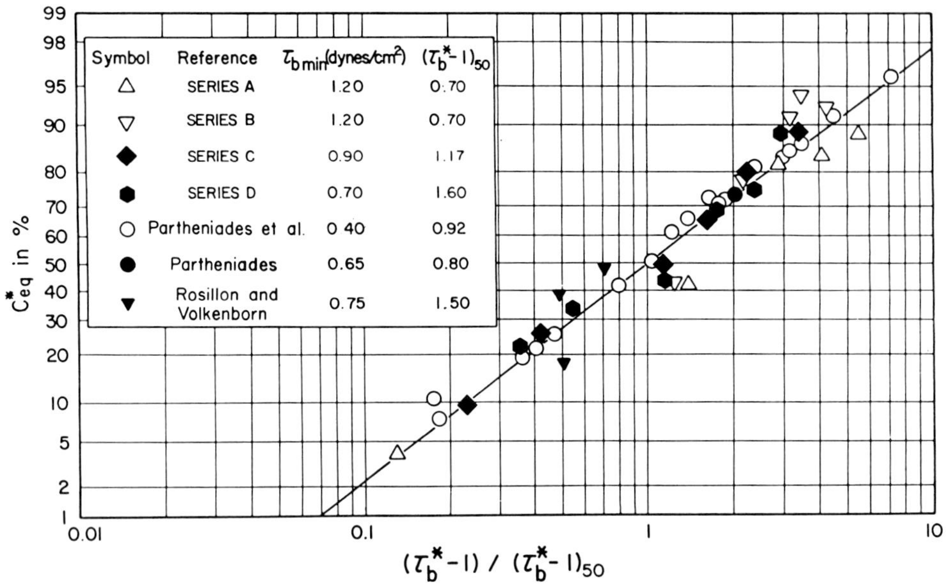

Figure 18.11. Relative equilibrium concentration versus $(\tau_b^* - 1)/(\tau_b^* - 1)_{50}$ for various sediment suspensions (Mehta and Partheniades, 1973, 1975, 1979, 1982; Partheniades, 1986a, b, c).

tion while the far-bed near-homogenous field contributes only to the formation of flocs and floc aggregates. It can reasonably be speculated that the floc strength, τ_{bmin}, increases with both its overall density and the size of clay packets since a higher number of clay platelets means more joints to break and heavier units to move (Partheniades, 1986b, c; Quinn, 1980). It follows that the density and erosive strength of beds formed by deposition of suspended fines of similar mineralogical composition increase with increasing bed shear stress during deposition. A stratified cohesive bed with respect to density and strength is thus developed whose resistance to erosion is expected to increase with depth of erosion below its original surface.

In a dispersive flow system, such as in an open channel, the properties of the deposited mud are expected to vary along the channel since the larger, denser, and stronger flocs deposit first. To investigate this variation, kaolinite clay was deposited in a 100-m-long straight flume under steady flow conditions and under various bed shear stresses, τ_0. The flume was divided into six different sections of about equal length defined as "reaches." Samples of deposited clay from each one of these reaches were tested in the annular channel-ring system (Dixit et al., 1982; Mehta et al., 1982). The results are shown in Figure 18.12. It is clearly observed that (1) in the same reach, τ_{bmin}, which is a

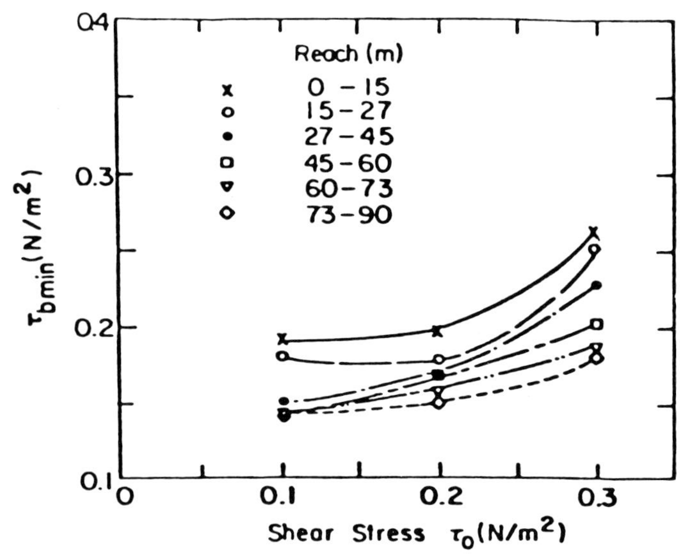

Figure 18.12. Variation of τ_{bmin} with shear stress, τ_0, during deposition in an open flume (Dixit et al., 1982; Mehta et al., 1982).

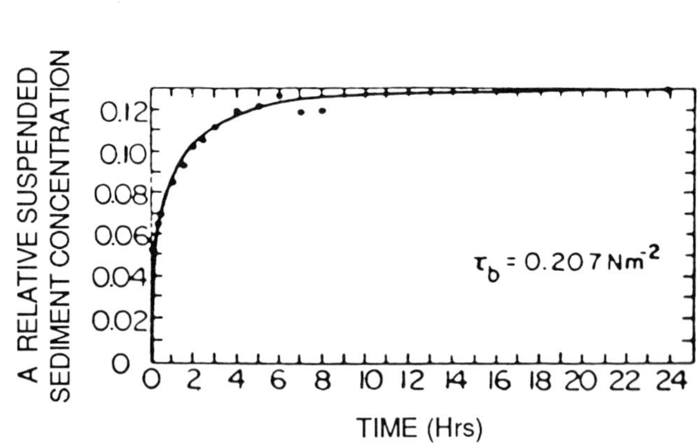

Figure 18.13. Relative suspended sediment concentration from erosion of stratified kaolinite bed (Mehta and Partheniades, 1979, 1982; Partheniades, 1986a, b, c).

CHAPTER 19

Fluidization of Soft Estuarine Mud by Waves

Mark A. Ross and Ashish J. Mehta

Introduction

The interaction between waves and soft muddy bottoms, a key process in governing estuarine and lacustrine cohesive sediment transport, is at present not well understood. What is quite well known, however, is that waves are significantly important in generating fluid mud, a high concentration near-bed slurry, which thereby becomes potentially available for transport by tidal currents. It follows that the precise mechanism by which fluid mud is formed by wave action over cohesive, porous solid beds is of evident interest in understanding and interpreting the microfabric of flow-deposited fine sediments in shallow waters. Results from preliminary laboratory tests described using known soil mechanical principles shed some light along these lines, and suggest that the fluidization process may be even more significant in generating potentially transportable sediment than previously realized.

Physical Considerations

Evidence of entrainment of mud by wave action over mud flats has been well documented, among others, by Kemp and Wells (1987), as illustrated in Figure 19.1. Of the four instantaneous (turbulence-mean), vertical concentration profiles for suspended sediment, one represents a prefrontal condition, two were obtained during frontal passage, and the last is postfrontal. The data were obtained during a 3-day period at a site on the eastern margin of the Louisiana chernier plain where the tidal range is less than 0.5 m. Wave height during the winter cold front passage was on the order of 13 cm and period 7 sec. Of particular interest is the development of a near-bed, high concentration layer by the frontal wind-generated waves (profiles 2 and 3), which was previously absent (profile 1). The postfrontal profile 4 further suggests that this layer may have persisted following the front, conceivably due to the typically low rate at which such a layer dewaters. The suspension concentration in the upper water column was higher following the front than that during the front, possibly due to advection from a neighboring area of higher turbidity.

The observed evolution of the suspension concentration profile during the front is characteristic of what occurs when progressive waves interact with a soft mud bed. In Figure 19.2, results based on a laboratory flume test on the resuspension of a soft estuarine mud bed by progressive waves are shown. The test conditions were similar to those in Run 1 mentioned later. Also shown are calculated profiles based on a solution of the vertical sediment mass transport equation for the particular test conditions. Details of this modeling effort have been given elsewhere (Ross, 1988). It suffices to note that the evolution of the concentration profile over the 4-hr test duration has evident similarities with the prototype data of Figure 19.1 during frontal passage. Near the bottom, due to entrainment of the mud–water interface resulting from the effects of hindered gravitational settling and upward, flow-induced diffusion strongly stabilized by the negative buoyancy of the high concentration mud, a significant concentration gradient, or lutocline, is observed to have developed. The height at which the lutocline stabilizes is largely determined by a balance between the rate of turbulent kinetic energy input and the buoyancy flux determined by the sediment settling rate (Hopfinger and Linden, 1982). In the water column above the lutocline diffusion due to the wave field was characteristically slow, so that the concentration increased to modest levels only.

For the formation of a lutocline the cohesive bed, essentially a porous solid, must first become fluidized. Fluidization of the cohesive bed, accompanied by measurable degradation in soil

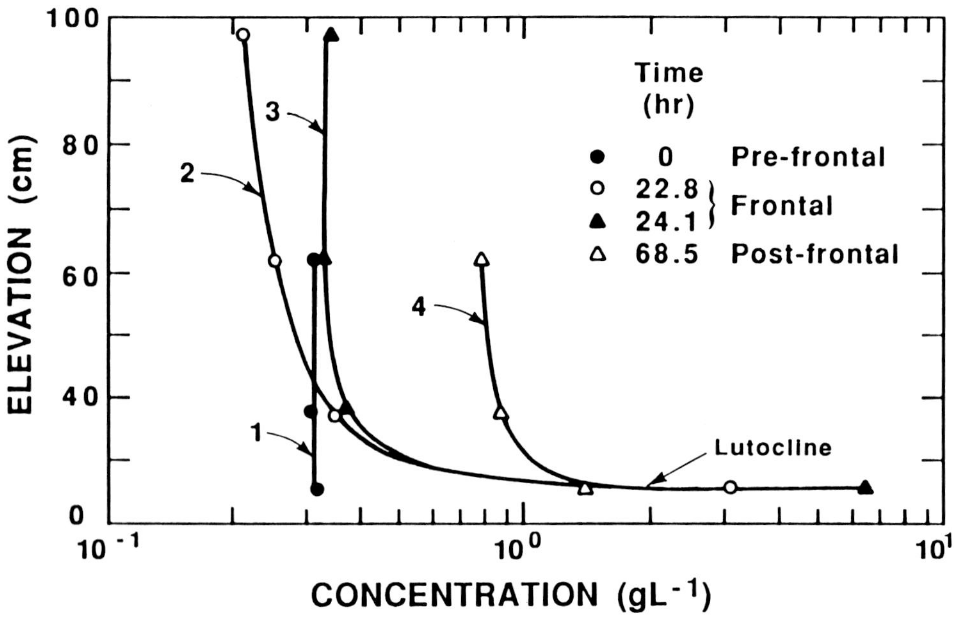

Figure 19.1. Vertical suspended sediment concentration profiles obtained before, during, and after the passage of a winter cold front at a coastal site in Louisiana. Times are relative to time of measurement of the prefrontal profile 1. (Adapted from Kemp and Wells, 1987.)

geotechnical properties, essentially proceeds via "shaking" and "pumping" (Liu et al., 1986; Pamukcu and Suhayda, 1987; Ross, 1988). Shaking refers to the action of oscillatory shear stress at the bed surface, which results in a visoelastic mud motion and eventual failure of the solid structure (Maa, 1986). Pumping relates to the buildup of excess pore pressure due to the dynamic pressure gradients under waves, which likewise breaks up the structural integrity of the solid mud matrix.

The process of bed fluidization can be examined by measuring the effective stress, since the cohesive bed is characterized by the occurrence of a measurable effective stress, while the overlying fluid has practically none (Parker, 1986). Figure 19.3 gives relevant definitions. In a two-phased system consisting of a deformable mineral skeleton filled with an incompressible fluid, the effective stress, σ', at any point is the difference between the total vertical stress, σ, and the pore water pressure, u_w, i.e.,

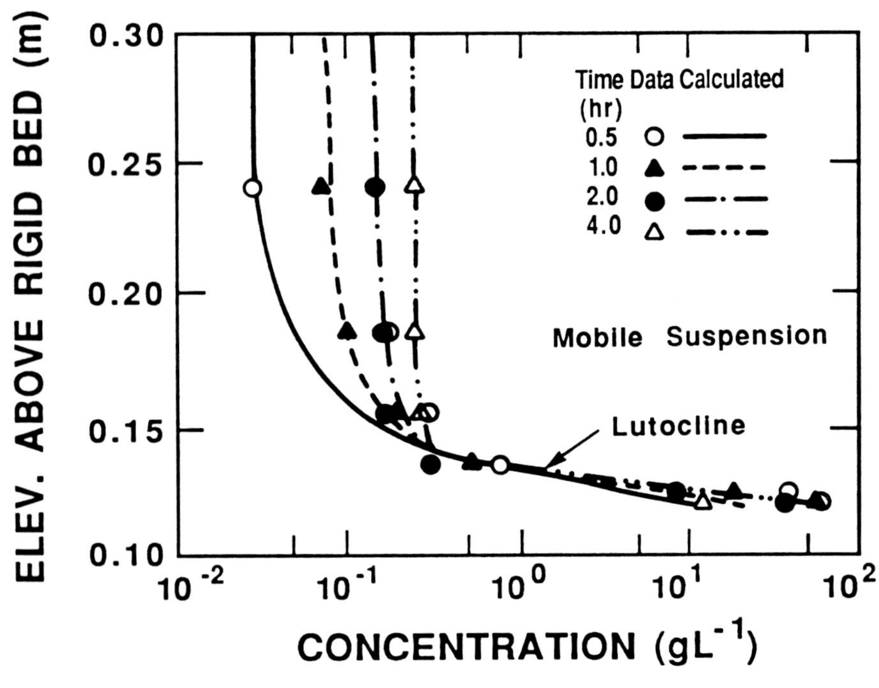

Figure 19.2. Vertical suspended sediment concentration profiles from a wave resuspension test in a laboratory flume.

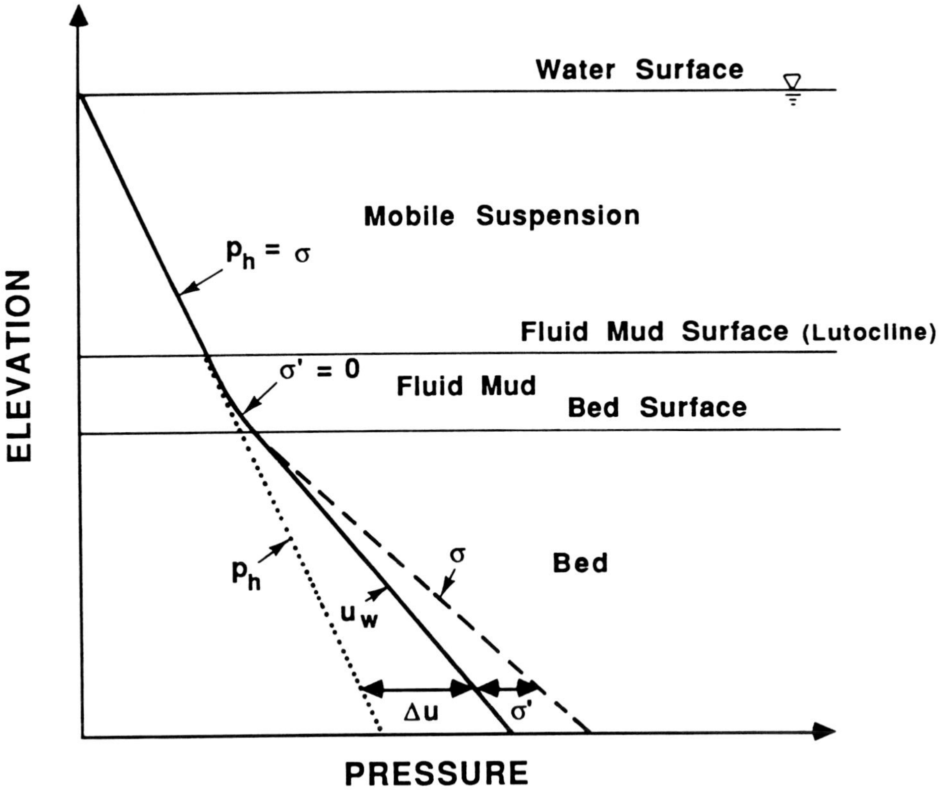

Figure 19.3. Definition diagram for the cohesive bed surface and fluid mud.

$$\sigma' = \sigma - u_w \tag{1}$$

Under dynamic conditions, the actual pore pressure may be higher than the hydrostatic pressure, p_h, by an amount Δu, the excess pore pressure. In this case if the sum of p_h and Δu approaches σ, liquefaction occurs (Perloff and Baron, 1976). The liquified material is typically much more amenable to transport by hydrodynamic forcing such as due to tidal currents, than is the bed.

With reference to Figure 19.3, the zero effective stress plane defines the bed surface. Above the fluid mud surface, or the lutocline, the hydrostatic pressure practically equals the total stress, since sediment concentration is relatively low in the mobile suspension. In the following sections, results from two experimental runs in a laboratory flume to track these two surfaces during wave action over the mud bed are described.

Experiments

The plexiglass wave flume used was 20 m long, 0.5 m wide, and 0.45 m high. Progressive, nonbreaking waves were generated at one end by a plunging-type wavemaker (Ross, 1988). A sloping, wave absorbing beach was installed at the other end to minimize wave reflection. The middle 8 m long reach, depressed 15 cm with sloping ends, contained the mud. A predominantly montmorillonitic mud of 3 μm dispersed particle (median) size from Tampa Bay, Florida was used. The mean slurry concentration was about 150 g/liter.

Total pressure and pore pressure in the mud were measured with Druck pressure gauges (models PDCR135/A/W and PDCR81, respectively). Pairs of these gauges were flush-mounted along the flume side wall at approximate elevations of 3, 5, 7, and 9 cm above the rigid flume bottom. Details of gauge calibration and data acquisition procedures have been reported by Ross (1988). The same reference may be consulted for measurement procedures for mud density and water surface profiles. The beds were prepared by introducing a mud slurry of known approximate density in the test reach, and then filling the flume with tap water to the desired level 7 days prior to the test. There was sufficient salt in the ambient fluid, in spite of tap water addition, to maintain the sediment in an aggregated state.

Figure 19.4 gives illustrative pressure profiles from Run 2 (water depth 32 cm, mud depth 13 cm, wave height 7 cm, period 1 sec). The reference hydrostatic pressure (p_h) line is obtained by integrating the density profile of the suspension from the free surface down to the "visual" mud surface, defined here as the lutocline, and, assuming constant density pore fluid (suspension), further down to the rigid bed. This would be the pore pressure if the sediment were composed of coarse, porous material. The total

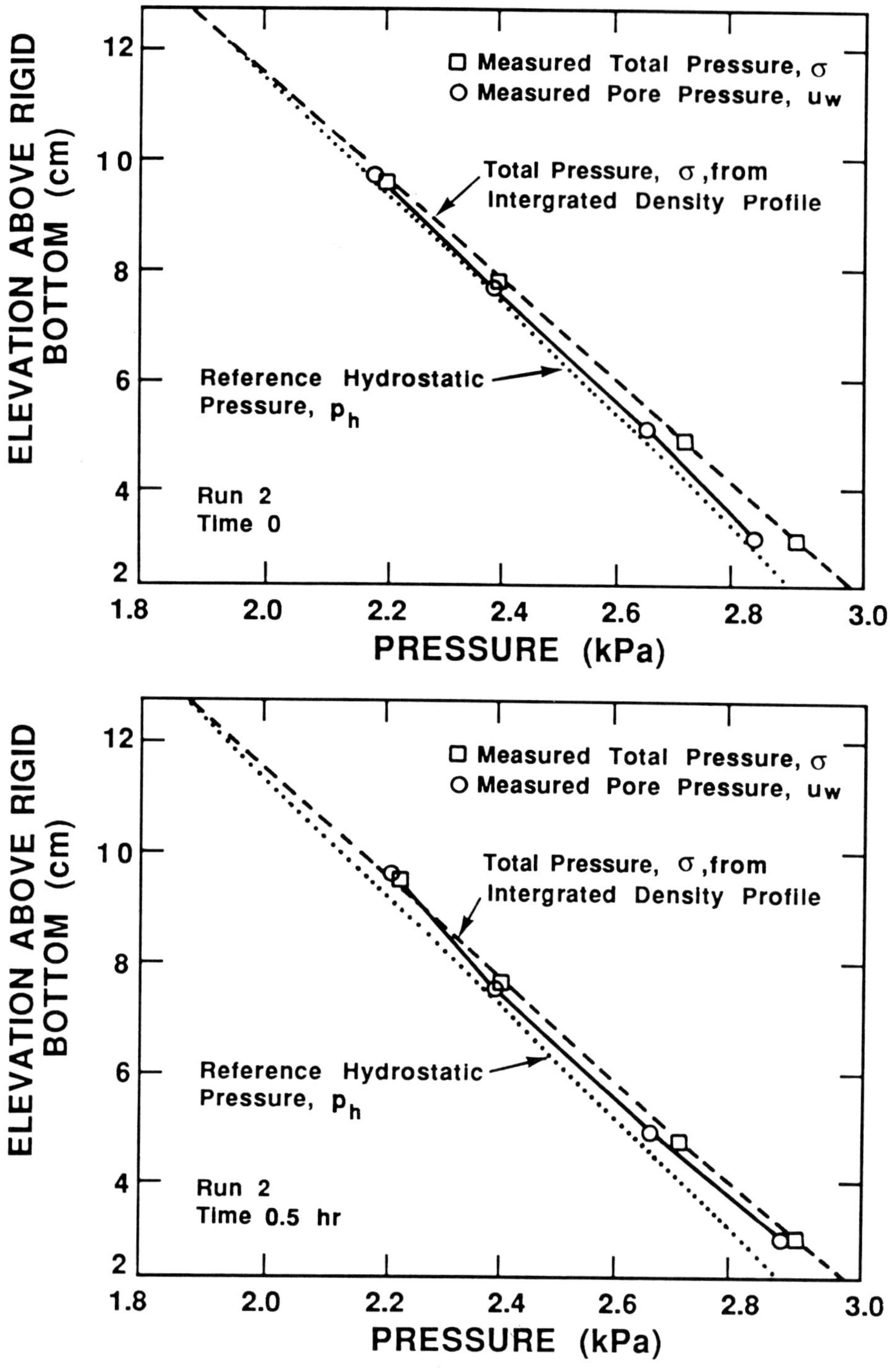

Figure 19.4. Total pressures and pore pressures in the mud from Run 2. (a) At the start. (b) After 30 min.

pressure, σ, at any level, z, is obtained by integrating $\rho(z)gdz$ from the free surface to any level z, where ρ is the bulk density of the mud–water mixture and g is acceleration due to gravity.

Examples of sediment dry mass concentration profiles below the mud surface from Run 1 (water depth 31 cm, mud depth 12 cm, wave height 6 cm, period 1 sec) at different times are shown in Figure 19.5. These can be converted easily to the corresponding bulk density profiles, given the sediment granular density of 2.65 g cm^{-3}.

In Figure 19.4a, which corresponds to the initial condition with practically no sediment in suspension above the mud–water interface, measured total pressures agree fairly well with the line obtained by integration of the bulk density profile. On the other hand, the measured pore pressures are slightly higher than the corresponding hydrostatic pressures. This suggests the presence of unreleased excess pore pressure even after 7 days of bed consolidation.

The cohesive bed surface was assumed to be represented by the 10^{-3} kPa effective stress level, a very small value, rather than the zero effective stress level, which was found to be difficult to identify. The 10^{-3} kPa level was at 9.2 cm elevation at the start (Fig. 19.4a). In Figure 19.4b it is seen, however, that through the buildup of pore pressure due to wave action after 30 min, the bed level dropped to 7.8 cm.

Based on these types of measurements, the development of the fluid mud layer with time is shown in Figures 19.6a and b for Runs 1 and 2, respectively. In both cases, there was an overall trend of dropping bed surface with continued wave action. Furthermore, in both cases, the mud–water interface exhibited a trend of initial entrainment followed by stabilization in the first 2 hr. In Run 2, which was continued for an additional 6-hr period, the mud surface dropped subsequently, mainly due to downstream advection of the fluidized material (Ross, 1988). As noted earlier, from the start the mud–water interface did not coincide with the bed surface, due to the presence of an underconsolidated layer of fluid mud (1.1 cm thick in Run 1 and 2.4 cm in Run 2).

Conclusions

Fluid Mud Evolution

The described tests demonstrate that the entrainment of the lutocline during wave-induced erosion (Figs. 19.1, 19.2) in itself does not provide a clue relative to the degree of availability of fluidized mud by a lowering of the bed level. Without the latter evidence, in fact, there may be a tendency to grossly underestimate the mass of fluid mud generated, based on observations of lutocline evolution alone. Related to this is the fact that, for example, in Run 1 fluid mud thickness increased from 1.1 to 5.3 cm. However, as the mud density (concentration) also increased with depth (e.g., Fig. 19.5), the mass of sediment that became fluidized increased quite significantly, from 0.17 kg/cm^2 of bed area at the beginning, to 1 kg/cm^2 after 2 hr.

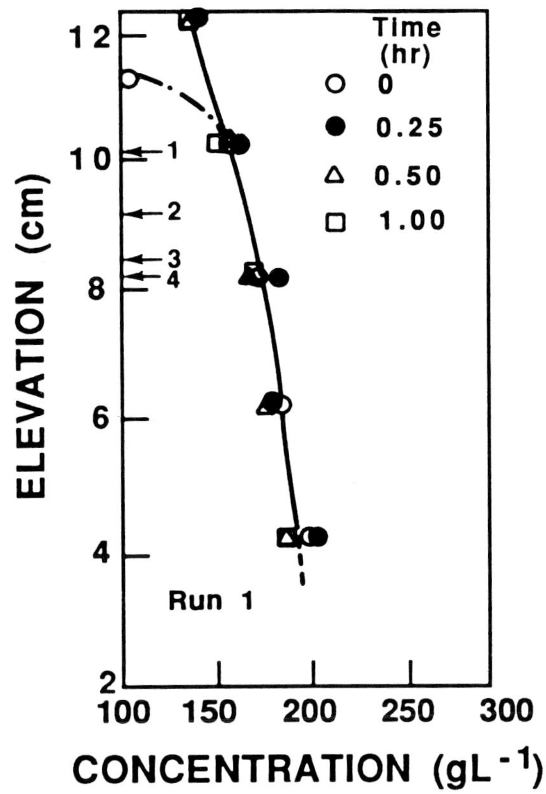

Figure 19.5. Representative mud concentration profiles from Run 1. Arrows indicate bed level at different times following test initiation.

It is worth noting that it may be unrealistic to relate a particular mud density as representing the bed surface during the fluidization process. Evidence presented in Figure 19.5, although of a rather tentative nature, suggests that the bed may have fluidized without a corresponding change in the density. The four enumerated elevations indicate the bed levels at the four times shown for Run 1. Notwithstanding an initial concentration change near the top, it is observed that the mud density did not change significantly during the first hour, in spite of the lowering of the bed level. Although a more conclusive evidence supporting this observation must await further research, the presented data seem to imply bed structural breakdown and dissipation of the effective stress during wave loading presumably via an internal rearrangement of the sediment aggregate matrix. Rheological data presented by, among others, Wright (1987) indicate that the strength of the cohesive bed is greater before it is sheared than after shearing begins. The fluidized material will therefore remain as such as long as wave action continues.

A logical follow-up question pertains to the process of bed recovery once wave action is arrested and the bed is reformed,

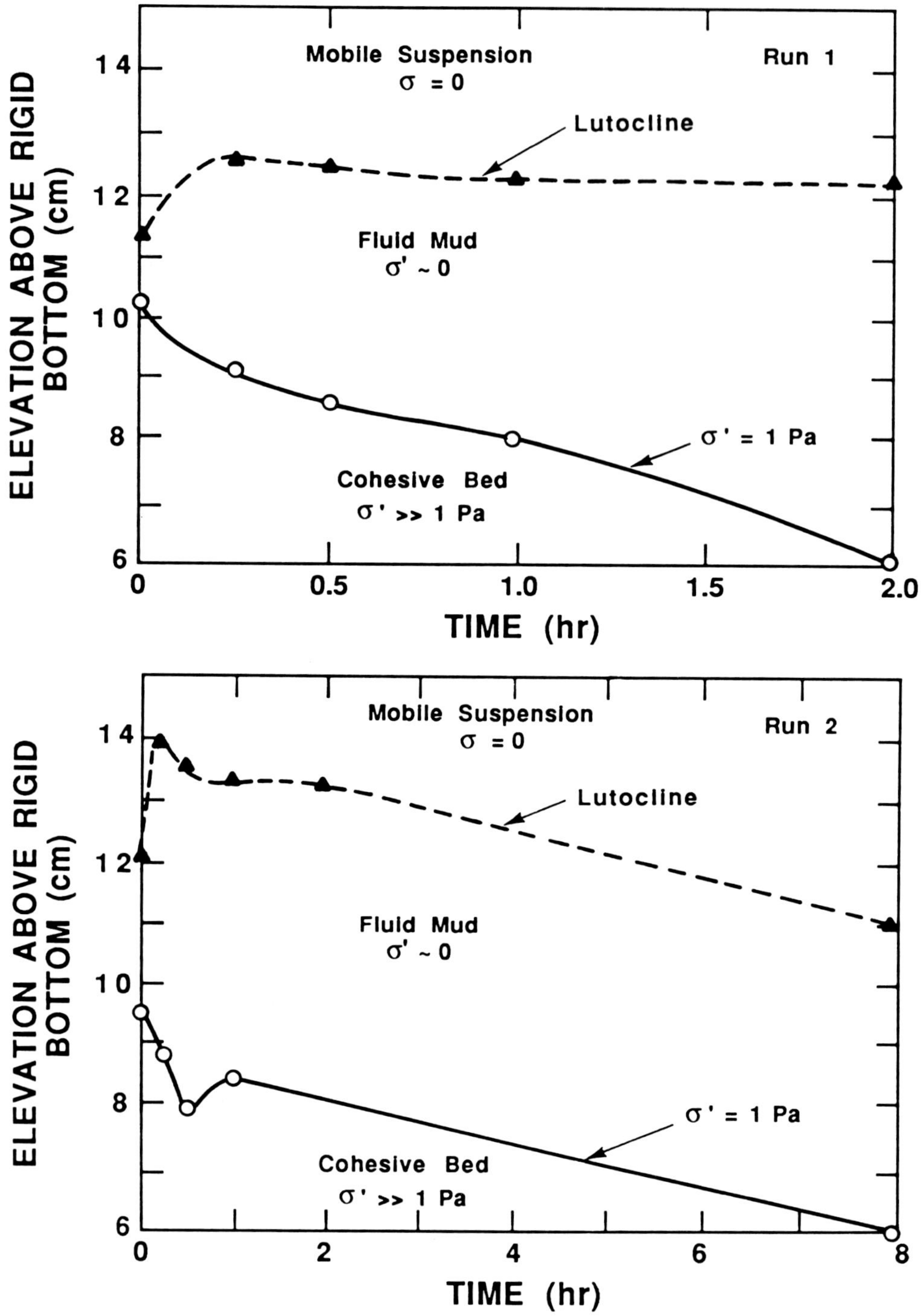

Figure 19.6. Generation of fluid mud from the cohesive bed under wave action. (a) Run 1. (b) Run 2.

a situation represented, for example, by the postfrontal profile in Figure 19.1. The likelihood that the cohesive bed is fully regenerated from the stationary fluid mud, without any measurable degree of dewatering, is an issue that needs to be elucidated experimentally.

Acknowledgment

Support provided by the U.S. Army Engineer Waterways Experiment Station through contract DACW 39-87-P-1064 is acknowledged.

References

Hopfinger, E.J., and P.F. Linden, 1982. Formation of thermoclines in zero-mean-shear turbulence subjected to a stabilizing buoyancy flux. Journal of Fluid Mechanics, v. 114, p. 157–173.

Kemp, G.P., and J.T. Wells, 1987. Observations of shallow-water waves over a fluid mud bottom: implications to sediment transport. Proceedings of Coastal Sediments '87, v. 1, American Society of Civil Engineers, New York, p. 363–377.

Liu, P.L-F, B.R. Turcotte, and F.H. Kulhawy, 1986. Wave propagation over a sediment trench. Journal of Waterway Port Coastal Ocean Engineering, v. 112(1), p. 64–77.

Maa, P.-Y., 1986. Erosion of soft muds by waves. Ph.D. thesis, University of Florida, Gainesville, 296 p.

Pamukcu, S., and J.N. Suhayda, 1987. High resolution measurement of shear modulus of clay using triaxial vane device. In: Cakmak, A.S. (ed.), Soil Mechanics and Liquefaction. Elsevier, Amsterdam, p. 307- 321.

Parker, W.R., 1986. On the observation of cohesive sediment behavior for engineering purposes. In: Mehta, A.J. (ed.), Estuarine Cohesive Sediment Dynamics. Springer-Verlag, Berlin, p. 270–289.

Perloff, W.H., and W. Baron, 1976. Soil Mechanics Principles and Applications. Wiley, New York, 745 p.

Ross, M.A., 1988. Vertical structure of estuarine fine sediment suspensions. Ph.D. thesis, University of Florida, Gainesville, 188 p.

Wright, V.G., 1987. Laboratory and numerical study of mud and debris flow. Ph.D. thesis, University of California, Davis.

CHAPTER 20

The Significance of Sediment-Flow Dynamics on Clay Microstructure Development: Riverine and Continental Shelf Environments

Huon Li and Richard H. Bennett

Introduction

The highly variable dynamics of coastal and continental shelf waters have a significant effect on the transport and deposition of sedimentary materials. The complex sediment transport mechanisms in continental shelf waters involve the interactions of multiphase flows of solids, liquid, and organics (dissolved and particulate) with multicomponent chemical and minerals species. Also biological processes are pervasive. Fine particles are transported as individual particles and floccules of various sizes and they interact with the fluid flow. The solid particle coagulation, breakup, deposition, erosion, and transport processes and mechanisms depend on the tidal cycles as well as the waves and currents. Particle associations at the sediment water interface are directly related to the particle associations in the suspension, and thus the microstructure at the sea floor is considered to be related to the floc sizes in the water column. The particles (individual particles and flocs) in suspension that are deposited eventually at the depositional interface are the fundamental "building blocks" of the bottom sediment.

The objectives of this chapter are to address (1) the significance and role of clay microstructure in flow dynamics (emphasis is on clay microstructure interactions with tide, wave, current, and turbulence), (2) the importance of scale in relation to microstructure and flow dynamics, (3) the role of particle-to-particle interaction, bonding, and shear that occurs near the sediment–water interface, and (4) the role of the "outer region" in sediment transport and microstructure development. Emphasis is on the physicochemical interactions rather than biochemical or biological processes, although these important processes can not be neglected in the final analysis microfabric development. A comprehensive treatment of the field of sediment transport and dynamics is beyond the scope of this chapter and thus limited reference material is discussed.

The surficial sediments are often reworked by organisms (bioturbation) that modify the physical and mechanical properties and microstructure of the sediment (Richardson and Young, 1980; Bennett and Nelsen, 1983). The clay microstructure of continental shelf sediment has been studied extensively by Bowles (1968a, b), Bennett (1976), and Bennett et al. (1977, 1979, 1981). Comprehensive studies of the clay microstructure and the physicochemistry of clays have been presented by Grim (1968), Mitchell (1976), and Bennett and Hulbert (1986). Collective papers on shelf sediment transport can be found in the book edited by Swift et al. (1972) and papers on cohesive sediment dynamics appear in the lecture notes edited by Mehta (1986).

Clay–Water Interactions

Clay Mineral Particles

Clay minerals are crystalline materials that are uniquely layered in plate-like structures. Clay minerals are classified as hydrous aluminum silicates (phyllosilicates). Their structures are commonly composed of composite layers built of components of tetrahedrally and octahedrally coordinated polyhedra. The very common clay minerals are kaolinite, smectite, illite, and mixed-layer clays. The clay mineral particles are usually less than 2 μm. Details of clay mineralogy and classification can be found in works by Grim (1968) and Mitchell (1976). Selected characteristics of the common clay mineral particles are listed in Table 20.1. The shape, size, cation exchange capacity, specific gravity, and specific surface are important properties of the clay minerals

Table 20.1. Characteristics of clay mineral particles.*

Mineral	Shape	Size (µm)	Cation exchange capacity (meq/100 g)	Specific gravity	Specific surface (m^2/g)
Kaolinite	Six-sided flakes	0.1–4 × 0.05–2	3–15	2.60–2.68	10–20
Montmorillonite (Smectite)	Flakes (equidimensional)	>10Å × up to 10	80–150	2.35–2.7	50–120 primary 700–840 secondary
Illites		0.003–0.1 × up to 10	10–40	2.6–3.0	65–100

*From Mitchell (1976).

that are important in clay microstructure–flow interaction behavior. For example, floculation and settling velocity are affected by the shape, size, specific gravity, cation exchange capacity, and specific surface of the interacting particles. Atoms of clay minerals are bonded into a three-dimensional structure, and the termination of this structure at a surface produces unbalanced force fields, positive or negative charges depending on the nature of the termination or break (Bennett and Hulbert, 1986). Isomorphous substitution within the crystal lattice also gives the clay particle a net negative charge. When clay mineral particles are 2 µm or less in size, the surface force fields become a significant factor in clay microstructure development. Due to the platy shape of the clay mineral particles and the fact that clay surfaces in general have a residual negative electrostatic charge, surface and colloidal forces exert important influences on particle-to-particle interactions. To preserve electrical neutrality, exchangeable cations are attracted and held on the surfaces and the edges, and in some clays between the unit cells. The quantity of these exchangeable cations necessary to balance the charge deficiency of a particular clay mineral is called the cation exchange capacity (CEC). The CEC is expressed as milliequivalents per 100 g of dry clay (Grim, 1968; Mitchell, 1976).

Suspended Clay Particles—Salt Water System

The behavior of the suspended clay particles in salt water is extremely complex. The flocculation behavior of clays in suspension is dependent on several chemical, physical, and environmental properties such as electrolyte concentration, dielectric constant, temperature, ion size, adsorbed water, pH, and clay mineralogy (Bennett and Hulbert, 1986). The particles may coagulate or remain stable depending on the balance between the electrostatic forces and the van der Waals forces and environmental conditions. The presence of organic matter is an important factor in the aggregation of clay particles. In addition, other complex variables include the bonding of clay particles by mucus

Table 20.2. Characteristics of two mud groups according to Ohtsubo and Muraoka (1986).

	First mud group	Second mud group
Type of cohesive sediment	Kaolinite Kaolinite + organic matter Montmorillonite Natural muds	Bentonite Bentonite + organic matter
Exchangeable cation	A^{3+}, Ca^{2+}, H^{+}	Na^{+}, Li^{+}
Resuspension behavior	1. Clear interface between bed surface and upper water 2. Cracks in the bed surface just before bed destruction 3. Bed destruction begins at the cracks 4. No flow of bed at bed destruction	1. Obscure interface between bed surface and upper water 2. Hazy streamwise streaklines consisting of tiny sediments 3. No cracks in the bed surface for any current condition 4. Flow of bed at bed destruction
Settling type	Collective subsidence	Free settling
Type of flow curve	Type A Existence of τ_{y1} (yield stress) Sharp bend in flow curve	Type B No τ_{y1} (yield stress) Smooth flow curve (power-law type)

*From Ohtsubo and Muraoka (1986).

secretions of algae, bacteria, diatoms, or other members of the benthic and epiflora and fauna common in coastal and oceanic environments (McCave, 1984). Even though neither the physicochemical nor the biochemical bonding forces can be simply predicted or easily measured at the present time, advances have been made in cohesive sediment dynamics in the last 30 years. Major contributions were made by Einstein and Krone (1962), Krone (1962, 1963, 1986), Partheniades (1962, 1965, 1977, 1986), Migniot (1968), McCave (1970, 1984), Mehta (1973, 1986), Owen (1977), and Hunt (1982, 1986).

More recently, Mimura (1989) presented a comprehensive review of cohesive sediment dynamics studies in Japan. Significant experiments on the critical shear stress and rate of resuspension were carried out by Ohtsubo and Muraoka (1986). Forty types of sediment were tested including clays (kaolinite, montmorillonite, and bentonite) and muds taken from different lakes and estuaries. To examine the organic effects, starch and agar were added, respectively, to the clay, and some of the natural muds were tested chemically. The most important result showed that cohesive sediment could be classified into two groups by the difference in the resuspension behavior, settling characteristics, and rheological properties over a wide range of sediment types and water contents. Kaolinite, kaolinite containing organic matter, montmorillonite, and natural muds were classified into one group, namely the first mud group. Bentonite and bentonite containing organic matter were classified as the second mud group. Some important characteristics of the two mud groups are listed in Table 20.2. The two different settling types, collective subsi-

Figure 20.1. Shear stress–strain rate relationship for two mud groups (A) Type A has a yield stress; (B) Type B has no well-defined yield strength. (From Ohtsubo and Muraoka, 1986.)

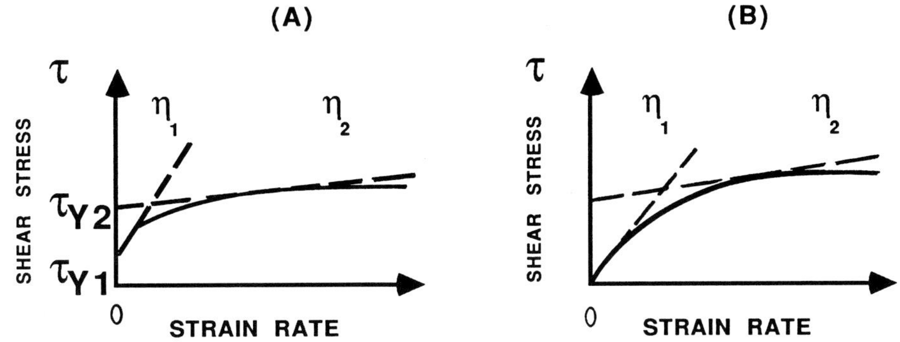

dence and free settling, indicate the difference in cohesive forces between the two mud groups. In the first mud group, collective subsidence settling occurs as the suspended particles aggregate and form a particular chain-like structure. The different types of relationship between shear stress and strain rate were obtained as shown in Figure 20.1. Type A curve, which has a yield stress τ_{y1} and a sharp bend indicative of material having a finite shear strength, was obtained for the first mud group. Type B curve, which has no well-defined yield strength, was representative of the second mud group. The existence of the yield stress τ_{y1} suggests that the first mud group has strong bonding and/or structure between particles.

The sizes of hydrated ions are expected to play an important role in the aggregation of suspended particles. Some of the important possible mechanisms for clay–saltwater interactions include (1) adsorbed hydrated cations which form the diffuse double layer, (2) hydrogen bonding, (3) hydration of exchangeable cations, (4) charged surface–dipole attraction, (5) attraction by van der Waals forces, (6) electroosmosis, (7) chemicoosmosis, (8) thermal diffusion, (9) external body forces, and (10) hydrodynamic forces.

Flow Dynamics

Water Mass Dynamics on the Continental Shelf

The transport of sediment in the water column and as bed load material is strongly influenced by both the fluid motion as well as the microstructure in suspension. The dynamics and microscale characteristics of the water masses also significantly affect the developmental history of fine-particle microstructure. On the continental shelf, a wide range of spatial and temporal scales of water motion exists. Since long-term changes involve deposition and low rate transport of clay materials, water motion of low intensity is important over geological time scales. Wave-induced water motion and currents, tidal-generated currents, and currents induced by fluid density differences are important for sediment transport under certain environmental conditions. A brief discussion of the various aspects of fluid flow important in sediment transport studies was presented by Weggel (1972) and a review of the coastal processes was given by Johnson and Eagleson (1966). According to Johnson and Eagleson, ocean currents can be classified into several groups: (1) currents generated by density distributions in the sea, (2) currents resulting from wind stresses, (3) tidal currents and those associated with internal waves, (4) currents caused by the activity of surface waves, and (5) currents induced by the intrusion of freshwater entering the ocean at river mouths. The various types of currents are depicted in Figure 20.2.

The flow regions in a vertical section of continental shelf water are shown schematically in Figure 20.3. The flow characteristics for the "wall" region are very significant in sediment transport and the dynamics are different from the main water mass body. The turbulence and dynamics of the "wall" region are of most importance in terms of sediment transport and suspended sediment microstructure.

Turbulence

Turbulence is the irregular motion of rotational flows. The thermodynamic and hydrodynamic variables of these flows depict the chaotic fluctuations in space and time coordinates. The following possible mechanisms of turbulence generation in the ocean were pointed out by Monnin and Ozmidow (1985):

1. Instability of vertical velocity gradients in drift flow—these velocity gradients are induced by direct effects of the wind on the upper mixed layer (UML).
2. Overturning of surface waves—this mechanism, and to a less extent the hydrodynamic instability of wave motions induced by surface waves, is the most powerful source of UML turbulence.
3. Instability of vertical velocity gradients in stratified large-scale oceanic flows—because of the large scale and low rate of turbulent energy generation, this mechanism is considered not important for shelf sediment transport.
4. Hydrodynamic instability of quasihorizontal mesoscale nonstationary flows—empirical evidence exists for the generation of small-scale turbulence.

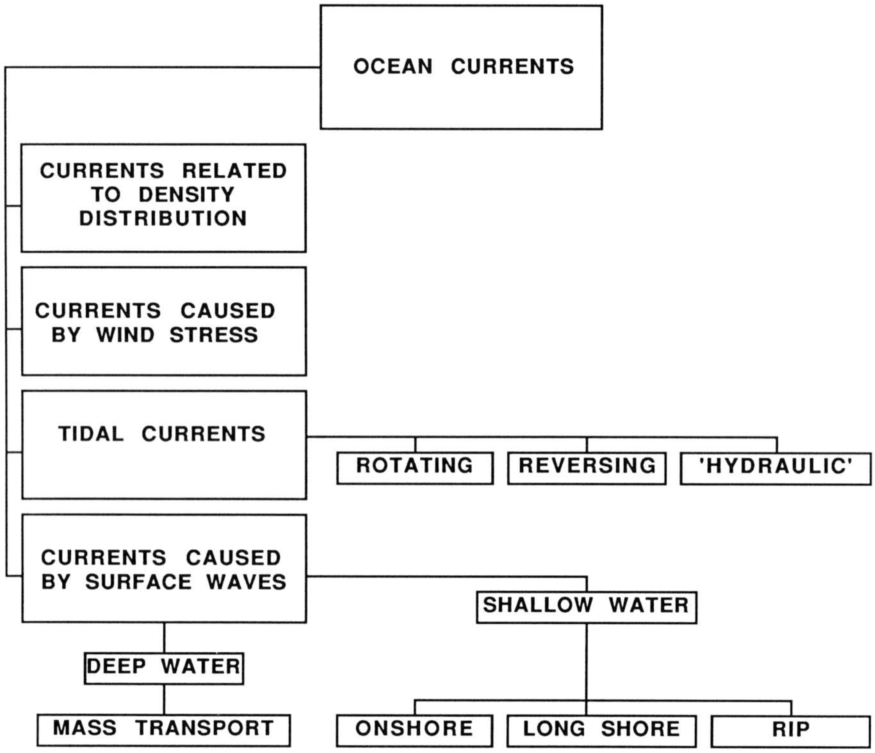

Figure 20.2. Classification of ocean currents. (From Johnson and Eagleson, 1966.)

5. Instability of local velocity gradients in internal waves—the instability of internal waves in the turbulent microstructure of the water increases the turbulent energy by a very small amount, which results in only slight dissipation.
6. Convection in layers with unstable density stratification—this mechanism can create turbulent (microstructure) layers.

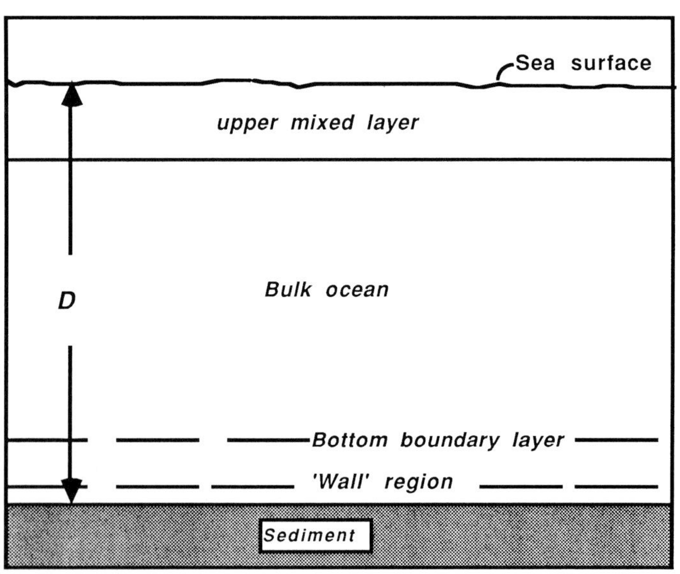

Figure 20.3. Flow regions in a vertical section (not to scale).

7. Instability of vertical velocity gradients in the bottom boundary layer (BBL)—this type of turbulence is created by Ekman layer instability.

In addition to the turbulence generation mechanisms listed above, the dynamics and turbulence characteristics in the "wall" region are important considerations in continental shelf water dynamics. The thickness of the "wall" region (Nychas et al., 1973) is defined by

$$\delta_w = 70\nu/u_* \qquad (1)$$

where u_* ($u_* = \tau_b^{1/2}\rho^{-1/2}$, here τ_b is the bed shear stress and ρ is the density of the water) is the friction velocity at the wall, and ν is the kinematic viscosity of the water. Turbulence in the "wall" region is *not isotropic and not homogeneous*. The turbulence, however, is termed isotropic if its statistical properties have no preferential direction, and it is said to be homogeneous if a random motion exists whose average properties are independent of position in the fluid. Classical works of turbulence were presented by Hinze (1959), Batchelor (1960), and Townsend (1976).

Dynamics in the "Wall" Region

A thin viscous sublayer exists next to the interface in the "wall" region for a smooth, rigid boundary. The ranges of the viscous sublayer and the "wall" region can be estimated as follows:

Table 20.3. Estimated thickness ranges for viscous sublayer and "wall" region.

	Friction velocity u_* (cm/sec)		
	0.1	1.0	10
Viscous sublayer (cm)	<0.5	<0.05	<0.005
"Wall" region (cm)	<7.0	<0.7	<0.07

$$\text{Viscous sublayer} \quad y < 5\nu/u_* \tag{2}$$

$$\text{"Wall" region} \quad y < 70\nu/u_* \tag{3}$$

where u_* is the friction velocity and ν is the kinematic viscosity of the water. Assuming $\nu = 0.01$ cm²/sec, the thickness ranges are estimated in Table 20.3. The condition for a boundary to be hydraulically smooth is that the dimensionless roughness Reynolds number

$$u_*k/\nu < 5 \tag{4}$$

where k is the roughness parameter (Schlichting, 1960). According to Einstein (1950), the total friction velocity (or shear stress) can be expressed as

$$u_{*_t} = u_* + u_{*_i} \tag{5}$$

where u_* is the "pure friction" component interacting between the fluid and the bed surface, irrespective of whether the surface is a flat plane or irregular, and u_{*_i} is the resultant of the drag forces acting on the irregularities. Discussions of the bottom friction were given by Yalin (1977). Based on Einstein's suggestion as stated in Eq. (5), the sea bottom is likely to be classified as smooth. Even for the rough boundary ($5 < u_*k/\nu < 70$, where k is the grain roughness parameter), the concept of the "wall" region is still valid.

The high shear and the dynamics of the viscous sublayer and the "wall" region are the key mechanisms that play a role in the deposition, resuspension, and floc breakup. From turbulent structure measurements in flat plate and pipe flow, the regions of maximum production and maximum dissipation are just outside the viscous sublayer (Hinze, 1959; Townsend, 1976). It is important to note that a dynamic viscous sublayer model was proposed by Einstein and Li (1955, 1956, 1958). The proposed model predicts a periodic growth and decay of the viscous sublayer. As a consequence of this model some estimates can be made for the generation of turbulence near the boundary. The unsteady and intermittent model was based on visual observations, pressure fluctuation measurements, and logic. Observations revealed that dye and sediment particles were exchanged between a smooth boundary and the turbulent flow through the viscous sublayer, thus making the assumption of a quasisteady sublayer impossible (Einstein and Li, 1958). Additional visual and quantitative studies of the viscous sublayer and "wall" region were carried out by Kline et al. (1967). An experimental investigation with hot-wire anemometers was performed by Gupta et al. (1971) to evaluate the spatial structure in and near the viscous sublayer. Later experiments (Nychas et al., 1973; Offen and Kline, 1975) confirmed that the production of the Reynolds stress is intermittent and is associated with a sequence of motions called "bursting" (Offen and Kline, 1974, 1975; Dyer, 1986).

Wave Boundary Layer

In continental shelf waters, the wave action may reach the sea bottom. It is well known that the irrotational theory can be applied to the entire gravitational wave motion except for a very thin layer adjacent to solid boundaries. Near solid boundaries the viscous effect cannot be neglected as compared with the inertia forces, and the thin layer can be treated according to the boundary layer theory. An experimental study was carried out by Li (1954) to investigate the wave boundary stability by oscillating a plate in still water with the range, amplitude, and period as computed by wave theory. According to Li (1954), the thickness of the oscillatory boundary layer is defined as

$$\delta_1 = 6.5\sqrt{\nu/\omega} \tag{6}$$

where ν is the kinematic viscosity and ω is the angular frequency. The thickness of the oscillatory boundary layer is very thin. For an oscillatory period of 10 sec, δ_1 has the thickness about 0.82 cm.

The critical Reynolds number at which the transition from laminar to turbulent takes place is found to be a constant

$$R = \omega^{1/2}d_1/\nu^{1/2} = 800 \tag{7}$$

where d_1 is the maximum amplitude of horizontal displacement at the boundary. The critical Reynolds number should be lower in the field (*in situ*) than the laboratory values stated above since the flow disturbances are much higher in the field.

Further information for wave boundary layers and their relation to sediment transport was provided by Teleki (1972) and Einstein (1972). Laboratory experiments were carried out by Kemp and Simons (1982) to investigate the interaction between waves and currents. Theoretical models were presented by Smith (1977), Grant and Madsen (1979), Grant et al. (1984), Herbich et al. (1984), and Shen (1986).

Detrital Particles and Flow Interactions

Fluid–Particle Dynamics

The transport of solid particles by random fluid motions that may include molecular motion and turbulence is diffusive in nature. For cohesive particles, the particle–particle interaction and the change of particle sizes make the problem very complex. Several scales of interaction are involved in the particle and flow dynamics: molecular scale, solid particle size, turbulent eddy

Table 20.4. Densities and shear strengths of aggregates in the capillary viscometer.*

Sample	Density (g/cm^3)	Shear strength (dyn/cm^2)	Cation exchangeable capacity† (meq/100 g)
Wilmington District	1.250	21.0	32
Brunswick Harbor	1.164	33.7	38
Gulfport Channel	1.205	45.9	49
San Francisco Bay	1.269	21.8	34
Delaware River	1.305	11	23
Potomac River	1.437	19	15
White River	1.212	48.6	60

*From Krone (1986).
†Cation exchange capacity of the $<10\ \mu m$ fraction.

size, and magnitude of the mean motion. Because of the random nature of sediment–fluid dynamics and the lack of complete understanding of the problems and microscale events, both statistical and semiempirical approaches are required to approach a solution to the sediment transport processes.

For clay particles to aggregate, mechanisms must exist to bring particles into collisions and the collisions must result in flocculation (coagulation). The flocculation process is due to the physicochemical interaction of the clay surfaces and surrounding electrolytic fluid. Common physical mechanisms involving particle collisions are Brownian motion, laminar and turbulent shear, and differential settling (see Bennett et al., this volume). A general equation for the particles undergoing flocculation including their size distribution was given by Friedlander (1977). The "coagulation kernels" were found for the Brownian motion, laminar shear, turbulent shear, and differential settling. Analogous to the work in the universal equilibrium concept of turbulence, a similar solution for coagulation was developed by Friedlander (1960) and later extended by Hunt (1982). Similar to flocculation, the equation for the particle size distribution undergoing breakup was given by Jeffrey (1982); however, the "breakup kernels" were not found. Additional data on the coagulation and breakup were presented by McCave (1984) and Hunt (1986). Based on known measurements, the size distribution can be estimated for shelf environments when the system is free of biological influences: (1) particle sized spectra are flat at low concentrations ($< 1 \times 10^{-5}\ g/cm^3$) in the size range 1–60 μm (Eisma and Kalf, 1979), and (2) at moderate concentration ($\sim 1 \times 10^{-4}\ g/cm^3$), a modal peak occurs between 5 and 20 μm (Kranck, 1973, 1981). Several modal peaks, however, may exist in the shelf where plankton are important. Once the size of the floc is known, the settling velocity of a single floc can be estimated theoretically or empirically.

Krone (1986) presented an hypothesis that described the orders of aggregates and laboratory measurement results of the capillary viscometer for density, shear strengths of aggregates, and cation exchange capacity (Table 20.4). The Bingham shear strength correlates with the cation exchange capacity as

$$\tau_B = 3.92 + 0.8447\ (\text{CEC}) \tag{8}$$

Table 20.5. Properties of sediment aggregates.*

Sediment sample	Order of aggregation	Density† (g/cm^3)	Shear strength (dyn/cm^2)	Porosity§ (n)
Wilmington District	0	1.250	21	0.862
	1	1.132	9.4	0.934
	2	1.093	2.6	0.958
	3	1.074	1.2	0.970
Brunswick Harbor	0	1.164	34	0.914
	1	1.090	4.1	0.960
	2	1.067	1.2	0.974
	3	1.056	0.62	0.981
Gulfport Channel	0	1.205	46	0.889
	1	1.106	6.9	0.950
	2	1.078	4.7	0.967
	3	1.065	1.8	0.975
San Francisco Bay	0	1.269	22	0.850
	1	1.179	3.9	0.905
	2	1.137	1.4	0.931
	3	1.113	1.4	0.966
	4	1.098	0.82	0.955
	5	1.087	0.36	0.962
	6	1.079	0.20	0.967
White River (in salt water)	0	1.212	49	0.885
	1	1.109	6.8	0.948
	2	1.079	4.7	0.967
	3	1.065	1.9	0.975

*From Krone (1986).
†Density in sea water = 1.025 g/cm^3.
§Solid particle density = 2.65 g/cm^3.

In Table 20.5, the results of the rotating cylinder viscometer are shown for the *order* of aggregation, density, and shear strength, as well as the porosity. Krone stated that

1. The *order* of the suspended aggregates depends on the history of the suspension: such as if the particles have been suspended in fluid zones having very high velocity gradients or if they are newly formed from dispersed mineral particles, the *order* would be zero. Aggregates that have formed in moderate velocity gradients were classified as one *order* higher.
2. Settling aggregates that adhere to the bed surface during deposition have a structure that at the bed surface is classified as one order higher than that of the depositing aggregate. The critical shear stress for deposition (τ_{cd}) also is considered one order of aggregation higher than the order of the depositing aggregate.

Flume studies of deposition of suspended cohesive sediment (Krone, 1986) showed that

$$dc/dt = -(W_s c/h)\,(1 - \tau_b/\tau_{cd}) \tag{9}$$

where c is the concentration, t is the time, W_s is the settling velocity, h is the water depth, τ_b is the bed shear stress, and τ_{cd} is the critical shear stress for deposition. The term $(1 - \tau_b/\tau_{cd})$ expresses the probability of the settling aggregates to adhere to the bed.

Table 20.6. Values of τ_{cd} for three sediments.*

Sediment	τ_{cd} (dyn/cm²)
Kaolinite	1.5
Bay mud	1.0
Maracaibo mud	0.8

*From Mehta (1986).

Mehta (1973, 1986) carried out extensive deposition tests under steady flows and reanalyzed some previous flume data. He found that the critical shear stress τ_{cd} depended on the sediment composition, and varied from 0.4 to 1.5 dyn/cm². The values of τ_{cd} are shown in Table 20.6 for three sediments.

Partheniades (1962, 1965, 1986) conducted experimental and theoretical studies for both erosion and deposition. The first experiments were conducted on a remolded bed at a comparable field density and of uniform consistency about 1 in. thick. At constant bed shear stresses above a threshold value, $\tau_{ce} = 0.96$ dyn/cm², the rates of erosion were constant and independent of the suspended sediment concentration (Partheniades, 1962, 1965, 1986). Data and observations indicated that no simultaneous erosion and deposition of sediment occurred during the experiments. Based on statistical considerations, models were developed by Partheniades for erosion, deposition, and sediment–flow interactions. The stochastic model for sediment–flow interaction was based on an extension of the Einstein (1950) approach for the bed-load function. The model indicates that simultaneous erosion and deposition of sediment particles may occur only in a limited range for cohesive sediments. Laboratory studies of erosion and deposition are very useful for studying and describing shelf sediment transport processes and clay microstructure development.

Turbulent Scale, Dissipation Rate, and Mean Shear Rate

Let us now address the subject concerning the scale of turbulence. For isotropic turbulence, Kolmogoroff made the following hypothesis (Hinze, 1959): "At sufficiently high Reynolds numbers there is a range of high wave-numbers where the turbulence is statistically in equilibrium and uniquely determined by the parameters ε and ν. This state of equilibrium is universal" (Hinze, 1959, p. 184). ε and ν describe the rate of turbulent energy dissipation and the molecular kinematic viscosity, respectively.

From dimensional reasoning, the characteristic scales can be obtained from the equilibrium range as

$$\text{Length scale: } \eta = (\nu^3/\varepsilon)^{1/4} \tag{10}$$

$$\text{Velocity scale: } \upsilon = (\nu\varepsilon)^{1/4} \tag{11}$$

$$\text{Shear rate scale: } G = (\varepsilon/\nu)^{1/2} \tag{12}$$

The rate of turbulent energy dissipation (or production), ε, in the ocean was estimated by Monin and Ozmidov (1985) as (1) in upper mixed layer, $\varepsilon \simeq 10^{-2} - 10^{-1}$ cm²/sec³; (2) in bulk ocean, $\varepsilon \simeq 10^{-5}$ cm²/sec³; and (3) in BBL, with turbulence created chiefly by Ekman layer instability, $\varepsilon \simeq 10^{-4}$ cm²/sec³. Based on these estimates, the characteristic scales were determined (Table 20.7). It is important to note that the assumption of local isotropy and equilibrium range is reasonable in all the three layers except in the "wall" region. In the "wall" region, the rate of energy dissipation and the mean shear rate are estimated based on the following relationships:

Table 20.7. Characteristic scales in equilibrium range for three layers of the ocean.

Region	Dissipation rate ε = (cm²/sec³)	Length scale η (cm)	Velocity scale υ (cm/sec)	Shear rate scale G (sec⁻¹)
Upper mixed layer	10^{-2}–10^{-1}	10^{-1}–0.56×10^{-1}	10^{-1}–1.8×10^{-1}	1–3.2
Bulk ocean	10^{-5}	5.6×10^{-1}	1.8×10^{-2}	3.2×10^{-2}
Bottom boundary layer	10^{-4}	3.2×10^{-1}	3.2×10^{-2}	10^{-1}

In the viscous sublayer, the viscous dissipation rate is

$$\varepsilon_v = u_*^4/\nu \tag{13}$$

and mean shear rate is

$$du/dy = u_*^2/\nu \tag{14}$$

At the edge of the "wall" region the turbulent dissipation rate is

$$\varepsilon = u_*^3/\kappa y \tag{15}$$

and mean shear rate is

$$du/dy = u_*/\kappa y \tag{16}$$

where κ is the von Karman's constant and has the value of 0.4 for clean fluids (may be lower with sediment).

The estimates of flow characteristics are given in Tables 20.8 and 20.9 for the viscous sublayer and for the edge of the "wall" region. When the friction velocity reaches the value of 1.0 cm/sec, high viscosity dissipation and high mean shear rates occur in the viscous sublayer (Table 20.8) and the turbulent dissipation rates at the edge of the "wall" region (Table 20.9) are higher than that of the dissipation rate in the upper mixed layer,

Table 20.8. Estimate of flow characteristics in the viscous sublayer.

	Friction velocity u_* (cm/sec)		
	0.1	1.0	10
Viscous dissipation rate ε_v (cm²/sec³)	10^{-2}	10^{2}	10^{6}
Mean shear rate $\frac{du}{dy}$ (sec⁻¹)	1	10^{2}	10^{4}

Table 20.9. Estimate of flow characteristics at the edge of "wall" region.

	Friction velocity u_* (cm/sec)		
	0.1	1.0	10
At $y = 70\ \upsilon/u_*$			
Turbulent dissipation rate ε (cm²/sec³)	3.6×10^{-4}	3.6	3.6×10^{4}
Mean shear rate $\frac{du}{dy}$ (sec^{-1})	3.6×10^{-2}	3.6	3.6×10^{2}

bulk ocean, and bottom boundary layer. Due to the fact that the turbulent dissipation rate and the turbulent production rate are much higher inside the "wall" region than outside, and because of the unsteady, intermittent, dynamic nature of this region, *the breakup of flocs takes place mainly in the "wall" region*. On the other hand, the flocs also may affect the turbulent characteristics, and the structure of the "wall" region, the fluid and solids interacting synergistically. The turbulence in the other layers contributes to the mixing, flocculation, diffusion, and deformation of the flocs, but at generally lower dynamic energies than in the "wall" region.

Description of the Particles and Flow Interactions on the Shelf and Riverine Environments: Significance to Clay Microstructure

Despite the complexity of the particle types and morphology and flow dynamics on the continental shelf, the mechanics of solid particle or floc movement can be described and approximated as follows:

1. The motion of any solid particle (or aggregate) is determined by Newton's laws. The forces acting on the particle include hydrodynamic forces, electrochemical particle–particle interaction forces, and gravitational force. Due to the various shapes of particles and the random nature of the force actions, the particle may reorientate and floccules may deform.
2. As an example, suppose we follow a solid particle (aggregate) in a location outside the "wall" region (i.e., $yu_*/\nu > 70$). Before a particle–particle collision, the motion of this particle can be considered as a single particle that is subjected to the actions of the steady current, periodic motions such as wave and tide, flow turbulence, Brownian motion, and gravitational force. This particle reorientates or moves more or less in a random manner. However, the motion of the particle has the tendency to follow a downward trajectory due to the gravitational force. Once particle-to-particle collision occurs, the compound particle may either bond or separate depending on the impact force and the physicochemical forces. The strength of the newly formed aggregate is probably dependent on (a) the number of molecular bonds at the particle–particle contact, (b) the type of available hydrated ions in the double layer, (c) the presence or absence of other molecules (organic and inorganic) in proximity to the points of contact, and (d) the geometry of the contact area and relative orientation of the particles [edge-to-edge (E-E) and edge-to-face (E-F) contact generally would be weaker than face-to-face (F-F) contacts]. After the particle-particle collision, the new aggregate will be considered as a single particle as before and the processes will continue in the flow regime outside the "wall" region.
3. As the aggregate drops inside the "wall" region and into the viscous sublayer, the particle may deform, break up, and eject outside the region due to the high shear and dynamics of this region. Because of the unsteady and intermittent nature in the "wall" region, the high lift force may prevent aggregate deposition. However, the aggregate may pass through the "wall" region and be deposited at the interface if the bottom friction velocity is below a critical value ($u_{*cd} \simeq 0.6$–1.3 cm/sec). Simultaneous erosion and deposition generally do not occur. The deposited sediment will consolidate, be reworked, and be altered by the mechanical, chemical, and biological forces and mechanisms driving microfabric development (Bennett et al., this volume). As the bottom friction velocity reaches another critical value ($u_{*ce} \simeq 0.6$–3.0 cm/sec) at another time, the deposited sediment will be resuspended.
4. Since the nature of the problem has virtually infinite degrees of freedom, the high variability in time, space, and scale, both a statistical concept and semiempirical approaches are required to develop meaningful predictive models of sediment transport and deposition and the processes and mechanisms in microfabric development.

Based on this study and the above descriptions and the discussions in previous sections, the suspended clay microstructure, can be described as follows: (1) clay mineral particle microstructure (fabric and physicochemistry) are the fundamental "building blocks"; (2) the available particle size distribution affects the aggregate size distribution, which commonly is flat in the range 1–50 μm (where plankton are important several modes can exist); (3) the density of the aggregates have the range 1.06–1.4 g/cm³; (4) after the initial contact, which is dominantly edge-to-face in the marine environment, stairstep structure (stronger and more stable structure) is preferred due to the high shear in the viscous sublayer and the "wall" region; and (5) chain-like and vortex-like structures may exist partially due to turbulence especially in the "wall" region.

Summary and Conclusions

The suspended clay particles in salt water may coagulate due to the effects of electrostatic forces, van der Waals forces, and biochemical activities. The flocculation–deflocculation behavior of clays in suspension is dependent on many physicochemical properties, especially the clay mineral types, their particle size distribution, and the ionic characteristics of the fluid. The particles and flow interactions such as Brownian motion, laminar and turbulent shear, and differential settling are of importance in the flocculation behavior.

A thin water layer near the bottom plays a very important role on the suspended clay particle and aggregates and on the particles at the depositional interface. This thin layer is called a "wall" region and has a thickness equal to $70\nu/u_*$, where ν is the kinematic viscosity of the fluid and u_* is the "pure" friction velocity. Outside the "wall" region, the turbulence and the mean shear play a role in flocculation. Inside the "wall" region, the particle aggregate may break up and be ejected outside the region due to the high shear and dynamics of the "wall" region and the microfabric of the floccules may be changed. The aggregate may pass through the "wall" region and be deposited on the bottom if the friction velocity is below a critical value (e.g., $u_{*cd} \simeq 0.6$–1.3 cm/sec).

The friction velocity due to the drag forces acting on the irregularities, u_{*i}, is a contribution mainly outside the "wall" region insofar as the particle–flow interaction is concerned. Due to the distinct roles of the u_* and u_{*i}, the application of the logarithmic law is not sufficient for the determination of the friction velocity and the roughness length.

The interaction of waves and currents is complex. For the friction velocity, a linear superposition is recommended as the first approximation.

References

Batchelor, G.K., 1960. The Theory of Homogeneous Turbulence. Cambridge University Press, London, 197 p.

Bennett, R.H., 1976. Clay fabric and geotechnical properties of selected submarine sediment cores from the Mississippi Delta. Ph.D. dissertation, Texas A&M University, College Station, TX, 269 p.

Bennett, R.H., and M.H. Hulbert, 1986. Clay Microstructure. International Human Resources Development Corporation, Boston, MA, 161 p.

Bennett, R.H., and T.A. Nelsen, 1983. Seafloor characteristics and dynamics affecting geotechnical properties at shelfbreaks. In: Stanley, D.J., and G.T. Moors (eds.), The Shelfbreak: Critical Interface on Continental Margins: Tulsa, OK, SEPM, 467 p.

Bennett, R.H., W.R. Bryant, and G.H. Keller, 1977. Clay fabric and geotechnical properties of selected submarine sediment cores from the Mississippi Delta. NOAA Professional Paper 9, U.S. Department of Commerce, 86 p.

Bennett, R.H., W.R. Bryant, and G.H. Keller, 1979. Clay fabric and related pore geometry of selected submarine sediments: Mississippi Delta. Scanning Electron Microscopy I, p. 519–524 and p. 424 (discussion).

Bennett, R.H., W.R. Bryant, and G.H. Keller, 1981. Clay fabric of selected submarine sediments: fundamental properties and models. Journal of Sedimentary Petrology, v. 51(1), p. 217–232.

Bowles, F.A., 1968a. Electron microscopy investigation of the microstructure in sediment samples from the Gulf of Mexico. Ph.D. dissertation, Texas A&M University, College Station, TX, 140 p.

Bowles, F.A., 1968b. Microstructure of sediments: investigation with ultrathin section. Science, v. 159, p. 1236–1237.

Dyer, K.R., 1986. Coastal and Estuarine Sediment Dynamics. Wiley, New York, 342 p.

Einstein, H.A., 1950. The bed-load function for sediment transportation in open channel flow. Technical Bulletin No. 1026, U.S. Department of Agriculture, Washington, D.C., 71 p.

Einstein, H.A., 1972. The basic description of sediment transport on beach. In: Meyer, R.E. (ed.), Waves on Beaches and Resulting Sediment Transport. Academic Press, New York, 462 p.

Einstein, H.A., and R.B. Krone, 1962. Experiments to determine modes of cohesive sediment transport in salt water. Journal of Geophysical Research, v. 67, p. 1451–1461.

Einstein, H.A., and H. Li, 1955. Shear transmission from a turbulent flow to its viscous boundary sub-layer. 1955 Heat Transfer and Fluid Mechanics Institute, UCLA, paper no. XIII, 16 p.

Einstein, H.A., and H. Li, 1956. The viscous sublayer along a smooth boundary. Journal of Engineering Mechanics Division, Proceedings, American Society of Civil Engineers, paper no. 945, 27 p.

Einstein, H.A., and H. Li, 1958. The viscous sublayer along a smooth boundary. Transactions of the American Society of Civil Engineers, v. 123, p. 293–313.

Eisma, D., and J. Kalf, 1979. Distribution of particle size of suspended matter in the southern bight of the North Sea and the Eastern Channel. Northerlands Journal for Sea Research, v. 13, p. 298–324.

Friedlander, S.M., 1960. On the particle-size spectrum of atmospheric aerosols. Journal of Meteorology, v. 17, p. 373–374.

Friedlander, S.M., 1977. Smoke, Dust and Haze: Fundamentals of Aerosol Behavior. Wiley, New York, 317 p.

Grant, W.D., and O.S. Madsen, 1979. Combined wave and current interaction with a rough bottom. Journal of Geophysical Research, v. 84, p. 1797–1808.

Grant, W.D., A.J. Williams III, and S.M. Glenn, 1984. Bottom stress estimates and their prediction on the northern California continental shelf during CODE-1: the importance of wave-current interaction. Journal of Physical Oceanography, v. 14, p. 506–527.

Grim, R.E., 1968. Clay Mineralogy, 2nd ed. McGraw-Hill, New York, 596 p.

Gupta, A.K., J. Laufer, and R.E. Kaplan, 1971. Spatial structure in the viscous sublayer. Journal of Fluid Mechanics, v. 50, p. 493–512.

Herbich, J.G., R.E. Schiller, R.K. Watanage, and W.A. Dunlap, 1984. Seafloor Scour. Marcel Dekker, New York, 320 p.

Hinze, J.O., 1959. Turbulence. McGraw-Hill, New York, 586 p.

Hunt, J.R., 1982. Self-similar particle-size distributions during coagulation: theory and experimental verification. Journal of Fluid Mechanics, v. 122, p. 169–185.

Hunt, J.R., 1986. Particle aggregate breakup by fluid shear. In: Mehta, A. J. (ed.), Estuarine Cohesive Sediment Dynamics. Lecture Notes on Coastal and Estuarine Studies 14. Springer-Verlag, New York, 473 p.

Jeffrey, D.J., 1982. Aggregation and breakup of clay flocs in turbulent flow. Advances in Colloid and Interface Science, v. 17, p. 197–211.

Johnson, J.W., and P.S. Eagleson, 1966. Coastal processes. In: Ippen, A.T. (ed.), Estuary and Coastline Hydrodynamics. McGraw-Hill, New York, 744 p.

Kemp, P.H., and R.R. Simmons, 1982. The interaction between waves and turbulent current: waves propagating with the current. Journal of Fluid Mechanics, v. 116, p. 227–250.

Kline, S.J., W.C. Reynolds, F.A. Scharub, and P.W. Runstadler, 1967. The structure of turbulent boundary layers. Journal Fluid Mechanics, v. 30, p. 741–776.

Kranck, K., 1973. Flocculation of suspended sediment in the sea. Nature, v. 246, p. 348–350.

Kranck, K., 1981. Particulate matter grain size characteristics and flocculation in a partially mixed estuary. Sedimentology, v. 28, p. 107–114.

Krone, R.B., 1962. Flume studies of the transport of sediment in estuarial shoaling processes. Hydraulic Engineering Laboratory and Sanitary Engineering Laboratory, University of California, Berkeley, 110 p.

Krone, R.B., 1963. A study of rheologic properties of estuarial sediments. Hydraulic Engineering Laboratory and Sanitary Engineering Laboratory, University of California, Berkeley, 91 p.

Krone, R.B., 1986. The significance of aggregate properties to transport processes. In: Mehta, A.J. (ed.), Estuarine Cohesive Sediment Dynamics. Lecture Notes on Coastal and Estuarine Studies 14. Springer-Verlag, New York, 473 p.

Li, H., 1954. Stability of oscillatory laminar flow along a wall. U.S. Army Corps of Engineers Technical Memorandum no. 47, 48 p.

McCave, I.N., 1970. Deposition of fine-grained suspended sediment from tidal current. Journal of Geophysical Research, v. 75, p. 4151–4159.

McCave, I.N., 1984. Erosion, transport and deposition of fine-grained marine sediments. In: Stow, D.A. and D.J.W. Piper (eds.), Fine-Grained Sediments: Deep-Water Processes and Facies. Blackwell Scientific Publications, Palo Alto, CA, 659 p.

Mehta, A.J., 1973. Depositional behavior of cohesive sediment. Ph.D. Dissertation, University of Florida, Gainesville, FL, 275 p.

Mehta, A.J. (ed.), 1986. Estuarine Cohesive Sediment Dynamics. Lecture Notes on Coastal and Estuarine Studies 14. Springer-Verlag, New York, 473 p.

Migniot, G., 1968. A study of the physical properties of various forms of very fine sediment and their behavior under hydrodynamic action. La Houille Blanche, v. 7, p. 591–620 (in French).

Mimura, N., 1989. Recent Japanese studies on cohesive sediment transport Journal of Coastal Research. Special Issue No. 5.

Mitchell, J.K., 1976. Fundamentals of Soil Behavior. Wiley, New York, 422 p.

Monin, A.S., and R.V. Ozmidow, 1985. Turbulence in the Ocean. D. Reidel, Boston, MA, 247 p.

Nychas, S.G., H.C. Hershey, and R.S. Brodkey, 1973. A visual study of turbulent shear flow. Journal of Fluid Mechanics, v. 61, p. 513–540.

Offen, G.R., and S.J. Kline, 1974. Combined dye-streak and hydrogen-bubble visual observation of a turbulent boundary layer. Journal of Fluid Mechanics, v. 62, p. 223–239.

Offen, G.R., and S.J. Kline, 1975. A proposed model of bursting process in turbulent boundary layer. Journal of Fluid Mechanics, v. 70, p. 209–228.

Ohtsubo, K., and K. Muraoka, 1986. Resuspension of cohesive sediments by currents. In: Wang, S.Y., H.W. Shen, and L.R. Ding (eds.), River Sedimentation, v. III, Proceedings of the Third International Symposium on River Sedimentation, p. 1680–1689.

Owen, M.W., 1977. Problems in the modeling of transport, erosion, and disposition of cohesive sediment. In: Goldbert, E.D. et al. (eds.), The Sea, v. 6, Wiley-Interscience, New York, 1048 p.

Partheniades, E., 1962. A study of erosion and deposition of cohesive soils in salt water. Ph.D. dissertation, University of California, Berkeley, CA, 181 p.

Partheniades, E., 1965. Erosion and deposition of cohesive soils. Journal of the Hydraulic Division, American Society of Civil Engineers, v. 91, no. HY1, Proceedings Paper 4202, p. 105–139.

Partheniades, E., 1977. Unified view of wash load and red material load. Journal of the Hydraulic Division, American Society of Civil Engineers, v. 103, no. HY9, Proceedings Paper 13215, p. 1037–1057.

Partheniades, E., 1986. A fundamental framework for cohesive sediment dynamics. In: Mehta, A.J. (ed.), Estuarine Cohesive Sediment Dynamics. Lecture Notes on Coastal and Estuarine Studies 14. Springer-Verlag, New York, 473 p.

Richardson, M.D., and D.K. Young, 1980. Geoacoustic models and bioturbation. Marine Geology, v. 38, p. 205–218.

Schlichting, H., 1960. Boundary Layer Theory, 4th ed. McGraw-Hill, New York, 647 p.

Shen, Y.P., 1986. Second order closure modeling of turbulent flow and sediment dispersion in coastal and estuarine waters. In: Wang, S.Y., H.W. Shen, and L.Z. Ding (eds.), River Sedimentation, v. III, Proceedings of the Third International Symposium on River Sedimentation, p. 1383–1396.

Smith, J.D., 1977. Modelling of Sediment Transport on Continental Shelves. In: Goldbert, E.D. et al. (eds.), The Sea, v. 6. Wiley-Interscience, New York, 1048 p.

Swift, D.J.P., D.B. Duane, and O.H. Pilkey (eds.), 1972. Shelf Sediment Transport: Process and Pattern. Dowden, Hutchinson & Ross, Stroudsburg, PA, 656 p.

Teleki, P.G., 1972. Wave boundary layers and their relation to sediment transport. In: Swift, D.J.P., D.B. Duane, and O.H. Pilkey (eds.), Shelf Sediment Transport: Process and Pattern. Dowden, Hutchinson & Ross, Stroudsburg, PA, 656 p.

Townsend, A.A., 1976. The Structure of Turbulent Shear Flow. Cambridge University Press, Cambridge, 429 p.

Weggel, J.R., 1972. An introduction to oceanic water motions and their relation to sediment transport. In: Swift D.J.P., D.B. Duane, and O.H. Pilkey (eds.), Shelf Sediment Transport: Process and Pattern. Dowden, Hutchinson & Ross, Stroudsburg, PA, 656 p.

Yalin, M.S., 1977. Mechanics of Sediment Transport. Pergamon Press, New York, 298 p.

CHAPTER 21

Silt Microfabric of Detrital, Deep Sea Mud(stone)s (California Continental Borderland) as Shown by Backscattered Electron Microscopy

Suzanne Reynolds and Donn S. Gorsline

Introduction

Mudstones constitute about 60% of the known sedimentary section (Potter et al., 1980), but papers dealing specifically with these sediments are rare. Muds contain few clues as to their origin and depositional history, so they are cursorily described and subsequently ignored, whereas the more unusual sand layers found interbedded with them are studied in detail. One reason for the lack of knowledge concerning mudstones may be a result of the inability to visually discern their individual components. One method of fine-scaled examination is backscattered electron (BSE) imaging, which has previously had only limited use in the investigation of shales (Krinsley et al., 1983; Pye and Krinsley, 1984) because high-resolution images only recently have been made possible due to the development of improved backscatter electron detectors. BSE imaging of polished thin sections yields micrographs at roughly the same scale of observation as regular optical microscopy. This scale, approximately 100–500×, yields information about the textural and structural arrangements of silt particles, termed herein "silt microfabric." Compared to optical micrographs, BSE images also provide (1) distinct delineation of micas, (2) higher resolution of clays, (3) strictly two-dimensional images, and (4) a more meaningful representation of microfabric.

This chapter delineates sedimentary structures and trace fossils seen in BSE images of deep-marine, detrital muds and mudstones from modern and ancient basins of the California Continental Borderland (Fig. 21.1).

Methods

In normal scanning electron microscopy, the primary electron beam interacts with the target material and transfers energy to it. This leads to the generation of secondary electrons (SE), which are collected by the SE detector and used to image the sample. Since the energy level of the secondary electrons is largely a function of the angle of incidence between the primary beam and the target material, the image produced in the SE mode reflects topography.

Some of the primary electrons, however, interact with the target material with little energy loss. In this case, the primary electrons are scattered through the material and eventually are reflected back out as backscattered electrons (BSE). The energy level of these electrons is a function of both the atomic number of the target material and the angle of incidence. If the topography of the specimen is minimized, the gray-level intensity of the image is almost wholly a function of atomic number contrast.

To prepare samples for BSE, thin (or thick) sections were polished on a lapping wheel, finishing with 600 grit polishing paper. Further polishing using a <5 μm alumina grit in an aqueous solution, and felt paper, did not result in a smoother surface. The alumina gouged out the resin and clay, leaving quartz, feldspar, and carbonate in positive relief. The 600 grit paper left some long gouges across the surface of the sample, which were sometimes apparent in the BSE image, but these did not significantly interfere with interpretation of the image.

Samples were very thinly coated with gold/palladium. Normal thicknesses of metal coatings decrease image quality significantly. The problem could be avoided by coating with carbon, although that method is more time-consuming. Best images are obtained using a 15 mm working distance and small spot size.

In these images, the bright spots are pyrite. Shell material is also quite bright, due to the inclusion of heavy metals. The white lathes are chlorite. Equant gray minerals are mostly quartz, dark gray, diffuse areas are clays, and black areas are voids and/or organic material.

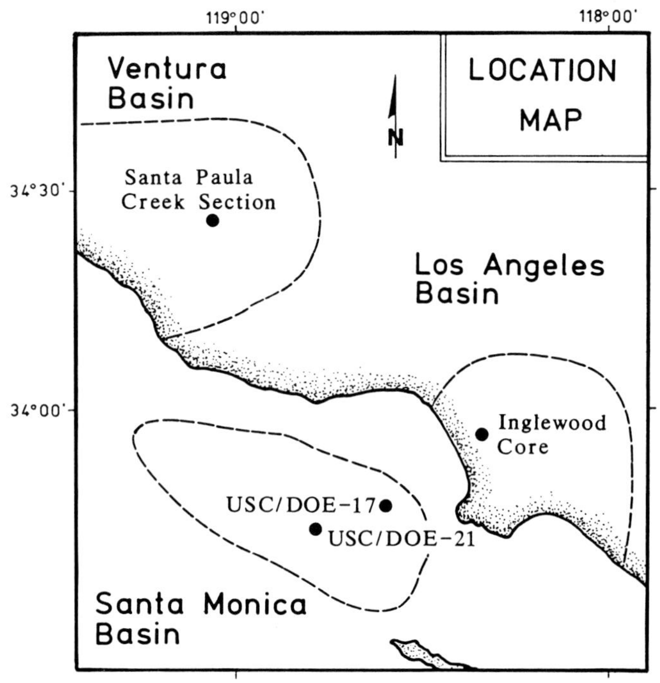

Figure 21.1. Sample location map.

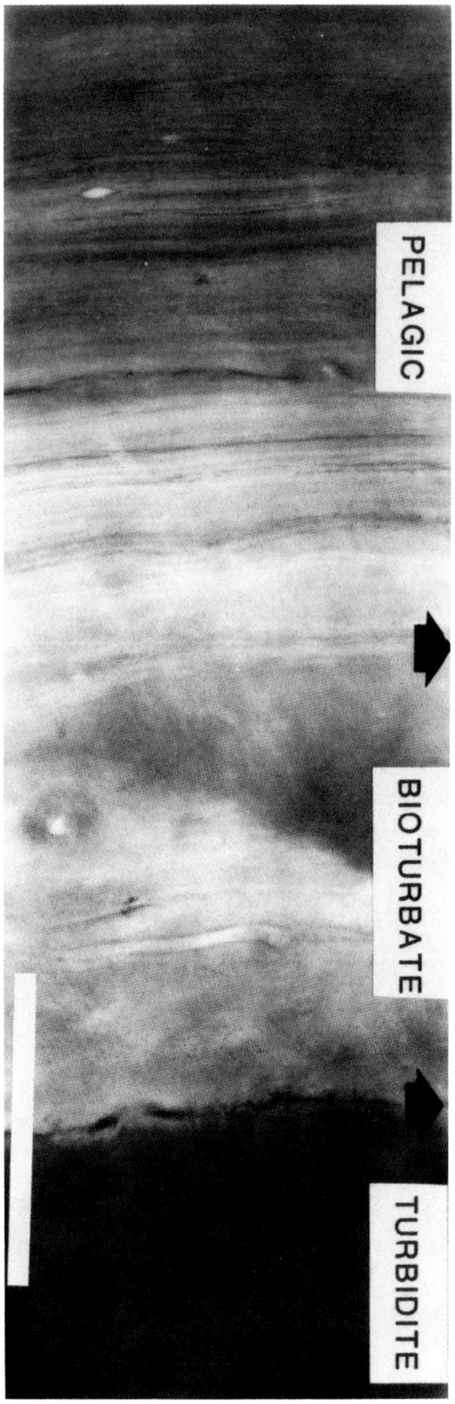

Figure 21.2. X-Radiograph of vertically-oriented core slab; scale bar is 5 cm. Arrows indicate boundaries between layers. The uppermost layer is characterized by pelagic laminations, the central layer has a bioturbate fabric, and the lowest layer is a turbidite.

Study Areas and Results

Santa Monica Basin

Santa Monica Basin, an inner basin of the California Continental Borderland, has been described in detail by Malouta et al. (1981). It lies directly off the narrow continental shelf near Los Angeles (Fig. 20.1). Sedimentation rates vary from 40 to 80 cm/1000 years. Maximum water depth is 960 m. Sediment is bioturbated to 800 m water depth; below 800 m, bottom waters are anoxic and primary laminae are preserved.

The largest amount of detrital input is from Hueneme Canyon to the north. A submarine fan complex dominates the northwest margin of the basin; the southeastern margin has less well-developed fans and mass movements. Turbidite layers are identified in the laminated zone by gray color, Bouma sequence sedimentary structures, and well-sorted, fining-upward grain size characteristics (Reynolds, 1986).

The central basin floor has surface sediments characterized by a mean grain size in the fine silt range, with 10–15% calcium carbonate and 8–12% organic material by weight. The clays present in these sediments are dominantly illite, montmorillonite, kaolinite, and chlorite; clay mineralogy does not vary significantly between gray turbidite layers and green hemipelagic or bioturbated sediments (Fleischer, 1970). Water contents of surficial material range from 50 to 70% of total volume.

Station USC/DOE-21 was cored on the central basin floor within the anoxic zone. On a freshly cut surface of this core, the upper 14 cm displayed green/black zebra-stripe laminations, which disappeared rapidly on exposure to oxygen. From 14 to 21 cm, the sediment was homogeneously green, and from 21 to 28 cm it was grey. The X-radiograph of this core (Fig. 21.2) indicated that the upper 14 cm was finely laminated, characteristic of pelagic deposition in the absence of bioturbation. From 14 to 21 cm, the sed-

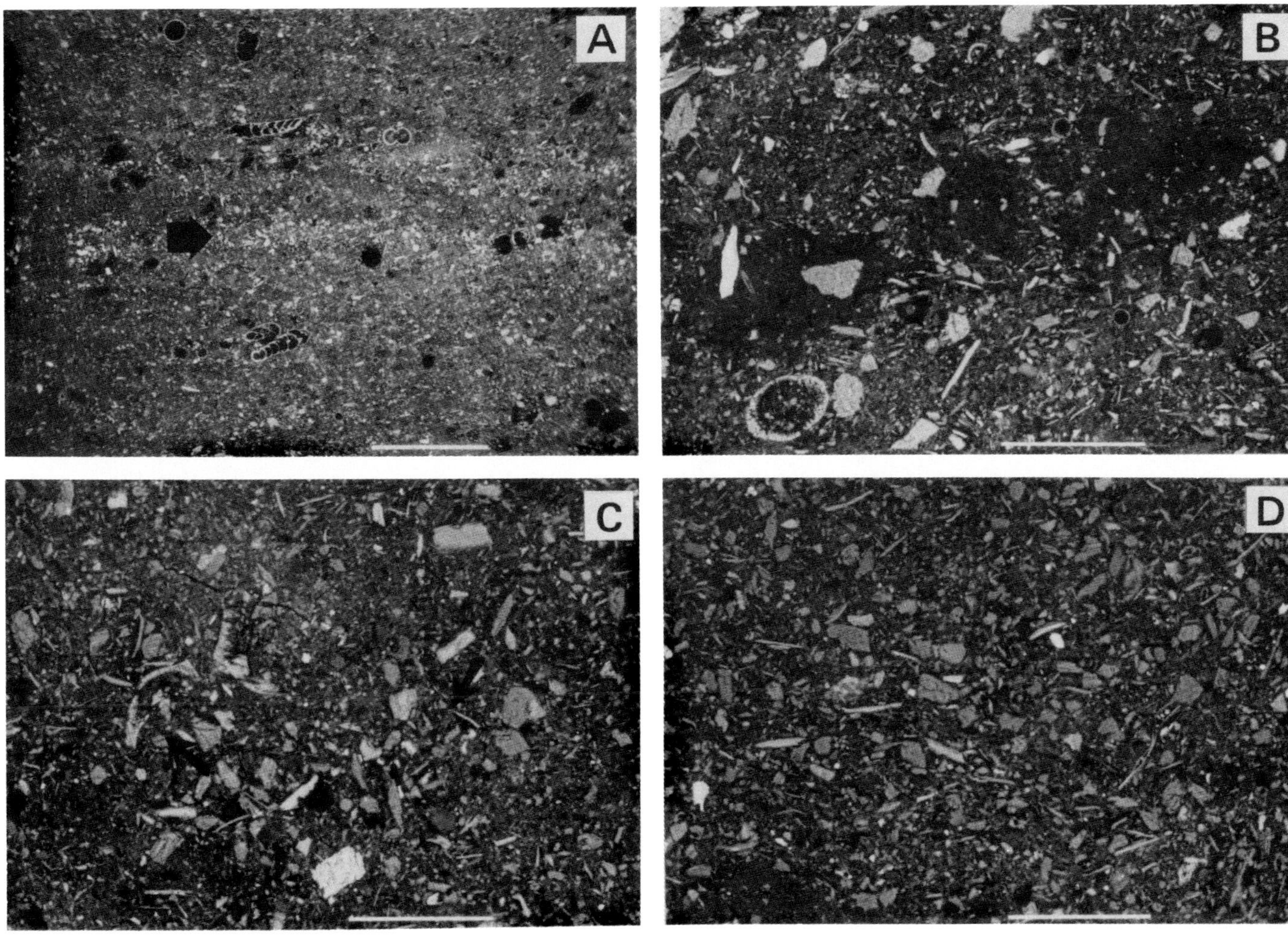

Figure 21.3. Backscattered electron (BSE) micrographs of recent muds from Santa Monica Basin; scale bar is 1 mm in A, 100 microns for B, C, and D. (A) Pelagic laminated mud; note silt and clay laminations that may result from nepheloid layer settling or flow. (B) Pelagic laminated mud; dark areas in the central portion of the micrograph are discontinuous organic laminations, probably resulting from deposition of organic bodies. (C) Bioturbated mud; the U-shaped distribution of silts is a meniscate backfill structure, which results from benthic deposit feeding. (D) Turbidite mud, near the base of the deposit. Coarse silt is abundant, but most grains are floating; note the preferred orientation of micas, a result of deposition during a waning current.

ment was bioturbated; individual horizontal burrows were apparent at 18 cm. Below 21 cm the opacity of the sample to X-rays markedly increased. Textural laminations were observed at 27 cm, and some individual burrows were present in the upper part of the bed.

Station USC/DOE-17 was cored on the lower slope of Santa Monica Basin at 750 m water depth. The X-radiograph displayed a mottled fabric characteristic of bioturbation.

BSE images of the laminated pelagic zone generally displayed clays with a small amount of silt. A regular alternation of a two-layer system was not present, as was observed through macroscopic examination. Instead, microscopic lamination occurred as irregularly spaced, thin stringers of sediment that had a composition different from the background material. These stringers differred in organic and textural content. Textural stringers were less than 1 mm thick, and were composed of silt overlain by clay, a silt layer alone, or a clay layer alone (Fig. 21.3A). Organic stringers were irregular and discontinuous and contained little detrital material (Fig. 21.3B).

The bioturbated mud had a higher silt content than the pelagic mud. This silt did not occur in thin horizontal lamina, but appeared irregularly through the sample in a mottled texture (Fig. 21.3C). Some of the mottling resembled the meniscate-backfill structure. Mica had random orientation throughout.

The turbidite mud also had a high silt content, but lacked the mottled structure of the bioturbated mud. Instead, silts were dispersed evenly across the sample. Silt grains were matrix supported. At the bottom of the bed, coarse silts

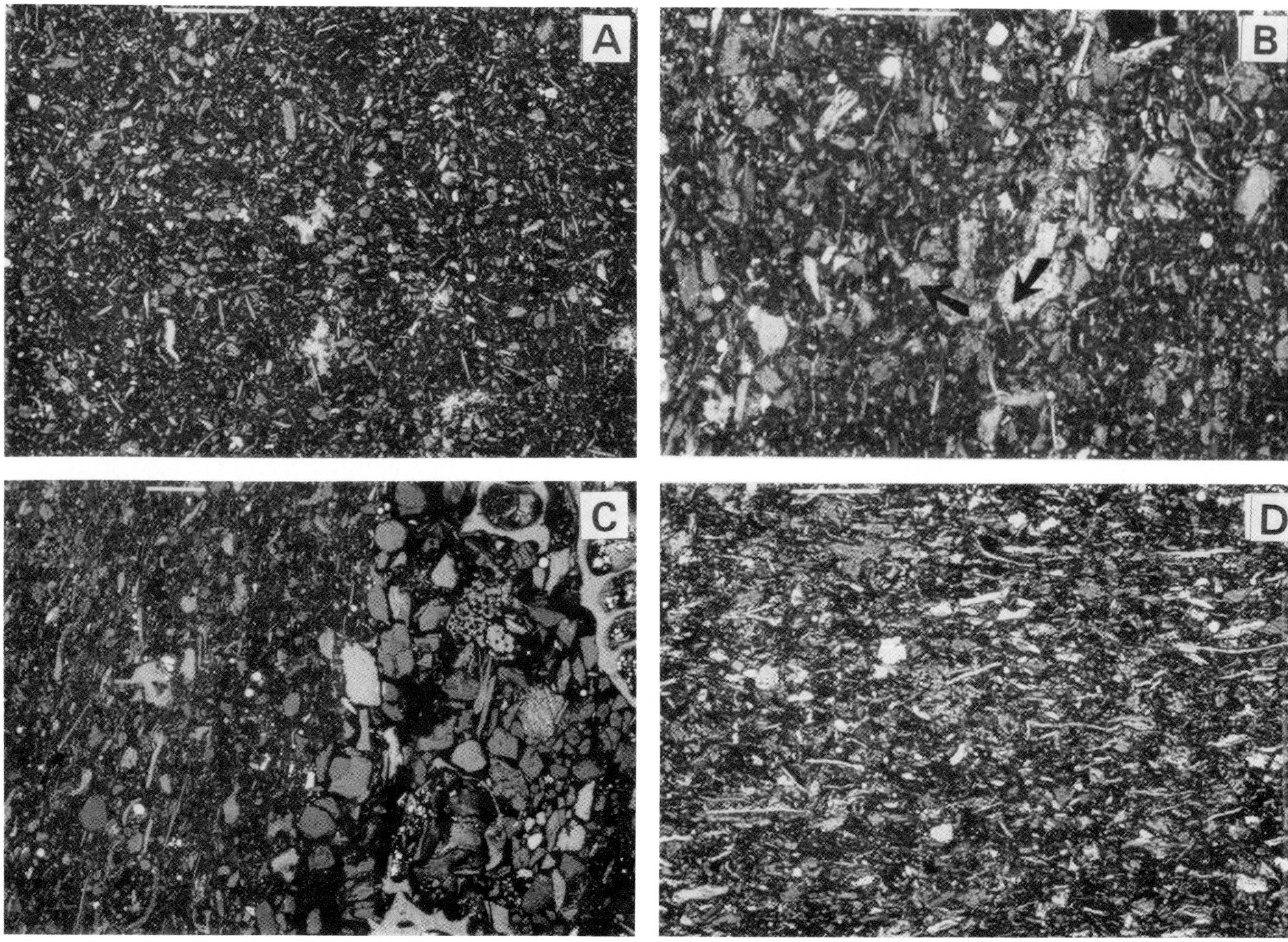

Figure 21.4. BSE micrographs; scale bar is 100 μm. (A) Turbidite mud, near the top of the deposit in Santa Monica Basin. Micas are in random orientation, a result of deposition from suspension; white dendritic patches are an artifact. (B) Bioturbated mudstone from the Los Angeles Basin, 276 m; note the meniscate backfill structure as shown by the arrow direction. (C) Contact between bioturbated (lower) and turbidite mudstone, Los Angeles Basin, 276 m; note the large foraminifera tests in the turbidite. (D) Turbidite mudstone, Los Angeles Basin, 276 m; this is near the base of the deposit, within Bouma division D. Note the preferred orientation of micas, a result of deposition during a waning current.

and micas were abundant and some micas exhibited preferred orientation (Fig. 21.3D). Upward in the deposit, silt size decreased and micas were in random orientation (Fig. 21.4A).

Los Angeles Basin

The Los Angeles Basin is one of the filled inner basins of the California Continental Borderland (Fig. 21.1). Its general geologic nature has been summarized by Yerkes et al. (1965). At depth within the basin lie Pliocene deep marine deposits that are roughly equivalent to modern deposition in the Santa Monica Basin. A continuous core was drilled through these strata in 1986 (Gidman et al., 1987) by Chevron Oilfield Research. A total of 427 m of Pliocene deep marine fan deposits was recovered. These are interpreted to have been deposited by turbidity currents on lobes and overbank areas (Schweller et al., 1987).

Three broad groups of mudstones were identified in these cores: (1) bioturbated mudstones—brownish green muds having no apparent sedimentary or bioturbate structures; (2) bioturbated sandy mudstones—brownish green muds having circular to ovoid, unlined, sand-filled burrows; and (3) turbidite mudstones—gray muds having no apparent sedimentary or bioturbate structures, found in association with a basal sandstone layer; the upper layers were bioturbated. The third mud type occurred only rarely, probably because bioturbation was intense and only the thickest deposition of mud, or very frequent turbidity current activity, allowed these to remain unbioturbated.

The descriptions of the studied intervals are as follows: *276 m* (depth below wellhead): turbidite mudstone with thin turbidite sandstone, interbedded with bioturbated mudstone (Fig. 21.5A);

639 m: an upper 2 cm of turbidite mudstone with a thin basal sandstone, underlain by bioturbated mudstone; *648 m*: bioturbated sandy mudstone with sand-filled, large diameter burrows (Fig. 21.5B).

Water contents ranged from 24% at 276 m to 17% at 648 m. The turbidite mudstones were less dense than the bioturbated mudstones. Density of all mudstones increased with burial depth. X-ray diffraction indicated that all of the samples contained chlorite, illite, and smectite. Kaolinite was a minor constituent in some samples.

The mean grain sizes of all samples varied from 6 to 7 phi. The bioturbated mudstones were composed of moderately to poorly sorted silts and clayey silts. The bioturbated sandy mudstones were poorly sorted clayey silts and sandy silts. The turbidite mudstones were moderately sorted clayey silts.

In BSE images of samples from 276 m, the bioturbated muds displayed spatial segregation of grain sizes, exhibited as lumps of clay-rich zones separated by zones of densely packed silts (Fig. 21.4B). This fabric was similar to the mottled texture of bioturbated muds in Santa Monica Basin. Some structures also resembled meniscate backfill structures. Preferred orientation of micas was only partly developed.

The contact between bioturbated mud and the overlying turbidite is shown in Figure 21.4C. At this scale of observation, the lower contact of the turbidite was not sharp. Silts intrude into the underlying, more clayey deposit. The lower portion of the turbidite was characterized by densely packed micas and coarse equigranular silts. Micas were wrapped around the silts as a result of consolidation. Silts were evenly dispersed across all images. This portion of the sample was grain supported, but floating grains increased in abundance upward in the deposit. The lower portion of the turbidite displayed preferential orientation of micas (Fig. 21.4D) that became more poorly developed upwards. Silt became progressively finer upward in the turbidite.

Near the top of the turbidite at 276 m, *Chondrites* burrows were present as blebs 1–2 mm in diameter (Fig. 21.6A). The interior of the burrow was dark due to low silt and mica content, and a high clay and organic content. High magnification BSE images of burrow fill and of the surrounding material indicated that the clay fabrics were roughly similar. Clays occurred singly or in face–face amalgamated packets. Particles and packets had an overall random orientation with respect to horizontal, but within some small areas, particles displayed a consistent orientation with respect to each other. This type of fabric is reminiscent of the turbostratic fabric defined by Aylmore and Quirk (1960).

At 639 m, preferred orientation was not well developed in the bioturbated mudstone (Fig. 21.6B). Packing was dense and homogeneous; floating silt grains were abundant. Spatial segregation of grain sizes was less apparent in these more deeply buried samples.

At the base of the mud turbidite at 639 m, preferred orientation of micas was well developed; imbrication of some silt grains was observed (Fig. 21.6C). The sample was matrix supported and densely packed. Upward in the sample, the degree of preferred orientation decreased, with some micas trending at high angles to horizontal. Overall, the difference in preferred orientation between the upper and lower portions of this turbidite was less than observed at 276 m.

In the bioturbated sandy mudstone, spatial segregation of silt and clay particles, in a mottled texture, was the dominant fabric element (Fig. 21.6D). Microscopic tests were abundant.

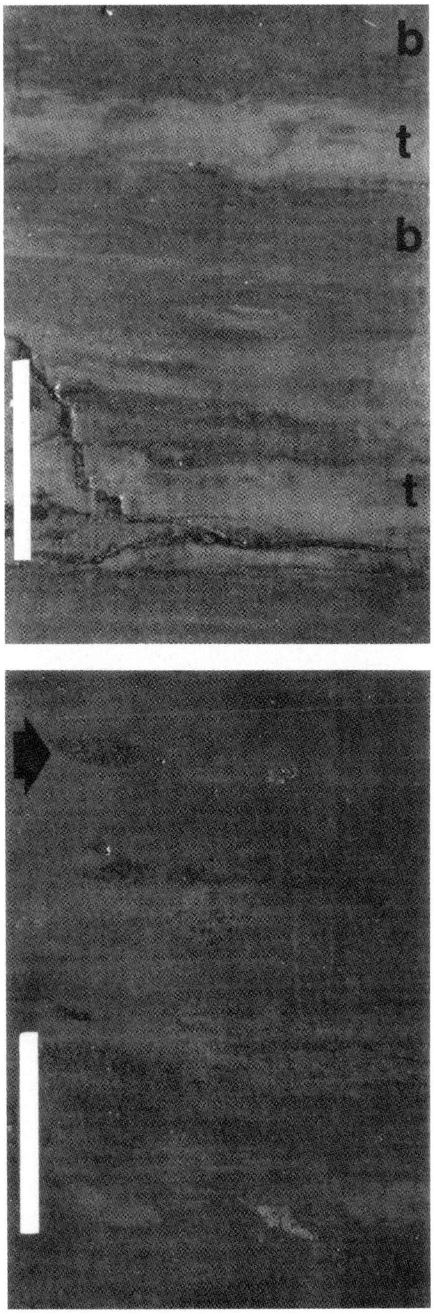

Figure 21.5. Photographs of typical mudstones within the Los Angeles Basin studied section; scale bars are 5 cm. (A) 276 m, interbedded grey turbidite (t) and green bioturbate (b) mudstones; crack in lower left and across lower part of sample. (B) 639 m, bioturbated sandy mudstone showing sand-filled burrow (arrow).

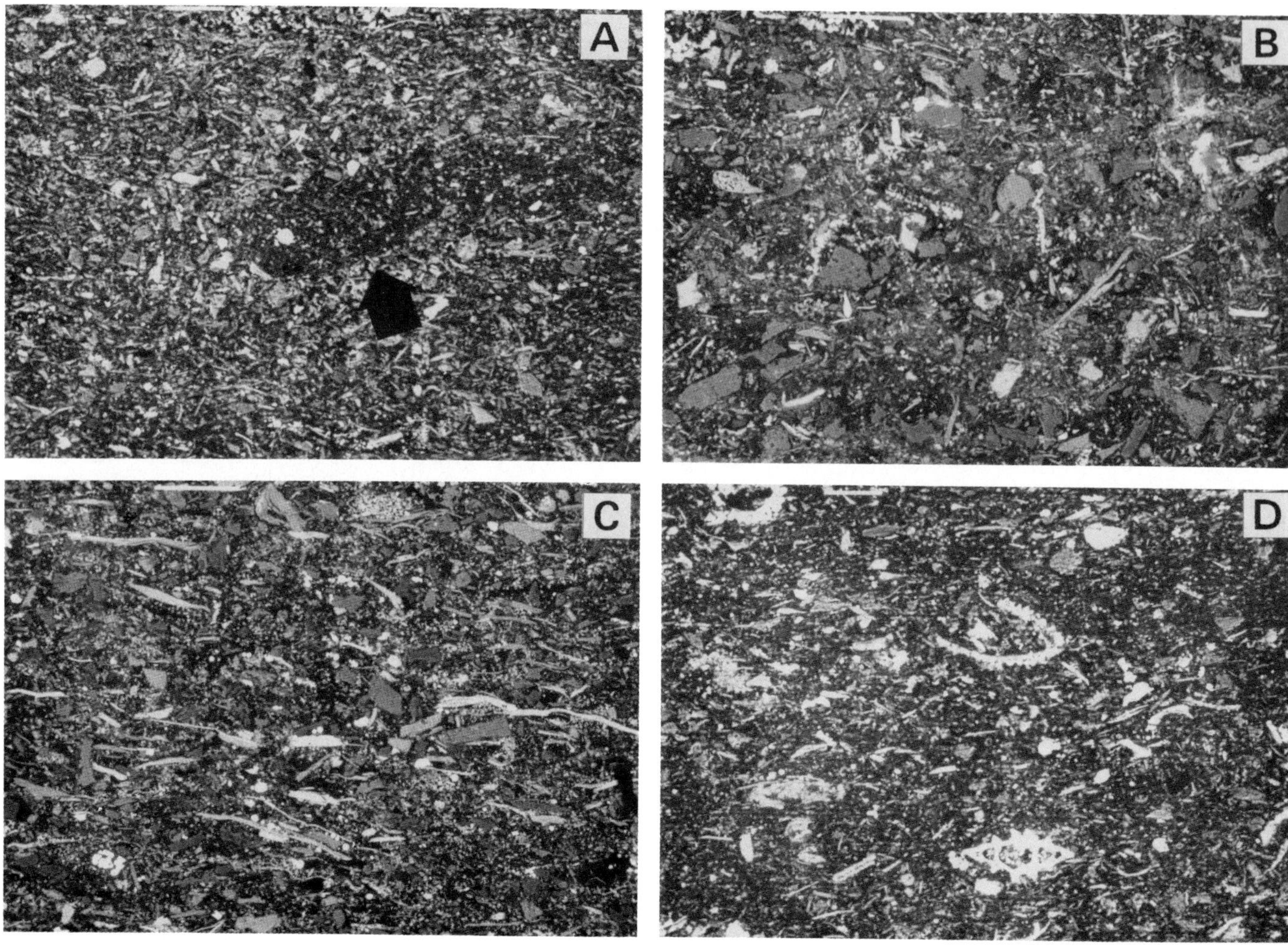

Figure 21.6. BSE micrographs of mudstones from the Los Angeles Basin; scale bars are 100 μm. (A) 276 m, isolated clay-filled *Chondrites* burrow within the upper portion of a turbidite mudstone; note random orientation of micas outside the burrow. (B) 639 m, bioturbated mudstone; note slight spatial segregation of clays and silts and random orientation of micas. (C) 639 m, turbidite mudstone; note abundance of pyrite, preferred orientation of micas. (D) 639 m, bioturbated sandy mudstone; note poorly developed preferred orientation of micas and prominent foraminifera tests; some textural mottling is apparent.

Meniscate backfill structures were common. A higher degree of preferred orientation was present in these bioturbated muds than in those at 276 m; however, the direction of orientation was locally variable.

Ventura Basin

The Plio-Pleistocene Pico Formation crops out (Fig. 21.1) in Santa Paula Creek near the city of Santa Paula, California (Johnson, 1978). It is comprised of deep-sea fan deposits of the Ventura Basin, and includes thick sequences of mudstones deposited under anoxic conditions. Exposures of these mudstones are excellent and sedimentary structures are prominent. Pelagic and turbiditic mudstones had distinctively different lithologies (Fig. 21.7). The pelagic mudstones were dense, mahogany-red, and finely laminated. The turbidite mudstones were soft, gray to tan, and exhibited textural grading and lamination characteristic of mud turbidites.

Because these samples were taken from an outcrop section, water contents were not determined, but is estimated to be 20%. The turbidites were clayey silts to sandy silts, with mean grain size in the medium to fine silt range. The pelagites were clayey silts with a mean in the fine silt range. X-ray diffraction indicated chlorite, montmorillonite, and illite in both types of strata.

In the BSE micrographs, pelagic and nepheloid-layer deposits were distinguishable by the presence of numerous pyrite cubes and framboids scattered evenly throughout the pelagic layers. However, the pyrite severely intefered with the interpretation of other fabric elements in the pelagic layers. Thin section images of the pelagites displayed thin millimeter to centimeter laminae resulting from kerogen layering; BSE images did not display a similar style of lamination. Rarely, very-fine-scale stringers of kerogen (Fig. 21.8A) or pyrite (Fig. 21.8B) were observed. Silt content was low, and samples were matrix supported.

In the nepheloid/turbidite layers (Fig. 21.8C), silt content was high, and silts were dispersed evenly across the sample; silt size decreased upward in the deposit. Textural lamination occurred as horizontal laminae of coarse silt. Preferential orientation of mica did not decrease upward in the turbidite deposits, as in the Santa Barbara and Los Angeles Basins samples.

The upper portion of one nepheloid/turbidite layer contained numerous clay-rich, oval-shaped areas about 0.5 mm in diameter (Fig. 21.8D). The shape and size of these features are consistent with copepod fecal pellets that frequently occur in recent anoxic basins.

Discussion

Pelagite Fabric

Sediment relations in the pelagites provide evidence as to their mode of deposition. Other studies of the clay microfabric of these sediments have shown that most material was deposited on the basin floor in the form of fecal pellets and large organic aggregates (Reynolds, 1988). Fecal pellet production occurs continuously throughout the year, but the rate of production increases during organic blooms. This leads to a relatively homogeneous composition through time, but a regular alternation of depositional rates. Sediment deposited during periods of rapid depositional rates will retain a high water content and therefore a low density; sediment deposited during periods of slower depositional rates will undergo more consolidation, and will therefore have a lower water content and a higher density. This regular alternation of depositional rates is reflected in X-radiographs as thin laminae of differing densities. Such regular alternations are not observed in backscattered images because these depict only compositional data. Occasional deposition of larger organic bodies results in randomly interspersed organic laminae.

Occasionally, the rain of pellets is interrupted by gravity flow deposition of detrital material, resulting in textural laminations. These thin gravity flows probably occur through high detrital runoff during major winter storms or deposition from nepheloid plumes in the water column.

Turbidite Fabric

The even distribution of silts across these samples results from the purely physical deposition of thick layers of sediment over a short span of time. Lower portions of the samples were deposited during a waning current; upper portions were deposited from suspension. This causes the decrease in silt size upward in the sample. Microfabric relationships suggest that the preferred orientation in the basal section of some mud turbidites resulted from depositional process—the basal sediment was deposited while the current was still moving, acting to orient the micas parallel to the depositional surface.

Figure 21.7. Photograph of representative Ventura Basin strata; scale bar is 5 cm. The section contains pelagic laminated (p) and turbiditic (t) mudstones.

Bioturbated Fabric

The bioturbate mud(stone)s are comprised of sediment that was originally deposited through either pelagic or turbiditic processes or both. Bioturbation disrupted the characteristic silt microfabric of the original depositional processes and produced textural mottling. Mottling results from particle size selection by organisms during particle feeding, as exemplified by the meniscate backfill structure.

Summary and Conclusions

Three types of mud(stone)s were examined, and distinctive microfabrics of each were delineated. Turbidites have an even distribution of silt across a sample; silt size decreases upward in

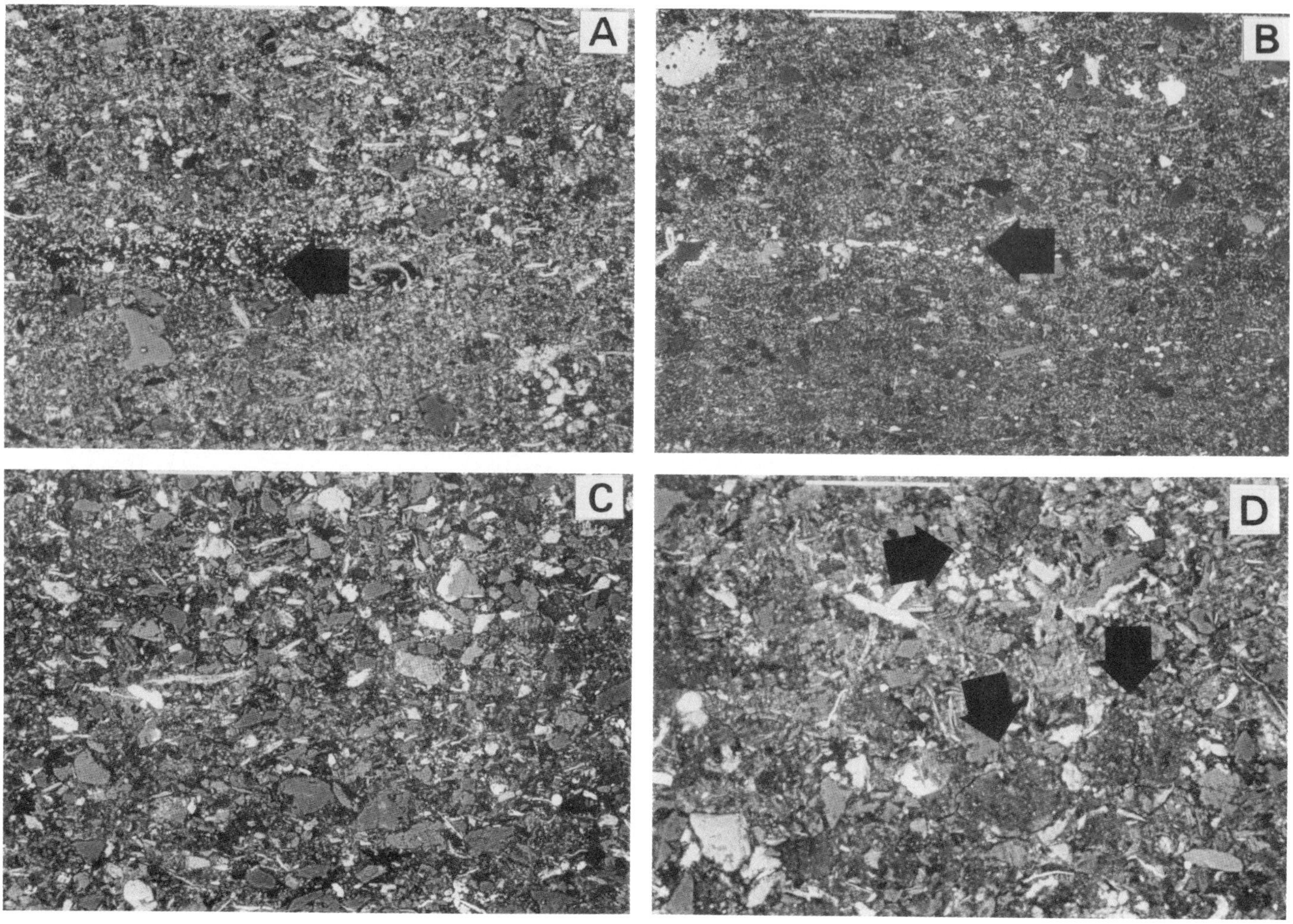

Figure 21.8. BSE micrographs of Ventura Basin mudstones; scale bars are 100 μm. (A) Pelagic mudstone with thin organic lamination; note the dominance of pyrite. (B) Pelagic mudstone with thin pyrite lamination. (C) Turbidite mudstone, near the base of the deposit; note the dominance of coarse silts and poorly developed textural laminations. Some preferred orientation of micas occurs. Grain-grain contacts dominate. (D) Upper portion of turbidite mudstone containing abundant fecal pellets as indicated by arrows. These are ovoid, clay-filled features.

the deposit. Lower portions of the deposit may be grain supported and exhibit a higher degree of preferential orientation of mica compared to upper portions.

Pelagites have a background matrix containing a low amount of silt that is distributed evenly across the sample. Thin horizontal stringers of material occasionally occur; these may be silt, clay, or organic layers.

Bioturbated muds have an uneven distribution of silt particles across a sample, resulting in a mottled texture.

This study has shown that microscopic examination of fine-grained sediments is an important source of environmental information. More work is necessary to verify and expand the model presented herein.

References

Aylmore, L.A.G., and J.P. Quirk, 1960. Swelling of clay-water systems. Nature, v. 183, p. 1752–1753.

Fleischer, P., 1970. Mineralogy of hemipelagic basin sediments, California Continental Borderland. USC-GEOL 70-7, Office of Naval Research, Washington, D.C., 208 p.

Gidman, J., A.E. Worthington, and G.A. Newman, 1987. Wellsite handling of poorly consolidated core, Inglewood Field, Los Angeles Basin. Bulletin of the American Association of Petroleum Geologists, v. 71, p. 560.

Johnson, B.J., 1978. Vertical sequence analysis of a deep sea fan system, Santa Paula Creek, California. M.S. Thesis, University of Southern California. Los Angeles, CA, 300 p.

Krinsley, D.H., K. Pye, and A.T. Kearsley, 1983. Application of backscattered electron microscopy in shale petrology. Geological Magazine, v. 120, p. 109–208.

Malouta, D.N., D.S. Gorsline, and S.E. Thornton, 1981. Processes and rates of recent (Holocene) basin filling in an active transform margin: Santa Monica Basin, California Continental Borderland. Journal of Sedimentary Petrology, v. 51, p. 1077–1096.

Potter, P.E., J.B. Maynard, and W.A. Pryor, 1980. Sedimentology of Shale. Springer-Verlag, New York, 306 p.

Pye, K., and D.H. Krinsley, 1984. Petrographic examination of sedimentary rocks in the SEM using backscattered electron detectors. Journal of Sedimentary Petrology, v. 54, p. 877–888.

Reynolds, S., 1986. A recent turbidity current, Hueneme Fan, California: reconstruction of flow properties. Sedimentology, v. 34, p. 129–137.

Reynolds, S., 1988. The fabrics of deep-sea detrital muds and mudstones: a scanning electron microscope study. Ph.D. Thesis, University of Southern California, Los Angeles, CA, 169 p.

Schweller, W.J., J. Gidman, A.A. Reed, and C.W. Grant, 1987. Topographically restricted deep-sea fan deposits, Inglewood field, Los Angeles Basin. Bulletin of the American Association of Petroleum Geologists, v. 71, p. 611.

CHAPTER 22

Physical Properties and Microstructural Response of Sediments to Accretion–Subduction: Barbados Forearc

Elliott Taylor, Patti J. Burkett, Jeri D. Wackler, and John N. Leonard

Introduction

Convergent margins are one of three major geological boundaries in terms of relative plate motion. These margins are sites where large piles of sediment are being physically transferred from one tectonic plate to another, often forming the roots of extensive mountain systems. Important global fluid and mass fluxes take place at these locations. The Barbados Ridge represents the growing pile resulting from sediment being scraped off the Atlantic plate and added to the Caribbean plate (Fig. 22.1). This transfer process results in accretion through lateral shortening and vertical thickening of the sediment package. Some of the sediment pile on the Atlantic plate is not initially offscraped. A major detachment fault, or decollement, separates material being accreted from that being subducted. Samples from this region were taken to determine the structural and physical changes associated with sediments undergoing accretion or subduction, and to understand the features that control the location of the slip plane, or decollement, between these two units.

Understanding sediment deformation and related physical changes at convergent margins has been the motive behind several Deep Sea Drilling Project (DSDP) and Ocean Drilling Program (ODP) coring programs. Sediment fabrics resulting from deformation during the accretion process are wide ranging and have been documented on scales varying from meters to microns. Small-scale visible structures include veins, scaly fabrics, scaly cleavage, folding, and faulting (Lundberg and Moore, 1986). Sediment physical properties may undergo significant to little change in these tectonically active areas, as described for the Barbados Forearc (Marlow et al., 1984), Middle America Trench (Shephard et al., 1982; Taylor and Bryant, 1985), Peru Trench (Suess et al., 1988), Cascadia margin (Carson, 1977), Aleutian Trench (Lee et al., 1973), Japan Trench (Carson and Bruns, 1980; Shephard and Bryant, 1980), and Nankai Trough (Bray and Karig, 1985, 1986; Johns, 1986). The relationship between the variability of physical change and related structural and tectonic features of these margins has been suggested through studies such as Carson et al. (1982), Shephard and Bryant (1983), and Carson and Berglund (1986). Cowan et al. (1984) described the development of strong, planar, preferred orientation in clayey mud related to a major thrust fault cored at the Barbados forearc. Chiou (1981) and Shephard et al. (1980) discussed some aspects of sediment microfabric reorientation and development of a preferred fabric direction in fine-grained sediments from the Middle America Trench. Knipe (1986) provided detailed scanning electron microscopy (SEM) and transmission electron microscopy (TEM) images of vein-related microfabric in Japan Trench sediments.

The Barbados convergent margin was cored during DSDP Leg 78A (Biju-Duval et al., 1984) and ODP Leg 110 (Mascle et al., 1988). Both cruises provided material from the Barbados forearc to study the process of offscraping and subduction. The Leg 110 coring program drilled at six sites located along a 22-km-long transect perpendicular to the deformation front (Fig. 22.1). In this chapter we focus on results from Leg 110 Sites 672 and 671. Site 671 penetrated 691 m below sea floor (mbsf), in 4932 m water depth, recovering cores from the offscraped accretionary prism and sediments being subducted below the decollement. Site 672, drilled 6 km seaward of the deformation front in 4982 m water depth, sampled 494 m of the sediment section resting on the Atlantic plate. In this study we document the microstructural changes in sediments in context of the observed physical properties, with emphasis on subcentimeter-scale features observed in X-radiography, SEM, and TEM.

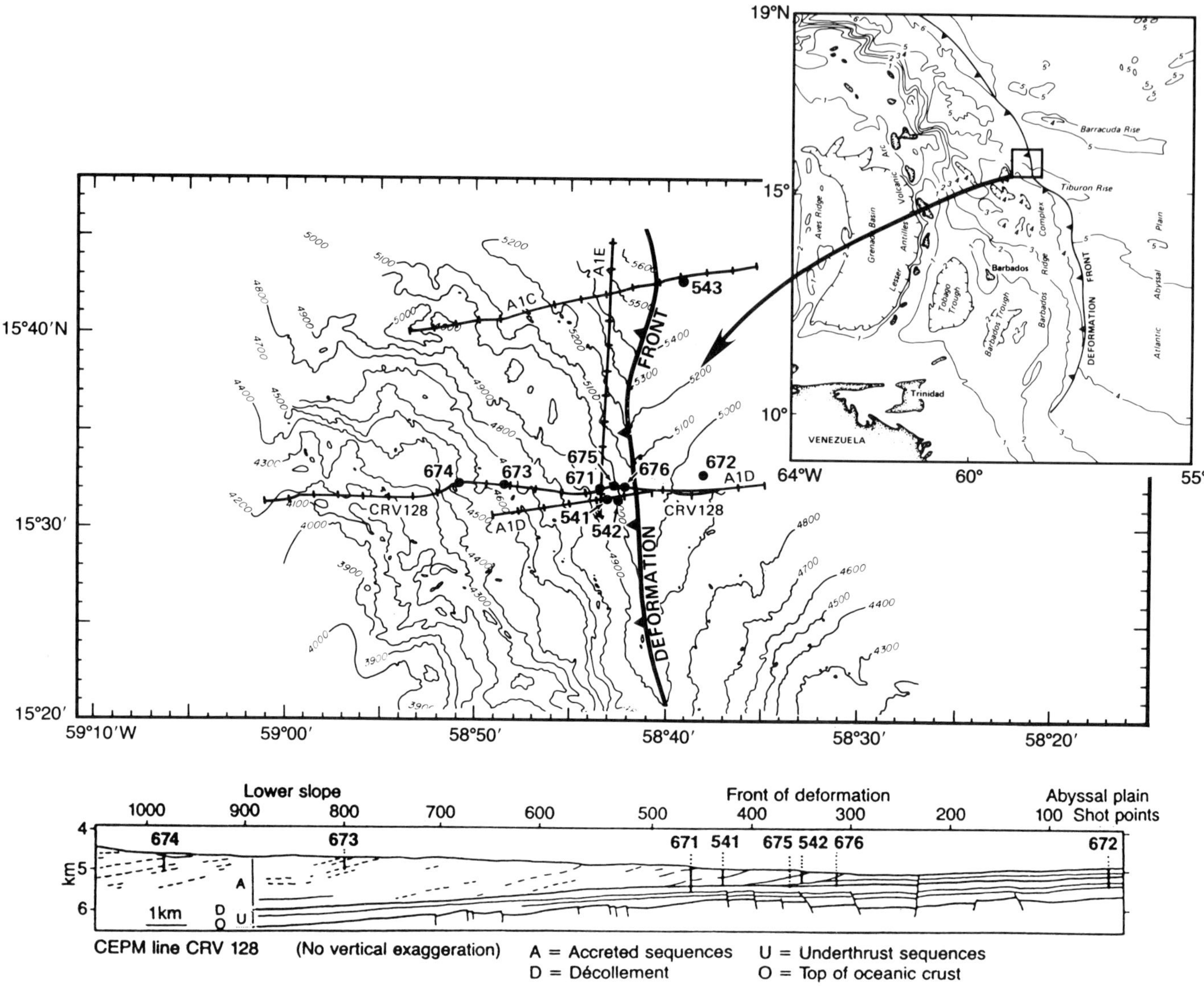

Figure 22.1. Location of Leg 110 drill sites across the Barbados Ridge.

Methods

Shipboard analyses of physical properties included index properties, GRAPE (indirect measure of bulk density through gammaray transmission), compressional wave velocity, vane shear strength, and thermal conductivity (Mascle et al., 1988). Shore-based studies covered both geotechnical and structural studies. Geotechnical testing included uniaxial consolidation (oedometer) tests (D2435 in ASTM, 1981; Lowe et al., 1964) and permeability (Lambe, 1951). Geotechnical samples were obtained to depths limited by sediment induration (approximately 200 mbsf at Sites 671 and 672).

Consolidation results are expressed as overconsolidation difference (OCD):

$$\text{OCD} = P'_c - P'_o$$

where P'_c is the preconsolidation stress and P'_o the effective overburden stress (Lambe, 1969). Permeabilities are expressed in terms of hydraulic conductivity. Values reported herein are interpolated for expected *in situ* void ratios and viscosity corrected to *in situ* temperatures using thermal gradients reported in Mascle et al. (1988).

Structural studies of the ODP Leg 110 cores include visual core descriptions complemented with shore-based X-radiographs, SEM, and TEM analyses. Samples for shore-based studies were selected to obtain the best coverage of major structural elements in the convergence process: offscraped, decollement, and underthrust sediments. Attempts were made to choose samples near

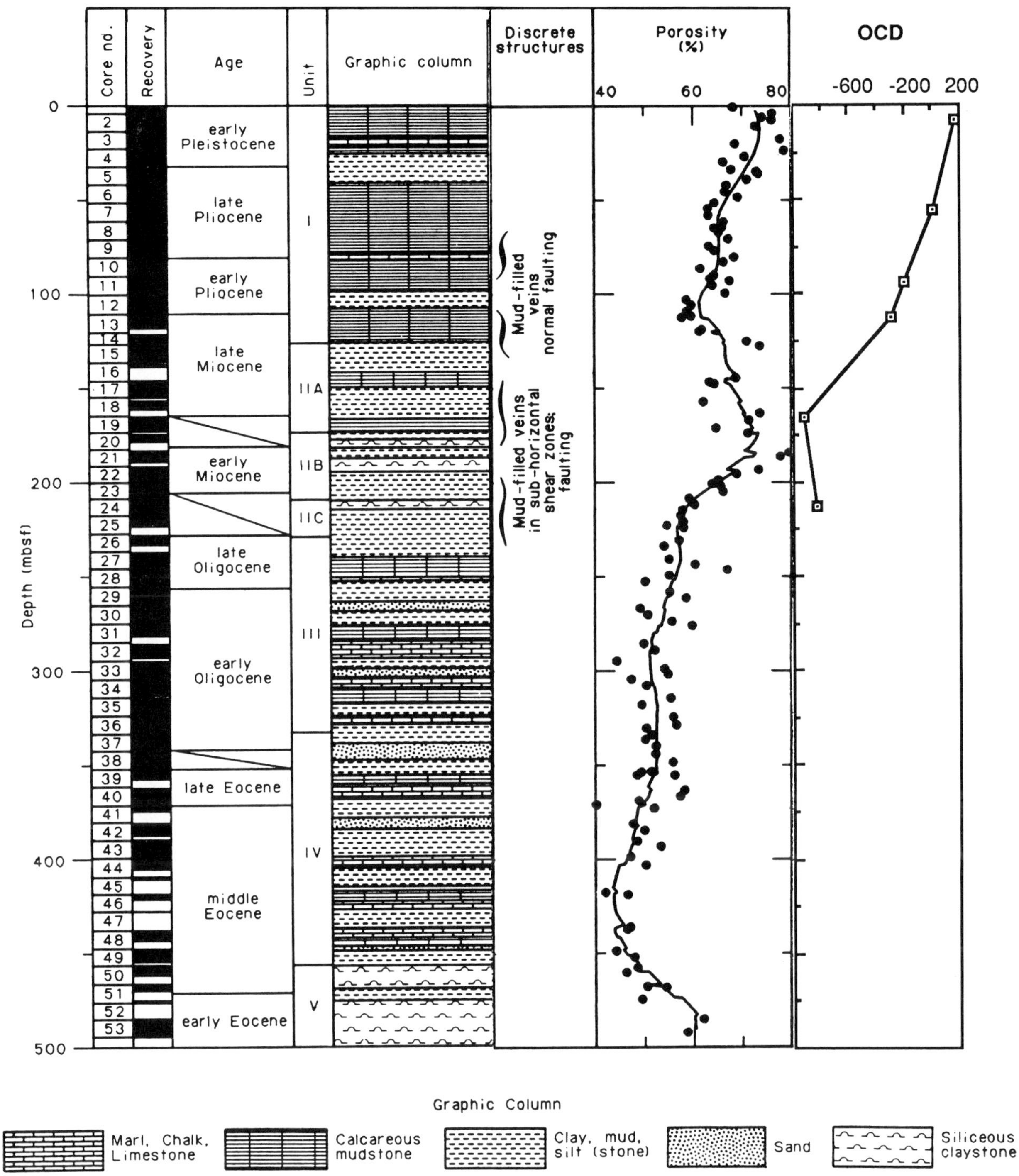

Figure 22.2. Lithology, structural features, porosity, and OCD of Site 672 sediments. Shallow overconsolidated sediments appear to be progressively underconsolidated with increasing depth.

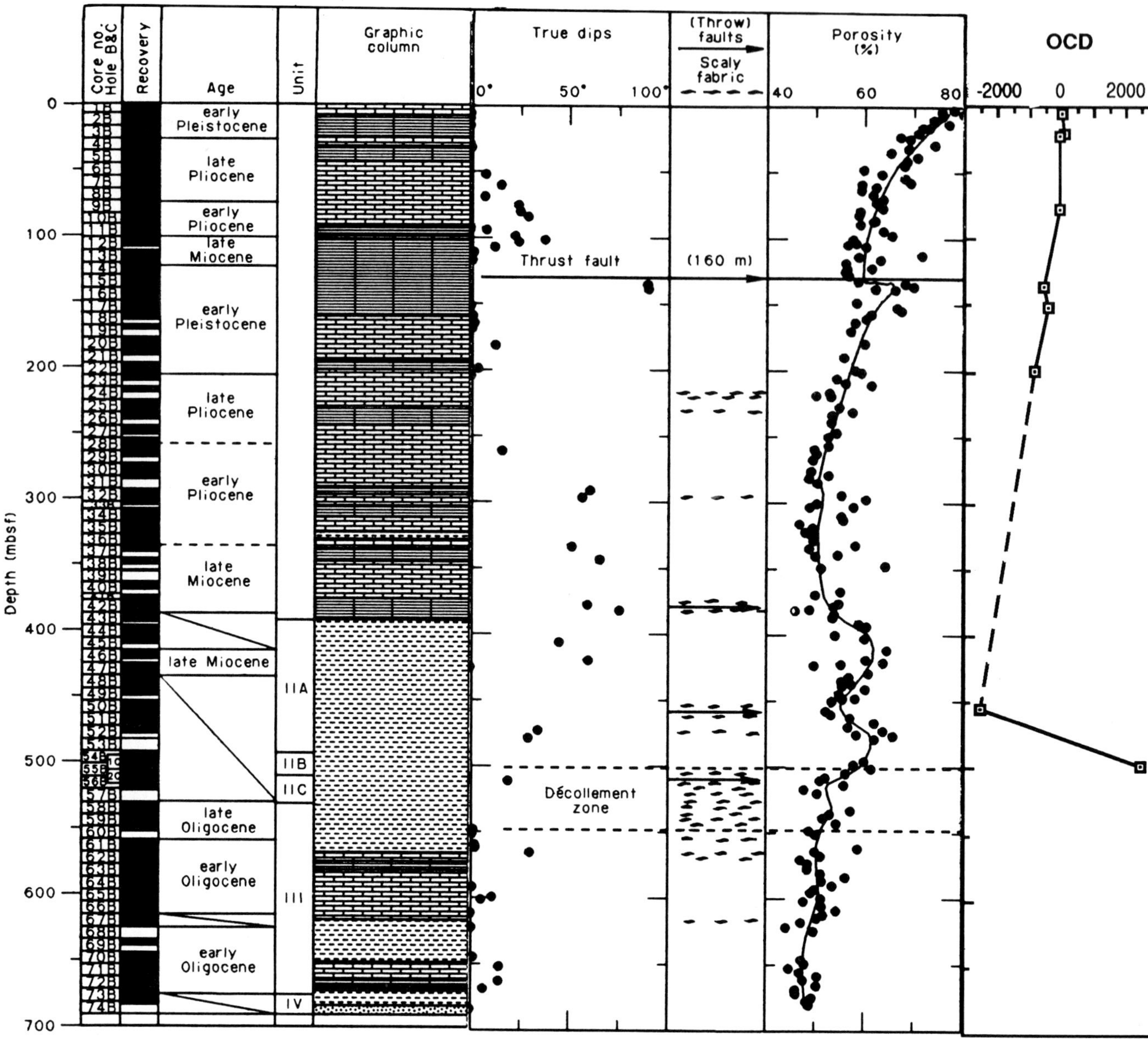

Figure 22.3. Lithology, structure features, porosity, and OCD of Site 671 sediments. The apparent underconsolidation of accreted sediments is strongly contrasted by the high OCD within the decollement (see Fig. 22.2 for key to symbols).

those for geotechnical analyses and from sites nearest the deformation front: Sites 671, 676, and 672.

X-Radiographic studies were carried out on selected 1-cm-thick slabs of core to obtain optimum resolution. Stereo pair X-radiographs were prepared to enhance structural interpretation. Exposures were obtained using methods described in Wackler (1988).

SEM and TEM samples were oriented blocks prepared for sputter coating (SEM) or epoxy impregnation (TEM) through critical point drying (Chiou, 1981; Burkett, 1987; Burkett et al., this volume). Most SEM samples were scanned at 15, 100–200, 450, and in cases at 1000× magnifications. TEM photomosaics were prepared to provide fine-scale coverage of six samples from Hole 671B. Hence, TEM samples represent material only from the offscraped sediment pile.

Sediment Physical Properties

The changes affecting the sediment column during convergence may be reflected in alterations of physical and geotechnical

Figure 22.4. Stereopair X-radiographs of millimeter-scale structural deformation in Barbados Ridge sediments. Top: Sample 672A-28X-3. Faint bidirectional faults with dips near 20° are located in the upper half (A). The large fracture in the middle is drilling induced (B). The light dendritic pattern at the top is interpreted as incipient dewatering structures (C). Bottom: Sample 671B-55X-5. Examples of microscopic subhorizontal scaly fabric are indicated by arrows.

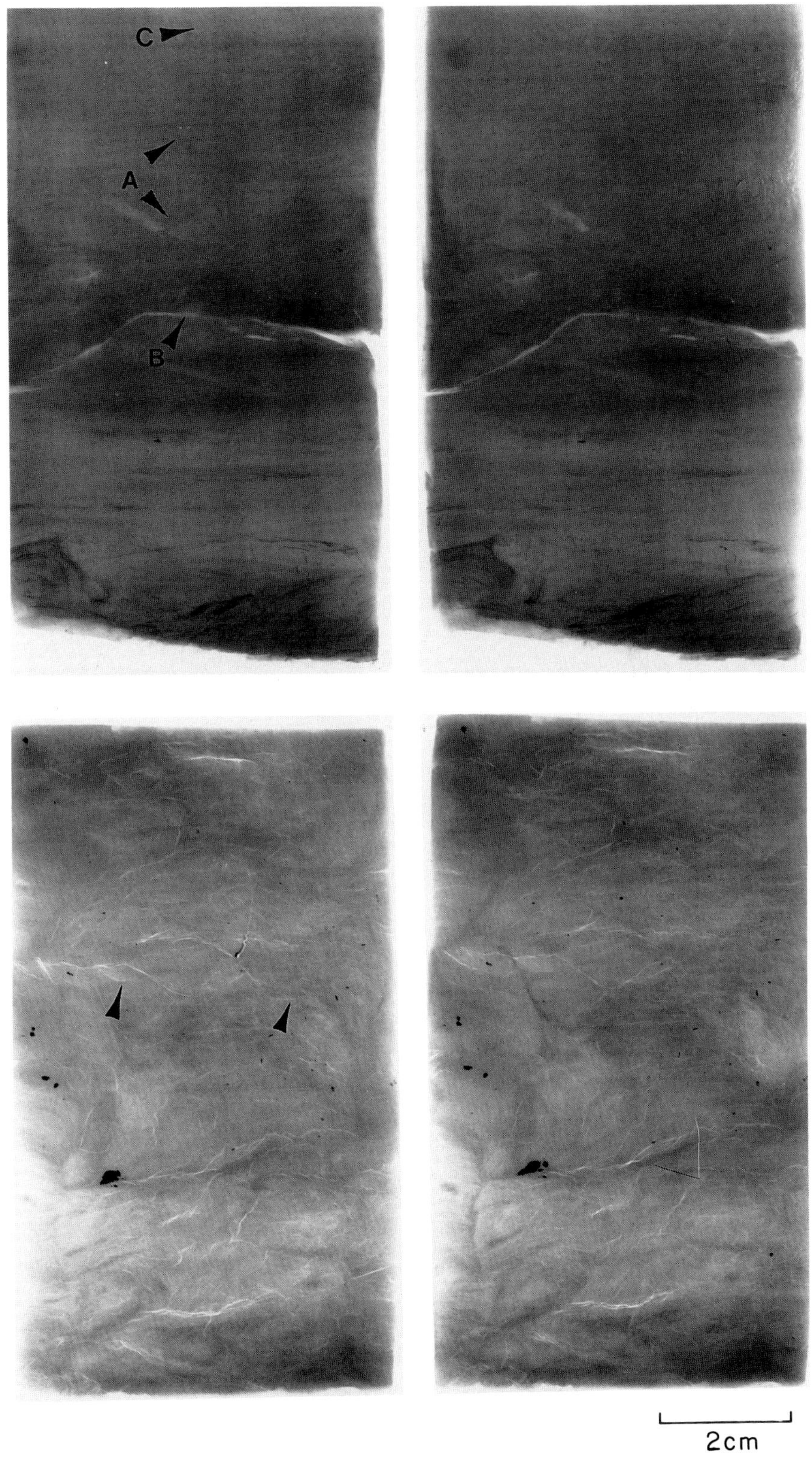

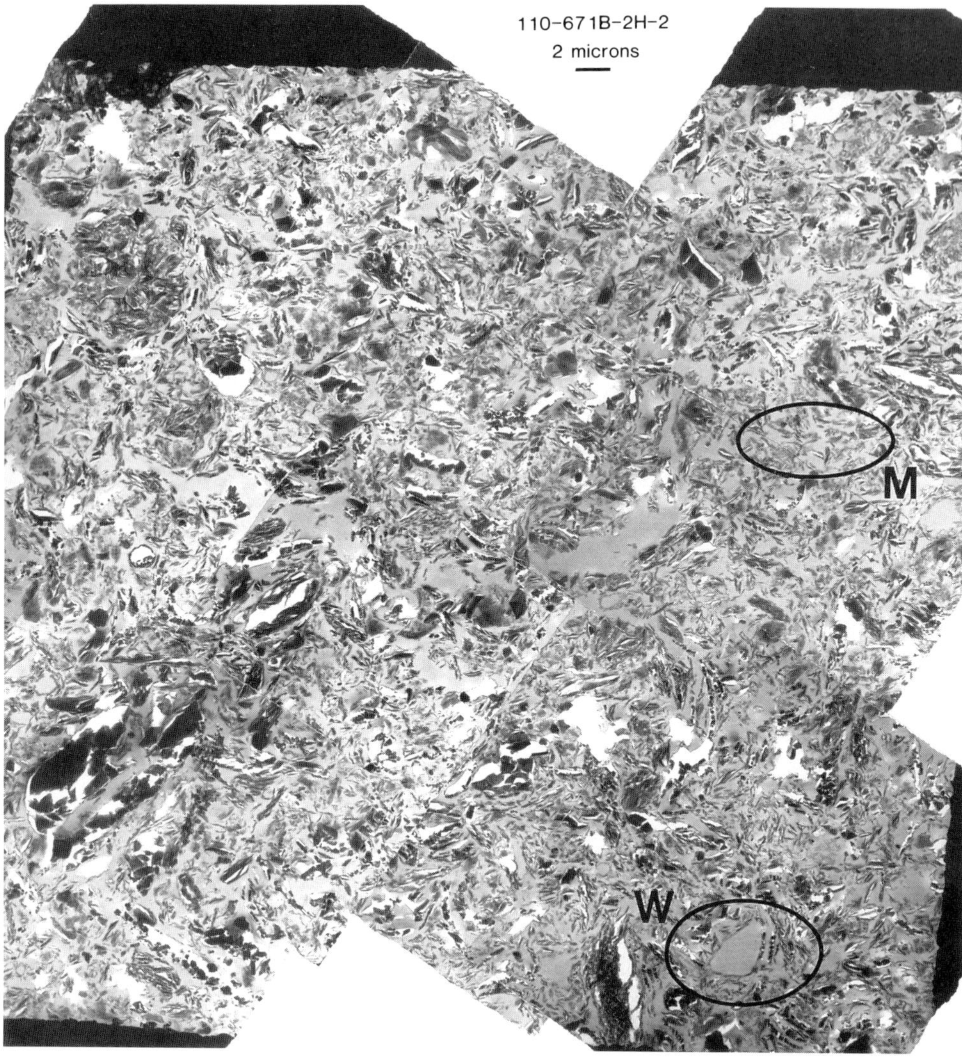

Figure 22.5(a–c). TEM photograph mosaics of Sample 671B-2H-2 (8.9 mbsf). Porosity: 76%. This unoriented sample exhibits an open structure of randomly oriented domains (M, matrix) surrounding biogenic particles (B), such as foraminiferal tests and nannofossils, and fractillites (F). Possible dewatering paths are indicated by channels (C) and by pore spaces with face-to-face arranged particles forming a rim around the pathway (W).

properties. Comparison of sediment characteristics from the seaward Site 672 with those of Site 671 in the prism provides an overview of the changes in the immediate vicinity of the deformation front. Site 672, drilled 6 km east of any evidence of convergence in seismic profiles, was intended to provide reference material against which sediments from within the arc could be compared. Structurally disrupted sediments and evidence of significant fluid flow within the cored section, however, suggest this site does not adequately serve as a reference (Mascle et al., 1988). Nevertheless, contrasting sediment properties between Sites 671 and 672 serve to illustrate the scale of structural and physical modification taking place during early convergence.

Site 672

The porosity versus depth profile of Site 672 sediments shows a pattern consistent with most marine sediments. A sharp decrease in porosity near surface proceeds downhole to a gentler gradient (Fig. 22.2). Superimposed on Site 672 is the porosity–depth function defined for terrigenous sediments (Hamilton, 1976). A clear departure for this trend occurs at about 120 mbsf, corresponding to a lithologic change from calcareous oozes/mud above to the smectitic and radiolarian mud(stone) of Unit 2 below. Maximum porosities of 78% occur at the top of the radiolarian-rich Subunit IIb and decrease sharply within Subunit IIc. Sediment composition appears to exert a major control on the physical properties at this site.

Geotechnical test results indicate near-surface sediments are overconsolidated above 7 mbsf, normally consolidated to about 50 mbsf, and underconsolidated below that depth (Fig. 22.2). Maximum apparent underconsolidation occurs near the future, or propagating, decollement within Subunit IIb. This underconsolidated state may be a result of low effective stress within the sediment column arising from excess *in situ* pore pressures at those depths. Below the propagating decollement, sediments are

Figure 22.5. (b)

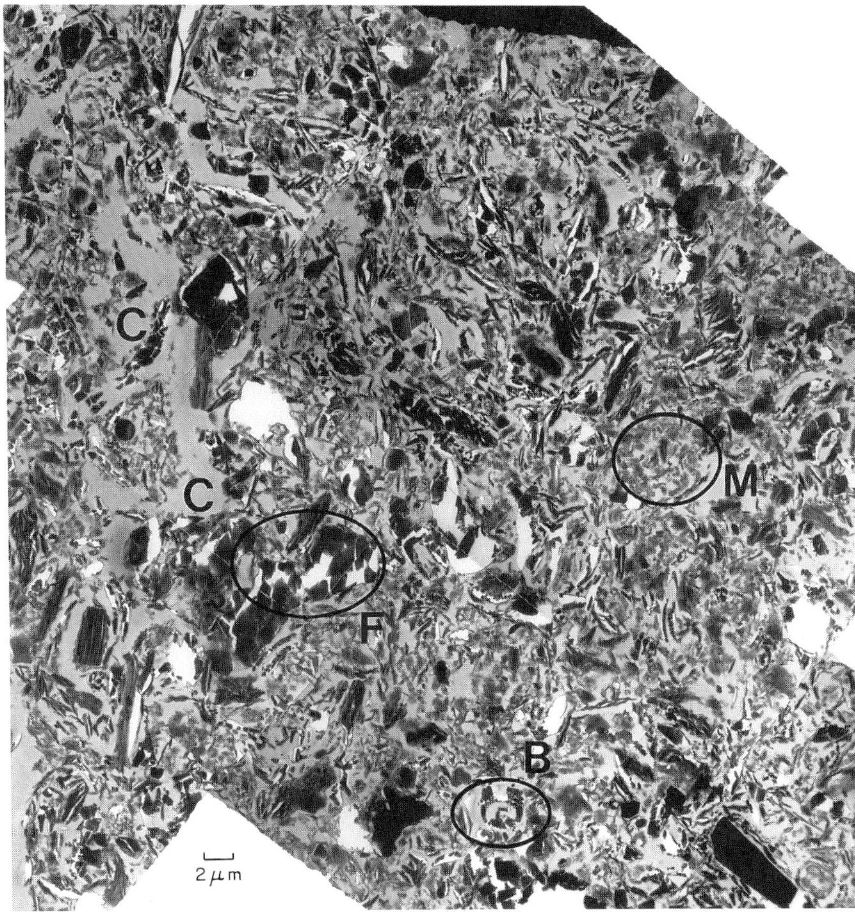

underconsolidated to the extent of those immediately above that major slip plane, thus disrupting the downhole trend of increasing underconsolidated behavior.

Permeabilities vary from about 10^{-6} to 10^{-8} cm/sec within the upper 200 m. These values are typical of hemipelagic sediments under similar burial depths (Bryant et al., 1981).

Site 671

The nearly 700-m-deep hole cored at this site sampled offscraped and underthrust sediment, coring through the decollement at 500–540 mbsf. Several thrust faults cut through the offscraped sediment section at this site, resulting in structural repetition of stratigraphic units (Fig. 22.3). Sediments in the uppermost 120 m exhibit a porosity–depth function similar to sediments at Site 672. A major thrust fault at 128 mbsf is evident in the nonuniform porosity profile. Below that depth, porosity decreases nonlinearly to near 200 mbsf where it resumes a fairly linear decrease with depth. Sediment composition, as at Site 672, strongly influences the index properties at Site 671. The repetition of porosity maxima at 410 and 475 mbsf is attributed to a probable thrust fault between these two depths. The decollement, stratigraphically located within a lower Miocene siliceous mudstone interval, is marked by an abrupt change in porosity. Values decrease from about 60 to 47% over the 40-m-thick interval.

Geotechnical results of Site 671 samples are similar to those obtained for Site 672. The OCD values are more negative with depth in the offscraped sequence. Maximum underconsolidation is probably located immediately overlying the decollement and, as at Site 672, may reflect lower effective overburden stress resulting from higher pore pressures at those depths. The distinct shift of the OCD within the decollement suggests sediments are effectively consolidating, probably as a result of shear-imposed deformation and/or the proximity to effective dewatering paths through more permeable fractured mudstone.

Permeabilities within Site 671 sediments are directly related to the porosity trends. The range of 10^{-7} to 10^{-9} cm/sec is slightly lower than that of Site 672. At comparable depths, Site 671 permeabilities are almost half an order of magnitude less than those of Site 672. This decrease across the arc, like porosity, indicates lateral consolidation is taking place within the accreted package.

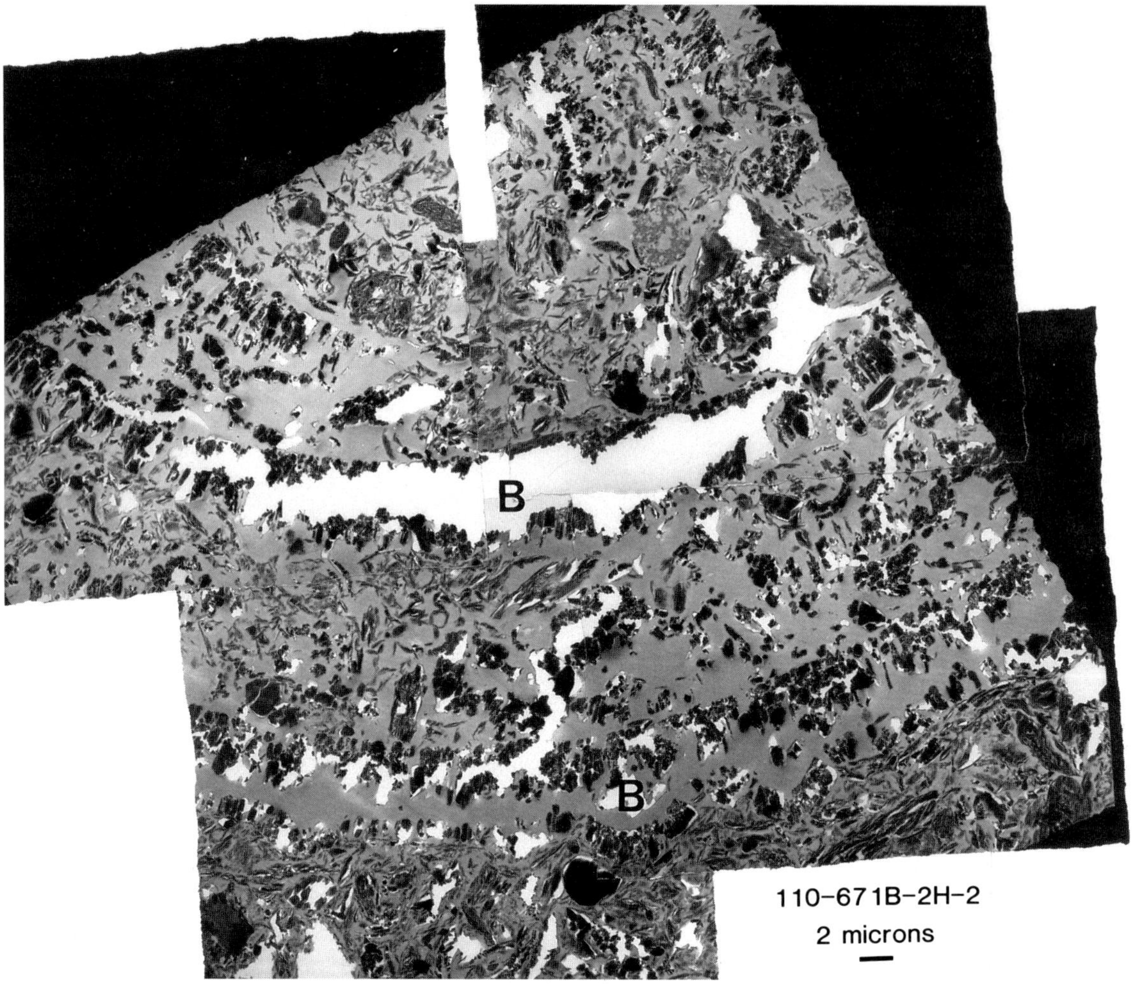

Figure 22.5. (c)

Sediment Macrostructural Features

Structural evidence of the convergent process is observable in sediments from both Sites 671 and 672. The decollement at Site 671 is perhaps the principal structural feature, characterized by oxidized, Mn-stained, siliceous mudstone with a pervasively subhorizontal scaly fabric. Evidence of fluid migration through scaly fabric is found in localized calcite or rhodochrosite-filled veins. Similar features are also described for sediments at Site 672, casting doubt on its usefulness as a "reference" site.

Zones of scaly fabric are found at both Sites 671 and 672, and not always associated with thrust faults. Scaly fabric and dewatering veins are characteristic of sediments overlying the decollement, whereas underthrust sediments appear undeformed or exhibit brittle cleavage. Other sites cored during Leg 110 also show sediments are progressively deformed as they are further incorporated into the accretionary wedge.

X-Radiography of select sediments serves to closely examine structural features on centimeter to millimeter scales. Wackler (1988) documented incipient development of scaly fabric, slickensided microthrusts, microfolding, and submillimeter dendritic patterns possibly reflecting fluid-escape structures (Fig. 22.4).

Sediment Microfabric

SEM and TEM photographs of selected samples from Site 671 were made to document fabric modification in accretion and subduction. The uppermost sediments have an open fabric of fine-grained sediments interspersed with silt- to sand-sized foraminifer tests and fractillites (fractured illites; Bryant and Bennett, 1988) (Fig. 22.5). The foraminiferal tests help maintain a high porosity in these sediments. Dewatering channels, rimmed by locally aligned domains, appear throughout these uppermost sediments. The edge-to-edge and edge-to-face clay particle arrangements in the shallow sediments of Hole 671B become progressively more oriented downhole, forming domains of face-to-face clay packets (Fig. 22.6). Core 671B-8H-4 exhibits numerous fractillites in TEM, and development of preferred subhorizontal alignment in localized fine-grained matrices.

Figure 22.6(a,b). TEM photographs of Sample 671B-8H-4 (68.9 mbsf). Oriented with upcore direction at top of photographs. Porosity: 63%. Fairly open fabric of particle domains show zones of preferred orientation (O). Some possible packets of remolded (R) fabric surround fractillites (F).

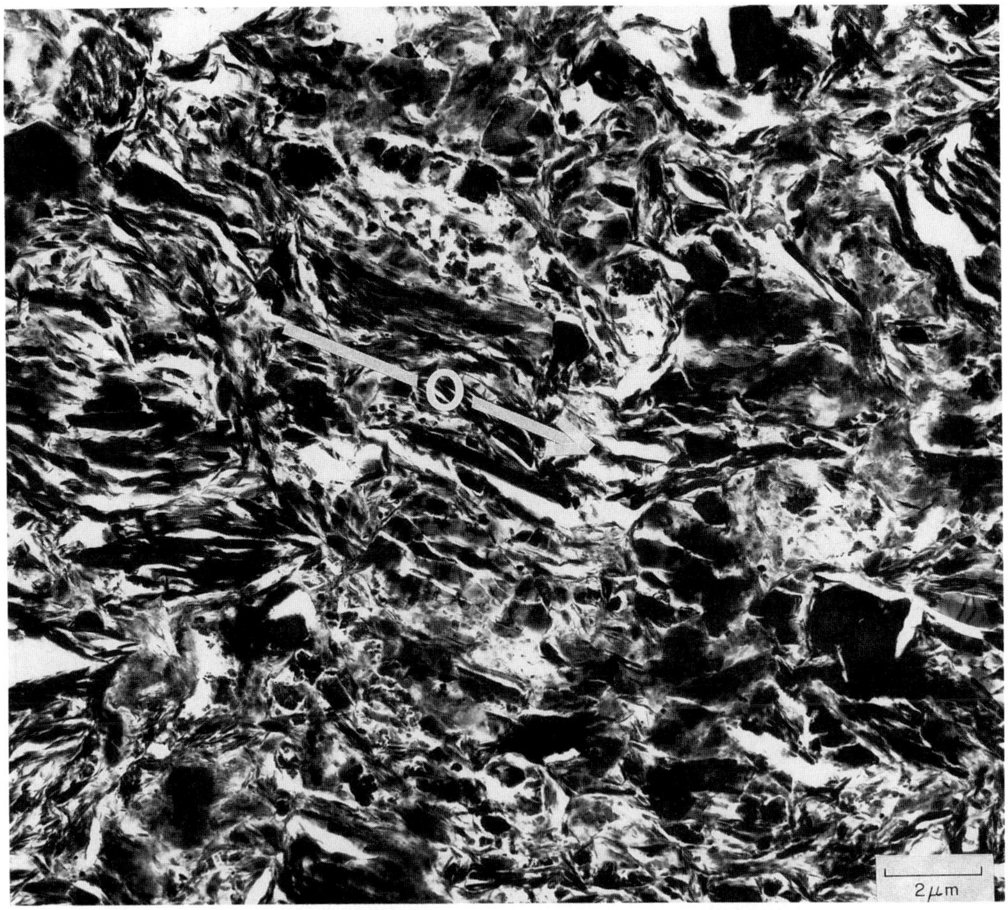

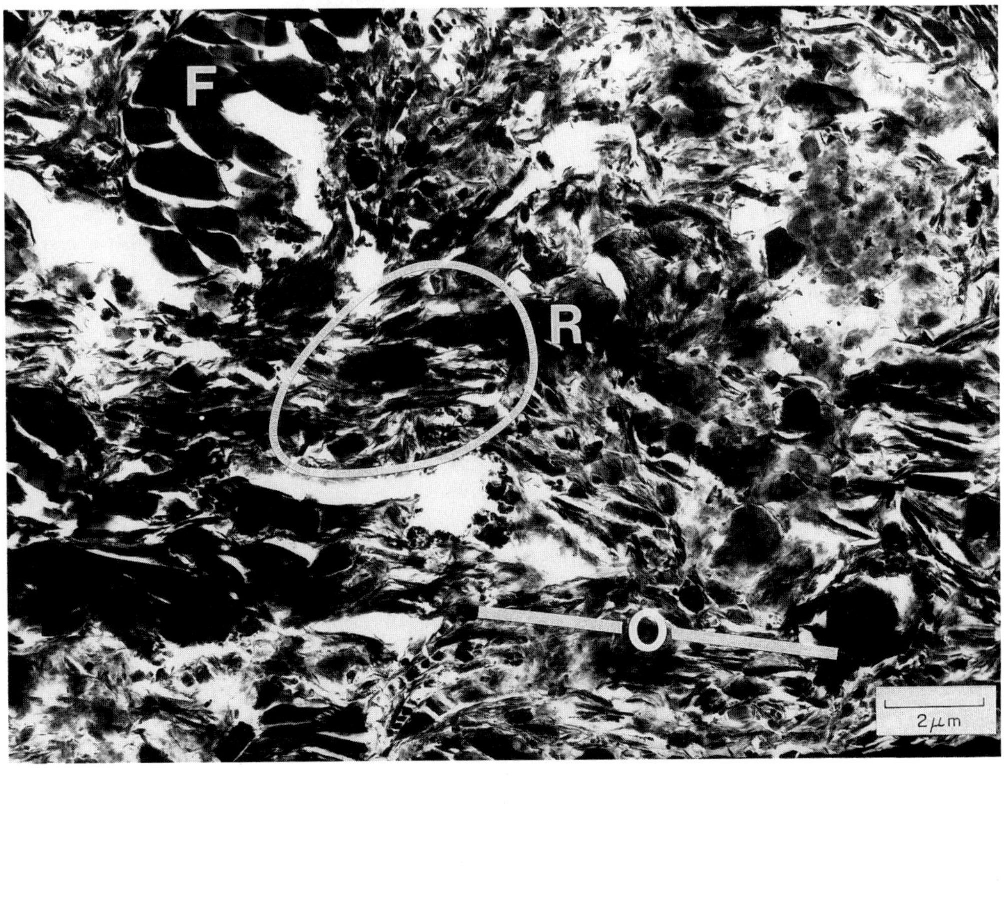

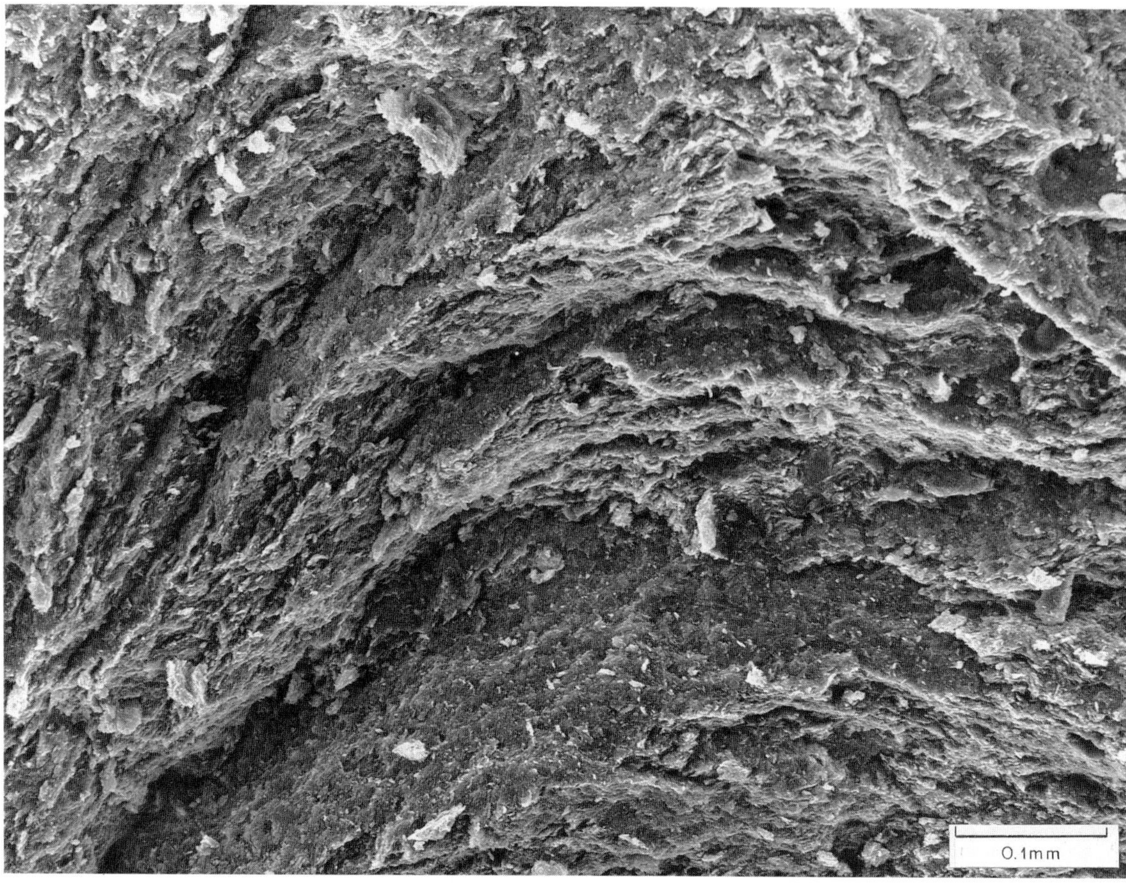

Figure 22.7. SEM photograph of Sample 671B-14X-4 (125.2 mbsf). Oriented with upcore direction at top of photograph. Porosity 58%. Sample from 3 m above the major thrust fault may reflect shear-related drag remolding of sediment fabric.

Localized remolding is suggested from swirled packets surrounding denser clays and fractillites. The lower porosity of this sample (63% compared to 76% for Sample 671B-2H-2) is associated with microfabric exhibiting local remolding and general development of a preferred orientation.

Sample 671B-14X-4 is located 2.8 m above the major thrust fault at 128 mbsf. The material in this sample represents the base of the uppermost thrust package. The porosity at the base of this package is nearly 50% compared to 72% porosity just below the fault. Figure 22.7 shows the relatively strong development of bedding parallel fabric with remolding possibly associated with near-field strain around the fault plane.

Sample 671B-28X-4 is from upper Pliocene calcareous mudstone lithologically equivalent to sediments sampled in Core 671B-8H. A major difference between these otherwise comparable samples is the much lower porosity of the deeper core: 50% instead of 60%. The lower porosity is evident in SEM and TEM images showing significantly less random fabric (Fig. 22.8). Zones of edge-to-edge aligned clay domains and particles are pervasive throughout this sample. The overall fabric within the oriented sample, however, does not appear to be preferentially aligned. Microshear planes form bent end-to-end chains. Lack of a general preferred orientation may account for no apparent fissility in these low porosity, calcareous mudstones.

Siliceous and smectitic mudstones of Unit II have a fine-grained, random, open fabric reflecting the porosities near 60% overlying the decollement (Fig. 22.9). The finer grain matrix does not exhibit dewatering channels or remolding evident in overlying sediments. The microfabric character of Sample 671B-45X-3 suggests little shear or compressive stress has altered this section of sediment.

Shear-related microstructural collapse of the open siliceous and smectitic mudstone fabric observed in Sample 671B-45X-4 takes place within, and below, the decollement. Sample 671B-56X-2, from the base of the decollement, has a very strongly oriented subhorizontal, bedding parallel fabric (Fig. 22.10). This type of fabric probably accounts for the scaly macrofabric described in visual core descriptions at this depth. The porosity loss through the 40-m-thick decollement is greater than 10% and accounts for the collapsed structure observed in this sample.

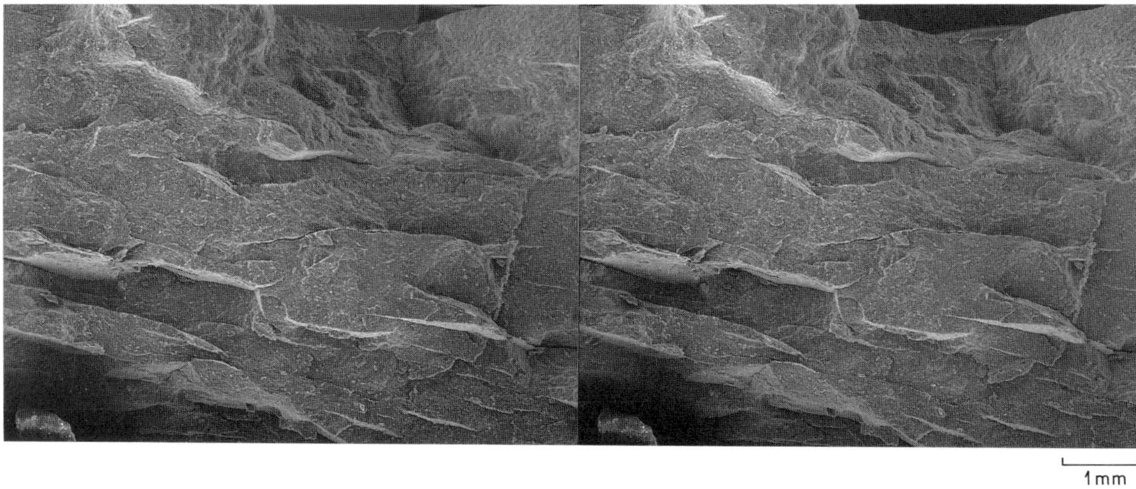

Figure 22.11. SEM stereo-photograph of Sample 671B-61X-2 (560.0 mbsf). Oriented with upcore direction at top of photograph. Porosity: 50%. The base of the decollement shows sediment deformation following shear-related consolidation. Dense, subhorizontal fabric creates low intergranular permeabilities but brittle scaly cleavage results in higher formation permeability.

Progressive consolidation, dewatering, and hardening of this fine-grained mudstone are apparent as bedding-parallel clay packets develop into mudstone with brittle, subhorizontal scaly cleavage (Fig. 22.11).

Microfabric and Geotechnical Characteristics

Although comparison of sediment physical properties between Sites 671 and 672 suggests little evidence of lateral consolidation effects on accreted sediments, structural deformation within the offscraped section testifies to stresses on the column. Triaxial test results from sediments overlying the decollement indicate rotation of major and minor stress directions is associated with accretion (Moran and Christian, 1990). However, Site 672 and 671 sediments are progressively underconsolidated with depth suggesting *in situ* excess pore pressures. Despite the complex stress field there, a major control on the geotechnical character of the sediments near the Barbados deformation front seems to be lithologic composition.

Sediment microfabrics observed in SEM and TEM seem to substantiate the compositional control on sediment deformation. The offscraped calcareous mud(stone) is deformed within tectonic packages and near fault zones. These sediments develop zones of highly aligned chains, distributed in an apparently random fashion through the overall sample. The result of applied stresses in the accreted, calcareous mud(stone) is a decrease in porosity and permeability that closely follows that of nonaccreted equivalent facies at Site 672. Except for localized zones, the calcareous mud(stone) has developed little to no fissility. Localized zones of bedding-parallel, preferential particle alignment may be associated with thrust-fault strains.

High porosity, fine-grained siliceous and smectitic mudstones in offscraped sediments above, and forming, the decollement show an open, random fabric immediately above this major displacement plane. The low permeabilities of overlying calcareous mudstone may impede pore fluid migration, resulting in excess pore pressure build-up. The increase of pore pressure, in turn, reduces the strength of this sediment (Taylor and Leonard, 1990). Lower in the decollement, shear-induced strains have developed a strongly oriented microfabric in these same sediments leading to a pronounced porosity decrease within the 40-m-thick zone. The bedding-subparallel fabric results in greater fissility, which under stresses in the decollement may lead to development of scaly fabric. This scaliness in turn enhances bedding-parallel permeability allowing dewatering and structural collapse of the fabric (Moore et al., 1988). Continued consolidation and increasing strength lead to the scaly cleavage found in sediments below the decollement.

The calcareous mud(stone) and smectitic/siliceous mudstones recovered from the Barbados Forearc exhibit contrasting physical properties and structural deformation. Sediment composition appears to play a key role as a focus for the developing decollement at this active margin. The microstructural deformation observed within the Leg 110 sediments suggests significant small-scale disruption takes place during the early stages of accretion–subduction, despite the apparently limited reaction implied by the physical properties record.

References

American Society for Testing Materials, 1981. Annual Book of Standards, Part 19: Natural Building Stones, Soil and Rock. Philadelphia (ASTM).

Biju-Duval, B., J.C. Moore, et al., 1984. Initial Reports DSDP Volume 78A. U.S. Government Printing Office, Washington, D.C., 603 p.

Bray, C.J., and D.E. Karig, 1985. Porosity of sediments in accretionary prisms and some implications for dewatering processes. Journal of Geophysical Research, v. 90, p. 768–778.

Bray, C.J., and D.E. Karig, 1986. Physical properties of sediments from the Nankai Trough, Deep Sea Drilling Project Volume 87A, Sites 582 and 583. In: Kagami, H., D.E. Karig, et al. (eds.), Initial Reports DSDP Volume 87. United States Government Printing Office, Washington, D.C., p. 827–842.

Bryant, W.R., and R.H. Bennett, 1988. Origin, physical, and mineralogical nature of red clays: the Pacific Ocean basin as a model. Geo-Marine Letters, v. 8(8), p. 189–249.

Bryant, W.R., R.H. Bennett, and C.E. Katherman, 1981. Shear strength consolidation, porosity, and permeability of oceanic sediments. In: Emiliani, C. (ed.), The Oceanic Lithosphere: The Sea, v. 7. Wiley, New York, p. 1555–1616.

Burkett, P.J., 1987. Significance of the microstructure of Pacific red clays to nuclear waste disposal. M.S. Thesis, Texas A&M University, College Station, TX, 79 p.

Carson, B., 1977. Tectonic modification of deep-sea sediments at the Washington-Oregon continental margin: mechanical consolidation. Marine Geology, v. 24, p. 289–307.

Carson, B., and T.R. Bruns, 1980. Physical properties of sediments from the Japan Trench margin and outer trench slope: results from Deep Sea Drilling Project Leg 56 and 57. In: Scientific Party, Initial Reports DSDP Volumes 56 and 57, Part 2. United States Government Printing Office, Washington, D.C., p. 1187–1199.

Carson, B., and P.L. Berglund, 1986. Sediments deformation and dewatering under horizontal compression: experimental results. In: Moore, J.C. (ed.), Structural Fabrics in Deep Sea Drilling Project Cores from Forearcs. Geological Society of America Memoir 166, p. 135–150.

Carson, B., R. von Huene, and M. Arthur, 1982. Small-scale deformation structures and physical properties related to convergence in Japan Trench slope sediments. Tectonics, v. 1, p. 277–302.

Chiou, W.A., 1981. Clay fabric of gassy submarine sediments. Ph.D. dissertation, Texas A&M University, College Station, TX, 248 p.

Cowan, D.S., J.C. Moore, S.M. Roeske, N. Lundberg, and S.E. Lucas, 1984. Structural fabrics at the deformation front of the Barbados Ridge complex. Deep Sea Drilling Project Leg 78A. In: Biju-Duval, B., J.C. Moore, et al. (eds.), Initials Reports DSDP Volumes 78A and 78B. United States Government Printing Office, Washington, D.C., p. 535–548.

Hamilton, E.L., 1976. Variations of density and porosity with depth in deep sea sediments. Jour. Sed. Pet., 46(2):280–300.

Johns, M.W., 1986. Consolidation and permeability characteristics of Japan Trench and Nankai Trough sediments from DSDP Leg 87, Sites 582, 583, and 584. In: Kagami, H.D., E. Karig, et al. (eds.), Initial Reports DSDP Volume 87. United States Government Printing Office, Washington, D.C., p. 827–842.

Knipe, R.J., 1986. Microstructural evolution of vein arrays preserved in cores from the Japan Trench. In: Moore, J.C. (eds.), Structural Fabrics in Deep Sea Drilling Project Cores from Forearcs. Geological Society of America Memoir 166, p. 135–150.

Lambe, T.W., 1951. Soil Testing for Engineers. Wiley, New York, 165 p.

Lambe, T.W., and R.V. Whitman, 1969. Soil Mechanics. Wiley, New York, 553 p.

Lee, H.J., J.W. Olsen, and R. von Huene, 1973. Physical properties of deformed sediments from Site 181. In: Kulm, L.D., R. von Huene, et al. (eds.), Initial Reports DSDP Volume 18. United States Government Printing Office, Washington, D.C., p. 897–901.

Lowe, J., P.F. Zaccheo, and H.S. Feldman, 1964. Consolidation testing with back-pressure. Journal of Soil Mechanics and Foundations Division ASCE, v. 90, p. 69–86.

Lundberg, N., and J.C. Moore, 1986. Macroscopic structural features in Deep Sea Drilling Project cores from forearc regions. In: Moore, J.C. (ed.), Structural Fabrics in Deep Sea Drilling Project Cores from Forearcs. Geological Society of America Memoir 166, p. 13–44.

Marlow, M.S., H.J. Lee, and A.W. Wright, 1984. Physical properties of sediment from the Lesser Antilles margin along the Barbados Ridge: Result from DSDP Leg 78A. In: Biju-Duval, B., J.C. Moore, et al. (eds.), Initials Reports DSDP Volume 78A. United States Government Printing Office, Washington, D.C., p. 549–558.

Mascle, A., J.C. Moore, et al., 1988. Proceedings Ocean Drilling Program, Part A-Initial Results Volume 110. Ocean Drilling Program, College Station, TX., p. 603.

Moore, J.C., A. Mascle, et al., 1988. Tectonics and hydrogeology of the northern Barbados Ridge: results from Ocean Drilling Program Leg 110. Geological Society of America Bulletin, v. 100, p. 1578–1593.

Moran, K., and H.A. Christian, 1990. Strength and deformation behavior of sediment from the Lesser Antilles forearc accretionary prism. In: Moore, J.C., A. Mascle, et al. (eds.), Proceedings of the Ocean Drilling Program Scientific Results, Volume 110. Ocean Drilling Program, College Station, TX., p. 279–288.

Shephard, L.D., and W.R. Bryant, 1980. Consolidation characteristics of Japan Trench sediments. In: Scientific Party Initial Reports DSDP Volumes 56 and 57, Part 2. United States Government Printing Office, Washington, D.C., p. 279–312.

Shephard, L.E., W.R. Bryant, and W.A. Chiou, 1980. Geotechnical properties of Middle America Trench sediments, Deep Sea Drilling Project Leg 66. In: Watkins, J.S., J.C. Moore, et al. (eds.), Initial Reports DSDP volume 66. United States Government Printing Office, Washington, D.C., p. 475–504.

Shephard, L.E., and W.R. Bryant, 1983. Geotechnical properties of lower trench inner slope sediments. Tectonophysics, 279–312.

Suess, E., R. von Huene, et al., 1988. Proceedings of the Ocean Drilling Program, Initial Reports, Volume 112. Ocean Drilling Program, College Station, TX., 1015 p.

Taylor, E., and W.R. Bryant, 1985. Geotechnical properties of sediments from the Middle America trench and slope. In: von Huene, R., J. Aubouin, et al. (eds.), Initial Reports DSDP Volume 84. United States Government Printing Office, Washington, D.C., p. 551–570.

Taylor, E., and J.N. Leonard, 1990. Sediment consolidation and permeability at the Barbados forearc. In: Moore, J.C., A. Mascle, et al. (eds.), Proceedings of the Ocean Drilling Program, Scientific Results Volume 110. Ocean Drilling Program, College Station, TX., p. 289–308.

Wackler, J.D., 1988. Microstructures and fabrics of sediments from the lesser Antilles accretionary complex. M.S. Thesis, Texas A&M University, College Station, TX, 159 p.

CHAPTER 23

Anomalous Stress History of Sediments of the Northwest Pacific: The Role of Microstructure

Kathleen A. Dadey, Margaret Leinen, and Armand J. Silva

Introduction and Background

This work is part of an extensive geotechnical investigation of a typical pelagic "red" clay deposit in the Northwest Pacific Ocean east of Shatsky Rise (Fig. 23.1). Details of this project and its results were presented by Dadey (1983) and Marine Geotechnical Consortium (1985). This chapter focuses on anomalous stress history observed in these sediments and considers the possible influences of microstructure. Standard piston and gravity cores are available from the area, but our work concentrated on hydraulic piston core DSDP 576A because of its length (66 m) and well-constrained coring conditions.

Geologic evidence indicates that the region is stable and that deposition has been slow and relatively continuous over the past 65 million years (Heath et al., 1985). Such a history should result in sediments that are normally consolidated, that is, have been subjected only to stresses resulting from the weight of the overlying sediment column and have consolidated completely under that weight. Consolidation is quantified with the overconsolidation ratio (OCR), calculated as the ratio of the preconsolidation stress to the effective overburden stress. The preconsolidation stress, defined as the maximum past stress (Lambe and Whitman, 1969), is determined from consolidation testing. OCRs greater than unity signify overconsolidation; ratios less than one signify underconsolidation. Normally consolidated sediments exhibit OCRs equal to one.

Our theory, however, is contradicted by the results of consolidation tests of DSDP 576A and other piston and gravity cores. Sediments below approximately 20 m are normally consolidated to slightly underconsolidated, but shallower samples exhibit significant degrees of overconsolidation (Fig. 23.2; Dadey, 1983; Marine Geotechnical Consortium, 1985).

In conventional soil mechanics, the presence of overconsolidation would imply that the upper portion of this deposit had borne stresses in the past greater than those it is now experiencing, producing a preconsolidation stress exceeding the present overburden (e.g., Lambe and Whitman, 1969). The most common cause of overconsolidation is removal of overburden by erosion. However, there is no evidence for large-scale erosion or mass wasting in the vicinity of DSDP 576A. Furthermore, any explanation for overconsolidation in shallow layers that relies on removal of overburden is problematical because of the existence of normally consolidated sediments below the overconsolidated zone.

A similar overconsolidation phenomenon, referred to as "apparent overconsolidation" (AOC), has been observed. Apparent overconsolidation has been described throughout the world ocean (e.g., Keller and Bennett, 1970; Bryant et al., 1981) and appears to be ubiquitous in shallow sediments, regardless of the consolidation state of deeper layers (Silva et al., 1976; Shephard et al., 1978).

The existence of the AOC layer has been attributed primarily to the inherent strength of flocculated, fine-grained marine sediments. The preconsolidation stress in these cases is an estimate of this strength, which is large relative to the overburden stress near the surface (e.g., Richards and Hamilton, 1967). Consequently, the extent, or depth of the AOC zone is limited by the steady increase in overburden and is generally confined to the upper 2–5 m of a deposit.

We believe that the persistence of overconsolidation to 20 m in these Northwest Pacific sediments represents an anomalous extension of the apparently overconsolidated layer. This theory is supported by porosities that remain high and shear strength trends that are relatively constant in the upper 20 m (Dadey, 1983; Marine Geotechnical Consortium, 1985). Similar phe-

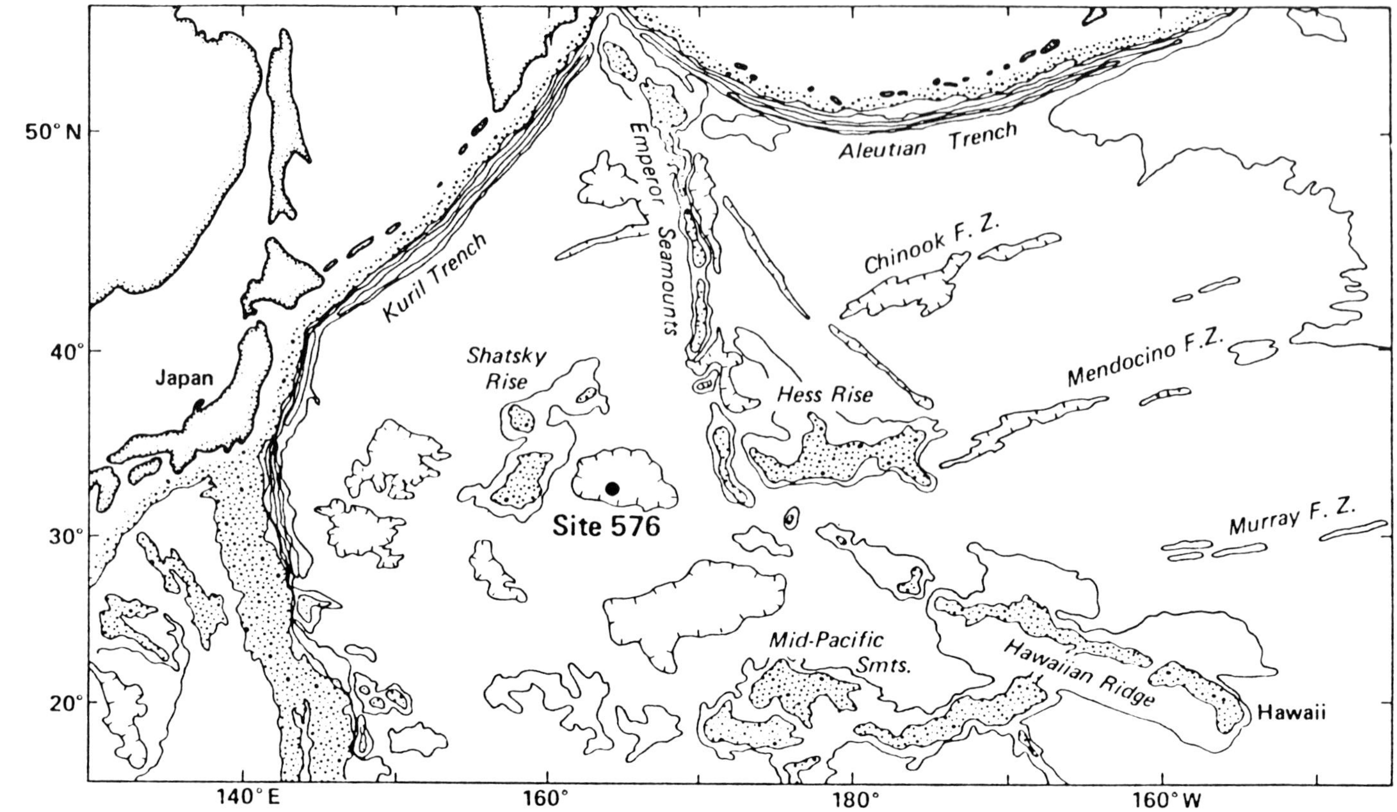

Figure 23.1. Map of the Northwest Pacific showing the location of Site 576. (From Heath, et al. 1985.)

nomenon have been noted in AOC layers of other deposits (e.g., Silva, 1979). Overburden stresses do not differ significantly from average values for the deep ocean (Bryant et al., 1981), suggesting that these sediments must be unusually strong. Because sediment microstructure is integral to all other properties, we examined the microstructure of DSDP 576A samples for clues to the anomalous strength and apparent overconsolidation.

Methods

Care was taken to preserve the specimens for microstructure analyses with as little disturbance as possible. Because surface tension effects and shrinkage are minimal with the critical point drying (CPD) method, it is the recommended drying method for high porosity, clayey marine samples (e.g., Bennett et al., 1977; Tovey and Wong, 1978) and was used in this study. Procedures used are similar to those reported by Burkett et al. (1987) and are described in detail elsewhere (Dadey, 1983).

After drying, specimens were fractured to expose fresh surfaces and mounted on microscope stubs with liquid graphite. Samples were coated with carbon and gold-paladium to ensure electrical conductivity. Samples used for energy dispersive X-ray spectrometry (EDS) were carbon coated only. A Cambridge Instruments S4 scanning electron microscope (SEM) and a Princeton Gamma Tech Model 3000 EDS system were employed.

Results and Discussion

Cementation has been proposed as an explanation for apparent overconsolidation in a number of other deposits (e.g., Bjerrum, 1967 Nacci et al., 1974). Cementing agents bind sediment particles together, forming a strong, rigid microstructure that results in a preconsolidation stress exceeding the overburden stress and an OCR greater than unity.

Our results suggest that some type of growth or recrystallization may exist at particle contacts in DSDP 576A sediments. For example, the EDS element spectrum of a contact point which appeared to be cemented (marked A in Fig. 23.3; note the fibrous connections) is shown in Figure 23.4. Relative to the average composition of the sample and the composition of a nearby silt particle (marked B in Fig. 23.3), the contact point is greatly enriched in iron and potassium (Fig. 23.4). The element

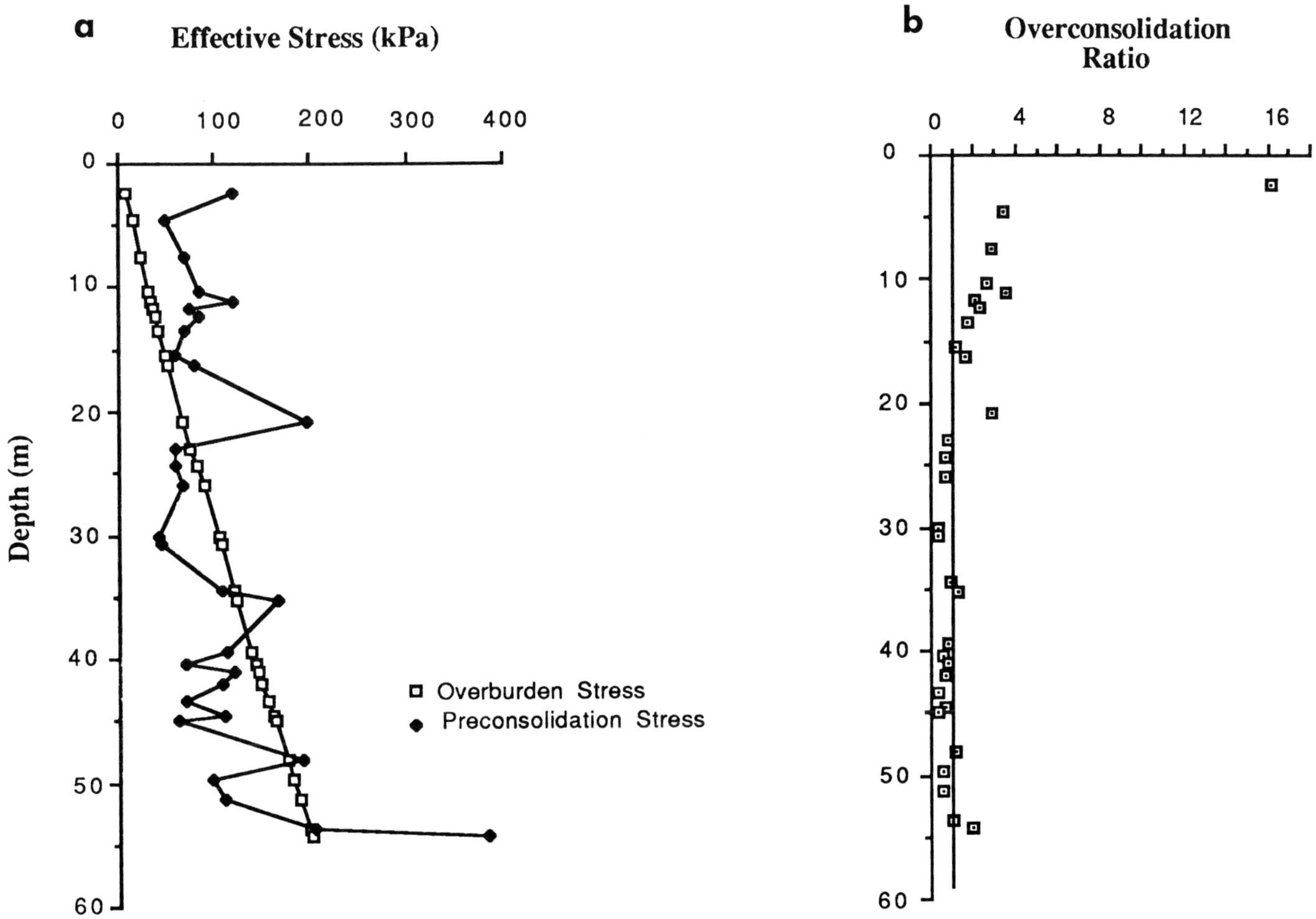

Figure 23.2. (a) Effective stress versus depth in Hole 576A. Open squares are calculated overburden stresses; closed symbols are preconsolidation stresses estimated from consolidation tests. (b) Overconsolidation ratio versus depth in Hole 576A. The line at OCR = 1 represents normal consolidation (see text).

spectrum resembles that reported for some types of authigenic illite (Welton, 1984). These results suggest that at least some contact points may serve as nuclei for precipitation or recrystallization of authigenic minerals.

In addition, the discovery of a correlation between trends in porosity and the percentage of silt is significant (Fig. 23.5) and unusual because it is inconsistent with the vast majority of published data. Higher porosity sediments tend to be fine grained and clay rich (e.g., Keller, 1974). Based on the SEM examinations, we believe that much of the silt-sized material is composed of clusters and aggregates of clay-sized particles (Fig. 23.6) and intergrown domains discussed below.

These aggregates are bound with such tenacity that they resist the rather severe chemical and mechanical deflocculation processes employed in sample preparation for grain size analysis (Folk, 1974). Consequently, they are identified as silt-sized material. An analogous situation was described by Heath (1969) in the Equatorial Pacific where an increase in grain size was correlated with opal cementation of clay plates.

Essentially, we believe that the aggregates act as clay-sized particles in contributing to the high porosity, but as silt-sized material in the grain size analyses. The resistance to deflocculation strongly suggests that these aggregates also contribute to the exception strength of these sediments.

We also observed regions of the fabric that we refer to as intergrown domains based on the definition of Smart and Tovey (1981). Distinct particle contacts are not visible and the fabric appears to be continuous over distances of several micrometers (Figs. 23.7, 23.8). One explanation for the intergrown domains is that they are artifacts of the SEM system. They may be anomalous images generated by regions in the fabric analogous to the "halos" described by Bennett et al. (1981). Observed in the transmission electron microscopy (TEM) mode, these halos consist of very thin, small clay particles that tend to cluster around

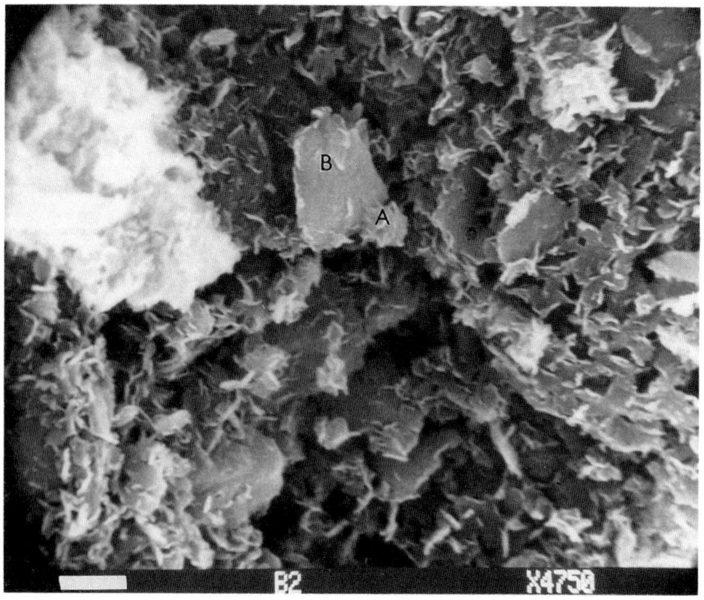

Figure 23.3. Micrograph of sample at 11.6 m in Hole 576A. A possible cemented contact is marked A. Note the fiberous nature of the fabric at the contact. Scale bar is 2 μm.

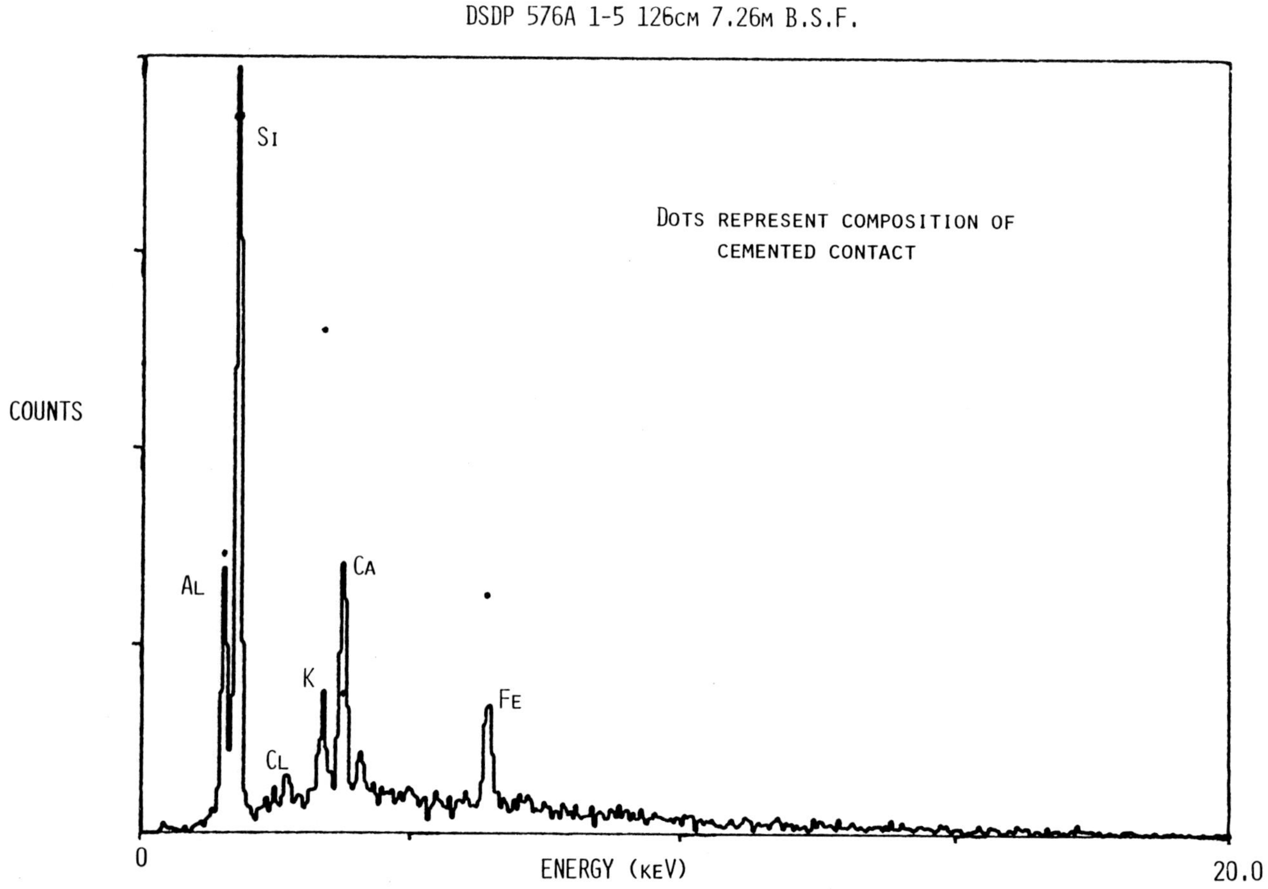

Figure 23.4. EDS spectrum of microfabric in Figure 23.3. The composition of the silt particle marked B in Figure 23.3 is considered average for the sample and a baseline. The composition of the contact point (A) is shown by dots.

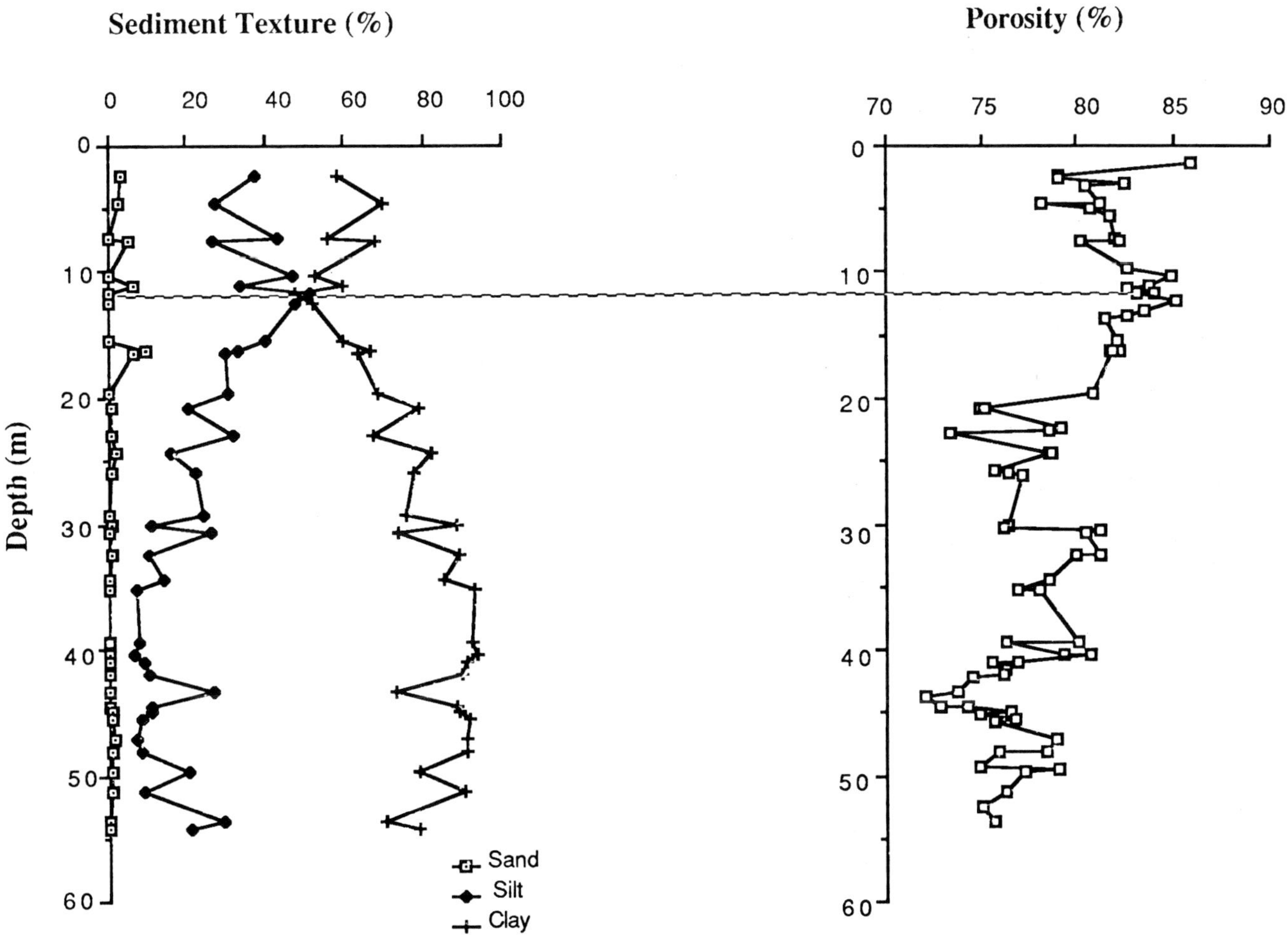

Figure 23.5. Sediment texture (grain size) and porosity versus depth in Hole 576A. Note the similarity between trends in silt content and porosity.

Figure 23.6. Micrograph of sample at 11.6 m in Hole 576A. A large cluster or aggregate of clay particles is visible in the center of the micrograph; a smaller one is to the left. Scale bar is 1 μm.

Figure 23.7. Micrograph of sample at 43.3 m in Hole 576A. An intergrown domain is visible in the lower right corner. Scale bar is 2 μm.

and between larger domains and particles. Because of their orientation and/or proximity to larger particles, the fabric in these areas may appear continuous (R. Bennett, personal communication, 1988). We believe this phenomenon accounts for some of our intergrown domains (e.g., region marked by the arrow in Fig. 23.9), but not all.

Many of the intergrown domains are visually similar to authigenic iron-rich semctite reported in the Eastern Pacific, which are believed to precipitate, with manganese micronodules, from amorphous Fe–Mn oxyhydroxides (Dymond and Eklund, 1978). In addition, the presence of relatively high percentages of iron and manganese oxyhydroxides (Schoonmaker et al., 1985) and Mn-micronodules (Heath et al., 1985) have been noted in these DSDP 576A sediments. Support for the occurrence of *in situ* mineral formation in this deposit is supplied by Schoonmaker et al. (1985) who attribute variations in clay mineralogy to diagenetic changes.

The EDS spectrum of an intergrown domain (Fig. 23.10) does show significant iron and manganese enrichment, but also depletion in both aluminum and silicon. Thus, some of the intergrown domains may actually be the smectite precursor.

Because intergrown domains were identified in nearly 80% of the samples examined, we hypothesize that they are the primary source of the unusual strength of this deposit. Their continuous structural framework is inherently stronger and less compressible than any arrangement composed of discrete particles.

We believe that both the clay aggregates and intergrown domains are distinct from the silt-sized, eolian-derived lithic fragments referred to as argillites by Burkett et al. (1987). Identification of argillites by SEM is not definitive; TEM is used. However, it is clear that argillite grains are a single unit, often sheathed in a layer of clay-sized plates. Neither the intergrown domain regions nor the clay aggregates/clusters fit this description. Nevertheless, the existence of argillites in our samples is

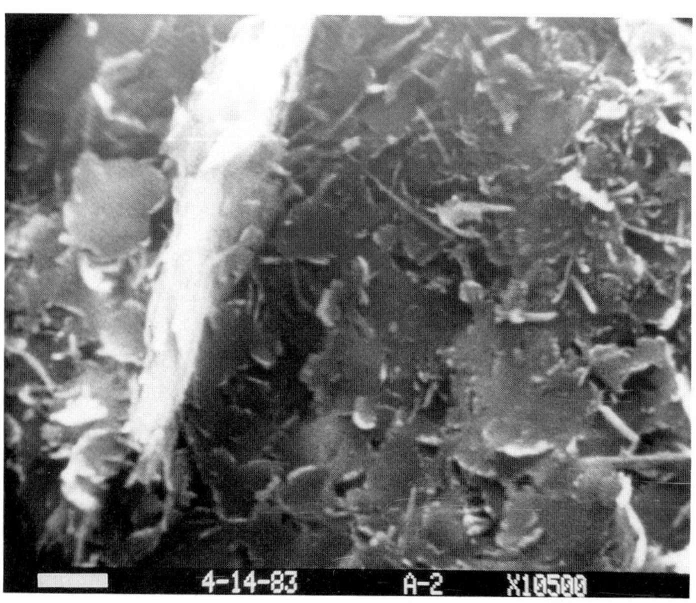

Figure 23.8. Micrograph of sample at 49.6 m in Hole 576A. A detailed view of an intergrown domain clearly showing the lack of visible particle boundaries. Scale bar is 1 μm.

Figure 23.9. Micrograph of sample at 15.56 m in Hole 576A. A large argillite, partially hidden by clay particles, is marked by A. A region of clusters of small clay plates on the argillite is marked by the arrow. Scale bar is 2 μm.

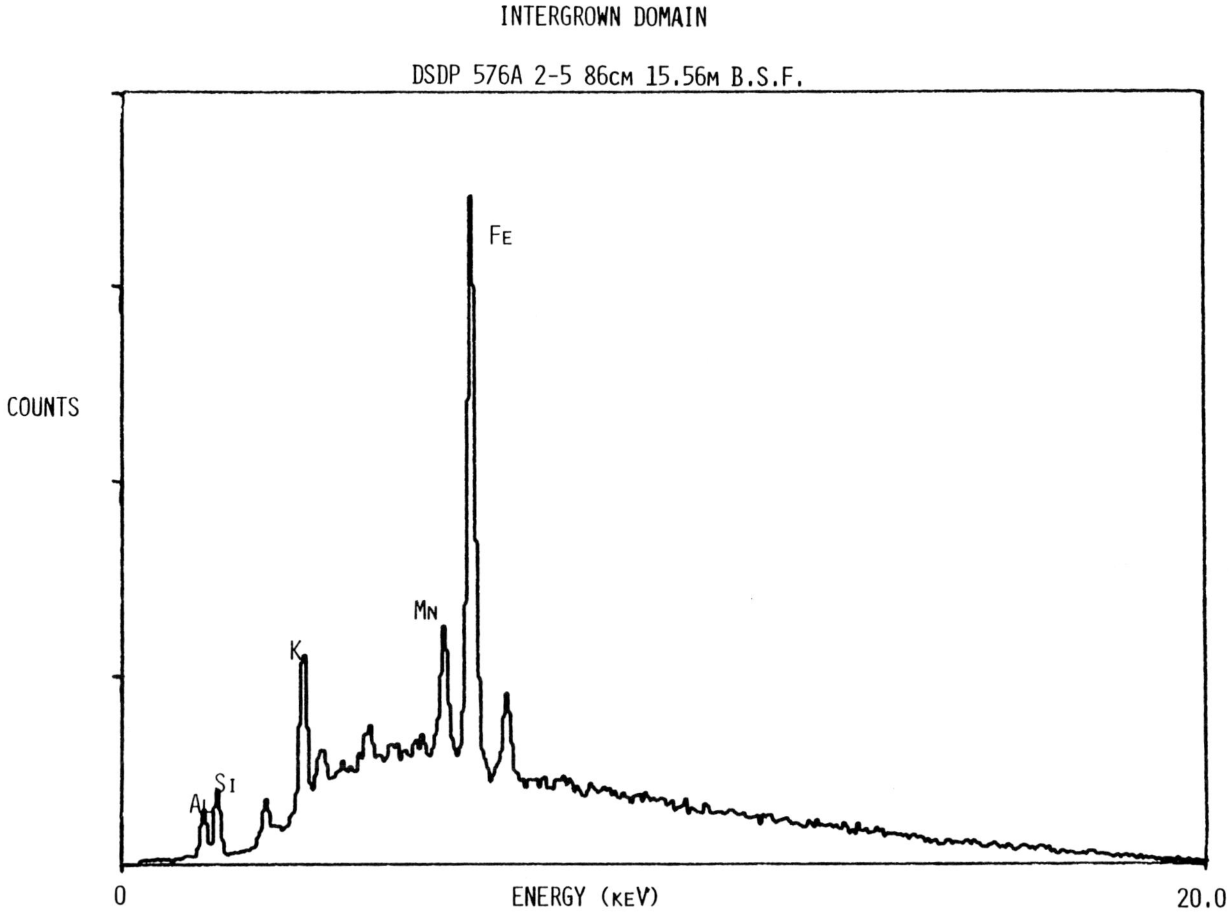

Figure 23.10. EDS spectrum of one intergrown domain.

likely, based both on their presence further downwind (Burkett et al., 1987) and visual evidence (Fig. 23.9; probable argillite marked A).

Conclusions

Overconsolidation discovered to 20 m in sediments of the Northwest Pacific represents an extension of the apparently overconsolidated layer. The origin of this overconsolidation is an unusually strong microstructure that is large, relative to the overburden stress in the upper 20 m. We identified three phenomena that contribute to this strong microstructure: cementation at particle contacts, the existence of strongly bound clay aggregates/clusters, and intergrown domains.

References

Bennett, R.H., W.R. Bryant, and G.R. Keller, 1977. Clay fabric and geotechnical properties of selected submarine sediment cores from the Mississippi Delta. NOAA Professional Paper 9, U.S. Dept. Commerce, Washington, 86 p.

Bennett, R.H., W.R. Bryant, and G.R. Keller, 1981. Clay fabric of selected submarine sediments: fundamental properties and models. Journal of Sedimentary Petrology, v. 51, p. 217–232.

Bjerrum, L., 1967. Engineering geology of Norwegian normally consolidated marine clays as related to settlements of buildings, Geotechnique, v. 17, p. 81–118.

Bryant, W.R., R.H. Bennett, and C.E. Katherman, 1981. Shear strength, consolidation, porosity and permeability of oceanic sediments. In: Emiliani C. (ed.), The Oceanic Lithosphere: The Sea v. 7. Wiley, New York, p. 1555–1616.

Burkett, P., R.H. Bennett, H. Li, F. Nastav, W.R. Bryant, L.E. Shephard, and W. Chiou, 1987. Microstructure of red clay from the Central Pacific deep-sea basin: significance to subseabed nuclear waste disposal. Sandia Report SAND86-2492, U.S. Dept. of Commerce, Washington, 87 p.

Dadey, K.A., 1983. Stress history of unlithified marine sediments in the Northwest Pacific. M.S. Thesis, University of Rhode Island, Kingston, RI, 150 p.

Dymond, J., and N. Eklund, 1978. A microprobe study of metalliferous sediment components. Earth and Planetary Science Letters, v. 40, p. 243–251.

Folk, R.L., 1974. Petrology of Sedimentary Rocks. Hemphill's, Austin, TX, 135 p.

Heath, G.R., 1969. Mineralogy of Cenozoic deep-sea sediments from the equatorial Pacific Ocean. Geological Society of America Bulletin, v. 80, p. 1997–2018.

Heath, G.R., D.H. Rea, and S. Levi, 1985. Paleomagnetism and accumulation rates of sediments at Site 576. In. Heath, G.R., L.H. Burckle, et al. (eds.), Initial Reports of the Deep Sea Drilling Project, v. 86. U.S. Government Printing Office, Washington, D.C., p. 459–502.

Keller, G., 1974. Marine geotechnical properties interelationships and relationships to depth of burial. In: Interbitzen, A. (ed.), Deep Sea Sediments Physical and Mechanical Properties. Plenum Press, New York, p. 215–223.

Keller, G.H., and R.H. Bennett, 1970. Variations in the mass physical properties of selected submarine sediments. Marine Geology, v. 9, p. 215–223.

Lambe, T.W., and R.V. Whitman, 1969. Soil Mechanics. Wiley, New York, 553 p.

Marine Geotechnical Consortium, 1985. Geotechnical properties of Northwest Pacific pelagic clays: Deep Sea Drilling Project Leg 86, Hole 576A. In: Heath, G.R., L.H. Burckle, et al. (eds.), Initial Reports of the Deep Sea Drilling Project, v. 86. U.S. Government Printing Office, Washington, D.C., p. 723–758.

Nacci, V.A., W. Kelly, M.C. Wang, and K. Demars, 1974. Strength and stress-strain characteristics of cemented deep-sea sediments. In: Interbitzen, A. (ed.), Deep Sea Sediments Physical and Mechanical Properties. Plenum Press, New York, p. 129–150.

Richards, A., and E.L. Hamilton, 1967. Investigations of deep-sea sediment cores. III. In: Richards, A. (ed.), Marine Geotechnique. University of Illinois Press, Urbana, p. 93–117.

Schoonmaker, J., F. Mackenzie, M. Manghani, R. Schneider, D. Kim, A. Weiner, and J. To, 1985. Mineralogy and diagenesis: their effect on acoustic and electrical properties of pelagic clays, Deep Sea Drilling Project Leg 86. In: Heath, G.R., L.H. Burckle, et al. (eds.), Initial Reports of the Deep Sea Drilling Project, v. 86. U.S. Government Printing Office, Washington, D.C., p. 549–570.

Shephard, L.E., W.R. Bryant, and W.A. Dunlap, 1978. Consolidation characteristics and excess pore water pressures of Mississippi Delta sediments. Proceedings of Annual Offshore Technology Conference, Paper No. 3167.

Silva, A.J., 1979. Geotechnical properties of deep sea clays a brief discussion. Proceedings 1st Canadian Conference on Marine Geotechnical Engineering Alberta, p. 262–269.

Silva, A.J., C.D. Hollister, E.P. Laine, and B. Beverly, 1976. Geotechnical properties of deep sea sediments: Bermuda Rise. Marine Geotechnology, v. 1, p. 195–232.

Smart, P., and N.K. Tovey, 1981. Electron Microscopy of Soils and Sediments: Examples. Claredon Press, Oxford, 154 p.

Tovey, N.K., and K.Y. Wong, 1978. Preparation, selection and interpretation problems in S.E.M. studies of sediments In: Whalley, B. (ed.), Scanning Electron Microscopy in the Study of Sediments. Geo-Abstracts, Norwich, p. 181–191.

Welton, J.E., 1984. SEM Petrology Atlas. American Association of Petroleum Geologists, Tulsa, 235 p.

II

Applied Clay Microstructure

A. Modeling—Past and Present: New Directions

CHAPTER 24

Modeling—Past and Present: New Directions—Overview

Sibel Pamukcu

The chapters presented in this section range from laboratory and *in situ* measurement of depositional and postdepositional properties of clay sediments and pure clay systems to numerical modeling of sediment transport and sediment stress deformation. This clearly reflects the diversity and the interdisciplinary nature of the theme of the section. The need to use physical and analytical models is borne out. Experimental methods are discussed that integrate concepts and principles of physicochemistry, microfabric, porometry, and the fundamental mass physical and mechanical properties of sediments to formulate reliable predictive models of microstructure and sediment behavior.

Main subject areas in this section are measurement and analytical methods. In each, either new techniques and/or the generation of new data and empirical correlations are presented. These are not necessarily exclusive of each other since the integrity of most predictive techniques is strongly dependent on measured data and observed behavior. Examples of new measurement techniques are the floc camera used *in situ*, and spectroscopy of absorbed dyes and the acoustic velocity of rock minerals, which are measured in the laboratory. These techniques involve quantitative measurement of physical, chemical, and mineralogical properties of clay suspensions. A number of chapters report new data and empirical information with respect to the influence of depositional and postdepositional environment on microfabric, and of microfabric and physicochemical properties of clay particles on their mechanical behavior. Other chapters address analytical methods, either on a macro- or microscale. One of the chapters reports a new analytical approach to characterize clay microstructure and two report numerical and theoretical findings on dewatering in accretionary sediments, incorporating *in situ* properties such as porosity, pore fluid density and pressure, mineralogy, and geological history.

New directions in modeling of mechanics of clay–water systems should incorporate an element of uncertainty due to variance of physical properties in time and space as ambient conditions change during depositional and postdepositional periods. Additional uncertainty also arises from the inability to acquire precise input data for analytical predictive tools. A more unified and probably more realistic approach in modeling and predictive methods can be achieved if physical and mechanical properties can be based on some fundamental properties of clay–water systems, such as microstructure, surface chemistry, dynamics of particulate matter, and pore fluid. Therefore, it is imperative that new measurement techniques be developed to acquire more realistic data or make use of existing techniques to widen the data base correlating micro- and macroproperties of these systems under a variety of environmental conditions.

The study by Pamukcu and Tuncan on the influence of contaminating agents such as municipal sludge, petroleum, and flyash on the mechanical behavior and microstructure of consolidated and sedimented systems of pure clays is the lead chapter of this section. Some of the physical properties such as permeability, low-strain shear modulus, and compressive strength are found to be altered by the incorporation of these agents into the pore fluid. These effects are attributed to microstructure changes in the presence of the chemicals, evidence of which was observed in SEM micrographs of typical samples. It was hypothesized that formation of clay flocs, coating of these flocs with organic material, and the addition of fibrous inclusions caused the observed increase in the low-strain shear modulus of specimens with sludge incorporated. Specimens mixed with fuel oil exhibited high water holding capacity and a corresponding loose, open microfabric.

Cenens, Schoonheydt, and Schryver studied the organization of clay particles in dilute aqueous solution as inferred from spectroscopy of adsorbed dyes. The results indicate that smectites occur as aggregates of individual clay platelets, even in the presence of Na^+ as the exchangeable cation. Only the extent of aggregation depends on the type of exchangeable cation. The results also suggest that if the aggregates consist of a regular packing of individual sheets face to face, then the average number of sheets is six and the distance between sheets must be at least two water layers. This study reflects the importance of detailed investigations of the fundamental physical and chemical properties of clay microstructure.

Nonclassical, temperature-dependent ion partitioning in the water viscinal to clay particles would influence significantly the properties of fine-grained sediments. Drost-Hansen discusses the effects of viscinal water on the sedimentation process, compaction, and the ultimate properties of sediments. The discussion indicates that with the exclusion of structure-making ions (such as Na^+ and Li^+) and the preferential retention of structure-breaking ions (such as K^+) in the interstitial water, a number of thermodynamic properties of sediments would change. These properties include specific heat, density, thermal expansion coefficient, isothermal and adiabatic compressibilities, and acoustic properties. Other behavior that might be influenced by the properties of viscinal water are the rheology of unconsolidated sediments and the permeability of consolidated sediments.

The investigation by Williams and Williams of the rheology and microstructure of concentrated illite suspensions includes study of the variation of yield stresses as a function of ionic strength, pH, and volume fraction of solids. The role of colloidal forces in determining microstructure, critical volume fraction, and rheological behavior is identified by analysis of possible interactions between illite particles as a function of ionic strength and pH. The results of this study indicate substantial influence of microstructure and physicochemical environment on the elastoviscous behavior of K-illite suspensions.

A computer controlled floc camera system consisting of a collimated flash, three cameras, and a depth sensor was assembled to obtain quantitative *in situ* data on suspended particulate matter in oceans during submersible operations. Syvitski, Asprey, and Heffler describe the analysis of photographs using an image analysis system linked to a mainframe computer. Instantaneous and absolute velocities of particles, their shapes, their volumes, and their minimum and maximum diameters can all be determined using the new instrumentation. The instrumentation and associated methods offer a significant advancement over existing approaches to obtaining ground truth information on suspended particulate matter.

A new approach for characterizing clay fabric is presented by Altschaeffl and Thevanayagam. A mathematical relationship is developed between fabric and strain, and the uniqueness of the relationship is examined. The theory was validated using SEM data on a natural clay. Results of the study indicate that development of the new tool will permit close modeling of sediment behavior using the strain space theory of plasticity.

The influence of microproperties on the macrobehavior of selected clay-rich sediments is investigated by Ferrell and Carpenter. Microtextural and microchemical observations are analyzed in an attempt to produce quantitative information for different sedimentary environments. Variables such as pore structure, revealed by the ratio of pore perimeter to pore area in digital images, may be characteristic of a particular environment. Particle size and shape may provide important evidence on postdepositional environment. Results of this study demonstrate that advances in microtextural studies will enhance our understanding of the geology and industrial utility of clay-rich sediments.

Brunton discusses a new technique for quantitative rock mineral analysis based on measurement of the acoustic velocity of a known volume of pulverized rock in suspension. The equation used is a function of measured acoustic velocity, the density and compressibility of each expected mineral, and the volume fraction of water. A nonlinear least-squares curve fitting is employed to obtain the volume fraction of each mineral. The tests were conducted at various temperatures. The report includes the equipment and the procedures of analysis.

Chang, Lennon, Pamukcu, and Carson developed a numerical model for coupled fluid expulsion/deformation of dewatering sediments. The model is a finite element scheme created to simulate the phenomena under lateral earth stresses. The mathematical formulation is based on Cam-clay deformation/distortion model and continuity of fluid mass combined with generalized Darcy's law. It allows temporal and spatial variations of the pore pressure, porosity, fluid density and temperature field. It was demonstrated that the fluid expulsion/deformation behavior of dewatering sediments could be predicted more closely using this model.

CHAPTER 25

Influence of Some Physicochemical Activities on Mechanical Behavior of Clays

Sibel Pamukcu and Mustafa Tuncan

Introduction

Clays are chemically active electrolyte–gas–solid systems. Their microfabric and physicochemical forces determine much of their physical properties and subsequent mechanical behavior. The prevailing physical and chemical conditions at the time of deposition influence the spatial arrangement of the clay particles. Changing environmental conditions such as temperature, pH, salinity, electrolyte concentration, and type, as well as the dielectric constant of the pore fluid may change the stability of the initially attained structures by inducing new particle associations and affecting pore sizes and pore fluid properties. Such changes subsequently may affect the ability of the clay to carry effective stresses as well as affecting its permeability, consistency, and volume change characteristics. An example of this situation is the ground migration or accidental spill of contaminants that may result in prolonged contact or permanent saturation of soils with chemicals. Prediction of long-term performance of clays under such potential circumstances needs to be based on all conceivable physical, chemical, and biological changes within the system.

Relating engineering properties and mechanical behavior of clays to their more fundamental characteristics, such as microstructure, mineralogy, chemistry, and surface morphology has received much attention in recent years. Increasing occurrences of soil contamination and the critical need to reclaim land and protect freshwater sources necessitate generation of fundamental information to better understand the long-term effects of contaminants. As clays and colloidal particles are particularly sensitive to the physicochemical conditions during deposition (and also to postdepositional changes in the environment) this study concentrates on these soil constituents.

Background

Soil is a mixture of inorganic and organic solids, gases, liquids, and various constituents, such as microorganisms. All of these components influence each other and react to achieve physical, chemical, and electrical stability within the soil system. The ongoing changes in the environment induce stresses that may alter this stability and create different products. One source of such stress that has gained much attention in recent decades is the human-induced pollution. Clays and colloidal particles are chemically active soil solids, the physical chemistry and microstructure of which have been shown to influence strongly their mechanical properties and behavior. Better understanding of the fundamental mechanisms of these influences would enable engineers and scientist to predict long-term soil properties and interpret any anomalies in the long-term behavior of engineered facilities with more confidence.

Clays exposed to environmental stresses (such as various chemicals in the pore fluid, change in pH or temperature of the pore fluid, and generation of gases through biological or chemical reactions) tend to display changes in their physical properties over time. A number of studies have addressed these phenomena (Whelan et al., 1975; Drost-Hansen, 1976; Faas and Wartel, 1977; Torrance and Pirnat, 1984; Acar et al., 1985; Evans et al., 1985; Acar and Ghosn, 1986; Du et al., 1986; Foreman and Daniel, 1986; Hettiaratchi and Hrudey, 1987; Madsen and Mitchell, 1987; Tuncan et al., 1988; Pamukcu et al., 1987, 1990). Introduction of various organic–inorganic ions into the pore fluid influences the colloidal stability of fine-grained soils (Sparks, 1986; Bohn et al., 1985). Activities such as adsorbtion and ion exchange, and the conditions such as pH, temperature, and degree of saturation under which these activities take place determine the type of changes in the structural and physicochemical properties of the soil. Thus, new particle associations, pore sizes, and pore fluid properties may be attained. This new structure and properties then determine the ability of soil to carry effective stresses, its mechanical response to various loading conditions, its permeability, consistency, and volume change behavior (Mitchell, 1976).

Theoretical Considerations

The structure of clay minerals has been studied extensively and is reasonably well understood (Grim, 1968; Mitchell, 1976; van Olphen, 1977; Sparks, 1986; Bennett and Hulbert, 1986). The ultimate clay mineral structure results from combinations of two fundamental units forming the atomic molecular lattice. One unit is an octahedral configuration and the other is a tetrahedral configuration. Both tetrahedron and octahedron units form continuous sheet-like structures. The basic tetrahedral and octahedral sheets are then layered in various configurations resulting in clay minerals with a wide range of behavior. The forces holding the atoms and molecules within a clay mineral can be classified as primary and secondary bonds. Primary bonds are mainly ionic and covalent type, whereas the secondary bonds, which are much weaker, are dipole bonds with specific dipole force identified as a hydrogen bond. In addition, van der Waals forces can result in molecular bonding.

Clay layers, made up of tetrahedral and octahedral sheets, often carry a net negative charge as a result of substitutions of certain cations within the sheet structure. The negative charge, resulting from the isomorphous substitution of cations by less positive cations, is often compensated by cations located on the layer surfaces. In the presence of water, these compensating cations have a tendency to diffuse away from the layer surface. Similarly, anions at higher concentrations, at a certain distance away from the clay surface, have a tendency to diffuse toward the layer. The tendency of the cations and anions to be, respectively, attracted and repelled by the clay surface, and diffuse away from or toward the surface gives rise to the term "diffuse double layer." In a static environment, this double layer has a constant charge that is largely determined by the type and degree of isomorphous substitutions and the resulting net negative charge in the mineral layers.

When the clay is dry, the adsorbed cations are tightly held to the negatively charged surfaces. When the clay is immersed in water, the "precipitated salt" associated with the cations on the particles goes into solution and tends to diffuse away. There are several theories on how the distribution of ions within the double layer originates. The most widely applied model in geotechnical engineering is the Gouy–Chapman model. This model of the diffuse double layer has been applied in attempts to explain the behavior of clay since the 1950s (Lambe, 1959). Subsequent refinements to the Gouy–Chapman model have been developed (van Olphen, 1977; Sparks, 1986), and these treatments have yielded a more detailed picture of the double layer charge distribution. However, most colloidal phenomena can be interpreted in a general way on the basis of the Gouy–Chapman model.

Based on the interaction of clay minerals and the accompanying diffuse double layer, definitions of different types of soil fabric have evolved. A "dispersed" soil structure is one in which the net electrical forces between adjacent particles (at the time of deposition) produced repulsion. If these interparticle forces are net attractive during deposition, the soil structure is said to be flocculated. In other words flocculated (or aggregated) structures are those in which particles tend to join together, whereas dispersed structures are those in which particles tend to move apart. The Gouy–Chapman model indicates that the tendency toward flocculation is increased by decreasing the double layer thickness as a decrease in the double layer thickness reduces the interparticle repulsion forces. Conversely, the tendency toward dispersion increases as the double layer thickness increases. To fully understand the clay–water system, it is necessary to examine in detail the variables within the Gouy–Chapman model and how they affect the clay structure and its associated behavior. The model predicts that an increase in electrolyte concentration, valence of ions, and temperature and a decrease in dielectric constant of the pore fluid increases the tendency toward flocculation. There are some other factors that are not taken into account by the Gouy–Chapman model. These factors have been either predicted or observed to influence clay structure and behavior. For example, a decrease in the size of the hydrated ions, pH, and anion adsorption increases the tendency toward flocculation.

Mineralogy and spatial arrangement of particles in a clay medium can usually provide significant information on the physical and mechanical behavior of the clay. For example, a flocculated "cardhouse" arrangement of clay particles suggests higher stiffness, a more unstable and thus more brittle mode of deformation, as well as higher permeability due to the open fabric and collapsed double layer. Conversely, a dispersed "parallel" arrangement of particles suggests lower stiffness, more stable and ductile mode of deformation, and lower permeability due to the swollen double layer. The size and distribution of pore spaces, size, type, and surface morphology of clay particles and other solid constituents, and particle assemblages all seem to have mechanical functions and contribute to the mechanical behavior of clay.

Literature Survey

There exists a number of studies conducted on fabric, structure, and property relationships since the 1950s. Increased attention to physicochemical aspects is relatively recent due to the increased need of specific knowledge about interactions between contaminants and soil constituents. A brief review of a few related studies are presented below.

Torrance and Pirnat (1984) investigated the effect of pH on rheology of marine clays at various degrees of salinity. Using a coaxial viscometer they found the yield stress to decrease as the pH increased. The yield stress increase with decreasing pH was attributed to a change in ion saturation. The influence of salinity on the remolded shear strength of these materials was found to be pH dependent. It has been observed that high pH tends to create dispersed clay systems (Mitchell, 1976). This occurs due to the dissociation of H ions, which results in increased negative charge on clay particles and increased repulsion between particles. At low pH positive charges dominate the edges of the particles leading to attraction and edge-to-face flocculation. Acid attacks clay particles causing release of ions, which enhance flocculation (van Olphen, 1977).

Drost-Hansen (1976) showed that the "vicinal water" or longer range, low-energy water extending over distances a few hundreds of molecular diameters may have significant effects on the depositional and overall properties of sediments. It was concluded that in this water, which may not be much different in structure than the bulk water, highly nonclassical, strongly temperature-dependent ion partitioning may occur, excluding from the bound water those ions that are strong water-structure makers, such as Li^+ and Na^+, while preferentially retaining structure-breaking ions, such as K^+. In such cases, the specific heat of sediment and other thermodynamic properties, including density, thermal expansion coefficient, and isothermal and adiabatic compressibilities affecting acoustic properties, would be influenced. It was also discussed that in unconsolidated sediments the rheological properties and in consolidated sediments the permeability would be affected by the anomalous viscosity of the vicinal water. Processes that were studied with respect to the properties of the vicinal water included sedimentation, aggregation, compaction, and consolidation.

Evans et al. (1985) through a series of long-term triaxial permeability tests showed that the permeation of clays by water saturated with some organic fluids causes significant increase in their permeability. When carbon tetrachloride and aniline were used, permeability increased between two to four orders of magnitude with respect to permeation with water alone. When the same materials were used as permeants at concentrations less than or equal to their solubility limits, no appreciable increases in the permeability were observed. Low-activity clays exhibited significantly less degradation of permeability than high-activity clays. The increase in permeability was attributed to the collapse or shrinkage of the diffuse double layer. From similar studies conducted by Acar and Ghosh (1986) and by Bowders (1985) similar conclusions were made. When clays were permeated with landfill leachate some changes in the permeability (Bowders et al., 1984; Fang and Evans 1988) and in the compressive and tensile strength were observed (Tuncan et al., 1988).

Lagaly (1985) reported that the adsorption of organic molecules by hydrophobic and hydrophilic clay minerals is more strongly governed by structural aspects than previously reported. Clay minerals bind organic compounds by adsorption, ion exchange, and intercalation, which may take place on the external or internal surfaces. The structural dependence of the interaction was demonstrated by the adsorption processes of methanol molecules, clustering polar molecules, and methanol–benzene mixtures on different types of smectites. It was also shown that organic molecules exert a large influence on the rheological properties of clays.

Gauffreau (1988) showed that soil particles coated with activated sludge develop some degree of hydrophobic character. The humic substances or organic compounds similar in chemical structure to that of humic substances present in activated sludge can be adsorbed onto clay mineral surfaces as macromolecular complexes bound together by di- and trivalent cations or hydrogen bonding. The polyvalent ions responsible for bonding these complexes in soils are Si^{4+} or Ca^{2+}, Fe^{3+}, and Al^{3+}. These complexes with polyvalent cations tend to make the humic substance less soluble. The water drop penetration time, a measure of water repellency, for the activated sludge mixed and heat-treated soil was observed to be about 500 times more than the original untreated soil. The height of the capillary rise was reduced by about a factor of 4 and the liquid–solid contact angle increased significantly. These findings indicated development of the hydrophobic or water repellency character of the soil due to the coating of the particles by the humic substances in the activated sludge.

Investigation

A number of laboratory studies on the effects of physicochemistry and microstructure on the engineering properties of clay systems have been conducted. Some of the results of these studies are presented and discussed below.

On Consolidated Clay

Variations of stiffness, shear deformation, and permeability of consolidated clay specimens were studied in the presence of increasing percentages of organic and inorganic additives. Scanning electron micrographs of various mixtures were analyzed to evaluate qualitatively their mechanical behavior based on the observations of the microstructure. Additives were selected among the waste materials commonly disposed by land or ocean. These were flyash, representing the inorganic waste, and activated sludge, representing the organic waste. Number 2 fuel oil (heating oil) was also used as a contaminating material in this set of tests.

The flyash was obtained from the Martins Creek Steam Electric Station of Pennsylvania Power and Light Company. The portion of the flyash finer than No. 200 sieve size was 60% by weight, which placed it at the lower range of typical flyash grain sizes. The sludge was obtained from the Allentown, Pennsylvania Waste Water Treatment Plant at Kline's Island. It was black

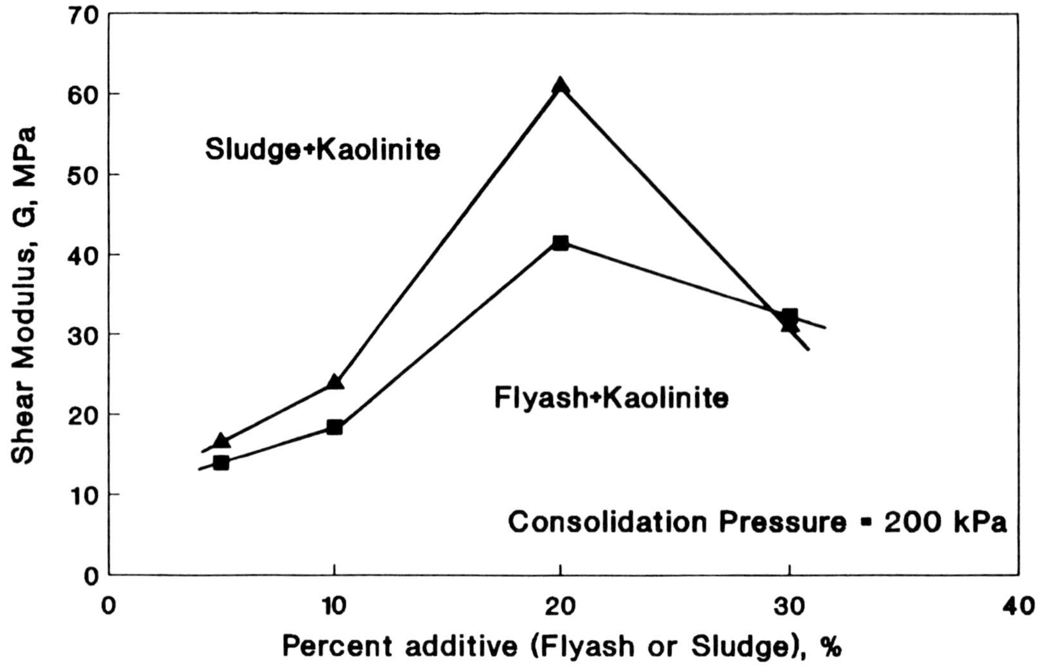

Figure 25.1. Variation of shear modulus with percentage additive for consolidated specimens of kaolinite and flyash and kaolinite and sludge.

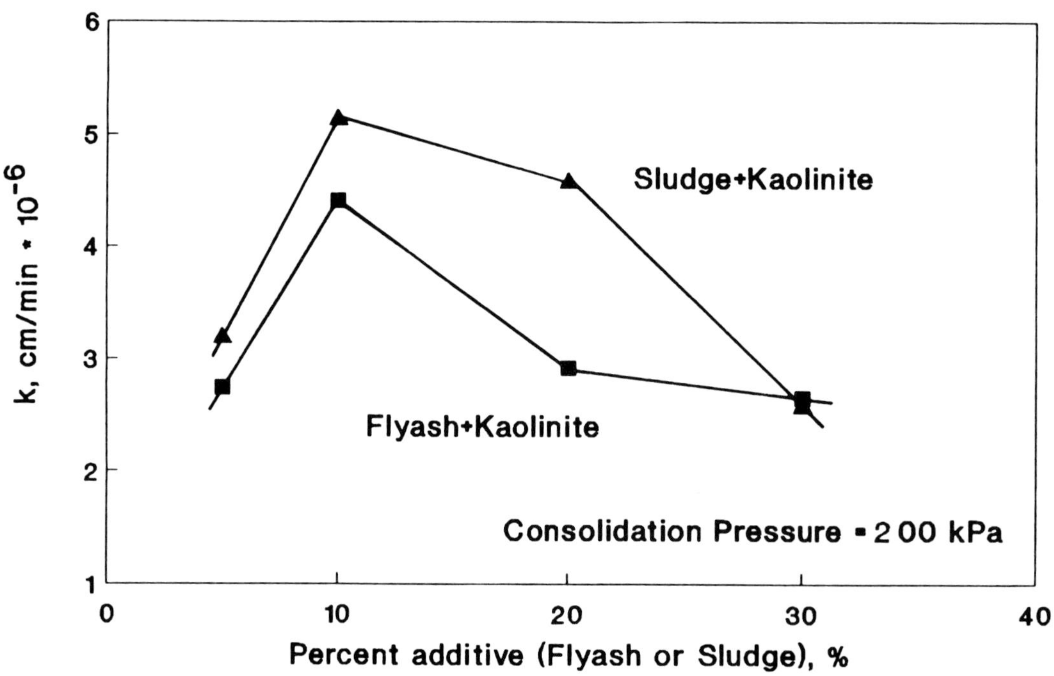

Figure 25.2. Variation of coefficient of permeability versus percentage additive for consolidated specimens of kaolinite and flyash and kaolinite and sludge.

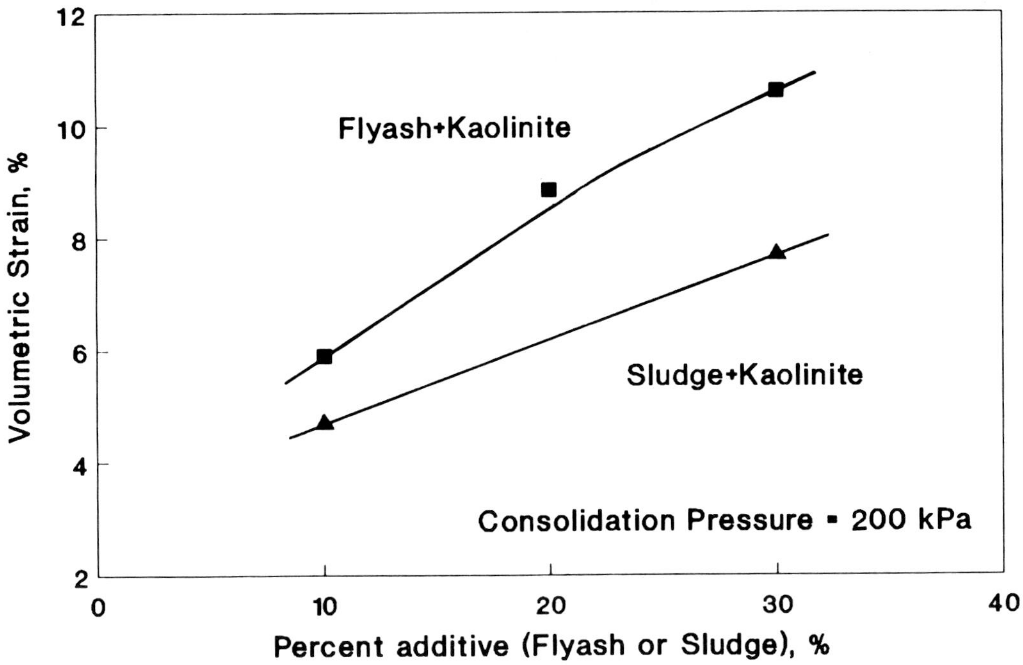

Figure 25.3. Variation of volumetric strain versus percentage additive for consolidated specimens of kaolinite and flyash and kaolinite and sludge.

in color, thick in consistency, and appeared to have fiberous solid particles in suspension. Its initial pH was 7.3, and water content 90.5% (based on wet weight, ASTM D2975). In the activated sludge process, a suspended aerobic microbial culture is used to treat wastewater. The microorganisms (mostly bacteria) metabolize organics. The waste products then settle from the aerated mixture. The settled material is what is known as activated sludge, which contains the organic compounds similar in chemical structure to humic substances. The specific gravity of No. 2 fuel oil was given as 0.85. The clays used were Georgia kaolinite with plasticity index of 39 and liquid limit of 70%, and a laboratory mixture of 4 clays named as "*marine clay*."

The clay samples were prepared by premixing with distilled water and the appropriate amount of additive in slurry form, and consolidating this slurry in large one-dimensional consolidation cells to 0.75 kg/cm^2 pressure. In a one-dimensional, or what is known as a K_o-consolidation cell, axial pressure is applied onto a cylindrical soil specimen that is constrained radially such that no lateral deformation can occur. At the completion of consolidation, each specimen was trimmed and mounted on the triaxial permeability devices and further consolidated to 2 kg/cm^2 under triaxial pressure. The Lehigh triaxial permeability systems of 11 units that were used in this study are described else where in detail (Evans and Fang, 1986). At this stage the permeabilities of the soil and additive mixture specimens were measured. The undestructed specimens were taken out of the permeability cells, trimmed, and mounted on a triaxial cell equipped with Bender Elements to measure the shear wave velocity and thus shear modulus. Bender Elements are piezoelectric chips that generate and receive shear wave signals through the column of soil. The incoming signal is collected via a Nicolet oscilloscope and the time of travel of the wave is measured. A number of specimens were left in the triaxial permeability cells and further tested for undrained shear strength at various confining pressures. In these tests the samples were not consolidated further at the increased confining pressures and the effect of induced pore pressure could be observed from the resulting data. The pH of the kaolinite and 5% additive slurries prior to consolidation was measured to be 6.4 for the kaolinite and sludge and 7.1 for the kaolinite and flyash.

Specimens of consolidated clay and additive mixtures were analyzed using a scanning electron microscope (SEM). The SEM specimens were prepared by the critical point drying technique. They were initially treated with ethyl alcohol and then acetone. A critical point drying apparatus was utilized to replace the acetone with CO_2. The specimens were then kept in a vacuum desiccator for extended periods of time to ensure removal of all the water from the pore space.

Figure 25.1 shows the variation of shear modulus with increasing percentages by weight of flyash and sludge. Both the shear modulus of the kaolinite and sludge mixture and kaolinite and flyash mixture increases with increasing percentage of additive up to around 20% and decreases from there on. This may indicate that the clay tends to acquire a flocculated structure at lower percentages of the additives and consequently the stiffness increases. Conversely, at higher percentages of the additives the

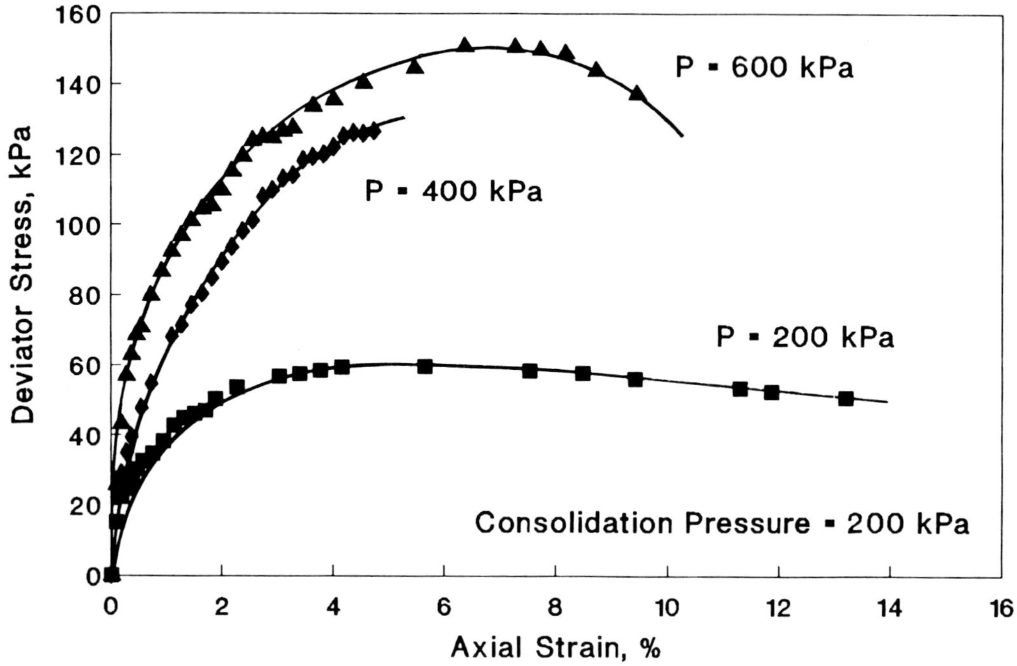

Figure 25.4. UU triaxial compression curves of 5% fly-ash-mixed consolidated kaolinite specimens at different confining pressures.

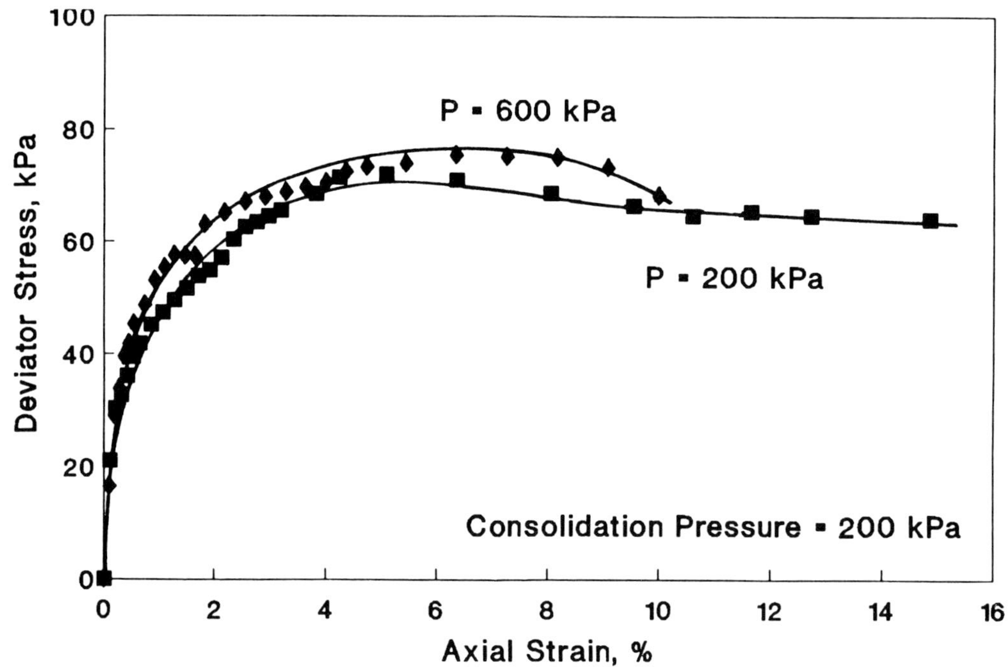

Figure 25.5. UU triaxial compression curves of 5% sludge-mixed consolidated kaolinite specimens at different confining pressures.

structure may be dispersed producing the effect of reduced stiffness. Trends that support similar reasoning were observed in permeability and volumetric strain behavior of the specimens, as illustrated in Figures 25.2 and 25.3, respectively. The coefficient of permeability increases to the left of the maximum, and decreases on the right of the maximum in Figure 25.2. The compressibility of the materials increases with the increasing amount of additive both for the sludge and the flyash mixed specimens as shown in Figure 25.3. These data agree with the assessment made for the behavior of shear modulus data. In other words, flocculated structures tend to have higher permeabilities and lower compressibilities, whereas dispersed structures display the opposite trends.

Figures 25.4 and 25.5 show, respectively, the triaxial compression curves for 5% flyash and sludge added kaolinite specimens. From Figures 25.1, 25.4, and 25.5, it is observed that the stiffness and the compressive strength values of the two normally consolidated specimens with 5% additives do not exhibit much difference. However, the compression data show that the effect of pore pressure development on the strength is more pronounced for the sludge mixed specimen than for the flyash mixed specimen. For the sludge mixed specimen there is little difference between the stress-strain curves of 2 kg/cm^2 and 6 kg/cm^2 ($\sim$200 and 600 kPa) confining pressures, which indicates more reduction in the effective stress due to excess pore pressure development. This behavior of sludge mixed clay as opposed to that of flyash mixed clay requires explanation. One possible explanation may be the agglomeration and flocculation of clay into "domain" structures with addition of sludge. Such a structure would be expected to exhibit higher stiffness when measured at strain levels below that at which the domains or agglomerations can be destroyed. The shear modulus measurements were made at approximately 10^{-3}% shear strain amplitude. It is hypothesized that the humic substances and organic mucus in the sludge coat the particles promoting flocculation and the fibrous particles act as reinforcing members between the flocculated domains. At higher strain levels, as in the triaxial compression test, these domains are crushed, giving way to a compressible, lower strength material. Existing pore pressures are expected to aggravate the situation, reducing the strength even further, as observed in here. Flyash mixed clay, on the other hand, showed higher strength due to the addition of silt sized particles that would increase its frictional resistance and reduce its compressibility.

The scanning electron micrographs of the 5% and 30% sludge mixed kaolinite specimens are inconclusive and do not confirm or refute the explanation given above. However, the flocculated, and more oriented, structure of the clay is evident in the 5% sludge mixed clay micrograph as predicted. Figure 25.6a and b shows the micrographs for the 5% and 30% sludge mixed clay, respectively. The fibrous inclusions that are expected to contribute to the higher stiffness can be detected in both of the pictures.

Figure 25.6. (a) SEM micrograph of 5% sludge-mixed kaolinite consolidated at 2 kg/cm^2 (scale bar 1 μm). (b) SEM micrograph of 30% sludge-mixed kaolinite consolidated at 2 kg/cm^2 (scale bar 1 μm).

Figure 25.7. SEM micrograph of 10% flyash-mixed kaolinite consolidated at 2 kg/cm² (scale bar 1 μm).

Figure 25.8. SEM micrograph of 20% fuel oil-mixed marine clay consolidated at 2 kg/cm² (scale bar 1 μm).

Table 25.1. Shear modulus test results on some consolidated marine clay samples.

Sample*	G† (MPa)	V_s‡ (m/sec)	Wet unit weight (kN/m³)	Water content (%)	Specific gravity
MC§ + distilled water	47.4	162.3	17.6	48.3	2.51
MC + saline water	48.1	165.0	17.3	46.4	2.27
MC + saline water 20% sludge	35.3	137.7	18.2	49.2	2.65
MC + saline water 20% fuel oil	14.3	112.6	11.1	56.3	2.00

*Consolidated at 0.75 kg/cm².
† Shear modulus.
‡ Shear wave velocity.
§ Marine clay.

The 30% sludge mixed clay appears to have a less oriented structure, and there is some evidence of agglomerations or flocculated domains.

The micrograph of 10% flyash-mixed kaolinite is given in Figure 25.7. The flyash particles cannot be detected in this micrograph. They are suspected to be embedded in the clay, which makes it difficult to identify them, or crushed and disintegrated under the compressive load. There is no evidence of agglomeration of particles, however the structure appears more compact than that of the sludge mixed specimens.

Table 25.1 presents the measured values of shear wave velocity, shear modulus, and some index properties of consolidated 20% No. 2 fuel oil, 20% sludge mixed, and control specimens of artificially prepared marine clay. The marine clay mixture was made up of 13% chlorite, 16% kaolinite, 50% illite, and 21% Ca-montmorillonite, which are the typical percentages of clay minerals found in Atlantic ocean sediments. The saline water specimens were prepared by mixing distilled water with a commercial product called "Instant Ocean" to obtain typical salt content of oceans. The fuel oil-mixed specimen shows a significant reduction in shear wave velocity and unit weight when compared to those of the control specimen. This specimen is also observed to retain the largest amount of water. The high water content, low density, and low shear modulus indicate a dispersed or non-flocculated matrix. Figure 25.8 shows the SEM micrograph of 20% fuel oil-mixed marine clay in which the open, loose structure is evident. However, the random orientation of the particles and the floc formations do not support the deductions made earlier. It appears that in this case the loose structure of the material dominates the mechanical properties rather than the fabric characteristics. Judging from the micrograph, fuel oil might have acted as a "gluing" agent to produce the flocs, but also might have trapped water resulting in reduced stiffness.

The unit weight of the sludge-mixed specimen is higher; however, the shear wave velocity is lower than that of the control specimen. Hence, the modulus reduction of the sludge mixed specimen is not as significant as observed with the fuel oil-mixed specimen. The reduced moduli values are attributed to the increased compressibility with the addition of organics. The increase in unit weight indicates retention of organics in the pore space of the clay mixture. Supporting this, the SEM micrograph of the sample, shown in Figure 25.9, indicates some orientation and a more compact matrix than that of the fuel oil sample.

Figure 25.9. SEM micrograph of 20% sludge-mixed marine clay consolidated at 2 kg/cm² (scale bar 1 μm).

On Sedimented Clay

Using large-scale sedimentation tanks (11 × 11 × 24 in.) equipped with sampling ports, mixtures of clay, sludge, and flyash were sedimented. Both distilled and saline water were used as sedimentation medium. After a period of sedimentation, the liquid above the sediment was drained and undisturbed specimens of the soil mixture were sampled. The soft and fluid-like sediment was stabilized by "slightly freezing" the samples using liquid nitrogen. Part of the specimens were obtained by allowing the sediment to consolidate under gravity for some time until it attained a moderately solid consistency for handling. Samples were then trimmed and mounted on the proper testing units and confining pressure applied before thawing or further remolding occurred. The shear modulus values for sedimented marine clay were measured to be on the order of 11.5 MPa for both the 10% fuel oil-mixed specimen, and the 10% sludge-mixed specimen under 0.5 kg/cm² confining pressure. The wet unit weight and water content of these specimens were 13.73 kN/m³, 132% and 18.63 kN/m³, 100%, respectively.

The materials used in the sedimentation tests were illite and laboratory-mixed marine clay (50% illite, 21% Ca-montmorillonite, 16% kaolinite, 13% chlorite). Sludge, flyash, and No. 2 fuel oil were used as additives. The pH values of various slurry mixtures were measured as illite and 5% sludge, 6.9; illite and 5% flyash, 7.5; marine clay and 10% sludge, 6.9; marine clay and 10% fuel oil, 7.7

Settlement and concentration curves were generated for illite and sludge, illite and flyash, marine clay and sludge, and marine clay and fuel oil mixtures. Figures 25.10 and 25.11 show the concentration versus depth measurements for illite and sludge and illite and flyash, respectively. Figure 25.12 shows the concentration versus time variation for the midsection of the sedimentation tanks. As observed, the settlement of the sludge-mixed clay is faster than the flyash-mixed clay initially. This supports the hypothesis that the flocculation and agglomeration may occur with the addition of sludge, which promotes the fast deposition. Figure 25.13 shows the actual settlement depth versus time, which also confirms this conclusion. SEM micrographs of sludge and flyash-mixed illite clay specimens are given in Figures 25.15 and 25.16, respectively. The inclusion of fibrous material of sludge and the rounded silt-sized particles of flyash is clearly seen in these micrographs. The coating of illite clay particles with fibrous texture can be detected in Figure 25.16. Due to the loosely deposited matrix, the flyash particles can also easily be seen.

The sedimentation results of marine clay mixtures are given in Figure 25.14. The sedimentation data of illite mixtures are superimposed on this figure for comparison. The additives used with the marine clay were fuel oil and sludge. The fuel oil sample had a higher rate of settling than the sludge sample. Initial slurry pH measurements indicated that the higher pH (7.7) of the fuel oil sample might have promoted dispersion and a slower rate of settlement. The higher rate of settlement, in this case, may be attributed partly to the weight of the fuel oil, part of which pooled on top of the sedimenting column, and probably contributed to sediment movement downward initially.

Conclusions

Variations of shear modulus, permeability, and compressibility were observed with increasing percentage of sludge or flyash in consolidated kaolinite specimens. Some of the behavior could be explained through clay microstructure changes and supporting evidence was gathered from SEM micrographs. It is hypothesized that the formation of clay floc and coating of these flocs with organic material and the stiffening effect of fibrous inclusions contribute to the increase in the low strain shear modulus

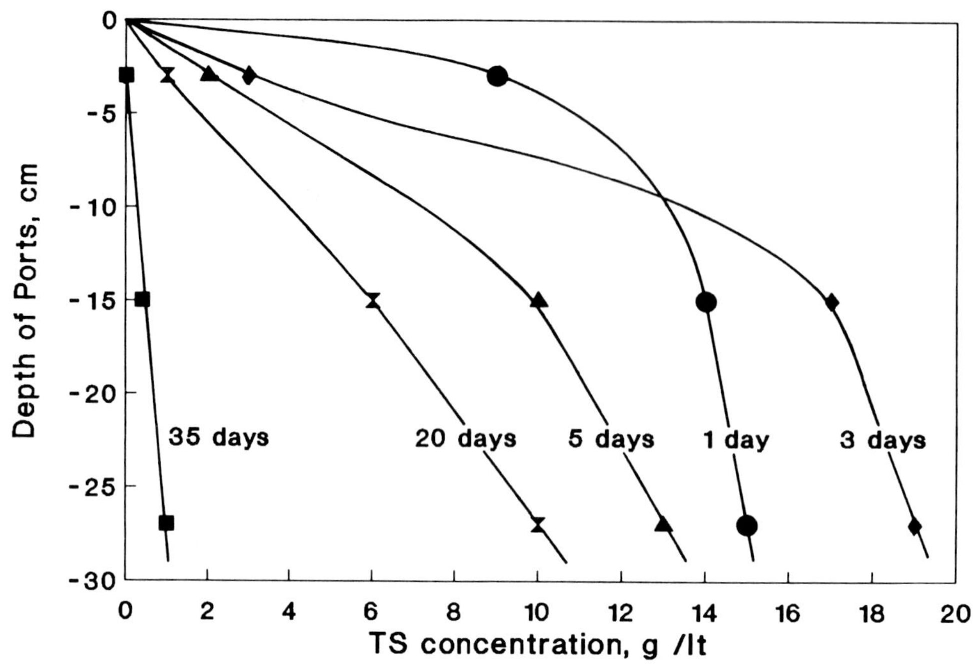

Figure 25.10. Concentration versus depth from the top of settling slurry for 5% sludge-mixed illite specimen.

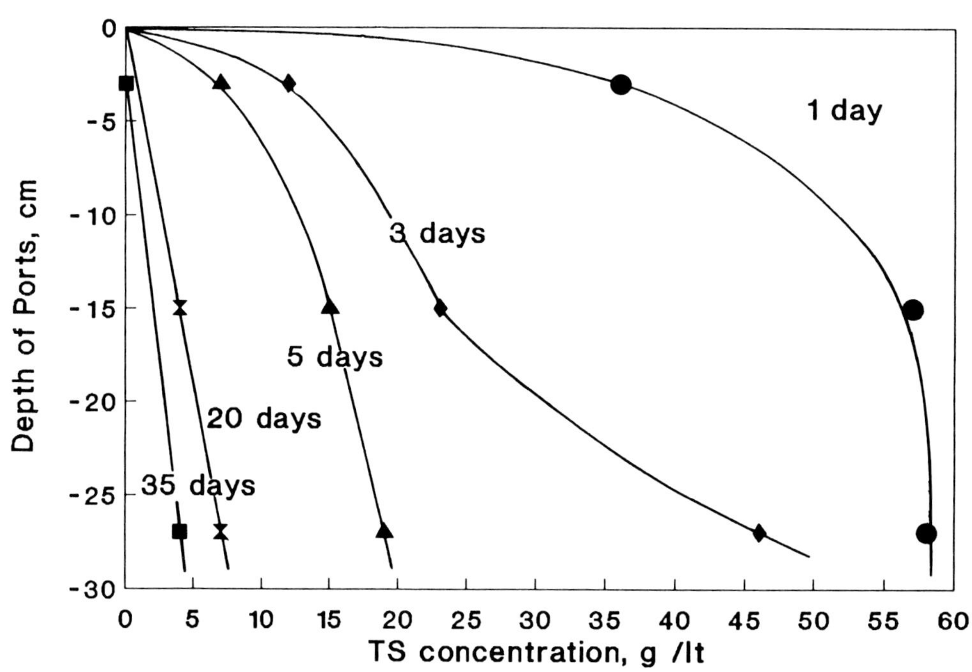

Figure 25.11. Concentration versus depth from the top of settling slurry for 5% flyash-mixed illite specimen.

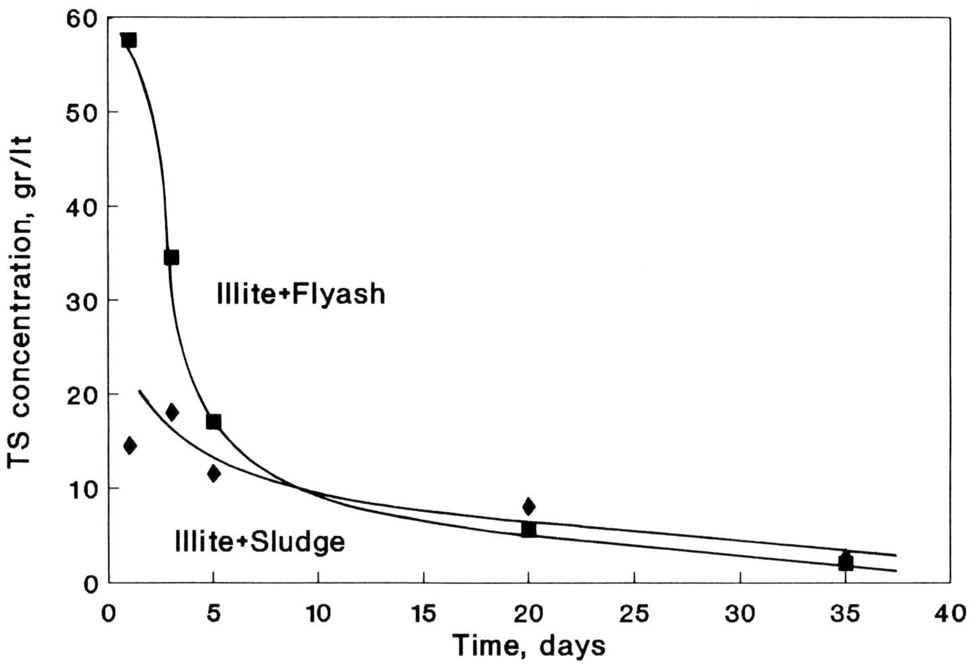

Figure 25.12. Concentration versus time variation for 5% sludge and flyash-mixed illite slurry specimens.

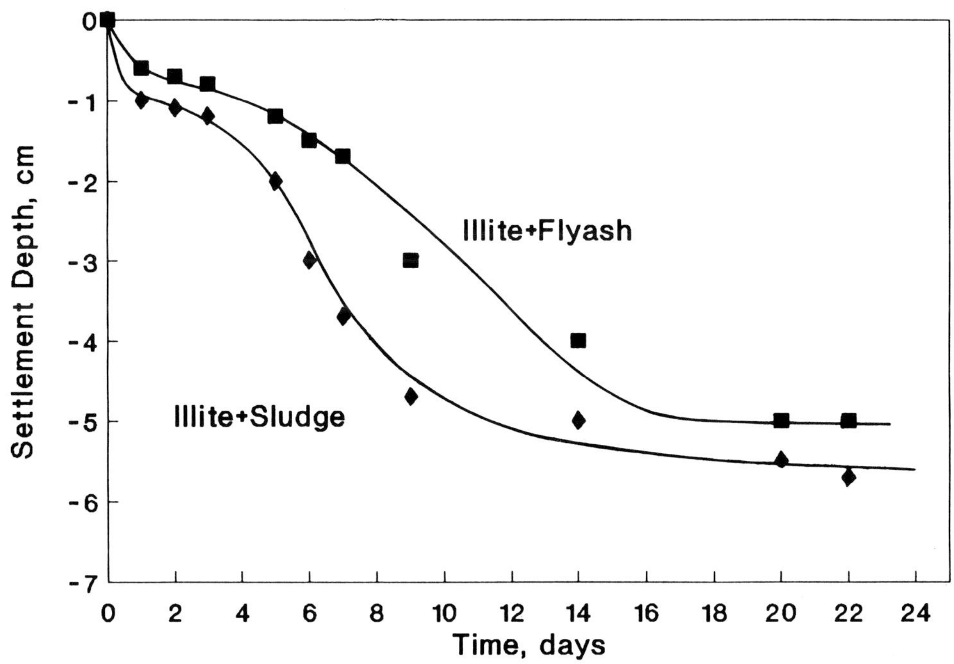

Figure 25.13. Settlement versus time variation for 5% sludge and flyash-mixed illite slurry specimens.

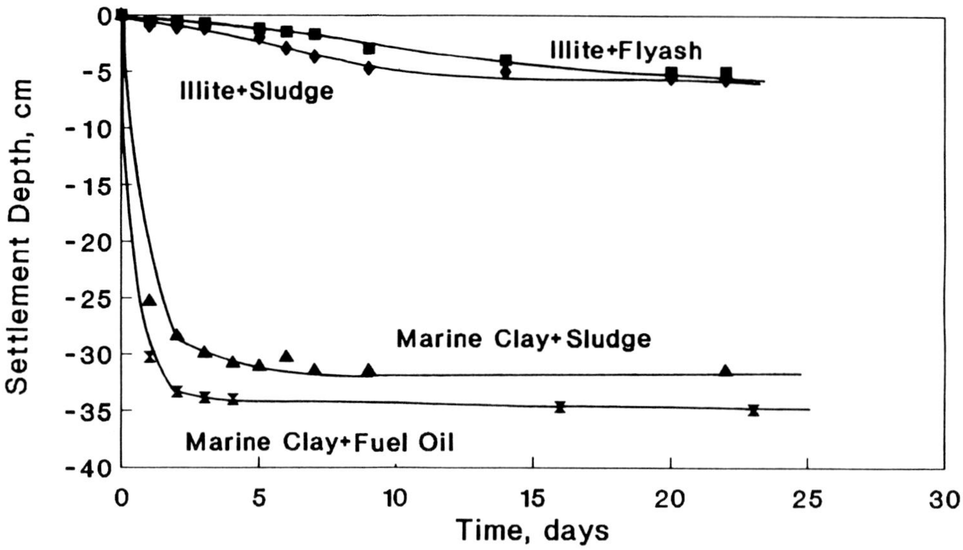

Figure 25.14. Settlement versus time variation for 10% sludge and fuel oil-mixed marine clay and 5% sludge and flyash-mixed illite slurry specimens.

Figure 25.15. SEM Micrograph of 10% sludge-mixed illite settled slurry specimen (scale bar 1 μm).

Figure 25.16. SEM micrograph of 10% flyash-mixed illite settled slurry specimen (scale bar 1 μm).

of the sludge-mixed specimens. Shear modulus and some index properties of sludge or fuel oil-mixed consolidated marine clay samples were also found to vary with the type of contamination. Fuel oil-mixed specimens exhibited low strength, high water holding capacity, and low unit weight. The porous, loose structure of these samples was evident in the micrographs. The rates of sedimentation of the clay mixtures were affected by the presence of contaminants, which influenced the microstructure of the sedimented material. The micrographs of these samples indicated these influences as well.

This study generated some information of a fundamental nature on the influence of contamination on physical properties of clay soils. However, it has also generated questions that can be answered only by further study in interpretation of the microstructure of the created mixtures as well as chemical interactions between contaminants and clays.

References

Acar, Y.B., and A. Ghosh, 1986. The role of activity in hydraulic conductivity of compacted soils permeated with acetone. In: Fang, H.Y. (ed.), Proceedings of Environmental Geotechnology Symposium. Envo Publishing Co., Bethlehem, PA, v. 1, p. 403–410.

Acar, Y.B., I. Olivieri, and S.D. Field, 1985. Transport of organic contaminants and geotechnical properties of fine grained soils. Proceedings of the XI International Conference on Soil Mechanics and Foundation Engineering, American Society of Civil Engineers, v. 3, p. 1237–1240.

Bennett, R.H., and M. Hulbert, 1986. Clay Microstructure. IHRDC Publishers, Boston, 161 p.

Bohn, H.L., B.L. McNeal, and G.A. O'Connor, 1985. Soil Chemistry. Wiley, New York.

Bowders, J.J., 1985. The influence of various concentrations of organic liquids on the hydraulic conductivity of compacted clay. Dissertation, University of Texas at Austin, 218 p.

Bowders, J.J., D.E. Daniel, G.P. Broderick, and H.M. Liljestrand, 1984. Methods for testing the compatibility of clay liners with landfill leachate. Hazardous and Industrial Solid Waste Testing: Fouth Symposium, American Society for Testing Materials, Special Technical Publication 886, p. 233–250.

Drost-Hansen, W., 1976. Effects of vicinal water on colloidal stability and sedimentation process. Journal of Colloid and Interface Science, v. 58(2), p. 251–262.

Du, B.L., G.K. Mikroudis, and H.Y. Fang, 1986. Effect of pore fluid pH on the dynamic shear modulus of clay. Hazardous and Industrial Solid Waste Testing and Disposal: Sixth Volume, American Society for Testing Materials, Special Technical Publication 933, p. 226–239.

Evans, J.C., and H.Y. Fang, 1986. Triaxial equipment for permeability testing with hazardous and toxic permeants. Geotechnical Testing Journal, v. 9(3), p. 126–132.

Evans, J.C., H.Y. Fang, and I.J. Kugelman, 1985. Organic fluid effects on the permeability of soil-bentonite slurry walls. Proceedings of The National Conference on Hazardous Waste and Environmental Emergencies, Environmental Protection Agency.

Faas, R.W., and S. Wartel, 1977. The effect of gas bubble formation on the physical and engineering properties of recently deposited fine-grained sediments. Geologie En Mijnbouw, v. 56(3), p. 211–218.

Fang, H.Y., and J.C. Evans, 1988. Long-term permeability tests using leachate on a compacted clay liner material. Groundwater Contamination: Field Methods, American Society for Testing Materials, Special Technical Publication 963, p. 397–404.

Foreman, D.E., and D.E. Daniel, 1986. Permeation of compacted clay with organic chemicals. Journal of the Geotechnical Engineering Division, American Society of Civil Engineering, v. 112(7), p. 669–681.

Gauffreau, P.E., 1988. Hydrophobic soil: A low cost alternative to clay lining materials. Thesis, Lehigh University, Bethlehem, PA, 97 p.

Grim, R.E., 1968. Clay mineralogy, 2nd ed. McGraw-Hill, New York.

Hettiaratchi, J.P.A., and S.E. Hrudey, 1987. Influence of contaminant organic-water mixtures on shrinkage of impermeable clay soils with regard to hazardous waste landfill liners. Hazardous Waste and Hazardous Materials, v. 4(4), p. 377–388.

Lagaly, G., 1987. Clay-organic interactions: problems and recent results. Proceedings of the International Clay Conference, The Clay Minerals Society, p. 343–351.

Lambe, T.W., 1959. Physico-chemical properties of soils: role of soil technology. Journal of Soil Mechanics and Foundation Engineering Division, American Society of Civil Engineering, v. 85, no. SM2.

Madsen, F.T., and J.K. Mitchell, 1987. Chemical effects on hydraulic conductivity and their determination. Open file report, Environmental Institute of Waste Management Studies, University of Alabama, Tuscaloosa.

Mitchell, J.K., 1976. Fundamentals of Soil Behavior. Wiley, New York, 422 p.

Pamukcu, S., M.K. Kavulich, and H.Y. Fang, 1987. A parametric sensitivity study on mechanical performance of flyash mixed soil. In: Evans, J.C. (ed.), Toxic and Hazardous Wastes, Proceedings of The 19th Mid-Atlantic Industrial Waste Conference, Technomic Publishing Co., Lancester, p. 589–599.

Pamukcu, S., M. Tuncan, and H.Y. Fang, 1990. Influence of some environmental activities on physical and mechanical behavior of clays. Physico-Chemical Aspects of Soil, Rock and Related Materials, American Society for Testing Materials, Special Technical Publication, STP 1095, p. 91–107.

Sparks, L.D., 1986. Soil Physical Chemistry. CRC Press, Boca Raton, FL.

Torrance, J.K., and M. Pirnat, 1984. Effect of pH on the rheology of marine clay from the site of south nation river, Canada, landslide of 1971. Clays and Clay Minerals, v. 32(5), p. 384–390.

Tuncan, M., L.I. Khan, and S. Pamukcu, 1988. The effect of leachate on geotechnical properties of clay liner. In: Varma, M.M., and J.H. Johnson (eds.), Hazardous and Industrial Waste, 20th Mid-Atlantic Industrial Waste Conference, Hazardous Materials Control Research Institute, Silver Spring, MD, p. 133–144.

van Olphen, H., 1977. An Introduction to Clay Colloidal Chemistry for Clay Technologists, Geologists and Soil Scientists, 2nd ed. Wiley, New York.

Whelan T., J.M. Coleman, J.N. Suhayda, and L.E. Garrison, 1975. The geochemistry of recent mississippi river delta sediments: gas concentrations and sedimentary stability. Preprint 2342, Proceedings of the 7th Offshore Technology Conference, Houston, p. 71–84.

CHAPTER 26

Organization of Clay Particles in Aqueous Suspension as Inferred from Spectroscopy of Organic Dyes

Jos Cenens, Robert A. Schoonheydt, and Frans C. De Schryver

Introduction

The clay–water system has attracted the attention of the scientific community from the very beginning of clay research because of its obvious agricultural and economic importance. The question of the swelling of smectites in water is 2-fold: (1) Do clays swell until they form a randomly distributed suspension of individual clay sheets? (2) How influential is the clay surface in determining the properties of adsorbed water?

These questions have been addressed mainly by studying the properties of the adsorbed water by a large range of techniques or by studying the thermodynamics of the clay–water system (Sposito and Prost, 1982).

Our approach is entirely different. The spectroscopic properties of the adsorbed dyes, methylene blue (MB) and proflavine (PF) (Fig. 26.1), depend on the same variables as those that determine the state of aggregation of the clay particles in aqueous suspension: the loading of the adsorbed species, the type of clay, the exchangeable cation, and the particle size (Cenens et al., 1987; Cenens and Schoonheydt, 1988). There must be a correlation between the spectroscopy of adsorbed dyes and the state of aggregation of the clay particles.

Principles

MB and PF are extremely strongly adsorbed by clays from aqueous solutions containing only the monomers, MBH^+ and PFH^+. As a consequence they do not discriminate between different types of sites, but adsorb on the first sites they encounter (Cenens and Schoonheydt, 1988). These sites may be on the edges of the crystals, on the planar surfaces, or in the interlamellar space, if it is available, that is, if the platelets are separated far enough from each other to allow free diffusion of MBH^+ or PFH^+ between the platelets. In any case, the dye molecules are concentrated around the clay particles, and aggregation of the dye molecules occurs. Dimers or even trimers are formed because of the strong hydrophobic interactions between the aromatic ring systems of the dye molecules (Bergmann and O'Konski, 1963; Haugen and Melhuish, 1964). The extent of the di- and trimerization reactions depend on the aggregation of the clay platelets: for a given loading, the smaller the available surface, i.e., the larger the aggregates, the more these reactions are pronounced (Cenens, 1988; Cenens and Schoonheydt, 1988). Thus, when di- or trimerization reactions can be followed quantitatively with absorption or fluorescence spectroscopies, evidences pertaining to clay particle organization in aqueous suspensions can be obtained.

Experimental

The clays used were hectorite (H) and Barasym SSM-100 (synthetic mica-montmorillonite) (BS), source clays of the Clay Minerals Society. Their properties, their preparation (saturated with Na^+), and the preparation of the dye–clay suspensions are described in detail elsewhere (Cenens et al., 1987; Cenens and Schoonheydt, 1988). They are essentially Fe^{3+}-free clays. Only the fraction < 0.3 µm of H and BS was used. BS is essentially a nonswelling clay (Wright et al., 1972). The initial concentration of the dyes is always 2.5×10^{-6} mol/dm^3. This value ensures the presence of only monomers in the starting solutions. The exchange, which is in all cases quantitative, is done in a carbonate buffer at pH 9 to avoid protonation of the dyes at the clay surface. All the loadings are expressed in percentage of the Na^+ CEC (cation exchange capacity) of the clays. Usually loadings do not exceed 25%. Absorption spectra were taken of the clay–dye suspensions and clay suspensions. They were subtracted from each other to obtain the spectra of the adsorbed dyes. Fluorescence

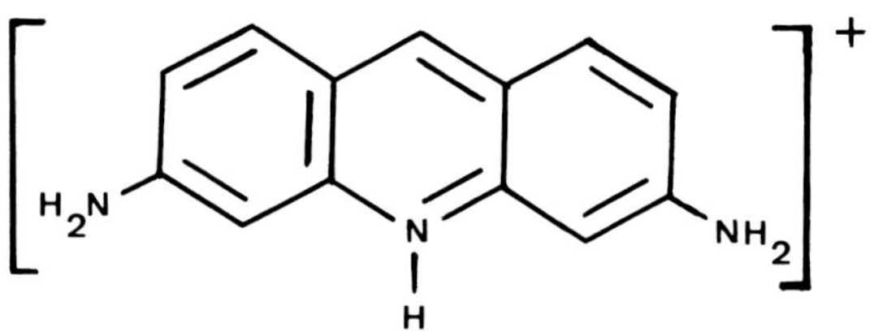

proflavine

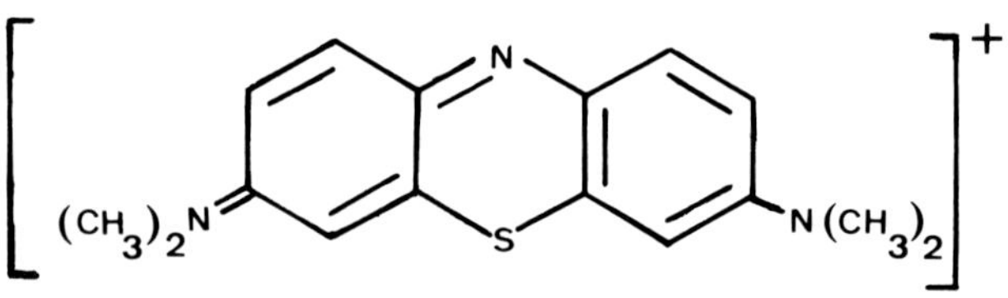

methylene blue

Figure 26.1. Structures of the dyes in aqueous solution.

spectra were taken of clay–PF suspensions and of a solution of 2.5 $\times$ 10^{-6} mol/dm^3 PF. The ratio of the intensities of the fluorescence spectra yields Φ_0/Φ, where Φ_0 is the quantum yield of the solution and Φ the quantum yield of the clay–PF suspension.

Results and Discussion

Typical absorption spectra of MB are shown in Figures 26.2 and 26.3. They show bands at 673 and 653 nm, respectively due to MBH^+ at the external surface and at the planar surface (Cenens and Schoonheydt, 1988). The latter may be the interlamellar surface, but we do not exclude the planar surface of completely delaminated clay sheets, if they occur in the suspension. The former is predominant on BS; the latter on H. As the loading increases, the band intensities of MBH^+ decrease and bands around 610 and 575 nm increase in intensity. These bands are ascribed to dimers [$(MBH^+)_2$] and trimers [$(MBH^+)_3$], respectively (Cenens and Schoonheydt, 1988). On H only dimers are formed, while on BS dimerization is followed by trimerization. In view of the fact that BS is nonswelling it is concluded that trimerization occurs only on the external surface of the clay particles.

The decrease of the intensities of the MBH^+ bands and the concomitant intensity increase of the bands of $(MBH^+)_2$ and $(MBH^+)_3$ with increasing loading suggest a reaction of the type

$$n\ MBH^+ \rightleftharpoons (MBH^+)_n$$

on the surface with $n = 2$ or 3. If this reaction is not hindered by specific surface interactions, then n, the dimerization constant K_d, and the trimerization constant K_t can be deduced from the spectroscopic data by the methods of solution chemistry (Cenens, 1988; Hida and Sanuki, 1970). Experimentally one follows the intensity decrease of the spectra with loading at 656 and at 673 nm for H and BS, respectively. The extinction coefficients of the different species at these wavelengths and the surface concentrations have to be known. The extinction coefficients of MBH^+, $(MBH^+)_2$, and $(MBH^+)_3$ on the surface were determined experimentally (Cenens and Schoonheydt, 1988). The surface concentration (mol/dm^3) can be estimated with two assumptions: (1) quantitative adsorption and (2) concentration of the molecules in a volume around the clay particles, made up of the surface area multiplied by 1 nm. This is the surface layer

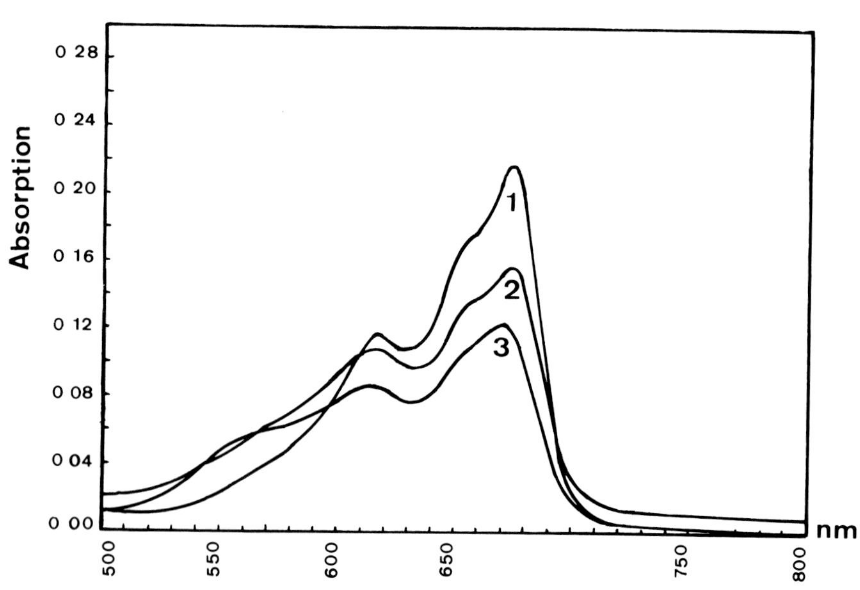

Figure 26.2. Spectra of MB on BS at pH 9 for loadings of (1) 0.38%, (2) 0.93%, and (3) 1.93% of the CEC.

Figure 26.3. Spectra of MB on H at pH 9 for loadings of (1) 0.14%, (2) 1.13%, and (3) 2.82% of the CEC.

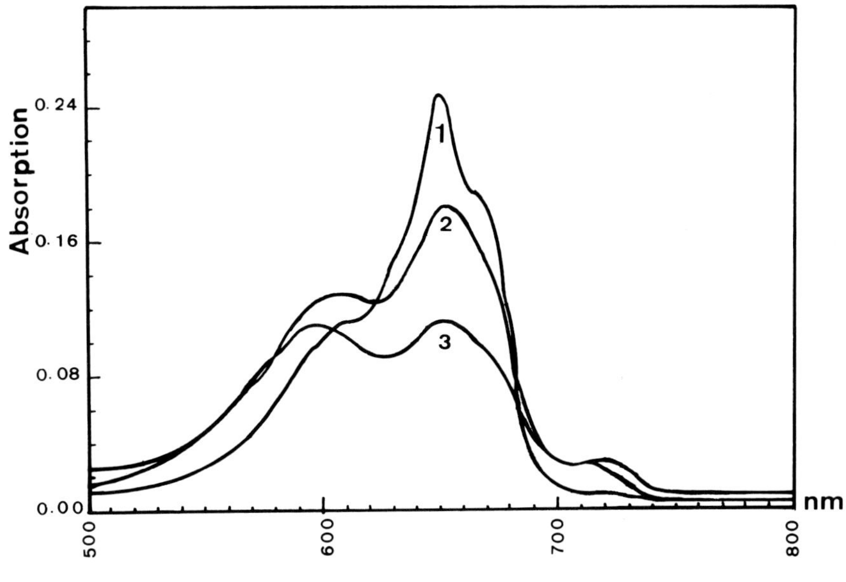

thickness in which the MB molecules are supposed to reside. The first hypothesis has been verified experimentally (Cenens and Schoonheydt, 1988). The second hypothesis is reasonable since MB is strongly adsorbed and the exchangeable cations reside preferentially in the Stern layer (Low, 1987). For H the surface volume is 775 $m^2/g \times 10^{-9}$ m or 7.75×10^{-7} m^3/g; for BS it is 1.33×10^{-7} m^3/g. The value of H is calculated with an average particle size of 300 nm; the value of BS is taken from van Olphen and Fripiat (1979).

Table 26.1 shows the dimerization and trimerization constants of MB on the surface of BS and Na^+–H, calculated from the best fit with the experimental data. The aggregation numbers are very close to 2 for H and to 3 for BS. Three conclusions can be drawn. (1) Almost pure dimerization occurs on H and almost pure trimerization on BS; (2) the successful use of 775 m^2/g in the calculation of the surface volume of H implies that dimerization occurs on the total surface or that the total surface of H is available for reaction. For BS only the N_2-BET surface area (van Olphen and Fripiat, 1979) has been used, indicating that trimerization occurs on the external surface of the aggregates of BS. (3) Trimerization and dimerization occur as in solution, but the values of the constants are smaller. In the literature solution values of dimerization constants for PF are 395–2700 mol^{-1} dm^3 and for MB 2000–5900 mol^{-1} dm^3 (Mataga, 1957; Bergman and O'Konski, 1963; Turner et al., 1972; Spencer and Sutter, 1979).

Dimerization on the interlamellar surface means that the clay sheets are far enough apart to allow free diffusion of MB^+. The minimum distance required is two water layers. Thus Na^+–H may occur as aggregates but the minimum distance between the clay sheets in the aggregate must be at least two water layers. Trimers cannot be formed in the interlamellar space probably because the clay sheets collapse once the dimers are formed in this region.

It is possible to quantify the available surface area in aqueous suspension by a study of the fluorescence of PFH^+, adsorbed on the clays. In the case of PF, trimerization does not occur, not even on BS, and the emission intensity in the region 500–600 nm decreases with the loading due to dimerization only (Cenens et al., 1987). When the natural logarithm of the ratio of the quantum yields of a 2.5×10^{-6} mol/dm^3 solution of PFH^+ and of the same amount of PF adsorbed on the clay are plotted against the surface concentration, as defined above, two observations can be made (Fig. 26.4). (1) The straight line behavior is indicative of a Perrin type of quenching with the dimer as quencher. Thus, the different species on the surface do not diffuse away from each other within the lifetime of the excited state. (2) For PF on Na^+–H two straight lines are obtained; for PF on BS only one line is obtained with the same slope as that of PF on Na^+–H above a loading of 1% or a surface concentration of 7 mmol/dm^3.

To explain these results, we advance the hypothesis that the mechanism of the quenching reaction is the same, independent of the location of PFH^+ on the surface. This means that the slopes of the straight lines would necessarily be equal.

This is the case for BS at all loadings and for H above loadings of 1% of the CEC (surface concentration of 7 mmol/dm^3). It indicates that the total surface area of H is available to PF above 1% of the CEC and that only the external surface area of the aggregates is available in the case of BS. This conclusion is the

Table 26.1. Aggregation constants of methylene blue on clays.

Clay	n	K_d (mol^{-1} dm^3)	K_t (mol^{-2} dm^6)	CEC (meq g^{-1})
BS	2.82		772	0.464
H	1.94	126		0.526

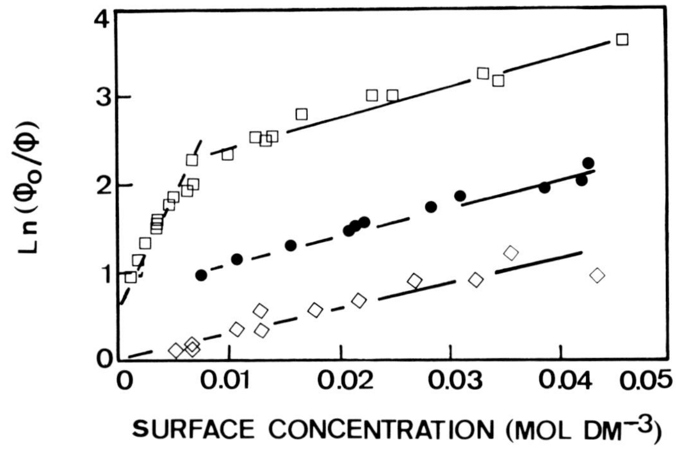

Figure 26.4. The natural logarithm of the ratios of the fluorescence quantum yields of a 2.5×10^{-6} *M* solution of PFH^+ (Φ_0) and of PF on clays (Φ) as a function of the surface concentration. □, H; ◇, BS; ●, external surface of H. Reprinted with permission from Cenens et al., ACS Symposium Series. Copyright 1990, American Chemical Society.

same as that derived above from the interpretation of the absorption spectra of MB.

For H with a loading below 1% of the CEC (surface concentration of 7 mmol/dm^3), our hypothesis dictates the same slope as for BS. The observation that the slope is higher indicates that the available surface area is smaller than 775 m^2/g or that the surface concentration in Figure 26.4 is underestimated. Equal slopes (black dots in Fig. 26.4) can be generated by reducing the surface area to 125 m^2/g. Our interpretation of that number is that below 1% of the CEC PF probes only part of the surface. That part can be interpreted as an external surface of aggregates of H in aqueous suspension.

The external surface of 125 m^2/g is only twice the N_2-BET surface area (van Olphen and Fripiat, 1979). This and our successful use of the N_2-BET surface area of BS suggest that the clay aggregates in aqueous suspension are very similar to those of the corresponding powdered materials. In the aggregates of H, however, the clay sheets are loosely bound, the average distance between them being at least two layers of water molecules. This is shown by the dimerization reactions, discussed above. It is an interesting hypothesis then that the aggregates of H, and possibly any clay, differ not so much in the size or number of individual sheets per aggregate, but more in the organization of the sheets in the aggregate.

Acknowledgments

The authors acknowledge the National Fund of Scientific Research (Belgium) for a research grant covering this research. R.A.S. is indebted to the same institution for a position as research director. J.C. acknowledges the I.W.O.N.L. (Belgium) (Instituut voor Wetenschappelijk Onderzoek in Nijverheid en Landbouw) for a Ph.D. grant.

References

Bergman, K., and C.T. O'Konski, 1963. A spectroscopic study of methylene blue monomer, dimer and complexes with montmorillonite. Journal of Physical Chemistry, v. 67, p. 2169–2177.

Cenens, J., 1988. Spectroscopy of dye molecules at clay surfaces: Ph.D. Thesis, Katholieke Universiteit Leuven. Leuven, Belgium, 150 p.

Cenens, J., and R.A. Schoonheydt, 1988. Visible spectrosopy of methylene blue on clays in aqueous suspension: hectorite, laponite B and barasym. Clays and Clay Minerals, v. 36, p. 214–224.

Cenens, J., R.A. Schoonheydt, and F.C. De Schryver, 1990. Probing the surface of clays in aqueous suspension by fluorescence spectroscopy of proflavine. In L.M. Coyne, S.W.S. McKeever, and D.F. Blake (eds.), ACS Symposium Series, v. 415, p. 378–395.

Cenens, J., D.P. Vliers, R.A. Schoonheydt, and F.C. De Schryver, 1987. Spectroscopic study of the surface chemistry of proflavine on clay minerals. In: Schultz, L.G., H. van Olphen, and F.A. Mumpton (eds.), Proceedings International Clay Conference, Denver, 1985, The Clay Minerals Society, Bloomington, p. 352–358.

Haugen, G.R., and W.H. Melhuish, 1964. Association and self-quenching of proflavine in water. Transactions of the Faraday Society, v. 60, p. 386–394.

Hida, M., and T. Sanuki, 1970. Studies of the aggregation of dyes. The scope of application of the maximum slope method. Bulletin of the Chemical Society of Japan, v. 43, p. 2291–2296.

Low, Ph.F., 1987. Structural component of the swelling pressure of clays. Langmuir, v. 3, p. 18–25.

Mataga, N., 1957. Note on the polymerization of dyes in solution. Bulletin of the Chemical Society of Japan, v. 30, p. 375–379.

Spencer, W., and J.R. Sutter, 1979. Kinetic study of the monomer-dimer equilibrium of methylene blue in aqueous solution. Journal of Physical Chemistry, v. 83, p. 1573–1576.

Sposito, G., and R. Prost, 1982. Structure of water adsorbed on smectites. Chemical Reviews, v. 82, p. 553–573.

Turner, D.H., G.W. Flynn, S.K. Lundberg, L.D. Faller, and N. Sutin, 1972. Dimerization of proflavine by the laser Raman temperature-jump method. Nature, v. 239, p. 215–217.

van Olphen, H., and J.J. Fripiat, 1979. Data Handbook for Clay Minerals and Other Non-Metallic Minerals. Pergamon Press, New York, 346 p.

Wright, A.C., W.T. Granquist, and J.V. Kennedy, 1972. Catalysis by layer lattice silicates. I. The structure and thermal modification of a synthetic ammonium dioctahedral clay. Journal of Catalysis, v. 25, p. 65–80.

CHAPTER 27

Some Effects of Vicinal Water on the Sedimentation Process, Compaction, and Ultimate Properties of Sediments

W. Drost-Hansen

Introduction

For many processes in aqueous systems the details of the molecular structure of water may safely be ignored. However, when dealing with dimensions in the range from colloidal particles to molecular dimensions (say 1 μm to 1 Å) it becomes necessary to pay careful attention to structural aspects of the aqueous environment. Thus, in dealing with the sedimentation of clays and other minerals as well as when describing fine-grained sediments both bulk and interfacial water structures play a role.

A vast literature exists dealing with the properties of bulk water and bulk aqueous solutions, yet the structure of these bulk phases is still not known with certainty. Even less is known about water at interfaces. Again a wealth of information exists on the properties of large surface-to-volume ratio systems but far less is known about the structure of interfacial water, specifically how proximity to an interface may change the bulk structure and how far into the bulk this effect is manifested.

Until fairly recently, the foundation of surface and colloid chemistry was the theory first developed about 45 years ago by Deryaguin and Laudau in the U.S.S.R. and, simultaneously, by Verway and Overbeek in the Netherlands, the so-called DLVO theory. The theory is especially applicable to the problem of colloid stability; it takes into account, semiquantitively, the van der Waals–London dispersion forces (leading to attraction between particles) and the electric double layer interactions, leading to repulsion. The theory was a truly major accomplishment; however, it does not represent a complete picture. Two other types of forces must be taken into account, namely solvation forces (resulting from surface-induced changes in the solvent structure) and, less generally, steric forces. The latter are sometimes also referred to as forces of entropic origin, and these may play a role when adsorbed polymers (or other flexible surface groups) are present on the interacting surfaces.

No one has contributed as much to unraveling the importance of solvation forces as Deryaguin and his hundreds of co-workers. (See for instance Deryaguin and Churaev, 1974, 1977, 1986 and Churaev and Deryaguin 1985.) Their investigations have provided a wide range of approaches to the quantitative study of solvation forces. However, over the past 10 or 15 years the most impressive tool for direct measurements of surface forces is the force balance apparatus. Deryaguin and Abricossova (1951, 1958) used a direct force balance method to investigate the attraction between macroscopic bodies, described by the Hamaker constant; Peschel and Adlfinger in Germany (1967, 1971) measured forces between large quartz surfaces in aqueous (and some nonaqueous) systems. The most successful approach has been that of Israelachvili and co-workers (Israelachvili, 1985, 1987b; Israelachvili and Adams, 1978; Israelachvili and Christenson, 1986), who have perfected a surface force balance capable of measuring forces ranging over more than four orders of magnitude (from macroscopic dimensions to within about 1 Å). Israelachvili (and co-workers) summarized the current state of affairs in surface forces and it appears that future efforts to describe particle–particle interactions must account both for the DLVO theory and the solvation and entropic effects. Such a theory will in this chapter be referred to as the DLVOI theory.

Among the reasons for Israelachvili's success is the fact that he perfected a technique for using mica surfaces. When properly cleaved such surfaces are smooth on the molecular level, thus allowing direct force measurements over distances from intimate contact (less that 1 Å) to microns (or longer). Israelachvili and co-workers (1978, 1985, 1986, 1987b) also demonstrated that near the mica surfaces many consecutive layers of solvent may be laid down with remarkable periodicity. The extent of these "ordered" layers may approach 6 to perhaps 10 molecules—or far beyond the two or three molecular layers previously assumed. However, while the forces due to the solvation effects may be

Table 27.1. Some vicinal water properties (comparison with bulk properties).

Property	Bulk	Vicinal	Reference*
Density (g/cm³)	1.00	0.97	a,b
Specific heat (cal/K-g)	1.00	1.25 ± 0.05	c,d
Thermal expansion coefficient (°C⁻¹) (25°C)	250×10^{-6} (25°C)	300–700 × 10^{-6}	e
(Adiabatic) compressibility coefficient (atm⁻¹)	45×10^{-6}	60–100 × 10^{-6}	f
Excess sound adsorption (a/υ^2) cm⁻¹/sec²	7×10^{-17}	~35 × 10^{-17}	g
Heat conductivity (cal/sec)/cm²/(°C/cm)	0.0014	~0.01–0.05	h,i
Viscosity (cP), Energy of activation	0.89	2–10	j
Ionic conduction (KCl) (kcal/mol)	~4	5–8	k
Dielectric relaxation frequency (Hz)	19×10^9	2×10^9	l

*References: (a) Etzler and Fagundus (1983); (b) Etzler and Fagundus (1987); (c) Braun and Drost-Hansen (1976); (d) Etzler (1988); (e) Drost-Hansen (1982); (f) Bruun et al. (1978a); (g) Bruun et al. (1978b); (h) Metsik and Aidanova (1966); (i) Drost-Hansen (1971) (see p. 155–159); (j) Peschel and Adlfinger (1970); (k) Schufle et al. (1976); (l) Alpers and Hühnerfuss (1983).

measured over distances approaching hundreds of molecular layers, this by no means implies that the water structure is affected over such large distances. Thus, Israelachvili (1987a) suggests that the extent of structure rearrangement may have dropped to about one-half the maximal extent over a distance of only, say, three or four molecular layers. Thus, while the solvation forces do correspond to partial structural rearrangements over distances 10 times greater than envisioned in the classical DLVO approach, these rearrangements are not particularly long range. It is at this point that the ideas of Israelachvili and the present author diverge in that I believe structural changes may indeed be induced in the vicinal water over very large distances—of the order of about 20–50 molecular diameters. Other differences also exist and these will be discussed below, especially the idea that the interfacially modified water responds in a highly nonclassical manner to temperature.

Vicinal Water

As discussed above most (or all) solid surfaces in contact with water or aqueous solutions modify the structure of the neighboring water and a vast literature exists on surface hydration effects due to the interaction of the water with such quantities as surface bound ions, dipoles, and electrical double layers. These interactions are usually highly energetic but of short range, generally involving only a few molecules, say 2 to 5 (at most). On the other hand, over the past 20 years evidence has been accumulating for the existence of low-energetics, longer range structural effects, apparently extending over distances of as much as 20 to 50 molecular diameters (say 50–130 Å) from the interface. These surface modified, aqueous layers are referred to as vicinal water layers (vw). It remains to be seen if the vicinal water structures discussed here are similar or identical to the structures responsible for the long-range hydration forces described by Israelachvili.

An important feature of the surface modified, long-range structure of the interfacial water (the vicinal water) is its ubiquitous nature: it appears that vicinal water occurs at most (or all) water–solid interfaces, regardless of the specific chemical nature of the surface. This is referred to as the paradoxical effect (PE) (see Drost-Hansen, 1976; Etzler and Drost-Hansen, 1983). Because of the PE it appears highly likely that one must allow for the effects of vicinal water in all sedimenting systems in nature and in all water-filled, porous sediments. The present chapter, therefore, addresses some of the likely processes and phenomena in natural sediment systems where vicinal water may be expected to play a role.

It is beyond the scope of this chapter to present the detailed evidence for vicinal water structuring. Instead, Table 27.1 gives some typical examples of estimates of vicinal water properties. It should be kept in mind that these estimates are notably affected by the distance from the solid surface inducing the vicinal structure. Very likely the maximal extent of vicinal water is found a few molecular layers from the interface (beyond the strongly bound water, such as the water of hydration of the ions in the surface.) The degree of structure modification likely decays away from the interface until unaffected bulk structure prevails, say 200 or so molecular diameters d_{H_2O} away from the surface.

For purposes of estimating the effects of vicinal water on thermodynamic or kinetic parameters a rather crude approximation is frequently made: instead of accounting for an exponentially decreasing vicinal water structuring on going from the surface into the bulk phase, an "effective thickness," Δr, of vicinal water is introduced. The data are treated as if all the water within Δr is vicinal water while the water beyond this thickness has its normal, bulk properties. Obviously, this approximation is made for arithmetical convenience only and does not reflect any physical "boundary" between vicinal water and bulk water. For extensive discussions of the depth of orientation of vicinal water see Hühnerfuss (1983) and Alpers and Hühnerfuss (1983).

Structure and Energetics of Vicinal Water

There exists relatively little specific information about the structure of vicinal water. In all probability the structure is not very different from the bulk. In the theory of Etzler (1983), the main difference is a slightly greater tendency for more complete four-coordination of the water molecules. By the same token, the energy difference between bulk water and vicinal water is probably small, say of the order of a few hundred calories per mole, i.e., less than RT at room temperature. For this reason many processes are relatively insensitive to vicinal water (namely if the process considered is highly energetic). On the other hand, phenomena involving rate processes such as diffusion may be strongly affected by the vicinal water. This is particularly true in cell physiology.

To illustrate those aspects that may be affected by vicinal water consider the processes of sedimentation, compaction, and consolidation into a dense sediment. Very likely vicinal water will influence several or many of the processes involved, as illustrated below. However, keeping in mind that vicinal water does not differ that much from bulk water it is possible that the net effect of the vicinal water structuring may be relatively small in some cases. Not enough is known about vicinal water to permit any but the crudest estimates to be made of the magnitude of the vicinal water influence.

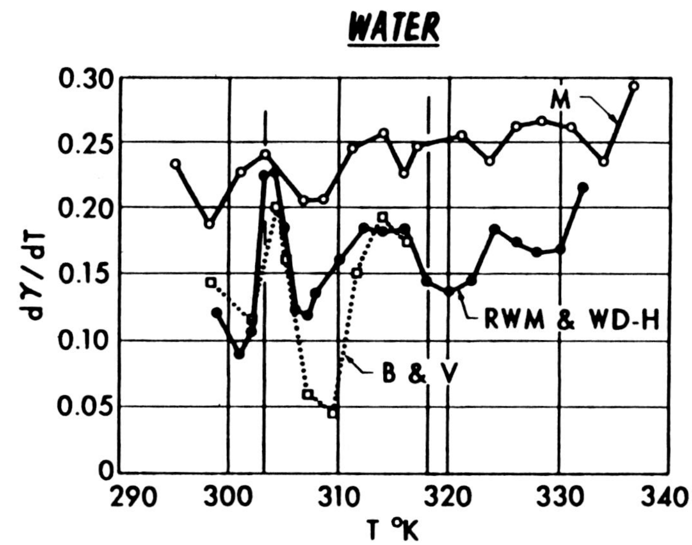

Figure 27.1. Derivative of surface tension with respect to temperature. Based on data by Moser, Bordi and Vannel, and Drost-Hansen and Myers (Drost-Hansen, 1965b).

Some Specific Aspects of Vicinal Water

One of the most surprising attributes of vicinal water is the occurrence of thermal anomalies in the properties of such interfacially modified water. As an example, Figure 27.1 shows the entropy of surface formation (ΔS_s) obtained by differentiating the apparent surface tension with respect to temperature (Drost-Hansen, 1965b). Our own data for surface tension were obtained using a double capillary method. A distinct peak near 30°C (303 K) is seen in all three sets of data; the results by Bordi and Vannel (1962) and those by Drost-Hansen and Meyer (1963) show a nearly 100% increase in ΔS_s around 30°C compared to both lower and higher temperatures. The effect is due to some type of structural transition in the vicinal water in the glass capillary/water/air region. (Note that there appear to be no anomalies in the surface tension of water measured in such a way as to eliminate or minimize the influence of a neighboring solid surface.)

The surface entropy data shown in Figure 27.1 are quite characteristic of the thermal anomalies of vicinal water. In fact more or less abrupt changes occur in the temperature coefficient of many aqueous surface properties, at or near 14–16, 29–32, 44–46, and 59–62°C. In general these temperature ranges are referred to as T_k.

Wiggins (1975) provided another example of abrupt changes in an aqueous surface property, namely the effect of vicinal water on interfacial ion distribution. Wiggins equilibrated samples of silica gel (with average pore diameter 140 Å) with solutions containing both Na^+ and K^+ in equimolar ratios. Three series of measurements were made using Na–K-sulfate, Na–K-iodide, and Na–K-chloride as the electrolytes. After equilibration was achieved Wiggins analyzed for Na^+ and K^+ concentrations and calculated the concentration ratio for each cation:

$$\lambda_K = \frac{[K^+]_i}{[K^+]_\sigma}$$

and

$$\lambda_{Na} = \frac{[Na^+]_i}{[Na^+]_\sigma}$$

where $[\cdot]_i$ and $[\cdot]_\sigma$ indicate, respectively, the cation concentration inside the pores and outside the pores (in the bulk solution). Wiggins then formed the ratio $\lambda_k/\lambda_{Na} = K^K_{Na}$. Finally the measurements were repeated at many closely spaced, different temperatures. The results are shown in Figure 27.2. In this figure note that the values for K all exceed 1.00. Note also that the three sets of data (for sulfates, iodides, and chlorides, respectively) are quite similar: it appears that the anions play no role in the cation distribution. Most impressive is the unexpected influence of temperature: well-defined peaks in the ion distributions are observed near 15, 30, and 45°C, rather precisely the temperatures at which thermal anomalies were reported as long as 20 years earlier (Drost-Hansen and Neil, 1955; Drost-Hansen and Lavergne, 1956).

Because of the great significance of Wiggins findings it was important to confirm these results. Using a similar type of silica gel, Wiggins experiments were repeated in my own laboratory (Hurtado and Drost-Hansen, 1978). Our results for Na^+–K^+ chloride were quantitatively identical to the results reported by Wiggins. Later Etzler and Liles (1986) performed similar experiments (using Li^+–K^+ ions) (in both water and D_2O) and observed peaks in the ion-partitioning ratio at the same temperatures as reported by Wiggins and by Hurtado and Drost-Hansen. Thus, there can be no doubt whatever that the effect is real and it must surely play a role in the natural sediments of interest to this conference. Thermal anomalies, similar to those described above,

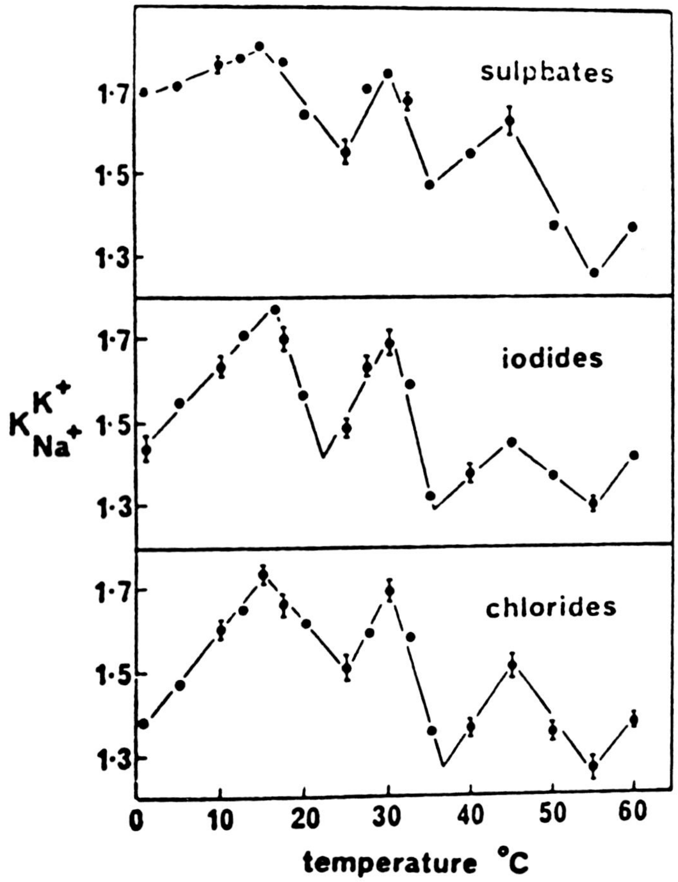

Figure 27.2. Potassium/sodium ion distribution in silica pores (from equimolar mixture of Na^+ and K^+) as a function of temperature. Top to bottom: sulfates, iodides, chlorides (Wiggins, 1975).

have been observed for many other properties of interfacial water (and aqueous solutions) (Drost-Hansen, 1965a, 1969a, b, 1978). The anomalies are most likely due to some type of phase transitions in the vicinal water. Among other properties that have revealed such anomalous temperature dependencies are viscosity, surface conductance, and disjoining pressure. These and other examples will be mentioned below.

Overview

In Table 27.2 are outlined some of the processes involved in sedimentation and compaction as well as some dynamic and equilibrium properties of the final sediment. Each of the steps in this table are discussed below.

Sedimentation

In classical theory we may use the Stokes fall velocity expression for a spherical particle settling in a homogeneous liquid phase:

Table 27.2. Some of the processes involved in sedimentation and compaction as well as some dynamic and equilibrium properties of the final sediment.

Process	Effects
Sedimentation	Fall velocity affected by vw viscosity and density characteristics
Aggregation	In principle, the DLVO theory plus the longer range hydration forces (Israelachvili) may be applied. Because of the vw the disjoining pressure between particles may be strongly influenced by temperature and pressure
Hindered settling	Counterflow affected by viscosity of vw
Compaction	In principle, DLVO theory plus Israelachvili hydration forces may be used to model the compaction process. vw affects ion redistribution and specific heat. Volume change due to elimination of overlapping vw shells
Consolidated sediment	Flow through sediments affected by vw. Ionic composition affected due to ion redistribution. Ionic conduction affected. Heat capacity and thermal conductivity affected. Sound velocity affected. Microbial activity influenced by effects of vw

$$v_s = \frac{2(\rho_p - \rho_0)r^2 g}{9\eta}$$

where v_s = particle velocity
ρ_p = particle density
ρ_o = solution density
r = particle radius
g = acceleration of gravity
η = viscosity of solution

For nonspherical particles it is customary to introduce a "form factor," f, that, in most cases, is not significantly different from unity.

In view of the likely existence of vicinal hydration of the particles, we propose (Drost-Hansen et al., 1987a) that the expression for v_s should be modified to take into account the change in effective diameter and effective density of the particle. r is replaced by $r + \Delta r$, where Δr is the "effective thickness" of the vicinal water layer. As the particle now appears larger than the "bare" particle, the net density is also affected. Thus, allowing for the effects of the vicinal water, the expression for the settling velocity becomes

$$v_{s,h} = \frac{2(\rho_{p,h} - \rho_0)(r + \Delta r)^2 g}{9\eta}$$

where $v_{s,h}$ is the fall velocity of the vicinally hydrated particle and $\rho_{p\cdot h}$ is the net density of the hydrated particle.

To a first approximation ($\Delta r \ll r$) $\rho_{p,h}$ becomes

$$\rho_{p,h} = \rho_p - \frac{3\Delta r}{r}(\rho_p - \rho_{vw})$$

Combining the two effects it is seen that the fall velocity is increased due to the increase in effective radius but the rate is decreased because of the lower effective density; for a particle

CHAPTER 28

Rheology and Microstructure of Concentrated Illite Suspensions

D.J.A. Williams and P.R. Williams

Introduction

The rheological properties of cohesive clay suspensions are of theoretical and practical interest in problems involving particulate transport at the sediment–water interface such as erosion, resuspension, and contaminant transfer and transport.

Cohesive suspension rheology shows a marked dependence on solids concentration ϕ. For $\phi \geq \phi_c$ where ϕ_c is a critical solids concentration, a volume filling network structure exists that gives rise to complex non-Newtonian behavior. In general these structured materials exhibit viscoelastic behavior and mechanical properties of interest in sediment dynamics such as elasticity and yield behavior are sensitive to physicochemical environment (Golden et al., 1982).

Illite, which is an important and common constituent of sediments, provides a useful model clay particle for systematic investigation of cohesive sediment rheology. This chapter describes the influence of pH, concentration of electrolyte, and ϕ on the viscoelastic behavior of well-characterized illite suspensions using the complementary techniques of shear wave propagation and applied stress rheometry. The use of these techniques permits meaningful estimates of elastic moduli and quasistatic yield stresses of these often weakly structured suspensions (James et al., 1987).

Experimental

Materials

A sample of raw illite was supplied by W. W. Emerson, CSIRO, Division of Soils, Urrbrae, Australia. The results of detailed characterization of this material in our laboratories (Beene, 1987) can be summarized as follows:

1. The clay is nonexpanding, dioctahedral, aluminous, potassic, and mica like containing K^+ as the interlayer cation and with a characteristic basal spacing of 10 Å. This description is in accord with the definition of an illite (see Srodon and Eberl, 1984).
2. Electron microscopic examination of the clay revealed particles with a predominantly platey habit, with platelet diameter of $\sim$ 0.043 μm and platelet thickness of $\sim$ 0.02 μm.
3. The solids density was 2648 kg/m^3 and the specific surface area (N_2 adsorption; BET) was 125.6 m^2/g.

Suspensions of required solids concentration were prepared by dispersing clay ultrasonically in KCl (Analar grade) solutions; the water used was twice distilled from an alkaline potassium permanganate solution in an all-glass still. Suspensions were allowed to equilibriate: changes in pH were negligible after 24 hr. Before making rheological measurements, suspensions were ultrasonicated for about 3 hr.

Shear Wave Propagation

The shear modulus G was measured using a modification (Williams and Williams, 1989) of the pulse shearometer (Rank Bros., Bottisham, Cambridge, U.K.), which consists of a pair of parallel discs housed in a cylindrical chamber containing the clay sample (50 ml) and which operates in the horizontal mode. Each disc is connected to a piezoelectric crystal and a shear wave is initiated by an electrical pulse to one of the crystals inducing rapid, small amplitude ($\sim 10^{-4}$ rad) torsional displacement of one disc relative to the other. The frequency of the wave generated in the instrument is $\sim$ 220 Hz. The velocity of the wave V is determined by measuring the time taken by the wave to travel the distance d between discs.

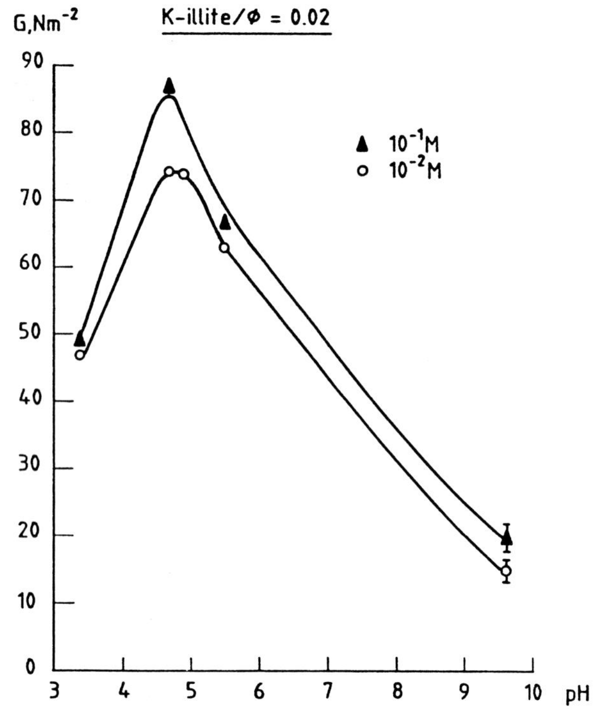

Figure 28.1. Variation of shear modulus with pH for K-illite suspensions at ϕ = 0.02 in 10^{-1} M KCl.

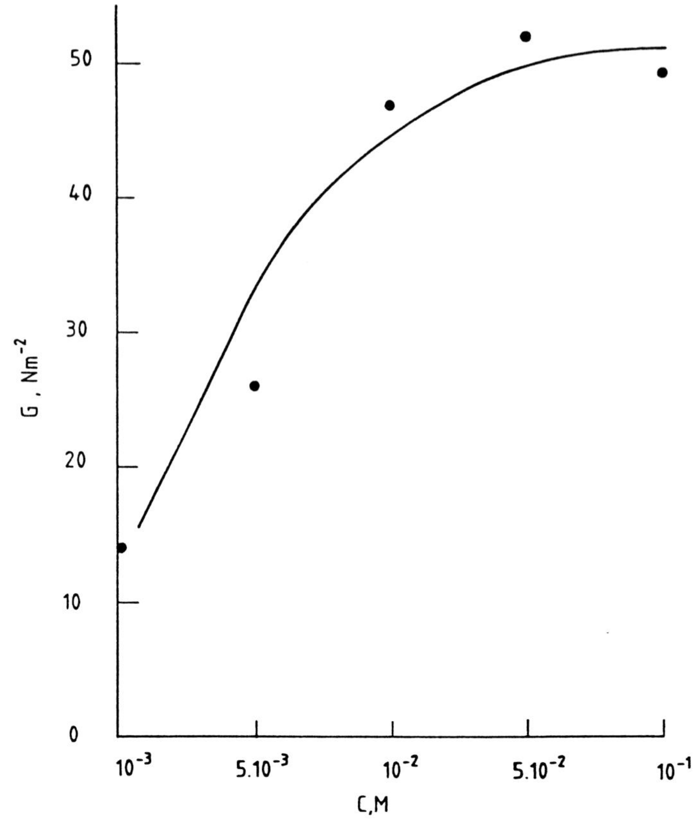

Figure 28.2. Variation of shear modulus G with KCl concentration for K-illite suspensions at ϕ = 0.02 and pH 3.4.

The dynamic modulus G^1 is related to the wave velocity by

$$G^1 = V^2\rho \frac{(1 - r^2)}{(1 + r^2)^2} \tag{1}$$

where ρ is the density of the sample and r is a damping factor given by

$$r = \frac{V}{\omega Z_0} \tag{2}$$

where ω is the angular frequency and Z_0 is the damping length. When r is small, i.e., when $\lambda_s/Z_0 << 1$ where λ_s is the wavelength of the shear wave, then G^1 is approximated by the shear modulus $G = \rho V^2$. This condition is met in the present work since the instrument is designed to work in the region where $\lambda_s/Z_0 \ll 0.1$, so that the shear modulus is the limiting value of the dynamic modulus reached at high frequency.

The time (t_e) required by a sample to recover its equilibrium structural level, following vigorous shaking in a flask and its insertion into the shearometer was recorded; t_e was then incorporated into applied stress rheometry.

Applied Stress Rheometry

Measurements of creep and yield stress were made with a constant stress rheometer (Deer; Series III) fitted with a smooth walled conicylindrical (Mooney) measuring geometry (bob radius 17.5 mm, height 6.5 mm, annular gap width 2 mm, cone angle 3°, gap under cone 76 µm. A detailed description of the theory and operation of this instrument pertinent to weakly structured suspensions is given elsewhere (James et al., 1987); appropriate shear stress (σ) and shear rate ($\dot{\gamma}$) relationships were used in calculations (Whorlow, 1980).

The deformation time $\gamma(t)$ (γ = shear strain) response of illite suspensions as a function of ϕ, pH, and electrolyte concentration C, was determined using the following procedure:

1. A test sample (16 ml) was loaded into the Mooney cell and subjected to a shear rate in excess of 100 s^{-1} for ~ 3 min and then allowed to stand for $t \geqq t_e$.
2. A constant stress was applied instantaneously to the sample and removed suddenly after 30 sec.
3. The sample was again allowed to equilibriate, i.e., stand for $t \geqq t_e$ before application of a slightly higher stress.
4. Repetition of (3) until the onset of thixotropic breakdown in the sample structure (increasing $\dot{\gamma}$ at constant σ and viscosity decrease, usually over several orders of magnitude, provides an operationally defined yield stress σ_y (James et al., 1987).

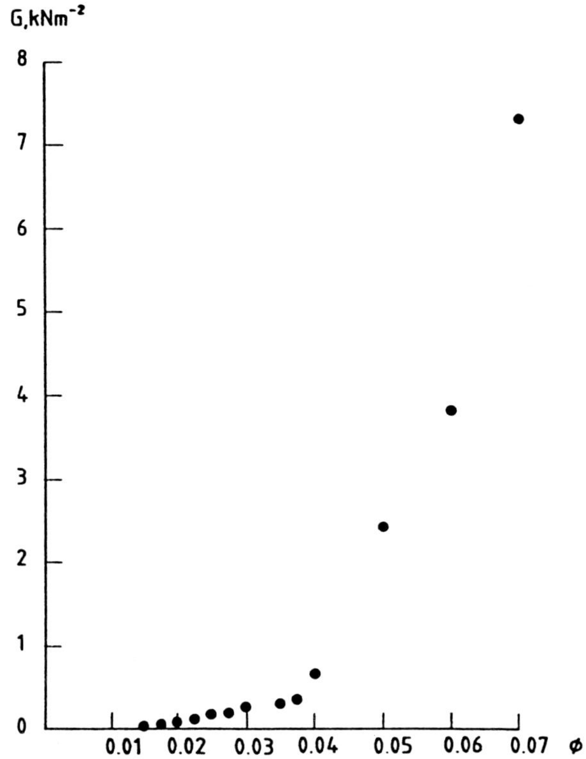

Figure 28.3. Variation of shear modulus G with ϕ for K-illite suspensions in 10^{-1} M KCl at pH 4.7.

Table 28.1. Rheological properties of illite suspensions at $\phi = 0.02$ in 10^{-2} M KCl.

pH	J_0^{-1} (N/m²)	σ_c (N/m²)	γ_c	E_c (J/m³)
3.4	38	0.03	8×10^{-4}	10^{-5}
4.7	87	0.07	8×10^{-4}	2.8×10^{-5}
5.5	26	0.03	10^{-3}	1.7×10^{-5}

Results and Discussion

The development of structure, which is associated with flocculation of illite particles, is an important factor influencing the rheological properties of these suspensions. Shear wave propagation and applied stress rheometry provide rheological information by perturbing the network structure from its equilibrium flocculated state; thus Figures 28.1, 28.2, and 28.3 reflect suspension structure and particle interaction forces associated with a given physicochemical environment.

Illite particles are dually charged with a net charge on the platelet edge, which depends on pH and electrolyte concentration, and a permanent negative charge on the face due to isomorphous substitution of Al^{3+} for Si^{4+} in the tetrahedral layer. At pH < PZC (point of zero charge) the edge possesses a net positive surface charge while a net negative charge exists at pH > PZC.

The development of structure is influenced not only by pH and KCl concentration but also by suspension treatment prior to the determination of G and creep compliance at equilibrium. Herein, this initially involves subjecting a suspension to a shear rate in which relative motion of particles occurs due to the velocity gradient, followed by a period t_e during which $\dot{\gamma} = 0$ and diffusion-limited agregation occurs.

Various modes of interaction between particles are thus possible depending on physicochemical environment, namely face–face (F-F) face–edge (F-E), and edge–edge (E-E, crossed or parallel modes). Face–edge interactions could lead to a card-house structure (van Olphen, 1963); however, the relative motion of two platelets in a velocity gradient could lead to a floc comprising overlapping plates in the F-E mode (James and Williams, 1982).

In contrast to other clays, e.g., kaolinite or montmorillonite, where the ratio of face diameter/edge thickness is typically 10:1 and 100:1, respectively, this ratio is ~ 2:1 for many of these illite particles. Thus pairwise interactions between platelets or small flocs formed in overlapping F-E or F-F modes will under diffusion-limited aggregation be much influenced by the physical and electrochemical properties of the edge.

The variation of G with pH at $\phi = 0.02$ is shown in Figure 28.1. Well-defined maxima occur at ~ pH 4.6 in the presence of 10^{-1} and 10^{-2} M KCl and emphasize the sensitivity of structure elasticity to chemical environment. For an E-E interaction model these data suggest a PZC for the edge of ~ pH 4.6;

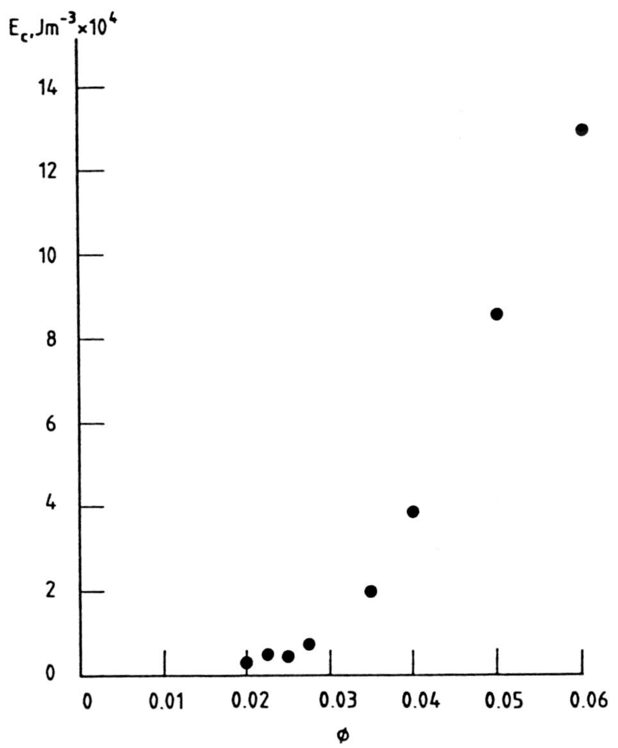

Figure 28.4. Variation of cohesive energy E_c with ϕ in 10^{-1} M KCl at pH 4.7.

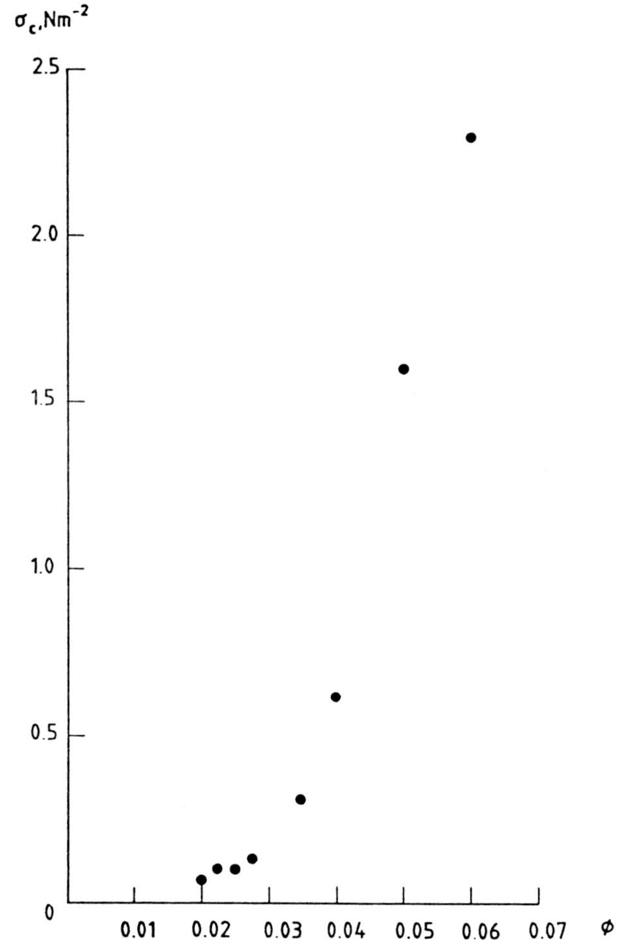

Figure 28.5. Variation in critical stress σ_c with ϕ in 10^{-1} *M* KC1 at pH 4.7.

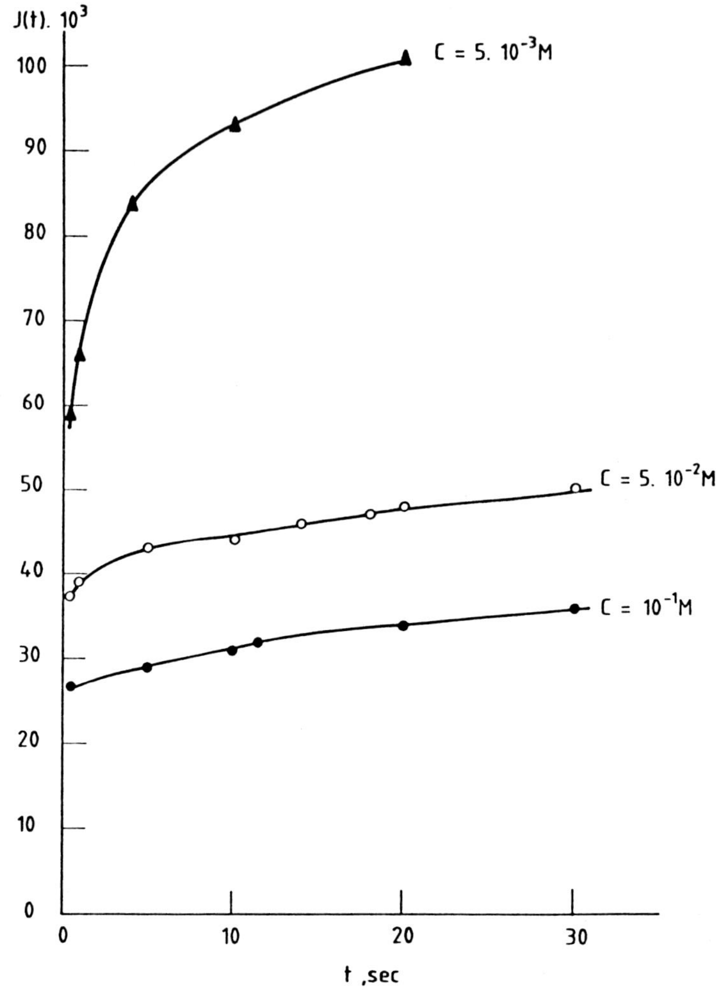

Figure 28.6. Creep compliance of K-illite suspensions at ϕ = 0.02 and pH 3.4 in 5×10^{-3} *M*, 5×10^{-2} *M*, and 10^{-1} *M* KC1.

excursions above and below this pH lead to a weakening of the structure due to increasing repulsion between similarly charged electrical double layers at constant ionic strength.

The marked increase in G with KC1 concentration for ϕ = 0.02 and pH 3.41 (Fig. 28.2) is consistent with an E-E interaction model. pH 3.41 is below the PZC and the net charge on the edge surfaces is positive and so coulombic repulsion between them will be reduced with increasing salt concentration. Note that under these conditions van der Waals attractive energy causes the network structure to become stronger with KC1 concentration. The limitation on electrical double layer compression on these edge surfaces with KC1 concentration is reflected in the behaviour of G at $\sim C \geq 10^{-2}$ M.

The dependence of G on solids concentration for $0.015 \leq \phi \leq 0.07$ shown in Figure 28.3 includes and considerably extends the $G(\phi)$ data of Williams and Williams (1989a, b) particularly near ϕ_c. The present data indicate that a volume filling network of flocculated particles is formed at $\phi_c \sim 0.01$ at pH 4.7 and in 10^{-1} *M* KC1. It is worth noting that although these suspensions are mechanically weak structures near ϕ_c, e.g., $G < 80$ N/m^2 typically, elasticity of the structure increases in a highly nonlinear manner with ϕ. The data are well represented by $G \sim \phi^n$ where $n = 3.9$ (least-squares fit; correlation coefficient 0.98). This power law dependence reflects a nonuniform flocculated structure and the value of n is within the range reported for other flocculated clay structures (see, e.g., Khandal and Tadros, 1988).

The cohesive energy E_c (or work required to disrupt the network structure) can be estimated with the relationship

$$E_c = \int_0^{\gamma_c} \sigma d\gamma \tag{3}$$

where γ_c is a critical strain corresponding (approximately) to the limit of linear viscoelastic behavior.

For these suspensions $J_0^{-1}(\sigma)$ $(= \sigma/\gamma_0)$ tends to a constant value for $\gamma_0 \leq \gamma_c$ so that γ_c and associated critical stress σ_c are readily estimated for a given physicochemical environment.

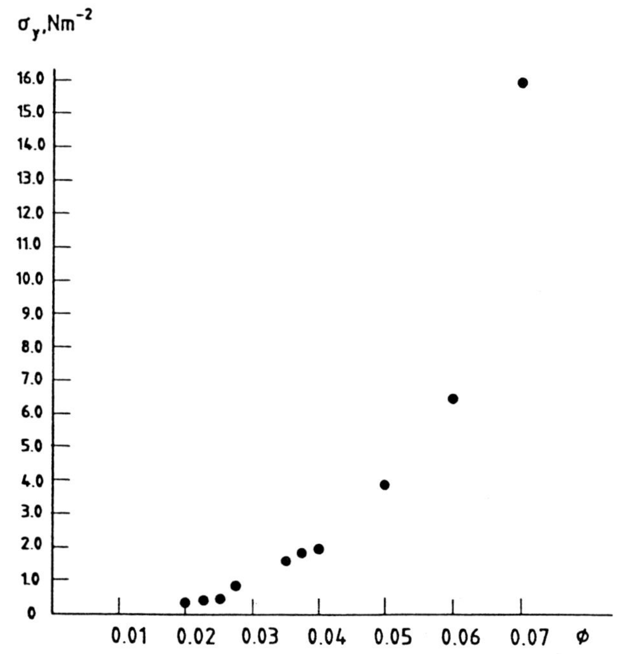

Figure 28.7. Variation of yield stress σ_y with ϕ for K-illite suspensions at pH 4.7 in 10^{-1} *M* KCl.

Values of E_c and concomitantly σ_c (Table 28.1) pass through a maxima at pH 4.7; this behavior is reflected in the values of J_0^{-1} (which approximates an elastic modulus) and supports the G (pH) data (Fig. 28.1).

Cohesive energy increases in a nonlinear manner with ϕ for a given chemical environment (Fig. 28.4). This behavior is also observed for σ_c (Fig. 28.5), which can be described by the relationship $\sigma_c \sim \phi^{3.4}$.

The evolution of the creep function $J(t)$ ($= \gamma/\sigma$) with time of application of a constant stress is typified in Figure 28.6 for $\phi = 0.02$ and pH 3.4. Analysis of $J(t)$ as $t \rightarrow 0$ shows that J_0^{-1} increases with C and confirms the behaviour of $G(C)$ shown in Figure 28.2.

The presence of a continuous network structure in these suspensions gives rise to an apparent yield stress σ_y that like σ_c shows a marked dependence on ϕ (Fig. 28.7) with $\sigma_y \sim \phi^{3.03}$. At stresses above σ_y the network structure collapses with a corresponding disappearance of measurable elastic response.

Conclusions

K-illite suspensions at $\phi \geq \phi_c$ exhibit viscoelastic behavior.

Shear moduli (G) at pH 3.4 increase with KCl concentration for $10^{-3}\,M \leq C \leq 10^{-1}\,M$ and G(pH) show well-defined maxima in 10^{-1} and 10^{-2} *M* KCl. This variation of G with C and pH is consistent with a structure in which interaction between particles and flocs is largely influenced by the physical and electrochemical characteristics of the edge surfaces.

Cohesive energy, critical stress, yield stress, and elastic modulus show a marked nonlinear dependence on ϕ in a given chemical environment.

References

Beene, G., 1987. Colloidal and surface properties of illite. Ph.D. Thesis, University of Wales.

Golden, S.P., J.W. Goodwin, and A.D. Olal, 1982. The dependence of the elastic properties of clay dispersions on the mode of interaction between particles. Transactions of the Journal of the British Ceramic Society, v. 81, p. 84–87.

James, A.E., and D.J.A. Williams, 1982. Particle interactions and rheological effects in kaolinite suspensions. Advances in Colloid and Interface Science, v. 17, p. 219–232.

James, A.E., D.J.A. Williams, and P.R. Williams, 1987. Direct measurement of static yield properties of cohesive suspensions. Rheologica Acta, v. 26, p. 437–446.

Khandal, R.K., and Th.F. Tadros, 1988. Application of viscoelastic measurements to the investigations of the swelling of sodium montmorillomite suspensions. Journal of Colloid and Interface Science, v. 125, p. 122–128.

Srodon, J., and D.D. Eberl, 1984. Illite. In Bailey, S.W. (ed.), Micas, Reviews in Mineralogy 13. Mineralogy Society of America, Washington, D.C., p. 495–544.

van Olphen, H., 1963. Introduction to Clay Colloid Chemistry. Interscience, New York, 301 p.

Whorlow, R.W., 1980. Rheological Techniques. Ellis Horwood, Chichester, U.K.

Williams, P.R., and D.J.A. Williams, 1989. Rheometry of concentrated cohesive suspensions. Journal of Coastal Research, Special Issue No. 5, p. 151–164.

Williams, D.J.A., and P.R. Williams, 1989b. Rheology of concentrated cohesive sediments. Journal of Coastal Research, Special Issue No. 5, p. 165–173.

CHAPTER 29

A Coupled Fluid Expulsion/Deformation Model of Dewatering Sediments

F. Tom Chang, G.P. Lennon, Sibel Pamukcu, and B. Carson

Introduction

The coupled dewatering and deformation behavior of sediments in an accretionary prism under horizontal tectonic loading at continental margins has been studied recently using the concepts of soil mechanics and flow through porous media. The dominant mechanism that contributes to fluid expulsion and deformation of sediments is high excess pore pressure that may be induced by mechanical loading, heating (Shi and Wang, 1985, 1986), or release of mineralogically bound water (Carson, 1987).

Recent advances in the numerical modeling of sediment deformation and dewatering in accretionary prisms provide preliminarily quantitative information for understanding the distribution of high excess pore pressures, and the rates and directions of fluid flows. However, the plastic behavior that characterizes major deformation of the sediments subjected to the tectonic loading was ignored.

In addition, the time-dependent distribution of porosity, and the effects of faulting due to tectonic loading on the hydraulic conductivity, rate of flow expulsion, and porosity should be included in such models. Existing coupled fluid expulsion deformation models in geotechnical engineering (e.g., Siriwardane and Desai, 1981; and Borja and Kavazanjian, 1985) do not accommodate these requirements.

In this paper, an updated two-dimensional, time-dependent finite element model is developed based on coupled behavior of flow through deforming sediments. The model calculates temporal and spatial distributions of high excess pore pressure, permeability, porosity, and rates of fluid expulsion for dewatering sediments under tectonic compression. The model also simulates thrust faults that occur in the sediments caused by tectonic compression. The effects of faults on the excess pore pressure and porosity distributions are also included.

The purpose of this study is not to duplicate the structural and excess pore pressure evolutions of any existing accretionary prisms, but to provide important information for the mechanisms of dewatering sediments under tectonic loading. To simulate the behavior of an actual accretionary prism using this model, the field data for the geometries of the boundary, the material properties, and the tectonic loading rate have to be attained first. The model is applied to a simpler test case as described later.

Theoretical Background Of the Present Model

The finite element formulation of the model is governed by the following two equations.

Displacement Equation

The displacement equation or equation of equilibrium can be expressed as

$$\frac{\partial \sigma'_{ij}}{\partial x_j} + \frac{\partial p}{\partial x_j} \delta_{ij} + F_i = 0 \tag{1}$$

where σ'_{ij} = components of the effective stress tensor, p = excess pore pressure, F_i = components of the body force vector, δ_{ij} = Kronecker delta, and $i,j = 1, 2$ represents the x, y directions, respectively, in two-dimensional Cartesian coordinates. Using the generalized Hooke's law, the constitutive law for the nonlinear (elastic–plastic) behavior of soil skeleton can be derived based on the incremental plasticity theory and the Modified Cam-clay model (Schofield and Wroth, 1968; Pamukcu

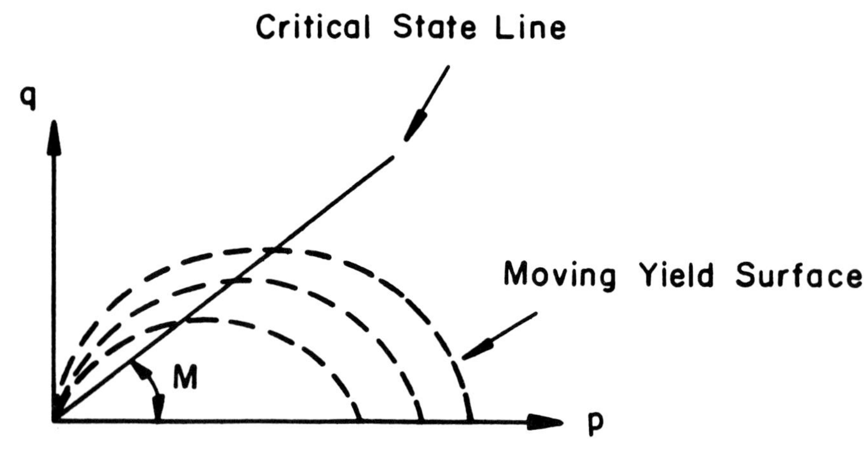

Figure 29.1. Yield locus in p–q space for the modified Cam-clay model.

and Suhayda, 1983; Desai and Siriwardane, 1984), which can be written as

$$d\sigma'_{ij} = C^{ep}_{ijkl} d\varepsilon_{kl}$$

where $d\sigma'_{ij}$ = components of the incremental stress tensor, $d\varepsilon_{kl}$ = components of the incremental strain tensor, C^{ep}_{ijkl} = material property tensor denoting elastic–plastic behavior = $C^{e}_{ijkl} - C^{p}_{ijkl}$, C^{e}_{ijkl} = elastic material property tensor, which can be determined by field seismic velocity data, C^{p}_{ijkl} = plastic material property tensor described by the Modified Cam-clay model, which can be determined by tests such as isotropic consolidation tests and triaxial tests (Desai and Siriwardane, 1984).

According to the Modified Cam-clay model, the loaded sediments behave like an elastic–plastic continuum and are subjected to successive yielding before reaching the critical state of stress, at which sediments flow as a frictional fluid without change in volume and shear stress. The successive ellipsoidal yield surfaces that correspond to the yielding of sediments at different states of stress after the elastic behavior, and the critical state line that represents the critical state of stress for a given type of sediment are shown in Figure 29.1 in p–q space (p = mean effective stress, q = deviatoric stress) for the Modified Cam-clay model.

Porous Media Flow Equation

The generalized Darcy's law is used to describe the slow flow behavior as follows:

$$q_i = -\bar{K}_{ij}\left(\frac{\partial p}{\partial x_j} + \rho_f g_j\right) \tag{3}$$

where q_i = specific discharge of fluid relative to solid grains, ρ_f = density of fluid, $\bar{K}_{ij}$ = permeability tensor, p = excess pore pressure, and g_j = gravitational potential vector. The permeability tensor, $\bar{K}_{ij}$, can be expressed in terms of hydraulic conductivity tensor, K_{ij}, and unit weight of fluid, γ_f, as K_{ij}/γ_f, or in terms of intrinsic permeability tensor, k_{ij}, and fluid viscosity, μ, as k_{ij}/μ.

Insertion of the generalized Darcy's law into the conservation of mass equation yields the porous media flow equation (Huyakorn and Pinder, 1983):

$$\frac{\partial}{\partial x_i}\left[\bar{K}_{ij}\left(-\frac{\partial p}{\partial x_j} + \rho_f g_j\right)\right] = \phi\beta\frac{\partial p}{\partial t} + \frac{\partial \varepsilon_{vol}}{\partial t} \tag{4}$$

where t = time, β = coefficient of fluid compressibility, ϕ = porosity, and $\partial\varepsilon_{vol}/\partial t$ = volumetric strain rate. The other parameters are defined as above.

Model Description

Galerkin's method (Zienkiewicz, 1977) and a weighted finite difference time-stepping scheme are used in deriving the finite element formulations for Eqs. (1) and (4), with three unknowns, u, v, and p, which are nodal displacements in the x, y directions, and excess pore pressure, respectively, for a two-dimensional model domain. The derived finite element equations in the incremental form coupling displacements and excess pore pressure can be written as

$$\begin{aligned}&[K_1]\{\Delta R_u(t_n)\} + [K_2]\{\Delta R_p(t_n)\}\\ &\quad = -\{\Delta F_1(t_n)\} + \{\Delta F_2(t_n)\} + \{\Delta P_1\} + \{\Delta L_r\}\end{aligned} \tag{5}$$

$$\begin{aligned}&[K_2]^T\{\Delta R_u(t_n)\} - (\alpha\Delta t[K_3] + [K_4])\{\Delta R_p(t_n)\}\\ &\quad = -\alpha\Delta t\{\Delta F_3(t_n)\} - (1-\alpha)\Delta t\{\Delta F_3(t_{n-1})\}\\ &\quad - \alpha\Delta t\{\Delta P_2(t_n)\} - (1-\alpha)\Delta t\{\Delta P_2(t_{n-1})\}\\ &\quad + \big((1-\alpha)\Delta t[K_3] - [K_4]\big)\{\Delta R_p(t_{n-1})\}\\ &\quad + [K_2]^T\{\Delta R_u(t_{n-1})\}\end{aligned} \tag{6}$$

where $[K_1]$, $[K_2]$, and $[K_3]$ are assemblage property matrices depending on C^e_{ijkl}, element shape functions, permeability tensor, and fluid density; $[K_4]$ is a matrix due to compressibility of fluid; $\{\Delta R_u(t_n)\}$ and $\{\Delta R_p(t_n)\}$ are the vectors of nodal displacements and excess pore pressures, respectively; $\{\Delta F_1(t_n)\}$ is a known load vector load due to initial stresses; $\{\Delta F_2(t_n)\}$ is a known load vector due to body forces; $\{\Delta P_1\}$ is a known load vector due to surface traction; $\{\Delta P_2(t_n)\}$ is a vector due to applied flow; $\{\Delta F_3(t_n)\}$ is a term contributed by material parameters such as permeability, fluid density, and viscosity; $\{\Delta L_r\}$ is a residual load vector due to plasticity (Chang, 1990); t = time; t_n and t_{n-1} are time levels at steps n and $n-1$, respectively; and α is a parameter that depends on the type of integration procedures. A six-node triangular element with quadratic variation for displacement and linear variation for excess pore pressure is adopted.

For given initial conditions of excess pore pressure, permeability, porosity, fluid viscosity, and density, and temperature distributions with proper boundary conditions, the displacements of sediments and excess pore pressure after the first time step are calculated throughout the domain. The porosity, permeability, fluid density, and temperature at each node in the finite element grid are then updated according to the newly computed excess pore pressure and displacements as explained below.

In the model, the initial porosity distribution of the sediments prior to being externally loaded is predicted based on the Athy's empirical exponential relation (Athy, 1930; Shi and Wang, 1986) between porosity and effective stresses.

$$\phi = \phi_0 \exp(-bp) \tag{7}$$

where ϕ_0 = initial porosity and p = mean effective stress. The value of b can be calculated from the observed porosity versus depth data, if available. Athy (1930) and Dickinson (1953) present b values ranging from 10^{-8} to 10^{-7} Pa^{-1} for shale. Porosity reduction is generally larger with the larger b value. After an external loading is applied to the sediments, the model has two options available for updating the subsequent porosity of the sediments: The first one is still based on the Athy's equation. The second option utilizes one of the unique features of the Modified Cam-clay model by updating the subsequent porosity according to the computed volumetric change (Chang, 1990). In addition, since the Modified Cam-clay model is an isotropic hardening model, the first option provides an alternative to better fit empirical data. For demonstration, the predicted subsequent porosity distribution presented in this chapter is based on the Athy's equation. Chang (1990) presents additional information on both techniques.

The fluid density and viscosity are assumed to be temperature dependent (Mercer et al., 1975). The approximate fluid density–temperature relationship used by this model is the first order Taylor expansions about a reference density (Voss, 1984), which can be written as

$$\rho(T) = \rho_0 + \frac{\partial \rho}{\partial T}(T - T_0) \tag{8}$$

where $\rho(T)$ = fluid density at temperature T, ρ_0 = reference fluid density at a base temperature T_0, and $\partial_\rho/\partial T$ = a constant value of density change with temperature. Temperature ranges in accretionary prisms are typically in the range of 20–100°C.

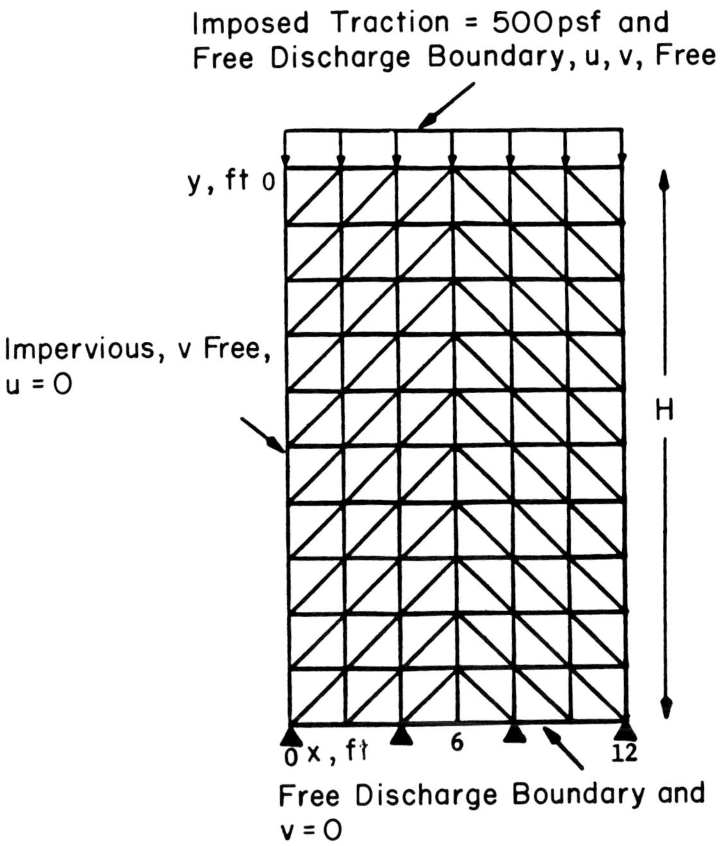

Figure 29.2. Finite element grid with imposed traction and boundary conditions of 1-D consolidation problem using triangular elements.

For the range 20–60°C, $\partial\rho/\partial T$ is approximately -0.375 [kg/(m^3/°C)]. The $\partial\rho/\partial T$ values for other temperature ranges will be obtained from other sources and updated in future versions of this model.

The temperature field is assumed time independent, but can be changed to vary with time if available data warrant it. Sediments "fail" due to tectonic compression when the state of stress at any individual finite element reaches critical state according to the Modified Cam-clay model. Faulting occurs along the boundaries between the failed and unfailed regions. The model determines which elements should have the initial value of permeability replaced with a specified higher value due to the newly formed conduit of fluid flow.

The model provides three options for updating the permeability distribution of the sediments. The first option uses the intrinsic permeability–effective stress relationship described by Morrow et al. (1984), the second uses the intrinsic permeability–porosity relationship described by Joseph et al. (1982); and the third method updates the hydraulic conductivity based on the hydraulic conductivity–porosity relationship described by Bryant et al. (1975). The intrinsic permeability can be obtained from the hydraulic conductivity.

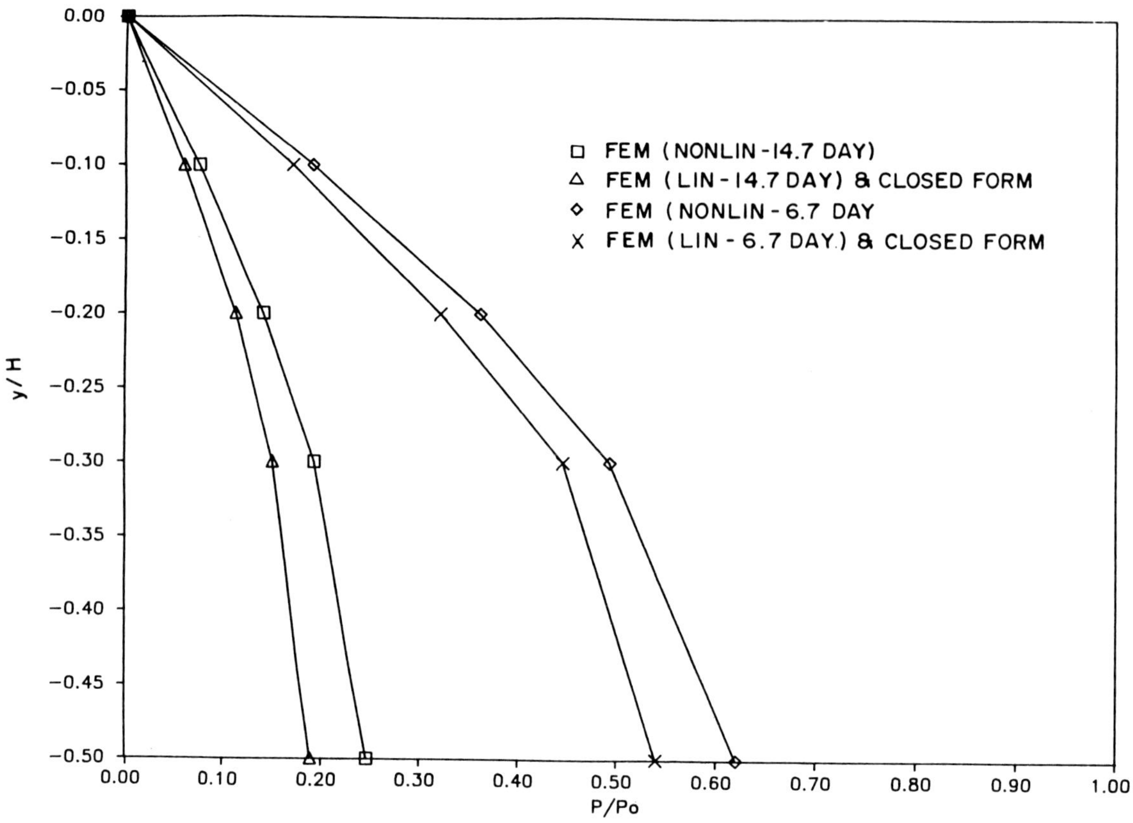

Figure 29.3. Comparison of excess pore pressure at different times.

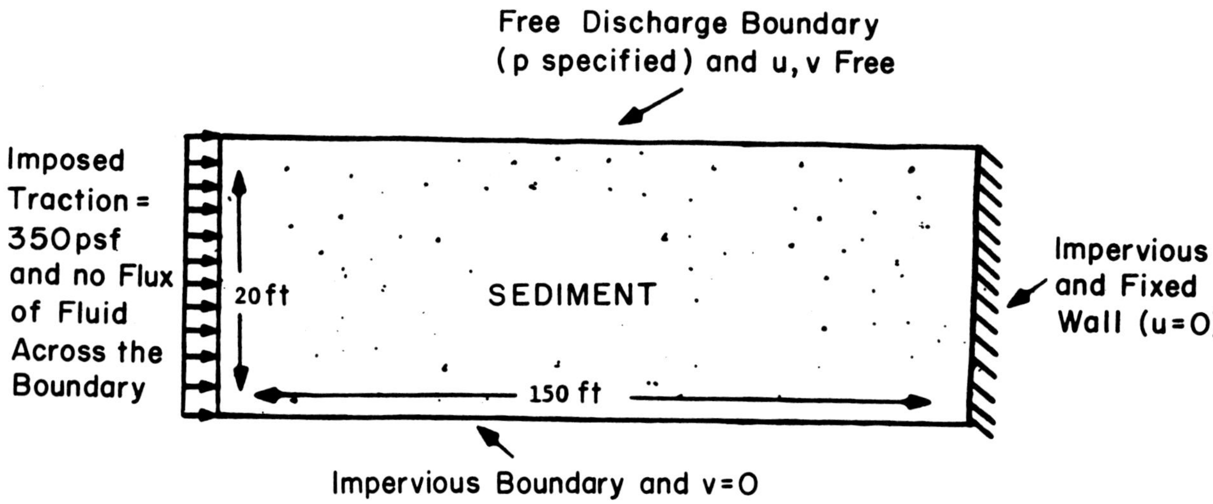

Figure 29.4. A 2-D plane strain problem of dewatering sediment under horizontal loading.

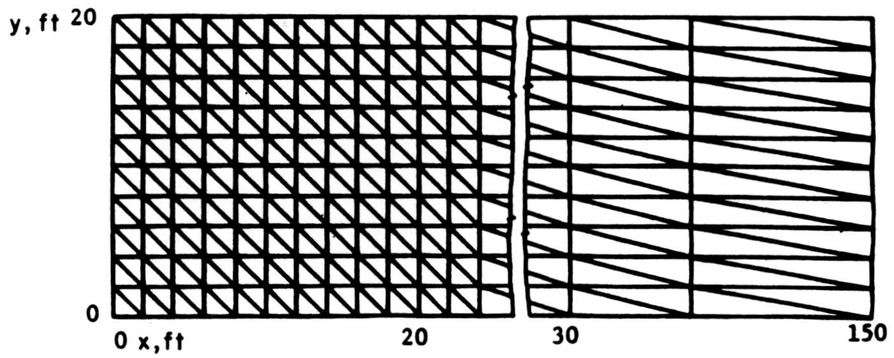

Figure 29.5. Finite element grid of 2-D plane strain problem using triangular elements.

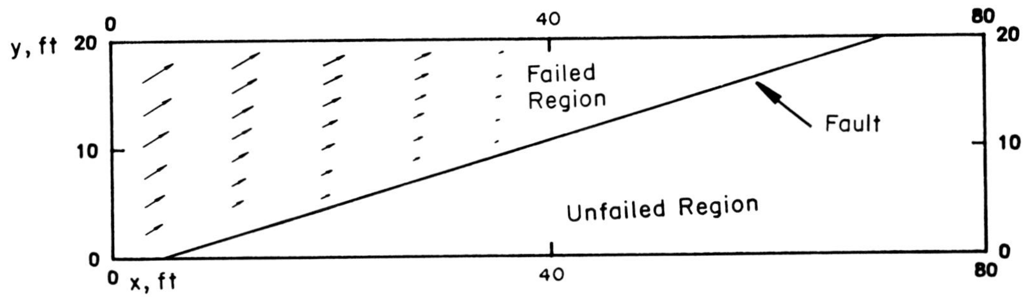

Figure 29.6. Predicted region of failure and location of fault. Arrows show plastic velocity distribution in the failed region.

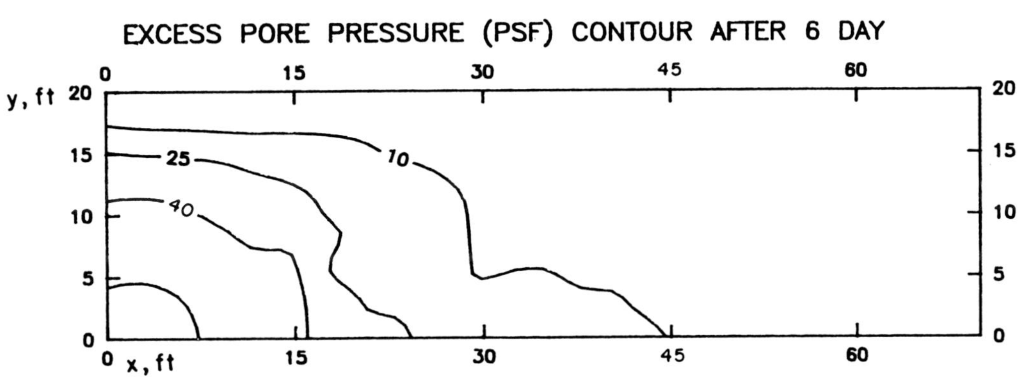

Figure 29.7. Excess pore pressure contour after 6 days with the development of fault.

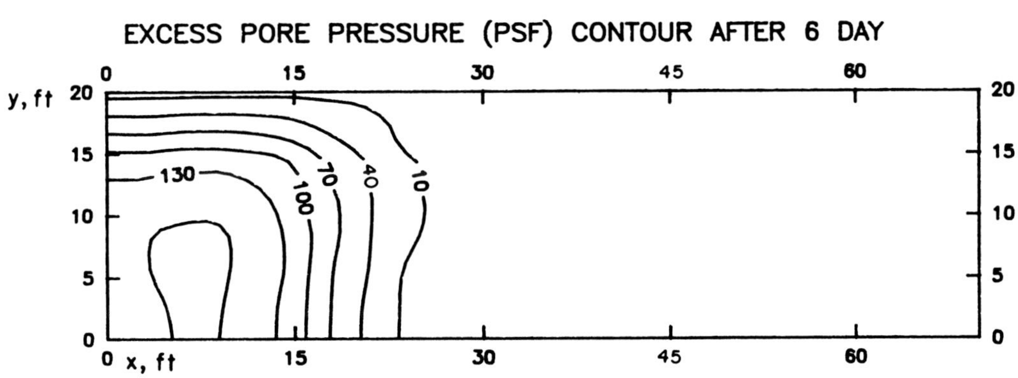

Figure 29.8. Excess pore pressure contour after 6 days ignoring the development of fault.

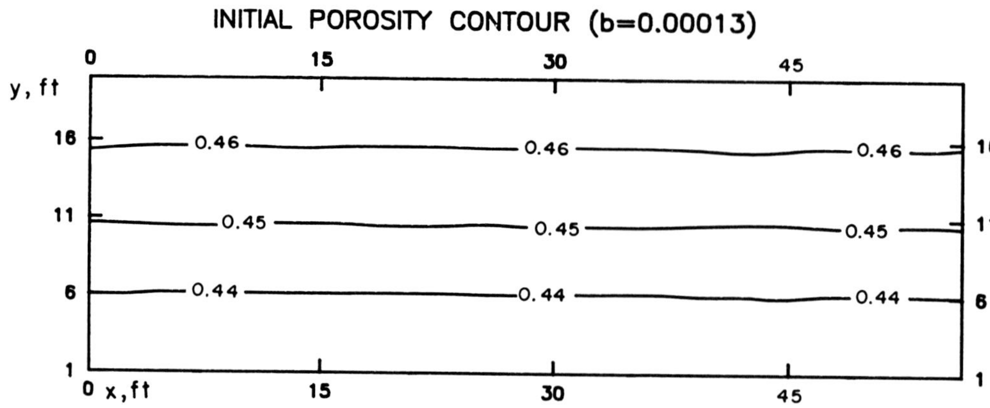

Figure 29.9. Initial porosity contour for sediment with material property parameter b = 0.00013 in Eq. (7).

This procedure is repeated for subsequent time steps and the evolution of displacements of sediments and excess pore pressures can be computed.

Verification of the Model

Figure 29.2 shows the finite element grid of a representative sample of saturated porous media under vertical loading and the imposed boundary conditions on excess pore pressure (p) and solid displacements (u, v, where u and v correspond to displacements in the x and y directions, respectively.) The material properties are as follows: hydraulic conductivity = 2.5×10^{-3} ft/day (7.62×10^{-4} m/day), unit weight of fluid and solids = 62.5 pcf (9.82×10^3 N/m^3) and 51.0 pcf (8.01×10^3 N/m^3), respectively, elastic modulus = 60,000 psf (2.87×10^6 Pa), Poisson's ratio = 0.4, slope of critical state line = 1.05, compression index (slope of the virgin compression curve) = 0.14, and recompression index (average slope of the recompression part of the compression curve) = 0.05.

Figure 29.3 shows the results of normalized excess pore pressure p/p_0 versus normalized depth y/H for the upper half of the domain, where p = excess pore pressure, p_0 = traction, y = depth from the surface, and H = depth of the sediment column. The lower half of the column has a symmetric pressure distribution for the simplified assumptions of constant hydraulic conductivity and porosity. The predicted results for dissipation of excess pore pressure from the linear elastic analysis agree with the closed form solution for one-dimensional consolidation (Terzaghi and Peck, 1951). The results from the nonlinear analysis are shown in the same figure for comparison. It can be seen that the inclusion of plastic deformation leads to a lower rate of dissipation of excess pore pressure. In addition, the model was also verified, with good agreement with other analytical solutions for other cases including two-dimensional plane strain consolidation (Chang, 1990).

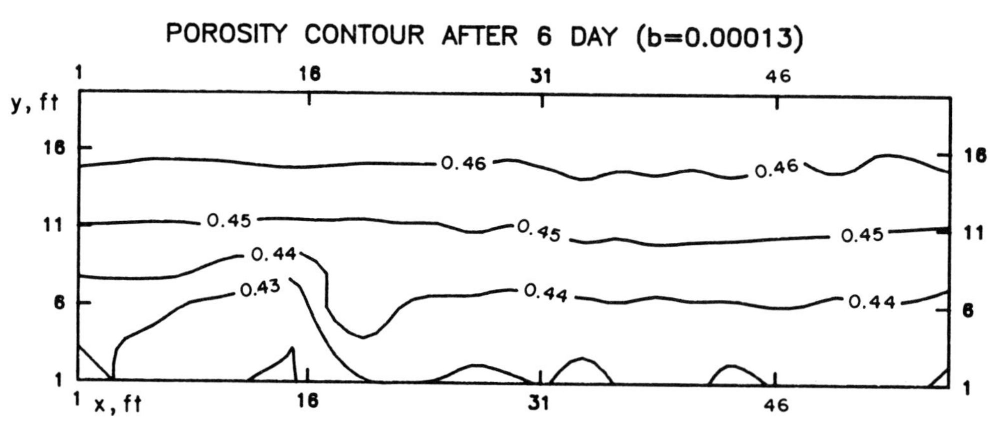

Figure 29.10. Porosity contour after 6 days with the development of fault for sediment with material property parameter b = 0.00013 in Eq. (7).

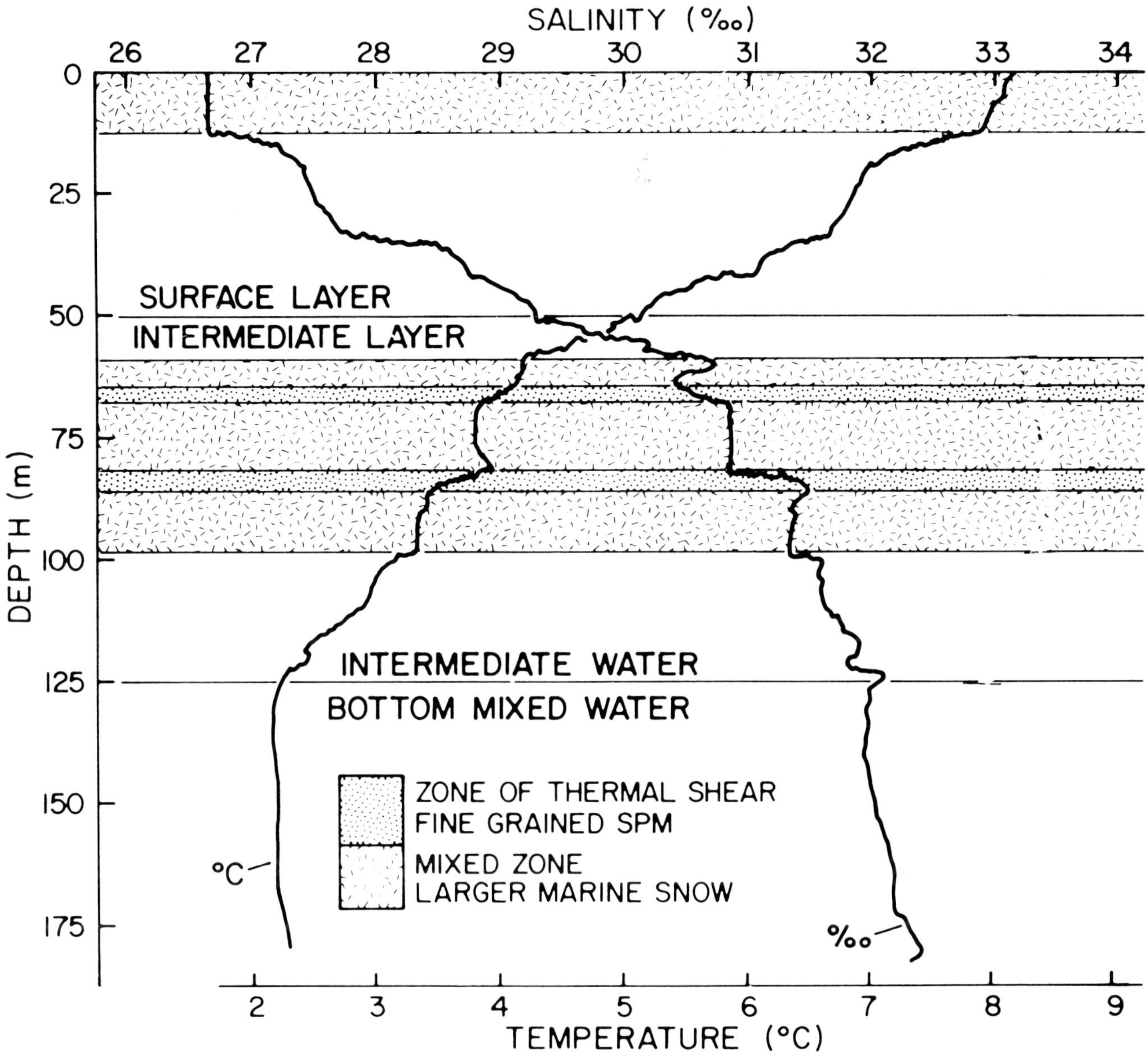

Figure 30.2. Fine structure in the intermediate layer of the lower St. Lawrence Estuary (cf. Fig. 30.1). Indicated are mixed zones within the water column containing large marine snow, and thermal shear zones containing only fine-grained SPM. Reprinted with permission from Syvitski et al., 1983, First observation of benthos and seston from a submersible in the lower St. Lawrence Estuary, Geographie Physique et Quaternaire, vol. 37, no. 3, Editor: Mme. Nicole Carette.

FCA Field Routine

The cameras are loaded with film and data backs are set in the time mode. A personal computer running a terminal emulator program is connected to the FCA computer. The program is loaded and initial parameters are set up. The FCA is lowered slowly ($\approx$ 0.25 m/sec). The descent of the FCA is traced on a 12-kHz sounder graphic recorder at a 1 sec sweep. Each time a photograph is taken, the pinger generates an additional four pings per second for 10 sec, thus providing a surface indication of the photo sequence on the graphic recorder. The pressure transducer used is accurate to within $\pm$ 1 m. The present flash unit needs 5 sec between photographs to recharge. Our normal mode of operation is to take photo profiles based on a linear or exponential depth interval, followed by a series of photos near the seafloor triggered by a time interval. We have also taken

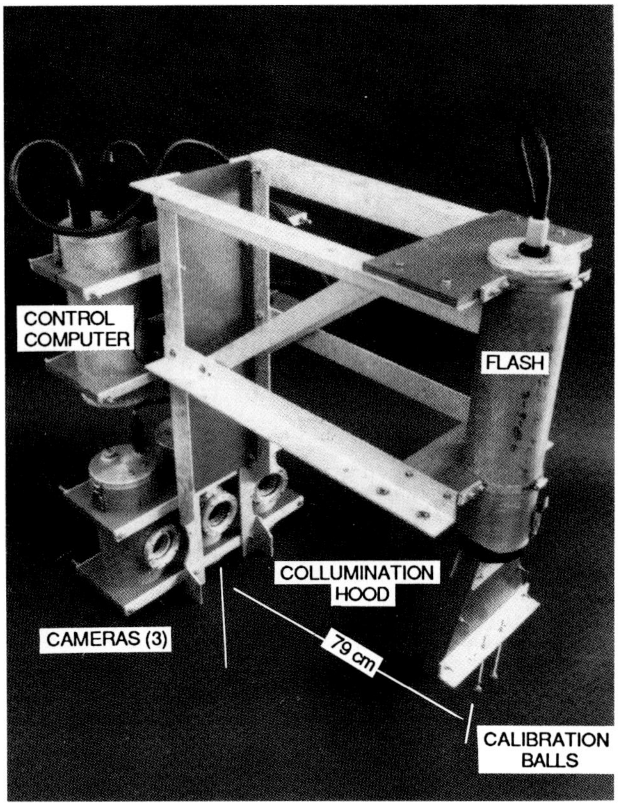

Figure 30.3. Photograph of the Atlantic Geoscience Center floc camera assembly (FCA) configured in the profiling mode.

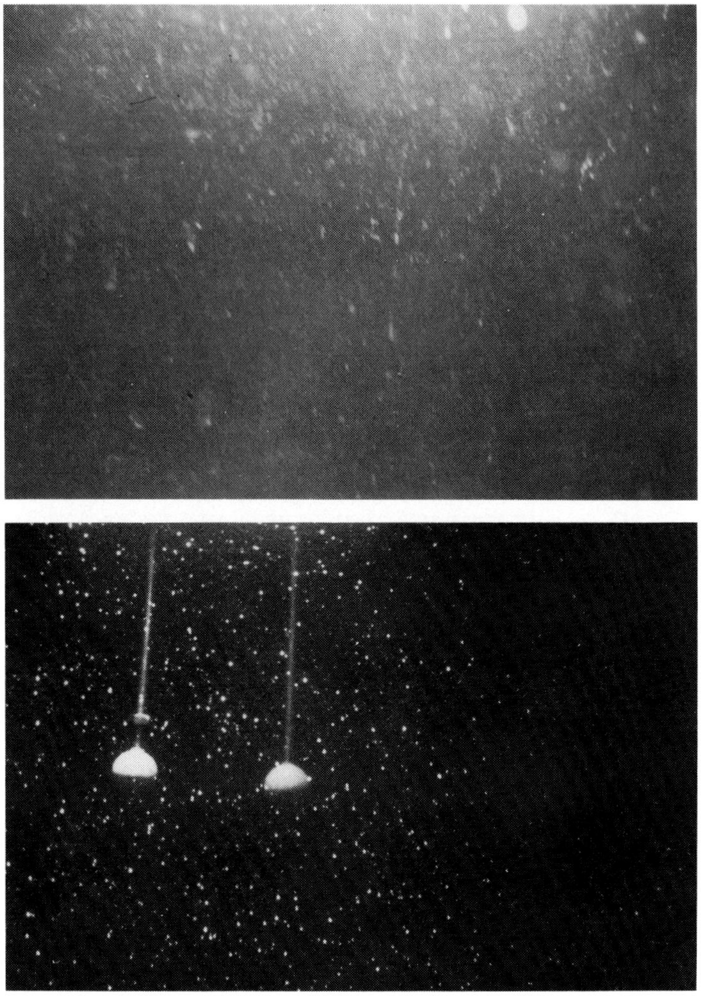

Figure 30.4. A comparison of (A) a submersible with (B) an AGC floc camera photograph of suspended particulate matter. The balls seen in the FCA photograph have a real diameter of 15 mm. The in-focus nature of suspended particles from the floc camera photo, compared to the submersible photograph, relates to the well-defined collimated nature of the illuminated water and the appropriate depth-of-focus of the camera. The photos were collected on separate cruises but at the same location within the St. Lawrence Estuary, Eastern Canada.

photos during both descent and ascent of the FCA. On retrieval of the FCA, the film is unloaded, and the data logged in the FCA computer are dumped.

For measurement of the absolute settling velocities of the individual particles, we use a modified FCA configuration. The FCA is mounted upside down on a large tripod so that particles are photographed as they settle ≈ 2 m above the sea floor. The flash collimator and illuminated volume are enclosed by a Lucite stilling tank with a closed bottom and a baffle of 10 cm squares on top. Particles entering the stilling tank are photographed directly through the walls of the tank. In this configuration, particles settle as if in a sediment trap and are photographed sequentially during their descent. The stilling tank attachment is necessary to eliminate the horizontal component of settling particles. Without the stilling tank, the particles would normally exit the illuminated slice of water column before the next picture could be taken.

FCA Lab Routine

The film is developed to produce uncut rolls of both negatives and positive prints. The negatives from the floc camera film are analyzed on a TAS PLUS, a Lietz automated image analysis system controlled by a LSI-11/2 microcomputer. A special Floc Camera program (FLOK) is used to analyze the negatives. Program FLOK scans the negatives by controlling the microscope stage recording the position coordinates (centroid), size, shape, and orientation of each particle captured on film. Program FLOK is set up to recognize the sediment particles by edge detection (based on noting sudden changes in the grey scale value between adjacent points).

The TAS system utilizes a programmable mechanical stage that moves new fields of view under the microscope and computer-linked television camera. Each 35 × 23 mm negative consists of 117 fields. The number of blocks analyzed depends

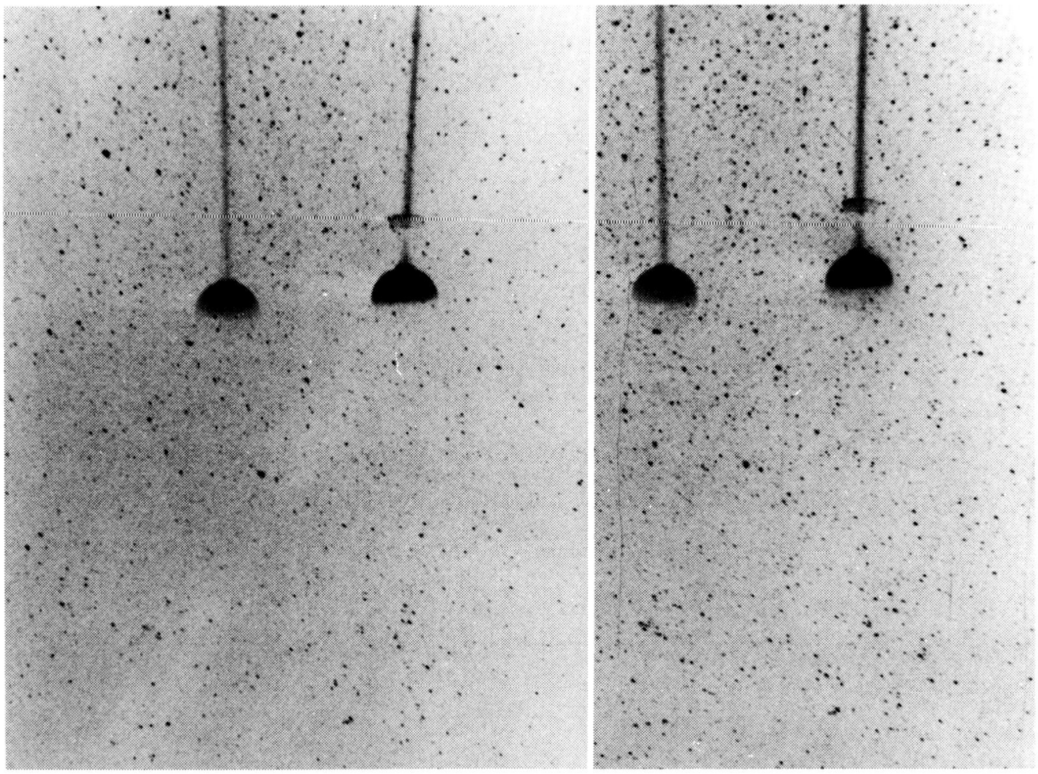

Figure 30.5. Photo pair from cameras 1 and camera 2 of particles suspended in the water at station 2 (48° 24.43′ N, 69° 09.61′) from the Gulf of St. Lawrence in Eastern Canada. The photo depth is 75 m below sea level and 175 m above the sea floor. The photos were taken on May 2, 1988.

on the concentration of particles within the water column. Data generated for this report came from 20 to 25 digitized fields. For a 20 field analysis, a 137.6 mm^2 area of the slide, which corresponds to 7740 mm^2 of the water column, is digitized. The thickness of the collimated flash is $\approx$ 25 mm. Thus the real volume of water digitized is 193,500 mm^3 or 0.2 liter. The lower limit of detection of a particle suspended in the water, based on the film grain we presently use, is 2 ∅ or 250 μm. Each negative takes about 30 min to analyze, recording information on hundreds or thousands of particles.

Data from the TAS FLOK program are eventually transferred to the BIO mainframe (Cyber 840) computer where the data are archived and analyzed by various programs including READY (Hackett et al., 1986) where the size frequency distribution of the suspended particles from each camera frame is ascertained. Stereo or three-dimensional (3-D) analysis need utilize information from only two cameras; the third camera provides useful quality control. Figure 30.5 is a stereo pair from adjacent cameras so that the 3-D image of the suspended particles may be seen by the reader using a standard stereoscopic viewer.

The shape, minimum and maximum diameters, and particle volume are normally determined for each particle. Where the FCA has been fixed rigidly within the water column, absolute settling velocities may be determined from changes in the vertical position of each particle between successive photographs.

Comparison of FCA Data to Conventional Data

Water was sampled synchronously with seven FCA profiles collected within the NW Gulf of St. Lawrence (see Syvitski, 1988, for location and sampling details). The concentration of SPM from the water samples was determined gravimetrically through filtration onto preweighed Nucleopore filters having a 0.45-μm nominal pore size. The number of marine snow particles (in units of number per litre) as determined with the FCA/TAS method can then be compared to the more conventional method for determining SPM concentration (in units of milligrams per liter). We find, for the Gulf of St. Lawrence, that the number of particles (per volume of water) of marine snow increases with the SPM concentration (Fig. 30.6A). The few exceptions include measurements within (1) the surface layer of river plumes, where the SPM concentration is high but the suspended particles had not all reached the stage of large floccules and agglomerates (i.e., as detectable using the FCA), and (2) the turbulent layer near the seafloor where suspended particles are

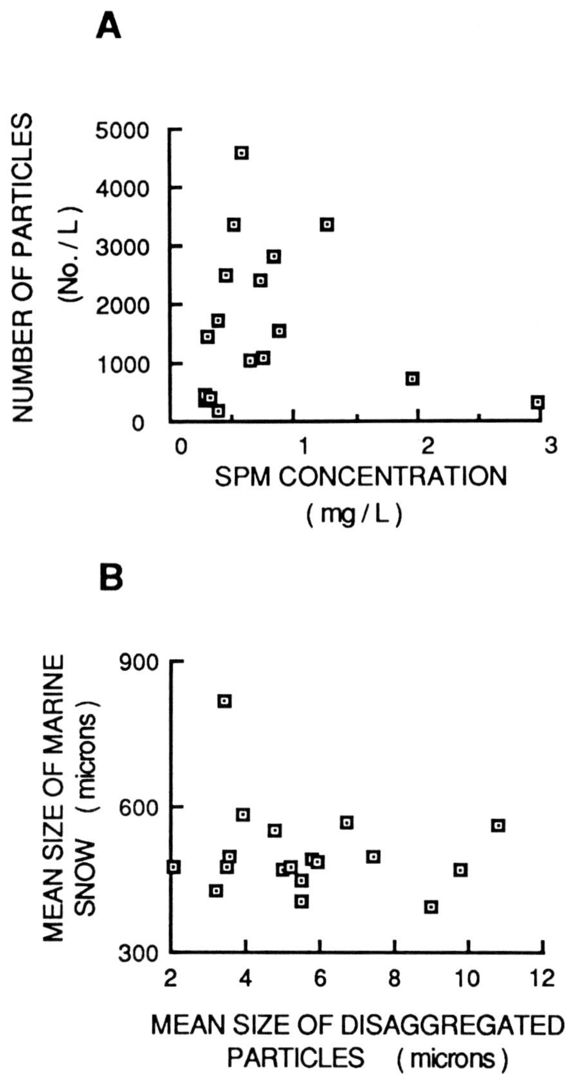

Figure 30.6. (A) Comparison of total particle concentration (in units of mg/liter or g/m^3) in the water, collected by 5-L Niskin bottles, with the number of particles > 250 μm (in units of No./liter) as observed using the FCA/TAS method. (B) Comparison of mean grain size of the deflocculated suspended particles with the mean grain size of the marine snow observed using the FCA/TAS system. Water samples were collected at the same time, depth, and station as was the FCA data. All samples and data were collected on BIO cruise DA88-008 to the N.W. Gulf of St. Lawrence, Eastern Canada (cf. Syvitski, 1988).

finely divided both from resuspension off the seafloor and from the break-up of the marine snow (Fig. 30.7).

Particles from the filtered water samples were also removed from the filter papers through initial soaking followed by low-intensity sonification. The individual particles composing the larger marine snow were dispersed in a sodium hexametaphosphate electrolyte solution. These disaggregated particles were then analyzed for mean grain size using a TAII computerized Coulter Counter involving different aperture tubes so as to cover the size range of 0.63–100 μm. No correlation was found between the floc components (i.e., the individual components that make up the marine snow) and the *in situ* size of the marine snow (Fig. 30.6B). This is expected from theory (Syvitski, this volume), where floc size is dependent on a particles residence time, biogeochemical interactions, and the distribution of turbulent shear within the water column.

FCA Station Profiles

We have chosen one station in the St. Lawrence Estuary from a recent cruise (BIO DA88-008) to demonstrate depth variations of the in situ properties of marine snow through a column of sea water (Fig. 30.7). Background information on the oceanography and sediment dynamics of this environment can be found in Syvitski et al. (1983). Figure 30.7 provides five representative FCA photos of the station 1 profile. The surface layer is dominated by finer flocs that increase in size dramatically with depth, especially just under the halocline (Fig. 30.1). Below 100 m the marine snow size is relatively constant until within the bottom mixed layer where the particle size decreases and the particle number increases. Five layers containing high numbers of large particles appear equally spaced within the water column and these layers are considered to represent tidal influence on the sediment delivery from the St. Lawrence River to the Estuarine basin. There appears an inverse correlation between the mean floc size and the number of particles of marine snow (Fig. 30.7F).

Size Frequency Distributions

Analysis of the particles observed using the FCA/TAS system reveals that particles suspended in the marine environment can be very large, i.e., in the 0.5–2 mm size range. Each individual marine snow particle is composed of many hundreds of finer particles. In the St. Lawrence Estuary, three types of size frequency distributions (SFD) of marine snow were recognized. A monotonically increasing SFD (Fig. 30.8A,D) typifies Surface Layer (Fig. 30.1) samples and suggests that much of the suspended particle population is of a size below the detection limit of the FCA/TAS. In other words, not all suspended particles are yet in the form of large agglomerates that normally need a lengthy residence time to form (Syvitski, this volume). A relatively flat SFD typifies samples from the intermediate layer (Fig. 30.1) samples and suggests most of the SPM load is in the form of marine snow but in a variety of sizes (Fig. 30.8B,D). Finally a Gaussian-shaped SFD (Fig. 30.8C,D) typifies samples from the bottom mixed layer (Fig. 30.1) whereby some of the marine snow particles are decreasing in their size due to the effect of turbulent shear within the layer. The change in the shape of the SFD is more dramatic in the FCA curves than is observed in the Coulter Counter produced curves (Fig. 30.8D,E).

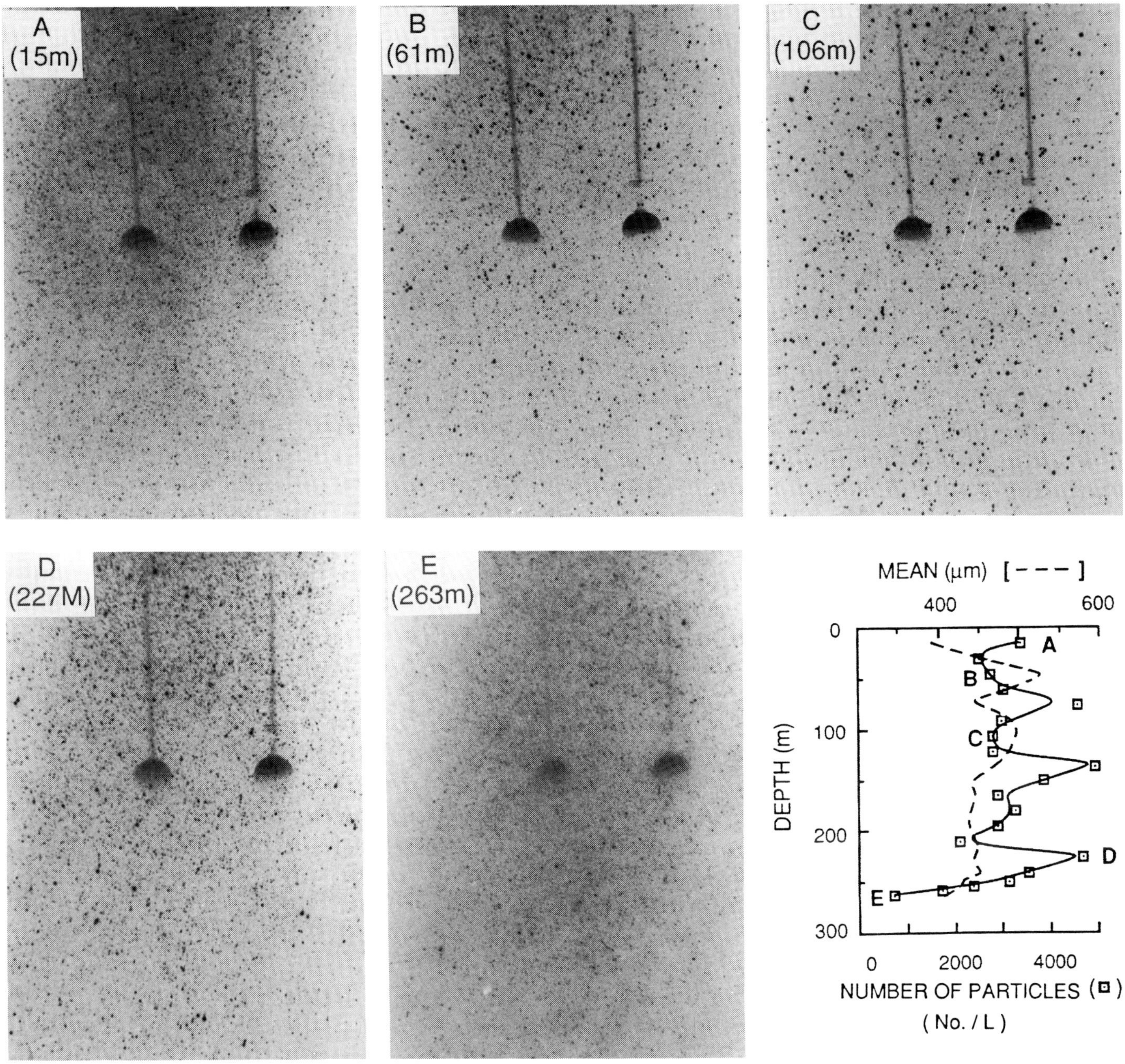

Figure 30.7. An example of a FCA profile with photographs and data from station 1 (BIO cruise DA88-008) situated at 48° 24.25′ N 69° 06.70′ W in the St. Lawrence Estuary, Eastern Canada. FCA photographs are from the middle camera (cf. Fig. 30.3) taken at depths of (A) 15 m, representing the surface layer, (B) 61 m, just below the surface layer, (C) 106 m, representing the intermediate cold water layer, (D) 227 m, representing the surface of the bottom mixed layer, and (E) 263 and collected 2 m above the sea floor. (F) Vertical profile of the mean diameter (μm) and particle density (No./liter) of marine zone photographed. These data may be compared with salinity-temperature profiles and submersible observations shown on Figure 30.3.

Settling Velocity Data

Using the configuration of the FCA for collecting photographs of suspended particles settling within a stilling tank, absolute velocities of various particles were measured using sequential photos taken 10 sec apart. The measurements were collected in Bedford Basin, part of the metropolitan harbor of Halifax, Nova Scotia, Canada, and at a depth of 25 m below the sea surface and 2 m above the sea floor. Figure 30.9 shows two photographs taken 20 sec apart. On each photo, eight of the same particles are

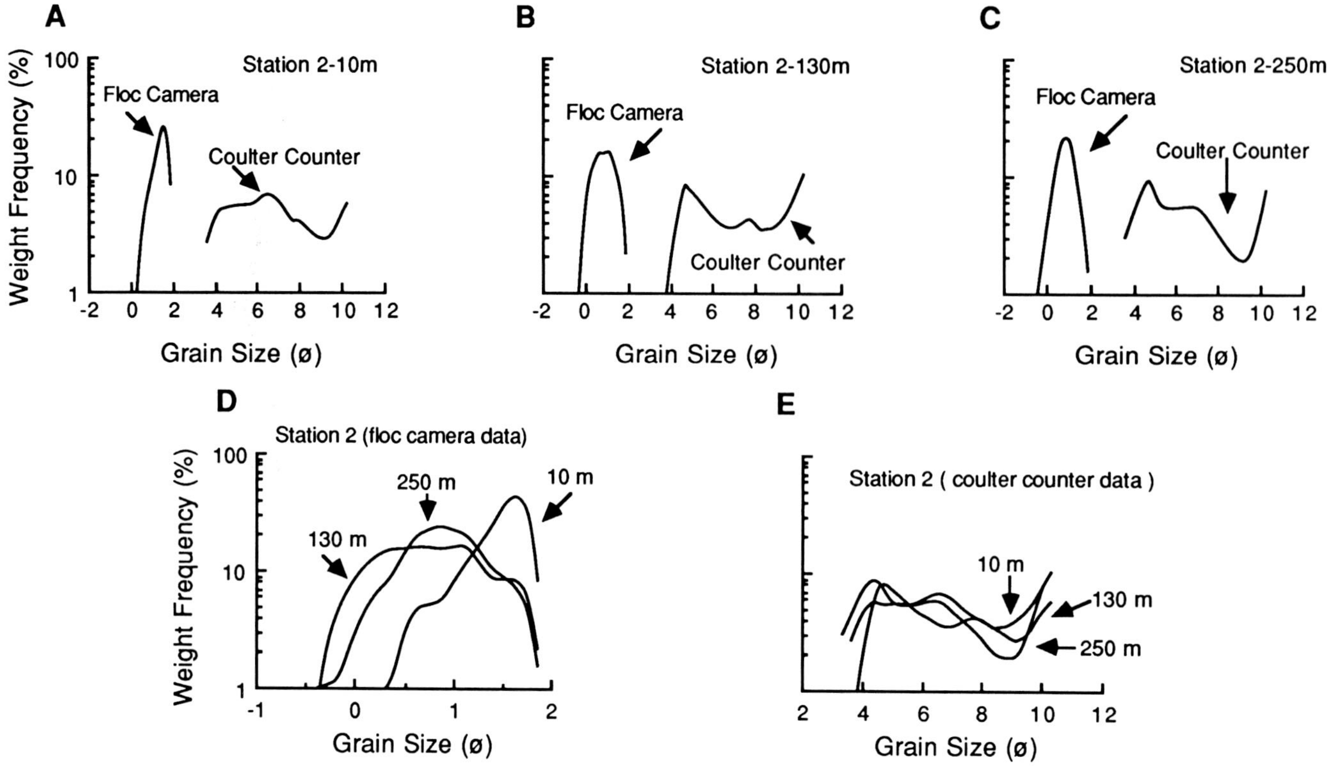

Figure 30.8. A comparison of Coulter Counter and FCA size frequency distributions from three depths representing the surface layer, the intermediate layer and the bottom mixed layer of the St. Lawrence Estuary (cf. Figs. 30.3 and 30.7): (A) 10 m sample, (B) 130 m sample, (C) 250 m sample, (D) the changing *in situ* size frequency distribution with depth, and (E) the relatively unchanging disaggregated or component size frequency distribution.

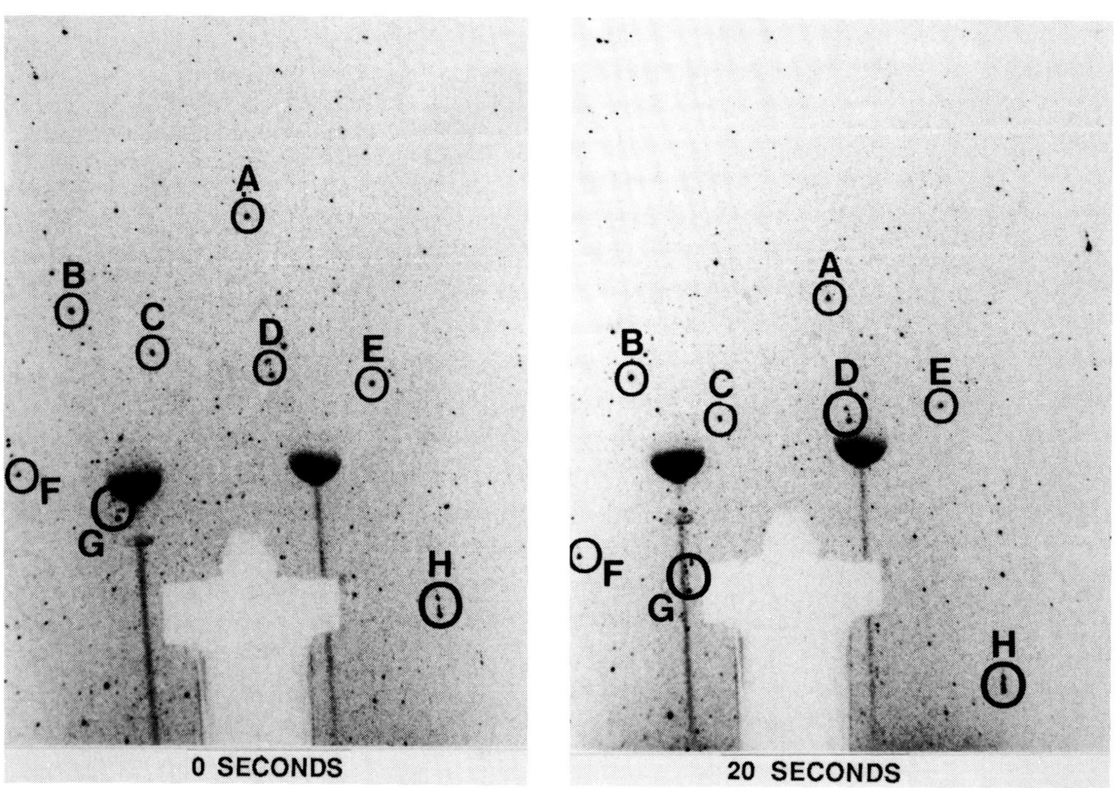

Figure 30.9. Sequential FCA photos taken 20 sec apart in the bottom mounted configuration. On each photo, eight of the same particles are identified, but at different positions with respect to the calibration balls.

identified, but at different positions as related to the calibration balls. Examining these particles over 8 to 10 photographs (i.e., 1.5–2 min of elapsed time), they were found to have individual settling rates between 0.6 and 1.3 mm/sec. Surprisingly the largest particles did not always have the largest settling velocity. The mean settling velocity of all particles analyzed was 1.05 mm/sec or 91 m/day. Using more complex sampling schemes and sediment traps, Syvitski et al. (1985) found that at similar depths in a British Columbia fjord, particles settled at 100 m/day. Their results were not based on individual particle settling rates; rather they reflected the mean settling rates for each size fraction. We believe that the settling rates reported here are the first of their kind.

Conclusions

A new floc camera assembly (FCA) has been constructed to take *in situ* stereo photographs of suspended sediment within a known volume of water. The photos collected are analyzed on a TAS image analysis system. The FCA/TAS configuration allows for the number of particles per volume of water to be ascertained, with individual information on each particle (identified through coordinates within a three-dimensional framework), including minimum, mean and maximum diameter, and particle area. Further data manipulation can produce the *in situ* size frequency distribution and, in the bottom mounted FCA configuration, absolute settling velocities of individual particles can be calculated. We expect that in very turbid environments, i.e., > 50 mg/liter, particle identification will be difficult under the present FCA configuration and possibly macrolenses should be used. The FCA combined with the TAS provides a viable means to obtain quantitative data on the *in situ* properties of suspended particulate matter in most marine environments.

Acknowledgments

This project was supported by the Geological Survey of Canada's Sedimentary and Marine Geology Branch under project 420-8141 "The physical behavior of suspended particulate matter (SPM) in natural aqueous environments." K.W.G. LeBlanc provided invaluable assistance in the preparation of this manuscript. Joe Kravitz (ONR) is gratefully thanked for support at the NORDA workshop. C.L. Amos and G.V. Winters reviewed an earlier version of the manuscript—thanks. Geological Survey of Canada contribution Number 43788.

References

Bishop, J.K.B., and J.M. Edmond, 1976. A new large volume filtration system for the sampling of oceanic particulate matter. Journal of Marine Research, v. 34, p. 181–198.

Farrow, G.E., J.P.M. Syvitski, and V. Tunnicliffe, 1983. Suspended particulate loading on the macrobenthos in a highly turbid fjord; Knight Inlet, British Columbia. Canadian Journal of Fisheries and Aquatic Sciences, v. 40, p. 100–116.

Folger, D.W., 1968. New particulate matter sampling devices and effects of techniques on marine suspensate recovery. Deep-Sea Research, v. 15, p. 657–664.

Hackett, D.W., J.P.M. Syvitski, W. Prime, and A.G. Sherin, 1986. Sediment Size Analysis System User Guide. Geological Survey of Canada Open File Report 1240, 25 p.

Hay, A.E., 1983. On the remote acoustic detection of suspended sediment at long wavelengths. Journal of Geophysical Research, v. 88, p, 7525–7542.

Hay, A.E., and D.G. Mercer, 1985. On the theory of sound scattering and viscous absorption in aqueous suspensions at medium and short wavelengths. Journal of Acoustical Society of America, v. 78, p. 1761–1771.

Lal, D., 1977. The oceanic microcosm of particles. Science, v. 198, p. 997–1009.

McCave, I.N., 1975. Vertical flux of particles in the ocean. Deep-Sea Research, v. 22, p. 491–502.

Shanks, A.L. and J.D. Trent, 1980. Marine snow: sinking rates and potential role in vertical flux. Deep-Sea Research, v. 27, p. 137–144.

Syvitski, J.P.M., 1988. CSS DAWSON 88-008 technical cruise report, May 01–May 17, 1988. Geological Survey of Canada Open File Report 1920, 61 p.

Syvitski, J.P.M., and D.E. Heffler, 1983. The floc camera. BIO Review' 83, p. 50–51.

Syvitski, J.P.M., and J.W. Murray, 1981. Particle interaction in fjord suspended sediment. Marine Geology, v. 39, p. 215–242.

Syvitski, J.P.M., N. Silverberg, G. Ouellet, and K.W. Asprey, 1983. First observations of benthos and seston from a submersible in the Lower St. Lawrence Estuary. Géographie physique et Quaternaire, v. 37, p. 227–240.

Syvitski, J.P.M., K.W. Asprey, D.A. Clattenburg, and G.D. Hodge, 1985. The prodelta environment of a fjord: suspended particle dynamics. Sedimentology, v. 32, p. 83–107.

Thorndike, E.M., 1975. A deep sea, photographic nephelometer. Ocean Engineering, v. 3, p. 1–15.

Winters, G.V., and D.E. Buckley, 1980. In situ determination of suspended particulate matter and dissolved organic matter concentrations in an estuarine environment by means of an optical beam attenuance meter. Estuarine and Coastal Marine Science, v. 10, p. 455–466.

CHAPTER 31

Characterization of Clay Fabric

A.G. Altschaeffl and S. Thevanayagam

Introduction

The behavior of clay soil has been described qualitatively in terms of its fabric and structure since Casagrande (1932) and Lambe (1958), among others, presented their concepts on the internal make-up of this soil. Fabric is the term ascribed to the internal arrangements and interrelationships among the constituent particles. Structure is the term that describes the combined effect of fabric, composition, and interparticle forces. The work of Rosenquvist (1959) provided a major step forward when he presented stereo photographs of the fracture faces of a clay soil, clearly showing an arrangement of particles; the arrangement differed for clays prepared differently. A number of investigators (e.g., Hodek, 1972; Mitchell, 1976; Arulanandan et al., 1973) made early efforts to describe the clay fabric and its anisotropy by various methods (e.g., optical transmission and electrical conductivity). Others (e.g., Sridharan et al., 1971 have used the distribution of the sizes of soil pores (pore-size distribution, PSD) as a measure of fabric. This PSD was then correlated with soil behavior successfully (Garcia Bengochea et al., 1979; Prapaharan et al., 1985). Rose diagrams have been used to describe the directional dependence on fabric. This chapter presents a novel approach to quantify soil fabric, its directional dependence and how fabric changes with subsequent deformation of soil.

Concept

Quantitative description of initial fabric would enhance characterization and forecasting of clay behavior under loading. On loading the fabric is continuously altered. Hence, it is necessary to develop techniques to quantify the changes in fabric as well. Whenever the material is distorted the fabric of the material is altered, and so is the strain or displacement field in the material. Hence, the strain and induced fabric of a material are inherently related to each other. By relating strain to changes in fabric, it appears possible to update the changes in initial fabric due to subsequent loading. On the other hand it allows prediction of strains based on fabric information as well. This idea requires a suitable choice of fabric descriptors that can be relatively easily measured.

As a preliminary attempt, let the pore network of clays be chosen as the clay fabric descriptor. The distribution of pore channels can be quantitatively described by a mathematical function that can be expanded in a tensorial form. The components in each tensor can be regarded as quantitative descriptor of fabric of the clay. The distortion of the pore channels due to deformation can be represented by a mathematical mapping of the initial form by an appropriate transformation matrix. This matrix can be related to strain and the fabric descriptors before and after deformation and hence develop the fabric–strain relation of the clay. With this as the context, a mathematical relationship for induced fabric and strain is developed and examined with preliminary experimental data in what follows. Tensor notation and summation convention over repeated indices are used.

Quantitative Determination of Fabric in Clays

Consider the distribution of pores in a vertical plane of a clay deposit shown in Figure 31.1 as a collection of infinitesimal line elements of length, dl, and thickness, dt, whose normal unit vector is $\mathbf{n}$. Choice of smaller values of dl and dt will improve the accuracy of measurement of distribution of pores. Draw arbitrary probing lines making angle α_i with the horizontal axis on this vertical plane in direction of a unit vector $\mathbf{q}$, where $q = (\cos \alpha_i, \sin \alpha_i)'$; the prime refers to the transpose of the matrix.

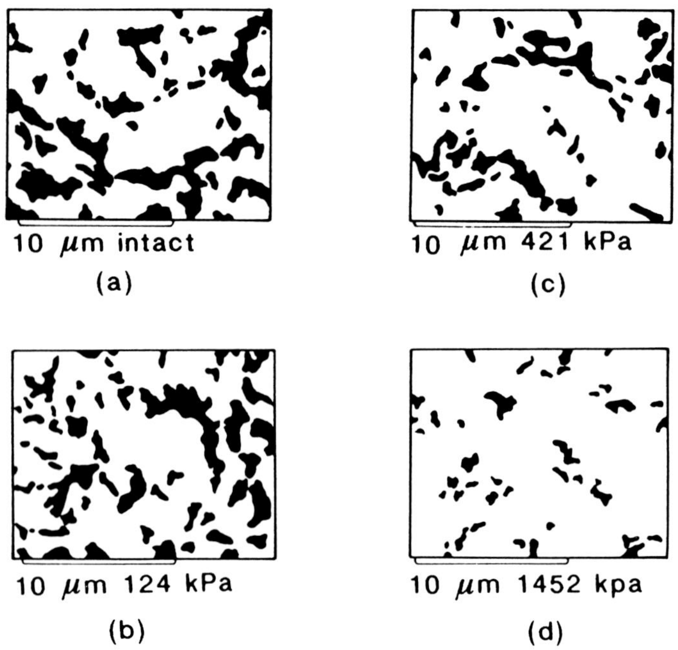

Figure 31.1. Graphic representation of porosity at different stress levels on a vertical plane. Pores are black and particles are white. (After Delage and Lefebvre 1984.)

Let the number of intersections of the pore elements (of thickness dt) per unit length of each probing line be denoted by $N(\alpha)$ [also denoted by $N(q)$]. $N(\alpha)$ is a periodic even function with a period, π [i.e., $N(\alpha) = N(\pi + \alpha)$]. If the clay is cross anisotropic, then $N(\alpha) = N(\pi - \alpha)$ as well.

The directional information $N(\alpha)$ and $N(q)$ can be expressed as a Fourier series expansion and another tensorial form in a Cartesian coordinate system as follows. The Fourier series expansion of $N(\alpha)$ is given by

$$N(\alpha) = \frac{C}{2\pi}\left[1 + \sum_{n=2}^{\infty} A_n \cos(n\alpha) + B_n \sin(n\alpha)\right] \quad (1a)$$

where

$$C = \int_0^{2\pi} N(\alpha)\, d\alpha \quad (1b)$$

$$A_n = \frac{2}{C}\int_0^2 N(\alpha)\cos(n\alpha)\, d\alpha$$

and (1c)

$$B_n = \frac{2}{C}\int_0^2 N(\alpha)\sin(n\alpha)\, d\alpha$$

$N(q)$ can also be expanded in a Cartesian coordinate system as

$$N(q) = \frac{C}{2\pi}\left[1 + D_{ij}q_iq_j + D_{ijkl}q_iq_jq_kq_l + \cdots\right], \quad i=1,2 \quad (2)$$

where, D_{ij}, D_{ijkl} . . . are termed as fabric tensors and can be expressed in terms of A_n and B_n (Kanatani, 1984). A_n and B_n can be obtained by numerical integration of Eqs. (1b) and (1c) and the data $N(\alpha)$ obtained from intersection counting.

Hence, the directional information regarding the porous network along the cross-sectional plane of a clay can be quantitatively expressed as tensors D_{ij}, D_{ijkl} . . . . Then it would be possible to relate this fabric information with soil parameters and characterize behavior of clays.

If higher order terms in Eqs. (1a) and (2) are neglected, the directional information $N(q)$ and $N(\alpha)$ can be further simplified as

$$N(q) = \frac{C}{2\pi}\left[1 + D_{ij}q_iq_j\right] \quad (3)$$

$$N(\alpha) = \frac{C}{2\pi}\left[1 + A_2\cos(2\alpha) + B_2\sin(2\alpha)\right] \quad (4)$$

where $D_{11} = A_2 = -D_{22}$ and $D_{12} = D_{21} = B_2$. The components of **D** provide a simplified quantitative description of initial fabric. The relationship between deformation and changes in fabric on subsequent loading is presented in the following.

Fabric–Strain Relationship

Consider the distribution of pore channel along the cross section of a clay specimen. Let $f(n)$ be the probability density function that describes the initial fabric of the material such that $f(n)\,dn$ is the length of the segments of fabric descriptors whose normal is oriented along the direction between unit vectors n_i and $n_i + dn_i$ per unit area along the cross section of the plane. Let the coordinate system of observation be X. Let the specimen experience and infinitesimal homogeneous strain ε_{ij} and a rigid body rotation r_{ij}. The coordinate axis of observation of fabric after the deformation shall be $\bar{X}$. The new probability density function after the above deformation shall be $f'(n')$, where n'_i is the corresponding unit vector in $\bar{X}$.

Let x_j be any point along the cross section of the clay specimen before it experiences ε_{ij}, r_{ij}. Let $\bar{x}_j$ be the coordinate of the same point after the above deformation. The distortion of x_j to $\bar{x}_j$ can be described by a linear mapping given by

$$\bar{x}_j = T_{ij}x_j$$
$$T_{ij} = l_{ik}(r_{kj} + \varepsilon_{kj} + \delta_{kj}) \quad (5)$$

where l_{ij} are the corresponding direction cosines between X and $\bar{X}$ and δ_{ij} is the Kronecker delta. Due to this mapping, the unit vector n_i in X is mapped as a vector m_i of length $L(n)$ in $\bar{X}$. Any small area in X is enlarged by determinant of the matrix T (det T) and mapped into $\bar{X}$.

Hence the unit vector n'_i in $\bar{X}$ is given by

$$n'_i = \frac{m'_i}{L(n)} \quad (6)$$

where

$$m_i' = T_{ij}\, n_j \quad \text{and} \quad L(n) = (T_{kl}\, n_l\, T_{kn}\, n_n)^{1/2} \tag{7}$$

The length of the corresponding fabric descriptors $f'(n')\,dn'$ per unit area in the deformed state (in $\bar{X}$) is given by

$$f'(n')dn' = \frac{L(n)\, f(n)\, dn}{det(T)} \tag{8}$$

Any vector y_i in X is mapped as y'_i in $\bar{X}$ given by

$$y'_i = \left[\frac{y_m y_m}{T_{kl} y_l T_{kn} y_n}\right]^{1/2} T_{ij} y_j \tag{9}$$

Taking the derivative (dy'_i/dy_i) at $y_i = n_i$, and substituting in Eq. (8), the relationship between $f(n)$ and $f'(n')$ is given by (Kanatani, 1985)

$$f'(n') = \frac{L^3(n)\, f(n)}{[\det(T)]^2} \tag{10}$$

Since T and L are related to strain, the final fabric descriptor $f'(n')$ is related to strain and initial fabric through the function $f(n)$. Hence Eq. (10) illustrates the inherent relationship between induced changes in fabric and strain. It also can be shown that the relationship between $f'(n')$, $f(n)$, and ε_{ij} is unique (i.e., for a given set of f and f' there is only one solution for ε_{ij}). With experimental evaluation of f and f', Eq. (10) leads to the solution of ε_{ij} and r_{ij} or vice versa. The functions f and f' can be related to initial and final fabric tensors as follows.

The probability density function $f(n)$ and $N(q)$ can be related by (Kanatani, 1984):

$$N(q) = \int |\, nq\,|\, f(n)\, dn \tag{11}$$

If $f(n)$ is known then the distribution of intersection count $[N(q)]$ can be obtained using the above relationship or vice versa. By taking inverse transformation of Eq. (11), and using Eq. (2), $f(n)$ can be expressed in terms of fabric tensors **D** as

$$f(n) = \frac{C}{8\pi}\left[1 + 3\, D_{ij}n_i n_j - 15 D_{ijkl} n_i n_j n_k n_l + \cdots\right] \tag{12}$$

If higher order terms are neglected Eq. (12) reduces to

$$f(n) = \frac{C}{8\pi}\left[1 + 3D_{ij}n_i n_j\right] \tag{13}$$

Similarly the functions after the deformation (denoted by a superscript prime) are given by

$$N'(q) = \frac{C'}{2\pi}\left[1 + D'_{ij}q'_i q'_j + D'_{ijkl}q'_i q'_j q'_k q'_l + \cdots\right] \tag{14}$$

$$f'(n') = \frac{C'}{8\pi}\left[1 + 3D'_{ij}n'_i n'_j - 15D'_{ijkl}n'_i n'_j n'_k n'_l + \cdots\right] \tag{15}$$

$$C' = \int_0^{2\pi} N'(\alpha)\, d\alpha \tag{16}$$

where $N'(\alpha)$ is the number of intersections per unit length of probing line after the deformation. Substituting f and f' in Eq. (10) one could obtain a general relationship for **D**, **D′** and $(\varepsilon_{ij}, r_{ij})$.

In the case of $X = \bar{X}$ (i.e., $l_{ij} = \delta_{ij}$), and no rigid body rotation takes place (i.e., $r_{ij} = 0$), substituting Eqs. (12) and (15) in Eq. (10), the solution for strain, ε_{ij} can be obtained. If the initial state is isotropic, Eqs. (2) and (12) reduce to

$$N(q) = \frac{C}{2\pi} \quad \text{and} \quad f(n) = \frac{C}{8\pi} \tag{17}$$

Substituting Eqs. (15)–(17) in Eq. (10)

$$C' = C\,(1 - 0.5\, \varepsilon_{kk}) \quad \text{and} \quad e_{ij} = -\, D_{ij} \tag{18}$$

where $e_{ij} = \varepsilon_{ij} - 0.5\varepsilon_{kk}\, \delta_{ij}$; e_{ij} is the deviatoric strain tensor. This technique to quantify initial and induced fabric is verified using available preliminary experimental data in what follows.

Example Solution

Data from the microstudies of Delage and Lefevbre (1984) have been examined. One-dimensional consolidation tests were performed on several specimens of St. Marcel clay with load increment ratio of 1.5 every 24 hr to final stresses of 23, 124, 421, and 1421 kPa. Then, stress was released at several stages and the sample was left at 4 kPa for 24 hr. Specimens were cut from samples and prepared for microscopy and porosimetry tests. The results of scanning electron microscopy observation were used to develop graphical representation of the porosity of specimens loaded to different stress levels on vertical planes and are shown in Figure 31.1.

For the purposes of the present investigation into the fabric–strain relationship and anisotropy, probing lines were drawn at regular intervals (20 or 30°), on magnified Figure 30.1 (vertical plane of the samples) for samples at (1) intact condition, and the specimens that were loaded to (2) 124 kPa and (3) 421 kPa.

Along each probing line the total pore length per unit probing line $[N(\alpha)]$ was estimated and this was presumed to be a suitable fabric descriptor, as a first estimate. To reduce the effort significantly, it was assumed that $N(\pi - \alpha) = N(\alpha)$ [i.e., the microstructure on the vertical plane is symmetric about the vertical axis (cross anisotropic)]. This assumption is reasonable for naturally deposited clays. The values of N for each sample are shown in Tables 31.1 and 31.2. Using Eqs. (1) and (2), the Fourier series expansion of $N(\alpha)$, the fabric tensors **D** were obtained. Neglecting higher order terms the distribution of $N(\alpha)$ is shown in Figure 31.2. The values of **D** for intact clay and those loaded to different stress levels are given by

Table 31.1. Observed values of $N(\alpha)$ for intact specimen and that loaded to 421 kPa.

		α (deg)						
		0	30	60	90	120	150	180
Intact	$N(\alpha)$	21.7	32.96	27.5	26.02	27.5	32.96	21.7
421 kPa	$N(\alpha)$	14.06	15.8	16.67	16.4	16.67	15.8	14.06

Table 31.2. Observed values of $N(\alpha)$ for specimen loaded to 124 kPa.

	α (deg)									
	0	20	40	60	80	100	120	140	160	180
$N(\alpha)$	25.6	28.23	29.0	25.5	22.6	22.6	25.5	29.0	28.23	23.7

Intact clay: $D_{11} = -D_{22} = 0.014$ and $D_{12} = D_{21} = 0.0$

124 kPa clay: $D'_{11} = -D'_{22} = 0.094$ and $D'_{12} = D'_{21} = 0.0$

421 kPa clay: $D'_{11} = -D'_{22} = 0.068$ and $D'_{12} = D'_{21} = 0.0$

These distributions were tested for possible data fluctuations using a statistical χ^2 test at a significance level of 0.005. The results indicate that the distribution $N(\alpha)$ for intact clay can be regarded as isotropic (i.e., any value of D_{ij} is due to the fluctuation of data and thus the fabric tensor can be set to be zero). For the case of 124 kPa specimen, the test showed that the fabric is in fact anisotropic and the values of D_{ij} indicate this anisotropy. The corrected values of D_{ij} are

Intact clay: $D_{11} = D_{22} = D_{12} = D_{21} = 0.0$

124 kPa clay: $D'_{11} = D'_{22} = 0.094$ and $D'_{12} = D'_{21} = 0.0$

Using Eq. (18) the vertical strain of the specimen loaded to 124 kPa is given by

$$\varepsilon_{11} = 2 \times 0.094 \times 100 = 18.8\%$$

This value compares very well with the measured value of 20% vertical strain for this specimen. In the case of 420 kPa, the χ^2 test showed that the distribution of $N(\alpha)$ can be considered as isotropic. This needs an explanation. It appears, as evident from Figure 31.2, that at low stress levels the strain induces distortion and compression of interaggregate pores, with very little amount of breakdown of the pores. At large stress levels, the interaggregate pores are broken and, therefore, these pores cannot be related to strain by a linear transformation as assumed in Eq. (5). In addition, the intraaggregate pores also may be affected. But the resolution of the microscopic data available from Figure 31.1 is not high enough to map these intraaggregate pores. This suggests that the chosen fabric descriptor was incomplete for the specimens subjected to high stress level. These reasons are expected to constitute the explanation on why the results obtained for the 421 kPa sample are different from that expected for small stress levels. This is a clear limitation at this stage.

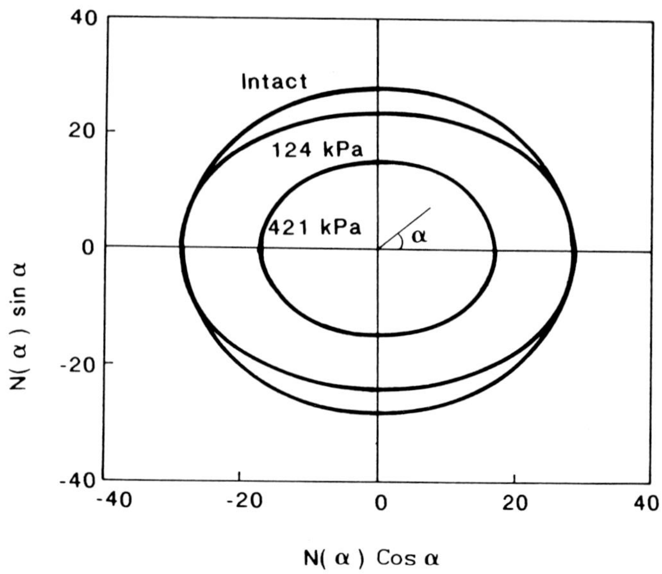

Figure 31.2. Distribution of $N(\alpha)$ (including up to second harmonics).

Despite the shortcomings, the example problem illustrates the viability of quantifying directional fabric information and the possibility of updating fabric upon further loading at small strains. The distributions of interaggregate pores appear to be fabric descriptors associated with mechanical behavior of clays and the material strain at small stress level. The attempt made here is only a first step for the first time to illustrate that it is possible to develop a quantitative measure of initial and load-induced fabric of clays. At this stage, the anisotropy only in the vertical plane was studied. This idea of quantifying the fabric can be extended to many other characteristics of clays that may be affected by mechanical response of clays and could be observed under microscopy. It is possible to extend the idea further and to quantify the 3-D anisotropy. Further research is needed in this direction.

References

Arulanandan, K., and S.S. Smith, 1973. Electrical dispersion in relation to soil structure. Journal Soil Mechanics and Foundations Division, ASCE, v. 99 (SM 12), p. 1113–1133.

Casagrande, A., 1932. The structure of a clay and its importance in foundation engineering. Contributions to Soil Mechanics, Boston Society of Civil Engineers, 1925–1940, p. 72–112.

Delage, P., and G. Lefebvre, 1984. Study of the structure of a sensitive Champlain clay and its evolution during consolidation. Canadian Geotechnical Journal, v. 21, p. 21–35.

Garcia Bengochea, I., C.W. Lovell, and A.G. Altschaeffl, 1979. Pore distribution and permeability of silty clays. ASCE, Journal of Geotechnical Engineering Division, v. 105, no. GT7, p. 839–856.

Hodek, R.J., 1972, Mechanism for the compaction and response of kaolinite. Ph.D. Thesis, Purdue University, 269 p.

Kanatani, K.I., 1984. Stereological determination of structural anisotropy. International Journal of Engineering Science, v. 22(5), p. 531–546.

Kanatani, K.I., 1985. Procedures for stereological determination of structural anisotropy. International Journal of Engineering Science, v. 23(5), p. 587–598.

Lambe, T.W., 1958. The structure of compacted clay. Journal of Soil Mechanics and Foundation Division, ASCE, v. 84(SM2), p. 34.

Mitchell, J.K., 1976. Fundamentals of Soil Behavior. Wiley, New York, 422 p.

Prapaharan, S., A.G. Altschaeffl, and H. Dempsey, 1985. Moisture curve of compacted clay: mercury intrusion method. Journal of Geotechnical Engineering Division, ASCE, v. 111(9), p. 1139–1143.

Rosenqvist, I. Th., 1959. Physio-chemical properties of soils: soil water systems. Journal of Soil Mechanics and Foundations Division, ASCE, v. 85(SM2), Part 1, p. 31–53.

Sridharan, A., A.G. Altschaeffl, and S. Diamond, 1971. Pore size distribution studies. Journal of Soil Mechanics and Foundations Division, ASCE, v. 97(SM5), p. 771–787.

CHAPTER 32

Microtexture and Microchemistry of Clay-Rich Sediments

R.E. Ferrell, Jr. and P.K. Carpenter

Introduction

The microtexture of clay-rich sediments is an important property of the material, reflecting its geologic history and controlling its behavior in industrial and environmental applications. Particle-to-particle arrangements that are the result of depositional and postdepositional conditions may promote the leaching of chemical constituents (as in "quick clays") or retard the flow of contaminants from landfills. In all situations it is essential to relate microcharacteristics to the macroscopic structures observed in the field.

The literature on microtexture is not very abundant and the references are scattered among journals in geology, civil engineering, agronomy, and related fields. The fundamentals of microtextural relationships were outlined in the works of Terzaghi (1925) and Casagrande (1932) more than six decades ago. They introduced the concept that clay mineral particles, especially in sensitive soils, were held together at their points of contact with sufficient strength to create a honeycomb structure with large interparticle voids. Much of the work that followed refined this model to account for differences in the sizes of the particles involved and the degree of parallelism of adjacent particles or flocculated aggregates. Bennett et al. (1977), Moon and Hurst (1984), and Bennett and Hurlbert (1986) have reviewed the literature on clay fabric. The microtexture of clay-rich sediments and soils varies from random to parallel domains composed of more than one clay particle and these features produce profound differences in the physical behavior of the materials. For example, O'Brien (1970) describes differences in domain orientation that are responsible for the geological distinction of mudstones (random) and shales (preferred). In this chapter some of the microtechniques and their application to geotechnical and geologic problems will be briefly described and illustrated. Most of the analyses were performed with a JEOL T-300 scanning electron microscope and a JEOL 733 electron microprobe.

Microtextural Analysis

Microtextural studies are useful in determining fluid flow mechanisms in water-sensitive sandstones and clay-rich sediments. Scanning electron microscope (SEM) studies of clays in sandstone reveal the changes in microtexture and the formation of gel structures as a result of varying salinity fluids passing through a simulated aquifer (Raffensperger, 1988). Chemical reactions associated with acidic wastes (Naymik, 1974) may cause clay-rich soils to develop a less dense microtexture. A microtextural study was used to determine whether fluids were moving through cracks and fissures or through the interparticle voids in consolidated sediments (Ferrell et al., 1989). It provides a good example of the application of combined micro- and macrotextural analysis methods.

In 1982, a train carrying large volumes of chemicals derailed near Livingston, Louisiana, and spilled the contents of several tank cars on the ground, causing much concern that the chemicals might contaminate shallow aquifers. The containment and clean-up procedures were effective in recovering most of the chemicals, but detailed sampling revealed traces of perchloroethylene in low permeability (10^{-7} cm sec^{-1}) clayey zones approximately 10 m below the surface. Ferrell et al. (1989) used X-ray radiographic and SEM analysis to determine that although the sediment microtexture was an effective barrier to chemical transport, open fractures provided a conduit for the rapid transport of the chemicals through the clayey sediments.

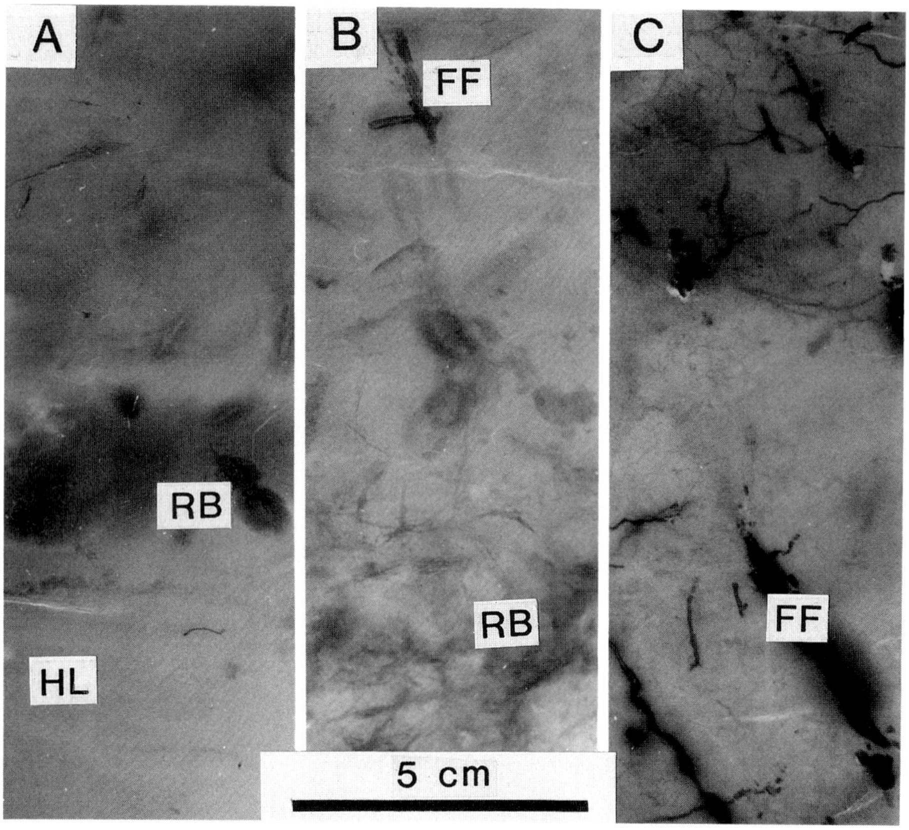

Figure 32.1. X-Ray radiographs (A–C) of slabs of cored sediments reveal structures that are not readily visible and provide guidance in the selection of samples for microanalysis. See text for an explanation of the symbols.

Macroscopic and Microscopic Correlation

X-Ray radiographs of typical soils or sediments near the Livingston site are illustrated in Figure 32.1. Bouma (1969) provides a general reference to the methods of X-ray radiography. These contact prints of the X-ray negatives record dense zones as dark areas and less dense zones as lighter areas. The clay-rich materials from three representative core sections exhibit heterogeneities and macrostructures that are the result of depositional and postdepositional soil-forming processes. An area of faintly stratified clay is present at HL in Figure 32.1A and iron oxide or hydroxide lined fractures or fissures are well developed at FF in Figure 32.1B and C. Fine-grained clays with iron-lined structures are the most abundant features in the radiographs. The structures at RB are probably roots and burrows while those at FF are due to shrinkage and ped formation during weathering. The important thing to realize is that these features are easily recognized in the radiograph even though they are not readily apparent during visual inspection. The X-ray radiograph clearly identifies three different macrotextural zones that are worthy of further study because they might have different permeabilities and microtextures. The information provided by these radiographs represents the essential first step in sample selection and the critical link between macro- and microtextural studies. It should be obvious that a random sampling of these cores should not be attempted because the observed microtextural data would be polygenetic. One sample might represent an original depositional fabric and another might be secondary. All microtextural studies should include a technique such as X-ray radiography to select representative areas for study and to link large-scale and microscopic features.

Microtextural Results

Typical microtextures of the sediments and soils illustrated in Figure 32.1 are presented in Figure 32.2A,B, and C. The SEM photomicrographs of Figure 32.2A and B are representative of the bulk of the cored materials. Figure 32.2C illustrates the iron-lined fractures and concretions that comprise about 2% of the core volume. All the clay-rich materials are mixtures of smectite, illite, kaolinite, and their interlayered varieties that formed as deposits in alluvial systems similar to those presently draining the area. Most of the soils formed on these sediments are characterized by a microtexture of essentially parallel bundles of clay flakes represented by Figure 32.2A and B. The oriented clays are especially evident in Figure 32.2B. The clays have very little intergranular void space (porosity) and their parallel oriented fabric

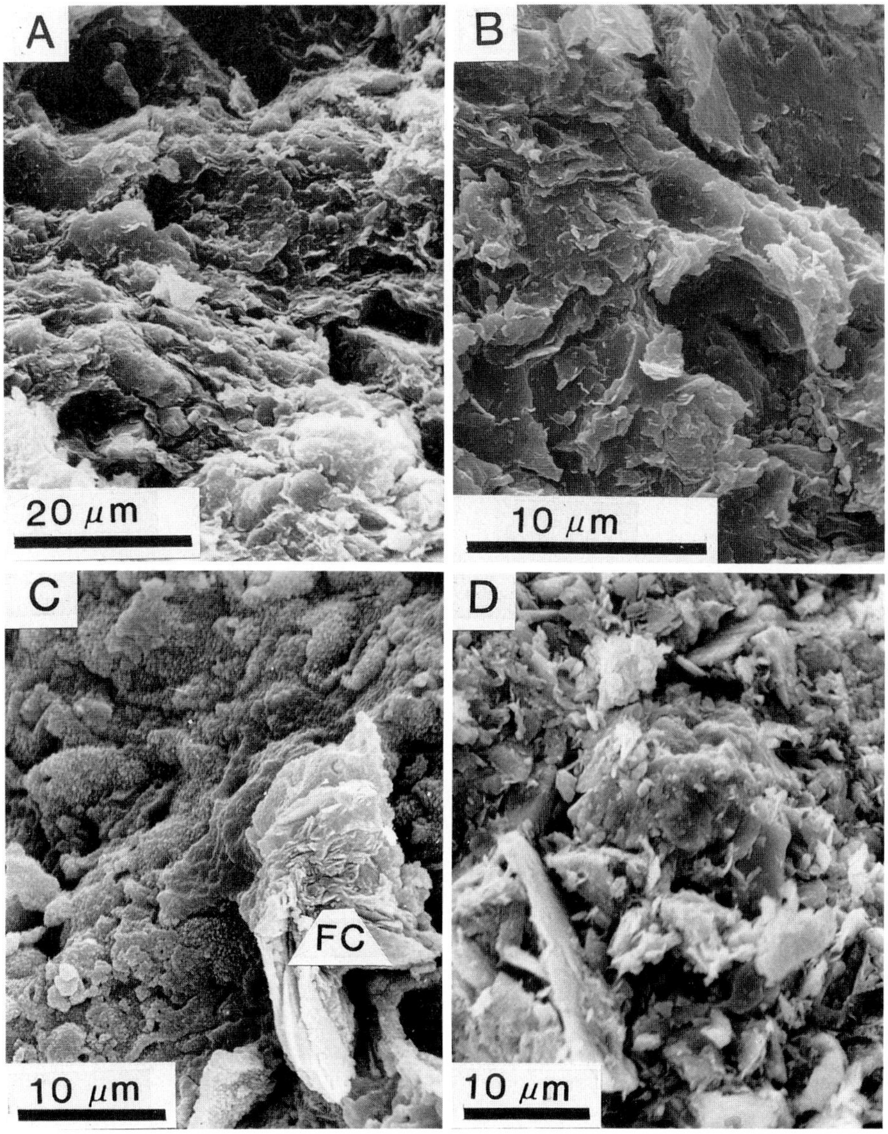

Figure 32.2. SEM micrographs of clay microtextures observed at Livingston site (A–C) and in the Oslofjord (D) reveal differences related to origin.

will make the permeability perpendicular to the layering much less than the horizontal permeability. These clays should act as barriers to fluid flow and intergranular permeabilities associated with this microtexture should be less than 10^{-7} cm sec^{-1}.

The principal mechanism of fluid transport through the Livingston sediments has been gravity flow along open fissures and cracks. The cracks formed as a result of sediment dewatering and pedogenesis and remained open even though several meters of overburden have accumulated. During the weathering associated with soil formation, the cracks were lined with iron oxide or hydroxide precipitates, which helped to maintain the openness of the fractures by cementation (Ferrell et al., 1989). Figure 32.2C reveals the microtexture of the materials lining a fracture. A large goethite or hematite aggregate of blade-shaped crystals is present at FC. Smaller crystals of the same composition coat the remaining grains of the photomicrograph. Their euhedral crystal form indicates that they grew into open pores in much the same way that crystals grow in a geode. Their well-developed form helps to support the argument that these crystal-lined fissures are the avenues for fluid transport.

Sediments of different origins often have microtextures that are different from the ones observed at the Livingston site. For example, illitic and chloritic clays derived from glacial debris and deposited in the Oslofjord, Norway, have a more open microtexture (Fig. 32.2D). The bundles of clay particles are arranged in a typically random fashion with many edge-to-face contacts (Ferrell, 1987). Black areas representing intergranular pore spaces are apparent and the permeabilities of these sediments should be much greater than the more compact arrangements discussed above. Clearly, origin and clay mineral

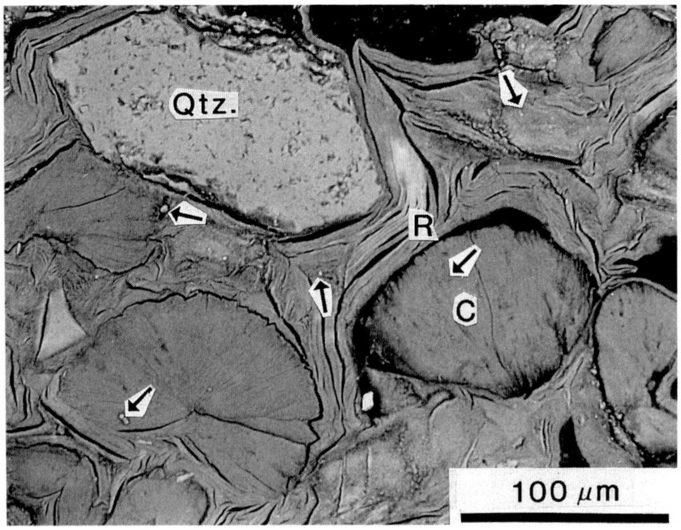

Figure 32.3. Backscattered electron image of vein-filling material composed of quartz (Qtz.), rectorite (R), and cookeite (C). Arrows indicate where quantitative chemical results were obtained. The analyzed volumes are about 15 μm in diameter.

composition influence the microtextures that form at different locations.

The microtextural studies provide a way to explain why the clay-rich materials at Livingston with low measured permeabilities were so rapidly by-passed by the perchloroethylene. The fluids were transported through an open fracture or fissure system lined with iron oxides or hydroxides. Perchloroethylene did not permeate the whole soil mass because the fractures and cracks were so readily available. This explanation provided a better understanding of the fluid flow and is an excellent example of the interpretation of physical behavior that can be accomplished with a combination of macro- and microtextural analysis techniques.

Microchemical Analysis

Microchemical studies provide important supplemental information for microtextural studies. Chemical determinations are often the only way to determine the identity of the particles forming a particular feature. For example, qualitative spot analyses of particles using an SEM with an energy-dispersive X-ray detector were used in the Livingston study described above to confirm that iron minerals were lining the fractures. Quantitative chemical determinations are usually obtained with an electron microprobe using wavelength-dispersive crystal spectrometers.

Spot Analyses

Two types of clays are present in a vein-filling material from Garland County, Arkansas. The typical appearance of the materials is illustrated in Figure 32.3. Cookeite, a chlorite, occurs as fan-shaped aggregates of mineral plates averaging less than 100 μm in diameter (C). The wrinkled material (R) surrounding the cookeite is rectorite, a regularly interstratified illite/smectite. Each of these minerals has a distinct average composition, which is summarized in Table 32.1. The reported means and standard deviations were calculated from the results obtained at widely scattered spots representing cookeite (11 positions) and rectorite (10 positions). The locations of some of the spot analyses are indicated by the arrows in Figure 32.3. The quantitative values were derived from X-ray intensities produced by a 10-μm-diameter beam and represent the composition of a domain about 200 nm in thickness. Oxide weight percentages were computed by the Bence–Albee procedure using hornblende, albite, and microcline as standard reference materials (Ferrell and Carpenter, 1990).

Table 32.1. Summary of quantitative chemical analyses of rectorite and cookeite illustrating the means and standard deviations (SD) obtained with an electron microprobe.

	Rectorite		Cookeite	
Oxide	Mean	SD	Mean	SD
SiO_2	53.92	1.10	43.08	1.43
Al_2O_3	40.59	0.87	55.63	3.46
FeO	0.93	0.46	0.45	0.07
MgO	0.53	0.11	0.31	0.11
CaO	0.04	0.03	0.01	0.01
Na_2O	3.86	0.37	0.10	0.01
K_2O	0.11	0.03	0.01	0.01

The results summarized in Table 32.1 are typical of those expected when clay minerals are analyzed. The actual values used in the calculations were normalized to 100% total determined oxides to eliminate point-to-point variations resulting from variable water contents and densities at the spots being analyzed. Relative total errors, including reproducibility and other analytical uncertainties, of the major oxides, SiO_2 and Al_2O_3, range from 2 to 6%. Minor oxides, such as Na_2O, may exhibit relative errors of 10% or greater if the oxide's abundance is near the limit of detectability of approximately 0.01 wt%. The results clearly define the chemical differences in the fan-shaped aggregates and the wrinkled materials. Some samples may not be as readily distinguished if they are composed of smaller or more intimately mixed crystals. The smallest analytical volume attainable with the standard operating conditions has a diameter of approximately 15 μm.

Image Analysis

Digital image processing methods provide a way to combine microchemical and microtextural observations. The brightness of an object in a backscattered electron (BSE) image is directly proportional to the mean atomic number of the material and the

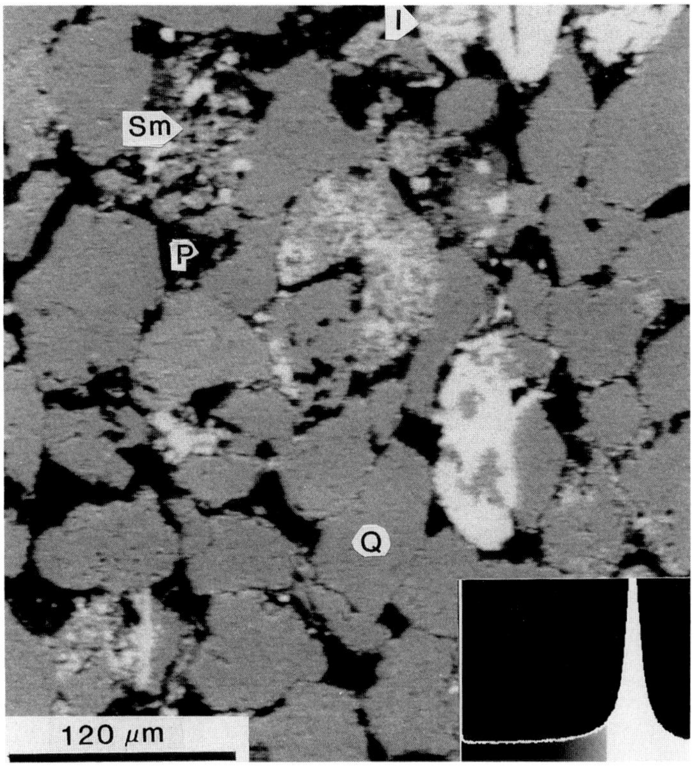

Figure 32.4. Backscattered electron image and gray-level spectrum (inset) of a sandstone illustrating the differences in brightness produced by atomic number differences in the grains and pore spaces. See text for an explanation of the symbols.

BSE images can be processed or segmented at selected gray levels to provide images of a selected phase's distribution or to provide quantitative estimates of its abundance. Figure 32.4 is an example of a sandstone from a petroleum reservoir that has experienced declining levels of production attributed to the presence of water-sensitive clays in the pore spaces. The chemical composition of the smectites in the pores produces a BSE gray level that is intermediate between the black pore spaces and the lighter gray quartz grains. Other illitic clays are the lightest features observed in the photomicrograph. The inset in Figure 32.4 illustrates a typical gray-level histogram for this sandstone. The illuminated bright area separates quartz and illite brightness values from those produced by the pore-filling epoxy and smectitic clays. The BSE gray-level spectrum can be separated at operator selected intervals to form binary images (black and white) depicting the distribution of the different phases. Calculating the area fraction (area % = volume %) of each gray-level segment reveals that porosity (P) accounts for 25 volume % of the sample and smectitic clays (Sm), quartz grains (Q), and illitic clays (1) represent 11, 54, and 7% of the sampled volume, respectively. The smectite percentage is almost double that found in adjacent rocks and accounts for the differences in production and permeability. Digital image processing methods can also determine the area, perimeter, diameter, and shape of features in the micrographs (Russ, 1986).

Summary

The three examples discussed above provide excellent demonstrations of the utility of microtextural and microchemical studies. The two techniques complement one another in unraveling geological, environmental, and materials science problems. The Livingston derailment is a particularly useful study with wide-ranging implications. The microtextural analysis in conjunction with the macrotextural information obtained from X-ray radiographs provides a unique insight into the mechanisms of fluid flow and an explanation for the rapid transport of perchloroethylene through impermeable strata. As these combined techniques become more widely used, the microtextural controls on the physical properties of clay-rich materials should be better understood.

References

Bennett, R.H., and M.H. Hulbert, 1986. Clay Microstructure. IHRDC Press, Boston, MA, 161 p.

Bennett, R.H., W.R. Bryant, and G.H. Keller, 1977. Clay fabric and geotechnical properties of selected submarine cores from the Mississippi Delta. NOAA Professional Paper 9, 86 p.

Bouma, A.H., 1969. Methods for the Study of Sedimentary Structures. Wiley, New York, 458 p.

Casagrande, A. 1932. The structure of clay and its importance in foundation engineering. Journal of Boston Society of Civil Engineers, v. 19, p. 168–221.

Ferrell, R.E., 1987. The microtexture of clay-rich sediments, Oslofjord. In: Schultz, L.G., H. van Olphen, and F.A. Mumpton (eds.), Proceedings of the International Clay Conference, Denver, 1985, p. 121–127.

Ferrell, R.E., and P.K. Carpenter, 1990. Applications of the EMPA-SEM-image analyzer in the study of clay. In: MacKinnon, I.A. (ed.), Electron Microscope and Microprobe Techniques in Clay Analysis. Clay Minerals Society, Bloomington, IN. In press.

Ferrell, R.E., A. Arman, and J.F. Grosch, 1989. X-Ray radiographic investigation of perchloroethylene migration at Livingston derailment site. ASTM Geotechnical Testing Journal, v. 12, p. 119–125.

Moon, C.F., and C.W. Hurst, 1984. Fabric of muds and shales: an overview. In: Stow, D.A.V., and D.J.W. Piper (eds.), Fine-Grained Sediments: Deep-Water Processes and Fabrics. Geological Society Special Publication, 15, p. 579–593.

Naymik, T.G., 1974. The effects of drying techniques on clay-rich soil texture. In: Arceneaux, C.J. (ed.), Proceedings of the 32nd Annual Meeting Electron Microscope Society of America, p. 466–467.

O'Brien, N.R., 1970. The fabric of shale—an electron microscope study. Sedimentology, v. 15, p. 229–246.

Raffensperger, J.P., 1988. Permeant-induced changes in the permeability and pore structure of unconsolidated sediments. M.S. Thesis, Louisiana State University, Baton Rouge, 185 p.

Russ, J.C., 1986. Practical Stereology. Plenum Press, New York, 185 p.

Terzaghi, K., 1925. Erdbaumechanik auf Bodenphysikalischer Grundlage. Franz Deuticke, Liepzig, Vienna, 399 p.

CHAPTER 33

Quantitative Rock Mineral Analysis

George D. Brunton

Introduction

Herzfeld (1930) derived a formula for the sonic velocity of a liquid suspension of small solid particles based on the assumptions: (1) the particles are small compared to the wavelength, (2) the amplitude of the waves is small compared to the wavelength, (3) the damping of the waves is negligible, and (4) the accompanying shear waves are negligible. His result (error in his original paper corrected by Randall, 1932) is

$$[k_1-k_2]/k_2 = [V_2^2-V_2^2]/BV_2^2-[V_2^2(p_1-p_2)/p_2]/^{V_2}$$
$$[1+(1/3)B(p_1-P_2)/p_2]$$

where

k_1 = adiabatic compressibility of the solid particles,
p_1 = density of the solid particles,
B = fractional part of the total volume of suspension occupied by the solid particles,
k_2 = adiabatic compressibility of the suspending liquid,
p_2 = density of the suspending liquid,
V_2 = velocity of sound in the suspending liquid,
V = velocity of sound in the suspension.

Herzfeld thought that results calculated from the formula would be valid even if the particles were nonspherical and anisotropic because of their random orientation in suspension. Randall (1932) used an ultrasonic interferometer to test Herzfeld's theory. He used a suspension of pyrex glass particles in water. The values of the ratios of the compressibilities for the solid–liquid for two different concentrations were negative. "Negative values for the compressibility are, of course, impossible. The broadening effect due to damping seemed to be of sufficient magnitude to account for the negative results obtained" (Randall, 1932).

Randall concluded that "There are three major limitations in the precision attainable in determining the compressibility of solids in powdered form by the velocity-suspension method: 1) The smallness of the ratio k_1/k_2, 2) the magnitude of the concentration B which it is possible to obtain, and 3) the accuracy of the velocities obtained for the liquid and for the suspension. It is estimated that a precision of 1% might be reached when the above three conditions are the most favorable, provided formula (1) [Herzfeld's] may be depended upon to that extent."

Urick (1947) reported that k_1/k_2 = approximately 1/40 for the mineral kaolinite and he measured the sonic velocity of suspensions with solid volume concentrations of "35–40% kaolin" by using sodium pyrophosphate deflocculant.

Wood (1944) assumed that for a mixture of two fluid media (air bubbles and water) that "the velocity [of sound] is the same as that in a homogeneous fluid of the same mean density p and the mean elasticity E as in the mixture. Let p_1E_1 and p_2E_2 represent the density and elasticity of the constituents 1 and 2, respectively. Then let x = the proportion of the first constituent by volume and $(1 - x)$ = the proportion of the second constituent by volume. The *mean density* p is therefore

$$p = xp_1 + (1-x)p_2$$

We must have also

$$1/E = x/E_1 + (1 - x)/E_2$$

whence the *mean elasticity*

$$E = E_1E_2/[xE_2 + (1-x)E_1]$$

and the *mean velocity*

$$c^2 = E/p = E_1E_2/[xE_2 + (1-x)E_1]\ [xp_1 + (1-x)P_2]$$

These expressions are applicable to a mixture of any two fluid media which do not react chemically."

Urick (1947) used a similar approach to the consideration of sonic velocity in suspensions. His formula for the velocity of sound in the mixture is

$$V_0^{2\prime} = 1/p_0k_0 = 1/[p_2B + p_1(1-B)]\ [k_2B + k_1(1-B)]$$

This equation is similar to that derived by Wood (1944) if compressibility is substituted for elasticity.

He used a one megacycle sonic interferometer to isothermally measure the sonic velocity in mixtures of kaolinite and water. The volume fractions of kaolinite varied from approximately 0.03 to 0.04; 90% of the particles were 0.4 and 5.0 μm. He used the dry grain density (2.60 g/cm³) of the clay to calculate volume fractions of his suspensions. His results for the ratios of the compressibilities of kaolinite to water at the three highest kaolinite concentrations were 0.016, 0.028, and 0.023 (mean = 0.022). Since the compressibility of water at 0°C is 50.92×10^{-12} cm²/dyn the mean ratio from Urick's results gives 1.1×10^{-12} cm²/dyn for the compressibility of kaolinite.

Shumway (1958) generalized Urick's equation (Urick, 1947) to

$$V_2 = 1/[F_wB_w + \text{Sum}\ (F_{ci}\ B_{ci})]\ [F_wR_w + \text{Sum}\ (F_{ci}\ R_{ci})] \quad (1)$$

where V = the acoustic velocity in the suspension described by the denominator, F_w = volume fraction of water, F_c = the volume fraction of each of the minerals i, R_w = the density of water, R_c = the density of each mineral i, B_w = the compressibility of water, and B_c = the compressibility of each mineral i in suspension. He measured the sonic velocity of distilled water, "fine quartz sand," continental shelf silt, calcareous ooze, a deep sea red clay, and a continental slope clay using a resonance method (Shumway, 1956) at a fixed volume fraction of the suspended material over a temperature range from approximately 5°C to approximately 80°C.

Analytical Technique

The sample is pulverized and examined for mineral content by means of X-ray diffraction. If one or more clay minerals are present, enough of these are separated from the coarse minerals by standard suspension techniques that their combined density and compressibility can be determined from sonic velocity measurements. After the densities and compressibilities of each mineral constituent of the sample are determined, a known volume fraction of the pulverized sample is suspended in a fluid (usually water) for determination of the individual volume fractions of each mineral. In addition to reduction of the particle size of the minerals by pulverization and deflocculation by peptization, the sample is countercirculated through the sonic cavity to ensure uniform suspension.

Equation (1) is used to determine the densities and compressibilities of minerals in a monomineralic suspension at constant temperatures because the volume fractions of the fluid and the mineral are adjustable constants, the density and compressibility of the fluid are constant at constant temperature, the mineral density and compressibility are independent variables, and the sonic velocity is the dependent variable. The density and compressibility results determined on a mineral at several temperatures are fit to a polynomial function of density vs. temperature and compressibility vs. temperature with standard curve fitting techniques. These polynomials are used to calculate the density and compressibility of each component at each temperature for the mineral analysis of the sediment sample.

In the mineral analysis, the densities and compressibilities of minerals and fluid are the constants of Eq. (1), and the volume fractions become the independent variables. The dependent variable, sonic velocity, is measured at a variety of temperatures. The constants are obtained for each temperature from the polynomial functions.

An accurate polynomial exists for calculating the velocity of sound in water from 0 to 100°C (Greenspan and Tsciegg 1957; Lovett 1969). A similar polynomial is fit to published data for the density of water over the range of temperatures that are covered. The compressibility of water, Bc_c, is calculated by use of Laplace's equation, $V^2 = 1/B_wR_w$, from the values of velocity and density determined from these polynomials. Hence, in Eq. (1) the volume fraction of the water, F_w, becomes the independent variable ($F_c = 1 - F_w$ for the case of one solid component), and the velocity of the suspension, V, becomes the dependent variable for a set of nonlinear simultaneous equations from the which the constants B_c and R_c are determined at constant temperature by a nonlinear least-squares fit of the data. Overdetermination of B_c and R_c values by this method provides greater precision for each value than can be obtained from single measurements.

References

Greenspan, M., and C.E. Tschiegg, 1957. Speed of sound in water by a direct method. Journal of Research of the National Bureau of Standards, 59C, p. 249–251.

Herzfeld, K.H., 1930. The propagation of sound in suspensions. Philosophical Magazine 9, p. 752–768.

Lovett, J.B., 1969. Comments concerning the determination of absolute sound speeds in distilled and seawater and Pacific SOFAR speeds. Journal of the Acoustic Society of America, v. 45, p. 1051–1057.

Randall, C.R., 1932. Ultrasonic measurements of the compressibility of solutions and of solid particles in suspension: Bureau of Standards Journal of Research, v. 8, p. 79–99.

Shumway, G., 1956. A resonant chamber method for sound velocity and attenuation measurements in sediments. Geophysics, v. 21, p. 305–319.

Shumway, G., 1958. Sound velocity vs. temperature in water-saturated sediments. Geophysics, v. 23, p. 494–505.

Urick, R.J., 1947. A sound velocity method for determining the compressibility of finely divided substances. Journal of Applied Physics, v. 18, p. 983–987.

Wood, A.B., 1944. A Textbook of Sound. G. Bell and Sons, London.

II

Applied Clay Microstructure

B. Measurements/Techniques/Sampling Strategy

CHAPTER 34

Applied Clay Microstructure: Measurements, Techniques, and Sampling Strategies for Clay Fabric Research—Overview

Peter Smart and W.A. Chiou

Engineers, soil scientists, and geologists have recognized that clay fabric plays an important role in determining soil behavior. Equally, the physical, chemical, and biological environments have a profound effect on clay fabric. Research of this subject has been in abeyance, while fabric research found using a variety of techniques (including X-ray diffraction, chemical analysis, and electron microscopy) has been reported sporadically. This is due principally to the extremely fine texture and complexity of natural sediments and soils, and to the lack of availability of high resolution instruments and their accompanying techniques.

Fortunately, the advancement of science and engineering and the emerging interdisciplinary studies in the past two decades have made the observation of clay fabric, both in the laboratory and in the field, more meaningful. Modern instrumentation, especially high-resolution transmission electron microscopes (TEM) and scanning electron microscopes (SEM), which facilitate resolutions of less than 2 and 15 Å, respectively, has made direct observations of individual clay particles attainable. Nevertheless, special techniques are required for observations ranging from field scale to molecular scale. An understanding of both the possibilities and limitations of these techniques and instruments is essential in extending our present qualitative and subjective descriptions to quantitative measurements. This type of comprehension is equally important in validating models of the physicochemical and micromechanical behavior of microfabric and is advancing understanding of the larger scale geological formations and deposits. New and improved sampling methods, numerical techniques, and statistical procedures are essential for a soundly based understanding of the behavior of fine-grained sediments. In this section, state-of-the-art techniques are presented and examined to quantify and classify the microstructure of suspended materials, unconsolidated and consolidated sediments, and rocks.

Baerwald, Burkett, and Bennett present a brief review and detailed synthesis of the technique of sample preparation for electron microscopic observation of clay microfabric. The techniques provide the reader with well-established, reliable methods for the preparation of soft clay samples for electron microscopic study. The structures of some clay minerals, iron oxides, and other constituents of sediments are changed when the samples are dried, yet high-resolution microscopy is necessary for examining the crystal structure of these constituents, which requires vacuum conditions. The chapter by Fukami, Fukushima, and Kohyama describes an advanced development of environmental cells that enables the samples to be wet (ranging from 100% relative humidity to liquid) while inside an electron microscope. This technique is expected to have an important role in future studies of marine sediments, aggregates, and marine snow.

The effect of biogenic methane on the microstructure is discussed by Chiou, Bryant, and Bennett. Sampling disturbance is an ever-present problem in geotechnical investigations. It is particularly acute in gassy sediments, since the bubbles of gas expand and remold the sediment when the pore pressure is released. Similar remolding also may occur in saturated sediments if the compressibility of the soil structure is less than that of the pore fluid. Ideally, the sample should be pressurized at the moment of sampling, and the pressures should be maintained through all subsequent stages of preparation and testing. The chapter by Chiou et al. describes an apparatus designed to meet this requirement.

Quantitative theories of the continuum behavior of sediments ought to be supported by quantitative observations of the microstructural phenomena involved. This has proved difficult to achieve and much past work had to rely on qualitative, semi-quantitative, or crudely quantitative methods of observing microstructural behavior. As image analyzers have become more readily available, more powerful, and faster, progress on developing quantitative methods has increased. Chapters by Ross and Ehrlich, Smart, Tovey, Leng, Hounslow, and McConnochie, Bhatia and Soliman, and Chiou, Bryant, and Bennett describe the state-of-the-art position. Ross and Ehrlich describe the petrographic image analysis technique and relate such data to physical properties. In addition to analyzing pore/throat associations from the point of view of permeability, they also begin to map and subdivide pores, grains, and clay assemblages. Smart et al. measure the overall anisotropy of clay sediments and begin to automatic-map differently oriented and randomly oriented patches of clay using an intensity gradient technique. Bhatia and Soliman discuss coarse-grained sediments using a thin-section technique with an image analyzer. Chiou et al. develop a simple approach to characterizing the orientation of clay fabric via analyzing transmission electron micrographs that provide a semiquantitative analysis of clay fabric.

Dilational and shear waves are both important in themselves and important tools for field measurements in geotechnical research. The propagation and measurement of these waves depend on various factors: composition, structure, pressure, etc. Unfortunately, field measurements are expensive and laboratory measurements are usually made under less than ideal conditions. These problems are addressed in several papers. Stoll and Ando and Yamamoto report laboratory studies aimed at explaining the physical basis of the phenomena. Both studies suggest that geoacoustic properties of marine sediments have a close relationship with their microstructure (fabric). Richardson, Muzi, Troiana, and Miaschi compare laboratory and field measurements and suggest that laboratory measurements of shear wave velocity can be calibrated to *in situ* values. Libicki and Bedford describe simulated laboratory work on the resuspendability of surface sediment in the field. They found the periodic nature of deposition, which, in relation to the fabric of sediments, becomes important in controlling the parameters of acoustic sensing.

Marine sediments collected from the Brazil Basin off the eastern coast of South America were examined by Orsi and Dunn and thus revealed a relationship between acoustic, elastic, and physical properties of sediment. A proposed clay fabric model was tested to deduce the probable mechanism for the changing of geotechnical properties of the sediments. The chapters in this section clearly demonstrate the importance of clay microstructure to a large number of scientific and engineering disciplines.

CHAPTER 35

Techniques for the Preparation of Submarine Sediments for Electron Microscopy

Roy J. Baerwald, Patti J. Burkett, and Richard H. Bennett

Introduction

The collection of high-quality sediment samples recovered from the sea floor is prerequisite to initiating laboratory investigations of sedimentological, microstructural, or geotechnical properties. These types of studies require virtually undisturbed material. Likewise, microfabric studies require the utmost care in the handling of samples and in the preparation of subsamples for electron microscopic observations. The literature is replete with discussions and techniques concerning submarine sediment coring and sample recovery in general. Richards (1962) presented a thorough review of the various types of coring devices and the significance of sample disturbance. In general, large-diameter thin-wall sampling devices provide high-quality sediment cores (Richards and Keller, 1961; Lambert and Merrill, 1979). High-quality sediment core samples now can be recovered from deep core drilling using specially designed sampling tools (DSDP, Vols. 64 and 68). Analyses and evaluation of sample quality were discussed by Bennett (1976) and Bennett et al. (1977). Discussion of the numerous sampling techniques is beyond the scope of this chapter and the interested reader is referred to the above references for details.

The purpose of this chapter is to present proven techniques for the preparation of sediment subsamples for conventional electron microscopy (EM) observations and study. Analytical procedures such as energy dispersive spectroscopy (EDS), wavelength dispersive spectroscopy (WDS), electron energy loss spectroscopy (EELS), and back-scattered electron analysis of polished specimens (Smart and Tovey, 1982) were considered beyond the scope of this chapter. Although many conventional EM techniques can be found in numerous technique manuals for the biologist, few exist for the geologist.

The critical element in the tedious process of sediment sample preparation for EM is to maintain and ensure microfabric integrity. The microfabric is defined as the orientation and arrangement or spatial distribution of the solid particles and their particle-to-particle relationships (Bennett, 1976; Bennett et al., 1977). The microfabric is intimately linked and related to the physicochemistry of the sedimentary material and the terms microfabric and physicochemistry are called, inclusively, the microstructure (Bennett et al., 1977; this volume). Because of the nature of the questions being addressed by scientists and engineers regarding the complex processes and mechanisms that drive the development of sediment microfabric, techniques and instrumentation must be continually pushed to their limits. In addition, new techniques must be developed to enlighten our perspectives and comprehension of the interactive environmental processes that produce the observed features. This chapter presents tested and proven techniques that may help to standardize future studies of microfabric and provide a common base for meaningful comparisons of data and observations among researchers. Some developing technologies and future requirements are discussed in a later section.

Techniques for Clay Sediment Preparation for TEM

Submarine sediment samples are comprised of organic material and assemblages of minerals, including clay, carbonates, and silica and many other rock-forming minerals. The intrinsic physical nature of the sediment types vastly differs, and is dependent on these various minerals comprising the sediment fabric that ultimately governs its physical properties and behavior under static and dynamic stresses. Cohesiveness of the sediment can vary greatly from weak to noncohesive as found in quartz sands and aragonitic and carbonate sediments (Bennett et al., 1990). Highly cohesive sediments are exemplified by smectite-rich Mississippi delta clays (Bennett, 1976) and Pacific red clays (Bryant and Bennett, 1988; Burkett et al., this volume). Knowledge of the mineralogy and intrinsic strength of the material is essential when determining the best technique for preparing a particular sediment to ensure integrity of the microstructure during the many steps of sample preparation.

A variety of methods for dehydration, infiltration, and embedding and cleaning of sediment soil samples for transmission electron microscopy (TEM) and scanning electron microscopy (SEM) have been described historically by Pusch (1966, 1967, 1968), Smart (1967a,b), Bowles (1968a,b), O'Brien (1971), Foster and De (1971), Hulbert and Bennett (1975), and Smart and Tovey (1982). Conventional methods of sample preparation for biological specimens employ "liquid dehydration" procedures with a graded series of fluids such as ethanol or acetone followed by a graded series of propylene oxide and epoxy during infiltration (Pease, 1964; Sjostrand, 1967; Hayat, 1970; Hayat and Zirkin, 1973). Often these conventional methods for biological sample preparation are not advantageous for marine clay sediment, characteristically having low coefficients of permeability ($k = 10^{-6}$–10^{-9} cm/sec) that require long dehydration and infiltration times because of the relatively high viscosities of the epoxy resins. Several specific methods are described in detail in this report. However, it is important to try several techniques simultaneously in the early stages until one method is found to be best for the particular sediment being studied. Bennett (1976) and Bennett et al. (1977) first described application of critical point drying (details of CPD method will be described later) and embedding techniques for submarine clay samples. Computer digitizing of TEM micrographs from samples that were critical point dried demonstrated no change in void ratio (ratio volume of the voids to the volume of the solids) indicating no disturbance of the microfabric (Bennett, 1976). This method has been preferred in subsequent studies (Bennett et al., 1977; Chiou, 1980; Chiou et al., 1983; Bennett and Hulbert, 1986; Burkett, 1987; Burkett et al., this volume). Figure 35.1 is a flow chart of techniques recommended for clay sediment sample preparation used in microfabric analysis. The initial six steps are identical for TEM and SEM, but are markedly different after critical point drying.

Procedures

Step 1. Using a small-diameter wire knife, rectangular-shaped subsamples from sediment cores are slowly cut, immersed in absolute ethanol (ethyl alcohol), and sealed to prevent dehydration. Care must be taken to minimize sampling disturbance during cutting. The rectangular shape of the sample orients the sample with respect to the bedding plane without etching or marking any of the surfaces that could degrade. The center of the sample is the highest quality region for the subsample because it is protected from disturbances such as compression or microshearing or from drag along the core wall. The subsamples should be trimmed to approximately 4 × 7 mm to ensure a proper fit later into BEEM capsules (Fig. 35.2b,c). Sometimes a smaller sample is needed if the sediment resists resin penetration, as do argillitic samples or shales; but, in general, small samples may be more susceptible to disintegration. Since visual recognition of samples is impossible after processing, careful labeling with a pencil, insoluble ink, on small paper labels is essential. Subsample orientation should be maintained to ensure that sectioning or trimming for SEM can be performed in a direction normal to bedding planes for standard observation procedures. Some studies may require precision and control for other directions of observation.

Step 2. The alcohol is carefully exchanged by pipetting or decanting and replacing with fresh alcohol stepwise until all the water is removed. To test for complete alcohol substitution, titration with a few drops of silver nitrate produces a white precipitate of Cl^-, i.e., sodium chloride remains in the interstitial water, thus incomplete replacement of pore water. Fresh water sediments will normally not have the Cl^- ion present, thus the silver nitrate test cannot be used as an indicator for exchange of interstitial water. The replacement step can be hastened by twice daily replacement of the ethanol, and is dependent on the permeability and sample size of the clay. Too much cutting and handling increases the possibility of sample disturbance. A significant advantage when preparing high water content sediment samples (soft or plastic in behavior) is to delay final trimming until after replacement with ethanol or complete dehydration because the clay often becomes more brittle.

Step 3. The absolute ethanol is replaced by amyl acetate, a transition fluid miscible with ethanol and with a lower dielectric constant than ethanol. An ethanol–amyl acetate–liquid CO_2 exchange electrochemically provides an optimal and gentle treatment of the fabric. Replacement of the ethanol with amyl acetate should take approximately the same amount of time as water replacement to ensure complete substitution. Following the replacement of ethanol with amyl acetate, the samples are wrapped individually in wetted lens tissue or placed in a perforated stainless-steel sample holder, and immersed in amyl acetate until the CPD step. The holder consists of eight internal compartments 9 × 5 mm, or configured into four compartments 9 × 10 mm for critical point drying (Fig. 35.2a).

Figure 35.1. Flowchart of the procedures for the preparation of sediment samples using SEM and TEM. Steps 1–6 are identical for TEM and SEM and describe dehydration and fluid replacement techniques. Steps 7T–14T describe impregnation, trimming, and ultrathin sectioning. Steps 7S–11S describe fracturing, mounting, and sputter-coating of sample for SEM. Extended from Bennett, 1976.

SAMPLE PREPARATION TECHNIQUES FOR MICROFABRIC ANALYSIS

DEHYDRATION

STEP	TECHNIQUE	INSTRUMENTATION OR DESCRIPTION
1	subsampling of clay specimen and trimming	wire knife
2	replacement of interstitial water with ethyl alcohol	silver nitrate test sealed glass containers
3	replacement of ethyl alcohol with amyl acetate	protected specimen in lens paper or perforated sample holder
4	replacement of amyl acetate with liquid CO_2	critical point apparatus
5	critical point drying with CO_2	critical point apparatus
6	placement and storage of individual specimen in desiccators	samples in vacuum desiccated or sealed jar with desiccant

TEM ONLY

STEP	TECHNIQUE	INSTRUMENTATION OR DESCRIPTION
7 T	placement of specimens into individual containers & into vacuum desiccator	plastic cup or Beem capsule protective lens paper removed
8 T	specimens impregnated under vacuum and removed	vacuum desiccator Spurr epoxy resin
9 T	specimens cured at 60° - 70° C	oven
10T	trimming of specimens prior to ultrathin sectioning	glass knife or manually with a razor blade
11T	ultrathin sectioning with microtome	diamond knife microtome
12T	placement of sections on copper grids	copper grid
13T	light carbon sputtering of ultrathin sections	vacuum evaporator
14T	ultrathin sections photomicroscopy	T.E.M.

SEM ONLY

STEP	TECHNIQUE	INSTRUMENTATION OR DESCRIPTION
7S	fractured specimen	razor blade
8S	mounted specimen	aluminum stub
9S	electrostatically clean fractured surface	CAB plastic tube; sample
10S	sputter coating with palladium/gold wire	vacuum evaporator
11S	Surface feature photomicroscopy	S.E.M.

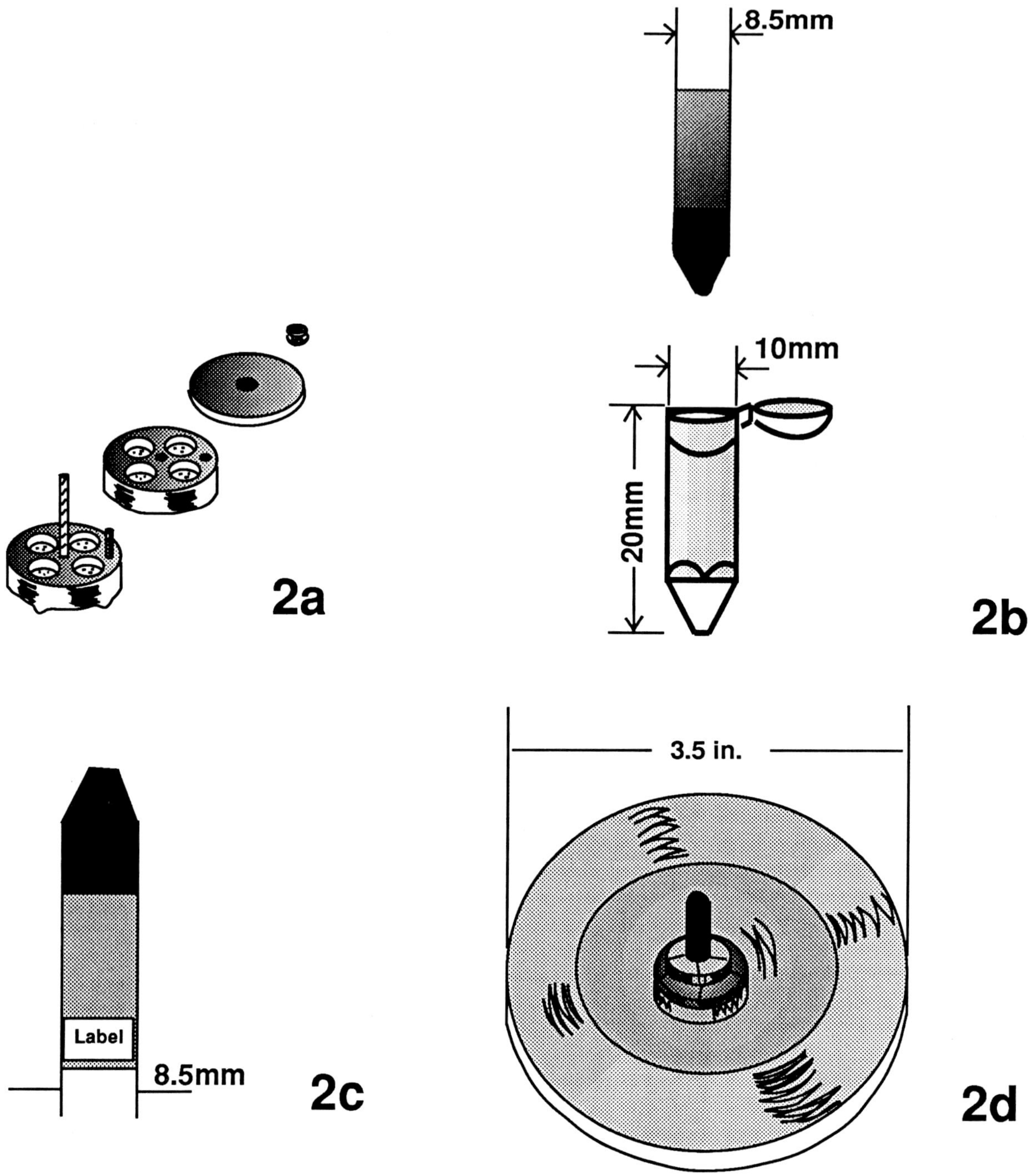

Figure 35.2. (a) Perforated specimen holder capable of containing eight small or four larger clay samples used in the CPD apparatus. (b) BEEM capsule mold for impregnating clay sample into Spurr low viscosity epoxy resin, and the embedded sample. (c) Enlargement of the sample showing embedded label. (d) Machined stainless-steel chuck for holding sample stationary for hand trimming under the binocular microscope.

Steps 4 and 5. Samples are critical point dried in a critical point apparatus using liquid CO_2 as the replacement fluid for the amyl acetate. At the critical temperature and pressure (31.3°C and 72.9 atm), no boundary exists between the liquid and the gas phase, because the density of the gas and liquid phase is identical (Anderson, 1951). Thus the sample can be dried without the deleterious effects of surface tension. Carbon dioxide is preferred because it is easily obtained, inexpensive, and can be incorporated into various types of critical point dryers at a safe and manageable pressure and temperature, for standard laboratory procedures. The amyl acetate is rapidly replaced with liquid CO_2 compared to previous transition steps because the liquid CO_2 is slowly purged through the sample chamber. This procedure continuously drives the chemical interface between amyl acetate and liquid CO_2 to "zero" (amyl acetate) by constant replacement with purging fluid. The time required for complete replacement of amyl acetate by liquid CO_2 varies between 20 and 60 min depending on sediment permeability (Bennett et al., 1977).

Step 6. After the samples are dry they are stored in a vacuum desiccator, either individually wrapped in lens paper, or contained in the perforated sample holder. It is critical that the samples remain completely dry for preservation of the structure and for complete resin impregnation.

Extended Procedures for TEM

Step 7T. BEEM capsules (Fig. 35.2b) are oven dried for moisture removal and stored in a desiccator. This is to prevent any moisture adsorption onto the clay surface or resin that may prevent complete polymerization.

Step 8T. The SPURR low-viscosity resin (Spurr, 1969) is mixed according to the manufacturer's hard cure formula consisting of

10.0 g vinyl cyclohexene dioxide (VCD)
26.0 g nonenylsuccinic anhydride (NSA)
4.0 g diglycidyl ether of polypropylene glycol (DER)
0.4 g dimethyl amino ethanol (DMAE) catalyst

Note: The resin should not come in contact with skin or be inhaled as the DMAE catalyst is reported to be a carcinogen. Using dust-free gloves, handle the uncured resin under a fume hood for maximum safety. We experimented with a nontoxic, lower viscosity resin, L.R. White, which required no mixing of components, but unfortunately distortion and warping occurred on curing. Therefore it is useless for fabric studies involving delicate, fine-grain, high porosity, clayey sediments. It could have some potential for embedding less sensitive materials such as well-cemented or shaley fabrics, but it is not recommended for precision studies of clay samples.

Individual samples are placed in BEEM capsules, placed in a tray, and positioned in a vacuum desiccator with proper plumbing to a rotary vacuum pump and to the funnel filled with SPURR (Fig. 35.3). Jim (1985) also advocated the use of SPURR for clay microstructure studies. A curved or bent glass tube with a dropper type end is attached through a rubber stopper and connected to a hose which in turn is connected to the funnel. A clamp on surgical tubing controls the flow of resin. A vacuum is slowly applied to the desiccator to reach optimum vacuum with a standard rotary pump. After about 0.5 hr under vacuum, the resin is slowly released to each BEEM capsule through the glass dropper tube and allowed to run along the side

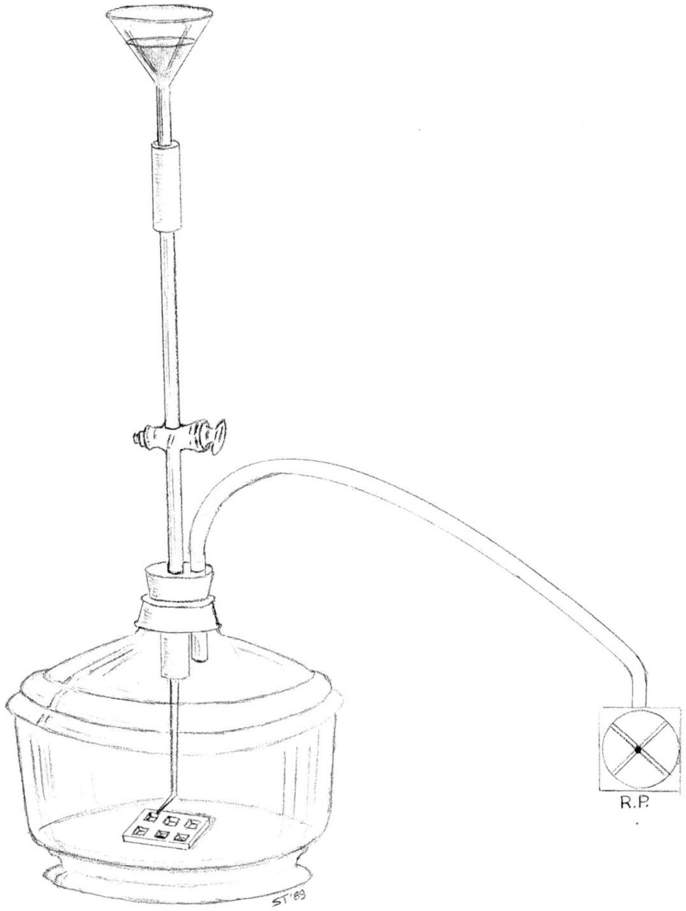

Figure 35.3. Overview of vacuum desiccator assembly for Spurr infiltration of sediment samples under vacuum.

of the BEEM capsule (do not drop resin directly on the sample). The BEEM capsule is filled with resin, the vacuum pump turned off, and the desiccator *very slowly returned* to atmosphere. After impregnation of the resin, a small probe (dental tool or bamboo stick) can be used to carefully adjust the final desired position of the sample and the label inside the capsule. Since SPURR components are toxic, it is safest to carry out all of the above activities in a fume hood with the exhaust fan running.

Step 9T. The samples are placed in a 70°C oven for 24 hr to polymerize the epoxy into a very hard plastic. The best procedure is to vent the oven through a fume hood during curing to avoid obnoxious fumes. The hardness of the resin can be adjusted by changing the relative portions of the four components, (see manufacturer's instructions) but the goal (though difficult to achieve completely) is to reasonably match the hardness of the sample material for consistency during ultrathin sectioning. Many marine clay sediments contain brittle illite, and the hard cure recipe is preferred. Waste resin should be cured and then discarded to avoid health risks. Also cure any of the leftover resin to prevent a sticky lab accident and alleviate any health

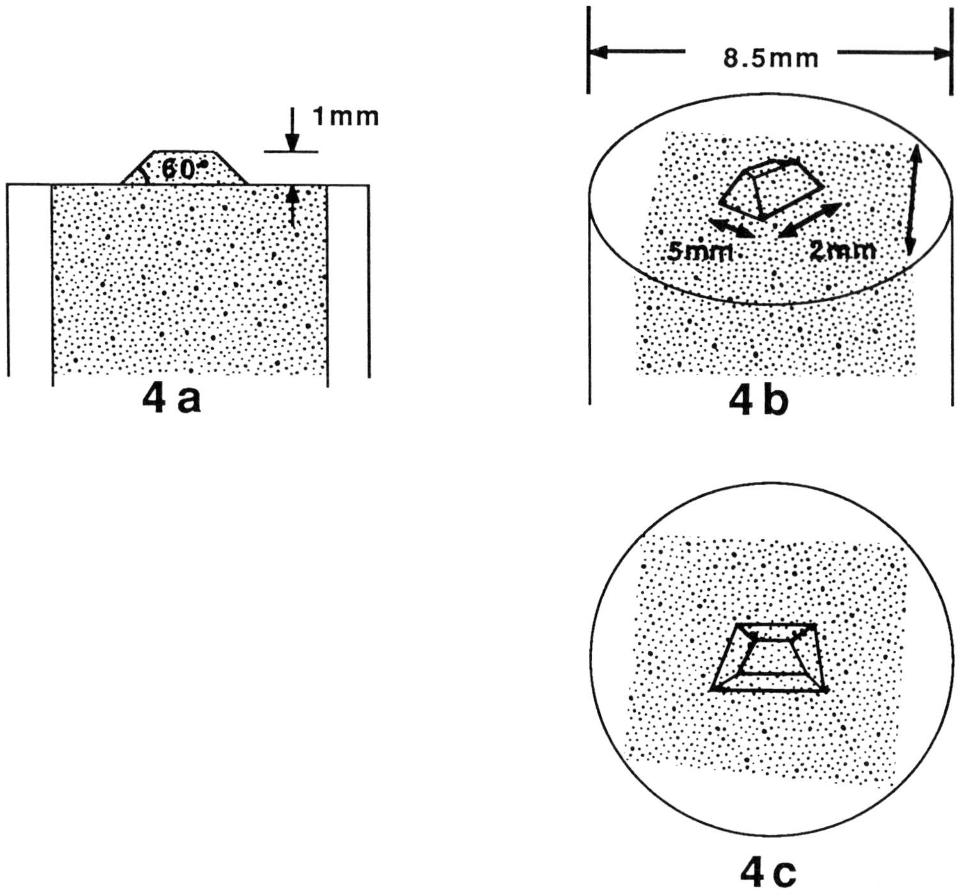

Figure 35.4. Diagram of the tip of the sample showing the geometry achieved by hand sectioning using razor blades. (a) Side view showing the 60° slope of the shoulders providing support during ultrathin sectioning. (b) Trapezoid is carved onto the block face. This shape aids in removal of sections from the knife edge during the follow-through stroke and adhesion of consecutive faces forming a ribbon. (c) Top view.

concerns. Remove cured samples from oven and immediately return to the desiccator to cool. Later remove the cured sample block (Fig. 35.2b,c) from the BEEM capsule using a razor blade. The inability to scratch the epoxy surface with your fingernail indicates a proper hard cure.

Step 10T. The sample is held stationary in a chuck (Fig. 35.2d) and viewed under a binocular dissecting microscope for the trimming steps. Optimal control of trimming the sample prior to ultrathin sectioning is achieved manually using razor blades. A new razor edge should be used for each stroke to prevent scarring and gouging of the block face. A fine file can also be employed usefully for trimming. Starting from the outside of the block, remove the four sides, and slowly work toward the center. As the clay is approached, angle the razor blade to cut about 60° (Fig. 35.4a) forming pyramid-shape sides that provides "shoulders" for stabilization during ultrathin sectioning. As the final size is approached, angle two opposite sides so that the desired final shape of the surface of the face is trapezoidal 0.5 × 1 mm (Fig. 35.4b,c). This shape is ideal for serial section analysis because adjacent edges of the sections adhere to each other and the shape facilitates orientation when viewing a ribbon under low magnification under the TEM. However, if serial sections are not required, a square or rectangle shape is preferable with the sides *not* parallel. This will ensure that the sections will not stick to each other and usually a maximum number of sections can be picked up with one try, since the sections will accumulate near the cutting area of the knife. As an alternate trimming method, a glass knife mounted in a microtome can be used but is more time consuming. The glass edge must be moved and realigned frequently to provide a clean sharp surface.

Step 11T. The trimmed sample block is placed in a microtome and ultrathin sectioned using a diamond knife mounted on a "boat" filled with water (Fig. 35.5a). The microtome incrementally draws the sample across the stationary knife and slices the ultrathin sections. The desired thickness of the ultrathin sections are approximately 800 Å, and appear gold because of their interference colors. The sections adhere to each other forming ribbons of floating trapezoid-shaped samples. The ribbons are then clustered using a single eyelash brush (Fig. 35.5b) (see Hayat and Zirkin, 1973).

Step 12T. A #300 mesh copper grid is held by its edge with forceps and lowered onto the sections, following through into the water, and turned over as it is brought out of the water. As an alternate technique, the grid may be immersed and brought up

Figure 35.5. Diagram of the diamond knife during ultrathin sectioning of processed, embedded clays. (a) The thin sections float on the water-filled boat (Redrawn after Bryant and Bennett, 1988.) (b) Ribbons can be easily seen through the binocular microscope with proper light and water adjustments. The eyelash brush is used to cluster and manipulate the sections prior to picking up on a copper grid. (c) Enlargement of the copper grid after attachment of the thin sections, by cohesion during wicking onto lens tissue. (d) Standard glass slide showing placement of copper grids onto the edges of thin strips of double stick tape prior to evaporation of carbon for stabilization.

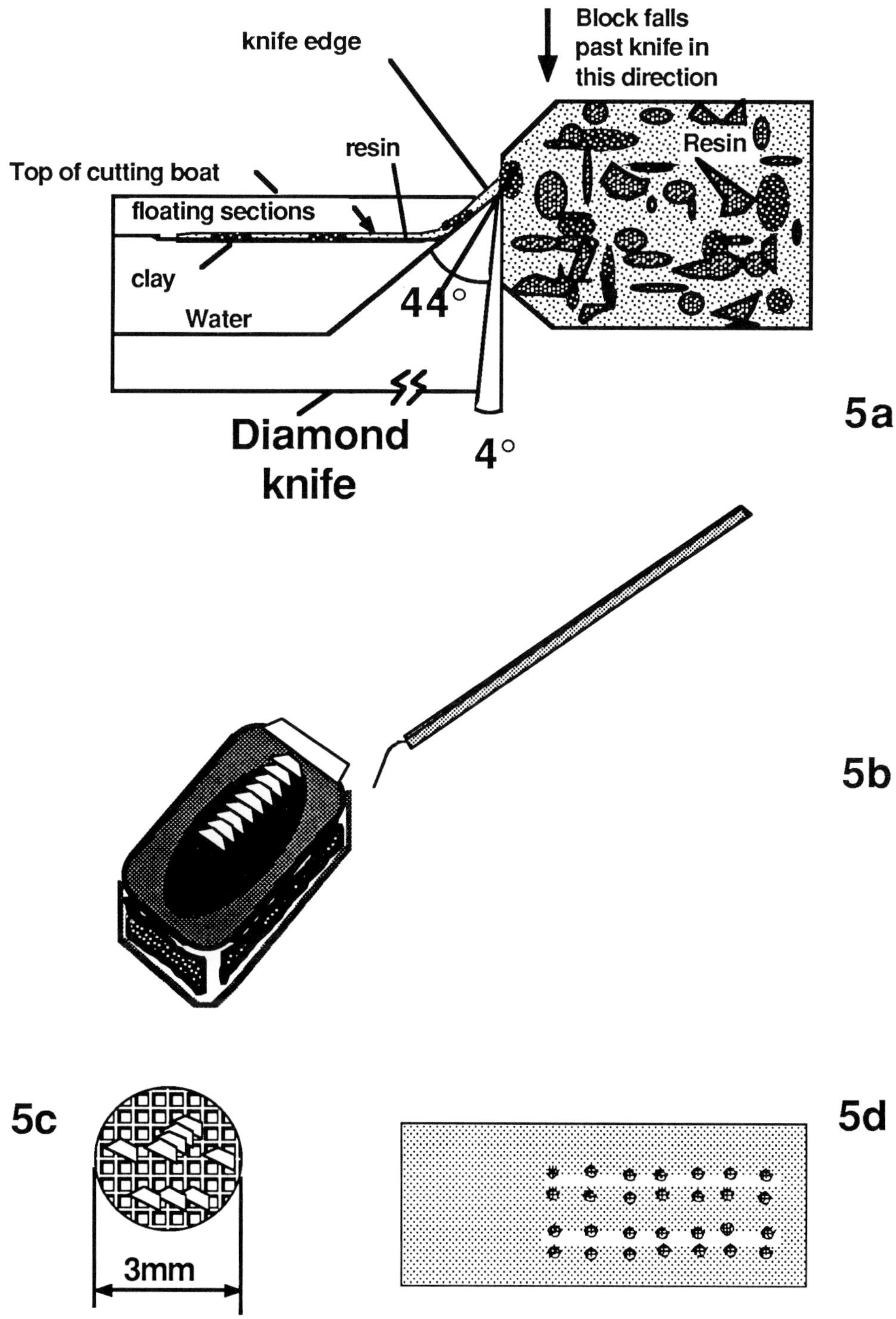

under the sections. The ultrathin sections are now positioned on top of the grid and the water is wicked onto filter paper (below the grid), causing the sections to adhere to the grid (Fig. 35.5c). The grids are stored in numbered grid boxes or clean BEEM capsules and carefully logged. A common but tragic occurrence is spilling, by opening the box too fast or upside down and subsequent mixing of grids.

Note: Grids vary in mesh size, commonly from #100 to #500. For adequate support of the sections and still enough "window" for viewing, a #300 is preferred.

Step 13T. Evaporation of a thin layer of carbon (pale gray as judged by a filter paper reference) unto the thin sections on the grids aids in stabilization by preventing charging of clay particles, etc. under the electron beam in the TEM. Strips of double-stick tape, about 4 mm wide are adhered to a glass slide (Fig. 35.5d). The copper grids are carefully placed face up on the tape so only the outer edge of the grid is adhering. Placing the grids close together allows even distribution of the vaporized carbon. The slide is placed in the vacuum evaporator, and current is generated through sharpened graphite rods resulting in deposition of carbon onto the copper grids. The grids are carefully removed from the glass slide after the chamber is cooled and vented, and stored in the grid box.

Step 14T. The samples are now ready for viewing and photographing in the transmission electron microscope. The photoprocessing commonly uses Kodak sheet film and chemicals with their recommended times. Other commercial sources such as Ilford or Agfa can also be used. Many TEM designs permit the use of 35-mm, 70-mm, as well as $3\frac{1}{4} \times 4$-in. film formats, each with a set of advantages and disadvantages depending on need.

The completed TEM procedure for subsampling and processing the sediment cores through developing and printing of the negatives and prints takes between 20 and 40 hr per sample, but fortunately several samples are processed simultaneously.

Extended Procedures for SEM

Initial preparation of the clay samples for scanning electron microscopy is the same as for TEM through the critical point drying step. The dried samples are stored in a desiccator prior to sputter coating.

Step 7S. The sample must be fractured to reveal a freshly exposed surface. A razor blade is pushed slightly into the sample and twisted causing fracturing of the sample.

Step 8S. The sample is then glued to the surface of an aluminum stub with a drop of carbon conductive paint or a glue, then allowed to thoroughly dry.

Step 9S. Following the procedure of Hulbert and Bennett (1975), a cellulose–acetate–butyrate (CAB) plastic tube rubbed with a lint-free cloth is passed above the clay surface resulting in cleaning with an electrostatic field and freeing the surface of loose particles. This cleans loose particles from the surface and avoids physical contact with the sample surface to be observed with the SEM.

Step 10S. The mounted stub is sputter coated with gold/palladium. The coating should be sufficiently heavy to reduce or prevent charging of the clay surface.

Step 11S. The sample is ready for viewing and photographing in the scanning electron microscope.

The procedures described above are for cohesive sediment samples typically comprised of clay minerals. Often samples predominantly of carbonate minerals are extremely fragile and disintegrate on drying because they lack cohesion. Alternative methods of preserving their structure have been developed and are described below.

Techniques for Carbonate Sediment Preparation for SEM

Often the surface of a sediment sample is too disturbed from the sampling procedure for reliable undisturbed structure studies by SEM. Several methods of exposing new or "fresh" surfaces by various fracturing techniques have been described by Smart and Tovey (1982). In addition, Tovey (1970) and Wong and Tovey (1975) have utilized "peeling" techniques consisting of silver glue, epoxy, plastic, or Sellotape for removal of thin layers of sediment for EM studies.

Scanning electron microscopy of cored, noncohesive carbonate sediment samples is difficult because of their extremely friable nature after critical point drying. Using standard techniques (Fig. 35.1 Steps 7S–11S) for clay samples usually results in disintegration of carbonate fabric under the electron beam or at various stages during preparation. Unstable specimens of any kind create erratic bursts of secondary electrons emitted from the specimen. The effect is known as charging. When charging occurs, the image is obscured on the CRT and photography becomes impossible due to the long photographic time exposures (45 sec to 1.5 min) necessary for low-noise, high-resolution negatives.

Special sample preparation techniques and microscope adjustments are required to preserve the integrity of the microfabric of carbonate samples (first reviewed in Bennett et al., 1990). After critical point drying, oolitic carbonate samples should be stored in a vacuum desiccator to prevent absorbed moisture. Friable samples will often explode when placed into the vacuum chamber if they are not completely free of water. Sample stability is greatly increased by using a low viscosity cyanoacrylate (Superglue) before metal coating. Enough glue should be applied to one side of the sample to penetrate approximately one-third of the sample.

The glue should be allowed to dry completely. Gently break open the sample to expose an undisturbed, glue-free interior surface held together by underlying hardened glue. The specimen should be mounted on a SEM specimen stub with cyanoacrylate and coated with a thick (deposit on sample should have a dark silvery-black color) layer of gold-palladium in the sputter coater. Too thin a coating results in charging, but too thick a layer will obscure structural details. Paint the lower edges of the sample and exposed aluminum stub with a layer of colloidal carbon ("TV TUBE KOAT") to alleviate charging.

In spite of these precautions, the sample remains fragile. Therefore, the SEM settings are important for high-resolution images. The kilovoltage setting should be moderately high (17–20 kV) to compensate for a necessary small probe size setting. Finally, a relatively high (200 μA) emission current is also required to compensate for the smallest possible probe size. A very small probe size will ensure high-resolution microscopy and at the same time reduce the likelihood of specimen damage or contamination resulting from heating by the scanning beam.

Preparation of Large-Area Ultrathin Films for TEM

Large area ultrathin films provide a substrate and support medium when attached to #300 or #400 copper grids. This technique has applications for clay and mineralogy studies such as selected area diffraction analysis for individual crystal identification, or serial section analysis for pore geometry studies. The various methods described in the literature for the preparation of ultrathin films for transmission electron microscopy (Pease, 1964; Sjostrand, 1967; Hayat, 1970) are all useful, but results are sometimes inconsistent depending on the researcher using the technique. Formvar and Parlodion are popular plastic used for making films. Very large ultrathin films may be prepared easily by applying drops of Parlodion dissolved in amyl acetate to the surface of a bowl of double distilled water. The solution quickly and quite evenly spreads out on the water surface. Rapid evaporation of the solvent leaves behind the pure plastic film that may be affixed to TEM grids in various ways. Unfortunately, ultrathin films produced in this manner are fragile, and can be easily broken mechanically or by heat produced by the electron beam. This problem is especially acute when large area films are required for single hole grids for serial section analysis. For best results, the attached films should be stabilized with a thin layer of carbon from a conventional carbon vacuum evaporator (Fig. 35.1, Step 13T) after the films have been attached to the grids.

Ultrathin films prepared from Formvar plastic are much stronger than Parlodion, and do not require carbon support. Properly prepared, these films are stable under the electron beam for timed exposures of 2 min or longer. This stability is achieved even with very large films stretched over single hole grids used for serial section analysis. However, consistency in preparation of ultrathin Formvar films is difficult. The fundamental problem is that the solvent, commonly ethylene dichloride, used to dissolve the Formvar does not spread evenly on an aqueous surface. The films must be dried onto a solid surface and then released, which can be difficult. Considerable sagging commonly occurs when attaching the film to large hold grids. Since 3-mm copper grids are thin, this sagging will inadvertently allow the film to attach itself to the solid substrate beneath as well as to the grids themselves. This results in destroyed or weakened films caused by the removal of grids from the substrate.

The University of New Orleans Biology laboratory has developed the following procedure that retains some conventional techniques while incorporating new methods. The significant improvements include better release of Formvar films from the glass slides and improved application of the films to the large hole grids. We have realized approximately a 90% success rate using the following techniques:

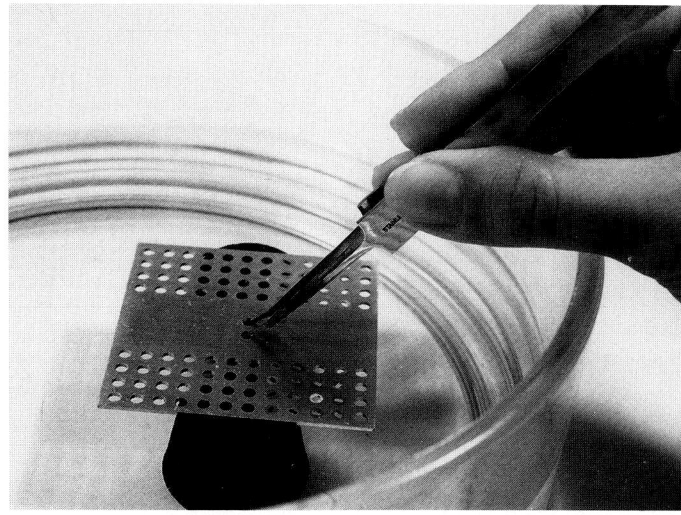

Figure 35.6. View of setup for attaching Formvar film to large hole grids. Resting palm of hand on edge of staining dish will ensure a steady hand during this process.

1. Rinse and fill a large staining dish (20 cm diameter × 8 cm high) with distilled water.
2. Place two No. 6 rubber stoppers as a base for an aluminum plate (8 cm × 8 cm × 1.5 mm thick) milled with 50 holes (2.8 mm D) plus two additional holes for lifting the slide with microforceps (Fig. 35.6). The holes must be slightly smaller than a standard 3-mm grid to support the entire circumference of the grid (Fig. 35.7).
3. Adjust the water level to approximately 0.5 cm above the aluminum plate and position new, single-slot, oval-shaped grids over the holes in the plate. Be careful to avoid air bubbles while placing the grids on the slide, shiny side up.
4. Wash several smooth glass slides with Comet or some other mildly abrasive detergent.
5. On a wet glass slide place a drop or two of liquid soap (7× cleaning solution works well) on both sides of the glass slide, diluted with a few drops of water.
6. Gently dry the glass slide with paper toweling or Kimwipes until slides look clear when held to a light source.
7. Place glass slides in a 60°C oven to dry for about 15 min. This will prevent excessive holes forming in the film.
8. Pipette several milliliters of Formvar solution (0.5% in ethylene dichloride) and flood approximately 75% of one side of the glass slide.
9. Drain excess liquid immediately from the slide with Whatman filter paper and then tap the end of the slide on the towel to ensure complete drainage of the excess. Air-dry the film for a few minutes.
10. Scrape the sides of the glass slide with a razor blade and make transverse cuts on the film near both ends of the attached film (Fig. 35.8).

Figure 35.7. Closeup view of 3-mm grids and slightly smaller holes drilled into aluminum support plate. Top two holes have films attached.

11. Position fluorescent light source and vary the angle of the slide with respect to the water surface so that interference colors can be seen.
12. Breathe on the film to create a layer of moisture.
13. Slowly touch the "distal" end of the slide to the water surface watching for the film to attach to the surface meniscus (Fig. 35.9). It may be necessary to assist this process by gently pulling the film with a pair of microforceps. If the film stops coming off at any point, remove the slide from the water, breathe on it again, and resume.
14. Silver to pale gold interference colors represent usable thicknesses that range from 70 to 120 nm (700 to 1200 Å) for routine fabric studies. High-resolution studies would require an excess of 125 kV since it is difficult if not impossible to cut clay microfabric in the "gray" interference color (>50 nm) thickness range.

    Note: If the films appear too thick as judged by blue, green, or pink interference colors, make new films by further diluting the Formvar solution by small increments.
15. Position a suitable floating Formvar film (push gently with microforceps) over the aluminum plate bearing the large hole grids. Carefully lift the aluminum slide straight up with the forceps (Fig. 35.6) and air dry in a standard size petri plate. The grids can be stored on the aluminum plate in a petri plate for later use. Lowering the water level by a siphon device works equally well.

Fixation of Clay–Organic Complexes

An exciting area of investigation is the role of organics in the formation of the microstructure of marine clay sediments. Biogenic components, including fecal pellets and marine snow, are the dominant carriers of organics through the water column. Biological and physical processes can result in resuspension of sediment, also important to sediment microfabric analysis. Complete preservation of *in situ* organics is fundamental for studying the relationships of clay minerals and associated organics. The standard fixation method for most biological specimens for TEM uses glutaraldehyde ($OCHCH_2CH_2CH_2CHO$) followed by osmium tetroxide (OsO_4). An aqueous sodium cacodylate [$Na(CH_3)_2AsO_2 \cdot 3H_2O$] buffer is often used to adjust the concentration of the osmolarity of the tissue. Eight percent glutaraldehyde can be purchased from several vendors (see appendix). The fixative is stored under nitrogen gas in 10 ml vials at 4°C to prevent polymerization of the aldehyde. Working solutions are prepared by making appropriate dilutions with

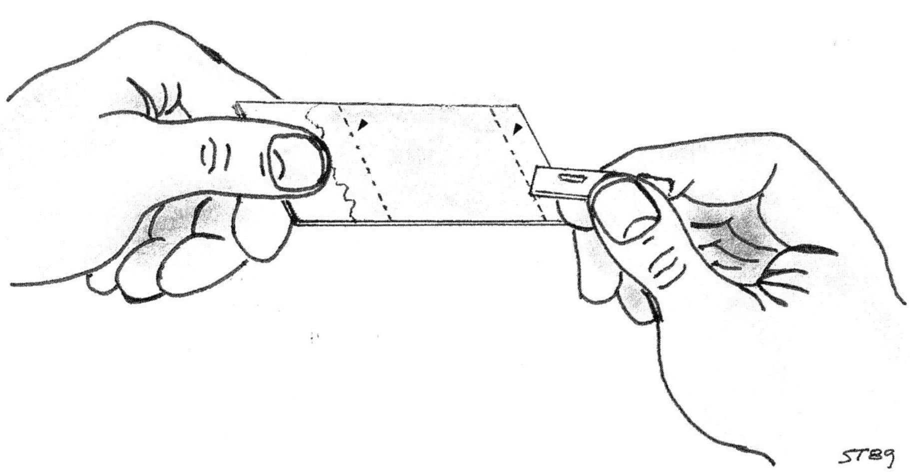

Figure 35.8. Diagrammatic representation of Formvar-coated slide. Positions where cuts in film should be made are indicated with arrows.

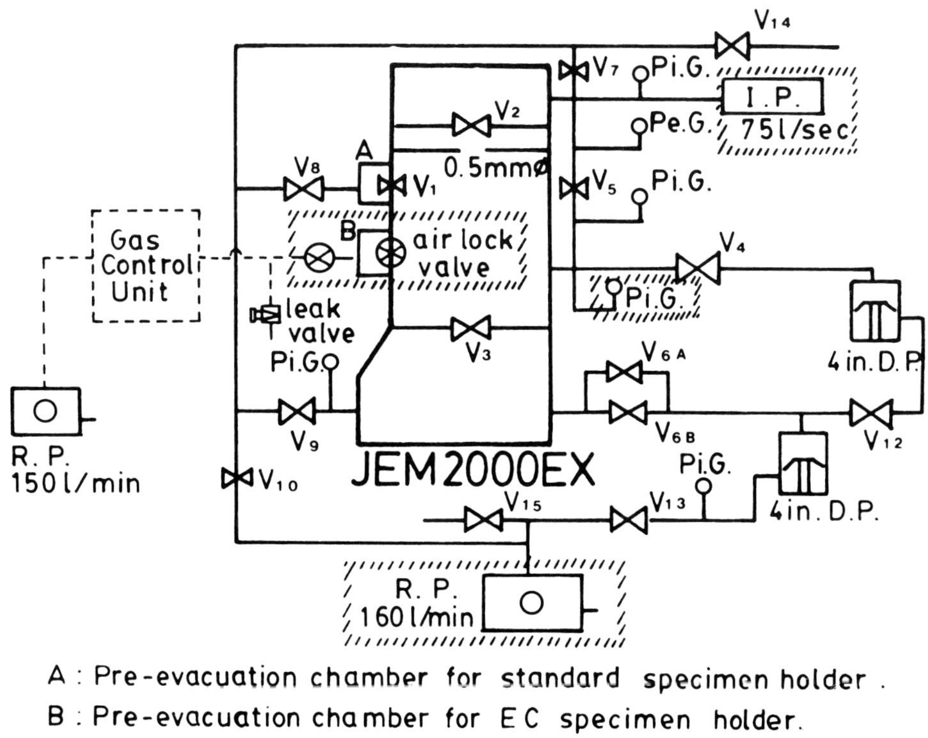

Figure 36.2. Modified pumping system of JEM-2000EX electron microscope for installation of EC device.

6. The specimen holder is then inserted into the electron microscope; the degree of vacuum in the electron microscope specimen chamber is also checked.

Preparation of Sealing Film

In the case of a film-sealed EC, to achieve high-resolution and high-contrast observation, the sealing film covering nine holes of the copper grid should be as thin and stable as possible. We have already developed the evaporated carbon film reinforced by a plastic microgrid with numerous minute holes (5 ~ 10 μm diameter), as the sealing film (Fukami and Fukushima, 1984). The thickness of this film is less than 20 nm, yet it will withstand pressure differences over 200 Torr between the inside and the outside of the EC. However, the microgrid in this type of sealing film restricts the visible field. To reduce this problem, we developed a new technique to make the sealing film. Preparation procedure for the new sealing film are as follows: (1) Make the plastic microgrid on a slide glass; (2) smooth the microgrid with ethyl acetate vapor; (3) carbon coat (about 20 nm in thick) the smoothed microgrid with rotary evaporation technique; (4) score the carbon film in 2 × 2 mm squares; (5) float off the scored carbon films in water; (6) pick up the floating film on the copper grid with nine holes; and (7) dissolve the plastic completely in ethyl acetate. A low-magnification image of the new

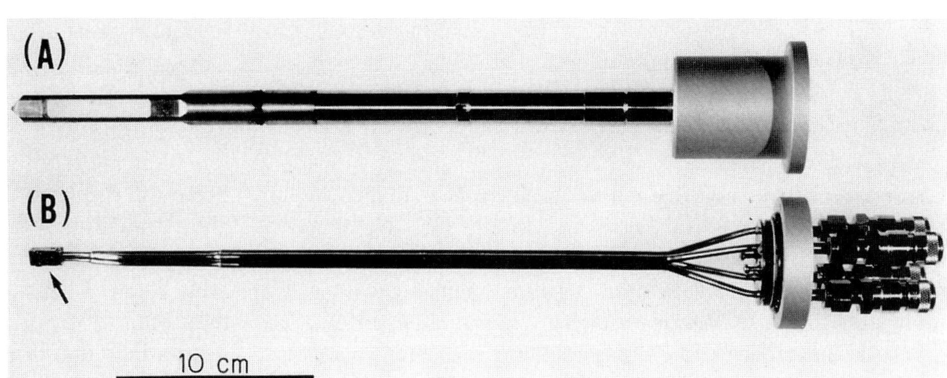

Figure 36.3. Side-entry type large specimen holder for EC observation. (A) Cover and (B) inner pipes and sealing block (arrow).

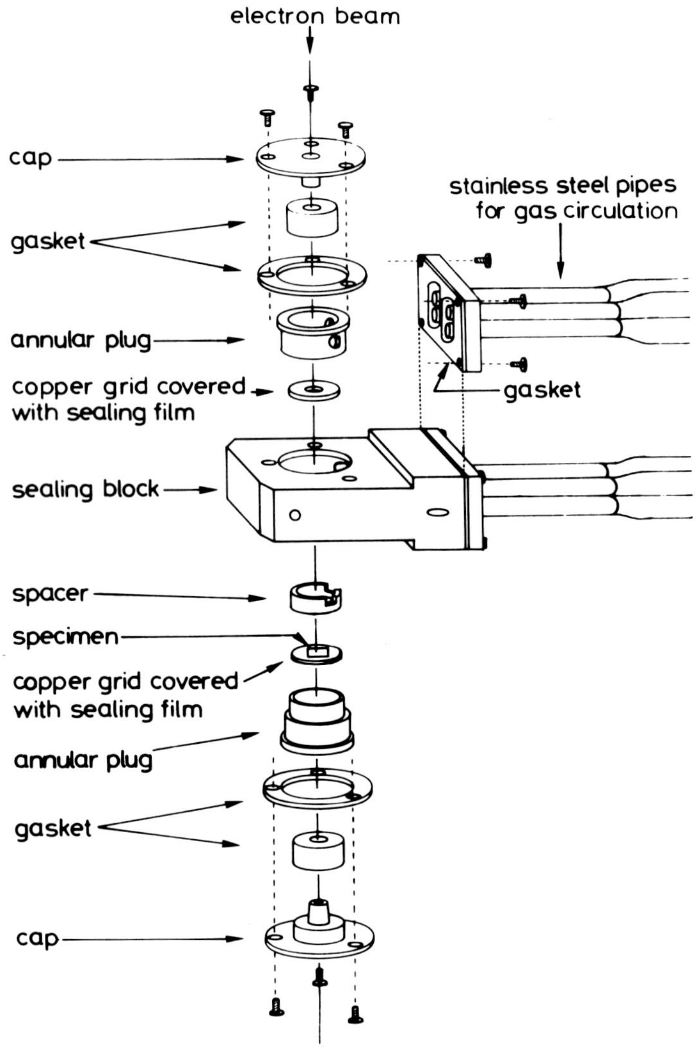

Figure 36.4. Arrangement of film-sealed EC constructed in sealing block.

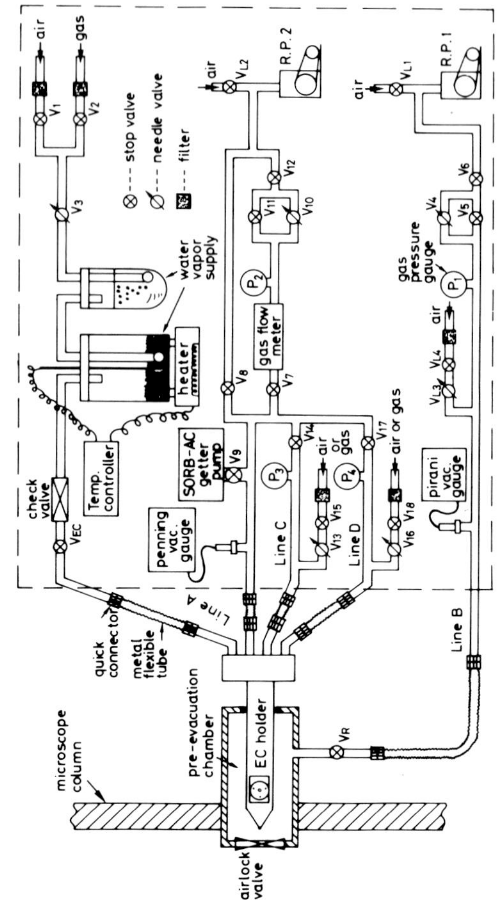

Figure 36.5. Diagram of gas environment control system.

type film broken intentionally is shown in Figure 36.6. It can be clearly seen that many dimples replicated the smoothed microgrid are arranged side by side. Figure 36.7 shows a comparison of the view fields with the new type of film (A) and with the old type of film (B).

Specimen Damage in the EC

Carbonaceous specimens, such as organic macromolecules and many biological materials observed in the EC, are severely damaged by the dissociated products, such as ions or radicals generated by the electron beam striking against gas or water molecules in the EC. A quantitative relation between electron dosage and morphological change of some specimens was measured in different gas environments of O_2, AR, N_2, He, air or water vapor, and in a vacuum (Fukushima et al., 1982). The decreasing rate of radius or width of the specimens was calculated as a function of the removal rate of their surface materials, against irradiating electron beam intensity. It was confirmed that the removal rate depends on the surrounding gas and the beam intensity, and that water vapor in the carrier gas accelerates the removal rate. Hydrated specimens, however, should be placed in a wet gas environment containing enough water vapor to avoid dehydration. Hydrated clay minerals are not as sensitive to such morphological damage in the EC in comparison with biological materials. However, raising the specimen temperature in the wet gas environment by exposure to the electron beam is an important factor in relation to the dehydration of the interlayer water. From the results of a previous study (Kohyama et al., 1978a) of phase transition of hydrated halloysite, it was confirmed that the specimen temperature could be kept below 65°C with the irradiating electron beam intensity of $10^{-4} \sim 10^{-3}$ A/cm² at the specimen plane for periods of 2 min exposure. Under such weak irradiating conditions of the electron beam, although the morphological damage can be neglected, the image becomes so dark

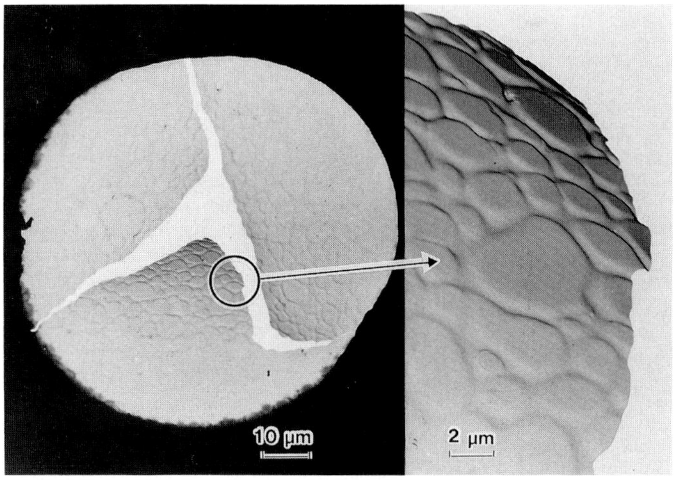

Figure 36.6. Electron micrographs of damaged new sealing film with rugged surface.

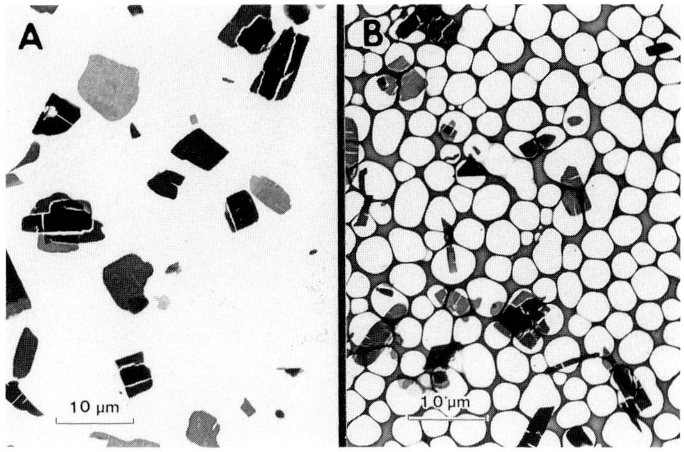

Figure 36.7. Images obtained with new type sealing film (A) and with old type (B).

that we must use highly sensitive photographic materials, such as X-ray film or an image intensifier for recording the image.

Observation of the Wet State of Clay Minerals

Some clay minerals, such as hydrated halloysite, halloysite (10 Å), smectite, interstratified minerals, vermicullite, and imogolite have the interlayer and/or adsorbed water as an essential component when they are in the natural wet state.

The interlayer or adsorbed water, however, is quite susceptible to dehydration in vacuum, in dried air, or at moderately high temperatures (even in wet air). Accordingly, we have observed only a "dried altered feature" instead of the "real feature" of wet and/or hydrated natural specimens when they are viewed by conventional electron microscopy.

To observe interlayer hydrates and complexes of clay minerals by electron microscopy, it is necessary to place the specimens in a special cell that controls water vapor or gas, as mentioned above. For this purpose, we applied the high-resolution EC to fully hydrated particles of halloysite (Kohyama et al., 1978a,b), hydrazine complexes of kaolin minerals including halloysite (Fukushima et al., 1980; Kohyama et al., 1982), and hydrated smectite (Fukushima et al., 1980; Kohyama et al., 1982). In this chapter the results found on fully hydrated halloysite (of tubular and spherical form) and smectite (as well as their layer lattice images) are reviewed.

Halloysite of Tubular and Spherical Forms

Halloysite is well known as a kaolin mineral composed of 1:1 layers, which occurs in different morphological forms (i.e., tubular and spheroidal forms). The main geological occurrence of halloysite is in soils and weathered rocks. The structure and morphology for dehydrated halloysite [halloysite (7 Å)] has been investigated with conventional electron microscopy, and the basic structure of halloysite (7 Å) was established. However, the structure of hydrated halloysite [halloysite (10 Å)] had not been examined directly until the authors studied it (Fukami and Fukushima, 1984; Fukushima et al., 1987) due to the susceptibility of its interlayer water to dehydration. We successfully observed the hydrated form of tubular halloysite in a transmission electron microscope equipped with an EC (Fukami and Fukushima, 1984; Fukami et al., 1987).

Specimen

The observed specimens were two kinds of hydrated halloysite of tubular and spherical forms from Kusatsu, Gunma Prefecture, and Utsunomiya, Tochigi Prefecture, Japan, respectively. Under natural room temperature and humidity, the halloysite (10 Å) showed a strong basal reflection of 10.1 Å by X-ray powder diffraction analysis. The spacing of the basal reflection ranges from 7 to 10 Å depending on the state of hydration. The strong basal reflection of 10.1 Å was largely lost when the interlayer water was dehydrated and the 7.4 Å reflection appeared clearly [halloysite (7 Å)]. This phenomenon also occurred at a temperature slightly higher than usual room temperature as well as in a vacuum of about 1 Torr at room temperature (for a short period). Once the interlayer water of halloysite (10 Å) was lost, it was not restored by treatment with water of ethylene glycol. That is, the dehydration is an irreversible reaction.

Experimental Procedures

A drop of the hydrated halloysite suspension was placed on a copper disk with nine holes covered with thin carbon film as sealing film. One minute was allowed for crystals to precipitate

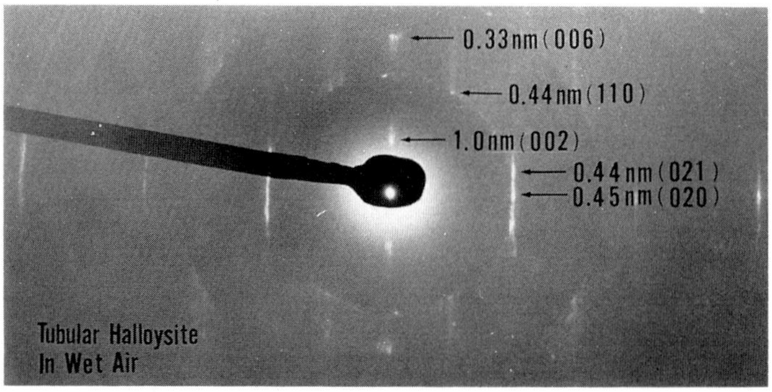

Figure 36.8. SAED pattern for hydrated state of tubular halloysite.

from suspension onto the surface of the film before the EC was inserted into the electron microscope.

Electron microscopic observation was performed by a conventional electron microscope (JEM-7A) equipped with the EC and operated at 100 kV. Generally, in this study, halloysite (10 Å) particles were observed under a water-saturated atmosphere at less than 100 Torr [EC(air)], which was then exhausted to vacuum [EC(vac)]. The same particles were observed in this vacuum. This process is called "dynamic observation."

Results and Discussion

SAED Analysis of Halloysite (10 Å) and Halloysite (7 Å)

SAED patterns of a single crystal halloysite (10 Å) could be obtained under the water-saturated air condition, EC(air), with a current density of approximately 10^{-4} A/cm^2 and exposed for a period of 2 min (Fig. 36.8). This pattern shows no change during exposure for a period of 10 min. Then, "dynamic observation" was applied and the same particle was observed in vacuum. The basal reflections of halloysite (10 Å) were observed clearly at 10.2 Å (Fig. 36.8) and coincided with that of the halloysite (10 Å) observed by X-ray powder diffraction analysis in the natural wet state, which indicated that the SAED analysis was performed without dehydration of the interlayer water of halloysite (10 Å). On the other hand, the SAED pattern obtained in the condition of EC(vac) shows 7.4 and 3.7 Å basal reflections (Fig. 36.9), which shows that the interlayer water in the halloysite (10 Å) was completely dehydrated. The (020) diffraction in both SAED patterns of halloysite (Figs. 36.8 and 36.9) appears at 4.45 Å, which indicates that b_0 = 8.90 Å, and the neighboring diffraction spot appears perpendicular to the b^* direction at about twice the spacing of the first-order basal reflection. The distance between the (020) diffraction spot and its nearest neighbor is about 20 Å for halloysite (10 Å) and is 14.6 Å for halloysite (7 Å). These values are just double the spacing of the first-order basal diffraction spot in each case. Therefore, we index the spot adjacent to (020) and (021) and the first basal diffraction spot as (002). From these results, we could confirm that halloysite (10 Å) and halloysite (7 Å) have monoclinic structures with two-layer periodicity.

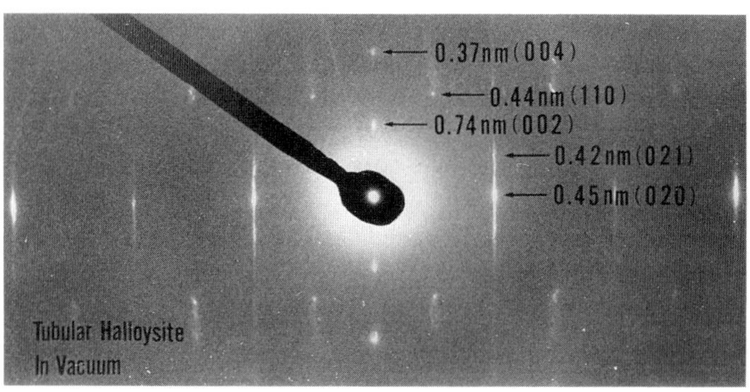

Figure 36.9. SAED pattern for dehydrated state of tubular halloysite.

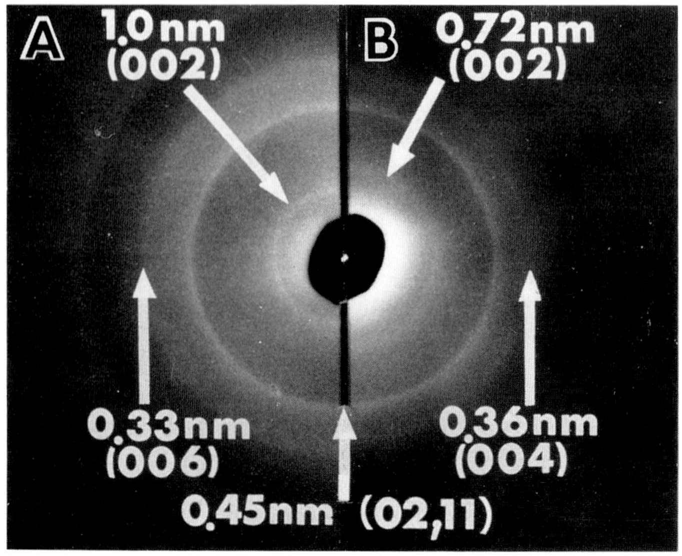

Figure 36.10. SAED pattern for the hydrated (A) and dehydrated (B) state of spherical halloysite.

Previously, two-layer periodicity had been reported by some investigators (Honjo and Mihama, 1954; Honjo et al., 1954; Souza Santos et al., 1965; Chukhrov and Zvyagin, 1966) for halloysite (7 Å). They were, however, all for the dehydrated form because the specimens had been dehydrated in the high vacuum of the electron microscope. In this study, we have obtained SAED analysis for tubular halloysite (10 Å) and have shown a monoclinic structure with two-layer periodicity in both halloysite (10 Å) and halloysite (7 Å). Parameter measurements for these specimens with a gold standard give the following values:

Halloysite (10 Å): $a = 5.14 \pm 0.04$ Å, $b = 8.90 \pm 0.04$ Å
$c = 20.7 \pm 0.1$ Å, $\beta = 99.7°$

Halloysite (7 Å): $a = 5.14 \pm 0.04$ Å, $b = 8.90 \pm 0.04$ Å
$c = 14.9 \pm 0.1$ Å, $\beta = 101.9°$

Figure 36.11. Tubular halloysite of the hydrated state in a wet air environment (A) and the dehydrated state in vacuum (B).

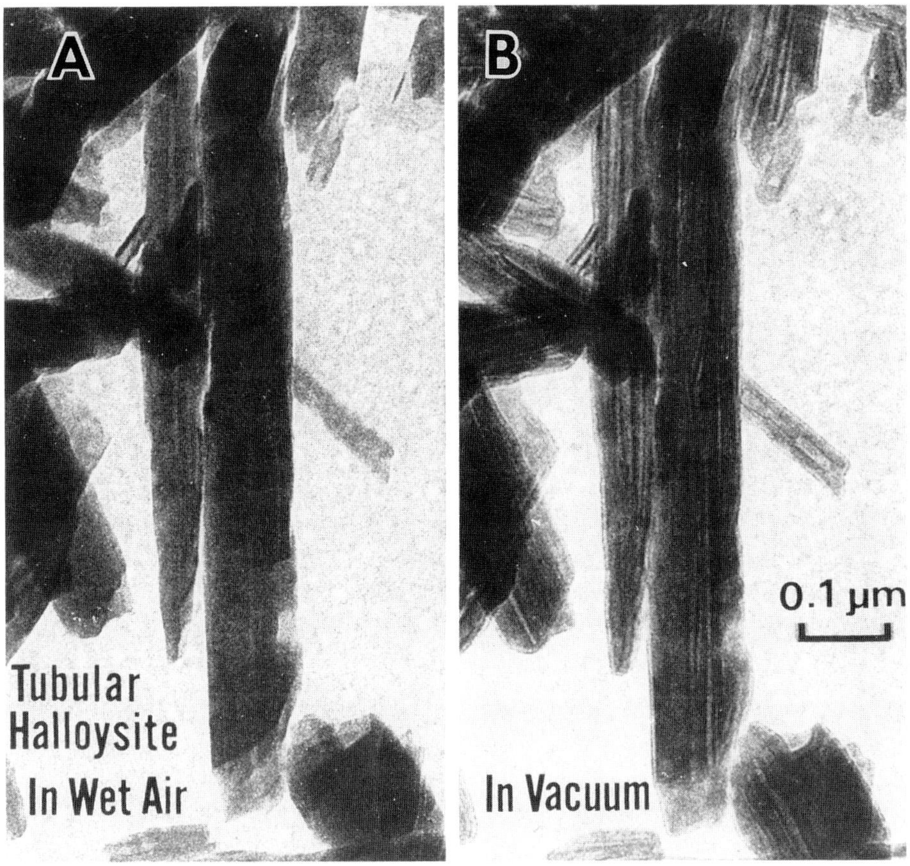

The analogous observation for the hydrated form of spherical halloysite (10 Å) was also observed in the EC(air), in which the 10-Å diffraction ring of the basal plane was detected (Fig. 36.10A) and then changed into the 7-Å ring in vacuum (Fig. 36.10B).

Morphological Changes due to Dehydration

Some morphological changes due to dehydration of interlayer water were clearly observed. Figure 36.11A was observed in wet air atmosphere [EC(air)] and Figure 36.11B in vacuum [EC(vac)] for tubular halloysite. By SAED analysis they were confirmed as the fully hydrated and the fully dehydrated forms of halloysite, respectively. The halloysite (10 Å) particles were observed as smooth and uniform contrast tubes with no recognized stripes except for a light center region. However, in the vacuum many dark and light stripes with widths of about 50–100 Å were observed along the tube axes.

The diameter of the halloysite tubes varies from 300 to 1200 Å with the maximum frequency around 500 Å. The diameter seems to increase with the dehydration of the interlayer water. The diameter in a wet air atmosphere compared with that in vacuum is shown in Figure 36.12. If the diameter did not change, the points would plot on the line of $y = x$ in Figure 36.12. They plot instead on the line of $y = 1.12x$. The apparent increase in the ratio of the diameter in vacuum against the diameter in wet air was estimated around 12%, which corresponds to the dehydration.

The analogous morphological change also was observed for spherical halloysite (10 Å) in this dynamic observation. That is, the clear zonal stripes appeared along the outer shape of the

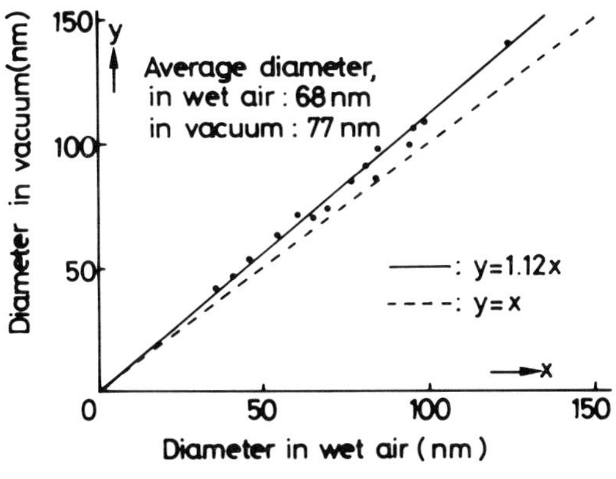

Figure 36.12. The comparison of diameters of tubular halloysite in a wet air environment and in vacuum.

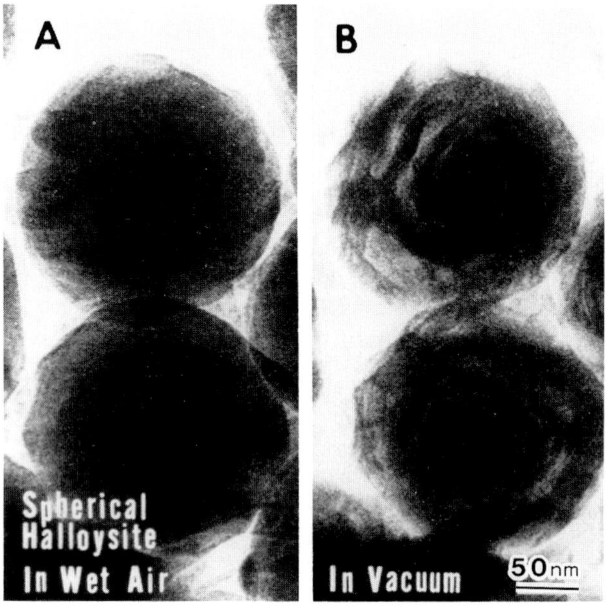

Figure 36.13. Spherical halloysite of the hydrated state in a wet air environment (A) and the dehydrated state in vacuum (B).

Figure 36.15. Layer lattice image of tubular halloysite (10 Å).

particles in vacuum (Fig. 36.13A,B). The diameter of the spherical halloysite (10 Å) however decreased about by 5% with dehydration (Fig. 36.14).

On the basis of these experimental results, the idealized model of morphological changes by dehydration is summarized as follows. When halloysite is fully hydrated, the crystal domains along the *c* axis are tightly connected with each other, but become separated as a result of dehydration, and then can be seen as stripes along the tube axis (tubular halloysite) or along the spherical outer shape showing a zonal contrast (spherical halloysite) in the transmitted image.

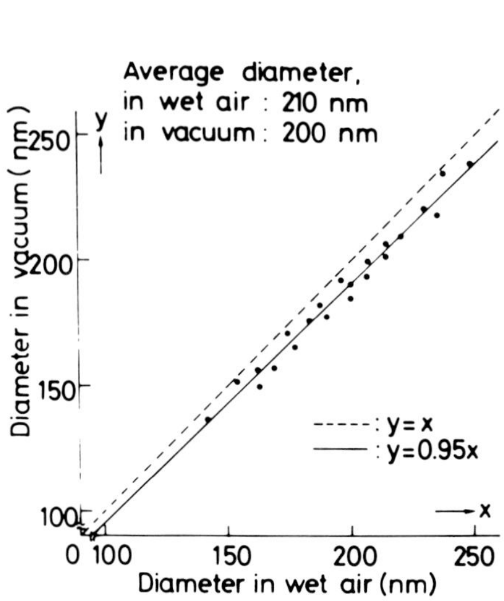

Figure 36.14. The comparison of diameter of spherical halloysite in a wet air environment and in vacuum.

This behavior may be explained by the appearance of many gaps in the tubes and the expansion of the rolled configuration caused by relaxation of inherent strain accompanying the segregation of each crystal domain. On the other hand, the antithetic result for the spherical halloysite showing that the diameter decreased could be caused by the structural differences between tubular and spherical morphologies.

Direct Observation of Layer Lattice Images

The layer lattice image of tubular halloysite (10 Å) was also observed in a wet air environment in the EC (Fig. 36.15). The lattice image of 10 Å was clearly seen to be parallel to the tube direction. The area around the center of the capillary is somewhat dark in contrast. This might reflect the presence of adsorbed water in the capillary in a wet air environment. In vacuum, the dark contrast disappeared and the central capillary was clearly seen in each tube.

Hydrated Smectite

The crystalline structure and morphology of smectite in the hydrated state had not previously been observed electron micro-

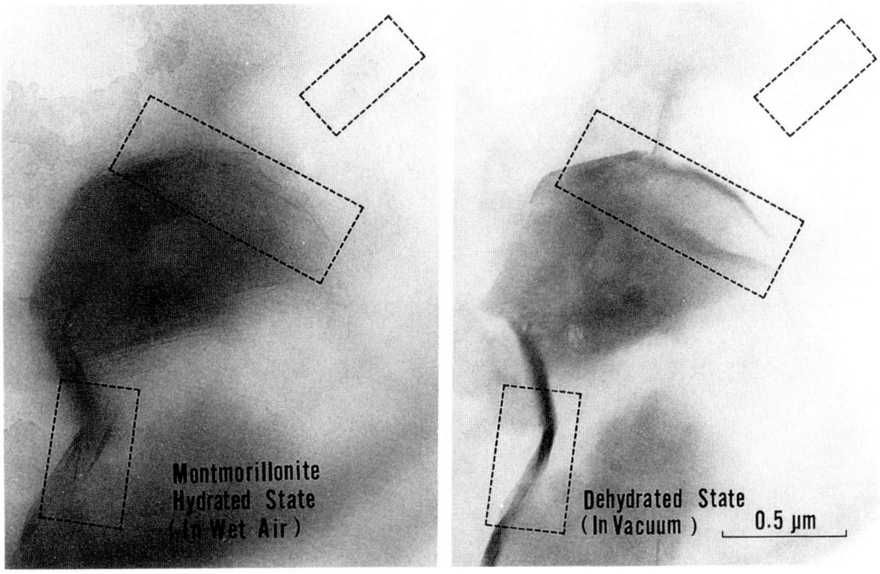

Figure 36.16. Curling of smectite particles due to dehydration.

scopically because the interlayer water is easily dehydrated on evacuation. Wyoming montmorillonite shows a typical aggregate structure under a conventional electron microscope. In these aggregates, thin but wide montmorillonite particles are mostly irregularly folded and have a foliated appearance.

When a smectite specimen with a thin water layer is carefully transferred to the EC, flat and unfolded lamellar particles are observed in a wet air environment (Fig. 36.16A). On evacuation, the flat edges of the lamellar particles gradually curl up and the curly edges that existed in a wet environment curl more as shown in Figure 36.16B.

Figure 36.17 shows basal spacings in wet air and in vacuum of SAED patterns of the smectite observed at the curled edges. The basal spacing changed from 17 to 10 Å during removal of interlayer water by evacuation. As the interlayer water is sensitive to the environment partial or full dehydration easily occurs through evacuation of wet air from the EC. SAED pattern of (hk0) in Figure 36.17, the so-called "net patterns," were not significantly affected by the difference in environmental conditions. The dehydrated smectite could be hydrated again within 30 min by introducing air saturated with water vapor into the EC. This is shown by the increase of the basal spacing to 16 Å

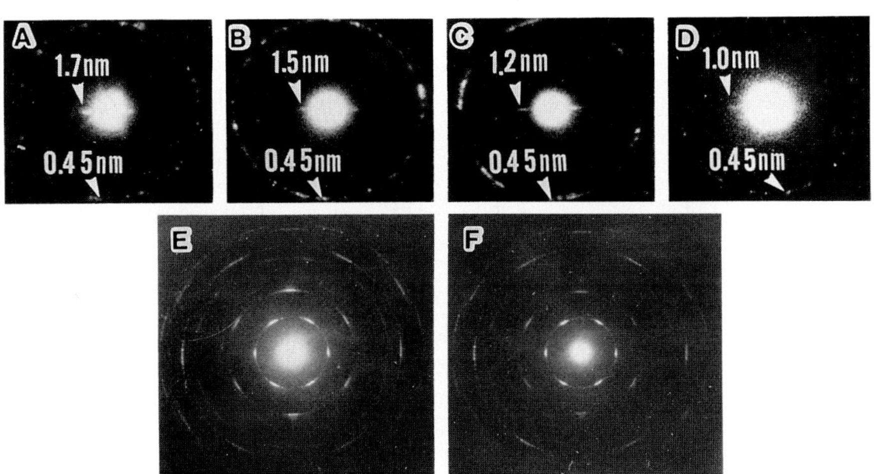

Figure 36.17. SAED pattern of various stages of hydration of smectite. (A–D) (00*l*) spots corresponded to various hydration stages; (E) and (F) net pattern in the hydrated state and in the dehydrated state.

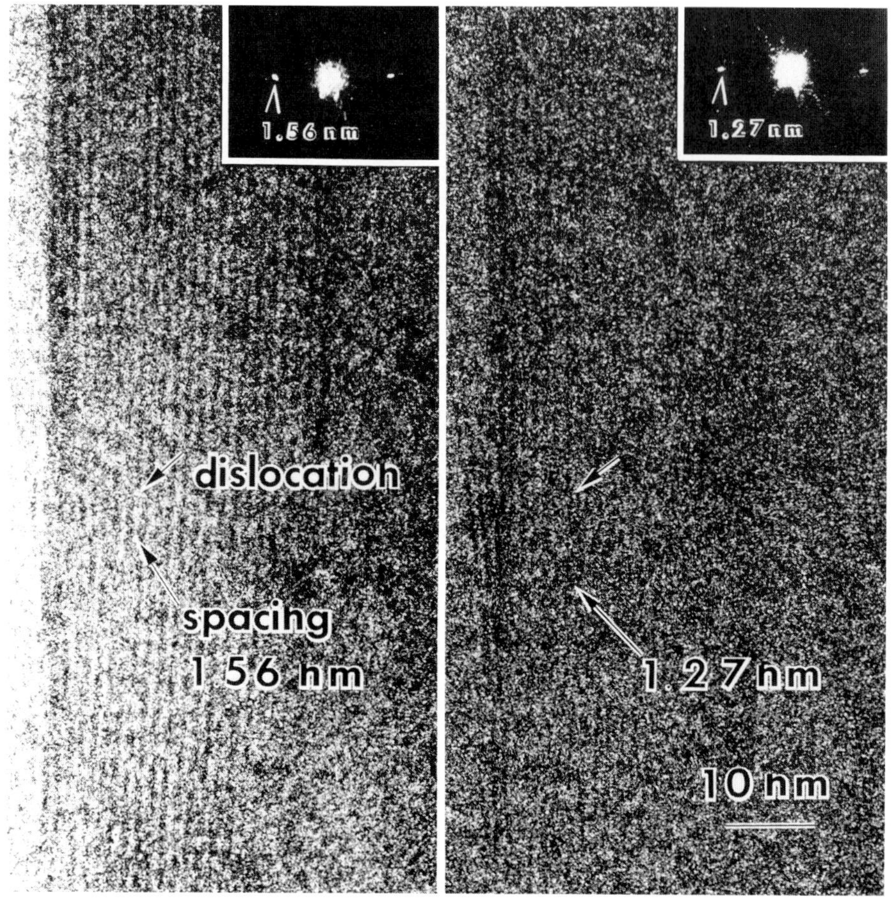

Figure 36.18. Layer lattice images of smectite in a wet air environment (A) and in vacuum (B).

in the SAED pattern and by the increase of the thickness of the curled edges.

Direct Observation of Layer Lattice Images of Hydrated Smectite

The basal diffraction spot of about 15 Å and that of the higher order were detected for the hydrated smectite. Since the EC has high resolution, we tried direct observation of the lattice images using the EC at an accelerating voltage of 100 kV.

The smectite specimen studied was a Ca-beidellite from the Seta mine, Hokkaido, Japan. The layer lattice image of the clay in the hydrated state was observed at the curled edges in a wet air environment in the EC as shown in Figure 36.18A. The optical diffraction pattern clearly shows a basal spacing of 15.6 Å. In vacuum, partial dehydration occurred resulting in a basal spacing of 12.7 Å as shown in Figure 36.18B. Moreover, a defect of lattice such as a dislocation is seen in both environments (arrowhead). This means the crystallinity was not greatly affected by the dehydration process.

Conclusion

Using a film-sealed EC attached to a conventional electron microscope, some hydrated clay minerals were observed. The hydrated state of halloysite showed a smooth contrast in the particles, but the dehydrated state showed a strange contrast along the tubular axis or spherical shape. These are the result of morphological changes by dehydration. The cell parameters of hydrated halloysite were first obtained by electron diffraction in the wet air environment, and the two-layer sequence was confirmed in halloysite (10 Å). Through dynamic observations, curling of smectite particles accompanying dehydration could be shown. Direct observation of layer lattice images was achieved in hydrated halloysite and smectite. The application of the EC to the study of hydrated clay minerals is very useful in examining the natural features of specimens.

Acknowledgment

The authors are grateful to Ms. Jill Yule of Northwestern University for her critical reading of the manuscript.

References

Chukhrov, F.V., and B.B. Zvyagin, 1966. Halloysite, a crystallo-chemically and mineralogically distinct species. Proceeding of the International Clay Conference, Jerusalem, v. 1, p. 11–25.

Fukami, A., and K. Fukushima, 1984. Environmental cell: observation of wet specimens. Proceeding of the 8th European Congress on Electron Microscopy, Budapest, v. 1, p. 71–80.

Fukami, A., K. Fukushima, A. Ishikawa, and K. Ohi, 1987. New side-entry environmental cell equipment for dynamic observation. Proceeding of the 45th Annual EMSA Meeting, Baltimore, p. 142–143.

Fukami, A., K. Fukushima, and A. Ishikawa, 1988. Imaging for chemical reaction process of wet specimens using new type of environmental cell with image processing system. Proceeding of the 9th European Congress on Electron Microscopy, York, v. 1, p. 145–146.

Fukushima, K., N. Kohyama, and A. Fukami, 1980. In-situ observation of clay minerals hydrated and intercalated with liquid chemicals using film-sealed environmental cell. Proceeding of the 38th Annual EMSA Meeting, p. 208–209.

Fukushima, K., M. Katoh, and A. Fukami, 1982. Quantitative measurements of radiation damage to hydrated specimens observed in a wet gas environment. Journal of Electron Microscopy, v. 31(2), p. 119–126.

Fukushima, K., A. Ishikawa, and A. Fukami, 1985. Injection of liquid into environmental cell for in situ observations. Journal of Electron Microscopy, v. 34(1), p. 47–51.

Honjo, G., and K. Mihama, 1954. A study of clay minerals by electron diffraction diagrams due to individual crystallite. Acta crystallo graphica, v. 7, p. 511–513.

Honjo, G., N. Kitamura, and K. Mihama, 1954. A study of clay minerals by means of single-crystal electron diffraction diagram-the structure of tubular kaolin. Clay Mineral Bulletin, v. 2, p. 133–141.

Kohyama, N., K. Fukushima, and A. Fukami, 1978a. Observation of the hydrated form of tubular halloysite by an electron microscope equipped with an environmental cell. Clays and Clay Minerals, v. 26(1), p. 25–40.

Kohyama, N., K. Fukushima, and A. Fukami, 1978b. Hydrated form of some clay minerals observed using a film-sealed environmental cell. Proceeding of the 9th International Congress on Electron Microscopy, v. 1, p. 72–73.

Kohyama, N., K. Fukushima, and A. Fukami, 1982. Interlayer hydrates and complexes of clay minerals observed by electron microscopy using an environmental cell. In: Olphen, H.V., and F. Veniale (eds.), International Clay Conference 1981. Elsevier, Amsterdam, p. 373–384.

Souza Santos, P. de, C.W. Brindley, and H. de Souza Santos, 1965. Mineralogical studies of kaolinite-halloysite clays: Part III. A fibrous kaolin mineral from piedade, Sao Paulo, Brazil, American Mineralogist, v. 50(5,6), p. 619–628.

CHAPTER 37

Clay Fabric of Gassy Submarine Sediments

W.A. Chiou, William R. Bryant, and Richard H. Bennett

Introduction

Objectives

The primary objective of this research was to delineate clay fabric microfeatures in sediment samples obtained by the use of a pressure core barrel. A secondary objective was to evaluate the difference in clay fabric between pressure core barrel samples and conventional core samples, and to enhance the understanding of the clay fabric of fine-grained gassy sediments.

Historical Review and Previous Works

Geologists have long been interested in the fabric of sediments, in particular the fabric of clastic sediments, but chiefly on large grain-sized particles such as sands, gravels, and tills. Study of the fabric of fine-grained argillaceous sediments has been in abeyance due to the extremely fine texture and complex composition of these sediments, and the necessity of high-resolution techniques to delineate individual particles. Although Henry Clifton Sorby, the father of sedimentology, presented the first microscopic study of rocks by means of thin section as early as 1851, it appears to be true, as Burnham (1970) stated, "Many geological authors . . . have been content to describe the large particles, and to dismiss the finer materials as a featureless intergranular paste of very finely crystaline materials."

The early interest in clay fabric study arose from attempts to determine the relationship between clay particle arrangement and the mechanical properties of soil. The early concepts of clay fabric were conceived by Terzaghi (1925), Goldschmidt (1926), and Casagrande (1932). Since then, clay fabric research has been carried out sporadically primarily by civil engineers, soil scientists, and geologists. A very comprehensive and informative literature review tracing the history of clay fabric study, from early concept through current knowledge to the future perspective, has been published by Bennett et al. (1977) and Bennett and Hulbert (1986). Moon (1972) also reviewed some of the early clay fabric concepts, particularly in the microstructure of fresh clay and compacted clays. A large collection of different soil fabrics accompanied with a comprehensive review and description of sample preparation techniques, including techniques for quantifying the results was published by Smart and Tovey (1981, 1982).

During the last 15 years, due to the realization that the fabric of a clay fundamentally affects a soil's mechanical behavior, and to the availability of the electron microscope as a tool to study clay microfeatures, research on clay fabric has been gaining momentum significantly. Excellent studies were presented at several international meetings, such as the Southeastern Asian Conference on Soil Engineering in 1970 and 1971, the Roscoe Memorial Symposium (Parry, 1971), the Third International Conference on Expansive Soils in 1973, the International Symposium on Soil Structure (Barden and Pusch, 1973), the Fourth International Working-Meeting on Soil Structure (Rutherford, 1974), a symposium held at the Third Meeting of Geological Societies of the British Isles (Whalley, 1978), and the 8th International Clay Conference (Schultz et al., 1987). Papers presented in these meetings demonstrated the significance of using electron microscope techniques, and the important role that clay fabric plays in the mechanical behavior of soils.

The published record contains few reports of studies dealing with the fabric of natural sediments in contrast to numerous studies of laboratory-prepared sediment samples. The use of laboratory-prepared pure clay specimens with different microfabric features could lead to misinterpretation of the mechanical behavior of natural soils (Collins and McGown, 1974). Furthermore, during the past few years, several geologists have begun to focus their attention on the relationships between clay fabric and depositional environments (O'Brien and Hisatomi, 1978; Bennett et al., 1977, 1980; Chiou et al., 1980; O'Brien et al., 1980; Shephard et al., 1980, 1982), particularly in solving sedimentary facies problems. Thus, it is important that specific, quantitative studies on natural sediments from different environments be carried out to fully understand the nature of the fabric in a natural environment. However, due to the limitations of both coring devices and laboratory instruments, undisturbed clay fabric has been difficult to obtain. This is especially true for deeply buried, gassy sediments in which clay fabric may have changed due to the release of hydrostatic pressure. Thus, it is important to determine the clay fabric of gassy sediments that have not been affected by the loss of downhole pressure, and to compare these results with those of studies performed using conventional methods.

Theoretical Considerations

Definitions and Terminology

As a result of the increasing number of clay fabric studies in different scientific disciplines, and the broad variety and complexity of fabrics that have been observed, there has been a proliferation of terms for the description of fabrics and fabric features. To prevent an overlap in meaning of terms with previous investigators, a brief explanation of a few key terms is given herein, and will be used in presentation of the results and interpretations in this report.

Clay fabric: Fabric, as used by sedimentary petrologists, refers to "the orientation in space of the elements of which a sedimentary rock is composed" (Gary et al., 1972). A fabric element of a sedimentary rock may be a single crystal, a detrital fragment, a fossil, or any component that behaves as a single unit with respect to an applied force (Fairbairn, 1949). Thus, fabric constitutes three-dimensional patterns in space, and includes factors such as packing, boundary relationships, discontinuities, grain size, or the presence or absence of a matrix, shape and roundness of particles, and orientation. In this study, the fabric elements are composed mainly of clay-size particles (<4 μm). Thus, clay fabric refers here to the spatial distribution, orientation, and particle-to-particle relationships of the <4-μm solid particles (mainly clay minerals) in the sediment (Bennett, 1976).

Domain: A stack of face-to-face or slightly stepped face-to-face parallel clay plates (Aylmore and Quirk, 1960, 1962; Bennett et al., 1977). It is essentially the same as the so-called book or packet structure of Sloane and Kell (1966). An array of such aggregates is referred to as turbostratic fabric (Biscoe and Warren, 1942; Olsen, 1962). Yong and Sheeran (1973) describe the grouping of domains or aggregates into larger fabric units as clusters.

Floccule or floc: A well-defined clay aggregate composed of several particles or domains having spatial arrangements and particle contacts that produce relatively large intravoids relative to the thickness of the individual particles that compose the floc (Bennett et al., 1977). In general a floc consists of a very porous network of randomly oriented clay flakes or clumps of flakes. Also, it may be composed of numerous face-to-face flocculated flakes arranged in a cluster in a stairstep fashion (O'Brien, 1971).

Chain: A series of clay particles or domains that link together in stepped face-to-face and/or edge-to-edge contact (Bennett et al., 1977). Chains normally appear to be long and continuous. A three-dimensional network of twisted chains of clay platelets having a stepped face-to-face association is termed stairstep cardhouse fabric (O'Brien, 1971).

Particle: A well-defined entity that is resolved by electron microscopy. It can be a single clay platelet or several platelets forming a domain. The particle can be thought of as the elementary unit or building block of clay fabric (Bennett et al., 1977).

All the terms described here and applied in this research are strictly descriptive and do not imply mode of formation.

Assumptions

Clay fabric research involves sampling, sediment drying, embedding, ultrathin sectioning, and the use of various electron microscopy techniques. Due to the soft gassy nature of the sediments in question and some inherent limitations of the techniques, several assumptions must be made:

1. No mineralogical changes result from the slight temperature but relatively low pressure change and the substitution of interstitial water with different fluids (ethyl alcohol and amyl acetate) during the replacement process prior to critical point drying. Studies reported by Range et al. (1969) and results of mineralogical studies by the author from Lee (1980) have demonstrated the stability of clay minerals under similar temperatures and pressures as used in this study.
2. No significant fabric changes result from the build-up of pressure in the critical point-drying chamber or high local pressure in the specimen. To prevent this type disturbance, the final stage of the critical point-drying procedure was performed very slowly as suggested by Tovey (1970) and should have little effect on fabric features (Tovey and Wong, 1973, 1978; Wong, 1975; Smart and Tovey, 1982).
3. No fabric disturbance results while substituting interstitial water with ethyl alcohol, amyl acetate, and liquid CO_2 before critical point drying.
4. Little compression occurs between the diamond knife and specimen (i.e., no plastic deformation occurred) while ultrathin sectioning the specimen. Any possible disturbance by compression has been minimized by adjusting the section condition (specimen position, cutting speed, diamond knife position, etc.) and by applying xylene vapor over the ultrathin sections to relieve any possible compression.

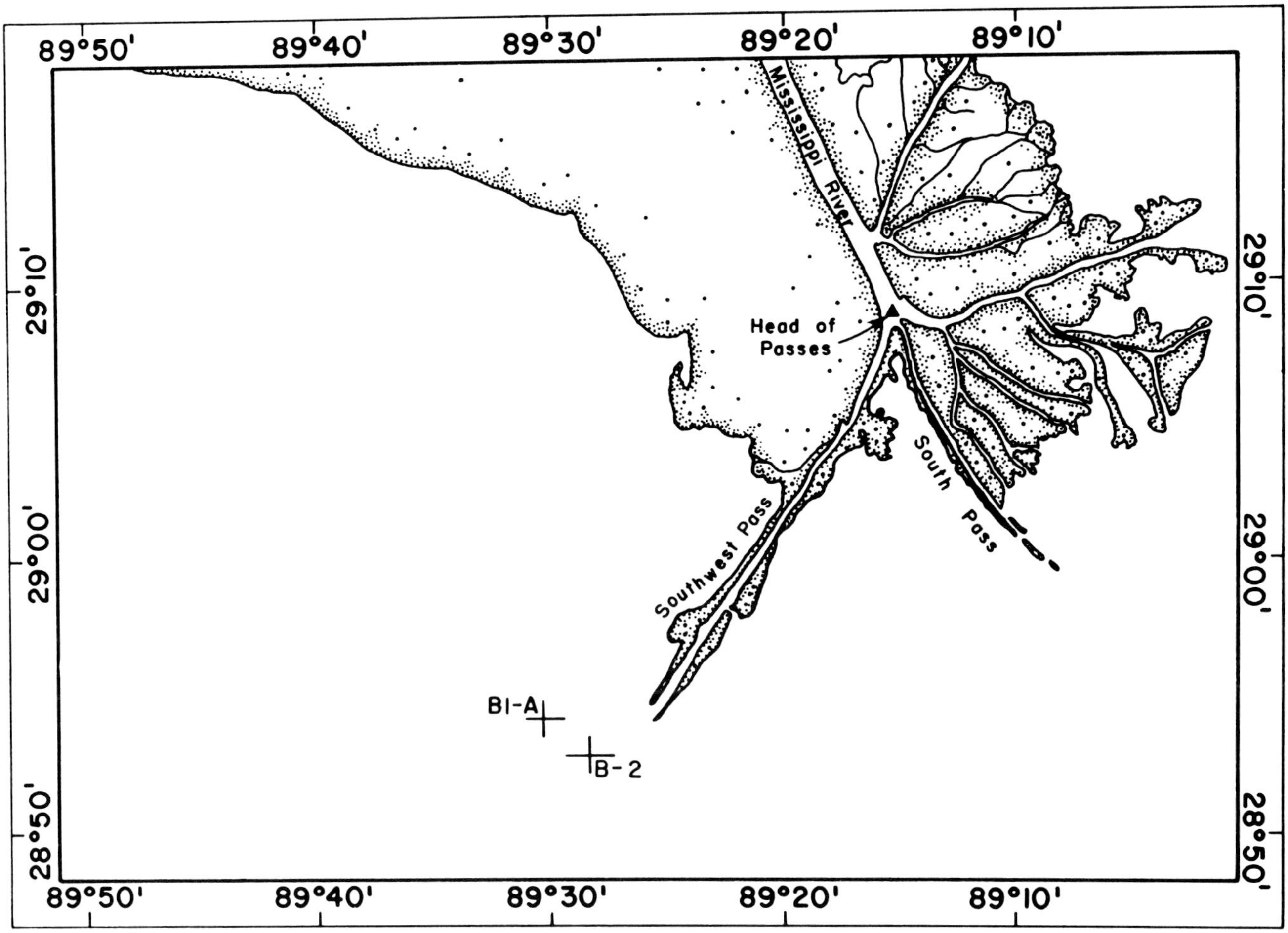

Figure 37.1. Index map showing sampling site locations.

5. No fabric disturbance results from the possible difference between the measured downhole pressure and actual *in situ* pore pressure if excess pore pressure existed. It has been assumed that the difference between *in situ* total pore water pressure and the measured downhole pressure was insignificant in terms of fabric changes.
6. No electron-optical or subsequent optical distortion of the image in the measurement and calculation of clay fabric orientation analysis exists.

Materials and Experimental Methods

Sample

General

Samples studied in this research were obtained using a "pressurized core barrel" sampling device that was developed at the Marine Geotechnical Laboratories, Department of Oceanography and Department of Civil Engineering of Texas A&M University. Detailed specifications and procedures for this sampling device are presented by Denk et al. (1981).

Essentially, this coring device seals sections of sediments in a core barrel and retains the sediment at its *in situ* downhole pressure. For further protection against pressure loss, downhole pressure is applied to the interior of the core barrel during withdrawal from the bore hole. This pressure is determined after the sample had been taken by applying helium gas pressure through a tube inserted into the interior of the core barrel just above the sample tube. Using a pressure regulator, the gas pressure is increased until a no-flow condition exists as indicated by a precise flow meter in the line. At this point the applied gas pressure just balances the downhole pressure (Denk et al., 1981). The sample should be maintained at the *in situ* pressure, but in reality, the pressure being maintained may well be that of the column of drilling fluid. The *in situ* downhole pressure on the cored sample was maintained during transport from the field to the laboratory and storage until laboratory analyses were made.

Table 37.1. Location and general information of sampling sites.*

Sample no.	Core no.	Location		Water depth		Sample depth below mudline		Measured downhole pressure	
		Latitude	Longitude	(m)	(ft)	(m)	(ft)	(kPa)	(psi)
1	B-1A	28°52′54″	89°28′39″	21	69	4.9	16	324	47
2	B-1A	28°52′54″	89°28′39″	21	69	7.9	26	365	53
3	B-1A	28°52′54″	89°28′29″	21	69	11.3	37	395	57
4	B-2	28°54′13″	89°30′10″	38	125	2.7	9	434	63
5	B-2	28°54′13″	89°30′10″	38	125	5.8	19	483	70
6	B-2	28°54′13″	89°30′10″	38	125	8.8	29	510	74
7	B-2	28°54′13″	89°30′10″	38	125	11.9	39	559	81

*The Lambert coordinates of Core B-1A: X = 2,594,001; Y = 82,970.
The Lambert coordinates of Core B-2: X = 2,585,823; Y = 90,832.

Location

The samples used in this study were obtained from an area located offshore Louisiana near the mouth of Southwest Pass within the Mississippi Delta complex (Fig. 37.1). The locations, water depth, sample depth, and measured downhole pressure of each sample are given in Table 37.1.

Geological Background

The birdfoot deltaic complex of the Mississippi River is situated on the continental shelf offshore Louisiana. The delta front represents the most recent area of deltaic sedimentation. Sediments are prograding seaward depositing clays and silty clays in the prodelta environment. During the past three decades, the geology, clay mineralogy, geochemistry, and geotechnical properties of these Mississippi deltaic sediments have been extensively studied. The present delta is probably one of the most thoroughly investigated marine environments in the world. Broad and extensive literature reviews of previous work in the Mississippi deltaic complex can be found in Shephard et al. (1979), Trabant and Bryant (1979), and references cited in these reports. Nevertheless, the only clay fabric research that has been carried out in the Mississippi Delta area was performed by Bowles, et al. (1969) and Bennett and associates (Bennett and Bryant, 1976; Bennett et al., 1977, 1979; Bohlke and Bennett, 1978, 1980).

Clay Fabric Analyses

Sample Preparation

The most critical steps in preparing samples for electron microscopic studies are the techniques employed in the dehydration of wet specimens and the process of embedding a specimen with an appropriate medium.

To accomplish this study, a special apparatus, "pressure vessel" (Fig. 37.2) for replacing interstitial water with intermediate fluid before critical point drying under equivalent *in situ* downhole pressure, was constructed. The detailed description of the special apparatus and procedures of drying specimens were presented in earlier papers (Chiou, 1980, 1981). Critical point-dried specimens were then embedded with a very low viscosity epoxy resin (SPURR) under vacuum, cured, and ultrathin sectioned (800–1000 Å thickness) with a diamond knife on a Sorvall MT-2 ultramicrotome for transmission electron microscopy (TEM) study. Detailed techniques of specimen embedding, thin sectioning, and ultramicrotoming, and electron microscopy [both TEM and scanning electron microscope (SEM)] observation were presented by Bennett et al. (1977), Baerwald et al. (this volume), Chiou (1981), and Chiou et al. (this volume).

To evaluate the difference in clay fabric between pressure core barrel samples and conventional core samples, sediments adjacent to those selected for the proposed pressurized clay fabric study were depressurized and then prepared at ambient pressure and analyzed by the same method (as that used for pressurized samples).

To ensure a more comprehensive and representative study of clay fabric sediment, specimens were cut as shown in Figure 37.3, so that the clay fabric of sediments could be viewed from different orientations of the sample relative to the core, i.e., side (thin sectioning parallel to core axis, A in Fig. 37.3), top (thin sectioning normal to core axis, B in Fig. 37.3), and random (C in Fig. 37.3). However, it was very difficult to maintain the precise orientation of ultrathin sections because of the difficulties encountered during sample preparation such as subsampling, embedding, and ultrathin sectioning. The orientation of ultrathin sections given here is thus rather rough, and is only an approximate orientation at best.

Orientation Analysis

The orientation of clay particles was based on the measured elongation direction of grain projection. To provide the simplest, fastest, and most accurate measurement of fabric orientation,

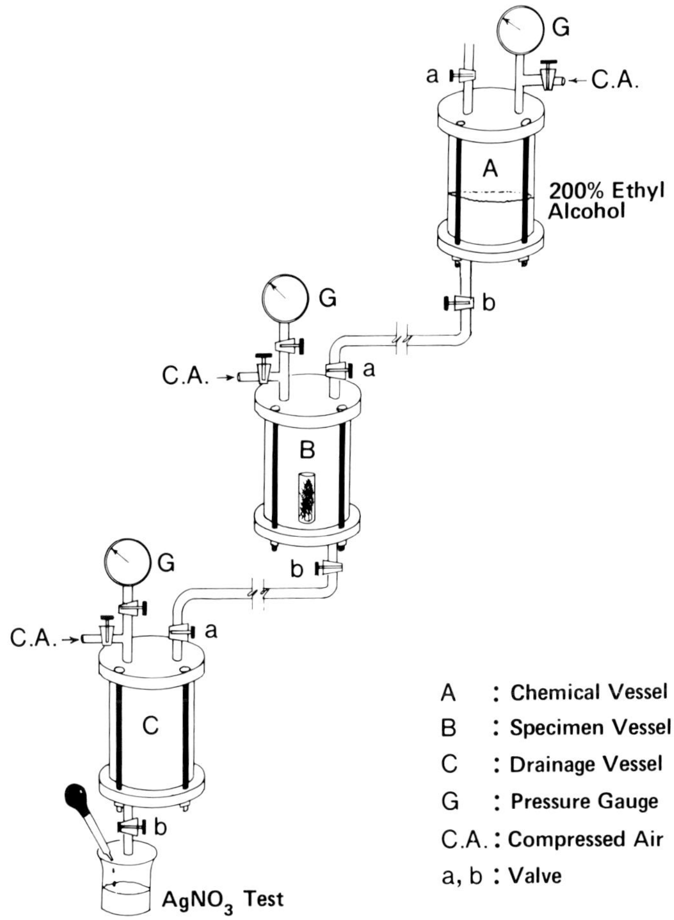

Figure 37.2. General arrangement of pressure vessels for substituting sediment interstitial water with intermediate fluid.

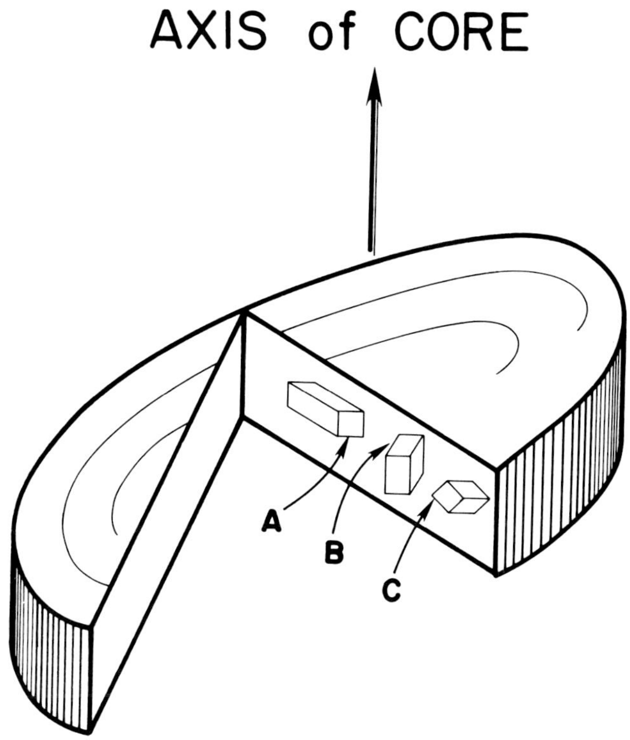

Figure 37.3. Schematic diagram showing the viewing orientation of embedded specimen. (A) Side or parallel to core axis view. (B) Top or normal to core axis view. (C) Random view.

the point counting method was used in this study. The detailed counting technique followed those of Chiou (1981) and Chiou et al. (this volume).

Designation of Sample Number

Different processing procedures and variations in sediment orientation for ultrathin sectioning necessitated a systematic designation of sample number. Samples designated with postfix (A) were prepared and dehydrated using the new technique described herein (i.e., sample dehydration was carried out entirely under the equivalent *in situ* downhole pressure). Sample postfix (B) was designated for samples that were dehydrated only partially under equivalent *in situ* downhole conditions, i.e., after changing the samples' interstitial water with absolute ethyl alcohol. The samples were then exposed to ambient pressures. Sample postfix (C) was designated for samples that were dehydrated after *in situ* downhole pressures were released, i.e., the conventional method as described by Bennett et al. (1977).

The capital letter without parentheses following the sample postfix indicates the orientation of the thin section observed in the transmission electron microscope. For example, sample #1 (A)-B means that this is the sample number 1 (Table 37.1) that was prepared following the new technique described in Chiou (1980, 1981) and the picture is viewed from the top of the core, i.e., the ultrathin section was cut perpendicular to the sediment column (or the axis of the core).

In addition, due to experimental failure, some samples may have the capital letter F behind A, i.e., (A-F), which means the proposed (A) method failed to work correctly due to an accident. Where the technique has not been completed successfully, the reasons for failure and the treatment after failure will be described.

Results

Lithology and Grain Size Analysis

The general lithology of the cores used in this investigation and the results of selected geotechnical property tests are shown in Figures 37.4 and 37.5.

The results of grain size analyses were performed at the same time the sample was fractionated for the clay mineralogy study,

Figure 37.8. Clay fabric of sample 1(A). Top: SEM micrograph showing the porous nature of the sediment in low magnification. The oval-shape depression is thought to be created by a large gas bubble. Bottom: SEM micrograph showing well-oriented clay fabric from the wall of the gas bubble-like depression.

study appears to be preserved fairly well. At low magnification clay particles are aligned in a general direction. Clay orientation and frequency distribution curves also indicate the material consists of fairly well-oriented clay particles over a relatively wide angle range. However, in some areas of the thin section, the clay fabric shows slightly random arrangement. Figures 37.15 and 37.16 illustrate these combined features. It also was found that clay fabric near large pore spaces seems to lie more randomly. In general, results of these two samples show mixed microfeatures of fairly oriented with some slightly randomly arranged clay fabric.

Samples 4(A-F) and 6(A-F) had been degassed to ambient pressure for 2 and 5 hr, respectively, during the dehydration of these specimens, although the original *in situ* downhole pressures were reapplied after degassing. Both TEM and SEM observations of these long degassed specimens show randomly arranged clay domains and clay platelets as compared to the others, even though some areas may consist of mixed microfeatures (Figs. 37.17–37.19). Fairly large void spaces and other swirl and channel-like features are also present. The framework of clay fabric in these long degassed samples is similar to those prepared by method (C).

Discussion

The Sediment

Based on the grain size, bulk mineralogy, and clay mineralogy analyses, the constituents of the sediments were the same throughout the two cored sections, although sample Nos. 1, 2, 4, and 6 showed a slightly higher weight percentage of coarse silt fraction than the others. Neither grain size analyses nor bulk and clay mineralogical analyses showed significant variations in these samples. The grain size distribution and mineralogical composition of the seven samples were very uniform, and the sediment can be classified as silty clay or mud (Folk, 1974).

There are several factors, both physical and chemical, that can influence fabric microfeatures of a clayey sediment. In this study, the physical factors such as grain size, bulk mineralogy, clay mineralogy, rate of deposition, and depth of burial (*in situ* downhole pressure or effective pressure) were taken into account. Because of sediment uniformity in the relatively short cores, and the high deposition rate during Pleistocene time, the sediment can be considered as being homogeneous. The clay fabric in the different samples discussed has been subjected to similar geological conditions and the relationship between the clay fabric and the effects of the *in situ* downhole pressure on the sediment can be compared.

Comparison of Clay Fabric Using Conventional and New Techniques

The clay fabric of samples prepared by method (B) or (C) of this research was similar to the previous studies of Mississippi Delta sediments (Bowles, 1969; Bowles et al., 1969; Bennett et al., 1977; Bohlke and Bennett, 1980). Bennett et al. (1977) found random arrangement of clay particles predominating over the greater portion of the ranges in void ratio in shallow buried sediments in the northwestern Gulf of Mexico and the Mississippi Delta. Bennett did not encounter noticeable preferred particle orientation in sediments buried less than 100 m in the Mississippi prodelta. These clay fabric observations appeared similar to those in sediments where the *in situ* downhole pressure was released before dehydrating the sample. Samples studied by Bennett had low to negligible gas content, thus, the comparison of results from this study to those of Bennett's is rather difficult.

In contrast, preferred orientation of clay particles was revealed by the new sample preparation method, although the

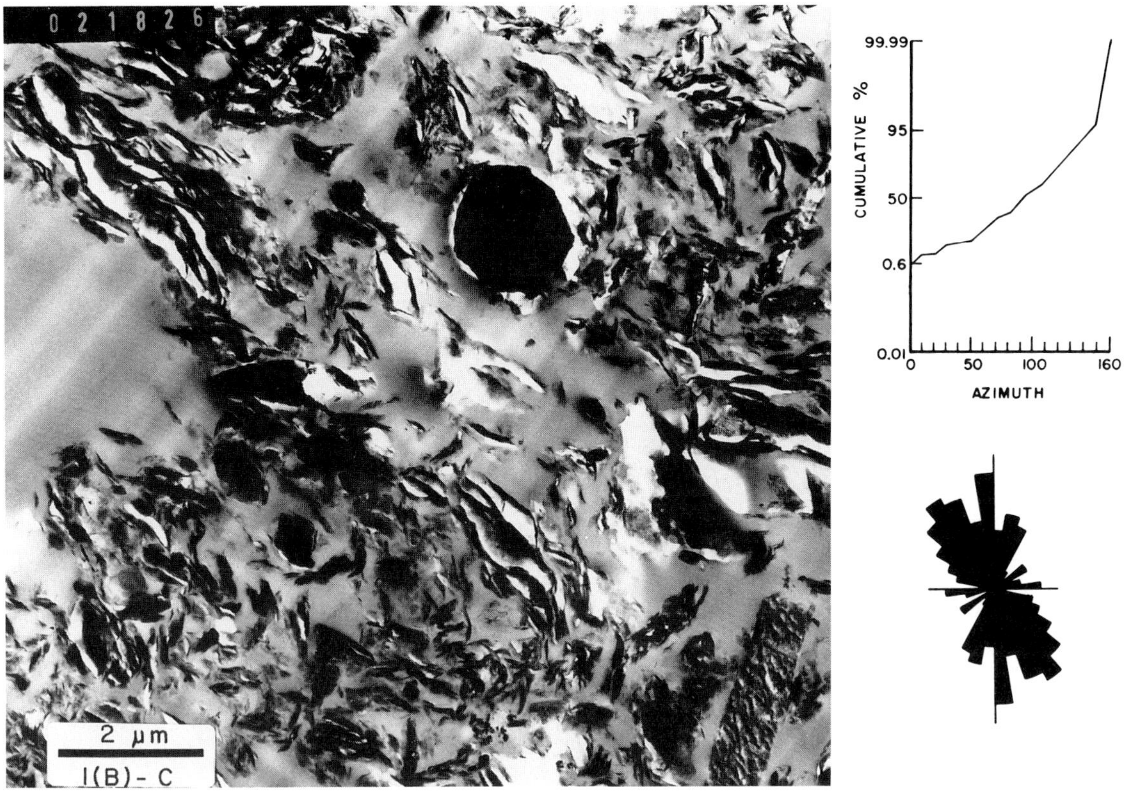

Figure 37.9. Clay fabric and orientation analysis of sample 1(B)-C. Left: Low magnification TEM micrograph depicting nonoriented clay fabric and large voids in the sediment. Upper right: Gentle slope of clay orientation distribution curve indicating randomly arranged clay fabric. Lower right: Rose diagram of clay fabric from this TEM micrograph (021826).

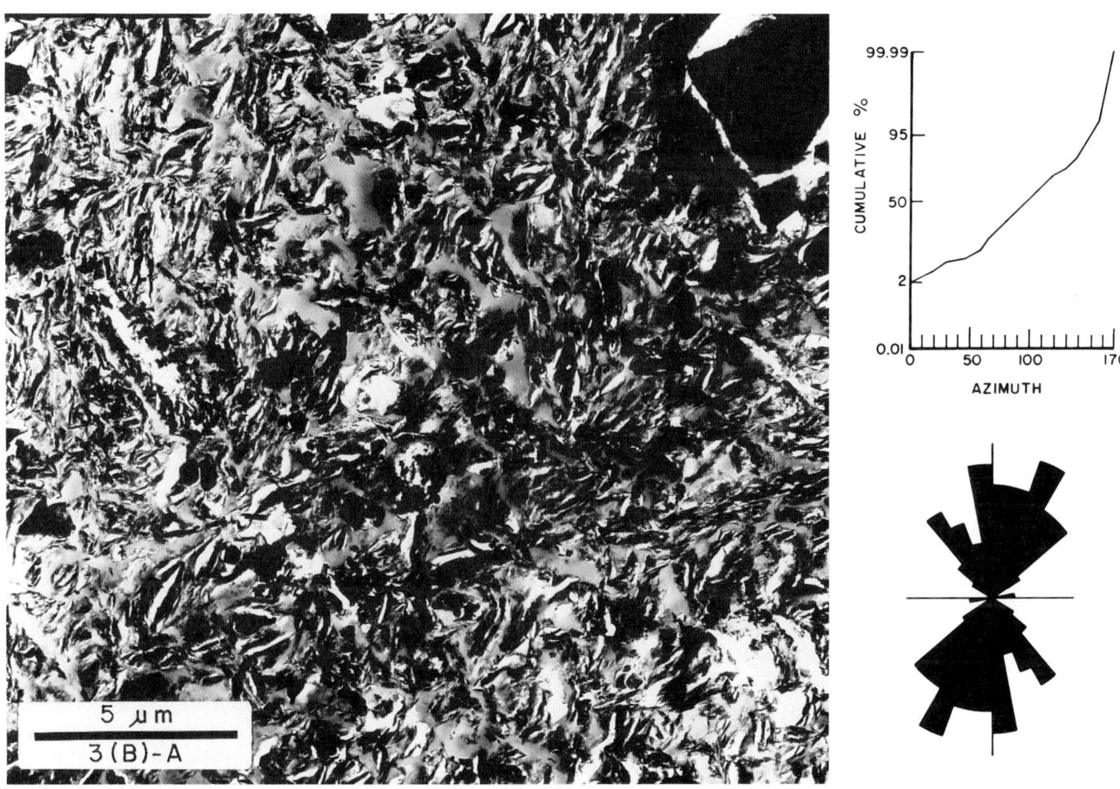

Figure 37.10. Clay fabric and orientation analysis of sample 3(B)-A. Left: Low magnification TEM showing the randomly arranged clay particles and domains spread through this specimen. Upper right: Clay particle orientation distribution curve revealing the typical randomly fabric. Lower right: Rose diagram showing clay orientation in the TEM micrograph.

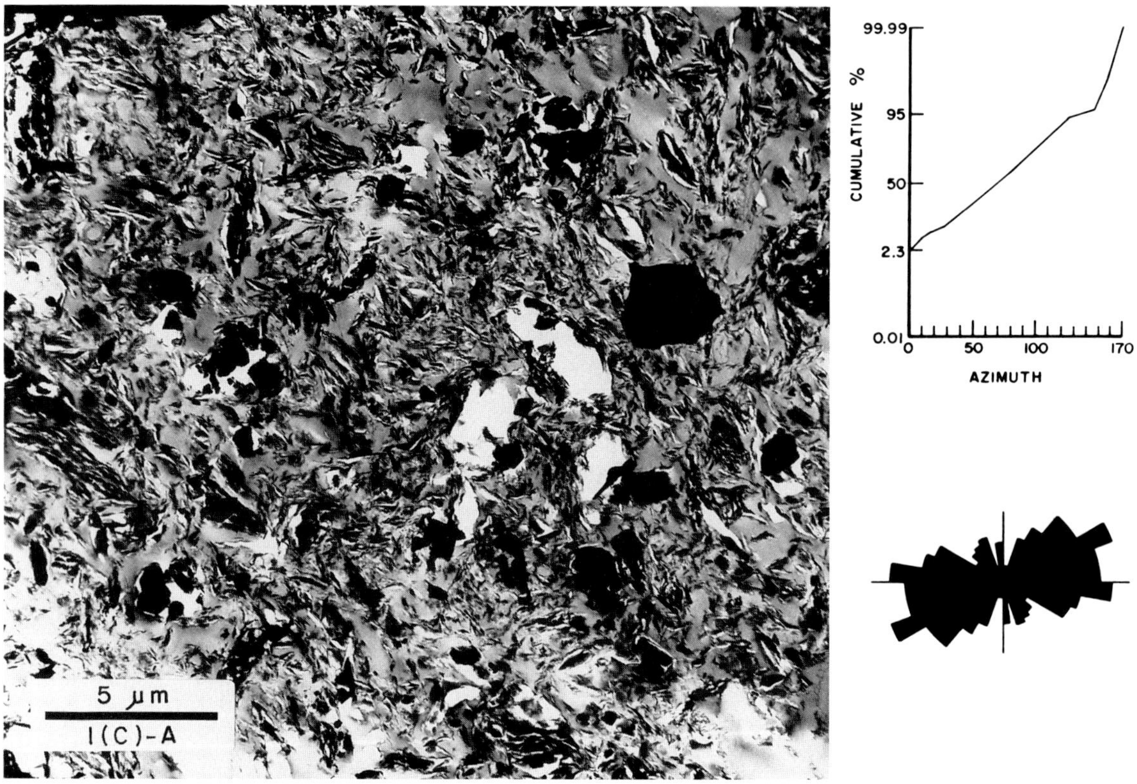

Figure 37.11. Clay fabric and orientation analysis of sample 1(C)-A. Left: Low magnification TEM micrograph revealing highly disoriented clay fabric. Upper right: Clay fabric analysis showing widely distributed clay orientation, i.e., poorly oriented fabric. Lower right: Rose diagram showing the same result of the same micrograph.

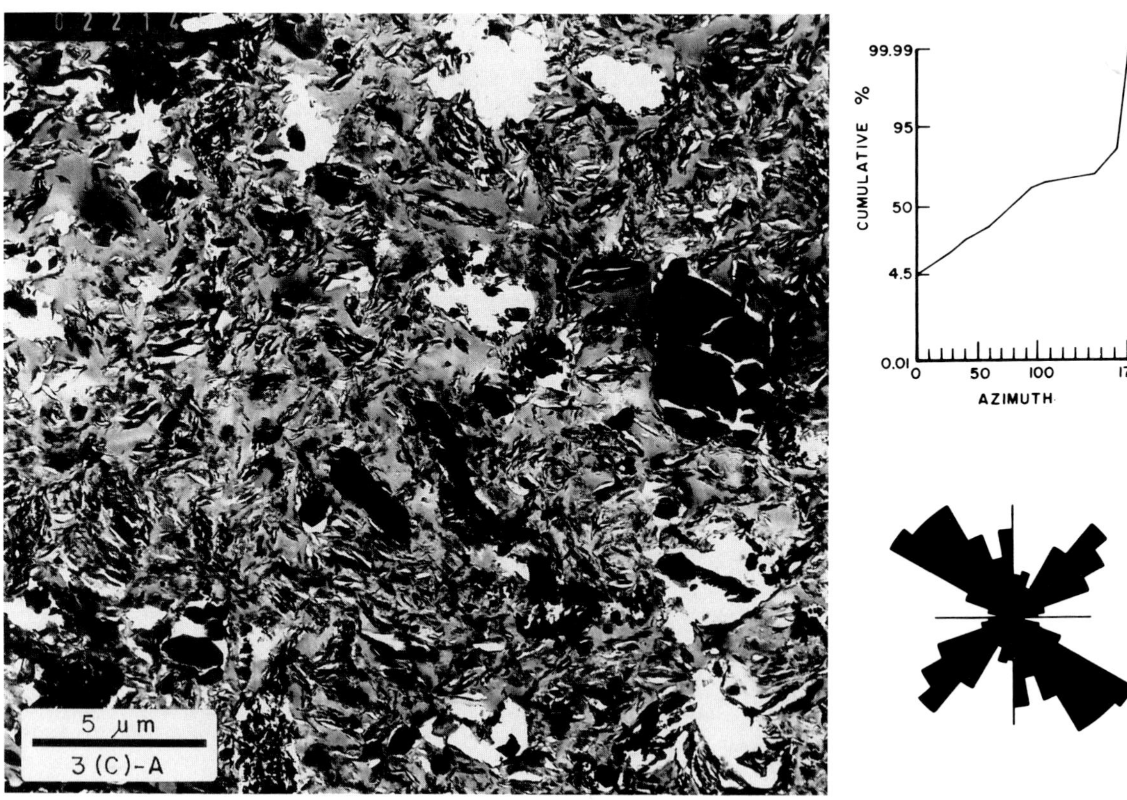

Figure 37.12. Clay fabric and orientation analysis of sample 3(C)-A. Left: Low magnification TEM micrograph depicting highly nonoriented and slightly swirled clay fabric. Upper right: Widely distributed clay orientation from the fabric measurement of TEM micrograph 022145. Lower right: Rose diagram showing randomly arranged clay particles in the TEM micrograph 022145.

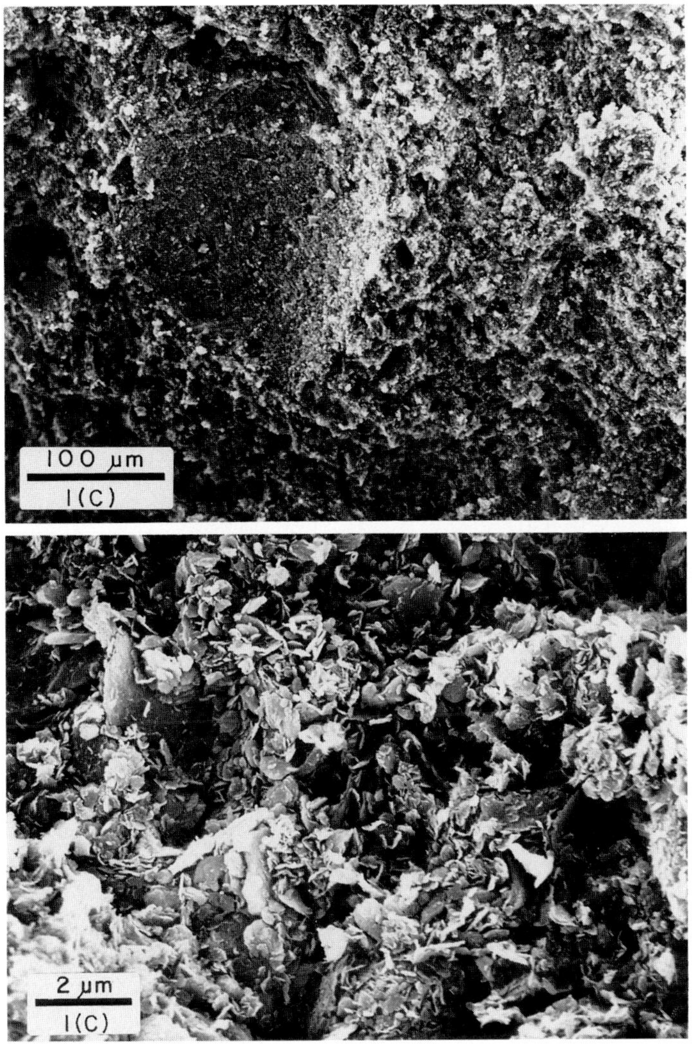

Figure 37.13. Clay fabric of sample 1(C)-A. Top: SEM micrograph of the porous sediment. No difference observed from Figure 37.8a. Bottom: Higher magnification of the "depression" area of the top micrograph revealing the randomly arranged clay fabric.

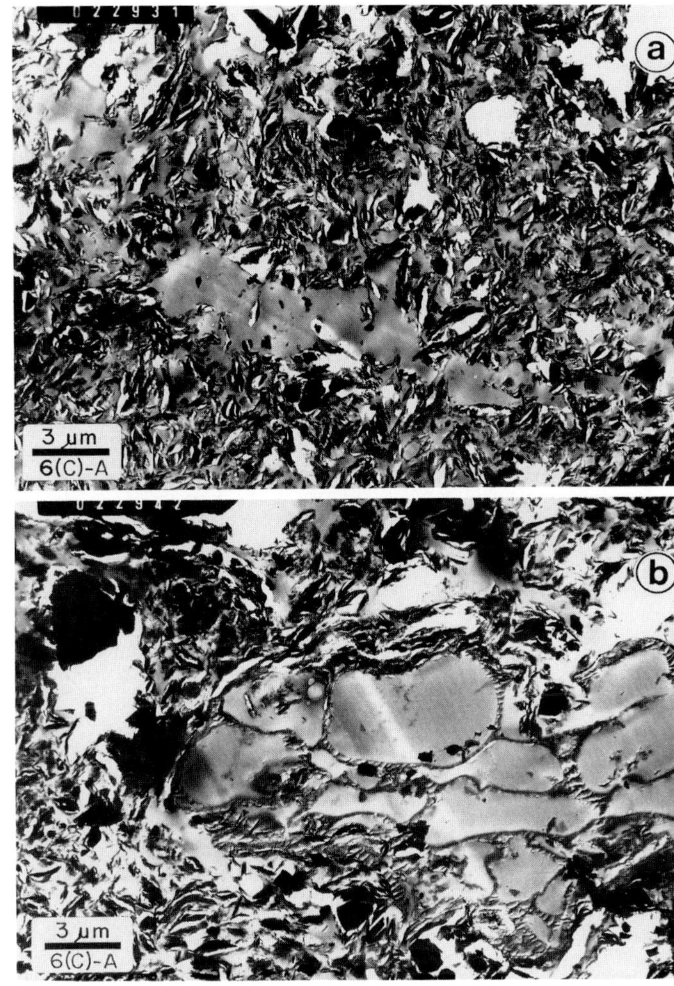

Figure 37.14. Clay fabric of sample 6(C)-A (TEM). Top: TEM micrograph showing large "pocket-like" pore space. It may be due to the degassing effect, i.e., gas expansion. Bottom: TEM micrograph showing large void spaces. The lines shown in the large void may be biogenic in origin.

preferred orientation was not always perfect in samples subjected to lower downhole pressures or effective overburden pressures. Since most of the clay particles are of plate-like shape, the clay particles tend to be aligned in a similar orientation due to the overburden pressure exerted across the maximum surface area of a clay particle (preferential orientation). While *in situ*, the overburden pressure (downward) was much greater than (or at least equal to) gas pressure (upward). The preferred clay orientation thus was maintained.

According to Whelan et al. (1981), methane concentration measured from the same pressurized core sediments used in this study ranged from 3450 to 137,140 ppm (μl CH_4(STP)/liter wet sediment). These values are generally higher than values found in companion samples taken with conventional wire-line equipment. Their results also showed that at least 98% of the methane was released from the sediment matrix within 3–5 hr after opening the pressure core barrel. The release of such a relatively high concentration of gas from the sediment would destroy the equilibrium system in the sediment matrix. Thus, the clay particles must undergo motion or rearrangement toward a new equilibrium system when exerting forces eased (i.e., when the gas or the overburden pressure was released).

Shear forces (stresses) exerted on clay particles must be created whenever the pressure changes (degassing) due to the release of *in situ* pressure. Consequently, clay particles will be realigned to another stable condition because of this force change. The mainly nonoriented but partially oriented clay fabric as shown in the results of method (B) and (C) probably resulted from the shearing stress and rearrangement of clay particles. Similar fabric relationships with regard to shear stresses and soil deformation have also been discussed by Sloane and Kell (1966), Morgenstern and Tchalenko (1967a,b), Smart (1967), Pusch (1970), Barden (1972), and Mitchell (1976).

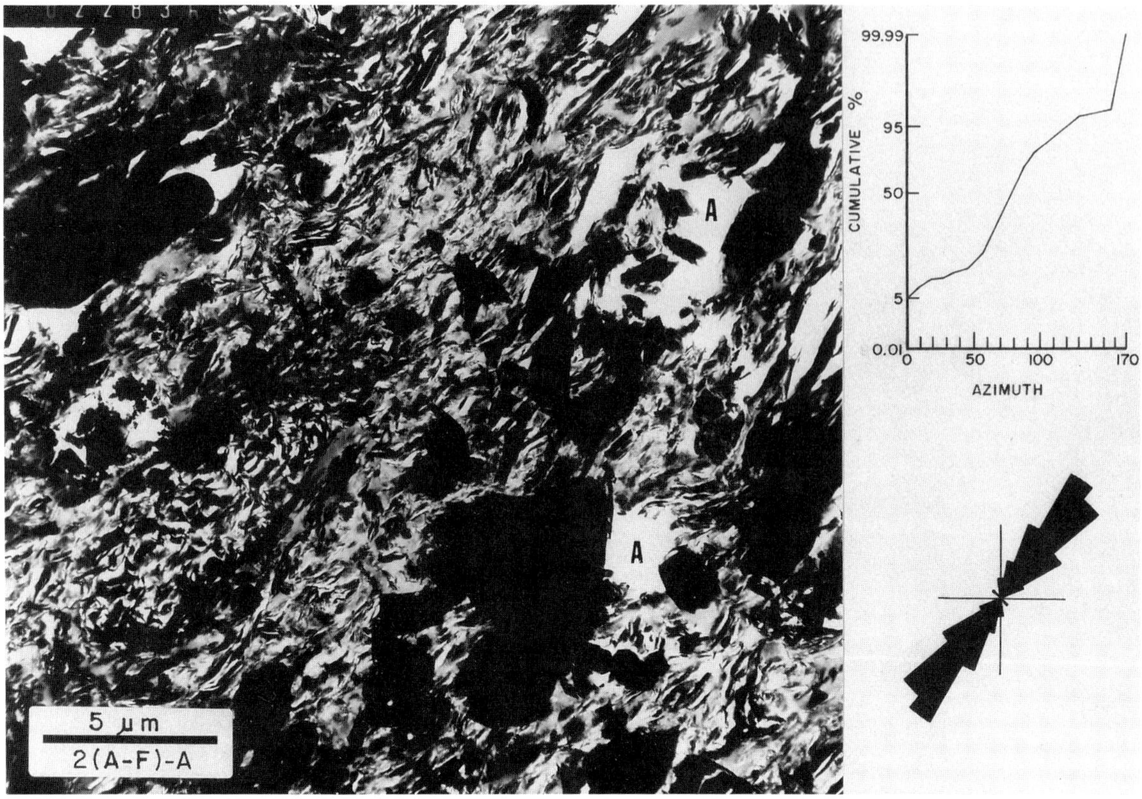

Figure 37.15. Clay fabric and orientation analysis of sample 2(A-F)-A. Left: Low magnification TEM micrographs showing the general direction of clay arrangement although slightly randomly arranged clay fabric occurs in some places. The abundant fine silts are also shown in the micrograph. The rather poor quality of this ultrathin section is mainly due to the large amounts of silt-size grains. It was also found that areas containing silts were not well impregnated with the embedding medium and were therefore torn out (areas A). Clay fabric in the central portion (from lower left to upper right) clay has been compressed a little as shown by elliptical-shaped clay aggregate (lower left). Thus, the orientation analysis of this micrograph is rather questionable. Upper right: Clay particle orientation distribution curve showing the mixed oriented and nonoriented fabric pattern. Lower right: Rose diagram showing clay orientation in the TEM micrograph (022836).

The highly nonoriented random microfeature depicted in samples prepared by method (C) probably represents disturbed microfeatures primarily due to the disequilibrium of sediment matrix by the release of *in situ* gas pressure. The channel-like void spaces with randomly arranged clay particles may reflect the avenues created by escaping gases.

Similar results between methods (B) and (C) suggest that clay fabric microfeatures of gassy sediment are rather delicate and that the fabric orientation can be altered if the *in situ* pressure is released before the sediment is completely dry regardless of drying time.

Clay Fabric vs. Degassing Time

During the course of this research, the accidental release of *in situ* downhole pressure in the exchange Plexiglas pressure vessel occurred during the testing of some samples. Although this was not the original intent, it did offer an opportunity to examine the effects of degassing as a function of time. Samples 2, 4, 6, and 7 offered such an opportunity.

Both sample 2(A-F) and sample 7(A-F) were accidentally degassed for approximately 10 sec and 45 min, respectively, and the clay fabric observed in TEM seems to be preserved fairly well. On the other hand, samples 4(A-F) and 6(A-F) had been exposed to ambient pressure for approximately 2 and 5 hr, respectively, and revealed highly random microfeatures showing the same fabric as from conventional samples 7(C) and 6(C). The degassing effect on disrupting clay fabric is time dependent. Although it was not possible to study this aspect in detail (i.e., degree of clay particle orientation versus length of time of degassing, rate of degassing and specimen size), it appeared that with a specimen size of 7 × 7 × 20 mm the clay fabric in the central portion of a specimen will not be disturbed if the length of degassing time is less than an hour. Of course, it also depends on the rate of leaking and gas concentration in the sample. This observation parallels the methane gas concentration studies made on the saline pressurized core barrel sediments (Whelan et al., 1981), which showed that at least 98% of the methane was released from the sediment matrix within 3–5 hr after opening the pressure core barrel.

The concurrence of well-oriented and highly random microfeatures in one sample examined [7(C)] is difficult to interpret. They may result from a partially disturbed sample or from an artifact created during sample preparation. Random microfeatures observed in sample 2(A-F) at high magnification probably indicate slightly degassed sediment near the sample edge. Clay fabric results of repressurized sediments also demonstrated the principle of clay fabric chemical irreversibility for marine sediment as proposed by Bennett et al. (1977).

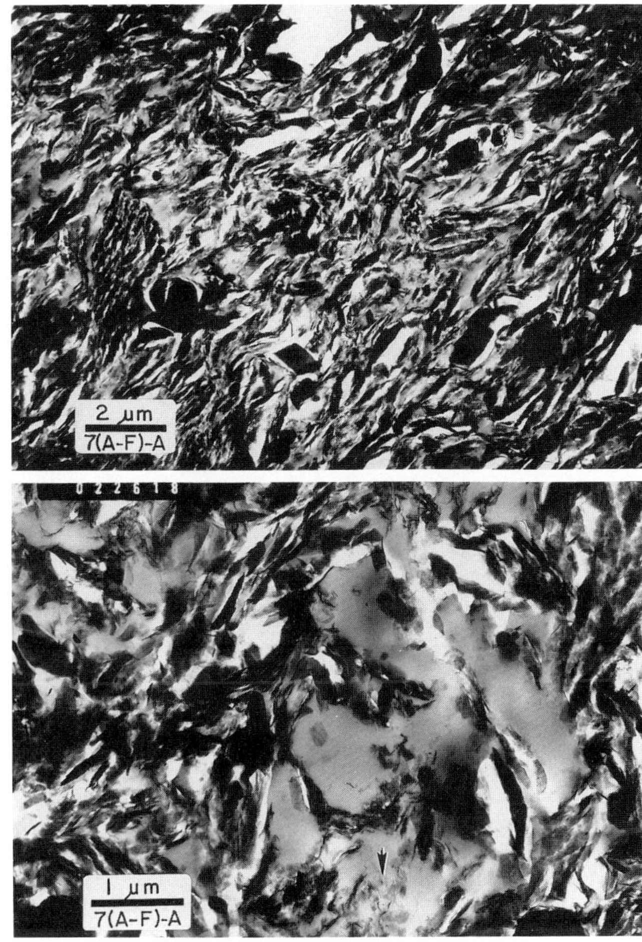

Figure 37.16. Clay fabric of sample 7(A-F)-A (TEM). Top: TEM micrograph showing that slightly swirled fabric pattern (center) that occurs in a generally well-oriented clay sediment. Bottom: TEM micrograph showing locally nonoriented clay fabric and large void spaces. Fluffy appearance (arrow) may be organic in origin.

Clay Fabric and Shear Strength

One of the major purposes of the pressure core barrel sampling project was to compare the *in situ* (pressurized) and conventional vane shear strengths. To accomplish this, vane shear equipment was taken into a hyperbaric chamber and shear tests were performed while at *in situ* downhole pressure. The same sample was then depressurized in the laboratory and the shear strength measured again using the same vane shear instrument.

By comparing both clay fabric and vane shear strength (1) in the hyperbaric chamber and (2) by the conventional method, the important relationship between these features is clearly shown (Fig. 37.20). Clay sediment with preferred orientation displays high shear strength, while clay sediment with random microstructure has lower shear strength. This observation is similar to the results presented by Matsuo and Kamon (1973) and Koff et al. (1973). Clay fabric with preferred orientation provides better sediment integrity and higher shear strength because of the greater surface area contact, and higher bonding force.

Conclusions

Based on the results of clay fabric studies on sediments recovered by the pressurized corer the following conclusions may be drawn:

1. The results presented in this research demonstrate that the pressure core apparatus and pressurized fabric techniques were successful. The new method, (method A) in this study, was able to remove gas and dehydrate the wet sediment samples by critical point-drying techniques without noticeably disturbing the clay fabric microfeatures. Nongassy sediment does not require the new method (A) described herein.
2. The results of detailed investigations of electron micrographs obtained by TEM and SEM observations, statistical calculations, and graphic analyses show that clay fabric microfeatures of gassy, deltaic sediments are affected by different sample dehydration techniques. The clay fabric reflected by these different methods of dehydration can be summarized as follows:

   a. The clay fabric of sediments prepared by the new method (A) was characterized by relatively well-oriented clay particles and domains, although random structures may occur locally. It appeared that the degree of preferred orientation increases with the overburden pressure. The size of domains was larger than those prepared by methods (B) and (C). Chains are well developed and relatively long. Void spaces are more elongate in shape. Orientation frequency curves are narrowly distributed and form a relatively steep slope.

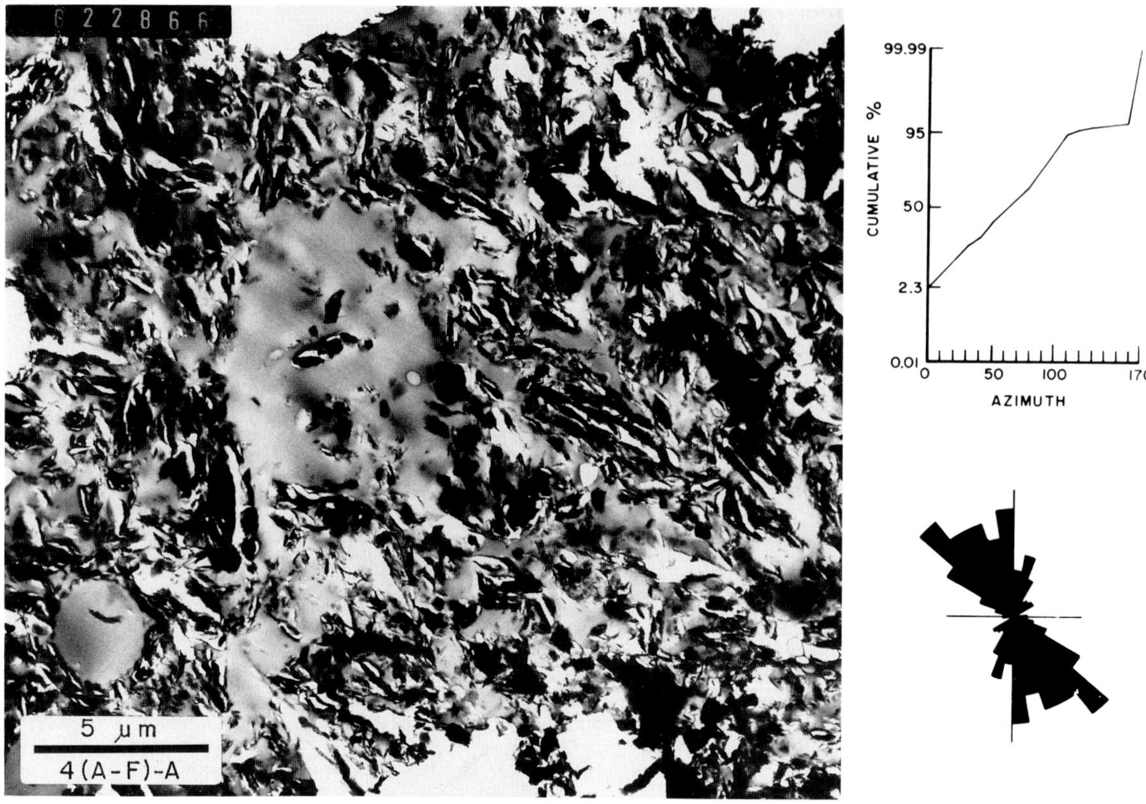

Figure 37.17. Clay fabric and orientation analysis of sample 4(A-F)-A. Left: Low magnification TEM micrograph revealing the high void spaces and the clay particles that seem to be floating in the spaces. Right: Clay particle orientation distribution frequency (top) and rose diagram (bottom) showing the mixed pattern of oriented and nonoriented clay fabric in the TEM micrograph (022866).

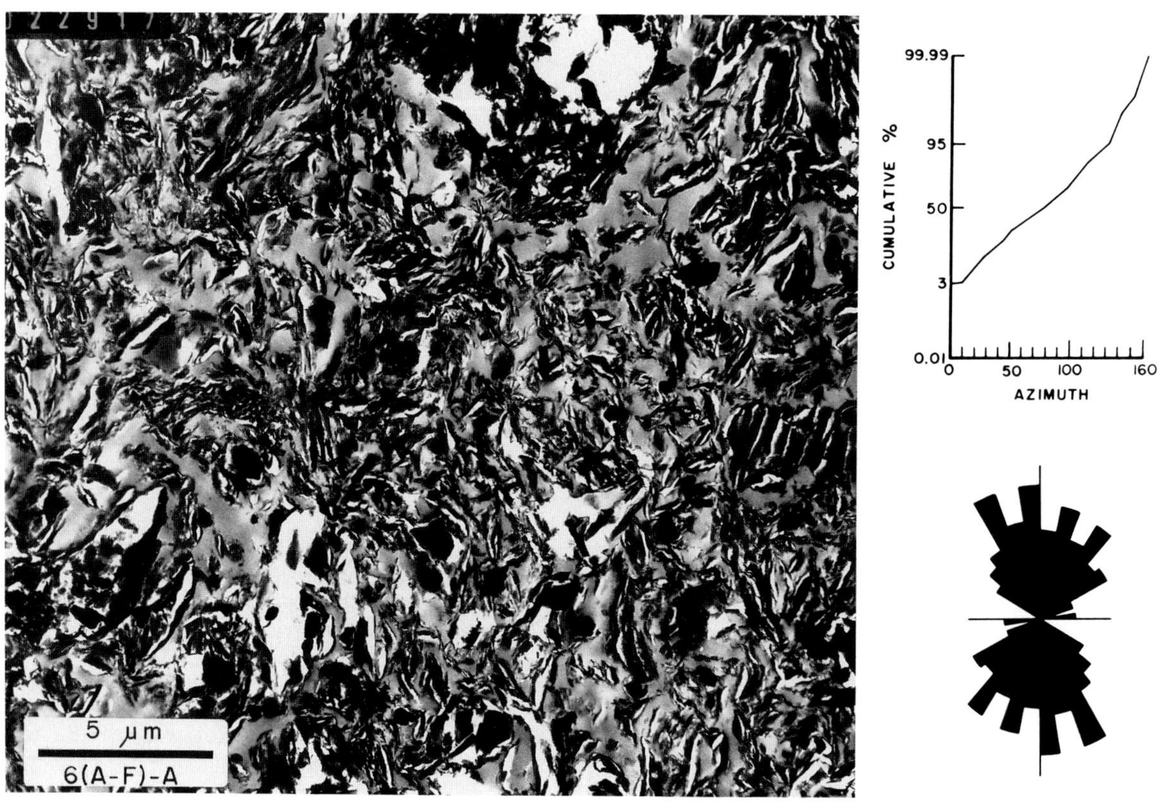

Figure 37.18. Clay fabric and orientation analysis of sample 6(A-F)-A. Left: Low magnification TEM micrograph revealing the randomly arranged clay particles and domains, and somewhat swirled pattern in the studied ultrathin sections. Right: Clay particle orientation analysis, as shown by orientation frequency distribution curve (top) and rose diagram (bottom) showing the highly randomly arranged clay fabric in this sample (TEM micrograph 022917).

Figure 37.19. Clay fabric of sample 4(A-F). A mosaic of SEM micrographs giving a general view over a relatively large area. Somewhat oriented clay arrangement is shown in upper portion of the micrograph whereas randomly arranged clay flakes associated with channel-like features (canyon-like in this micrograph) meandering through the central portion of the micrograph.

b. The clay fabric of sediments prepared by method (B) depressurized gassy sediment was typified by a random arrangement of clay particles. Degree of randomness seems to decrease with overburden pressure. The size of domains was smaller as compared to those of method (A). Clay particles appear to be floating in irregular void spaces. Clay particle orientation frequency curves are widely spread with gentle to low slopes.
c. The clay fabric of gassy sediments prepared using method (C) is primarily characterized by highly nonoriented clay particles (except one case in sample no. 7). The fabric appears to result in more randomness than method (B). Clay particle orientation frequency curves are widely distributed with a flat or slightly concave shape toward the lower frequency.

3. The comparison of clay fabric microfeatures in different stages of sediment degassing shows that the degassing (depressurizing) effect on the clay fabric texture in a sediment matrix is time dependent.
4. The relationship of clay fabric microfeatures and vane shear strengths in both *in situ* pressure and ambient pressure conditions illustrates the close correlation between these two important geotechnical properties. Sediment with preferred clay orientation at *in situ* pressures has higher shear strength than degassed nonoriented clay fabric at ambient conditions.

Suggested Areas for Future Research

This investigation represents only an initial step toward an understanding of the mechanisms that may influence the complex clay fabric microfeatures in gassy, clayey sediments. The present research has attempted to evaluate the effect of only the physical pressure factor (release of pressure on gassy sediment) on the many complex variables that may interfere with the clay fabric of a sediment. During the course of this research several areas for future study were identified.

1. How deep or to what depth will clay fabric microfeatures of a gassy sediment still be affected by the release of in situ pressure?
2. What is the detailed relationship between clay fabric and degassing time with different sampling (or burial) depth?
3. What are some statistical and mathematical technique(s) or formula(s) that can be developed to describe the relationship between clay fabric and shear strength or other geotechnical properties?

Based on the observation of thousands of TEM micrographs, the writers found that if the clay particles in the ultrathin sections can be identified easily and correctly, clay petrology, analogous to clastic sedimentary petrology, sandstone, or carbonate petrology, can be developed and applied to the understanding of fine grained particles. The most important aspect

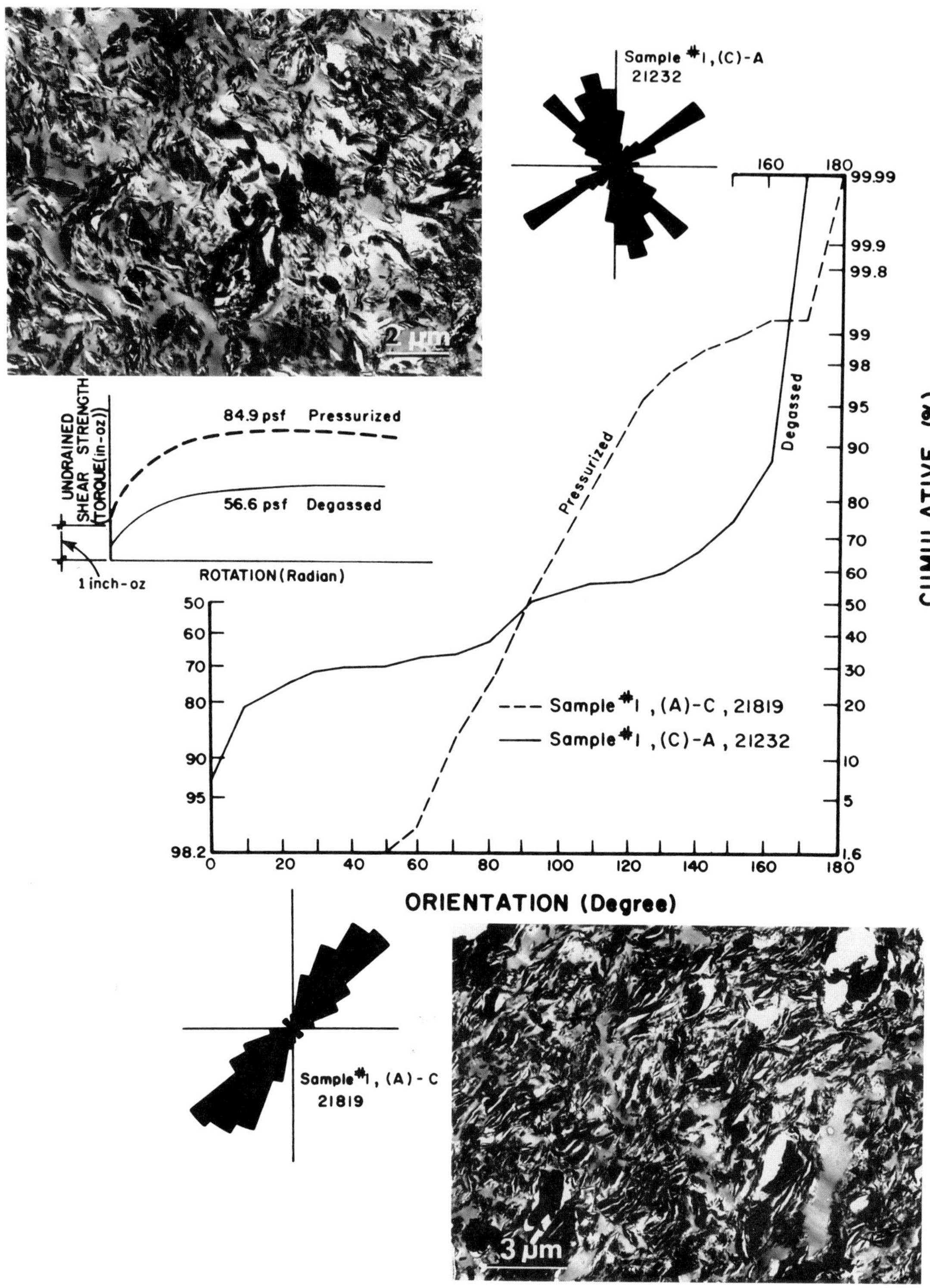

Figure 37.20. Comparison of clay fabric and shear strength. Clay fabric and shear strength between pressurized (new technique developed in this study) and degassed (conventional technique) sediment samples. TEM micrograph and orientation analysis (cumulative orientation frequency curve and rose diagram) indicating well-oriented clay fabric from the pressurized sample whereas the degassed sample reveals randomly arranged clay fabric.

of such a future research area is to improve the software and hardware of the X-ray (EDS) system to become a highly efficient particle composition analyzer, so that the texture, morphology, mineralogical, and chemical composition can be studied at the same time.

Acknowledgments

This chapter constitutes a portion of the senior author's doctoral dissertation under the direction of the co-authors. The authors are grateful to Dr. Hilton Mollenhauer (Veterinary Toxicology and Entomology Research Laboratory, U.S. Department of Agriculture) who graciously permitted free access to the transmission electron microscopy laboratory and helped in many ways with the electron microscopy study. The authors wish to thank the members of Geotechnical Properties Laboratories at Texas A&M University, Drs. Wayne Dunlap, Les E. Shephard, Michael Regan, Elliott Taylor, and Mark Johns, for their generous assistance in providing and preparing pressurized core samples for this research. This chapter has benefitted from discussions and reviews with Drs. Les E. Shephard, Louis J. Thompson, Thomas R. McKee, U. Grant Whitehouse (deceased), Richard Rezak, and David D. McGrail (deceased). This research was supported by the National Oceanic and Atmospheric Administration (NOAA), U.S. Department of Commerce, and the U.S. Geological Survey (Texas A&M Research Foundation Project #4076 and 3997).

References

Aylmore, L.A.G., and J.P. Quirk, 1960. Domain or turbostratic structure of clay. Nature, v. 187, p. 1046–1043.

Aylmore, L.A.G., and J.P. Quirk, 1962. The structural status of clay systems. Clays and Clay Minerals, v. 9, p. 104–130.

Barden, L., 1972. The influence of structure on deformation and failure in clay soils. Geotechnique, v. 22, p. 153–163.

Barden, L., and R. Pusch (eds.), 1973. Proceedings of the International Symposium on Soil Structure, Gothenburg, Sweden. Swedish Geotechnical Society and Swedish Society for Clay Research, 251 p., with 44 p. Appendix.

Bennett, R.H., 1976. Clay fabric and geotechnical properties of selected submarine sediment cores from the Mississippi Delta. Ph.D. dissertation, Texas A&M University, College Station, TX, 269 p.

Bennett, R.H., and W.E. Bryant. 1976. Clay fabric and selected geotechnical properties (abs.). 25th Annual Clay Minerals Conference, Corvallis, Oregon, p. 32.

Bennett, R.H., and M.H. Hulbert, 1986. Clay Microstructure. International Human Resources Development Corporation, Boston, MA, 161 p.

Bennett, R.H., W.R. Bryant, and G.H. Keller, 1977. Clay Fabric and Geotechnical Properties of Selected Submarine Sediment Cores from the Mississippi Delta. NOAA Professional Paper 9, U.S. Department of Commerce, NOAA, Rockville, MD, 86 p.

Bennett, R.H., W.R. Bryant, and G.H. Keller, 1979. Clay fabric and related pore geometry of selected submarine sediment. Mississippi Delta. In: O'Hare, A.M.F. (ed.), Scanning Electron Microscopy, Vol. 1, AMF O'Hare, Chicago, IL., Scanning Electron Microscopy, Incorporated, p. 519–524.

Bennett, R.H., D.N. Lambert, M.H. Hulbert, H. Schmoll, and B.M. Bohlke, 1980. Clay fabric of sediments from various depositional environments (abs.). 29th Annual Clay Minerals Conference, Clay Minerals Society, Waco, TX, p. 22.

Biscoe, J., and B.E. Warren, 1942. An x-ray study of carbon black. Journal of Applied Physics, v. 13, p. 364–371.

Bohlke, B.M., and R.H. Bennett, 1978. Mississippi prodelta crusts: a clay fabric and geotechnical analysis: Phase I report for the U.S. Geological Survey. NOAA, AOML, Miami, FL, 67 p.

Bohlke, B.M., and R.H. Bennett, 1980. Mississippi prodelta crusts: a clay fabric and geotechnical analysis. Marine Geotechnology, v. 4, p. 55–82.

Bowles, F.A., 1969. Electron microscopy of the microstructure in sediment samples from the Gulf of Mexico. Ph.D. dissertation, Texas A&M University, College Station, TX, 113 p.

Bowles, F.A., W.R. Bryant, and C. Wallin, 1969. Microstructures of unconsolidated and consolidated marine sediments. Journal of Sedimentary Petrology, v. 39, p. 1546–1551.

Burnham, C.P., 1970. The micromorphology of argillaceous sediments: particularly calcareous clays and siltstones. In: Osmond, D.A., and P. Bullock (eds.), Micromorphological Techniques and Applications. Soil Survey Techniques Monograph, No. 2, p. 83–96.

Casagrande, A., 1932. The structure of clay and its importance in foundation engineering. Journal Boston Society of Civil Engineering, v. 19, p. 168–208.

Chiou, W.A., 1980. A new technique in preparing *in situ* marine sediments for clay fabric study. Proceedings of the 38th Electron Microscopy Society of America Annual Meeting, p. 204–205.

Chiou, W.A., 1981. Clay fabric of gassy submarine sediments. Ph.D. dissertation, Texas A&M University, College Station, TX, 248 p.

Chiou, W.A., L.E. Shephard, and W.R. Bryant, 1980. Clay fabric and geotechnical properties of trench sediments (abs.). 29th Annual Clay Minerals Conference, Clay Minerals Society, Waco, TX, p. 28.

Collins, K., and A. McGown, 1974. The form and function of microfabric features in a variety of natural soils. Geotechnique, v. 24, p. 223–254.

Denk, E., W.A. Dunlap, W.R. Bryant, L.J. Milberger, and T.J. Whelan, 1981. A pressurized core barrel for sampling gas-charged marine sediments. Proceedings of the 13th Annual Offshore Technology Conference, p. 43–52.

Fairbairn, H.W., 1949. Structural Petrology of Deformed Rocks. Addison-Wesley, Cambridge, 344 p.

Folk, R.L., 1974. Petrology of Sedimentary Rocks. Hemphill Publishing Co., Austin, TX, 182 p.

Gary, M., R. McAfee, Jr., and C.L. Wolf, eds., 1972. Glossary of Geology. American Geological Institute, Washington, D.C., 805 p.

Goldschmidt, V.M., 1926. Undersolkelser over lersedimenter. Nordisk jourbrugsforkning, No. 4–7, p. 434–445.

Koff, C.L., A.S. Polyakov, and E.M. Sergeev, 1973. Changes in microtexture of marine clay sediments during their diagenesis. In: Barden L., and R. Pusch (eds.), Proceedings of the International Symposium on Soil Structure, Gothenburg, 1972. Swedish Geotechnical Society and Swedish Society for Clay Research, p. 227–242.

Lee, H.T., 1980. Compressibility and permeability of clays at high pressure. M.S. Thesis, Texas A&M University, College Station, TX, 148 p.

Matsuo, S.I., and M. Kamon, 1973. Microscopic research on the consolidated samples of clayey soils. In: Barden, L., and R. Pusch (eds.), Proceedings of the International Symposium on Soil Structure, Gothenburg. Swedish Geotechnical Society and Swedish Society for Clay Research, p. 197–199.

Mitchell, J.L., 1976. Fundamentals of Soil Behavior. Wiley, New York, 422 p.

Moon, C.F., 1972. The microstructure of clay sediments. Earth-Science Reviews, v. 8, p. 303–321.

Morğenstern, M.R., and J.S. Tchalenko, 1967a. Microscopic structure in kaolin subjected to direct shear. Geotechnique, v. 17, p. 309–328.

Morgenstern, M.R., and J.S. Tchalenko, 1967b. Microstructure observation on shear zones from slips in natural clays. Proceedings of Geotechnical Conference, Oslo, Norway, Norwegian Geotechnical Institute, v. 1, p. 147–152.

O'Brien, N.R., 1971. Fabric of kaolinite and illite floccules. Clays and Clay Minerals, v. 19, p. 353–359.

O'Brien, N.R., and E. Hisatomi, 1978. Sedimentological study of a turbidite cycle, Kii Peninsula, Japan. Memoirs of the Faculty of Science, Kyoto University, Series of Geology and Mineralogy, v. XLL(2), p. 177–186.

O'Brien, N.R., K. Nakazawa, and S. Toluhashi, 1980. Use of clay fabric to distinguish turbiditic and hemipelagic siltstones and silts. Sedimentology, v. 27, p. 47–61.

Olsen, H.W., 1962. Hydraulic flow through saturated clay. Clays and Clay Minerals, v. 9, p. 131–161.

Parry, R.H. (ed.), 1971. Stress-strain behavior of soils. Proceedings of the Roscoe Memorial Symposium, Cambridge University, Oxfordshire, England, G.T. Foulis and Company, Ltd., 752 p.

Pusch, R., 1970. Microstructural changes in soft quick clay at failure. Canadian Geotechnical Journal, v. 7, p. 1–7.

Range, K.J., A. Range, and A. Weiss, 1969. Fire-clay type kaolinite or fire clay mineral? Experimental classification of kaolinite-halloysite minerals. In: Heller, L. (ed. in chief), Proceedings of the International Clay Conference, Tokyo, Japan. Israel Universities Press, Jerusalem, p. 3–13.

Rutherford, C.K. (ed.), 1974. Soil microscopy. Proceedings of the 4th International Working Meeting on Soil Micromorphology, Kingston, Ontario, Canada, Aug. 27–31, 1973. The Limestone Press, Kingston, Ontario, Canada, 857 p.

Scafe, D.W., 1968. A clay mineral investigation six cores from the Gulf of Mexico. Ph.D. dissertation, Texas A&M University, College Station, TX, 75 p.

Schultz, L.G., H. van Olphen, and F.A. Mumpton (eds.), 1987. Proceedings of the International Clay Conference, Denver, 1985. The Clay Minerals Society, Bloomington, IN, 456 p.

Shephard, L.E., W.R. Bryant, and W.A. Dunlap, 1979. Geotechnical properties and their relation to geologic processes in South Pass Outer Continental Shelf Lease Area, Block 28, 47 and 48, Offshore Louisiana. Technical Report 79-5-T, Dept. of Oceanography, Texas A&M University, 125 p.

Shephard, L.E., W.A. Chiou, and W.R. Bryant, 1980. Consolidation characteristics and clay fabric of Japan Trench sediments. Geological Society of America Annual Meeting, Atlanta, GA, Abstracts with program, v. 12, p. 521.

Shephard, L.E., W.R. Bryant, and W.A. Chiou, 1982. Geotechnical properties of Mid-American Trench sediments. In: More, J.C., and J.S. Watkins (eds.), Initial Reports of the Deep Sea Drilling Project, v. 66. U.S. Government Printing Office, Washington, D.C., p. 475–504.

Sloane, R.L., and T.R. Kell, 1966. The fabric of mechanically compacted kaolinite. Clays and Clay Minerals, v. 14, p. 289–296.

Smart, P., 1967. Particle arrangements in kaolin. Clays and Clay Minerals, v. 15, p. 241–254.

Smart, P., and N.K. Tovey, 1981. Electron Microscopy of Soils and Sediments—Examples. Oxford University Press, Oxford, 176 p.

Smart, P., and N.K. Tovey, 1982. Electron Microscopy of Soils and Sediments—Techniques. Oxford University Press, Oxford, 264 p.

Sorby, H.C., 1851. On the microscopic structure of the calcareous grit of the Yorkshire coast. Quarterly Journal Geological Society, London, v. 7, p. 1–6.

Terzaghi, K., 1925. Erdbau Mechanik auf Bodenphysikalischer Grundlage. Franz Deutiche, Leipzig und Wien, 399 p.

Tovey, N.K., 1970. Electron microscopy of clays. Ph.D. Dissertation, University of Cambridge, Cambridge, England.

Tovey, N.K., and K.Y. Wong, 1973. The preparation of solid and other geological materials for the SEM. In: Barden, L., and R. Pusch (eds.), Proceedings of the International Symposium on Soil Structure, Gothenburg, Sweden. Swedish Geotechnical Society and Swedish Society for Clay Research, p. 59–67.

Tovey, N.K., and K.Y. Wong, 1978. Preparation, selection and interpretation problems in scanning electron microscope studies in sediments. In: Whalley, W.B. (ed.), Scanning Electron Microscopy in Study of Sediments. Geo Abstracts, Norwich, England, p. 181–199.

Trabant, P.K., and W.R. Bryant, 1979. Submarine geomorphology and geology of the Mississippi River Delta front. Technical Report 79-1-T, Dept. of Oceanography, Texas A&M University, 118 p.

Whalley, N.B. (ed.), 1978. Scanning Electron Microscopy in the Study of Sediments. A symposium held at the 3rd Meeting of Geological Societies of the British Isles, University College of Wales, Swansea, Sept. 24–25, 1977, Norwich, England, Geo Abstracts, 414 p.

Whelan, T. III, W.R. Bryant, and N.A. Dunlap, 1981. Methane concentrations and distribution in pressure core samples from Mississippi Delta sediments. (abs.), American Association Petroleum Geologists Bulletin, v. 65, 1981 AAPG-SEPM Annual Meeting, San Francisco, CA, p. 1007.

Wong, K.Y., 1975. Microfabric changes during the deformation of clays. Ph.D. dissertation, Cambridge University, Cambridge, England.

Yong, R.N., and D.E. Sheeran, 1973. Fabric unit interaction and soil behavior. In: Barden, L., and R. Pusch (eds.), Proceedings of the International Symposium on Soil Structure, Gothenburg, Sweden. Swedish Geotechnical Society, p. 176–183.

CHAPTER 38

Objective Measurement and Classification of Microfabrics and Their Relationship to Physical Properties

Cynthia M. Ross and Robert Ehrlich

Introduction

Assessment of microstructure requires a series of steps beginning with sampling and careful sample preparation followed by characterization. Characterization can be done to some extent physically by measuring such properties as permeability, diffusivity, and shear strength. However, the results of such tests are always ambiguous in that more than one microstructure fabric can yield similar values and conversely similar appearing fabrics may yield very different physical measures. The only way to resolve these ambiguities is to observe the microstructure directly. In almost all cases, this amounts to visual inspection of fabric elements intersecting a planar thin section through the sediment. One of the many reasons to assess petrographic images is determination of relationships between microstructure and physical properties. Another set of objectives involves classifying the microfabric to understand the process history of the sediment. A third objective would be to relate microstructure to an easily observable external condition such as side scan sonar or bottom photographs.

These three goals can be achieved poorly if at all by simple visual inspection of the sections. The very efficiency of the human eye–mind system makes such a procedure highly subjective and nuances easily observed are impossible to sufficiently characterize with precision and conciseness. Computer-assisted image analysis procedures provide a means of achieving the goals of fabric characterization. An image analysis system lacks the flexibility and breadth of the human eye–mind system, but has the advantages of consistency and virtually limitless memory. If the sediment fabric were relatively simple there would be no barrier for rapid penetration of image analysis into the field – that is, if sediment fabric consisted of black circles on a white background or grids of various sizes, programming the image analyzer would be a simple task. However, sediment microfabric, especially in muds and shales, is complex with elements of highly variant size and geometry. Therefore, sediment fabric characterization demands generalized image analysis methods that can be applied to complex shapes and textures.

For the past 10 years, we have engaged in developing a series of procedures termed petrographic image analysis (PIA), which is designed to uniquely characterize and classify complex fabrics. Our work has been concerned primarily with patterns of porosity in sandstones and limestones. However, inspection of TEM photomicrographs indicates patterns comparable to those observed in our rock studies.

Design Parameters of Petrographic Image Analysis

The PIA system was designed from its conception to characterize images of sediment fabric elements. Analytical goals included precise characterization of size, shape, and connectivity of fabric elements and their objective classification. However, an equally important character is the ability to process hundreds or thousands of images per day. Only with such a processing rate can we analyze enough fabric samples to ensure the generality of our observations.

The Binary Image

The first step in image analysis is the acquisition of digitized images, which initially involves the digitization of an analog video scanner signal. The digitized image consists of an array of pixels ("picture elements"), each of which has an associated intensity or brightness value. With transmitted light petrography,

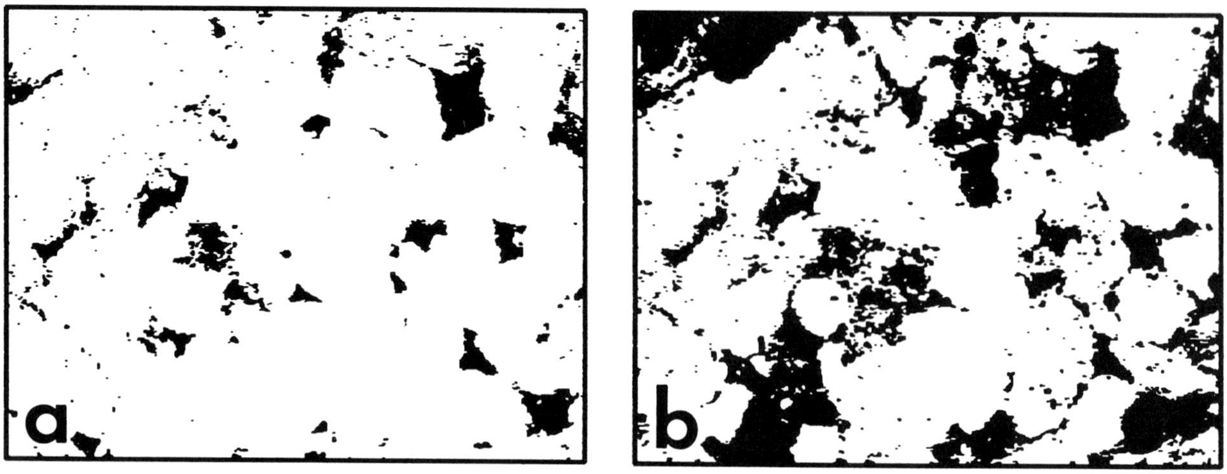

Figure 38.1. Binary images from petrographic image analysis of standard thin sections. (a) Porels (pore elements), blue-dyed epoxy-impregnated porosity that intersects thin section surface, are black. (b) Fabrels (fabric elements), all nonquartz elements including porosity, are black. (From Ferm et al., 1989.)

each field of view is digitized three times through red, blue, and green filters (R, B, and G, respectively). By combining the information of these three color planes, the color of each pixel as well as its intensity can be determined. A binary image is then derived using a digital filter. In porosity studies using blue-dyed epoxy-impregnated thin sections, intensity ranges and differences in the intensities from one color plane to another are used to segment blue pore pixels from nonpore pixels. The digital filter accepts or rejects pixels based on such criteria and constructs a binary image wherein the intensity of pixels of one sort are reset to a value of unity and pixels of the other sort are set to an intensity of zero. Any image may be treated algorithmically as the RBG images obtained by transmitted light microscopy where any color or intensity range may be segmented from the rest of the image.

Pixels, Porels, and Fabrels

The next step in the image analysis of textural elements is spatially relating individual pixels within the binary image to each other. For example, the binary image in Figure 38.1a consists of black pore pixels and white nonpore pixels. By determining those pore pixels that are adjacent to one another, the algorithm is able to identify discreet groups of pore pixels. Each group of neighboring pore pixels is termed a pore element or "porel" where a porel represents a two-dimensional slice of a three-dimensional pore. Using the same criteria, fabric elements or "fabrels" consist of adjacent black pixels that, in this particular case, represent any nonquartz portion of a thin section (Fig. 31.1b). Once individual elements are distinguished, the size, shape, and connectivity of these pore or fabric elements can be analyzed.

Processing of Porels and Fabrels

Size, shape, and connectivity of objects, in this case porels and fabrels, are easily measured if the objects vary only in one property. A simple system consists of constant-shaped objects varying only in size, shape, or connectivity. In such a system, a simple set of descriptors (e.g., area and perimeter) would uniquely characterize all the variation within the system and physical properties would be readily modeled with a basic descriptor set. However, fabric and pore elements are quite complex forming lace-like doilies and branching structures, and varying over several orders of magnitude in size.

To characterize highly variable porosity and texture patterns, binary images are subjected to an erosion–dilation differencing algorithm (Crabtree et al., 1984), which measures the size, area, perimeter, connectivity, and shape of each porel or fabrel. The erosion–dilation algorithm strips off the outermost layer of pore or fabric pixels (erosion), then expands any remaining object (porel or fabrel) one pixel layer (dilation). This process is repeated for a two-pixel layer, a three-pixel layer, and so on, progressively smoothing the object. The erosion–dilation procedure stops once all object pixels have been eroded leaving no pixels for subsequent dilation.

Throughout the erosion–dilation process, the algorithm records the area lost after each erosion–dilation cycle. Area lost on the last erosion–dilation cycle is the smooth component of the object while areas lost during previous erosion–dilation cycles are various sized roughness elements. Each object, field of view, thin section, or photomicrograph can be characterized by a frequency distribution of rough and smooth area components. The size (diameter) of smooth and rough area components equals twice the number of erosion–dilation cycles multiplied by pixel dimensions (which are dependent on microscope magnification). The erosion–dilation differencing algorithm also records the number and sizes of "thoats" (two-dimensional constrictions in fabrels or porels) as well as the perimeter attributed to each cycle.

The variable set produced by the erosion–dilation algorithm sufficiently defines microfabrics such that physical properties can be accurately modeled. Variables related to permeability (Ehrlich and Horkowitz, 1984) and mercury porosimetry (McCreesh, 1987) include the ratio of smooth to rough pore area and the proportion of two-dimensional pore throats 6–10 μm wide. Although these variables are strongly related to physical properties, they are not directly observable in thin section. To increase the genetic nature of petrographic image analysis output, the resultant data are subjected to a classification procedure.

Objective Classification of Porosity and Fabric Elements

A pattern recognition algorithm, SAWVEC (Ehrlich and Full, 1988), derived from EXTENDED CABFAC (Klovan and Meisch, 1976) and the QMODEL family of algorithms (Klovan and Meisch, 1976; Full et al., 1981, 1982), determines the number of distinct pore or fabric types, the characteristics of each type, and the relative proportion of each fabric or pore type for each thin section. The rough and smooth area distributions combined using maximum entropy criteria (Full et al., 1984) are the basis of the classification system.

Petrographic Image Analysis Applications

Numerous authors have demonstrated that PIA data are strongly related to physical properties such as mercury porosimetry (McCreesh, 1987; Etris et al., 1988), permeability (Ehrlich and Horkowitz, 1984; Yuan, 1987), log response (Ross et al., 1986), and fracture toughness (Ferm et al., 1989). To demonstrate PIA methodology, texture classification, and the relation to physical properties, a summary of fracture toughness relationships (Ferm et al., 1989), and an example of clay microfabric classification will be presented.

Fabrels and Fracture Toughness

Fracture toughness is a measure of a material's resistance to the growth of tensile fractures. Previous studies outlined by Ferm et al. (1989) indicate that microfractures selectively propagate through pores, matrix, and along grain boundaries and that fracture toughness is controlled by grain size and shape. For the 15 study sections from five sandstones, Ferm et al. (1989) determined petrographic characteristics via PIA that are statistically related to fracture toughness. For each field of view, two binary images were collected; one binary image consists of porosity (porels) while the other contains any nonquartz constituent including porosity (fabrels). PIA-derived character of porosity and nonquartz grain area (NGA) is then compared with fracture toughness values.

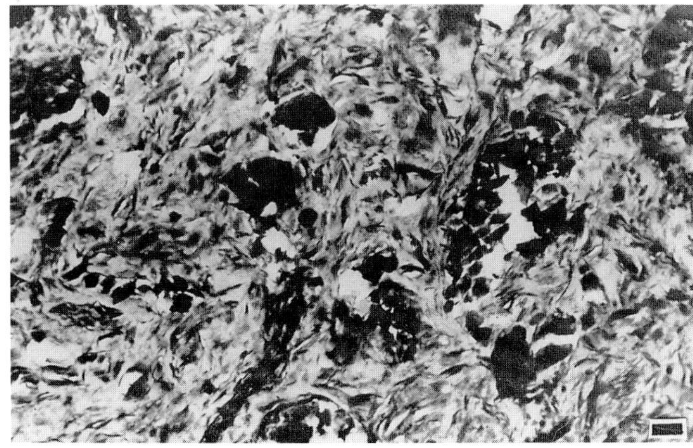

Figure 38.2. TEM photomicrograph from the "crust" zone of Mississippi prodelta muds in BH-3A. Scale bar = 1 μm. (From Bohkle and Bennett, 1980.)

The fracture toughness relationship with PIA measures was determined for two distinct groups. Group 1 consists of quartz arenites with point grain contacts and highly interconnected NGA. Fracture toughness of these samples is accurately predicted ($r^2 = 0.88$, $p < 0.006$) by total optical porosity (TOP) alone. TOP differs from whole rock or bulk porosity in that isolated pores and porosity connected by throats small enough to prevent epoxy impregnation are excluded from TOP. In this sample set where the TOP area and NGA area are almost equal, stress is transmitted freely through connected pore space accumulating at point grain contacts where rock failure occurs.

The second set, Group II, consists of lithic arenites with sutured or interpenetrating grain contacts and discrete, patchy NGA. Fracture toughness, in this case, is predicted by NGA perimeter (minus perimeter contributions from bubble trains) and the perimeter of the largest NGA components ($r^2 = 0.77$, $p < 0.01$). TOP is not related to fracture toughness in this sample suite, since TOP contributes much less area to NGA than in the previous group. Due to extensive sutured grain contacts, much more stress can accumulate in Group II samples. However, stress can be dissipated through large discrete patches of NGA.

Fabrels and Clay Microstructure

To evaluate the ability of PIA procedures to classify clay microstructures, 25 fields of view were taken from a TEM photomicrograph of a Mississippi prodelta mud sample (Fig. 38.2). Each field of view represents a 30 μm² area of the clay sample. A gray level threshold segmented black and dark gray portions of the TEM photomicrograph from lighter areas. The resultant binary textures were then subjected to the erosion–dilation

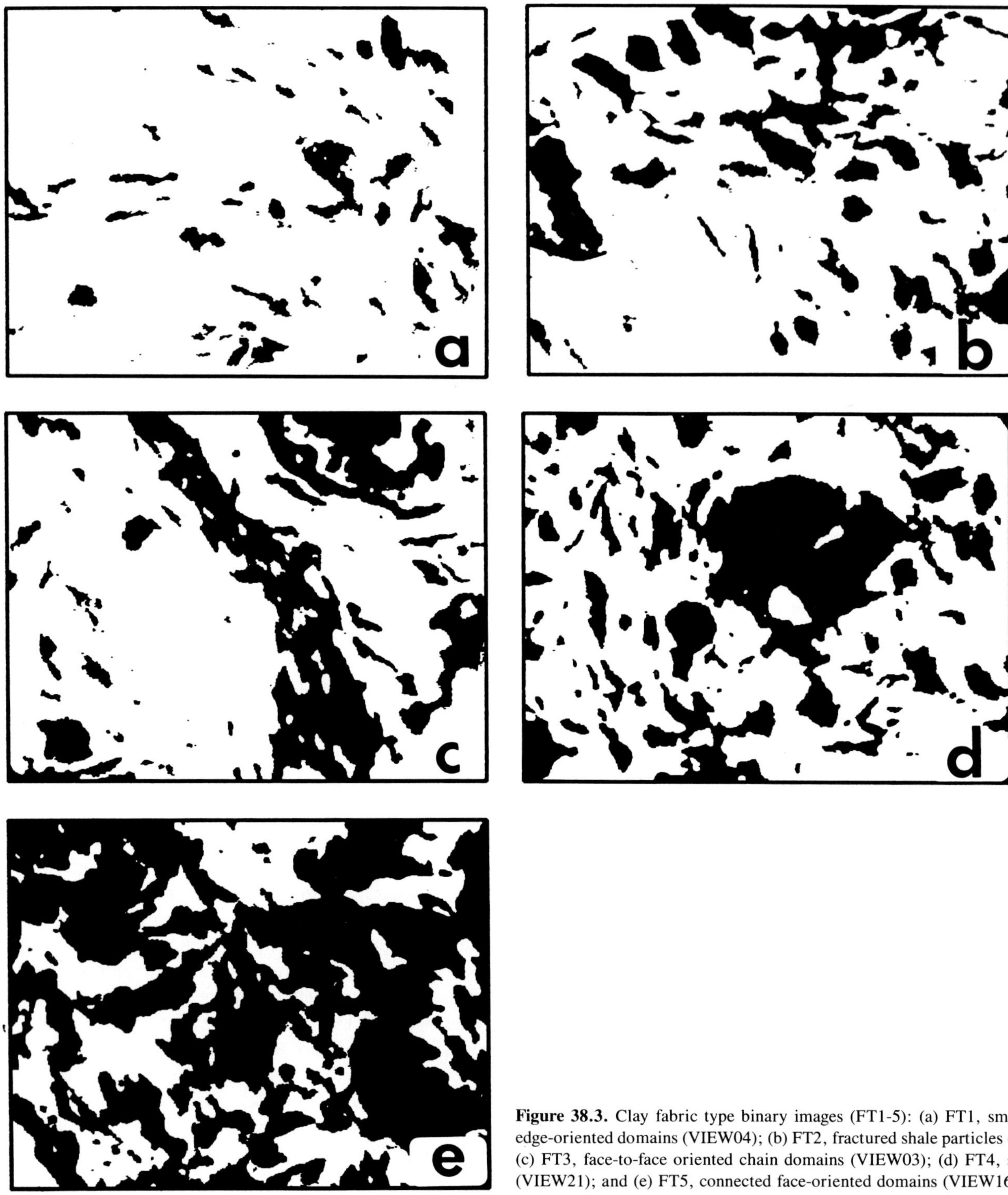

Figure 38.3. Clay fabric type binary images (FT1-5): (a) FT1, small discrete edge-oriented domains (VIEW04); (b) FT2, fractured shale particles (VIEW20); (c) FT3, face-to-face oriented chain domains (VIEW03); (d) FT4, shale clasts (VIEW21); and (e) FT5, connected face-oriented domains (VIEW16).

differencing procedure and classified according to the size, shape, and complexity of the segmented microfabric.

Within the TEM photomicrograph, five distinct fabric types were defined including face-to-face oriented chain domains and shale clasts (Fig. 38.3) based on the analysis of the rough and smooth fabric component distributions by the pattern recognition algorithm, SAWVEC. Fabric type 1 (FT1, Fig. 38.3a) consists of small discrete edge-oriented domains. Like FT1, fabric type 2 (FT2, Fig. 38.3b) is composed of small discrete domains. However, these domains are larger and are usually components

Table 38.1. Fabrel type end member compositions with percentage fabrel area attributed to smooth and rough fabrel diameters for each fabrel type.

Diameters (μm)	Percentage fabrel area				
	FT1	FT2	FT3	FT4	FT5
Smooth					
0.000–0.294	42.18	11.94	0.00	2.20	0.14
0.295–0.791	1.19	24.89	0.00	10.21	0.00
>0.791	0.00	0.00	14.89	43.10	10.05
Rough					
0.000–0.113	31.15	9.56	3.89	5.30	0.00
0.114–0.181	18.17	10.13	7.51	7.75	0.00
0.182–0.271	3.35	17.85	5.46	9.31	4.80
0.272–0.339	0.00	11.78	7.50	5.93	0.00
0.340–0.475	3.21	11.34	15.78	9.04	0.00
0.476–0.656	0.00	2.70	28.14	0.00	0.06
0.657–0.972	0.76	0.00	17.04	3.25	18.92
>0.972	0.00	0.00	0.00	4.10	66.22

of "fractillites," fractured shale particles as identified by Bryant and Bennett (1988). Fabric type 3 (FT3, Fig. 38.3c) consists of face-to-face oriented chain domains. Fabric type 4 (FT4, Fig. 38.3d) is composed of shale particles that have remained (for the most part) intact through the sectioning process. These particles may be compositionally different from the "fractillite" type shale clasts or perhaps their shape or size preserved them during the sectioning procedure. The largest and most complex fabric type, fabric type 5 (FT5, Fig. 38.3e), consists of connected face-oriented domains.

The end member compositions in the SAWVEC (Ehrlich and Full, 1988) output contains the distribution and sizes of the rough and smooth fabric components for each fabric type (Table 38.1). For instance, fabric type 1 (FT1) is composed predominantly of small smooth objects (42.18%; 0–0.294 μm) with small roughness elements (49.32%; 0–0.181 μm). The smooth fabrel dimensions (0–02.94 μm) are basically equivalent to the diameter range of the largest inscribed circle for FT1 and the percentage (49.32%) indicates the proportion of the fabrel area included in this circle. Any elongations or projections (which are common in FT1; Fig. 38.3a) are classified as roughness elements.

By knowing the sizes and relative area proportions of the smooth and roughness elements, one can infer the general size and shape of fabrel type and directly observe them in binary images. For example, fabric type 4 (FT4) consists of relatively large (> 0.791 μm) smooth fabrel diameters representing 43.10% of the fabrel area (Table 38.1). Roughness elements occur over a wide size range (0 to > 0.972 μm) with most of the roughness area (37.33%) occurring between 0 and 0.475 μm (Table 38.1). This indicates that the fabrel type is predominantly circular with comparatively small-scale irregularities. To verify this, one must observe the binary image with the greatest proportion of FT4.

The fabric type end member proportions (Table 38.2) from the pattern recognition algorithm, SAWVEC, indicate the relative proportion of each fabric type within a field of view. The greatest proportion of fabric type 4 (FT4) is in VIEW21 (Table 38.2). This field of view (Fig. 38.3d) consists predominantly of FT4 (56.97%) with 12.43% of FT1, 23.73% of FT2, 4.20% of FT3, and 2.63% of FT5 (Table 38.2). The large dark fabrel in Figure 38.3d is a shale clast representative of fabric type 4 (FT4). This single fabrel comprises 56.97% of the fabrel area in VIEW21.

Table 38.2. Fabric type end member proportions for each field of view.

VIEW	FT1	FT2	FT3	FT4	FT5
VIEW01	0.0997	0.4915	0.3212	0.0795	0.0081
VIEW02	0.1539	0.1465	0.0811	0.2289	0.3896
VIEW03	0.1278	0.1286	0.6341	0.0978	0.0118
VIEW04	0.8270	0.1996	−0.0395	0.0084	0.0044
VIEW05	0.1989	0.6474	0.2532	−0.0871	−0.0124
VIEW06	0.1310	0.1270	0.2820	0.3598	0.1003
VIEW07	0.0264	0.1964	0.3927	0.2445	0.1400
VIEW08	0.0491	0.5086	0.3993	0.0026	0.0404
VIEW09	0.0551	−0.0992	0.3267	0.1827	0.5347
VIEW10	0.1648	0.3225	0.2696	0.2379	0.0052
VIEW11	0.1772	0.0300	0.0771	0.5284	0.1873
VIEW12	0.2740	0.3153	0.0582	0.2206	0.1319
VIEW13	0.2163	0.1641	0.1514	0.3571	0.1111
VIEW14	0.3052	0.2432	0.0716	0.3277	0.0524
VIEW15	0.0873	0.4742	0.2397	0.1001	0.0988
VIEW16	0.0083	0.0240	0.1300	0.0998	0.7379
VIEW17	0.0208	0.1535	0.3661	0.2124	0.2472
VIEW18	0.2010	0.2263	0.0980	0.4192	0.0555
VIEW19	0.2668	0.7262	0.0368	−0.0249	−0.0049
VIEW20	0.1148	0.8654	0.0390	−0.0187	−0.0006
VIEW21	0.1243	0.2378	0.0420	0.5697	0.0263
VIEW22	0.0229	0.2310	0.4967	0.0029	0.2466
VIEW23	−0.0016	0.4752	0.2571	0.2102	0.0591
VIEW24	0.0655	0.1605	0.2297	0.1898	0.3545
VIEW25	0.1611	0.4080	0.2318	0.1466	0.0524

The relative proportion of each fabric type along with other PIA variables such as percentage fabrel area can readily be related to physical measures such as porosity, permeability, and shear strength. Additional photomicrographs (at the same magnification) of clay, mud, or shale samples with corresponding physical measurements are necessary for this type of PIA application. This procedure depends on the availability of consistently high-quality samples and subsequent images. The existence of sampling and processing artifacts such as cracks, plucked grains, and grinding grit will degrade the relationship between PIA data and physical measurements. These problems must be rectified before PIA is performed.

Future microfabric studies should include the bright areas associated with fractured shale clasts (Fig. 38.2) in the digitized image. These bright areas are caused by plucking of fractured shale particles during the sectioning procedure. By including both the dark fractured shale pieces with the bright area associated with the "fractillites," the original shape of the shale clast can be analyzed and related to physical test results. Addition of an orientation subroutine to the erosion–dilation differencing algorithm would provide further enhance clay, mud, and shale microfabric studies by distinguishing between parallel and swirling fabric patterns.

Conclusions

Petrographic image analysis (PIA) provides an optimal means of objectively measuring and classifying microfabrics. PIA information derived from petrographic thin sections is sufficient to precisely model physical response such as fracture toughness. PIA procedures have successfully determined the number, characteristics (size, shape, and connectivity), and proportion of fabric types for each field of view within a TEM photomicrograph of Mississippi prodelta muds. We have demonstrated that PIA can be applied to microfabric studies. Relationships between fabric character and physical measures can subsequently be determined in a consistent, objective manner.

Acknowledgments

The authors wish to extend special thanks to Dr. Richard H. Bennett for his discussions of the clay microfabric section. Also, we express our appreciation to J. Barclay Ferm for his review of the fracture toughness section.

References

Bryant, W.R., and R.H. Bennett, 1988. Origin, physical, and mineralogic nature of red clays: the Pacific Ocean basin as a model. Geo-Marine Letters, v. 8(4), p. 189–249.

Bohlke, B.M., and R.H. Bennett, 1980. Mississippi prodelta crusts: a clay fabric and geotechnical analysis. Marine Geotechnology, v. 4(1), p. 55–81.

Crabtree, S.J., Jr., R. Ehrlich, and C. Prince, 1984. Evaluation of strategies for segmentation of blue-dyed pores in thin sections of reservoir rocks. Computer Vision, Graphics, and Image Processing, v. 28, p. 1–18.

Ehrlich, R., and W.E. Full, 1988. Sorting out geology—unmixing mixtures. In Mann, J., R. McCammon, and T. Jones (eds.), Use and Abuse of Statistical Methods of the Earth Sciences. Oxford University Press, Oxford, in press.

Ehrlich, R., and K.O. Horkowitz, 1984. A strong transfer function links thin-section data to reservoir physics. Proceedings of the 59th Annual Technical Conference and Exhibition, Society of Petroleum Engineers of AIME, 13263, 7 p.

Etris, E.L., D.S. Brumfield, R. Ehrlich, and S.J. Crabtree, Jr., 1988. Relations between pores, throats and permeability: a petrographic/physical analysis of some carbonate grainstones and packstones. Carbonates and Evaporites, v. 3(1), p. 17–32.

Ferm, J.B., R. Ehrlich, R.L. Kranz, and W.C. Park, 1989. Predicting the fracture toughness of sandstone from petrographic thin sections. Bulletin for the Association of Engineering Geologists, submitted for publication, in press.

Full, W.E., R. Ehrlich, and J.E. Klovan, 1981. EXTENDED QMODEL—objective definition of external end members in the analysis of mixtures. Mathematical Geology, v. 13(4), p. 331–344.

Full, W.E., R. Ehrlich, and J.C. Bezdek, 1982. FUZZY QMODEL—a new approach for linear unmixing. Mathematical Geology, v. 14(3), p. 259–270.

Full, W.E., R. Ehrlich, and S.K. Kennedy, 1984. Optimal configuration and information content of sets of frequency distributions. Journal of Sedimentary Petrology, v. 54(1), p. 117–126.

Klovan, J.E., and A.T. Meisch, 1976. EXTENDED CABFAC and QMODEL computer programs for Q-mode factor analysis of compositional data. Computers and Geosciences, v. 1, p. 161–178.

McCreesh, C.A., 1987. Determination of pore type-throat size relationships in sandstones and their implications in terms of diagenesis and petrophysics. Ph.D. Thesis, University of South Carolina, Columbia, South Carolina, 110 p.

Ross, C.M., R. Ehrlich, J.A. Gellici, and B.O. Richardson, 1986. Relationship between wireline log response and quantitative pore geometry data in Shannon sandstone. Hartzog Draw Field, Wyoming. (abs.) American Association of Petroleum Geologists Bulletin, v. 70(5), p. 641–642.

Yuan, L.-P., 1987. The petrography of permeability: determination of relationships between thin section data and physical measurements. Ph.D. Thesis, University of South Carolina. Columbia, South Carolina, 125 p.

CHAPTER 39

Automatic Analysis of Microstructure of Cohesive Sediments

Peter Smart, N.K. Tovey, X. Leng, M.W. Hounslow, and I. McConnochie

Introduction

Qualitative interpretation of electron micrographs of soils and sediments is both useful and necessary (see, e.g., Smart and Tovey, 1981; Grabowska-Olszewska et al., 1984; Bennett and Hulbert, 1986). However, quantitative instrumental interpretation is needed:

1. to express more precisely that which can be seen by eye;
2. to present data in a form suitable for statistical testing;
3. to retrieve information that cannot be obtained by eye;
4. to release expensive staff for productive work.

The object of this chapter is to give a concise summary of work that has just started in the Universities of Glasgow and East Anglia to develop digital image processing for these purposes. Although optical image processing can retrieve hidden information and represent it in a form suitable for qualitative interpretation by eye (Smart and Tovey, 1982), this method would be more difficult to develop for the present purposes and will not be discussed here. The following sections will discuss various topics in turn, and the necessary historical references will be cited at appropriate points in these sections.

Figure 39.1 illustrates the scheme of analysis that is being developed. Routines shown in upper case have been tested on an IBM-compatible PC, but a transputer was preferred for Topcontour. For many micrographs, two or sometimes three parallel streams of calculation are recommended in the early stages. The most interesting part is the right-hand branch, some of which is entirely novel; however it will be easier to explain the left-hand branch first.

Instrumental Factors

Scanning electron microscopy of ground cross sections of impregnated samples is being developed. Success depends on the adequacy of the microscope; and interpretation must allow for some contribution to the signal from particles below the surface. This method has the advantage that the samples are robust, can be examined extensively, and can be stored for future use. Two archives of transmission electron micrographs of ultrathin sections are also available. These include both studies of deformation under stress and microstructures of undisturbed soils and sediments. Further work on natural soils is in hand using this technique. The advantages of TEM are that the micrographs cover a large area at high resolution, and they may be rescanned at different magnifications or orientations. Some conventional scanning electron microscopy is also in hand.

The proper choice of magnification appears to be critical. When a picture is digitized, the sampling frequency must be at least twice as great as the frequency of detail that is to be resolved; if not, closely spaced but separate points and lines will "fall between pixels" and merge into a wholly undiscernible mess. Some vendors of image analyzers recommend, and some of the formulas in use presuppose, that the sampling frequency is approximately five times the frequency of detail. However, the higher the sampling frequency, the less the field of view. For example, the magnification should be high enough to resolve individual clay plates within domains and low enough to include many domains within the field of view. On this basis, digitization should be just within the empty magnification based on the smallest detail to be resolved. (The argument here has a purpose different from the original discussion of resolution in the scanning electron microscope given by Oakley, 1972; Smart and Tovey, 1982, Section 6.5).

Digitization to 256 gray levels is recommended, especially for studying orientation, which involves subtracting the gray level of one pixel from that of another.

In the present work, the pixel raster is square (apart from calibration errors). Rectangular rasters are less suitable for

DIGITISE

mask

adaptive threshold

		DIFFERENTIATE, U
dilate particles	erode particles	HISTOGRAM, U
erode particles	dilate particles	DIFFERENTIATE, U.A
dilate particles	erode particles	ENCODE A
measure e′	measure silt	TOPCONTOUR
calculate e″		

restore large voids and silt

histogram, all features

measure features, external

measure features, internal

Figure 39.1. Flow chart for analyzing electron micrographs of soil.

orientation studies, but they are surprisingly popular with vendors, and some trouble was experienced in avoiding them. Consideration was also given to hexagonal rasters such as in the Leitz TAS system, but the formulas in use at present came out more neatly in the square system. To minimize spurious directional effects, many of these formulas are based on almost circular local arrays of pixels; for example, by discarding the four corner pixels in a 5×5 array, the radius from the center to the boundary is constant to within $\pm 6\%$. For convenience, this will be called the 20-array, because the central pixel has 20 near neighbors.

Masking

Although unnecessary in the work done so far, it may be expedient to use hand masking to remove rare features from the area of analysis.

Phase Discrimination

At its simplest, phase discrimination is a matter of distinguishing between light gray particles and dark gray voids in negatives of transmission electron micrographs of ultrathin sections. In a perfect section, there is a relatively minor problem of choosing the exact gray level as a threshold for separation; but in a real section, spurious local variations of mean gray level often pose a major problem. Climpson and Taylor (1976) avoided this by dissolving the particles from the ultrathin sections before microscopy; and it would not arise in scanning electron microscopy of cross sections. When these spurious variations of mean gray level cannot be avoided, adaptive thresholding may give some improvement. In this technique, the gray level of each pixel is compared with the mean gray level in the surrounding area. In an analogous technique of image enhancement, which we call declouding, the gray levels are adjusted to remove the spurious local variations and to give a better picture on the monitor.

Void Ratio

In general, total void ratio, e, cannot be safely measured in a microscope lest there be too many voids too small to resolve. Assume that the saturated moisture content, m, and the specific gravity of the particles, s are known. Then

$$e = ms \tag{1}$$

Consider such a case in which micrographs show that the particles are in groups, such as domains or random clusters, and that the void ratio may be divided into inter- and intragroup void ratios, e' and e'', respectively:

$$e = e' + e'' \tag{2}$$

Also, define the apparent intergroup porosity, n', as

$$n' = e'/(1 + e' + e'') \tag{3}$$

Then, n' is given approximately by the ratio of area of intergroup voids to total area of micrograph; this may result in a slight underestimation when thin sections are used. Thence

$$e' = n'(1 + e) \tag{4}$$

and

$$e'' = e - e' \tag{5}$$

Proceeding in this way, it has been shown that the intradomain void ratio of one kaolin remained constant during consolidation under low pressures, and decreased during consolidation under higher pressures (Smart and Tovey, 1982, Figure 1.43). This suggests that in general we have to consider at least four possibilities:

1. Rigid domains slip without themselves deforming.
2. Domains remain entire but deform by interparticle slips, etc., within themselves.
3. Domains break into smaller domains.

4. Domains grow into larger domains by capturing particles from adjacent domains.

These points will be discussed further below. In another example, taking average data from Pusch (1971) and assuming $s = 2.6$ leads to the results that brackish illitic clay (under 4.4 N/cm²) had $e' = 0.55$ and $e'' = 1.66$, whereas marine illitic clay (under 9.8 N/cm²) had $e' = 1.31$ and $e'' = 0.71$. These results suggest a preliminary approach to numerical classification of soil microstructures, in which the effect of nonclay particles must also be considered.

The two studies mentioned above used hand-mapping and blacking-in, respectively. The present strategy is to use adaptive thresholding to get a black-and-white picture of voids and particles, dilate the particles just enough to obliterate the intragroup voids while merely decreasing the size of the intergroup voids, erode the particles, i.e., dilate the remaining voids by the same amount, to restore the intergroup voids to their former size apart from some smoothing of their boundaries, check that the intragroup voids have not reappeared, and measure n' (also the size, shape, and orientation of each void). Figure 39.1 permits extra erosion followed by dilation, but this may be unnecessary.

If possible, each erosion and dilation should be done in a single pass using a large circular array. However, if the processing power is limited, it may be necessary to use either the 8-array, i.e., a square of 1.4 pixels "radius," or the 4-array, i.e., a diamond of 1 pixel "radius." Several successive dilations using either of these small arrays will expand an isolated point into a large square or diamond, respectively. Some commercial packages appear to do this; but alternate use of these arrays would be equivalent to using the 20-array, which is a circle of 2.2 pixels radius, and this would be a preferable expedient. However, even IBM PCs using simple Fortran are capable of 7.4 pixels radius.

Silt

Examination of scanning electron micrographs of ground cross sections of impregnated samples suggests that silt particles might be identified by reversing the procedure for intergroup voids, as indicated in the center of Figure 39.1.

Orientation of Clay Plates

Since there is an hypothesis that anisotropic behavior results from anisotropic packing of anisotropic clay plates, considerable attention has been given to measuring this anisotropic packing, even though the hypothesis may itself require elaboration.

Our approach has stemmed from the introduction of the intensity gradient approach (Unitt, 1975). Consider a micrograph, with image coordinates, x, y and intensity, i.e., gray level, I. Let

$$\mathbf{U} = \mathrm{grad}(I) \tag{6}$$

$$U = \mathrm{mod}(\mathbf{U}) \tag{7}$$

and

$$A = \arg(\mathbf{U}) \tag{8}$$

i.e., at any point, U is the greatest rate of change of intensity and A is the direction in which this greatest change occurs. In an ideal micrograph, I will change only at the edges of particles, and the histogram of the corresponding values of A will indicate preferred orientations. For example, a vertical section of consolidated kaolin gave

horizontal	45.5%
inclined right	26.5%
inclined left	16.4%
vertical	11.6%

A theoretical method of developing formulas for U and A is given by Smart and Tovey (1988) with further comments by Tovey et al. (1988). Formulas may be developed for arrays of various sizes, shapes, and orientations, and for various degrees of freedom.

The examples here used a scaled version of the 20,14-formula (see either paper quoted) equivalent to

$$\begin{aligned}1000\frac{\partial U}{\partial x} = {} & 13I_{i-1,j-2} - 13I_{i+1,j-2} + 77I_{i-2,j-1} \\ & - 207I_{i-1,j-1} + 207I_{i+1,j-1} - 77I_{i+2,j-1} \\ & - 70I_{i-2,j} - 280I_{i-1,j} + 280I_{i+1,j} \\ & - 70I_{i+2,j} + 77I_{i-2,j+1} - 207I_{i-1,j+1} \\ & + 207I_{i+1,j+1} - 77I_{i+2,j+1} + 13I_{i-1,j+2} \\ & - 13I_{i+1,j+2}\end{aligned} \tag{9}$$

and a corresponding form for $\partial U/\partial y$, where I_{ij} is the intensity at the ith pixel in the jth row.

Present opinions are as follows:

1. Arrays larger than the 20-array are usually unnecessary.
2. The 20,14-formula is the most suitable for new work. This formula uses all 20 neighboring pixels in the array to calculate U and A at the center; it assumes implicitly that the edge of a particle is diffused over about 5 pixels, and the micrograph must be taken accordingly.
3. Unitt's original 9-point formula is faster and might be used for on-line work. This formula also assumes that the edge of a particle is diffused over about 5 pixels.
4. To salvage information from old micrographs that had been taken at too small a magnification, either Unitt's original 5-point formula, which is the smallest central difference formula, or even the 3-point forward difference formulas might be necessary.

5. Any form of preprocessing of the original data will serve only to enhance errors (except that missing points from blemished CCD cameras and missing scan lines must be patched).

To illustrate a problem that can arise when a small formula is applied to a high contrast micrograph, consider the 3-point forward difference formula applied to a micrograph in which I is either 0 or 255 (voids or particles respectively). Let L be the pixel spacing. Then

$$\frac{\partial I}{\partial x} = \frac{1}{L}(I_{i+1,j} - I_{i,j}) \tag{10a}$$

and

$$\frac{\partial I}{\partial y} = \frac{1}{L}(I_{i,j+1} - I_{i,j}) \tag{10b}$$

whence

$$\mathrm{A} = \operatorname{atan}\left[\frac{I_{i,j+1} - I_{i,j}}{I_{i+1,j} - I_{i,j}}\right] \tag{11}$$

With the values given for I, A can only be a multiple of 45°.

There is another problem that might arise in a high contrast micrograph. Consider Unitt's 9-point formula in a one-dimensional case:

$$\frac{\partial I}{\partial x} = \frac{1}{12L}(I_{i-2,j} - 8I_{i-1,j} + 8I_{i+1,j} - I_{i+2,j}) \tag{12a}$$

$$\frac{\partial I}{\partial y} = 0 \tag{12b}$$

The sign of $I_{i+2,j}$, although theoretically correct, is contrary to intuition; and if only $I_{i+2,j}$ is high, A will be in error by 180°. Fortunately, in many analyses, nondirected directions are used, i.e., A is reduced into 0–180°, in which case the error would be immaterial. However, it might become necessary to consider alternative formulas in which the partial derivatives are calculated as weighted averages of central difference pairs all of which are "the right way round."

In the present procedure, U is calculated first in one pass over the image. Then the histogram of U is found, and the lower quartile mode obtained by counting up the histogram; this is faster than calculating the lower quartile per se and probably sufficiently accurate here. In a second pass over the image, A is calculated if U is greater than the lower quartile mode, otherwise A is given a dummy value which cannot be mistaken for an angle.

Figure 39.1 also shows a coding routine, which simplifies the data by reducing the angle to horizontal, inclined right or left, vertical, or undecided, and displays the results as green, blue, yellow, red, or black pixels, respectively, as a visual check.

Random Structure

If clay particles are arranged in random clusters or face-on, then many different values of A will occur within a small area. Alternatively, where there is local preferred orientation, most of the values of A will be similar.

Several methods of analysis were considered. Some preliminary work was done on finding domains by direct searching. The strategy is calculate A by the intensity gradient method and select all values of A close to any desired direction. When the corresponding pixels are displayed, there will be relatively thick clusters in the locations of the domains that are being sought, and relatively thin sprinklings elsewhere. Several methods of cluster analysis exist for combining the clusters and discarding the outliers. One method would be to calculate interpixel distances, to group those that were close together, and to discard the others. Another method would be to dilate all the pixels strongly, so that closely spaced clusters merged together. An equal erosion would reduce all the isolated pixels to their original size. Further erosion would eliminate the outliers. Finally, a second dilation equal to the second erosion would restore the clusters to their proper sizes. Since the outermost pixels in clusters would have been derived from the edges of the outermost particles in the domains, the final clusters should correspond with the domains exactly. Consideration was also given to using the principal component of the Fast Fourier Transform of a small square around each pixel to characterize that pixel ready for a virtually instantaneous further analysis. This is probably the most powerful method available, but it is also probably the most expensive. Alternatively a simple sine wave could be fitted directly to each area. The equation is

$$I = S \cos\left[\frac{2\pi}{W}(x \cos A + y \sin A) + P\right] \tag{13}$$

where there are four quantities to find:

S = amplitude (replaces u),
W = wavelength (possibly not needed),
A = direction perpendicular to the wave (corresponds exactly to the previous use of A),
P = phase angle (probably not needed).

A quicker procedure, which owes something to Stafford and Ambler (1988), would be to consider four local traverses through each pixel, in each raster direction and in the two bisecting directions. Measures of the mean wave lengths in these four directions would yield the strength and direction of any preferred orientation. Another attractive method would be to characterize each pixel by calculating the consistency ratio for a small circular area around each pixel (see Smart and Tovey, 1982). The method used here, which we call topcontouring, is almost equivalent to an approximation to the consistency ratio method.

In topcontouring, a circular area around each pixel is examined. If there is a preferred orientation, that orientation is ascribed to the pixel, otherwise the pixel is coded as random unless there are too few real values of A within the area, in which case the pixel is coded as undecided. This procedure is new, and some tuning is required to optimize the size of the circle and the criteria for preferred orientation and for indecision; but experience so far is

Figure 39.2. Unconsolidated kaolin, originally 512 × 512 pixels, cropped to about 450 × 450, see text.

encouraging. In one way, this procedure is an elaboration of dilation; both procedures expand particles into voids. In the future, therefore, the intergroup voids will be found first, stored separately, and reinserted into the picture after topcontouring.

Mapping of Domains

Present experience suggests that provided the magnification and other criteria are chosen correctly, topcontouring will map domains and small random clusters (and large voids and silt). In a region containing large domains that are oriented randomly, the domains will usually be mapped separately, and their randomness will then be evident from the histogram of their orientation.

Three examples, chosen to test the technique, are shown here. These were all differentiated, encoded, and topcontoured; then the boundaries of the regions were superimposed on the original micrographs. Vertical and horizontal refer to the micrographs as printed not to the original sample.

Figure 39.2 shows unconsolidated kaolin (cf. Smart and Tovey, 1981). The picture width is 4 μm approximately, 512 pixels. The magnification is much larger than would normally be used to show some points more clearly. The radius for topcontouring was 32 pixels; this choice will be discussed below. A 32-pixel border has been left unprocessed. Just above the center, a large

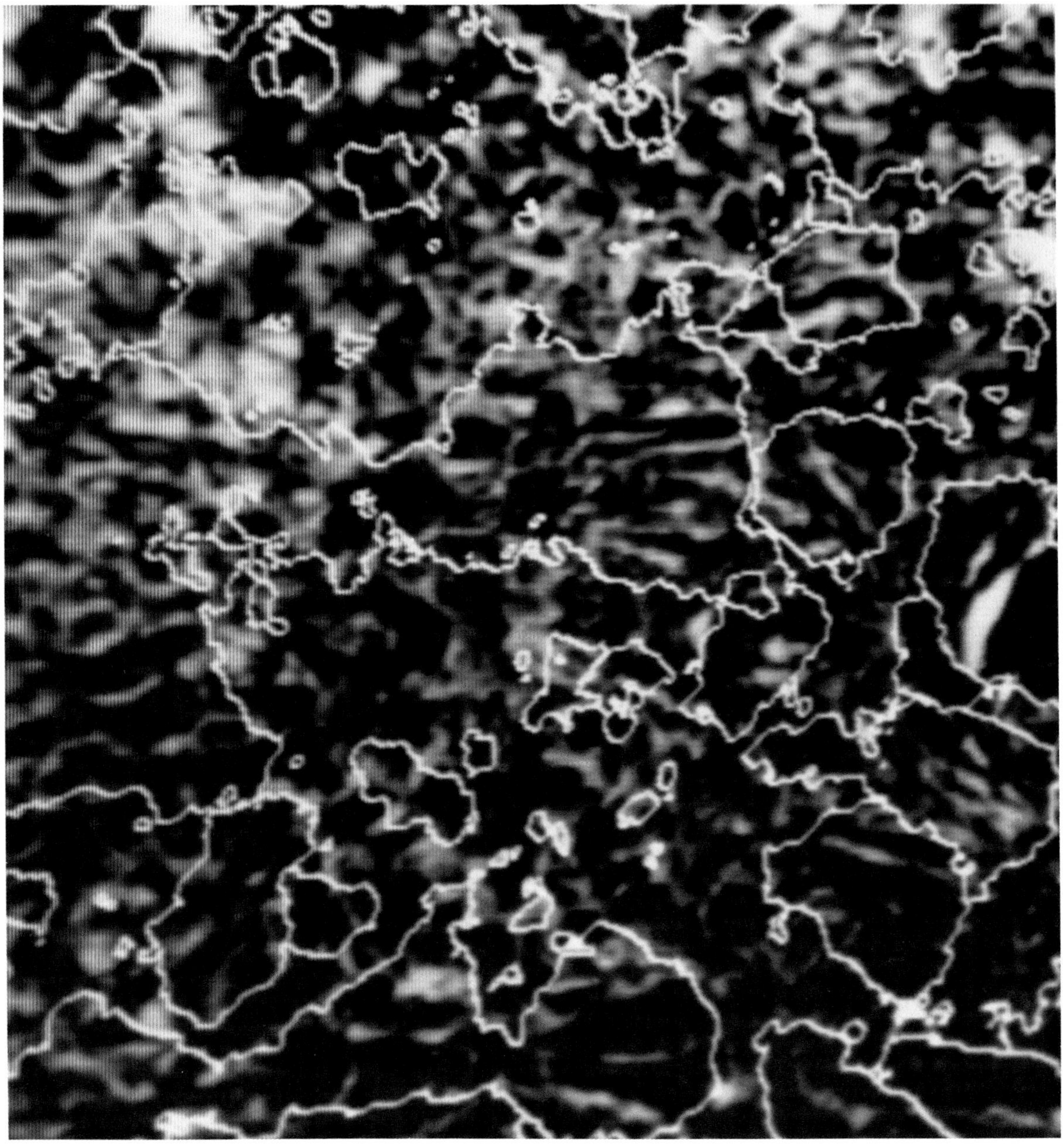

Figure 39.3. Horizontal section through turbostratic structure with preferred horizontal orientation.

domain sloping down to the right has been mapped reasonably well. At the top of this domain, the particles are horizontal, and these have been distinguished correctly. At the lower left, there is a feature that had been mapped by hand as a single vertical domain, and which the automatic method has divided into three regions, two vertical and one inclined. Inspection of the original micrograph showed that the automatic method has responded to some fine-scale structure that the hand-mapper had ignored. In some regions of the micrograph, e.g., just above a large particle on the right-hand side, the automatic method has incorrectly responded to variations of intensity within the voids. Separate mapping of the large voids and large particles is expected to give some improvement here.

Figure 39.3 is a (true) horizontal ultrathin section through consolidated kaolin; the picture width is 20 μm approximately, 352 pixels, with topcontoured using 20 pixels radius. No attempt was made to select a good micrograph, because one objective is to handle poor-quality images. In the original analysis, this micrograph had been used merely to support a conclusion from (true) vertical ultrathin sections that the structure was turbo-

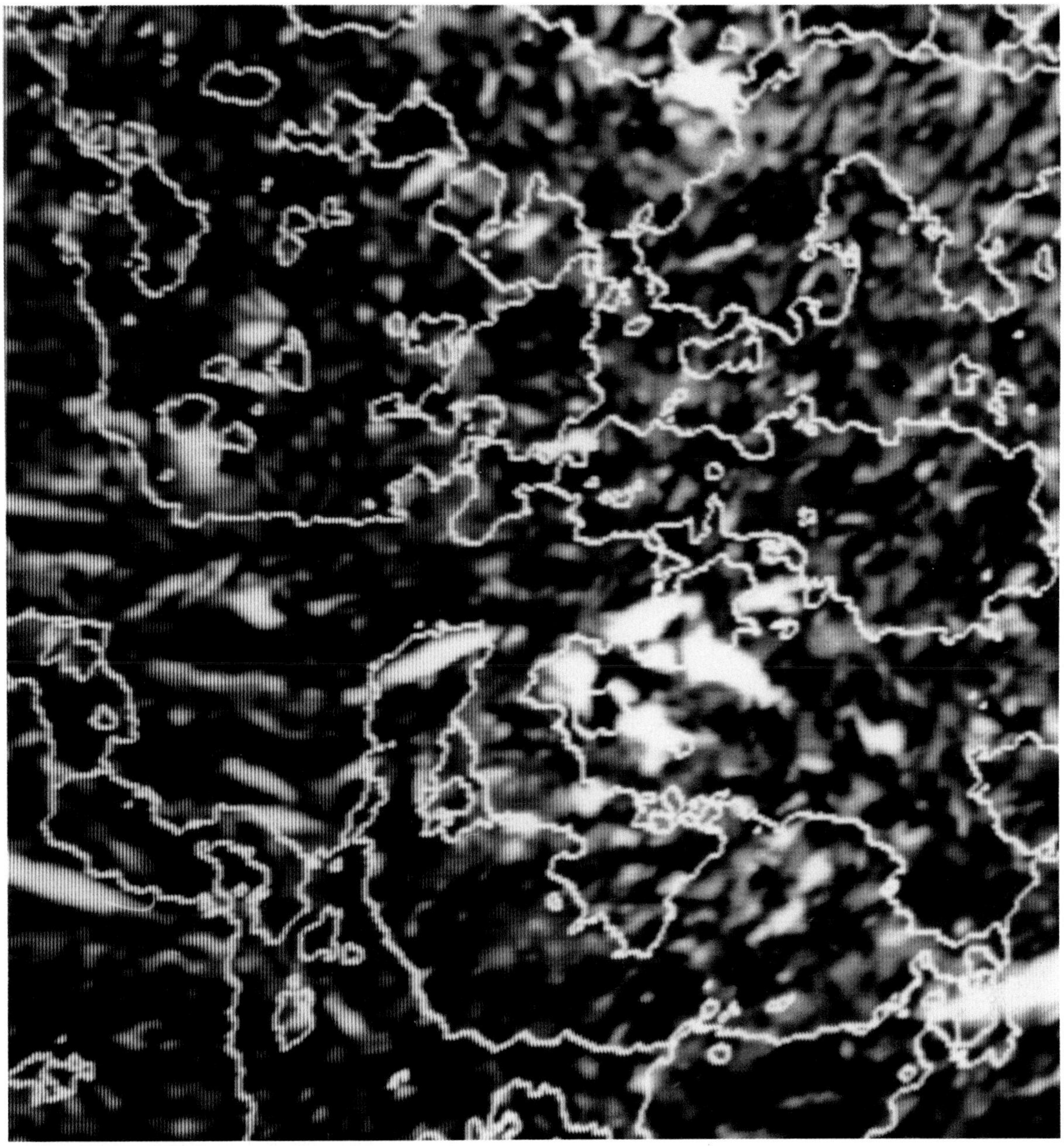

Figure 39.4. Random structure.

stratic with preferred (true) horizontal orientation. Despite some fussiness, the automatic mapping focuses attention on regions such as horizontal orientation in the center, down-right orientation just to the right of this central region, and random orientation above it. Overall, the percentage areas in the regions are undecided 0%, vertical 1.6%, down-right 9.5%, horizontal 36.0%, down-left 5.9%, and random 47.0%.

Figure 39.4 is an ultrathin section of kaolin, which had originally been classified merely as random structure; picture details are as in Figure 39.3. The automatic mapping has subdivided the micrograph into approximately equal areas of horizontal, down-left, and random, with negligible areas of vertical and down-right. In doing so, the automatic method brings to the attention of the analyst features that require consideration.

In the automatic method of mapping described above, the operator can control the definition of an undecided direction, the definition of randomness, and, through the choice of radius when topcontouring, the scale of discrimination. All three of the examples discussed above were topcontoured at various radii up to 128 pixels and the "best" results selected subjectively. Subjec-

Table 39.1. Percentage areas in regions of Figure 39.4 mapped using various radii.*

	Code					
Radius	0	1	2	3	4	5
128	0	0	0	5.4	0	94.5
64	0	0	0	25.6	9.8	64.4
32	0	0	0.4	27.5	23.0	48.9
20	0	1.3	2.9	29.4	26.2	39.9
16	0	3.8	4.5	31.7	27.3	32.5
8	0	11.7	12.6	32.6	27.2	13.7
0	6.9	20.2	19.2	27.7	25.7	0

*Radius in pixels; codes: 0, undecided; 1, vertical; 2, down right; 3, horizontal; 4, down left; 5, random.

tive control seems essential here. Table 39.1 shows results for Figure 39.4. The encoded micrograph, radius = 0, had 6.9% undecided. In this micrograph a small radius sufficed to smooth away the undecided pixels. The encoded micrograph also had approximately equal percentages in the four directions, thus supporting the original classification as random. On a large scale, radius = 128, almost the whole micrograph was mapped as random, again supporting the original classification. By using intermediate radii, more detail was brought out.

Intragroup Analyses

Once domains and random clusters have been mapped, either automatically or by hand, in addition to measuring their external attributes (e.g., area, length, breadth, orientation), the perfection of packing can be studied using, for example, the deviations of the intensity gradients from the mean direction, or by analyzing the components of the Fast Fourier Transform. Here, only one transform is required for each domain, so the problems of expense raised above no longer apply.

Conclusions

Established methods of image analysis, such as adaptive thresholding, dilation and erosion, and intensity gradient analysis, are already available and suitable for the measurement of intergroup void ratio and hence the calculation of intragroup void ratio, and for the measurement of overall orientation distributions. Newer methods are now being rapidly developed to map and measure random structure and domains. Methods already exist to permit intragroup analyses to start as soon as isolated groups are presented for analysis. Taken together, digital image analysis of electron micrographs is rapidly becoming a powerful tool for studying structural changes during deformation and for classifying natural soils of known and unknown geological provenance.

Acknowledgments

The work is being supported by AFOSR Grant 87-0346, SERC Transputer Initiative, and the Universities of Glasgow and East Anglia. We thank Prof. G. Petrie, Mr. D.A. Tait, and Dr. J.W. Kay, for various suggestions, and members of our own departments for help.

References

Bennett, R.H., and M.H. Hulbert, 1986. Clay Microstructure. International Human Resources Development Corporation, Boston, MA, 161 p.

Climpson, N.A., and J.H. Taylor, 1976. Pore size distribution and optical scattering coefficients of clay structures. TAPPI, v. 59, p. 89–92. (TAPPI, the journal of the Technical Association of the Pulp and Paper Industry).

Grabowska-Olszewska, B., V. Osipov, and V. Sokolov, 1984. Atlas of the Microstructure of Clay Soils. Państwowe Wydawnictwo Naukowe, Warsaw.

Oakley, C.W., 1972. The Scanning Electron Microscope. Cambridge University Press, Cambridge.

Pusch, R., 1971. The influence of stress on clay microstructure. National Swedish Building Research Summary R13. Svensk Byggtjänst, Stockholm.

Smart, P., and N.K. Tovey, 1981. Electron Microscopy of Soils and Sediments: Examples. Oxford University Press, Oxford.

Smart, P., and N.K. Tovey, 1982. Electron Microscopy of Soils and Sediments: Techniques. Oxford University Press, Oxford.

Smart, P., and N.K. Tovey, 1988. Theoretical aspects of intensity gradient analysis. Scanning, v. 10, p. 115–121.

Stafford, J.V., and B. Ambler, 1988. Seedbed characterisation by video image analysis. Presented at the 11th Conference of International Soil Tillage Research Organisation, Edinburgh.

Tovey, N.K., P. Smart, and M.W. Hounslow, 1988. Quantitative orientation analysis of soil microfabric. Presented at the 8th International Working-Meeting on Soil Micromorphology.

Unitt, B.M., 1975. A digital computer method for revealing directional information in images. Journal of Physics E, Series 2, v. 8, p. 423–425.

CHAPTER 40

The Application of Image Analysis Techniques to Microstructure Studies in Geotechnical Engineering

Shobha K. Bhatia and Aly Soliman

Introduction

The microstructure (fabric*) of granular soils is defined as the spatial arrangement of particles and associated voids (Brewer, 1964; Oda, 1972). Research has shown that the microstructure of granular soils play important roles in the soils' engineering behavior and properties. It is also believed that the behavior of granular materials can be evaluated by studying their microstructures. Techniques for studying microstructure of granular soils include nonoptical and optical methods. The nonoptical methods are mainly used to determine the pore size distribution and coordination number of granular materials (see Juang, 1981; Oda, 1977). For the optical methods, a microscope, either light or electron, is used to extract information from thin sections or photographs of thin sections. Quantitative analysis is performed manually using a microscope equipped with a mechanical stage and a point counter (Oda, 1976; Mitchell et al., 1976).

In this paper, optical techniques for extracting quantitative information from thin sections using an automated image analysis system are discussed. The advantage of using the image analyzer is the reduction of processing time and of the amount of manual work involved in extracting quantitative information from thin sections and the increased accuracy of such measurements.

Microstructure and Engineering Properties of Granular Soils

The dependence of the engineering properties and behaviors of soils on their microstructure has been evaluated over the last three decades. Several microstructure elements were proposed, and engineering properties and behaviors of soils were related to these elements.

Oda (1972, 1976) suggested that the mechanical behavior of granular soils can be determined by considering the following four microstructure elements: (1) preferred direction of the long axes of nonspherical particles, (2) intensity of the preferred direction, (3) three-dimensional distribution of normals to tangential planes at contacts, and (4) average number of contacts per particle, which is closely related to the void ratio of the material. According to Oda, the first two fabric elements refer to orientation, while the latter elements refer to packing. In addition, he proposed experimental techniques for determining the three fabric elements, (1) to (3), by means of thin sections and a microscope equipped with a mechanical stage. According to Oda, the void ratio and its distribution can be used instead of the coordination number to define the packing characteristics. He also proposed a technique for determining the void ratio and its distribution from thin sections.

Mitchell et al. (1976) studied the influence of microstructure on liquefaction behavior of sand using the techniques proposed by Oda (1972, 1976). They studied the following elements: (1) the orientation of grains and (2) the three-dimensional distribution of normals to tangential planes at contacts. In addition, they hypothesized that other elements, such as pore sizes, pore size distribution, and distribution of voids within samples may have an important influence on the liquefaction behavior of sand. However, these elements were not investigated in their study because of time constraints.

One of the fundamental relationships between microstructure and engineering properties of granular soils is the relationship between the pore size distribution and permeability (Juang, 1981; Nye 1985). According to Juang (1981), methods for determining the pore size distribution include (1) probabilistic theory

*The term fabric is used to define the microstructure of granular soils.

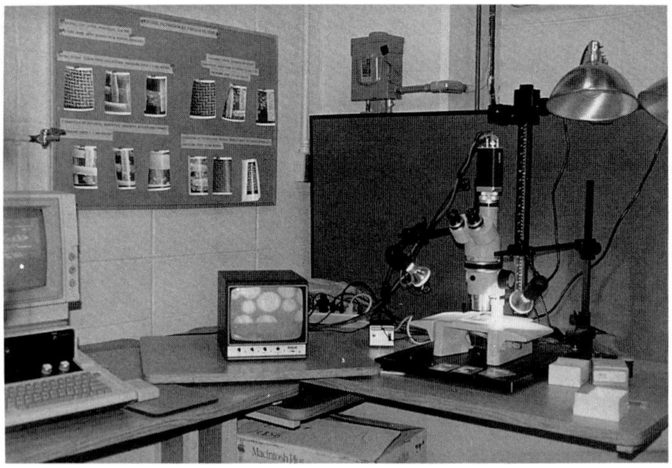

Figure 40.1. Image processing system.

and representative grain size distributions (indirect), (2) probabilistic theory and scanning electron microscope techniques (indirect), (3) capillary suction technique, and (4) mercury intrusion technique. He determined the pore size distribution using the latter technique and predicted permeability by applying a probabilistic theory to the measured pore size distribution. His results were in accord with the measured values of permeability.

Nye (1985), on the other hand, determined the pore size distribution of Ottawa sand by drawing random lines on printed binary images generated by an image analyzer for thin sections and manually measuring the length of each portion that intercepts the pore space. The distribution of the intercept lengths was regarded as the pore size distribution, and was used to predict permeability using Marshall's equation (Marshall, 1958). The predicted values of permeability were overestimated. According to Nye (1985), the quality of the analyzed image was poor due to problems with magnification and this consequently affected the results. The image analyzer, currently available at Syracuse University, can produce high-quality binary images that could be used to determine the pore size distribution with greater accuracy than Nye (1985).

Thin Section Analysis to Study Microstructure

In the past 15 years, quantitative analyses of fabric elements were performed on thin sections cut from laboratory-prepared

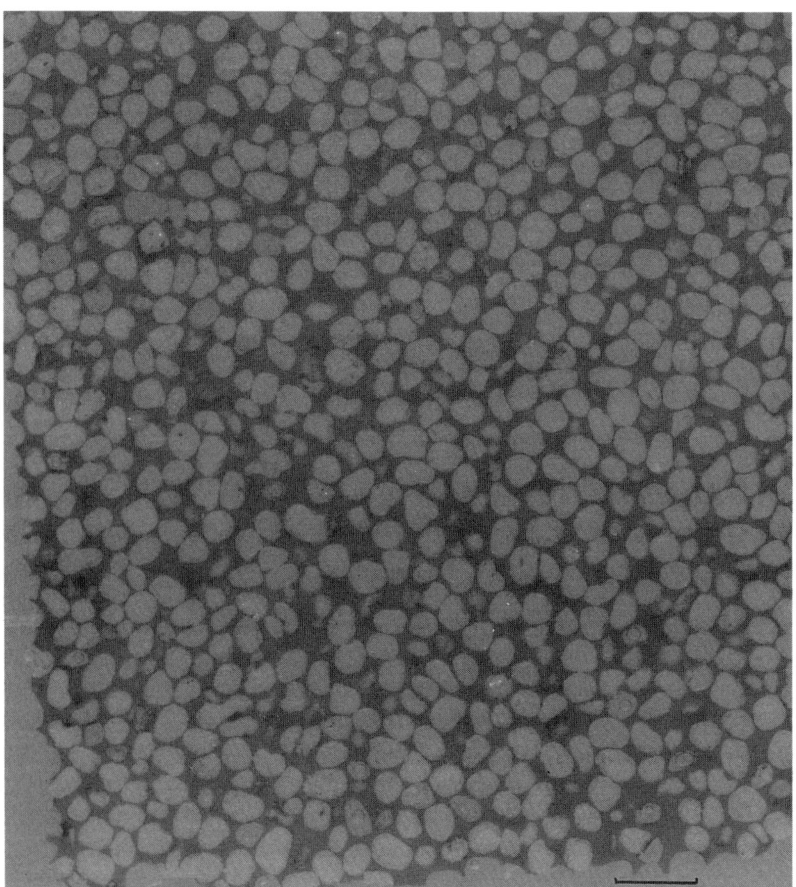

Figure 40.2. Enlarged photograph of a thin-section of Ottawa sand (C-190) in medium dense condition.

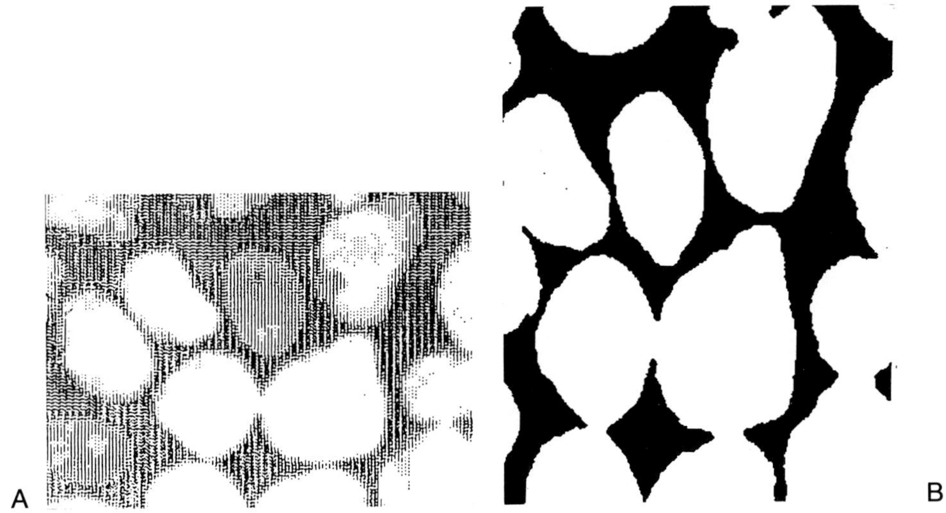

Figure 40.3. A gray-scale image produced by the image analyzer for the outlined part shown in Figure 40.2.

three-dimensional samples. The method included (1) impregnation of samples with polyester–resin having a very low viscosity, (2) preparation of thin sections using a diamond saw and a polishing machine, and (3) analysis of the thin section.

From the thin section, most of the fabric elements previously described can be measured. In the past, most measurements from thin sections were made either by projecting an image on a screen and manually taking measurements (Oda, 1976) or by using a microscope equipped with mechanical stage (Oda, 1972; Mitchell et al., 1976). In general, these techniques were subject to systematic errors from the operator and were time consuming. In addition, manual measurement is tiring work for the operator's eyes.

With the recent development of automated image analyzing computers, such measurements on thin sections or photographs can be quickly and easily performed with greater accuracy.

Principles of the Image Analyzing Computer

Image analyzers are digital computers that make it possible to extract and process quantitative information from thin sections or photographs. The image analyzer used for this study is the Image Plus system (Fig. 40.1) supplied by Dapple Systems of Sunnyvale, California. The system consists of the following components:

1. a video camera for scanning images, i.e., thin sections or photographs;
2. a 9″ monochrome monitor for viewing images;
3. an Apple IIe computer for data extraction and analysis;
4. a 12″ monochrome monitor, which is the screen on which the Apple IIe computer displays its text, graphics, and the digitized images; and
5. a printer.

To be able to analyze very fine microstructures and to avoid taking photographs of thin sections, it is recommended that the television camera be attached to a light microscope or that there be a direct input from a scanning electron microscope.

The process of image analysis can be summarized as follows:

1. The thin section or photograph is placed underneath the camera and thus appears on the camera monitor. Figure 40.2 shows a typical enlarged photograph for a thin section from a sample of Ottawa sand C-190, which was analyzed by using the image analysis system.
2. The image is acquired by the camera. This step converts the original image into another digitized image and stores it in the computer memory. The digitized image is an array of squares, whereas each square (a pixel) is defined by three values; two spatial coordinates (X,Y) and a gray level intensity value. The gray level is a measure of brightness and is restricted to integral values. The digitized image is called the gray scale image. The Image Plus system distinguishes 64 different gray level ranges from 0 (black) to 63 (white). It is important to note that more expensive image processing systems have 256 gray level options. In our system, the image is digitized as a grid of 256 × 256 pixels and the light signal at each pixel is coded as a 6 bit word. Many advanced automated Image Analysis systems have $(512)^2 \times (1024)^2$ or $(512)^2 \times (4096)^2$ pixel density and 8 or higher bits of gray level resolution. (Fig. 40.3a shows a typical gray-scale image for the outlined part of the photograph in Fig. 40.2).
3. A binary (black and white) image can then be created from the stored gray-scale image by selecting upper and lower limits for the gray levels available in the gray-scale image. In the binary image, all pixels on the gray-scale image whose numerical gray level lies between the selected limits will

appear, in white, on the computer's graphics display, while all other pixels on the gray-scale image will appear in black. The white region of the binary image represents the features to be measured and analyzed. Figure 40.3b shows a typical binary image obtained by the image analyzer for the gray-scale image shown in Figure 40.3a. Note that in this figure, particles are shown in black and voids are shown in white.

Image analyzers provide accuracy and reliability in measurements and reduce the processing time as well. The overall accuracy of the analysis depends mainly on the quality of the thin section or the photograph. A high-quality thin section or photograph has a sufficient contrast between the different phases of which the material is composed. The contrast of an image can be achieved either in the preparation of the thin section (by mixing the epoxy with black or blue dye to provide contrast with the grains) or in the processing of the photograph (by using high-contrast paper and/or special films). Note that the better the quality of the thin section, the better the quality of its photograph, and less time is consumed in editing and processing the binary image for analysis.

Analysis of the binary image is the central task of automated image processing systems. Various automated and semiautomated systems are available to perform different tasks. The tasks to have a number of distinct stages:

1. to separate the objects of interest from the background;
2. to extract the appropriate information from each object;
3. in some cases to relate the objects to each other or to some general features of the scene;
4. to prepare the resulting analysis in a manner convenient for the operator to use, and
5. to accomplish these separate actions under the control of a procedure of suitable flexibility.

Distinguishing the objects from the background is usually an easy task for the human observer but one that is peculiarly difficult to program in an automated image analysis system. A major division of systems is between those that isolate objects entirely automatically, by the program, and those that involve the operator's interaction. The isolation of objects invariably relies exclusively on the large-scale (low spatial frequency) pattern of light over the pixel grid. Whether the identification and specification of the object present serious programming difficulties depends largely on the further information that may be required; it may, for example, be relatively simple to determine that an object is present, but much more difficult to delineate its outline exactly. Based on our experience, the systems that allow operator intervention are more suitable. Operators can use a light pen to point to an object or draw its outline on the monitor screen. Once the objects are within a picture from the background, an immense number of further analytical procedures may be attempted. Basically, objects may be counted, measured, classified, and spatially related. Systems where such analytical procedures are built in are more useful.

A variety of general purpose image analysis systems are available today. These systems, in principle, can accomplish automatically all the stages of acquiring the picture, digitizing it, separating objects from background, and then analyzing them. In using the complete system, the operator is usually required to select a field of view, but, thereafter, the system takes over to an extent dependent on the program employed. However, in most systems, the operator may intervene at will or may be called on, in the running of the program, to declare procedure options at various points in the analysis routine. Table 40.1 lists seven such systems commercially manufactured. It is important to point out that there are many other systems available today that are not included in this list. The purpose of this list is to point out important differences between specific commercially available systems.

All of the systems except the one manufactured by Carl Zeiss, Jena, use television camera input, scanning in the image plane reimaged into the camera (Rosen, 1984). The systems significantly differ in their pixel density and gray level resolution in acquiring the image. Higher pixel density and gray level resolution mean better digitized images; more expensive is the general purpose image analyzer, and higher is the pixel density and gray level resolution. The majority of the systems work with monochrome detection, except the system from Carl Zeiss (Rosen, 1984). In addition, most of the systems have some form of operator's interaction. The cost of a completely automated image processing system depends to a large extent on the power and range of its analytical capabilities. All the systems listed in Table 40.1 are sold with a variety of options. As a rough guide, most versatile systems fall in a price range between \$15,000 to \$120,000.

Applications of Image Analysis in Geotechnical Engineering

In this section, improved techniques to characterize the fabric of granular soils are presented. The techniques utilizing powerful routines provided in the Image Plus system were designed to reduce the amount of manual work needed to extract data from thin sections or photographs. Both elements of fabrics and specific application are discussed in the following sections.

Orientation of Nonspherical Particles

Investigators (Mitchell et al., 1976) used the orientation of grains to study the influence of sample preparation techniques on engineering properties of granular soils. The orientation of nonspherical particles has been represented by the inclination of the longest axis of the grains with respect to a fixed reference (Oda, 1972, 1977). Figure 40.2 is an enlarged photograph (on high contrast paper) of a thin section of black colored resin, medium dense Ottawa sand impregnated with C-190. To extract information about the orientation of grains with respect to a fixed reference, the image analysis system must separate each grain from the background. As it can be seen in Figure 40.2, the gray level

Table 40.1. General purpose automated image analysis sytems.*

Manufacturer	Instrument	Scan plane	Pixel density	Gray level resolution	Light detection†	Operator interaction§
Bausch and Lomb	Omnicon 3000	Image	625 × 1066	9 bits	M	K,L
Cambridge Instr.	Quantimet 900	Image	704 × 896	6 or 8 bits	F	K,L,P
Carl Zeiss, Jena	Morphoquant	Object	512 × 512	7 bits	C	K
Dapple Systems⅜	**Image Plus**	**Image**	**256 × 256**	**6 bits**	**M**	**K,T**
Joyce Loebl	Magiscan 2	Image	$512^2 \times 1024^2$	6 or 8 bits	F	K,L
Leitz	TAS Plus	Image	512 × 512	8 bits	F	K,L
Reichert Kontron	Ibas II	Image	$512^2 \times 4096^2$	8 bits	C,F	K,T

† M, monochrome; C, color; F, monochrome detection with false color display.
% K, keyboard; L, light pen; T, digitizing tablet; P, potentiometer knobs.
⅜ Used in our study.

Figure 40.4. Two binary images for different areas shown in Figure 40.2.

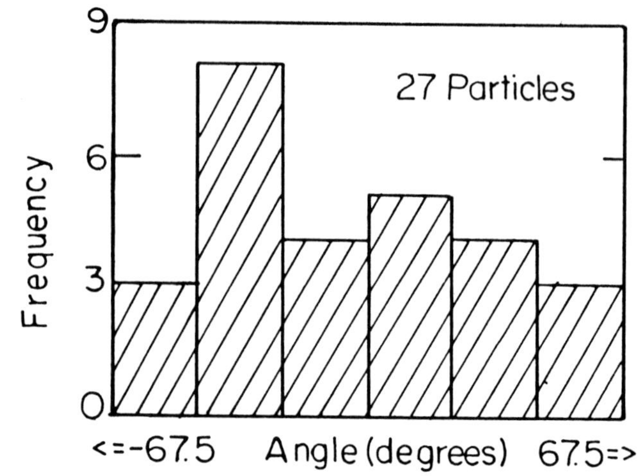

Figure 40.5. Orientation angle histogram for the binary images of Figure 40.4.

of the background and grains is not too difficult. The black pigment mixed in the resin was not enough to create much distinction between grains and pores.

After acquiring the image, binary image thresholding was carried out be setting an upper and lower level directly on the gray scale brightness histogram to isolate the grains. Binary image editing operations such as erosion–dilation was also used. Once the grains are clearly identified from the background, the manual mode was used to separate each grain, which the image analysis system would not otherwise do. Therefore, systems without a manual mode cannot be used for this kind of study.

The Image Plus system can automatically determine the longest axis of each feature, i.e., grain and measure its orientation angle with respect to the horizontal axis provided that all grains on the binary image are not touching. Figure 40.4 shows two binary images of a section of the photograph shown in Figure 40.2. These two images were used to produce the distribution of the orientation angle, $\bar{\theta}$, shown in Figure 40.5. The preferred direction of the long axes of the particles, θ can be determined using the expression (Oda, 1976)

$$\tan(2\bar{\theta}) \frac{\sum_{i=1}^{n} \sin(2\theta i)}{\sum_{i=1}^{n} \cos(2\theta i)} \tag{1}$$

where n is the number of particles measured and the intensity of the preferred orientation can be calculated using the following expression (Oda, 1976):

$$\text{V.M.} = \left\{ \left(\sum_{i=1}^{n} \sin\left[(2\theta_i)\right]^2 + \left(\sum_{i=1}^{n} \cos\left[(2\theta_i)\right]^2 \right\}^{1/2} \tag{2}$$

The value V.M., which is called vector magnitude, varies from 1 to 100. A complete random distribution of the orientation of the long axes gives a vector magnitude of 0%. On the other hand, a 100% value of the V.M. means that all the observed orientations are exactly parallel to the calculated preferred direction $\bar{\theta}$. For the binary images shown in Figure 40.4, the orientation angle, $\bar{\theta}$, was found to be -1.0 and the intensity of the preferred orientation was 10%.

The orientation angle measurements are done by analyzing 25–30 grains each time and accumulating this information for the entire section. The advantage of analyzing fewer grains each time is that measurement accuracy is higher and the editing of the binary image is easier.

The Pore Size Distribution

Extensive work has been conducted on the influence of pore size distribution on permeability. A variety of equations (Childs and Collis-George, 1950a,b; Marshall, 1958; Millington and Quirk, 1961) have been developed to calculate permeability of soil based on the pore size distribution of soil. It has been pointed out that all of the above equations are similar in performance. Nye (1985) used Marshall's equation to predict the permeability of clean sand, whereas he measured pore size distribution from thin sections using a Model 2000 Image Analyzer from Image Technology Corporation. He obtained the pore size distribution by drawing many parallel lines on printed binary images generated by an image analyzer and by manually measuring the length of each portion that intercepted the pore space. Nye considered the intercept as a void diameter, and used these measured void diameters in Marshall's equation to calculate permeability. The physically measured values of the permeability were much smaller than the calculated values. These results should not be a surprise, because coarsely spaced intercept lines cannot represent the pore size distribution.

The technique used by Nye (1985) to characterize the pore size distribution can be done automatically using the Image Plus system. A technique to determine the pore size distribution from thin sections using the Image Plus system is proposed:

Figure 40.6. (a) A binary image containing vertical lines only. (b) Applying the line intercept method to the binary image of Figure 40.4b. (c) Intercept lengths derived from subtracting Figure 40.4b from 40.6b.

1. Prepare representative binary images from the thin section as the ones shown in Figure 40.4a.
2. A binary image containing random straight lines is prepared and stored. For this case, vertical lines are selected for simplicity (Fig. 40.6a).
3. Add the binary image created in Step 2 to any of the binary images created in Step 1. This step results in an image such as the one shown in Figure 40.6b.
4. The image shown in Figure 40.6b is then subtracted from the original image shown in Figure 40.4b. This step results in the image shown in Figure 40.6c, which contains the portion of the vertical lines that intercept only the pore space. The length of each portion is automatically measured by the image analyzer, and the measured quantities stored on a file for later analysis.
5. Repeat Steps 2 through 4 for all other binary images prepared in Step 1. The measured intercepts obtained from each case are then added to the file created in Step 4.
6. The distribution of the measured intercepts is then obtained, which is the desired pore size distribution. Figure 40.7 shows the distribution for the two binary images shown in Figure 40.4a,b.

It must be pointed out that this technique can be fully automated and a very fine grid of vertical lines can be used for intercept measurements. It is yet to be seen whether the use of such calculated intercepts in Marshall's equation would lead to better prediction.

The Void Ratio and Its Distribution

The void ratio in any binary image can be easily calculated as the ratio of the area of voids (black) to the area of solid particles (white). The distribution of the void ratio in the thin section can be determined by using the method suggested by Oda (1976).

Oda (1976) proposed an experimental technique to determine the frequency distribution of void ratios of granular materials by means of thin sections and a microscope. The procedure suggested by Oda to determine the distribution of void ratios follows:

1. Divide the thin section into polygons enclosed by straight lines, connecting the centers of gravity of the particles as shown in Figure 40.8.
2. Measure the areas occupied by the solid particles A_{si} and by voids A_{vi} for each polygon.
3. Calculate the void ratio e_i for each polygon as follows:

$$e_i = \frac{A_{vi}}{A_{si}} \tag{3}$$

4. Repeat Steps 1 through 3 for at least two sections (vertical and horizontal sections) to account for the effect of fabric anisotropy on the frequency distribution of e_i.

The computation of the quantities of A_{vi} and A_{si} can be performed using a planimeter or an image analyzer.

Oda's method can be utilized by the Image Plus system. Since this technique does not require all centers of gravity of the particles to be connected as shown in Figure 40.8, it cannot be fully

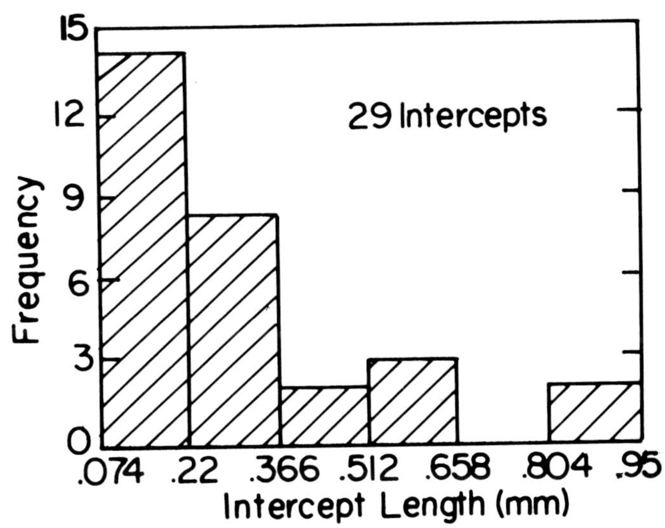

Figure 40.7. Intercept length histogram produced by the image analyzer for the two binary images shown in Figure 40.4a and b.

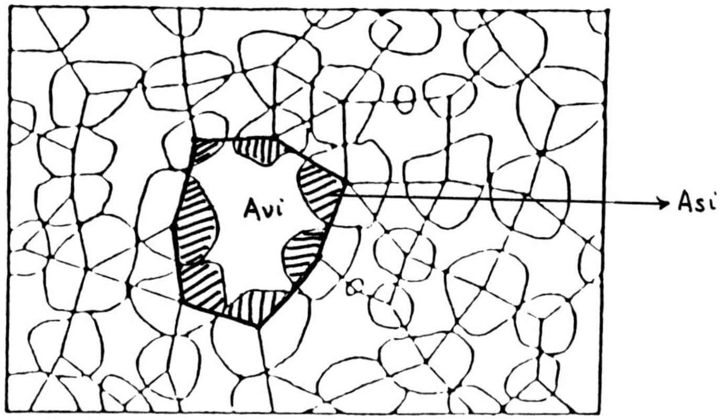

Figure 40.8. Determination of the frequency distribution of the void ratio using Oda's method (Oda, 1976).

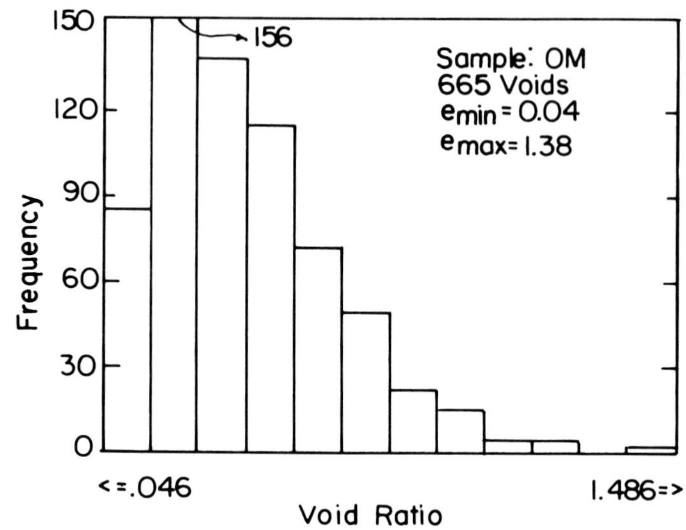

Figure 40.9. Histogram of the frequency distribution of void ratio for Ottawa sand C-190 (medium density).

automated on a computer. The measurement can be performed manually on video images:

1. A video image containing approximately 8 to 12 particles is first viewed on the screen.
2. A polygon is drawn on the screen by marking all the centers of gravity of all particles that surround a void. The area of this polygon, A_i, can thus be calculated by the computer.
3. The void associated with the polygon is then outlined, and the area of the void, A_{vi}, can thus be calculated by the computer. The void ratio, e_i, is then calculated by expression 1.
4. Steps 1 through 3 were repeated for a sufficient number of voids to obtain a continuous distribution of e_i. This distribution was regarded as the frequency distribution of void ratio.

Figure 40.9 shows a typical frequency distribution of void ratios obtained from four thin sections for a medium dense sample of Ottawa sand C-190, in which 665 polygons were used. Extensive data on a variety of granular soil have been presented by Bhatia and Soliman (1990), showing the influence of particle shape on the frequency distribution of void ratios. It is important to note that the void ratios obtained from the thin sections were lower than those of the samples as calculated from phase relationships. In Table 40.2, the comparison is given in terms of void ratio, e, and porosity, n, for the three different materials. As evident, the optical porosity or the void ratio is consistently lower than the overall sample. Similar observations were made by Etris et al. (1988).

However, a theory by Delesse (see Underwood, 1970) formulates that the void ratio of the thin section should be equal to the void ratio of the sample. Assuming that the theory is correct, the image analyzer underestimated the void ratio of the sample.

In general, it is known that the thin section underestimates the void ratio of the sample due to its finite thickness (Underwood, 1970; Oda, 1976; Russ, 1985). But it was found that this error accounts for only 2–3% of the error in volume fraction. Accordingly, it seems that this is not the major source of error in this current study. Possible sources of errors are listed below:

1. Error due to manual measurement. This source of error may cause a decrease or an increase in void ratio. However, outlining a void bias due to human tendency to draw around the boundary of the particle rather than on the boundary may cause a considerable decrease in the void ratio. Also, the thickness of the tracing line may have an effect on the accuracy of the measurements.
2. The drawing device was very sensitive to any disturbance, so it was very difficult to accurately trace the boundaries of the particles. In some cases, the measurement was repeated until a good estimate was arrived at.
3. Distortion due to deficiencies in the video acquisition system. This may account for the apparent enlarging of particles. However, this has not been verified at this stage of our research.

Two procedures were used to analyze the thin sections. For the glass beads the sections were viewed directly by the computer, while for Ottawa and Crystal silica sands, enlarged photographs were taken because the particle size was too small. However, the same amount of error was encountered for all sections. More precise analysis is needed to investigate this important observation.

Research is in progress at Syracuse University to examine the source of the differences between optical and physically derived values of void ratios.

Finally, this error in void ratio has been assumed to have a low significance on the shape of the frequency distribution of void ratios. Accordingly, Figure 40.9, which represents the effect of particle shape on the frequency distribution of void ratios, is a valid one.

Table 40.2. Comparison of porosity void ratio obtained from thin sections to those derived from physical samples.*

Sample description	Sample porosity (%)	Optical porosity (%)	Sample void ratio	Optical void ratio
Glass beads (dense)	35.0	28.0	0.54	0.39
Glass beads (medium)	38.2	30.5	0.62	0.44
Glass beads (loose)	41.8	32.4	0.72	0.48
Ottawa sand (dense)	34.2	28.0	0.52	0.39
Ottawa sand (medium)	38.2	31.0	0.62	0.45
Ottawa sand (loose)	42.5	33.3	0.74	0.50
Crystal silica (dense)	44.4	34.1	0.80	0.50
Crystal silica (medium)	46.5	37.1	0.87	0.59
Crystal silica (loose)	50.4	39.7	1.02	0.66

*From Bhatia and Soliman (1990).

Microstructure of Clogged Geotextiles

In geotechnical engineering, there is a great need to evaluate the filtration performance of clogged geotextiles. Prediction of filtration performance can be evaluated by determining the void granulometry in which the trapped particles reduce the pore spaces between fibers (Masounave et al., 1980). The analysis of cross sections of geotextiles by an image analyzer must be preceded by chemical treatment by which a sample of virgin clogged fabric is encased in a synthetic resin inert to the fiber material. When the voids have been filled, a controlled temperature cooling period is necessary to ensure hardening of the resin without formation of gas bubbles. Finally, the hardened resin block containing the geotextile is cut and polished to obtain a well-defined section.

An image analyzer can be used to analyze the cross-section. Many times, virgin geotextiles (Fig. 40.10a) are analyzed to evaluate their structure parameters, such as maximum and minimum pore sizes and pore size distributions of geotextiles. The study of clogged cross sections with an image analyzer is critical to the study of clogging mechanisms in geotextiles. The clogging could be due to particles trapped permanently in the geotextiles at various locations, bacterial fungi and algae growth, blocking of pore spaces by coarse particles, and blinding of geotextiles (Rollin and Lombard, 1988).

In Figure 40.10a,b the magnified cross section of unclogged and clogged geotextile is shown, and Figure 40.10c a magnified cross section of the soil (sil-co-sil 75) is shown, which was used for filtration tests to study the clogging behavior. With an image analyzer one can easily measure pore space and pore size distribution of the geotextile (Fig. 40.10a) and to analyze the clogged cross section, various filters in the microscope can be used to differentiate between geotextile fiber and the soil. Therefore, acquiring two images separately (geotextile fiber and the soil particle that has clogged the geotextile) and superimposing these images will provide an estimate of the magnitude of clogging. This technique is being used at Syracuse University to study the clogging mechanism of geotextile with gap-graded silts and sands.

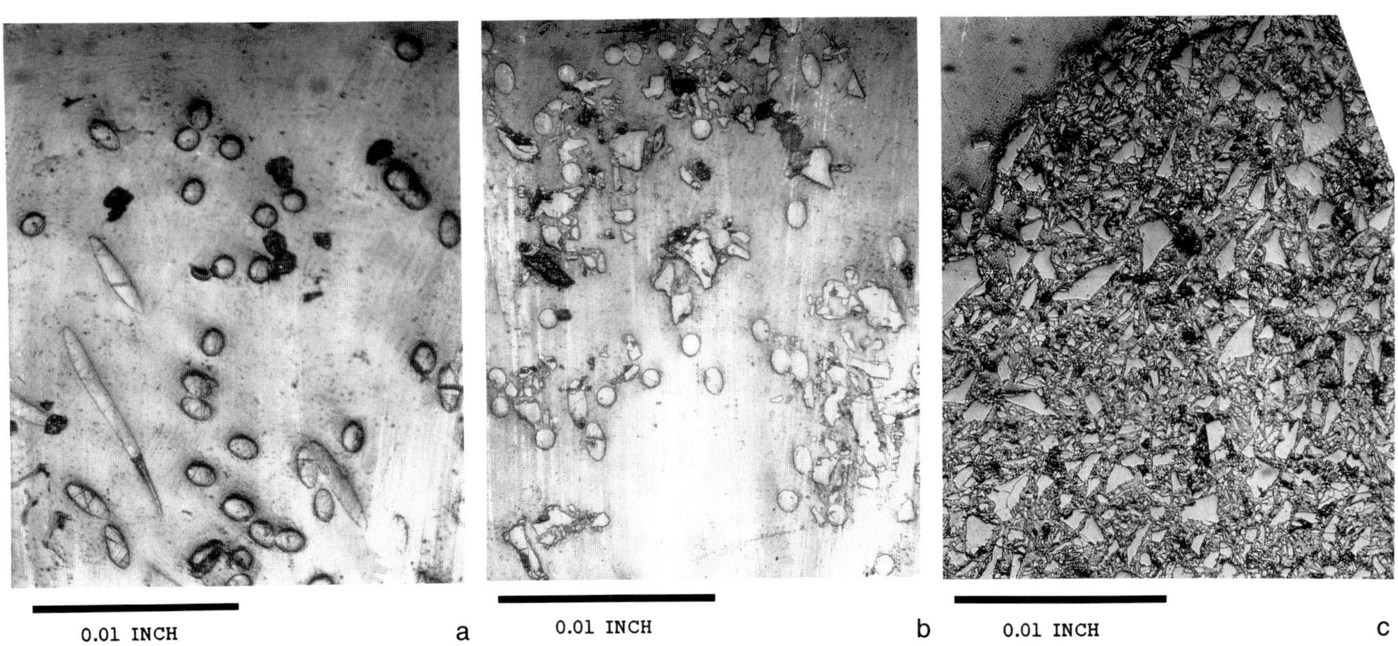

Figure 40.10. (a) Magnified cross section of unclogged nonwoven geotextiles. (b) Magnified cross section of partially clogged nonwoven geotextiles. (c) Magnified cross section of the soil used for the clogging tests.

Figure 40.11. (a) A binary image of a polished rock surface (Ehrlich et al., 1984). (b) A binary image of deighton soil (Murphy et al., 1977). (c) A binary image of medium dense Ottawa sand (Bhatia and Soliman, 1990). (Note: In all of the above figures, the black space represents pores. Figures are not drawn to scale.)

Discussion

Geologists (Ehrlich et al., 1984) and soil scientists are often interested in the measurement of pore spaces, size, orientation, pore perimeter, and pore roughness present in rocks and compacted soils. In Figure 40.11a, a typical binary image of a rock sample is shown that was analyzed by an image analyzer to calculate porosity and pore perimeter. As can be seen from this figure, the pore spaces (black areas) are generally discontinuous. In Figure 40.11b, a binary image of the Deighton soil (compacted) is shown that was analyzed by the Image Analyzer, Quantimet 720 by Murphy et al. (1977). Again the pore spaces (black areas) are more discontinuous. Measurement of such discontinuous area is an easy task by any Image Analyzer system and in such measurements the operator's input can be minimalized.

However, the geotechnical engineers are not only interested in knowing the pore size, void ratios, and pore size distribution, but also the geometry of each grain, its center of gravity, and its orientation (see Fig. 40.11c). In addition, to calculate the void ratio distribution and the orientation of grains, the operator has to edit the image. For a better editing and manipulation of the binary image, only a few grains can be analyzed at one time. Therefore, it is believed that semiautomated image analysis systems that can add and subtract the image will be more useful as compared to a fully automated system.

The Image Plus system used for our investigation has digital density of 256×256 and has only 64 different gray levels. There are many moderate priced image analysis systems that have a much higher pixel density and more levels of gray (up to 256). It is our opinion that some of the errors that occurred in our measurements are due to fewer gray levels available in our system. We are in the process of upgrading the system. As pointed out by Berryman (1985) a digitized image with 512×512 pixel resolution and 256 gray levels gives sufficiently good gray image quality for analysis.

We believe that for the measurements required to study the fabric of granular soils, the image analysis system should have the option of edge enhancement. Edge enhancement is a form of filtering, accomplished on the binary image to enhance the edges of the object. Many large automated image analysis systems have some built-in filters. From our limited experience with using such general filters for our study, it proved to be ineffective (see Fig. 40.12a). We feel that for our samples, we had to design our filter. Therefore, image analysis systems that have options to add filters for specific problems, are more versatile.

Certain image processing software programs have options of performing image texture calculations. The result of such analysis can enhance the objects (see Fig. 40.12b). By defining thresholding levels on the textured image, the outer boundaries of the grain can be defined. It is yet to be concluded that the use of a filter and texture analysis will improve our calculations.

Finally, we found that instead of analyzing thin sections, analyzing the polished surfaces of the samples will reduce some of the errors.

Conclusion

The microstructure studies of granular soils have been a topic of great importance for geotechnical engineers and researchers because engineering properties of granular soils depend on their microstructures. Thin section techniques have been widely used to study microstructures. In the past, measurement from the thin sections was done manually. Use of an image analysis system can greatly improve such measurement. In this chapter, the use of image analysis for a variety of geotechnical problems, particu-

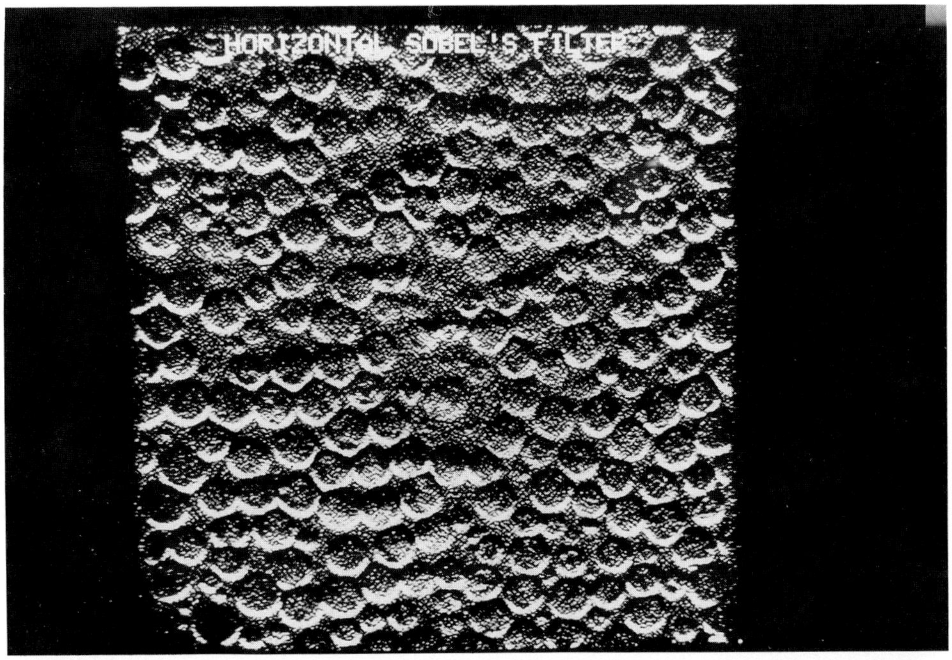

a

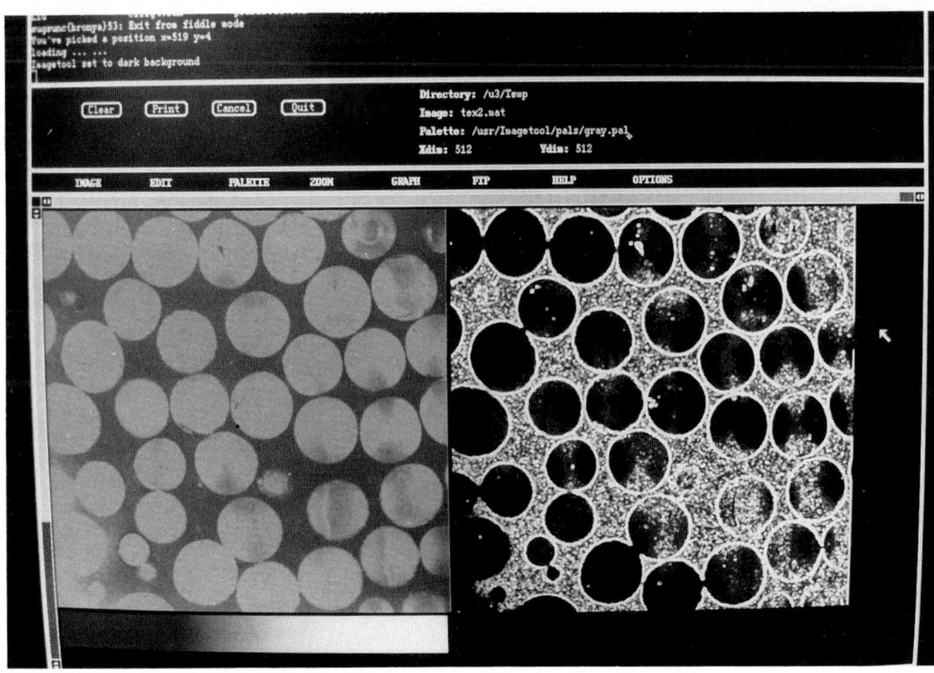

b

Figure 40.12. (a) Use of a high frequency filter (horizontal SOBEL filter) to highlight the edges. (b) Use of texture analysis to highlight the edges. (Left side of the figure is the acquired image and right side is the textured image).

larly dealing with microstructures, was discussed. In each case, typical data and results were shown to demonstrate the successful application of image analysis in microstructure studies of granular soils.

The focus on automatic processing systems for the analysis of granular soils has brought greater understanding of a variety of features. Smaller, less automated image analysis systems present some advantages over the larger systems. Because of the varied nature of significant features in geotechnical engineering, the ability to manipulate the data is crucial, less automated systems, such as the Image Plus system discussed in this chapter, allow for this ability, which in turn can increase accuracy. Also, the cost of smaller image analysis systems is obviously lower than the fully automated ones, which is often prohibitive for many researchers.

Acknowledgments

The authors gratefully acknowledge the support provided by the National Science Foundation and Syracuse University to establish the image analysis laboratory in the Civil and Environmental Engineering Department at Syracuse University. The authors appreciate the comments provided by Dr. P. Smart, Glasgow University, England and continuous assistance provided by Dr. R. Rao, SUNY-College of Environmental Science and Forestry, Syracuse, NY.

References

Berryman, J.G., 1985. Measurement of spatial correlation functions using image processing techniques. Journal of Applied Physics, v. 57(7), p. 2374–2384.

Bhatia, S.K., and A. Soliman, 1990. Frequency distribution of void ratio of granular materials determined by an image analyzer. Soil and Foundation, March, v. 30, No. 1, p. 1–16.

Bhatia, S.K., E. Nye, and A. Soliman, 1986. Image analysis application for soil fabric study. South Asian Regional Conference, December 3–6, Bangkok, Thailand.

Brewer, R., 1964. Fabric and Mineral Analysis of Soils. Wiley, New York, 470 p.

Childs, E.C., and N. Collis-George, 1950a. The permeability of porous materials. Proceedings Royal Society Series, A, v. 201, p. 392–405.

Childs, E.C., and N. Collis-George, 1950b. Movement of moisture in unsaturated soils. Transactions, International Congress Soil Science, Amsterdam, I, p. 1–4.

Duda, R., and P. Hart, 1973. Pattern Classification and Scene Analysis. Wiley, New York, 482 p.

Ehrlich, R., S.K. Kennedy, S.J. Crabtree, and R.L. Cannon, 1984. Petrographic image analysis, image analysis of reservoir pore complexes. Journal of Sedimentary Petrology, v. 54(4), p. 1365–1378.

Etris, E.L., D.S. Brumfield, R. Ehrlich, and S.J. Crabtree, 1988. Relations between pores throats and permeabilities: a petrographic/physical analysis of some carbonate grainstone and packstones. Carbonates and Evaporates. v. 3(1), p. 17–32.

Juang, C., 1981. Pore size distribution of sandy soils and the prediction of permeability. Ph.D. Thesis, Purdue University, West Lafayette, IN.

Marshall, T.J., 1958. A relationship between permeability and size distribution of pores. Journal of Soil Science, v. 9, p. 1–8.

Masounave, J., R. Denis, and A.L. Rollin, 1980. Prediction of hydraulic properties of synthetic non-woven fabrics used in geotechnical work. Canadian Geotechnical Journal, v. 17, p. 517–525.

Millington, R.J., and J.P. Quirk, 1961. Permeability of porous solids. Transactions, Faraday Society, v. 57, p. 1–8.

Mitchell, J.K., J.M. Chatoian, and G.C. Carpenter, 1976. The influence of sand fabric on liquefaction behavior. Report to U.S. Army Waterways Experiment Station, Vicksburg, MS.

Murphy, C.P., P. Bullock, and R.H. Turner, 1977. The measurement and characterization of voids in soil thin sections by image analysis. Part I, Principles and Techniques. Journal of Soil Science, v. 28, p. 498–508.

Nye, E.S., 1985. An investigation into structure-permeability relations of an Ottawa sand. M.S. Thesis, Syracuse University, Syracuse, NY.

Oda, M., 1972. Initial fabrics and their relations to mechanical properties of granular materials. Soils and Foundations, v. 12(1), p. 17–36.

Oda, M. 1976. Fabrics and their effects on the deformation behaviors of sand. Special Issue, Department of Foundation Engineering, Faculty of Engineering, Saitama University, Japan.

Oda, M., 1977. Co-ordination number of its relation to shear strength of granular material. Soils and Foundations, v. 17(2), p. 29–42.

Rollin, A.L., and G. Lombard, 1988. Mechanisms affecting long-term filtration behavior of geotextiles. Journal of Geotextiles and Geomembranes, v. 7, p. 119–145.

Rosen, D., 1984. Instruments for optical microscope-image analysis. In: Barer, R., and V.E. Cosslett, (eds.), Advances in optical and electron microscope, v. 9. Academic Press, London and New York, p. 323–354.

Rosenfeld, A., and A.C. Kak, 1976. Digital picture processing. Academic Press, New York.

Russ, C.J., 1985. Practical stereology, special printing for Dapple Systems, Inc., Sunnyvale, CA.

Underwood, E.E., 1970. Quantitative Stereology. Addison-Wesley, Reading, MA.

CHAPTER 41

Quantification of Clay Fabric: A Simple Technique

W.A. Chiou, William R. Bryant, and Richard H. Bennett

Introduction

It has always been a desire on the part of most geotechnical engineers or scientists to quantify studies whenever possible. Fabric measurements are no exception. A number of quantitative techniques have been reported by many geologists to quantify fabric measurements in sedimentary rocks or deposits (Krumbein, 1939; Dapples and Rominger, 1945; Pincus, 1953, 1969; Curray, 1956; Harrion, 1957; Durand and Greenwood, 1958; Schmoll and Bennett, 1961; Potter and Pettijohn, 1963, 1977; Lafeber, 1965; Watson, 1966). However, most of these quantitative studies were made on tills, gravels, sands, sandstones, and fossils.

Quantitative study of fine-grained clay particles did not begin until the mid-twentieth century when X-ray technology had become popular. A quantitative X-ray diffraction technique involving peak ratios was developed by Brindley (1953) and Martin (1962). Martin (1970) also described the use of a texture goniometer diffractionmeter that produced three-dimensional information. Other X-ray techniques have been proposed, including those of Fairbairn (1943), Ho (1947), Kaarsberg (1959), Silverman and Bates (1960), and Odom (1967), although they have not been widely applied. The applications of the polarizing optical microscope for determining clay fabric are limited. Morgenstern and Tchalenko (1967) have developed a technique for two-dimensional analysis, whereas Lafeber (1967) presented a three-dimensional analysis technique using the optical polarizing microscope.

Recently, quantification of clay fabric data using electron microscopy has become important. Pusch (1962) and Conley (1965) have applied quantitative methods to micrographs of dispersed clay particles. Transmission electron micrographs of intact clay samples have also been examined quantitatively by Pusch (1966, 1970) and Smart (1966, 1973). Their techniques have been further developed by McConnachie (1971, 1974). However, no sensitive mathematical techniques had been established for the rigorous description of the fabric of fine-grained soil at the electron microscopic level. Later Foster and Evans (1971), Foster (1973), Bennett et al. (1977, 1989), Smart and Tovey (1982), and Smart et al. (this volume) developed methods using computer image analysis techniques.

Stereoscopic measurements on transmission electron micrographs were first reported by Helmcke (1954), and later by Haanstra (1966) and Boyde (1967). Boyde (1967) described measurements on stereoscopic scanning electron micrographs using a floating spot, and in 1970 he employed a technique to assist in making such measurements. Lane (1969, 1972) described the geometry involved in fabric measurements. Although his approch was more general than most, he still made some approximations. Tovey (1970) described a method for the measurement of a third dimension that did not require using micrographs in a stereo pair. Tovey (1973a) and Tovey and Wong (1974) made the modification of this analysis more general, and developed an analysis procedure for measuring the third dimension (Tovey, 1973b).

Although the quantification of clay fabric is a subject still developing, quantitative analysis can be a valuable tool when combined with statistical analysis. Use of a specific quantification procedure makes the interpretation of fabric data less arbitrary and less subject to interpretation by personal preconception. To date, all the proposed techniques for quantification of clay fabric are engineering oriented and require some instrumentation to perform the analysis. In geology, a considerable volume of literature has accumulated on the quantitative analysis of geologic structures of rocks by direct measurements of their constituent elements in the field or on optical micrographs of thin sections. Many of the methods described can,

Table 41.1. Results of clay fabric analyses by statistical calculation.

Sample no.	Method* and thin section orientation†	TEM micrograph no.	Grains of count	Slope‡ or $\tan^{-1}$	Degree of orientation (degree)	Sample no.	Method* and thin section orientation†	TEM micrograph no.	Grains of count	Slope‡ or $\tan^{-1}$	Degree of orientation (degree)
1	(A)-A	21218	300	1.4516 55.4	20	4	(C)-B	22822	300	0.8491 42.3	35
	(A)-A	21221	100	1.0976 47.7	21		(C)-B	22883	100	0.7500 36.9	40
	(A)-C	21819	300	1.6667 59.0	18		(C)-B	22890	100	0.6429 32.7	41
	(A)-C	21810	100	1.6364 58.6	17	5	(A)-A	22515	300	1.2676 51.7	22
	(A)-C	21787	100	1.4516 55.4	20		(A)-A	22517	100	1.7308 60.0	15
	(B)-C	21826	100	0.7826 38.7	36		(A)-A	22523	100	1.2329 51.0	23
	(B)-C	21826	100	0.6081 31.3	42		(B)-A	22564	300	1.2329 51.0	21
	(C)-A	21227	300	0.6522 33.1	40		(B)-A	22567	100	0.9783 44.4	27
	(C)-A	21232	100	0.5590 29.2	61		(B)-A	22579	100	0.7143 35.5	42
	(C)-A	21888	100	0.7258 46.0	39		(C)-A	22538	300	0.5623 2.94	48
2	(A-F)-F	22836	300	1.0227 45.6	22		(C)-A	22542	100	0.6122 31.5	43
	(A-F)-A	22840	100	1.0588 46.6	22		(C)-A	22945	100	0.8108 39.0	34
	(A-F)-A	22844	100	0.7692 37.6	35		(C)-B	22555	100	0.6000 31.0	37
	(C)-A	22851	300	0.6475 32.9	40	6	(A-F)-A	22917	300	0.7895 38.3	36
	(C)-A	22855	100	0.6000 31.0	43		(A-F)-A	22920	100	0.7500 36.9	35
	(C)-A	22857	100	0.7438 36.6	39		(A-F)-A	22922	100	0.6716 33.9	42
3	(A)-A	22011	300	1.4516 55.4	21		(C)-A	22930	300	0.6718 33.9	42
	(A)-A	22016	100	1.6071 58.1	17		(C)-A	22931	100	0.7895 38.3	35
	(A)-A	22018	100	1.7647 60.5	15		(C)-A	22932	100	0.5590 29.2	52
	(B)-A	22489	300	0.5625 29.4	55	7	(A-F)-A	22586	300	1.4286 55.0	20
	(B)-A	22494	100	0.5625 29.4	54		(A-F)-A	22583	100	1.2857 52.1	20
	(B)-A	22510	100	0.7087 35.3	38		(A-F)-A	22585	100	1.5254 56.8	17
	(C)-A	22145	300	0.5625 29.4	54		(C)-A	22599	300	1.1688 49.5	21
	(C)-A	22147	100	0.5625 29.4	49		(C)-A	22600	100	1.0588 46.6	20
	(C)-A	22150	100	0.5921 30.6	48		(C)-A	22601	100	1.3636 53.8	19
4	(A-F)-A	22866	300	0.9184 42.6	31		(C)-B	22622	300	0.5478 28.7	54
	(A-F)-A	22867	100	0.9677 44.1	30		(C)-B	22624	100	0.5882 30.5	50
	(A-F)-A	22870	100	0.9575 43.8	31		(C)-B	22631	100	0.5921 30.6	46

*Method (of sample dehydration): (A) Sample dehydrated entirely under the equivalent *in situ* downhole pressure. (B) Sample dehydrated only partially under the equivalent *in situ* downhole pressure. (C) Sample dehydrated by conventional technique, i.e., dehydrated after *in situ* downhole pressure was released. (A-F) Sample failed to dehydrate correctly as proposed by method (A).

†Thin section orientation: A, Side or parallel to core axis view. B, Top or normal to core axis view. C, Random view. Detailed description of sample preparation method and thin section orientation was given by Chiou et al. (this volume).

‡Slope: the slope of particle orientation frequency curve

$$\text{Slope} = \frac{95\text{-}5}{\Delta 95\text{-}\Delta 5}$$

with minor modifications, be applied to fine-grained sediments provided the specimens are in the form of ultrathin sections.

This chapter describes a simple approach, analogous to other geologic fabric research, to characterize the orientation of clay fabric using transmission electron micrographs.

Method

General

Transmission electron micrographs used in this study were obtained from the "pressurized core barrel" project, which was developed at the Marine Geotechnical Laboratories of Texas A&M University. Detailed specifications and procedures for sampling were presented by Denk et al. (1981).

Samples prepared for transmission electron microscopy study (including sample dehydration, embedding, and ultramicrotomy) follow the methods of Chiou (1980, 1981) and Chiou et al. (1983). Basically, clayey submarine sediment samples obtained from the Mississippi Delta were critical-point dried, embedded by SPURR epoxy and ultrathin sectioned with a DuPont diamond knife on a Sorvall MT-2 ultramicrotome to a thickness of approximately 800–1000 Å (Bennett et al., 1977).

In this study, a total of 56 measurements were made from seven samples with different orientations of the ultrathin sections. Three types of sample preparation techniques, methods (A), (B), and (C), were employed and are described in Table 41.1.

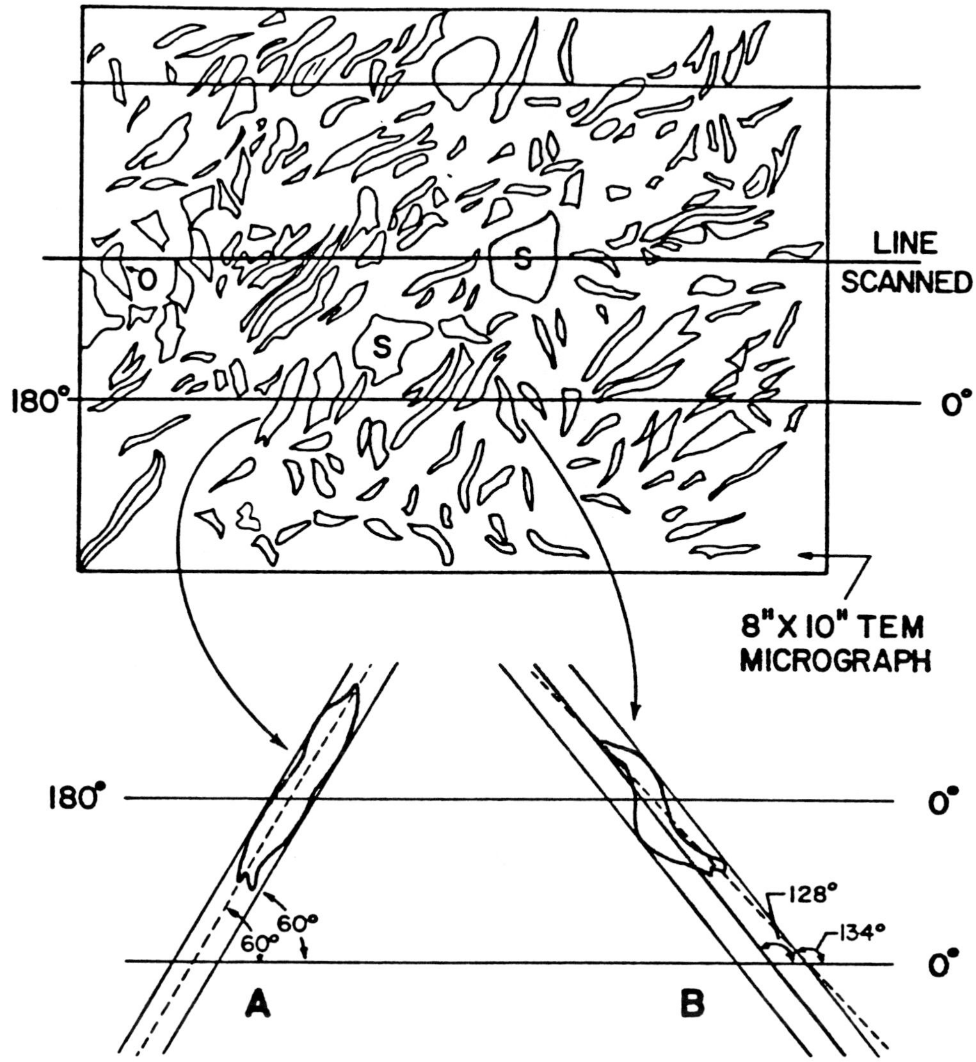

Figure 41.1. Schematic diagram showing how the orientation of clay paticles was measured from a TEM micrograph. Enlarged particles (A) and (B) show the relation between particle shape and the procedures of measuring particle elongation direction. Under condition (A), the direction of long dimension elongation (dashed line) is the same as that of the least projection elongation (solid lines). Under condition (B), the direction of long dimensional elongation (134°) is different from the direction of the least projection elongation (128°). Nonclay particles, such as silts (S), and organism tests (O) are not counted in the orientation analysis.

Electron Microscopy

All ultrathin sections were viewed in a Philips transmission electron microscope, Model 300, at 100 kV accelerating potential. To examine the fabric features, the specimens were viewed under the lowest magnification (approximately 300×) to seek the best and most representative areas, i.e., to prevent viewing possible artifacts. The specimen was then examined under increasingly higher magnifications. At least 8 to 10 observations were made on different portions of each specimen examined. The images were recorded on 8.3 × 10.2 cm (3¼ × 4 in.) Kodak electron image film 4463 (cat. 122 2140).

Orientation Analysis

The orientation of clay particles was determined to be the measured elongation direction of grain projections. The direction of elongation was assumed to be that of two parallel lines with minimum separation that can be drawn tangent to the grain projection.

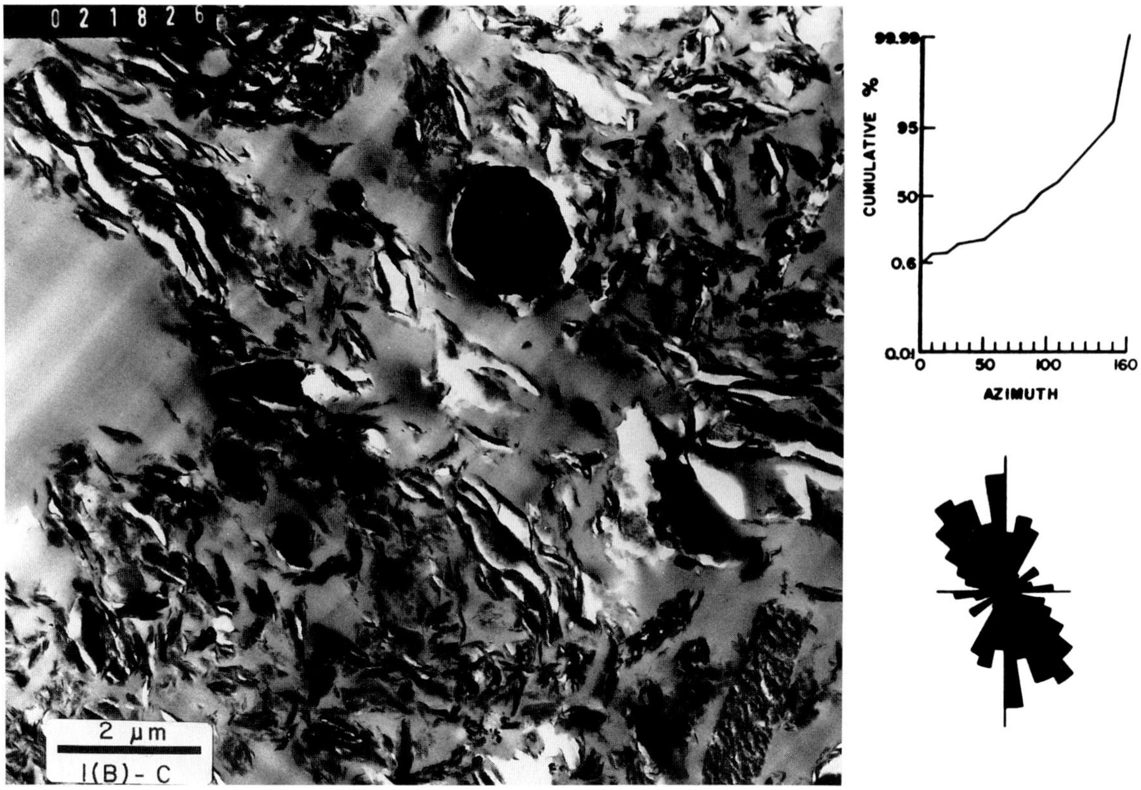

Figure 41.2. Clay fabric and orientation analysis of sample 1(B)-C showing the difference between "apparent" and "true" clay orientation. Left: Low magnification TEM micrograph depicting "apparently" oriented clay fabric and large voids in the sediment. However, clay fabric orientation analysis by the proposed technique reveals the random arrangement of clay particles. Upper right: Gentle slope of clay orientation distribution curve indicating randomly arranged clay fabric. Lower right: Rose diagram of clay fabric from this TEM micrograph (021826).

However, due to the elongated flaky nature of clay (mineral) particles, the direction of elongation is the same or very close to the longest dimension (i.e., the longest line that can be drawn on a particle projection). Only grains showing platy or flaky morphology on the transmission electron micrographs were measured. The measurement was carried out using an enlarged 8 × 10 in. (20.3 × 25.4 cm) micrograph.

To provide the simplest, fastest, and most accurate measurement of fabric orientation, the point counting method was used in this study. A scan line or grid system was set up on a clear plastic transparency so the entire micrograph was covered (Fig. 41.1). A transparent plastic protractor with a cross-hair at the center was used to overlay on the plastic sheet which in turn overlaid the enlarged 8 × 10 in. electron micrograph. Every grain that came under the cross hair was measured while the protractor was moved along a line until a predetermined number of measurements had been reached. The measured orientations were recorded and then tabulated for further analysis. Point counting results are unbiased regardless of the closeness of the measurements.

The principal problem in grain orientation analysis is obtaining a sufficiently large number of reasonably accurate measurements. In petrography studies, most investigators measure between 200 and 500 grains per thin section (Krumbein, 1935; Krumbein and Pettijohn, 1938; Friedman, 1958; van der Plas, 1962; Griffiths, 1967). Possibly as few as 100–200 measurements are necessary for a single analysis to satisfy specified confidence limits of the mean (Rosenfeld et al., 1953). In this study, the orientation of at least 300 grains (at lower magnification, e.g., 10,000× on an 8 × 10 in. micrograph) and 100 grains (at higher magnification, e.g., >25,000× on an 8 × 10 in. micrograph) were measured for each sample.

Specific orientation of each particle was assigned to one of the eighteen 10° intervals between 0 and 180°. Results of particle orientation measurements were then expressed by polar coordinate (rose) diagrams for the purpose of comparison. Furthermore, to express the degree of orientation (L = degree) of clay fabric in the sediments and for the purpose of a more absolute rather than arbitrary comparison from one sample to another, a simple mathematical analysis also has been performed in this search.

The number of counts within each of the 18, 10°, intervals were summed and assigned as the total number in an interval. In addition to having a polar diagram for comparison, the number of counts in all orientation classes was summed and adjusted to a total of 100%. Thus, a cumulative frequency curve can be constructed on a probability percentage coordinate diagram to show

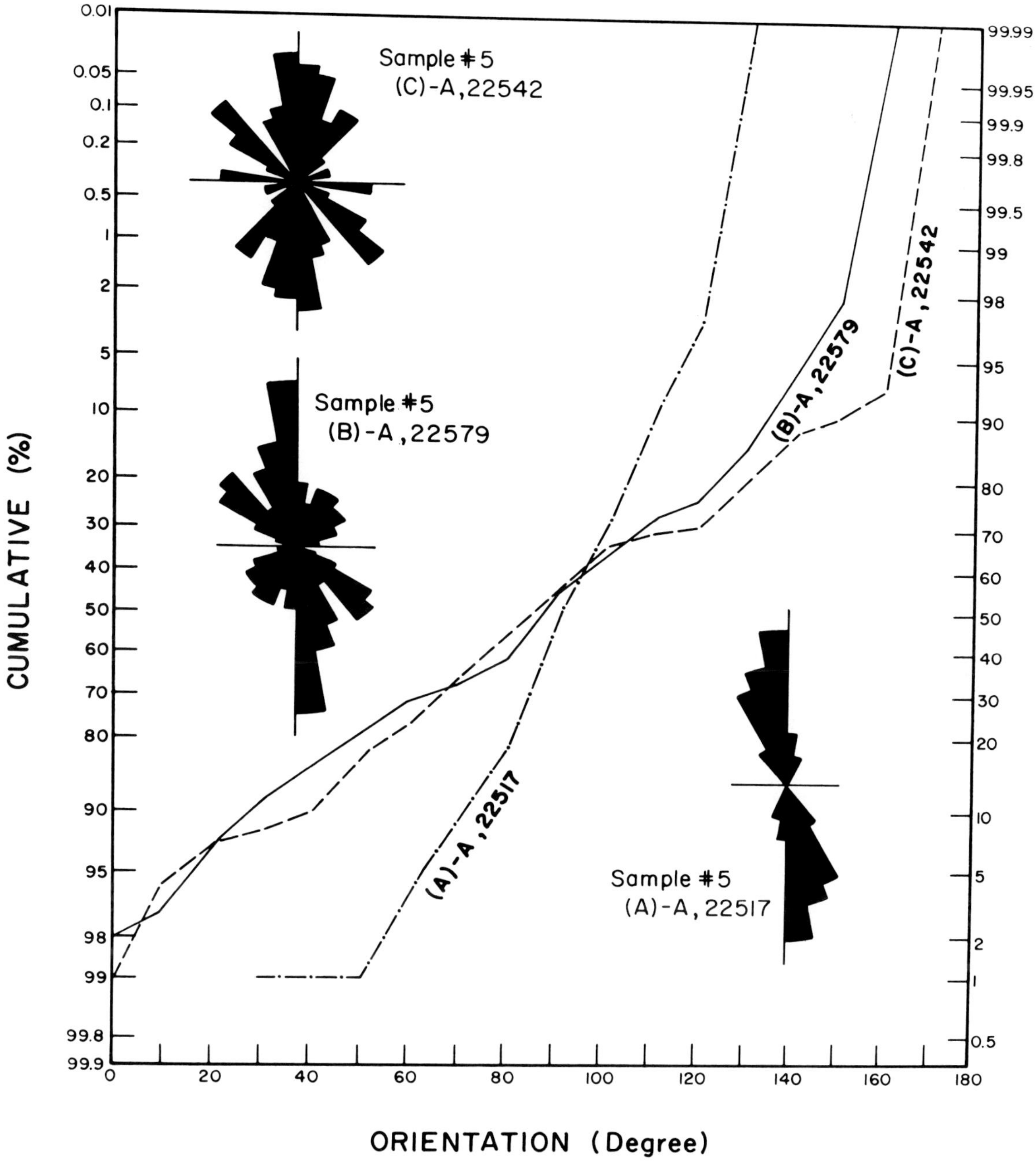

Figure 41.3. Comparison of results of clay particle orientation analyses between different sample preparation methods, (A), (B), and (C), from Mississippi Delta sediments clay fabric; results obtained by transmission electron microscopic study of the corresponding samples are shown in Figure 41.4.

distribution of grain orientation. The maximum occurrence (the mode) of fabric orientation was assigned to 90–100° intervals to create an absolute frame for the comparison of clay fabric in the different samples.

Percentages were accumulated beginning from the origin, or 0° end of the abscissa, and the cumulative number percent (not weight percent) was then plotted at corresponding degree class boundaries. The data points may be connected by either a curve or straight-line segments (as was the case in this study). The use of straight-line segments assumes that samples are normally distributed between orientation classes.

Graphic computation techniques for characterizing degree class orientations were used to avoid the lengthy calculations required by moment statistics. The formula developed by Folk

Figure 41.4. TEM micrograph showing different clay fabric microfeatures by different sample preparation techniques; clay fabric orientation analysis of the corresponding specimens is illustrated in Figure 41.3.

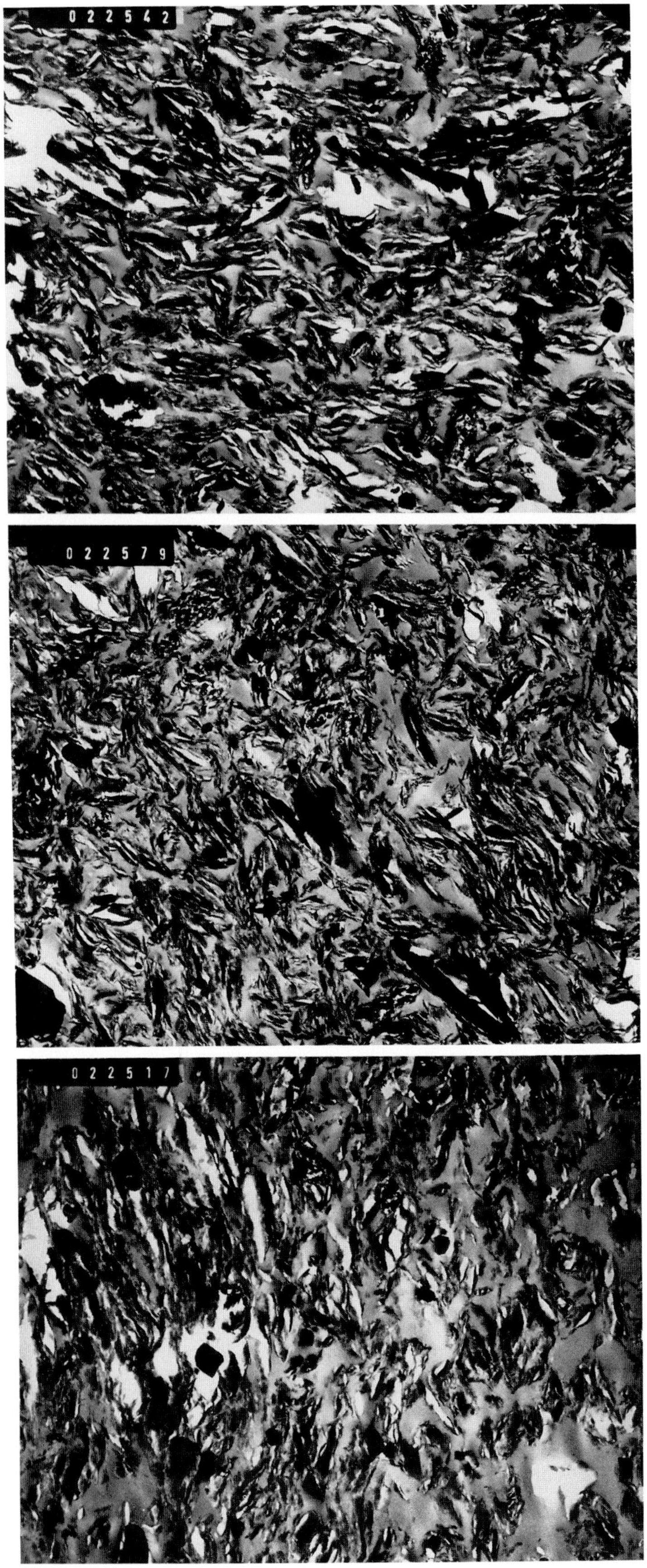

(C) - A

Degassed

2 µm

(B) - A

Degassed

2 µm

(A) - A

Pressurized

3 µm

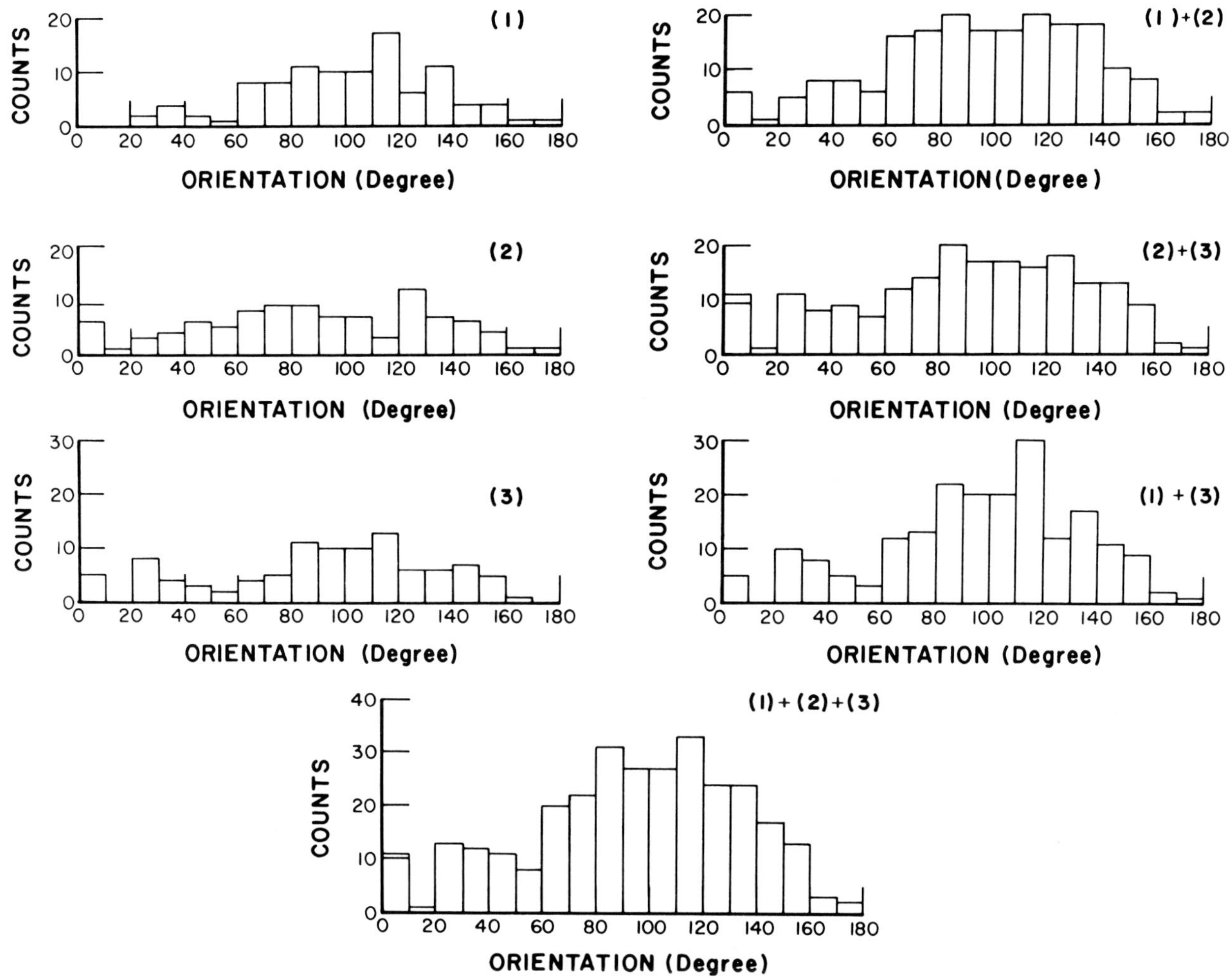

Figure 41.5. Histograms showing statistical results of clay particle orientation analysis. Histograms 1, 2, and 3 represent 100 counts each from the same TEM micrograph. Histograms of 1 + 2, 2 + 3, and 1 + 3 represent 200 counts each by adding any two of the 100 counts. Histogram of 1 + 2 + 3 represents the results of 300 counts by combining the three 100 counts.

and Ward (1957) for computing sorting (standard deviation, or a measure of spread around the mean) was applied here as "degree of orientation," or "sort the particle orientation." The degree of orientation equals

$$\frac{\angle 84 - \angle 16}{4} + \frac{\angle 95 - \angle 5}{6.6}$$ where $\angle 84$, $\angle 16$, $\angle 95$, $\angle 5$

represent the degree class of orientation at 84, 16, 95, and 5%, respectively, in the orientation frequency curve. The principle of using this formula is essentially similar to the concept of sorting. That means that if the fabric of a sediment is well oriented, then the orientation of clay particles will fall in a narrow range (i.e., more parallel to each other), thus the sorting will be better or the slope of orientation frequency curve will be steeper.

Results

Rose diagrams and orientation frequency curves illustrating the clay particle orientation of a clayey sediment are the first steps toward a quantitative approach (Fig. 41.2). To be more specific in comparing fabrics in sediments, the degree of orientation was used as described in the previous pages.

Based on the results of clay particle orientation measurements from TEM micrographs (Table 41.1), a good correlation exists between clay particle orientation and sample preparation method (also see Chiou et al., this volume). The relationship between the degree of clay particle orientation and the simple statistical results is clearly shown also in Figures 41.3 and 41.4, and can be categorized as follows:

1. If a sediment is composed of well-oriented clay particles, e.g., clay fabric revealed by method (A) in this study, the slope (of orientation frequency curve) generally has a value of greater than 1.00 or the degree of orientation is approximately smaller than 25°.

   The lower the value for the degree of orientation, the better the preferred orientation of the clay fabric.
2. If a sediment is composed of highly randomly arranged clay particles [e.g., clay fabric revealed by method (C) in this study] the slope generally has a value of less than 0.75 or the degree of orientation is approximately greater than 35°.
3. If a sediment has a slope value between 0.75 and 1.00 or the degree of orientation is somewhere between 25 and 35°, then the clay fabric of the sediment is neither a very well oriented nor highly nonoriented random arrangement.

From the results of this study, it is suggested that at least two or three traverses of 300 counts of clay particles at lower magnification are necessary to provide a general view of clay fabric, and about three or four traverses of 100 counts on medium or higher magnification are required for greater confidence. Otherwise, results based on a single counting may lead to misinterpretation of the real picture of clay fabric in the sediment matrix (Fig. 41.5).

It also must be noted that the result given above was based on TEM micrograph studies only. The classification was therefore on a two-dimensional basis. Nevertheless, it demonstrates the possibilities of a quantitative approach to clay fabric analyses by simple mathematical techniques, and provides the prospect for further development of quantitative analysis of clay fabric. Description of clay fabric is rather arbitrary and very dependent on the person who describes the microfeatures. For example, at first glance, clay fabric of Figure 41.2 seems to be oriented preferentially in a NW–SE direction. However, the calculation of actual measurement of clay particles showed that the overall particle orientation is random. The apparent clay fabric orientation is probably due to visual error resultant of a few large clay particles and domains in the NW–SE direction. Therefore, a more quantitative approach, as demonstrated in this chapter, for clay fabric analysis is highly suggested. This very inexpensive technique (essentially no instrumentation cost required) can provide reliable data within a relatively short period of time, although the computer image enhancement and quantitative image analysis may be one of the best approaches to use in the future.

Acknowledgments

This chapter represents a portion of the first author's doctoral dissertation. It has benefitted from reviews and suggestions from Dr. Les E. Shephard. We wish to acknowledge Dr. Hilton H. Mollenhauer of the Veterinary Toxicology and Entomology Research Laboratory, U.S. Department of Agriculture, who graciously permitted free access to the transmission electron microscopy laboratory. This work was supported by the National Oceanic and Atmospheric Administration, U.S. Department of Commerce, and the U.S. Geological Survey (Texas A&M Research Foundation Project #3997 and 4076).

References

Bennett, R.H., W.R. Bryant, and G.H. Keller, 1977. Clay fabric and geotechnical properties of selected submarine sediment cores from the Mississippi Delta. NOAA Professional Paper 9, NOAA, U.S. Dept. of Commerce, 86 p.

Bennett, R.H., K.M. Fischer, D. Lavoie, W.R. Bryant, and R. Rezack, 1989. Porometry and fabric of marine clay and carbonate sediments: determinants of permeability. Marine Geology, v. 89, p. 127–152.

Boyde, A., 1967. A single stage carbon-replica and some related techniques for the analysis of electron microscope images. Journal Royal Microscopical Society, v. 86, p. 359–370.

Boyde, A., 1970. Practical problems and methods in the three-dimensional analysis of scanning electron microscope images. Proceedings of the 3rd SEM Symposium, IIT Research Institute, Chicago, p. 105–112.

Brindley, G.W., 1953. An x-ray method for studying orientation of micaceous minerals in shales, clays, and similar materials. Mineralogical Magazine, v. 30, p. 71–78.

Chiou, W.A., 1980. A new technique in preparing in situ marine sediments for clay fabric study. Proceedings of the 38th Electron Microscopy Society of America Annual Meeting, p. 204–205.

Chiou, W.A., 1981. Clay Fabric of Gassy Submarine Sediment. Ph.D. dissertation, Texas A&M University, College Station, TX, 248 p.

Chiou, W.A., L.E. Shephard, W.R. Bryant, and M.P. Looney III, 1983. A technique for preparing high water content clayey sediments for thin and ultra thin section study. Sedimentology, v. 30, p. 295–299.

Conley, R.F., 1965. Statistical distribution pattern of particle size and shape in the Georgia kaolins. Clays and Clay Minerals, v. 14, p. 317–330.

Curray, J.R., 1956. The analysis of two-dimensional orientation data. The Journal of Geology, v. 64, p. 117–131.

Dapples, E.C., and J.F. Rominger, 1945. Orientation analysis of fine-grained clastic sediments. A report of progress. The Journal of Geology, v. 53, p. 246–261.

Denk, E., W.A. Dunlap, W.R. Bryant, L.J. Milberger, and T.J. Whelan, 1981. A pressurized core barrel for sampling gas-charged marine sediments. Proceedings of the 13th Annual Offshore Technology Conference, p. 43–52.

Durand, D., and J.A. Greenwood, 1958. Modification of the Rayleigh test for uniformity in analysis of two-dimensional orientation data. The Journal of Geology, v. 66, p. 229–238.

Fairbairn, H.W., 1943. X-ray petrology of some fine-grained foliated rocks. American Mineralogist, v. 28, p. 246–256.

Folk, R.L., and W.C. Ward, 1957. Brazos River bar, a study on the significance of grain size parameters. Journal of Sedimentary Petrology, v. 27, p. 3–27.

Foster, R.H., 1973. Analysis of soil microstructure, In: Barden, L., and R. Pusch (eds.), Proceedings of the International Symposium on Soil Structure. Swedish Geotechnical Society and Swedish Society for Clay Research, p. 5–13.

Foster, R.H., and J.S. Evans, 1971. Image analysis of clay fabric by quantimet. The Microscope, v. 19, p. 377–401.

Friedman, G.M., 1958. Determination of sieve-size distribution from thin-section data for sedimentary petrological studies. The Journal of Geology, v. 66, p. 349–416.

Griffiths, J.C., 1967. Scientific Method in Analysis of Sediments. McGraw-Hill, New York, 508 p.

Haanstra, H.B., 1966. Stereophotography with the electron microscope. Phillips Technical Review, v. 27, p. 231–237.

Harrion, P.W., 1957. New techniques for three-dimensional fabric analysis of till and englacial debris containing particles from 3 to 4 mm in size. The Journal of Geology, v. 65, p. 98–105.

Helmcke, J.G., 1954. Theories und praxis der electron mikrosko-pischen stereoaufnamen. Optik, v. 2, p. 201–225.

Ho, T.L., 1947. Petrofabric analysis by means of x-rays. Geological Society of China Bulletin, v. 27, p. 389–398.

Kaarsberg, E.A., 1959. Introductory studies of natural and artificial aggregates by sound propagation and x-ray diffraction methods. The Journal of Geology, v. 67, p. 447–472.

Krumbein, W.C., 1935. Thin-section mechanical analysis of indurated sediments. The Journal of Geology, v. 43, p. 482–496.

Krumbein, W.C., 1939. Preferred orientation of pebbles in sedimentary deposits. The Journal of Geology, v. 47, p. 673–706.

Krumbein, W.C., and F.J. Pettijohn, 1938. Manual of Sedimentary Petrology. Appleton-Century-Crofts, New York, 549 p.

Lafeber, D., 1965. The graphic representation of planar pore patterns in soils. Australian Journal of Soil Research, v. 3, p. 143–164.

Lafeber, D., 1967. The optical determination of spatial orientation of platy clay minerals in soil thin-sections. Geoderma, v. 1, p. 359–369.

Lane, G.S., 1969. The application of stereographic techniques to the scanning electron microscope. Journal Scientific Instruments (Journal Physics E), series 2, p. 565–569.

Lane, G.S., 1972. Dimension measurements. In: Hearle, J.W., J.T. Sparrow, and P.M. Cross (eds.), The Use of the Scanning Electron Microscope. Pergamon Press, New York, p. 219–238.

Martin, R.T., 1962. Research on the Physical Properties of Marine Soils. M.I.T., Dept. of Civil Engineering, Soil Div. Publ. No. 127.

Martin, R.T., 1970. Fabric of Consolidated Kaolinite. M.I.T., Dept. of Civil Engineering, Soil Div. Publ. No. 254, 73 p.

McConnachie, I., 1971. Electron microscopy of the consolidation of a kaolin. Ph.D. Thesis; University of Glasgow, Glasgow, Scotland.

McConnachie, I., 1974. Fabric changes in consolidated kaolin. Geotechnique, v. 24, p. 207–222.

Morgenstern, N.R., and J.S. Tchalenko, 1967. The optical determination of preferred orientation in clays and its application to the study of microstructure in consolidated kaolin, I and II. Proceedings Royal Society London, Series A, v. 300, p. 218–250.

Odom, I.E., 1967. Clay fabric and its relation to structural properties in mid-continent Pennsylvania sediments. Journal of Sedimentary Petrology, v. 37, p. 610–623.

Pincus, H.J., 1953. The analysis of aggregates of orientation data in the earth sciences. The Journal of Geology, v. 61, p. 482–509.

Pincus, H.J., 1969. Sensitivity of optical data processing to changes in rock fabric, Part I, Geometric patterns, Part II, Standardized grain patterns, Part III, Rock fabrics. International Journal Rock Mechanics and Mineral Science, v. 6, p. 259–276.

Potter, P.E., and F.J. Pettijohn, 1963. Paleocurrents and Basin Analysis. Springer-Verlag, Berlin, 296 p.

Potter, P.E., and F.J. Pettijohn, 1977. Paleocurrents and Basin Analysis, 2nd Corrected and Updated Edition. Springer-Verlag, Berlin, 420 p.

Pusch, R., 1962. Clay particles—their size, shape, and arrangement in relation to some important physical properties of clays. Statens rad for byggn adsforskning Handle. No. 40, Trans., Stockholm, Sweden.

Pusch, R., 1966. Investigation of Clay Microstructure by Using Ultra-Thin Sections. Geotechnical Institute Reprints and Preliminary Report, No. 15, Stockholm, Sweden.

Pusch, R., 1970. Microstructural changes in soft quick clay at failure. Canadian Geotechnical Journal, v. 7, p. 1–7.

Rosenfeld, M.A., L. Jacobsen, and J.C. Ferm, 1953. A comparison of sieve and thin-section techniques for size analysis. The Journal of Geology, v. 61, p. 114–132.

Schmoll, H.R., and R.H. Bennett, 1961. Axiometer—a mechanical device for locating and measuring pebble and cobble axes for macrofabric studies. Journal of Sedimentary Petrology, v. 31, p. 617–622.

Silverman, E.N., and T.F. Bates, 1960. X-ray diffraction study of orientation in the Chattanooga Shale. American Mineralogist, v. 45, p. 60–68.

Smart, P., 1966. Soil structure, mechanical properties and electron microscopy. Ph.D. dissertation, University of Cambridge, Cambridge, England, v. 1, 187 p.

Smart, P., 1973. Statistics of soil structure in electron microscopy. In: Barden, L., and G.R. Pusch (eds.), Proceedings of the International Symposium on Soil Structure. Swedish Geotechnical Society and Swedish Society for Clay Research, p. 69–76.

Smart, P., and N.K. Tovey, 1982. Electron Microscopy of Soils and Sediments—Techniques. Oxford University Press, Oxford, 264 p.

Tovey, N.K., 1970. Electron microscopy of clays. Ph.D. dissertation, University of Cambridge, Cambridge, England.

Tovey, N.K., 1973a. Quantitative analysis of electron Micrographs of soil structure. In: Barden, L., and R. Pusch (eds.), Proceedings of the International Symposium on Soil Structure. Swedish Geotechnical Society and Swedish Society for Clay Research, p. 50–57.

Tovey, N.K., 1973b. A general photogrammetric method for the analysis of scanning electron micrographs. Proceedings of SEM Conference, Institute of Physics, London, p. 82–87.

Tovey, N.K., and K.Y. Wong, 1974. Some aspects of quantitative measurements from electron micrographs of soil structure. In: Rutherford, G.K. (ed.), Proceedings, 4th Soil Microscopy Conference. The Limestone Press, Kingston, Ontario, Canada, p. 207–222.

van der Plas, L., 1962. Preliminary note on the granulometric analysis of sedimentary rocks. Sedimentology, v. 1, p. 145–157.

Watson, G.S., 1966. The statistics of orientation data. The Journal of Geology, v. 74, p. 786–797.

CHAPTER 43

Geoacoustic Properties of a Marine Silt

R.D. Stoll

Introduction

In the unlithified sediments just beneath the seafloor the propagation characteristics of both dilatational and shear waves play key roles in choosing a realistic geoacoustic model. In this near-bottom region the dynamic moduli of the sediments are functions of frequency and overburden pressure. Moreover, at low frequencies there appears to be a fundamental difference between the viscous damping that occurs in coarse sediments such as sand, owing to overall fluid motion relative to the skeletal frame, and the damping observed in fine sediments, which is attributed mainly to local fluid motion near each intergranular contact.

We have carried out laboratory experiments on a micaceous silt of glacial origin to measure the complex shear and Young's modulus over a wide range of frequencies extending from about 1 Hz to over 1.5 kHz. The experiments were performed by exciting torsional, flexural, and extensional modes of motion at very low amplitudes. At high frequencies the response at several different resonant frequencies was studied using a variant of the resonant column method while at low frequencies, the phase difference between the driving force and resulting motion was measured to determine intrinsic damping.

The results of our experiments have been incorporated into the Biot theory by describing the stiffness of the sediment frame using a simple viscoelastic model, with parameters chosen on the basis of our experimental data. As a result the curves showing Q or logarithmic decrement for both compressional and shear waves are markedly changed with the overall response at lower frequencies controlled largely by the damping due to "local" fluid motion and the effects of overall permeability observed only at very high frequencies.

In this chapter we describe the experimental work that has been done and show how the results are incorporated into the Biot theory. In the following discussion we assume that wave amplitude and logarithmic decrement are small so that different measures of damping may be related in a particularly simple way. For example, the quality factor Q, the logarithmic decrement δ (for both traveling waves and stationary vibrations), the attenuation coefficient α, and the ratio of the imaginary to real parts of any appropriate complex modulus, $\overline{M} = M_r + iM_i$ are related approximately by the expression

$$\frac{1}{Q} = \frac{\delta}{\pi} = \frac{\alpha V}{\pi f} = \frac{M_i}{M_r}$$

where V is phase velocity and f is frequency in Hz.

Experimental Work

During the course of our laboratory experiments, which are designed to measure the geoacoustic properties of marine sediments, we have used several different experimental setups in an effort to study the effects of frequency over the widest possible frequency range without tampering with specimen boundary conditions. A schematic of the experimental setup used in our first series of tests (Stoll, 1979) is shown in Figure 43.1 along with an example of the kind of output that is recorded. The sediment used in these experiments was confined in a thin, cylindrical shell composed of a spiral of steel wire embedded in latex rubber. The wire was 0.25 mm in diameter and the spacing between adjacent loops was also approximately 0.25 mm. The diameter of the specimen was 4.9 cm and the length was about 27 cm. The shell offers very high resistance to radial deformation, but virtually no resistance to twisting or to small amounts of axial compression or extension. The bottom of the reinforced

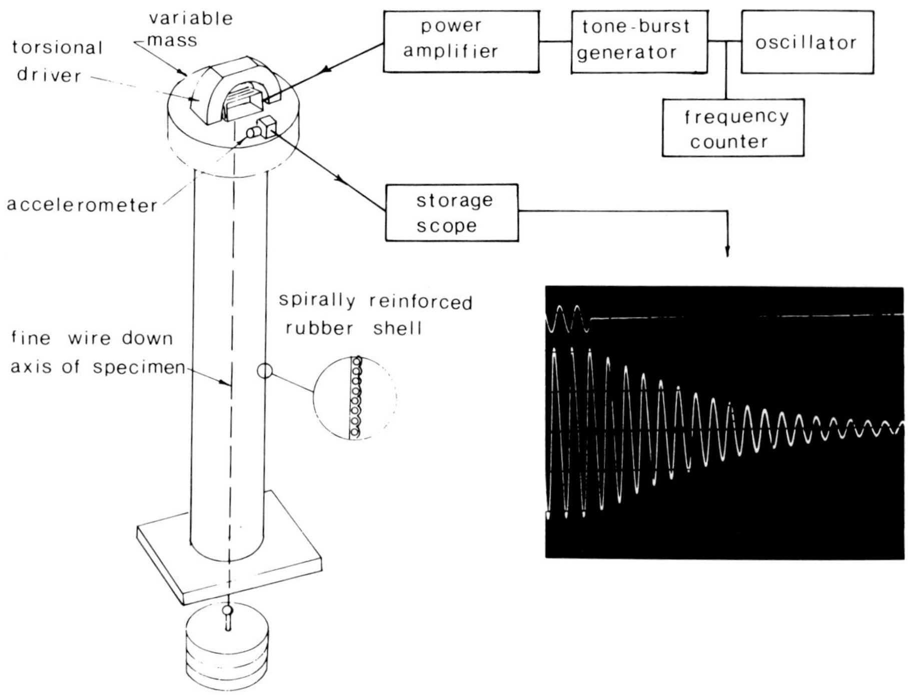

Figure 43.1. Schematic of experimental setup for studies of torsional vibration. (After Stoll, 1979.)

shell is sealed to a cylindrical pedestal fixed to a rigid base, and a tiny hole through the pedestal containing an "O" ring seal allows a thin stainless-steel wire to pass through the pedestal without leakage of pore water. The axial wire, which is attached to the top cap of the specimen, is used to apply static, axial stress to the specimen. Different axial stress levels are attained by placing weights on a platform attached to the lower end of the wire beneath the rigid base.

The upper mass, which moves with the top of the specimen, is composed of a cap with rough bottom (sand particles embedded in epoxy cement) in direct contact with the sediment and a circular plate, which supports several transducers and a lightweight driver coil. In addition one or more annular rings made of brass may be attached to the plate to vary the mass moment of inertia of the top assemblage. Thus the specimen and top mass form a torsional pendulum with a resonant frequency that depends on the stiffness of the specimen and the mass moment inertia of the top.

The use of an articulated shell and a mechanical system for static loading in lieu of the more commonly employed resonant column apparatus with pressure chamber and rubber membrane was adopted because it allowed much easier preparation of fully saturated specimens and stable static loading over long periods of time. Dynamic torsional excitation of the specimen was effected by placing one or more coils mounted on the top of the specimen in a magnetic field. In our early experiments with long, thin specimens the field was produced by a large horseshoe magnet with special pole pieces and iron core positioned inside the coil in the same manner as in a galvanometer with a D'Arsonal movement. The current flowing in the driver coil was controlled by a variable frequency oscillator, a power amplifier, and a tone burst generator. The rotational motion of the specimen was measured by a pair of horizontally oriented accelerometers mounted on the outer edge of the top mass.

In the resonant column tests the specimen is driven to resonance at several different power settings. At each resonance the frequency is adjusted to obtain maximum amplitude after which the power is switched off electronically and the decay of the wave form is recorded using a storage oscilloscope or a digital data acquisition system installed in a microcomputer. A typical record showing the driver signal and output of one of the displacement transducers is shown in Figure 43.1. The logarithmic decrement is determined from the rate of decay of the rotational motion after the driver is shut down. At each resonance the low-amplitude response is measured first and the maximum amplitude is limited to a predetermined value to avoid disturbance to the specimen.

In our more recent tests, particularly those on shorter specimens of larger diameter, we have used pairs of coils and magnets located on the periphery of the upper mass as drivers. In addition the motion of the specimen is measured with noncontacting capacitive displacement transducers that have the same sensitivity at all frequencies of interest. These transducers utilize a signal frequency of several megahertz and so can be used to measure the displacement of a target oscillating at frequencies in excess of 1 kHz. Moreover, they can resolve accurately displacements of the order of 10^{-7} cm so that it is possible to monitor

Figure 43.2. Experimental setup for measurement of phase difference between driving torque and resulting motion of top of specimen. (After Stoll, 1985a.)

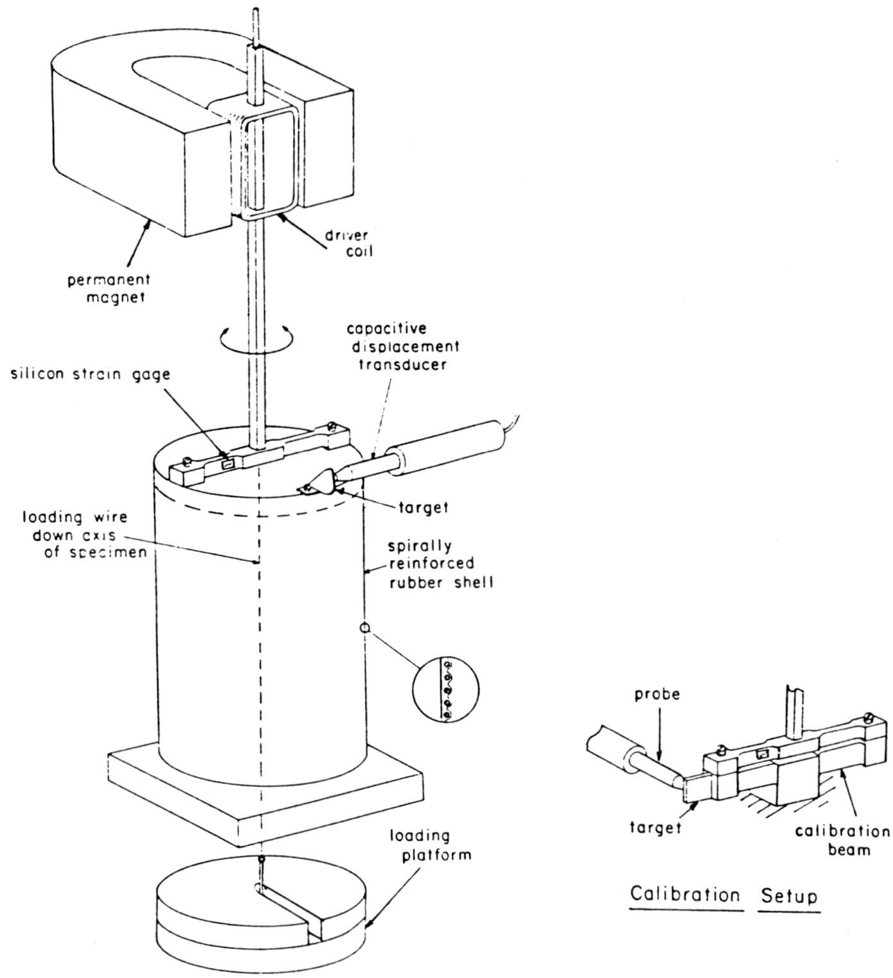

motion corresponding to maximum shear strains of less than 10^{-6} in the specimen. At higher levels of strain, the response of the specimen tends to become amplitude dependent so that the results are no longer representative of an acoustic wave propagating far from the source.

In addition to the resonant column tests, an experimental technique has been developed to measure modulus and damping at very low frequencies (Stoll, 1985a,b). The experimental setup for the low-frequency measurements is shown in Figure 43.2. In these experiments the phase difference between the driving couple and the resulting torsional motion is used to measure the shear modulus and Q of the sample. The experiments are performed with the aid of a microcomputer, which controls the experiment and obtains a robust sample of many cycles of loading. The microcomputer is also used to display and process data as well as to calculate the phase difference based on Fourier analysis.

The torque applied to the top of the specimen is measured by sensing the strain in a torque beam attached at two points to the top plate of the specimen. A silicon strain gage bridge and an instrumentation amplifier provide the necessary electrical output for sampling by an A/D converter. In most cases a robust sample consisting of more than 100 cycles of motion is taken after which Fourier transformation is used to determine the phase of the torque and resulting displacement. The phase difference measured in this manner is then corrected for any phase shift caused by the electronic circuitry based on a set of calibration data obtained by loading a high-Q aluminum beam proportioned to give displacements of the same order as those expected during the real tests (see Fig. 43.2). The calibration is important since we are measuring real phase differences in the specimen of sediment that are of the order of one degree or less!

In this chapter we present the results of tests on an inorganic, micaceous silt obtained from a glacial deposit in Long Island Sound, New York. The mean grain diameter of the silt is 0.025 mm and the maximum and minimum sizes are 0.1 and 0.005 mm, respectively. A typical set of data points for a resonant column test on this material is shown in Figure 43.3. The data

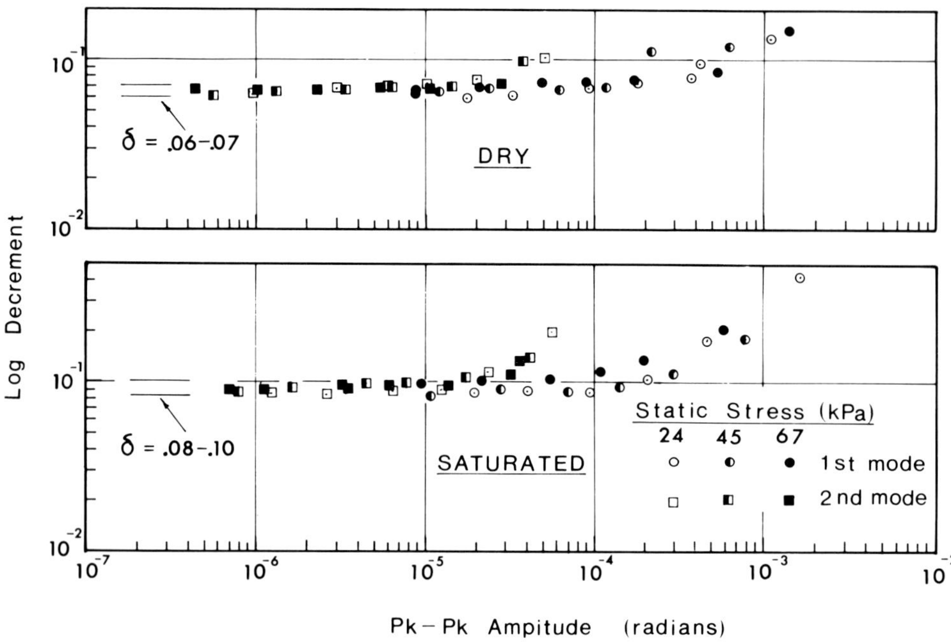

Figure 43.3. Logarithmic decrement versus amplitude for water-saturated and dry silt at three different levels of static axial stress.

are based on the first and second modes of torsional vibration at three different levels of static axial prestress. Both "room dry" and fully saturated samples were tested, and, as can be seen from the figure, there is a small but significant increase in the logarithmic decrement in the saturated tests. The effect of increasing amplitude of motion is also clearly evident in this figure. When the peak-to-peak motion at the top cap of the specimen (which is what is shown) is translated into maximum shear strain amplitude based on an analysis of the mode shapes (see Stoll, 1979), the horizontal portions of these curves are found to correspond to shear strain amplitudes of about 10^{-6} or less.

At different times during the period from 1978 to the present we have performed tests on the same silt material using the equipment shown in Figures 43.1 and 43.2. The results of this series of tests are given in Figure 43.4, which shows a very consistent picture of how the damping varies over a range of frequencies from 2 Hz to over 1 kHz. All the data shown in this figure are for one value of confining stress (overburden pressure) equivalent to about 1 m of embedment in the water-saturated sediment and for the lowest level of shear strain amplitude attainable in our tests. There is a continuous increase in the logarithmic decrement with increasing frequency up to the limit

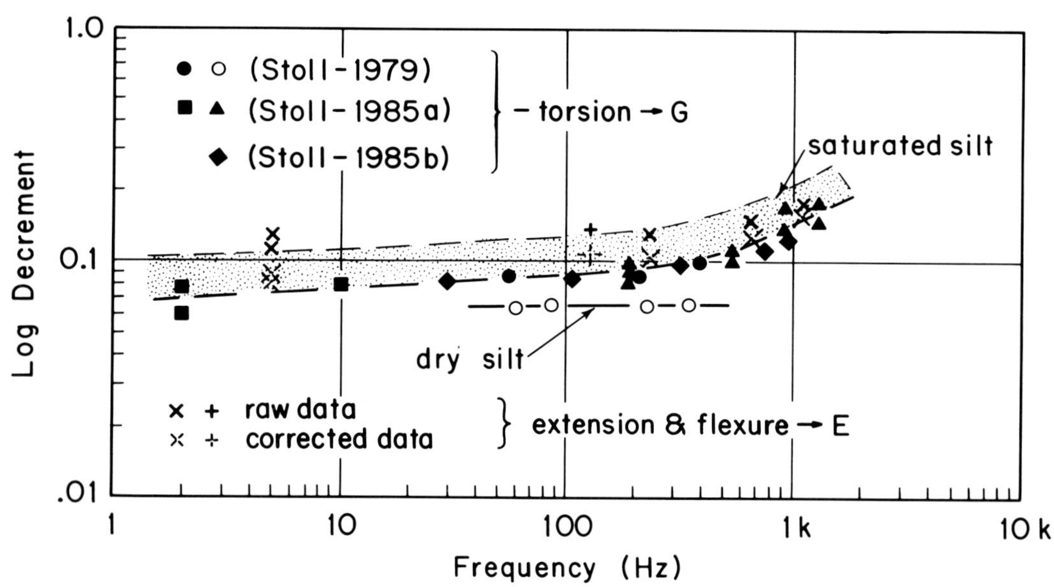

Figure 43.4. Logarithmic decrement versus frequency for water-saturated and dry silt. E is Young's modulus and G is shear modulus (used interchangeably with μ).

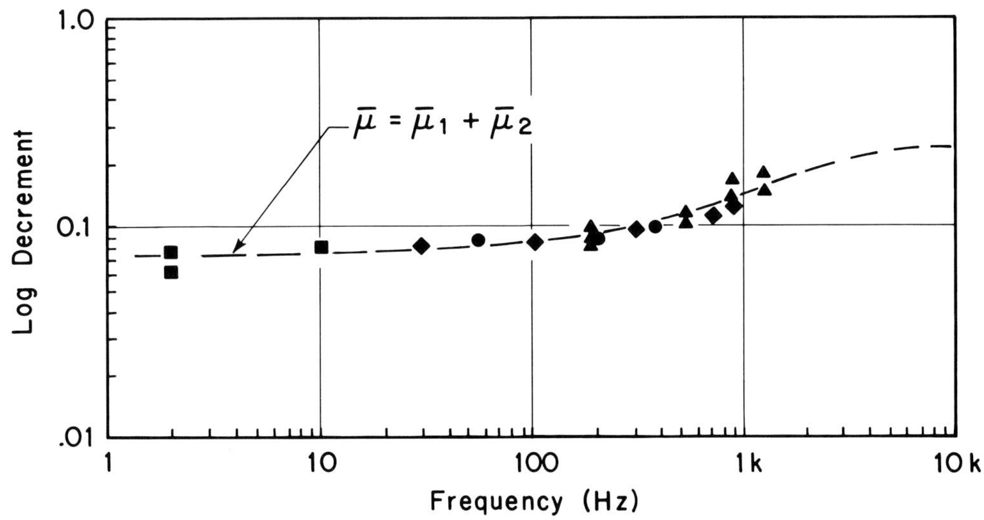

Figure 43.5. Fit of logarithmic decrement predicted by model based on overall complex modulus $\bar{\mu}$ to torsional data of Figure 43.4.

of our data in the low kilohertz range. Above about 1.5 kHz resonances in our equipment began to affect the data; however, the trend suggests that the damping will increase even more. We expect the maximum to occur somewhere in the low kilohertz range analogous to the results for water-saturated rock of high porosity given by Spencer (1981) and Murphy (1982).

The effect of frequency that is seen in Figure 43.4 cannot be modeled by the Biot theory if the moduli of the skeletal frame are complex constants and a single coefficient of permeability is specified. For the silt used in our experiments, the permeability measured in a constant head test was found to be 2.5×10^{-10} cm^2 at a porosity of 50%. When this permeability and a constant complex modulus are used in the Biot theory, no appreciable increase in log decrement is predicted until the frequency is over 10 kHz (this will be shown in a later figure—see also Stoll, 1986). In recent theoretical studies (Stoll, 1980, 1985b) we described a physical mechanism for damping that is based on the notion of local fluid flow near the intergranular contact points or in regions of close proximity between particle surfaces in fine grained sediments. The effect of a viscous fluid in these restricted regions is similar to the effect of "squeeze film" motion, which is well known in lubrication theory—the result is an intergranular force proportional to the rate of approach of the particles. Thus if we make the moduli of the skeletal frame strain rate dependent or viscoelastic, with a distribution of relaxation times appropriate to experimental results, the Biot theory can be used to reflect the effects of both "local" fluid motion governed by the *strain rates* of the frame and overall motion of the fluid field relative to the frame produced by the *average pressure gradient* in the fluid. Since these two mechanisms of energy loss are fundamentally different, with phase differences between the controlling motions, their effects cannot be combined in a single, generalized permeability operator.

A model formulated in the manner described above may be termed a "poroviscoelastic model" and it should be noted that the response will not be the same as if the overall response was based on an ordinary viscoelastic model. Moreover, there are compelling reasons to keep the effects of global relative motion of the fluid separate from the effects of local fluid motion by using a Biot type formulation, even though the resulting model may be more complex and involve the determination of more parameters. For example, in the case of a partially saturated sediment, the strain rate effects of local motion may still be largely applicable whereas the effect of relative global fluid motion will be markedly different.

The basic moduli that are input to both the Biot and the Gassmann theories are the complex shear modulus $\bar{\mu}$ and the complex bulk modulus $\bar{K}$ determined in a water environment under conditions wherein the pore water pressure remains constant. Unfortunately measurement of the bulk modulus, or equivalently Young's modulus, involves volume change so that the compressibility and mobility of the fluid and therefore the lateral boundary conditions play an important role in the interpretation of the experimental results if undrained, dynamic tests are used. To avoid this problem we have run a few tests involving extension and compression as well as flexure on partially saturated specimens so that most of the effect of fluid compressibility is suppressed. Under these conditions local viscous effects still affect the overall response, although possibly in a somewhat modified manner because of capillary effects. Although the tests involving only torsional motion are relatively easy to interpret, those involving axial extension–compression or bending necessitate corrections for the effects of shear and rotatory or lateral inertia, particularly if the diameter-to-length ratio of the specimen is not very small. Both the observed response and data corrected for the effects of our experimental configuration are shown in Figure 43.4. Although there are only a few data points, they follow quite closely the trend of the data from all of our torsional experiments suggesting that, until we have more data, a reasonable assumption may be to take the log decrement

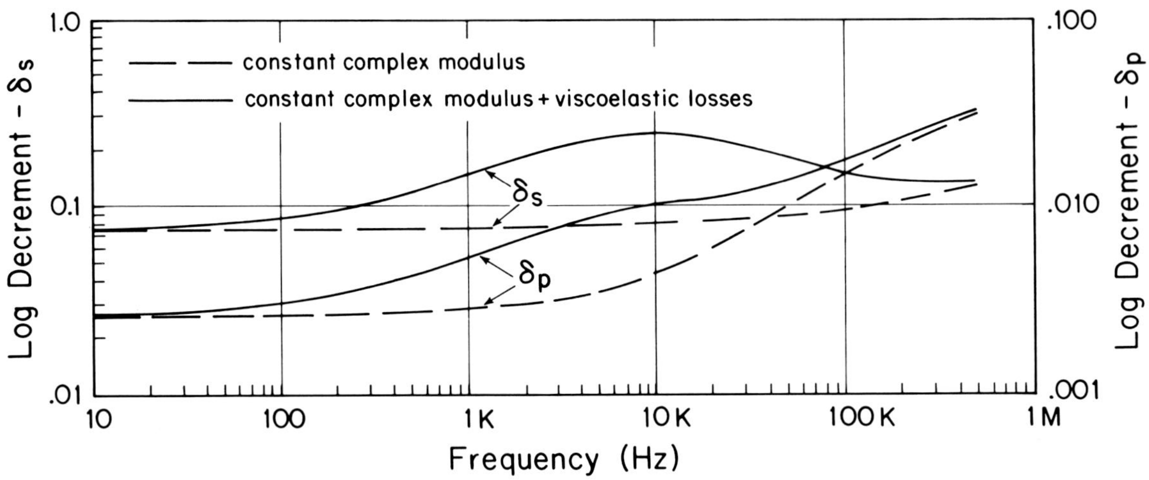

Figure 43.6. Overall response of sediment based on Biot theory with and without the effects of "local" viscous losses. δ_s is log decrement for shear waves and δ_p is log decrement for dilatational waves in water-saturated sediment.

corresponding to extensional motion as governed by Young's modulus, $\bar{E}$, as equal or slightly larger than the decrement for shearing motion as governed by $\bar{\mu}$, with the same frequency dependence. Clearly this is an area of experimental research that needs further work!

To incorporate into the Biot model the kind of frequency dependence shown in Figure 43.4, it is convenient to choose a simple viscoelastic model to describe that part of the frequency-dependent response owing to the effects of local fluid motion. The simplest choice would be a two parameter "Maxwell" model with a single relaxation time (e.g., see Flugge, 1967). However, with a view toward matching future data obtained at higher frequencies, we have chosen to model the viscoelastic response with a "Cole–Cole" model (Cole and Cole, 1941) with three free parameters. This phenominological model, which has been used in a somewhat different way by Spencer (1981) to fit the extensional response of waterbearing sedimentary rocks, is based on an assumed distribution of relaxation times that leads to a complex modulus with real and imaginary parts, N_1 and N_2, given by

$$N_1 = N_\infty = \frac{(N_0 - N_\infty)\,[1 + (\omega\tau_0)^{1-\alpha} \sin \tfrac{1}{2}\alpha\pi]}{1 + 2(\omega\tau_0)^{1-\alpha} \sin \tfrac{1}{2}\alpha\pi + (\omega\tau_0)^{2(1-\alpha)}}$$

$$N_2 = \frac{-(N_0 - N_\infty)(\omega\tau_0)^{1-\alpha} \sin \tfrac{1}{2}\alpha\pi}{1 + 2(\omega\tau_0)^{1-\alpha} \sin \tfrac{1}{2}\alpha\pi + (\omega\tau_0)^{2(1-\alpha)}} \qquad (1)$$

where N_0 and N_∞ are the relaxed and unrelaxed elastic moduli, τ_0^{-1} is the frequency at which N_2 is a maximum, and α is a parameter that determines the sharpness of the peak of the response curves. $\alpha = 0$ corresponds to a single relaxation and therefore a distribution of relaxation times that is a delta function (i.e., a single Maxwell element), whereas α greater than about 0.4 corresponds to a broad, bell-shaped distribution of relaxation times (see Spencer, 1981 for an example of the effect of a changing α). Since this portion of our model reflects only the changes in stiffness due to local viscous losses, we choose N_0 to be zero since there is vanishing influence of these viscous losses as the frequency approaches zero. This leaves three free parameters, α, τ_0, and N_∞, which may be chosen to fit our experimental data. It should be noted that the variable N in Equation (1) may be used to represent part or all of any complex modulus (i.e., shear, extensional or bulk) for which experimental data are available.

The final value of the overall complex shear modulus is the sum of a constant complex modulus $\bar{\mu}_1$, chosen to match the data at very low frequencies where viscous losses are negligible, plus a frequency-dependent complex modulus $\bar{\mu}_2$ determined from Equation (1) so that

$$\bar{\mu} = \bar{\mu}_1 + \bar{\mu}_2$$

where the real and imaginary parts of $\bar{\mu}_2$ are given by N_1 and N_2 of Equation (1). An example of such a model fitted to the torsional data of Figure 43.4 is shown in Figure 43.5. Since our data do not extend above about 1.5 kHz, the location and amplitude of the maximum logarithmic decrement are rather arbitrary. Many other combinations of α, τ_0^{-1}, and μ_∞ will undoubtedly fit the limited data set equally well so that more data from the high-frequency range is necessary to make the best final choice. Nevertheless, with the values used in the example, we are able to describe $\bar{\mu}$ quite satisfactorily in the range from 10 Hz to 1.5 kHz, which is of prime interest in geoacoustic modeling.

When the complex shear modulus described above is introduced into the Biot theory, with the logarithmic decrement for shear and dilatation assumed to be equal, we obtain response curves such as those shown in Figure 43.6 labeled "constant complex modulus + viscoelastic losses." The curve labeled "constant complex modulus" is the response that would have

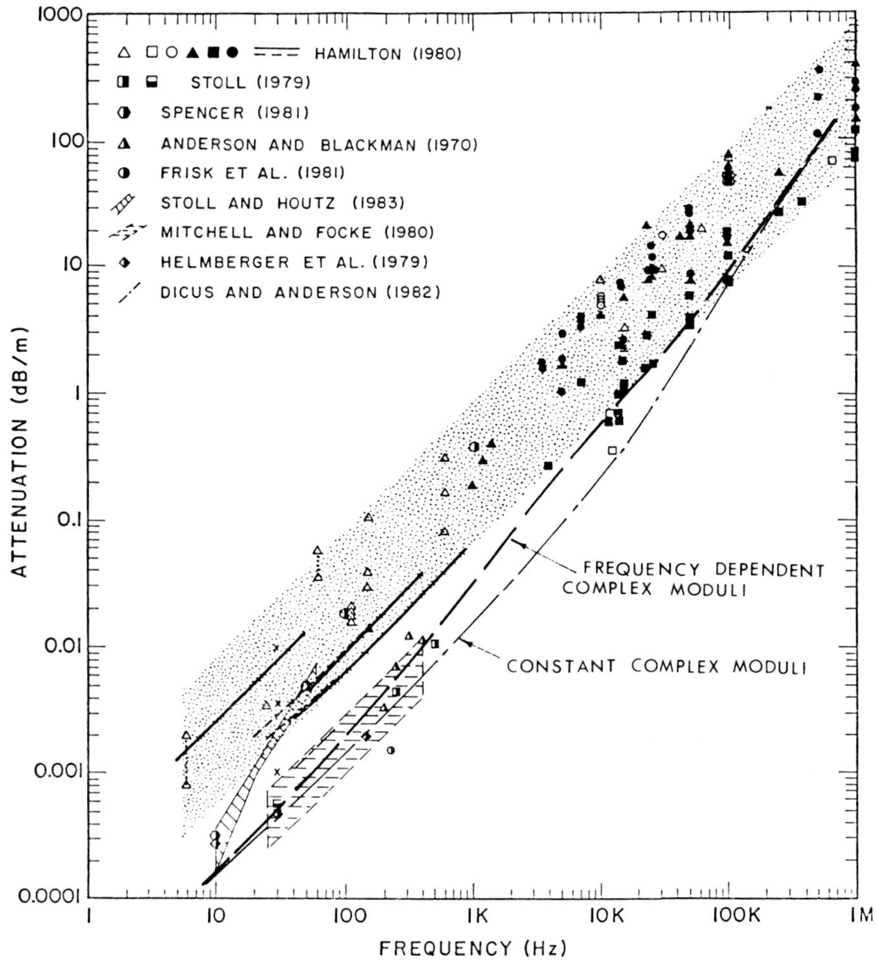

Figure 43.7. Attenuation of dilatational waves based on the Biot theory—comparison of model predictions with various data.

been predicted had we chosen the complex modulus of the skeletal frame to be a constant, thus neglecting the effects of "local" fluid losses but including frictional losses and the usual viscodynamic losses resulting from overall motion of the pore water relative to the skeletal frame as controlled by the permeability and associated parameters. In the latter case we see that the logarithmic decrement is nearly constant until we approach frequencies of 2–3 kHz in the case of dilatational waves and even higher for the case of shear waves. Thus it appears that in the lower frequency range local viscous losses (and intergranular friction, which determines the asymptotic value of the logarithmic decrement at low frequencies) are of prime importance in determining the overall intrinsic attenuation to be expected in the fine-grained material under scrutiny.

Conclusions

When the results for dilatational waves shown in Figure 43.6 are converted to attenuation, we obtain the curves shown in Figure 43.7. This figure shows experimental data from many different sources, and in the low-frequency range from 10 Hz to 1 kHz it can be seen that much of the more recent field data falls well below the older data that had been collected by Hamilton (1980) (the speckled band). The predictions of the model, based entirely on the laboratory results described above, agree quite well with the newer field data in the low-frequency range and, when the effects of local viscous losses are included, also match the experimental data at much higher frequencies (the data for clays, silts, and other fine-grained sediments fall near the bottom of the speckled band at high frequencies). Thus the Biot model, which has proved to be extremely useful in predicting the frequency-dependent response of coarse sediments, also provides a link between intrinsic damping at high and low frequencies for fine-grained materials provided the proper strain rate dependence of the skeletal frame is included.

Acknowledgments

The work described in this chapter was sponsored by the Office of Naval Research, Code 11250A, under contract No.

N00014-87-K-0204. Lamont-Doherty Geological Observatory Contribution No. 4504.

References

Anderson, R.S., and A. Blackman, 1970. Attenuation of low frequency sound waves in sediments. Journal of the Acoustical Society of America, v. 49, p. 786–791.

Cole, K.S., and R.H. Cole, 1941. Dispersion and adsorption in dielectrics. Journal of Chemical Physics, v. 9, p. 341–351.

Dicus, R.L., and R.S. Anderson, 1982. Effective low frequency geoacoustic properties inferred from measurements in the northeast Atlantic. Naval Ocean Research and Development Activity report No. 21.

Flugge, W., 1967. Viscoelasticity. Blaisdell, Waltham, MA.

Frisk, G.V., J.A. Doutt and E.E. Hays, 1981. Bottom interaction of low frequency acoustic signals at low grazing angles. Journal of the Acoustical Society of America, v. 69, p. 84–89.

Hamilton, E.L., 1980. Geoacoustic modeling of the seafloor. Journal of the Acoustical Society of America, v. 68, p. 1313–1340.

Helmberger, D.V., G. Engen and P. Scott, 1979. A note on velocity, density and attenuation models for marine sediments determined from multi bounce phases. Journal of Geophysical Research, v. 84, p. 667–671.

Mitchell, S.K., and S.K. Focke, 1980. New measurements of compressional wave attenuation. Journal of Acoustical Society of America, v. 67, p. 1582–1589.

Murphy, W.F., III, 1982. Effects of partial water saturation on attenuation in Massilon sandstone and Vycor porous glass. Journal of the Acoustical Society of America, v. 71, p. 1458–1468.

Spencer, J.W., 1981. Stress relaxations at low frequencies in fluid saturated rocks: attenuation and modulus dispersion. Journal of Geophysical Research, v. 86, p. 1803–1812.

Stoll, R.D., 1979. Experimental studies of attenuation in sediments. Journal of the Acoustical Society of America, v. 66, p. 1152–1160.

Stoll, R.D., 1980. Theoretical aspects of sound transmission in sediments. Journal of the Acoustical Society of America, v. 68, p. 1341–1350.

Stoll, R.D., 1985a. Computer-aided studies of complex soil moduli. In: Woods, R. (ed.), Measurement and Use of Shear Wave Velocity for Evaluating Dynamic Soil Properties. American Society of Civil Engineers, New York, p. 18–33.

Stoll, R.D., 1985b. Marine sediment acoustics. Journal of the Acoustical Society of America, v. 77, p. 1789–1799.

Stoll, R.D., 1986. Acoustic waves in marine sediments. In: Akal, T., and M. Berkson (eds.), Ocean Seismo-Acoustics. Plenum, New York, p. 417–434.

Stoll, R.D., and R.E. Houtz, 1983. Attenuation measurements with sonobuoys. Journal of the Acoustical Society of America, v. 73, p. 163–172.

CHAPTER 44

Sediment Shear Waves: A Comparison of *In Situ* and Laboratory Measurements

Michael D. Richardson, Enrico Muzi, Luigi Troiano, and Bruno Miaschi

Introduction

In recent years, scientists from such diverse fields as geophysics, seafloor engineering, sedimentology, soil mechanics, and underwater acoustics have devoted considerable attention to the measurement of sediment shear wave velocity and/or sediment dynamic modulus. These fundamental sediment properties are important to predicting the stability of sediment slopes, the consolidation behavior of sediments, the strength of marine foundations, and the conversion of water-borne energy to sediment shear wave energy at the seabottom, to give just a few examples.

Sediment shear wave velocity has been measured *in situ* using probes deployed by scuba divers, submersibles (Hamilton et al., 1970), and remotely from surface ships (Bennell et al., 1982). Shear wave velocity has also been measured in and between boreholes using explosive and various vibratory techniques (Warrick, 1974). Scholte waves and Love waves have been used to estimate shear wave velocities in surficial sediments by numerous investigators (Rauch, 1986; Akal et al., 1986; Snoek, 1990).

Hamilton (1976, 1980, 1987), in recent reviews of *in situ* measurements of shear wave velocity, found that the relatively few good measurements had such a wide range of values as to make the prediction of shear wave velocity in surficial sediments difficult and tenuous. Hamilton reported typical velocities of 50–150 m/sec in the upper meter of clays increasing to 100–200 m/sec at 10 m depth. Sands had similar values for the upper meter of the sediment increasing to 200–300 m/sec at 10 m.

Numerous attempts have been made to measure shear wave velocity of natural and artificial sediments in the laboratory. Many of these measurements have been based on the ceramic bender transducer technology pioneered by Shirley (1978). Shear wave velocities have been measured on freshly cut cores (Richardson, 1983; Richardson et al., 1987; Schultheiss, 1985; Lavoie 1988). Shear wave velocities have also been measured on artificial sediments at atmospheric pressure (Horn, 1980; Brunson and Johnson, 1980) and under confining pressures meant to represent consolidation under several meters of sediment (Schultheiss, 1981). Lovell and Ogden (1984) measured shear wave velocity gradients on both surficial and naturally consolidated sediments under confining pressures representing 0–400 m overburden pressure. Laboratory measurements of shear wave velocity have also been made using the resonant column test (see Hardin and Richart, 1963 for a review of these techniques). Shear wave velocities as low as 2 m/sec have been reported for artificial sediments created from settled kaolinite (Shirley and Hampton, 1978) and typical velocities of 20 m/sec (silts and clays) and 50 m/sec (sands) have been reported for surficial sediments collected with cores (Richardson et al., 1987).

Seismic refraction techniques (Danbom and Domenico, 1987) have also been used to determine shear wave velocities in marine sediments, but these techniques integrate shear wave velocities over profiles kilometers long and hundreds of meters thick. More short-range seismic experiments, such as those reported by Stoll et al. (1988), are required to determine sediment geoacoustic properties in the upper few meters of sediments. Recent advances in deep-towed seismic sources and receivers will also increase the vertical resolution of these techniques (Fagot, 1986).

Shear wave velocity can be estimated using the empirical relationships of Hamilton (1971, 1976, 1987) and Bryan and Stoll (1988) or calculated from physical models such as the Biot–Stoll Model (Ogushwitz, 1985; Biot, 1962; Stoll, 1980). Both models (given appropriate depth dependent input parameters), as well as empirical relationships, can be used to estimate shear wave velocity with depth in the sediment. The relatively few shear wave measurements, differences in measurement techniques, and a controversy about the actual physical mechanisms that

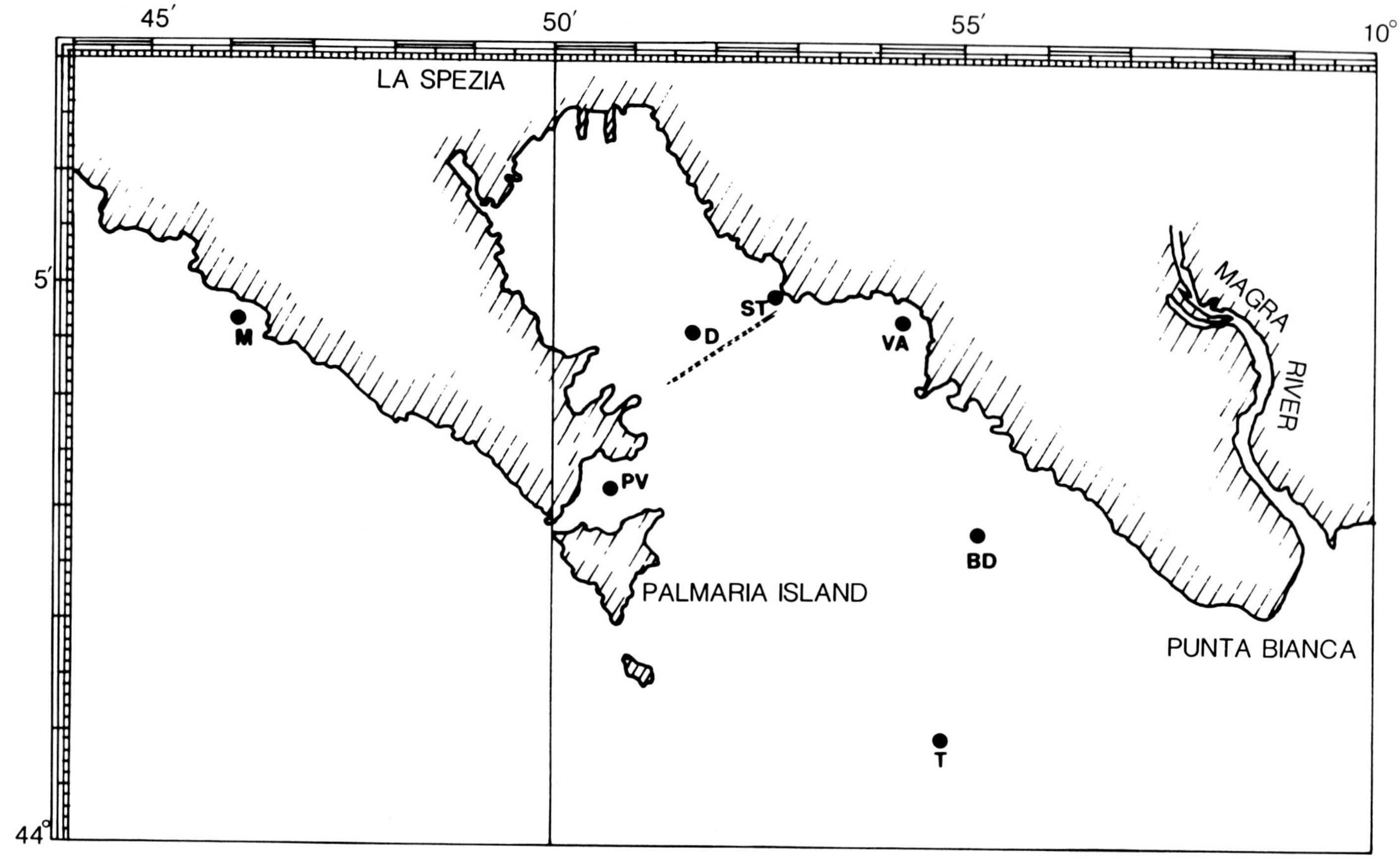

Figure 44.1. Location of sampling sites: Diga (D), Venere Azzura (VA), Santa Teresa (ST), Portovenere (PV), Turf (T), Boa Dragaggio (BD), and Monasteroli (M). Viareggio site (43° 48.62′N, 10° 07.16′E) was 33 km southeast of Palmaria Island.

control this type of low strain acoustic propagation have led to a rather confused picture as to the actual velocities of shear waves in surficial marine sediments.

It is the purpose of this chapter to compare values of shear wave velocity obtained both *in situ* and in the laboratory using similar measurement techniques. The existence of an empirical relationship between *it situ* and laboratory shear wave velocity is explored. Empirical relationships between *in situ* shear wave velocity and easily measured sediment physical properties are examined. Hamilton (1987) laments the lack of *in situ* measurements in modern marine sediments. The data we present and measurement techniques we develop will help fill this void and lead to an improved fundamental understanding of the propagation of acoustic waves through marine sediments.

Materials and Methods

General

Eight sites were chosen to represent a wide selection of sediment types within diving depths (Fig. 44.1). Several of the sites have been the locations for saclantcen acoustic and geoacoustic experiments conducted over the last 6 years (Rauch, 1980, 1986; Akal et al., 1984, 1986; Richardson, 1986; Schmalfeldt, 1986; Snoek et al., 1986; Snoek and Rauch, 1987; Snoek, in press).

Sediments were collected using a 12.0-cm-inside diameter PVC hand-operated corer. At least three cores were collected at each site. Nearbottom temperature and salinity were measured by scuba divers using hand-held probes. *In situ* shear wave velocity measurements were made with the probes described in the next section. At least three deployments were made at each station. Sediment cores were carefully transported to the laboratory and kept under refrigeration at 4°C until laboratory shear wave velocity measurements were made. After the acoustic measurements, sediment samples were collected from each core for mass property determination. All data, reported herein, were measured from sediments at or collected from 30 cm below the sediment-water interface.

A summary of environmental conditions for each station occupancy is given in Table 44.1. During our study measured salinities ranged from 37.5 to 38 ppt and are not reported for each deployment.

Table 43.1. Summary of environmental conditions for the eight sampling sites (some sites sampled more than once).

Site	Depth (m)	Date	Temperature (°C)	Sediment type	Porosity (%)	Density (g m^{-3})
Diga	7	6 October 87	26.0	Silty-clay	69.2	1.54
Diga	6	14 March 88	12.4	Silty-clay	68.9	1.54
Venere Azzura	7	15 March 88	12.5	Sand	47.1	1.88
Santa Teresa	10	17 March 88	12.5	Silty-clay	67.5	1.54
Portovenere	12	18 March 88	12.5	Silty-clay	63.4	1.63
Turf	18	27 April 88	14.5	Silty-sand	50.8	1.83
Diga	7	28 April 88	14.5	Silty-clay	–	–
Boa Dragaggio	14	30 April 88	14.5	Sand/silty-clay	57.9	1.71
Venere Azzura	7	25 July 88	24.1	Sand	43.7	1.91
Monasteroli	16	28 July 88	19.5	Sand	43.7	1.91
Turf	18	28 July 88	18.8	Sand/silty-clay	52.6	1.77
Viareggio	22	29 July 88	19.5	Silty-clay	61.9	1.60

In Situ Measurements

Sediment shear wave velocity was measured using a pulse technique. Transmitters and receivers were identical 1.25 in. (31.75 mm) square × 0.019 in. (0.48 mm) thick bimorph ceramic benders (Fig. 44.2). The ceramics were potted in a stainless-steel ring with silicone rubber (Hardness = 35 Shore A) to allow relatively unrestricted bender movement. A thin covering of much harder polyurethane resin (Hardness = 80 Shore A) holds the ceramics in place and provides a tough coating to protect the ceramics during insertion into the sediment. The received signals were amplified using a 40-dB gain amplifier located in the head of the receiver probes. A block diagram of the shear wave measurement system is presented in Figure 44.3. Shear waves are generated as a 6-cycle sine wave pulsed every 10 msec. Driving frequency (135–1120 Hz) and driving voltage (150–230 V p-p) varied depending on coupling characteristics, sediment shear wave velocity and attenuation, and the pathlength between receiver and transmitter. Transmitted and received signals were recorded with a digital waveform recording oscilloscope.

In October 1987 three isolated probes were used to test the system at the Diga site. The transmitter was placed by hand at 30 cm depth in the sediment and two receivers were inserted to 30 cm depth 200 cm on either side of the transmitter. The probes were inserted by hand to eliminate any electrical or mechanical connection between probes. After time-delay measurements were made, the receivers were moved successively in 25-cm intervals closer to the receiver. The resulting 17 distance vs. time delay measurements were plotted (Fig. 44.4) to determine the shear wave velocity (25.4 m/sec) and offset at nominal zero distance (0.013 cm). Receivers were then rotated 180° to demonstrate phase reversal of the received signal, a characteristic of shear and not compressional wave received signals.

The beam pattern of the combined transmitter–receiver system was investigated by rotating the receivers in a semicircle (50 cm radius) around the transmitter. The resultant 1.0 and 12.0 dB loss of signal at 45 and 90° suggests a wide beam pattern in the horizontal axis. A wide beam pattern in the vertical axis was demonstrated in a similar manner. These October trials proved that the shear wave probes could be used to measure accurately shear wave velocity up to distances of 200 cm, and that, because of the relatively wide beam pattern, the probes were insensitive to small changes in relative orientation.

In the March trials, the shear wave probes were rigidly attached to a 200 cm long stainless steel frame. The receivers were placed at 30 and 70 cm distance from the transmitter. A small amount of energy passed through the frame complicating the time-delay measurements. We were able to visually separate the frame-borne and sediment-borne signals by making time-delay measurements over a wide range of frequency (100–5000 Hz). In April the shear wave transmitter was potted in a 70 × 190 mm cylinder of silicone rubber that eliminated most of the energy transmission through the frame. For added isolation the receivers were potted for the July trials. The frame used for the April and July measurements was triangular (100 cm on a side) and held four compressional wave probes in addition to the shear wave probes (Fig. 44.5). Examples of received signals are presented in Figure 44.6.

Divers were required to deploy the current frame to avoid damage to the delicate probes. The next generation frame has been designed to operate independently of divers, and will contain probes to measure sediment temperature and electrical resistivity in addition to sediment shear and compressional wave velocity and attenuation. In this chapter we restrict ourselves to the presentation of *in situ* shear wave velocity.

Laboratory Measurements

Laboratory shear wave velocity measurements were made using the pulse technique described by Richardson et al. (1987). Shear waves were generated and received by bimorph ceramic bender elements cantilever mounted on a brass mass (Fig. 44.7). Transducers were electrically and mechanically isolated from each

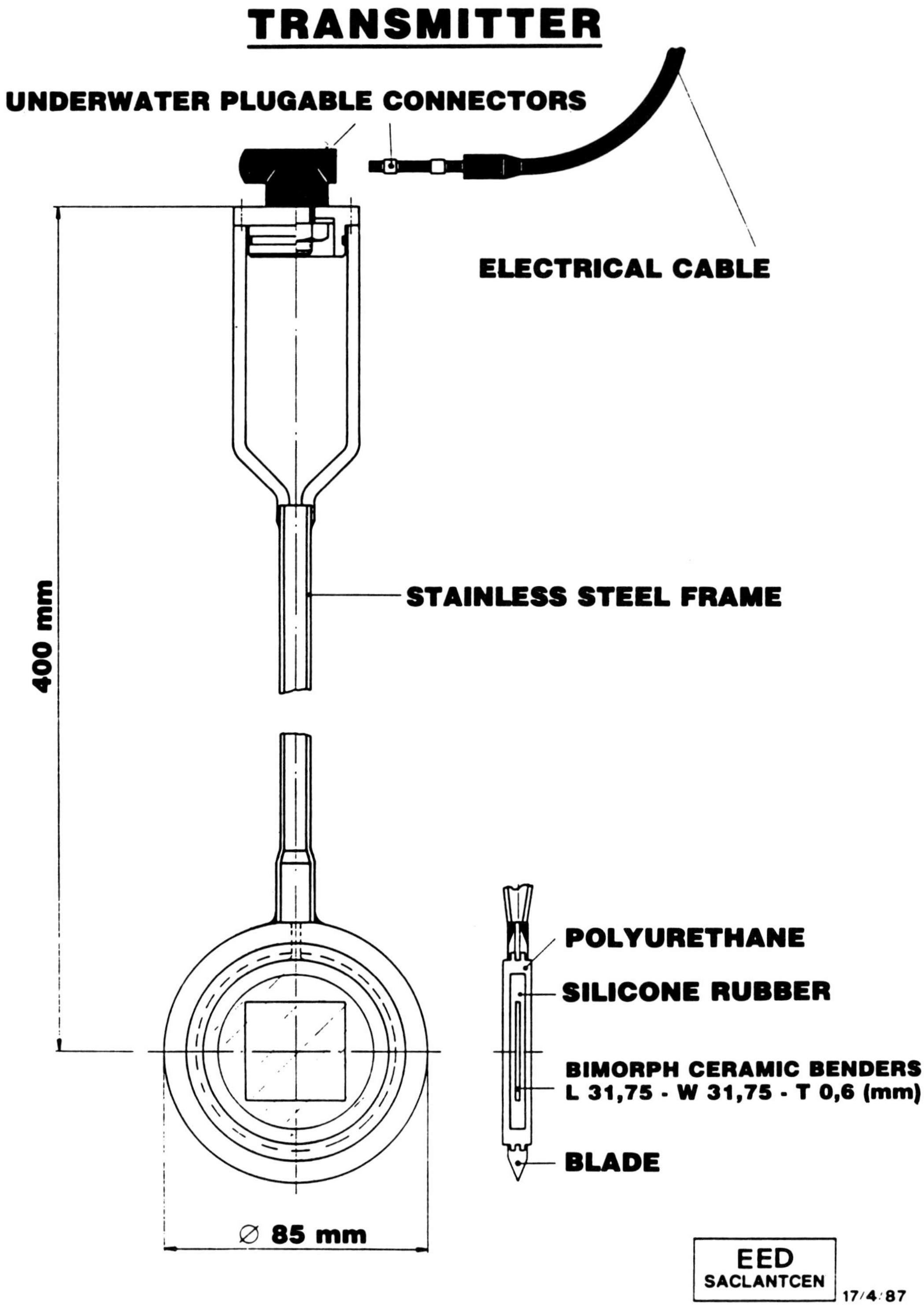

Figure 44.2. *In situ* shear wave transmitter (a) and receiver (b).

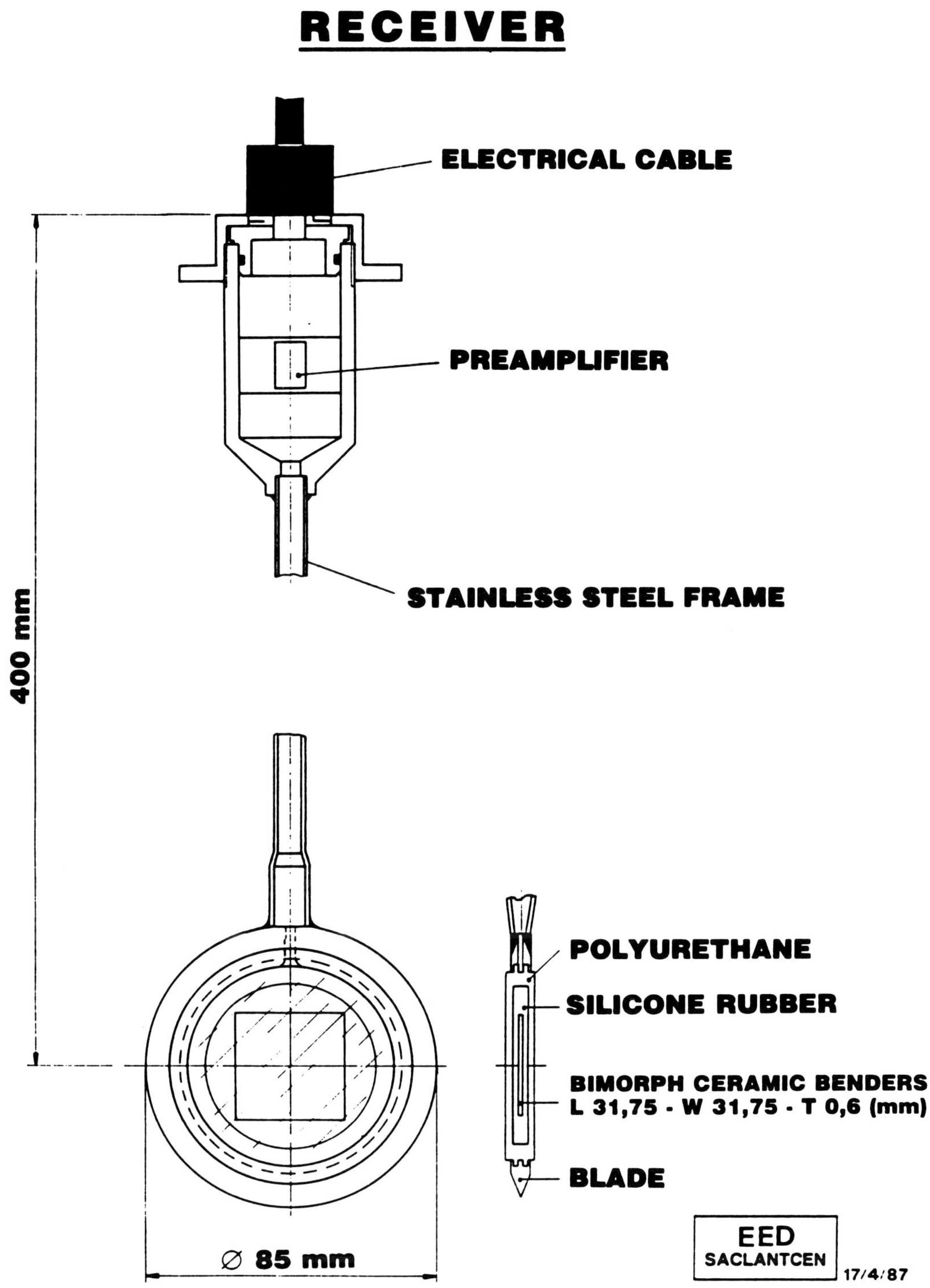
RECEIVER
ELECTRICAL CABLE
PREAMPLIFIER
STAINLESS STEEL FRAME
400 mm
POLYURETHANE
SILICONE RUBBER
BIMORPH CERAMIC BENDERS
L 31,75 - W 31,75 - T 0,6 (mm)
BLADE
∅ 85 mm
EED
SACLANTCEN
17/4/87

Figure 44.2. (b)

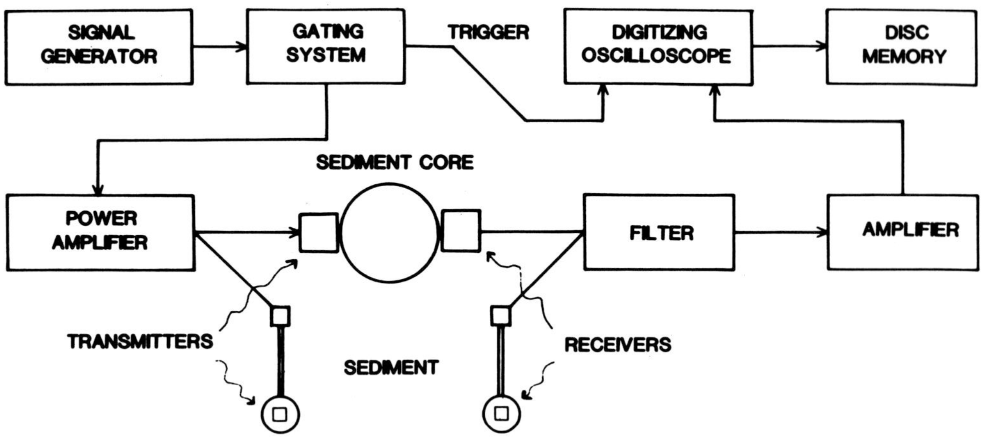

Figure 44.3. Block diagram of *in situ* and laboratory shear wave measurement system electronics. Preamplifiers (40 dB gain) located in the receivers are not shown.

other with rubber foam (Fig. 44.7) and the sediment was grounded to the electronics to eliminate electromagnetic feedover. The transmitter was driven by a 150–200 V p-p pulsed sine wave. Driving frequencies ranged from 150 to 1500 Hz depending on sediment type. The same electronic instruments were used to generate and record signals for *in situ* and laboratory shear wave measurements (Fig. 44.3). Examples of transmitted and received signals are presented in Figure 44.8.

Most time-delay measurements were made on sediments that remained in the 12 cm PVC cores. We drilled 3-cm-diameter holes in opposite sides of the core liner, snugged the transducers against the sediment suface, and recorded both time delay and distance between transmitter and receiver. Received signals were observed over a wide frequency range to separate shear wave signals transmitted through the sediment from those signals propagating along the core–sediment interface. Signals propagating along the core–sediment interface had lower amplitudes, much narrower bandwidths, and shorter time delays than shear wave signals transmitted through the sediment. Values of shear wave velocity measured on sediments removed from the cores were not significantly different from sediments remaining in cores. This suggested that we had successfully separated these signals. A time delay was subtracted from each measurement to account for the transit time of the signal through the electrical and mechanical system. This correction factor, measured with transducers touching, ranged from 2 to 14% of the sediment time-delay measurements (Fig. 44.8).

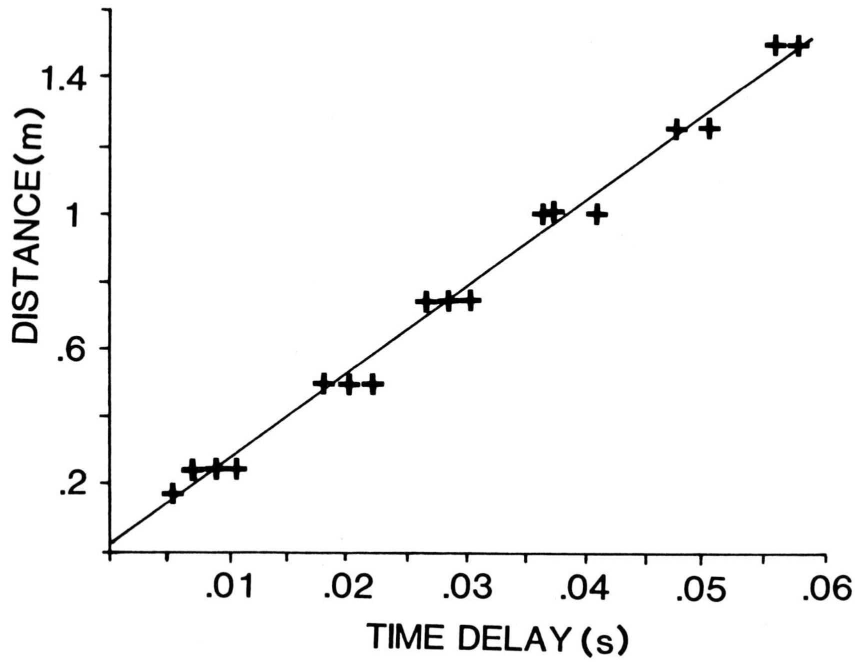

Figure 44.4. Shear wave velocity (25.4 m/sec) calculated from repetitive distance and time-delay measurements.

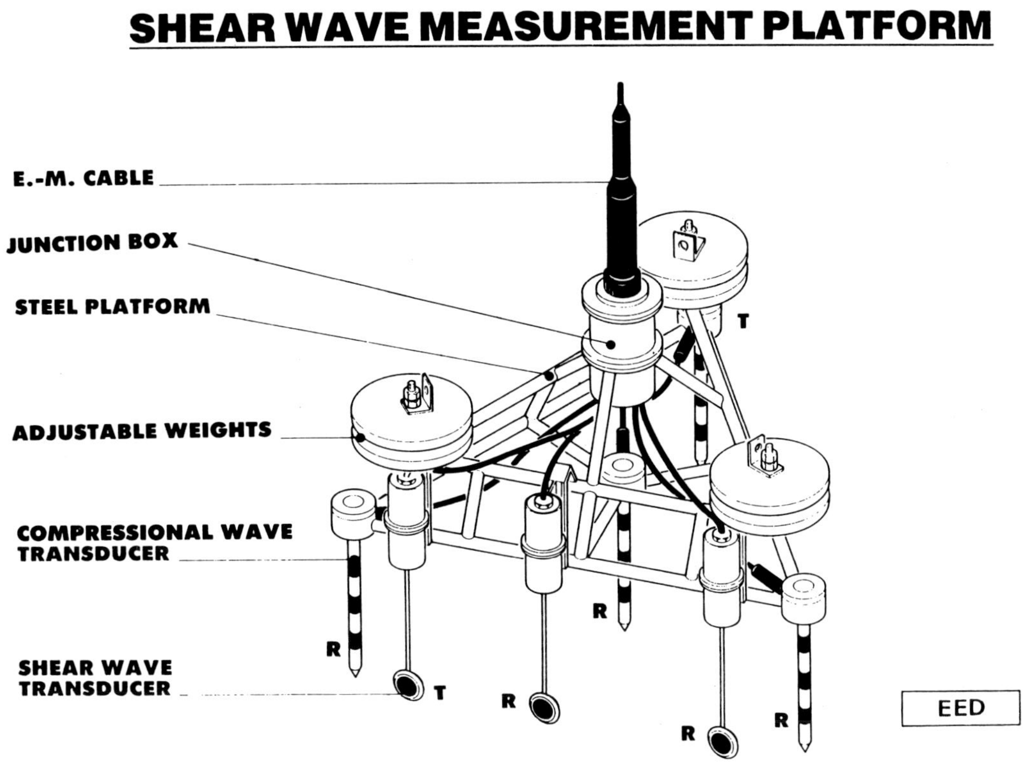

Figure 44.5. Acoustic shear wave measurement system as deployed in April and July 1988.

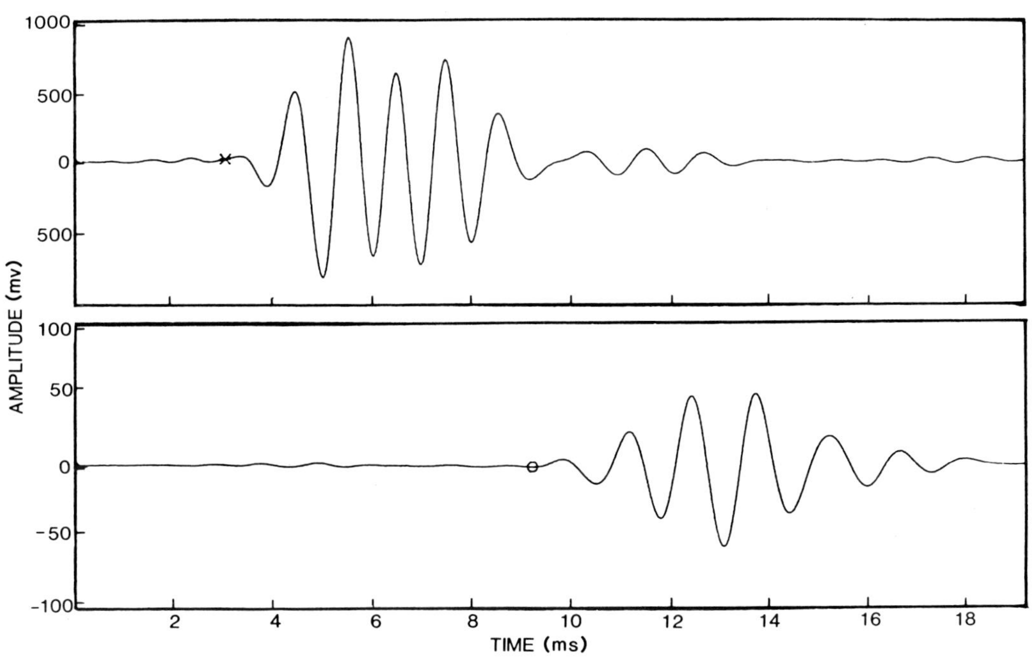

Figure 44.6. Examples of signals recorded from two shear wave receivers at the Venere Azzura site. Calculated shear wave velocities were 88.2 m/sec at 33 cm (top) and 82.4 m/sec at 71 cm (bottom) distance between probes.

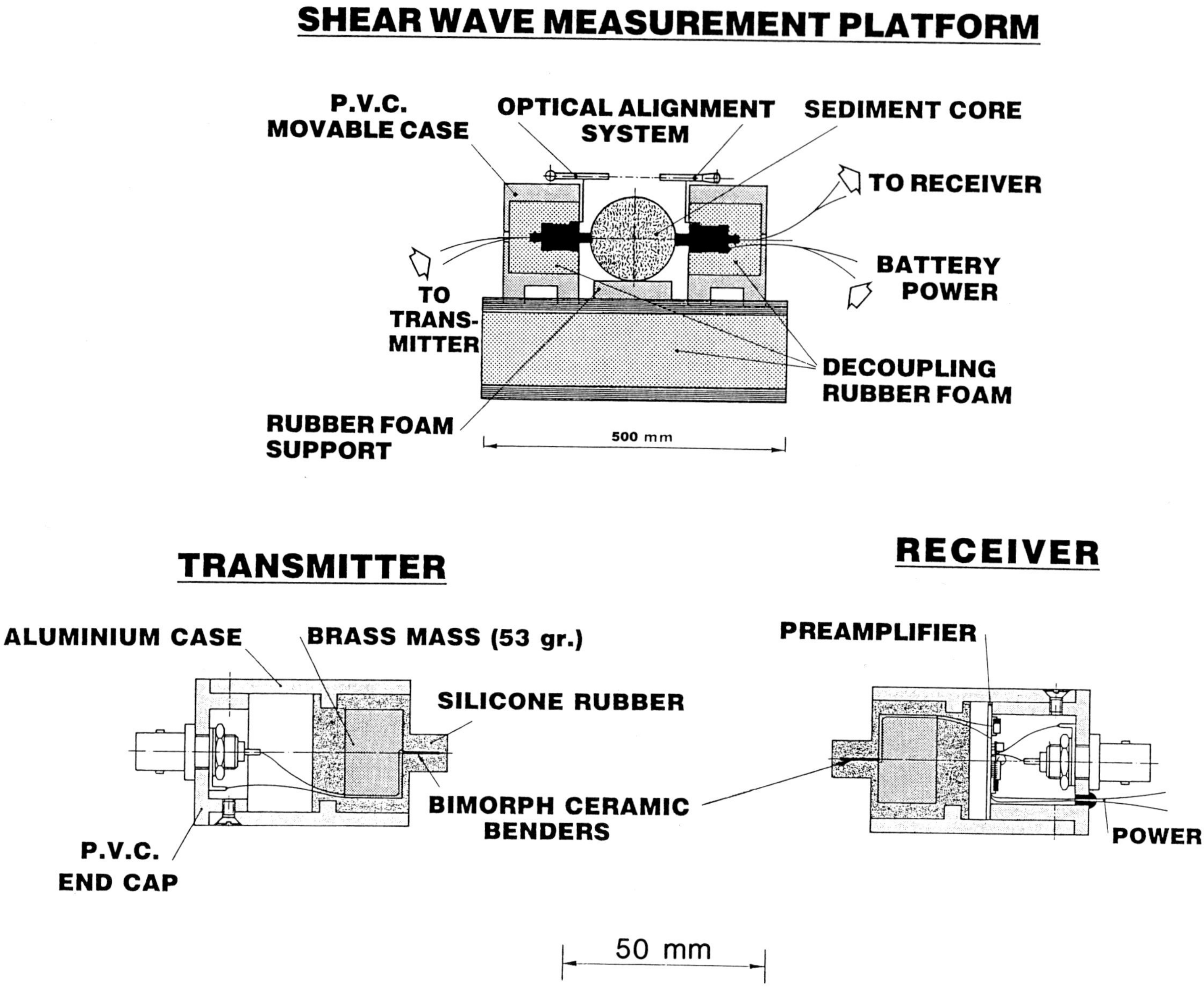

Figure 44.7. Laboratory shear wave measurement system.

Sediment subsamples were collected from the cores after laboratory shear wave measurements were completed. Dry-sediment density was determined with a helium pycnometer. Sediment porosity, void ratio, and wet density were calculated from weight loss of the sediment dried in an oven at 105°C for 48 hr, and from the measured dry density (Kermabon et al., 1969).

Results

Values of *in situ* sediment shear wave velocity ranged from 16.4 m/sec in the silty-clay sediments of Santa Teresa to 90.5 m/sec in the hard packed fine sands at Monasteroli (Table 44.2). Mean values of shear wave velocity measured on core sediments collected from the same locations were 6.5–22.1 m/sec less than mean *in situ* values. Shear wave velocity (*in situ* and laboratory) was negatively correlated with porosity and void ratio and positively correlated with sediment wet density (Table 44.3).

Discussion

The graphic relationship presented in Figure 44.9 suggests that laboratory values of shear wave velocity, measured at atmospheric pressure, can be corrected to *in situ* conditions using the following formula:

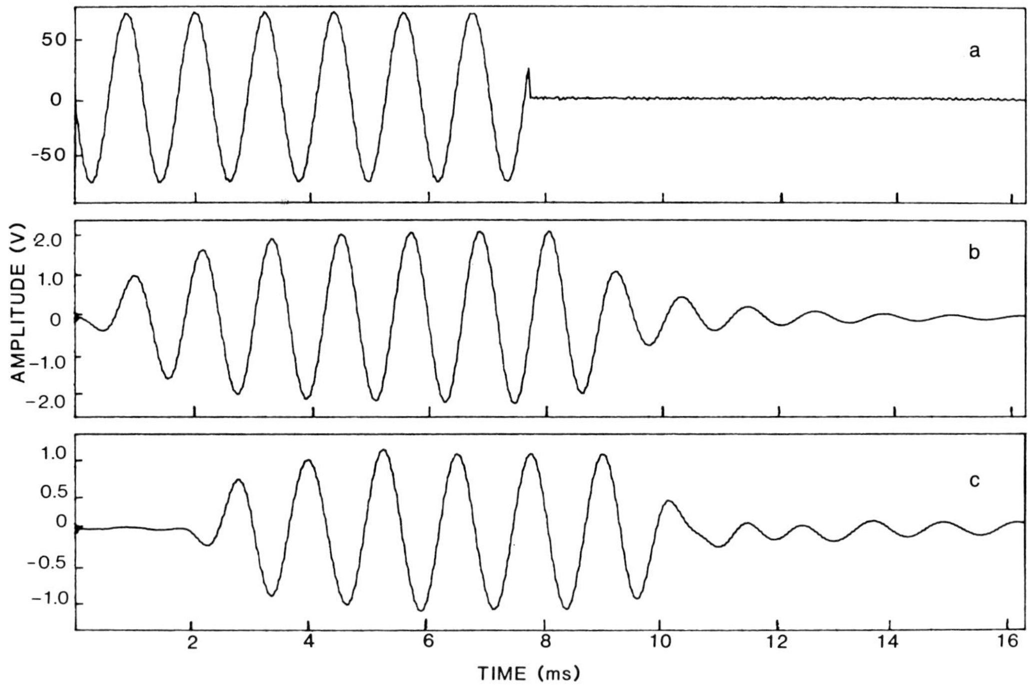

Figure 44.8. Examples of transmitted (a) and received signals for cored sediments collected at the Monasteroli site. The time delay with transducer and receiver touching (0.14 msec in b) was subtracted from time delay measured across 11.5 cm of sediment (1.86 msec in c) to calculate a shear wave velocity of 66.9 m/sec for this sandy sediment.

$$V_s\ (in\ situ) = 10.43 + 1.17\ V_s\ (\text{Lab}) \quad (1)$$

In spite of the high correlation between these two measurements ($R^2 = 0.975$), this formula should be applied with caution to other data sets. The relation applies only to surficial sediments and should not be extrapolated outside of the limited range of the data set.

Richardson et al. (1987) listed several factors that might contribute to the differences in laboratory and *in situ* measured values of shear wave velocity. These included (1) disturbance of sediments during collection, handling and measurement, (2) changes in pore pressure and/or physical characteristics that result from the release of confining pressure when sediments are removed from the bottom, (3) differences in frequencies used for measurements, (4) differences in techniques used to measure shear wave velocities or shear modulus, (5) poor measurement techniques, and (6) natural variability of shear wave velocity in sediments. We can add (7) changes in sediment temperature, (8) differences in strain values used for measurements, (9) disturbance of sediments during insertion of probes, (10) creation of excess pore pressure during insertion of probes, and (11) the possibility of strong vertical gradients of shear wave velocity in near surface sediments.

We can dismiss eight of these factors for the current comparisons. *In situ* and laboratory shear wave velocity measurements were made with the same type of transducers at approximately the same frequencies and strain levels. Both laboratory and *in situ* transmitters were driven with a 150–230 V p-p pulsed sine wave. The resultant behavior of the sediments under these low strains (<0.00001%) is to be considered purely elastic, yielding

Table 44.2. Summary of values of shear wave velocity measured *in situ* and from core samples in the laboratory.

Site	Date	V_S (*in situ*) (m/sec)		V_S (lab) (m/sec)	
		Mean	Range	Mean	Range
Diga	6–7 October 87	25.4	22.0–27.0	15.6	13.7–18.1
Diga	14 March 88	27.0	25.8–28.2	16.2	11.9–19.8
Venere Azzura	15 March 88	78.8	65.7–89.9	61.4	60.5–62.9
Santa Teresa	17 March 88	19.7	16.4–23.3	13.2	10.2–15.4
Porotovenere	18 March 88	29.3	24.8–37.4	14.3	10.5–16.8
Turf	27 April 88	41.7	33.7–57.9	24.4	22.6–25.6
Diga	28 April 88	23.6	19.5–28.1	–	
Boa Dragaggio	30 April 88	40.2	37.0–45.3	22.4	19.2–29.1
Venere Azzura	25 July 88	77.4	69.7–88.6	–	
Monasteroli	26 July 88	83.4	75.7–90.5	61.3	55.6–70.8
Turf (sand)	28 July 88	74.0	72.1–75.7	53.7	46.6–61.1
Turf (mud)	28 July 88	41.6	34.6–46.8	21.1	16.8–25.1
Viareggio	29 July 88	27.1	24.1–33.0	14.9	12.9–17.9

Table 44.3. Pearson product-moment correlation coefficients (r) calculated between values of *in situ* and laboratory shear wave velocities (m/sec) and sediment physical properties.

	Porosity (%)	Void ratio (%)	Wet density (g cm^{-3})
V_S (lab)	−0.85	−0.82	0.85
V_S (*in situ*)	−0.91	−0.87	0.92

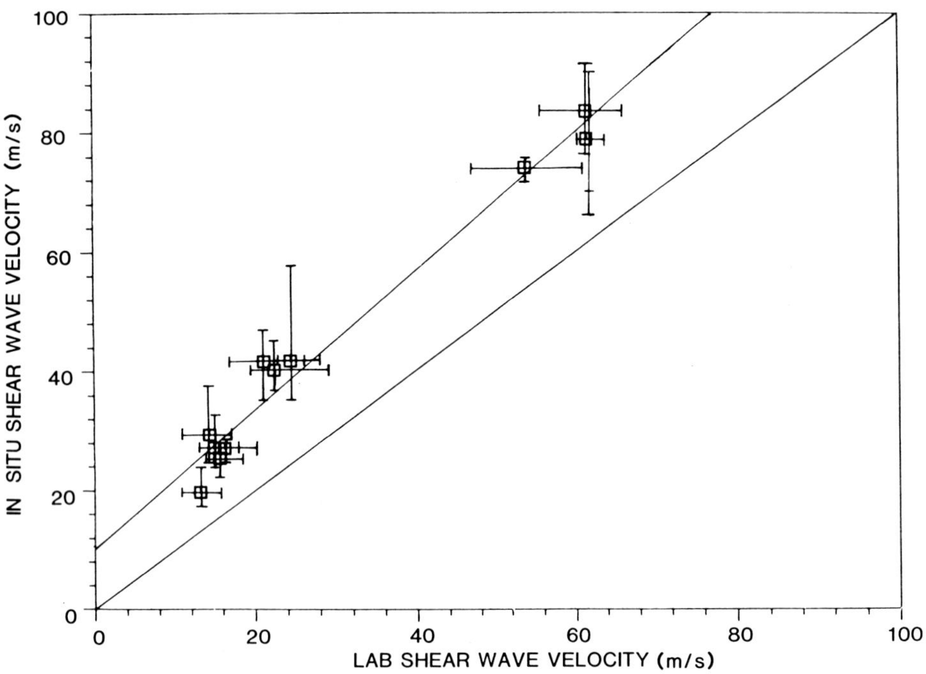

Figure 44.9. Comparison of laboratory and *in situ* shear wave velocities.

the maximum values of dynamic shear modulus and of shear wave velocity (Davis and Bennell, 1986). The resonant frequency of the pulsed sine wave, for both *in situ* and laboratory probes, ranged from 135 to 1500 Hz, depending on the mechanical impedance of the sediment. This frequency was generally lower for muds (135–430 Hz) and higher for sands (300–750 Hz). The time-delay measurements for single sediment specimens varied less than 5% over a wide frequency band (100–3000 Hz). The range of natural variability of values of shear wave velocity is presented in Figure 44.7 and preserves the basic relationships as reported. Values of shear wave velocity measured at sites sampled more than once were not significantly different in spite of differences in sediment temperature. The area disturbed by insertion of probes into sediments was small compared to pathlengths over which *in situ* shear wave measurements were made. At two sandy sites, *in situ* shear wave velocities made both immediately after probe insertion and after a 3-hr time delay. Shear wave velocities were the same, indicating excess pore pressures, created by insertion of the probes into the sediment, dissipated rapidly in these highly permeable, sandy sediments. Although great care was used to develop accurate measuring techniques, we cannot rule out systemic errors caused by poor techniques. The most likely causes for the lower laboratory shear wave velocities are sediment disturbances during collection, transportation, storage, and measurement both by mechanical manipulations and by changes in sediment confining pressures.

Numerous comparisons between values *in situ* and laboratory shear wave velocity have been made for terrestrial sediments using cross-hole and/or down-hole seismic techniques and laboratory resonant column tests (Cuny and Frey, 1973; Anderson and Woods, 1975; Anderson et al., 1978; Arango et al., 1978; Stokoe and Richart, 1975). Care must be taken in comparing these results to ours because strain amplitude, effective stress, time, and frequency of vibration must be accounted for (Davis and Bennell, 1986). Laboratory resonant column tests were run on sediments that had been subjected to effective confining pressures of up to 100 m, brought to the surface then compressed to *in situ* pressures. This can result in permanent changes in sediment microstructure. Our samples had no such stress history and no attempt was made to return sediments to *in situ* surficial conditions. In spite of these major differences in techniques, our results are in general agreement with comparisons of *in situ* and laboratory values of shear wave velocities reported for terrestrial sediments. Stoll et al. (1988), in a summary of these studies, reported values of *in situ* dynamic shear modulus to be 1.3–2.5 times the laboratory values. *In situ* dynamic shear moduli, calculated from values of shear wave velocity and density for this study, were 1.7–4.5 (mean 2.8) times calculated laboratory shear moduli (Table 44.4).

Akal et al. (1984, 1986) reported velocities of ducted Love waves from four of the sites occupied during this study. Measurements were made at short ranges (< 25 m) using stacked received signals from up to five ocean-bottom seismometers in series. Values of Love wave velocity (considered by Akal to be equivalent to values of shear wave velocity) at the Santa Teresa (16 m/sec), Portovenere (30 m/sec), Venere Azzura (65 m/sec), and Monasteroli (90 m/sec) sites were similar to *in situ* shear wave velocity values reported here. Akal's measurements at the

Table 44.4. Comparison of calculated and measured values of sediment dynamic modulus.*

Site	Date	Void ratio	Dynamic shear modulus (atm)		
			Calculated	Lab	*In situ*
Diag	6–7 October 87	2.23	11.1	3.7	9.8
Diga	14 March 88	2.23	11.1	4.0	11.1
Venere Azzura	15 March 88	0.90	103.8	70.0	115.2
Santa Teresa	17 March 88	2.08	14.2	2.6	5.9
Portovenere	18 March 88	1.75	24.5	3.3	13.8
Turf	27 April 88	1.04	81.4	10.8	31.4
Boa Dragaggio	30 April 88	1.41	43.3	8.5	27.3
Monasteroli	26 July 88	0.78	128.0	70.9	131.1
Turf (sand)	28 July 88	0.75	75.4	54.1	102.7
Turf (mud)	28 July 88	1.13	73.5	7.8	30.3
Viareggio	29 July 88	1.63	29.9	3.5	11.6

Dynamic shear modulus was calculated from the empirical relationship between void ratio, confining pressure (effective stress), and shear modulus given by Bryan and Stoll (1988).

Monasteroli site were for sandy-gravel sediments, in contrast to the sandy sediments we collected. The depth of propagation of Love waves in the sediment was estimated to be between 0 and 3 m, complicating comparisons between techniques.

Bryan and Stoll (1988) summarized the effects of mean effective stress (p') and void ratio (e) on sediment dynamic modulus (μ) with the following relationship:

$$\mu = \mu_0 (p')^n \exp(\tau e) \quad (2)$$

where $\mu_0 = 2526$ atm, $n = 0.50$, and $\tau = -1.5$. The formula was based on 494 concurrent laboratory measurements of dynamic shear modulus, confining pressure (range 14–700 kPa), and void ratio (range 0.35–2.5) compiled from the literature. Overburden pressure (p_0) at 30 cm sediment depth was calculated from the void ratio and sediment wet density. Mean effective stress was set equal to $0.67p_0$, after Stoll et al. (1988), and the sediment dynamic modulus was calculated for each of our sample sites (Table 44.4). Sediment dynamic modulus (μ) was also calculated from mean values of sediment shear wave velocity (V_s) and wet density (ρ) using the following:

$$\mu = V_s^2 \rho \quad (3)$$

Values of calculated sediment dynamic shear modulus were nearly equal to or higher than *in situ* measured values and much higher than laboratory measured values. The calculated mean effective stress was low (less than 3 kPa) and probably quite variable at 30 cm depth in the sediment. This variability and the rapid increase in predicted dynamic shear modulus in the upper meter of sediment make more exact comparison difficult.

In the marine environment biological, chemical and physical processes alter surficial sediment (upper 1 m) properties (Richardson and Young, 1980; Richardson et al., 1983; Richardson, 1983). These processes can increase sediment dynamic rigidity by compacting the sediment or by increasing chemical

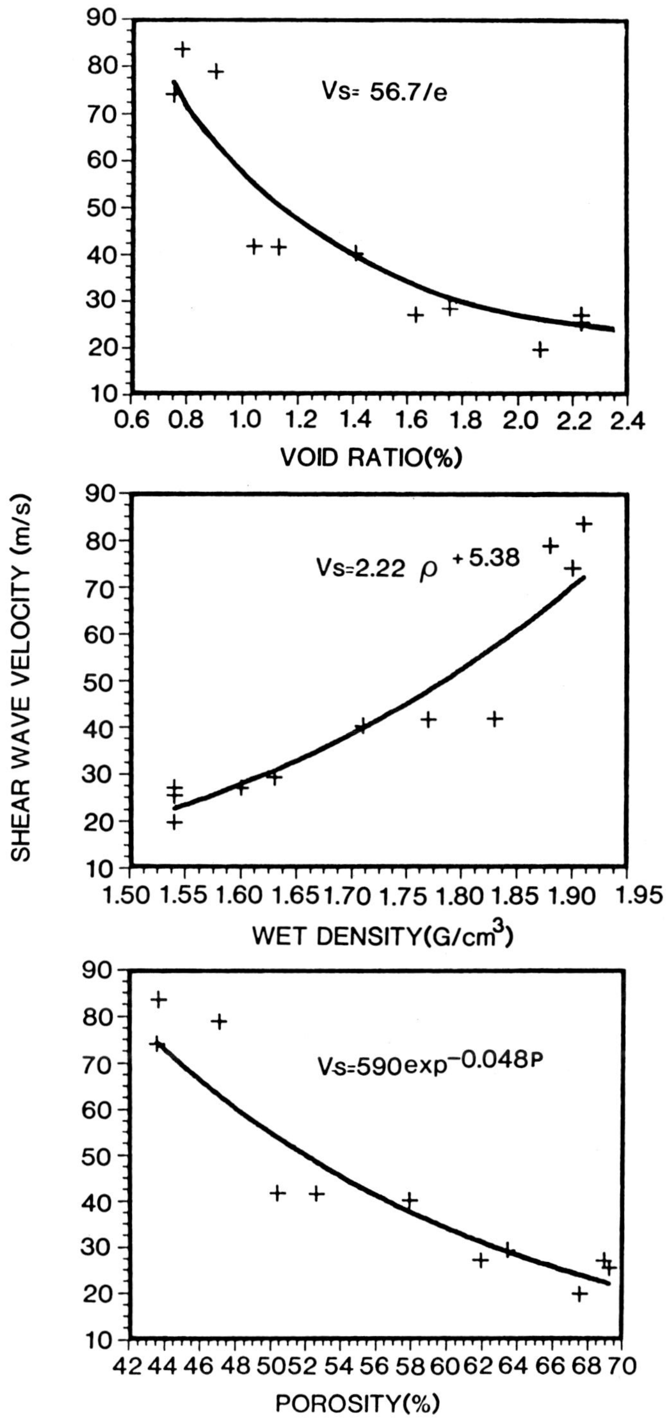

Figure 44.10. Empirical relationships between *in situ* shear wave velocity and sediment physical properties.

bonding between particles. Sediment dynamic rigidity can be reduced by the feeding activities of bottom animals. Marine sediments can therefore either be underconsolidated or overconsolidated with respect to overburden pressures. Modification of the formulations of Hamilton (1987) and Bryan and Stoll (1988) may be needed to predict sediment shear wave velocity gradients in the upper meter of sediment.

The empirical relationships between *in situ* sediment shear wave velocity and easily measured sediment physical properties (Fig. 44.10) provide reasonable estimates of surficial shear wave velocities for most marine sediments. Additional concurrent measurements are required to refine and extend this relationship to other sedimentary provinces. Hamilton (1971) suggests shear wave velocities should be highest in very fine sands with porosities of 45–55%. Sediments that are coarser or finer should have lower values of shear wave velocity because of a reduction in dynamic rigidity. The empirical relationship presented by Bryan and Stoll (1988) predicts an increase in sediment dynamic rigidity with increasing void ratio values over the range of 0.35–2.5%. Additional measurements of shear wave velocities are required to extend our empirical relationships to coarse sand and gravel sediments.

The rapid increase in shear wave velocity predicted for the upper few meters of sediment in reviews by Bryan and Stoll (1988) and Hamilton (1976, 1980, 1987) complicates comparison and predictions of sediment shear wave velocity. These empirical predictions were based on laboratory measurements of artificial, terrestrial and marine sediments and extrapolation of *in situ* seismic measurements to the upper few meters. Very few data are available on the gradients of *in situ* shear wave velocity in the upper few meters of marine sediments. An extensive measurement program is, therefore, required to define the variability and vertical gradients of shear wave velocity in marine sediments.

Acknowledgments

We would like to thank Captain A. Spairani and the crew of the R.V. Manning for their excellent support. Diving support was provided by M. Paoli (Saclantcen), and Proteco Sub. The support of the Electronic and Acoustic Engineering Department under the direction of A. Barbagelata is gratefully acknowledge. Sediment physical properties were determined by F. Turgutcan and B. Tonarelli. This chapter benefited from the helpful comments of T. Akal, J. Hovem, T. Muir, and R. Thiele.

References

Akal, T., P. V. Curzi, and E. Michelozzi, 1984. Geoacoustic and geotechnical properties of sediments: a pilot study. Societa Geologica Italiana Memorie, v. 27, p. 411–421.

Akal, T, H. Schmidt, and P. V. Curzi, 1986. The use of Love waves to determine the geoacoustic properties of marine sediments. In: Akal, T., and J. M. Berkson (eds.), Ocean Seismo-Acoustics. Plenum Press, London, p. 841–852.

Anderson, D. G., and R. D. Woods, 1975. Comparison of field and laboratory moduli. In: *In Situ* Measurement of Soil Properties, Raleigh, North Carolina. ASCE, New York, p. 69–92.

Anderson, D. G., C. Espana, and V. R. McLamore, 1978. Estimating *in situ* shear moduli at competent sites. In: Proceedings of the Specialty Conference on Earthquake Engineering and Soil Dynamics, Pasadena, California. ASCE, New York, p. 181–197.

Arrango, I., Y. Moriwaki, and F. Brown, 1978. *In-situ* and laboratory shear velocity and modulus. In: Proceedings of the Specialty Conference on Earthquake Engineering and Soil Dynamics, Pasadena, California. ASCE, New York, p. 198–212.

Bennell, J. D., P. D. Jackson, and P. Schultheiss, 1982. Further development of sea-floor geophysical probing. In: Oceanology International 82, Brighton, U.K. Spearhead Publications, Kingston-on-Thames, p. 4–8.

Biot, M. A., 1962. Generalized theory of acoustic propagation in porous dissipative media. Journal of the Acoustical Society of America, v. 34, p. 1254–1264.

Brunson, B. A., and R. K. Johnson, 1980. Laboratory measurements of shear wave attenuation in saturated sand. Journal of the Acoustical Society of America, v. 68, p. 1371–1375.

Bryan, G. M., and R. D. Stoll, 1988. The dynamic shear modulus of marine sediments. Journal of the Acoustical Society of America, v. 83, p. 2159–2164.

Cunny, R. W., and Z. B. Frey, 1973. Vibratory *in situ* and laboratory soil moduli compared. Journal of the Soil Mechanics and Foundations Division, American Society of Civil Engineers, v. 99, p. 1055–1076.

Danbom, S. H., and S. N. Domenico, 1987. Shear-wave Exploration. Society of Exploration Geophysicists, Tulsa, 275 p.

Davis, A. D., and J. D. Bennell, 1986. Dynamic properties of marine sediments. In: Akal, T., and J. M. Berkson (eds.), Ocean Seismo-Acoustics. Plenum Press, London, p. 501–510.

Fagot, M. G., 1986. Development of a deep-tow seismic system: a new capability for deep-ocean acoustic measurements. In: Akal, T., and J. M. Berkson (eds.), Ocean Seismo-Acoustics. Plenum Press, London, p. 853–862.

Hamilton, E. L., 1971. Elastic properties of marine sediments. Journal of Geophysical Research, v. 76, p. 579–604.

Hamilton, E. L., 1976. Shear wave velocity versus depth in marine sediments: a review. Geophysics, v. 68, p. 985–996.

Hamilton, E. L., 1980. Geoacoustic modelling of the sea floor. Journal of the Acoustical Society of America, v. 68, p. 1313–1340.

Hamilton, E. L., 1987. Acoustic properties of sediments. In: Lara-Saenz, A., C. Ranz-Guerra, and C. Carbo-Fite (eds.), Acoustics and Ocean Bottom. II FASE Specialized Conference. Consejo Superior de Investigaciones Cientiticas, Madrid, p. 3–58.

Hamilton, E. L., H. P. Bucker, D. L. Keir, and J. A. Whitney, 1970. Velocities of compressional and shear waves in marine sediments determined *in situ* from a research submersible. Journal of Geophysical Research, v. 75, p. 4039–4049.

Hardin, B. O., and F. E. Richart, 1963. Elastic wave velocities in granular soils. Journal of the Soil Mechanics and Foundations Division, American Society of Civil Engineers, v. 89, SM1, p. 33–65.

Horn, I. W., 1980. Some laboratory experiments on shear wave propagation in unconsolidated sands. Marine Geotechnology, v. 4, p. 31–54.

Kermabon, A., C. Gehin, P. Blavier, and B. Tonarelli, 1969. Acoustic and other physical properties of deep-see sediments in the Tyrrhenian Abyssal Plain. Marine Geology, v. 7, p. 129–145.

Lavoie, D., 1988. Geotechnical properties of sediments in a carbonate slope environment: Ocean drilling project site 630, Northern Little Bahama Bank. In: Proceedings of the Initial Reports of the Ocean Drilling Project Leg 101, Vol. B. U.S. Government Printing Office, Washington, D.C., in press, p. 315–326.

Lovell, M. A., and P. Ogden, 1984. Remote assessment of permeability/thermal diffusivity of consolidated clay sediments. Final Report to Commission of the European Communities, Nuclear Science and Technology, Luxembourg, 168 p.

Ogushwitz, P. R., 1985. Applicability of Biot theory; I. Low-porosity materials,

II. Suspension, III. Wave speeds versus depth in marine sediments. Journal of the Acoustical Society of America, v. 77, p. 429–464.

Rauch, D., 1980. Experimental and theoretical studies of seismic interface waves in coastal waters. In: Kuperman, W. A., and F. B. Jensen (eds.), Bottom-Interacting Ocean Acoustics. Plenum Press, London, p. 307–327.

Rauch, D., 1986. On the role of bottom interface waves in ocean seismo-acoustics: a review. In: Akal, T., and J. M. Berkson (eds.), Ocean Seismo-Acoustics. Plenum Press, London, p. 623–641.

Richardson, M. D., 1983. The effects of bioturbation on sediment elastic properties. Bulletin Societe Geologique de France, v. 25, p. 505–513.

Richardson, M. D., 1986. Spatial variability of surficial shallow-water sediment geoacoustic properties. In: Akal, T., and J. M. Berkson (eds.), Ocean Seismo-Acoustics. Plenum Press, London, p. 527–536.

Richardson, M. D., and D. K. Young, 1980. Geoacoustic models and bioturbation. Marine Geology, v. 38, p. 205–218.

Richardson, M. D., D. K. Young, and K. B. Briggs, 1983. Effects of hydrodynamic and biological processes on sediment geoacoustic properties in Long Island Sound, U.S.A. Marine Geology, v. 52, p. 201–206.

Richardson, M. D., P. V. Curzi, E. Muzi, B. Miaschi, and A. Barbagelata, 1987. Measurements of shear wave velocity in marine sediments. In: Lara-Saenz, A., C. Ranz-Guerra, and C. Carbo-Fite (eds.), Acoustics and Ocean Bottom. II FASE Specialized Conference, Consejo Superior de Investigaciones Cientiticas, Madrid, p. 75–84.

Schmalfeldt, B., 1986. A comparison of seismic and hydroacoustic measurements at very low frequencies in different shallow water areas. In: Akal, T., and J. M. Berkson (eds.), Ocean Seismo-Acoustics. Plenum Press, London, p. 653–662.

Schultheiss, P. J., 1981. Simultaneous measurements of P & S wave velocities during conventional laboratory soil testing procedures. Marine Geotechnology, v. 4, p. 343–367.

Schultheiss, P. J., 1985. Physical and geotechnical properties of sediments from the Northwest Pacific: Deep sea drilling project leg 86. Initial Reports of the Deep Sea Drilling Project, 86. U.S. Government Printing Office, Washington, D.C., p. 701–722.

Shirley, D. J., 1978. An improved shear wave transducer. Journal of the Acoustical Society of America, v. 63, p. 1643–1645.

Shirley, D. J., and L. D. Hampton, 1978. Shear wave measurements in laboratory sediments. Journal of the Acoustical Society of America, v. 63, p. 607–613.

Snoek, M., 1990. An investigation of seismo-acoustic interaction with the seafloor in shallow water using explosive sources. SACLANTCEN SR, SACLANT Undersea Research Centre, La Spezia, Italy, in press.

Snoek, M., and D. Rauch, 1987. Anisotropic behavior of interface wave propagation for near surface sediments. In: Lara-Saenz, A., C. Ranz-Guerra, and C. Carbo-Fite (eds.), Acoustics and Ocean Bottom, II FASE Specialized Conference. Consejo de Investigaciones Cientiticas, Madrid, p. 201–210.

Snoek, M., G. Guidi, and E. Michelozzi, 1986. Interface waves studies on the Ligurian shelf using an OBS array. In: Akal, T., and J. M. Berkson (eds.), Ocean Seismo-Acoustics. Plenum Press, London, p. 663–672.

Stokoe, K. H., and F. E. Richart, 1975. *In situ* and laboratory shear wave velocities. Proceedings of the Eighth International Conference on Soil Mechanics and Foundation Engineering, v. 1, Moscow, p. 403–409.

Stoll, R. D., 1980. Theoretical aspects of sound transmission in sediments. Journal of the Acoustical Society of America, v. 68, p. 1341–1350.

Stoll, R. D., G. M. Bryan, R. Flood, D. Chayes, and P. Manley, 1988. Shallow seismic experiments using shear waves. Journal of the Acoustical Society of America, v. 83, p. 93–102.

Warrick, R. E., 1974. Seismic investigation of a San Francisco Bay mud site. Bulletin of the Seismological Society of America, v. 64, p. 375–385.

CHAPTER 45

Geoacoustic Properties in the Near-Surface Sediment in Response to Periodic Deposition

Charles Libicki and Keith W. Bedford

Introduction

The problem of sediment deposition is complex, encompassing the effects of turbulent forcing and dissipation, erosion, and deposition. Boundary layers involved extend from planetary in scale to nearly microscopic. Both model studies and laboratory studies have indicated that the history of the depositional process can have a profound impact on later erosion, by determining the availability of erodible material and its degree of consolidation. Yet traditional field observations have been of the broad survey type, in which observation of system dynamics is uncertain, given the difficulty of observing exactly the same site on subsequent passes (Parker and Kirby, 1982). There is a growing body of work, however, in which field experiments are conducted from a fixed location to minutely observe the evolution of boundary layers at a single site. This method of operation presents its own opportunities and problems. The opportunity is to study in detail, and synoptically, the sediment erosion and deposition process as it is happening. With the instrumentation trained on a particular site, calibration can be made fairly exact and the investigation can concentrate on the dynamics of the process. In addition, since the instrumentation is fixed, small excursions in the sediment surface, or changes in its composition or morphology, can be accurately noted. The problem is, can this small scale activity be sensed acoustically? To address this problem, the following scenario is constructed: A fine sediment is deposited to a depth of 5 or 10 cm and allowed to consolidate for 24 hr. At that time, a second deposition of the same magnitude occurs and the consolidation is observed by a vertically incident acoustic beam for another 24 hr. The fine sediment is chosen, not only for its direct applicability in dredged material dynamics, but also because we can reasonably expect soft sediments to undergo a significant amount of consolidation due to selfweight in that amount of time. Sands, by contrast, tend to compact largely due to the influence of waves and turbulence (Krone, 1974) by a process that is much less well understood. Periodic deposition is modeled to investigate the possibility that it will generate acoustically visible interfaces.

To this situation we apply models (Fig. 45.1) of self-weight consolidation, acoustic transmission, and reflectivity models (and other relations as needed) and pose the following questions: In order to form a sufficient model of the consolidation and scattering process, what field data are necessary, and what can be inferred? How thick must a layer be to be resolved? How does a newly formed unconsolidated layer appear acoustically?

Consolidation Model

The consolidation model chosen for this exercise was first presented by Gibson et al. (1967). Unlike Terzaghi's classic model of consolidation, it is not limited to small strains, and thus can function in primary consolidation of soft soils, where strains of 50% or more may be observed (Cargill, 1985). Permeability and effective stress can be specified as arbitrary functions of the void ratio, e, which is related to the porosity, β by $e = \beta/(1 - \beta)$ (effective stress may be a function not only of present void ratio, but also of the history of the void ratio, in the case of swell or rebound from primary consolidation). Resistance to flow is modeled according to Darcy's Law, but is based on the fluid velocity *relative to the soil frame* rather than on that of the fluid alone. The governing equation is then

$$\left(\frac{\gamma_s}{\gamma_w} - 1\right)\frac{d}{de}\left[\frac{k(e)}{1+e}\right]\frac{\partial e}{\partial z} + \frac{\partial}{\partial z}\left[\frac{k(e)}{\gamma_w(1+e)}\,\frac{d\sigma'}{de}\,\frac{\partial e}{\partial z}\right] + \frac{\partial e}{\partial t} = 0 \tag{1}$$

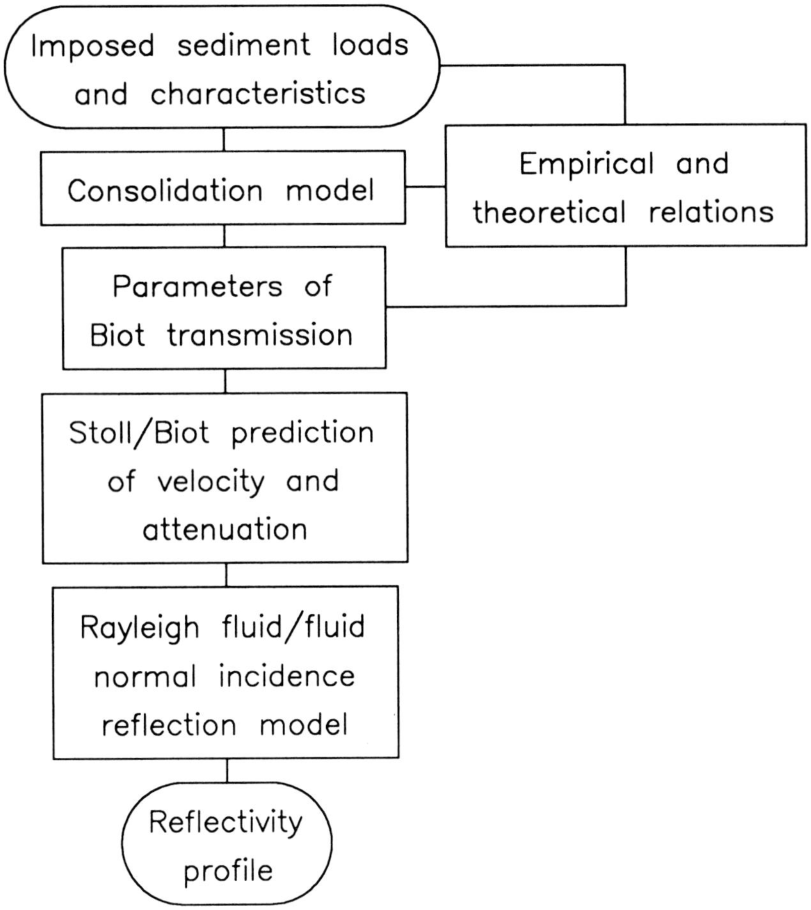

Figure 45.1. Models flow chart.

In this equation, the controlling dependent variable is e, the void ratio. The independent variables are time, t, and a transformed distance from the established base, z. Since vertical exchange of soil material (that is, intermixing of soil strata) is not modeled, a Lagrangian description of the consolidation process is adopted. Each differential stratum of sediment is assigned a height based on only the volume of sediment within the stratum. This height remains constant for each stratum, as sediment is not permitted to leave, although water does. In terms of true height above the base x,

$$z = \int_0^x \frac{d\xi}{1 + e(\xi)} \tag{2}$$

Other parameters include k, the permeability, and σ', the effective stress, both of which are specified as functions of the void ratio. These parameters may also depend on the history of the deposition and consolidation process, but that additional complexity is avoided here. Finally, the unit weight of the sediment and water are given by γ_s and γ_w, respectively.

The consolidation equation is nonlinear in form, and unless special restrictions are placed on the functions $k(e)$ and $\sigma'(e)$, no analytical solutions are known to exist. In the original paper, a Runge–Kutta numerical solution was performed. Monte and Krizek (1976) employed a finite element scheme, and Cargill (1985) used an explicit finite difference algorithm. The solution chosen here is an implicit, second-order correct procedure, with iteration progressing by the Newton–Raphson method. This is preferred due to its flexibility, its favorable stability criteria compared to explicit solutions, and the ability to incorporate the desired order of differentiability in the functions $k(e)$ and $\sigma'(e)$.

The time and space grid (keeping in mind that the space coordinates are transformed to preserve the Lagrangian solution) has the solution centered about a point at grid node i in space and between time steps j and $j + 1$ in time, at a remove of $\theta\Delta t$ from the old time step, where θ is assigned a value of 0.55. Numerical stability requires a value greater than 0.5 when using large time steps, but a much greater value of θ tends to introduce phase shifts into the solution. According to this scheme, for any parameter g, the value itself is approximated as

$$g \approx \theta g_i^{j+1} + (1 - \theta) g_i^j \tag{3}$$

the first derivative is

$$\frac{\partial g}{\partial z} \approx \frac{\theta}{2\Delta z}\left(g_{i+1}^{j+1} - g_{i-1}^{j+1}\right) + \frac{1-\theta}{2\Delta z}\left(g_{i+1}^{j} - g_{i-1}^{j}\right) \tag{4}$$

the second derivative is

$$\frac{\partial^2 g}{\partial z^2} \approx \frac{\theta}{\Delta z^2}\left(g_{i+1}^{j+1} - 2g_i^{j+1} + g_{i-1}^{j+1}\right) + \frac{1-\theta}{\Delta z^2}\left(g_{i+1}^{j} - 2g_i^{j} + g_{i-1}^{j}\right) \tag{5}$$

and the time derivative is

$$\frac{\partial g}{\partial t} \approx \left(g_i^{j+1} - g_i^{j}\right) \Big/ \Delta t \tag{6}$$

Applying the appropriate discretizations yields a set of simultaneous nonlinear algebraic equations, one per grid node, which are then solved by Newton–Raphson iteration.

The permeability/void ratio function has been found by experiment to follow a log/linear form over a range of several decades of permeability (Been and Sills, 1981)

$$k = A \exp(Be) \tag{7}$$

The log-linear relation can be incorporated directly into the numerical scheme to supply all the orders of differentiability required. Effective stress/void ratio relations tend to follow a more sigmoid curve (Monte and Krizek, 1976; Been and Sills, 1981); and so a cubic spline is passed through experimentally determined data points, which supplies the second-order continuity required by the Newton–Raphson procedure.

Boundary conditions at the soil base depend on the permeability of the substratum, whether it is free-draining, impermeable, or semipermeable. For this study, the impermeable boundary condition was chosen, in which case (Cargill, 1985)

$$\frac{\partial e}{\partial z} = -(\gamma_s - \gamma_w)/\frac{d\sigma'}{de} \tag{8}$$

where the denominator on the right-hand side is taken from the empirical relation of effective stress vs. void ratio.

Because there is no applied stress at the surface, the void ratio at the surface ought to remain constant throughout the consolidation process, and this is the upper boundary condition employed in the "no overburden" simulations discussed below. Experimental evidence, however (Been and Sills, 1981; Hayter, 1986), indicates that some consolidation will take place over time, even in the absence of applied stress, with decreases in the void ratio even at the sediment–water interface. To model this, Lee and Sills (1981) introduced the notion of an artificial overburden, which they modeled by setting an effective upper boundary at some distance (in terms of the Lagrangian coordinates) above the true surface. In the simulations performed here, the artificial overburden was constructed by including extra layers of deposited material. Whenever new material is to be deposited, this artificial layer is removed, the new deposit is placed on the "real" surface, and new artificial layers are added to the very top of that. This is the method employed in the "Been and Sills overburden" simulations discussed below.

Geoacoustic Model

The model employed for acoustic transmission within the saturated sediment is an adaptation of Biot's (1956) theory, with numerical solution and extensions by Stoll and Bryan (1970). The model has found application in a wide range of conditions from suspensions to porous rock (Ogushwitz, 1985). In each case, good agreement has been found between measured and predicted velocity, attenuation, and dispersion (variation in wave speed with frequency). Biot theory is based on the motion of the included water, the motion of the soil frame, and the motion of the coupled mass. The resulting equations predict three types of traveling waves: fast compressional, transverse, and slow compressional. The latter is subject to high attenuation and has only rarely been observed. The transverse or rotational wave is of minor importance in the high porosity, vertical incidence simulations employed here. The remaining fast compressional wave is found by assuming a solution for the pressure wave of the form

$$p = A \exp[i(\omega t - \ell x)] \tag{9}$$

where ω is the imposed acoustic frequency, the attenuation (in nepers of pressure per meter, or some compatible unit of length) is given by $\mathrm{Im}(\ell)$, and the phase velocity is given by $\omega/\mathrm{Re}(\ell)$. The value of ℓ (for both compressional waves) is given by the solution to the determinant equation

$$\begin{vmatrix} H\ell^2 - \rho\omega^2 & \rho_f\omega^2 - C\ell^2 \\ C\ell^2 - \rho_f\omega^2 & m\omega^2 - M\ell^2 - \dfrac{i\omega F\eta}{k} \end{vmatrix} = 0 \tag{10}$$

where

$$H = \frac{(K_r - K_b)^2}{D - K_b} + K_b + \frac{4}{3}\mu$$

$$C = \frac{K_r(K_r - K_b)}{D - K_b}$$

$$M = \frac{K_r^2}{D - K_b}$$

$$D = K_r[1 + \beta(K_r/K_f - 1)]$$

$$F = \frac{1}{4}\,\frac{\kappa T(\kappa)}{1 - 2T(\kappa)/i\kappa}$$

$$T(\kappa) = \frac{\mathrm{ber}'(\kappa) + i\,\mathrm{bei}'(\kappa)}{\mathrm{ber}(\kappa) + i\,\mathrm{bei}(\kappa)}$$

$$\kappa = a(\omega\rho_f/\eta)^{1/2}$$

Table 45.1. Parameters taken from Been and Sills (1981).

Variable	Name	Value
k	Permeability	$1.8738 \times 10^{-6} \exp(0.5233e)$ cm/sec
σ'	Effective stress	Cubic spline through graphed data
e_0	Initial void ratio	10.55
d	Particle diameter	8.5×10^{-4} cm
de/dz_∞	The ultimate slope of the void ratio	1.0 cm^{-1}
$e_{0,\infty}$	Ultimate void ratio at zero effective stress	6.5

and ber and bei are the real and imaginary parts of the Kelvin function (Abramowitz and Stegun, 1964) and ber′ and bei′ are, respectively, their derivatives. The remaining parameters are

ρ_r = density of the sediment particles
K_r = bulk modulus of the sediment particles
ρ_f = density of the fluid
K_f = bulk modulus of the fluid
η = viscosity of the fluid
β = porosity
k = permeability
a = pore size parameter
m = tortuosity
μ = complex shear modulus of the frame
K_b = complex bulk modulus of the frame

Of these parameters, some values are well established (for instance, density and viscosity of water), some are sediment properties that may be determined in the laboratory (density and bulk modulus of the sediment particles), some are parameters of consolidation (porosity and permeability), and still others are determined empirically. The process of establishing values for each of the parameters is covered in the simulations section.

The acoustic reflection model chosen is that of a one-dimensional scattering from discontinuities or gradients in acoustic impedance according to the Rayleigh fluid/fluid reflection model (Hamilton, 1974). Reflectivity at vertical incidence is given by

$$R = (\rho_2 v_2 - \rho_1 v_1)/(\rho_2 v_2 + \rho_1 v_1) \tag{11}$$

where ρ is the bulk density, v is the sound speed, and 1 and 2 refer to adjoining layers. Acoustic response is then given by the reflectivity, diminished by attenuation of the beam in its two-way transit through the overlying sediment. Response is a unitless quantity, as it is scaled to the output power of the transducer.

In practice, determination of the response profile proceeds from the received signal by two broad classes of "inverse scattering" procedures: dynamic deconvolution, in which each layer is recursively "peeled back" to reveal its own properties and the transformations effected on the continuing signal, and a linear equation approach, in which the scattering process is posed as a set of integral equations, or in the discrete case matrix-based equations, and in which the system as a whole is solved simultaneously. The recursive approach tends to offer the greatest simplicity of derivation, while retaining the numerical stability and efficiency with noisy data of the linear equation formulation (Bruckstein and Kailath, 1987). As a simulation, or "forward" rather than "inverse" problem, this study has as its end product the response profile itself. Inverse scattering becomes necessary when actual field data are employed with the models to infer sediment composition.

Simulations

This study simulates two cycles of instantaneous deposition, each followed by 24 hr of consolidation. The instantaneous deposition consists of either 10 or 5 cm of sediment in an unconsolidated, stress-free state. The simulations are further subdivided into those where the sediment surface remains stress free, and therefore at the initial void ratio, for all time, and those where an artificial overburden is imposed to mimic the laboratory observation that even the sediment at the very surface tends to consolidate over time.

The input data is taken from Been and Sills (1981). The sediment for their laboratory consolidation experiments consisted of estuarine mud from Combwich in Somerset, which was wet sieved and passed through a 75-μm sieve. The resulting material was a uniformly graded silt with 30% clay. The specific data, which were taken from Been and Sills' experiments (experiment 15), were initial void ratio, permeability vs. void ratio, effective stress vs. void ratio, particle size, and overburden parameters de/dz_∞ and $e_{0,\infty}$. The latter parameters are the ultimate slope (in Lagrangian coordinates) of the void ratio profile, and the ultimate void ratio at zero effective stress; Been and Sills estimate them by fitting the linearized form of Gibson's equation (discussed below) to their experimental results. The values are given in Table 45.1. The spline fit to the effective stress vs. void ratio is given in Figure 45.2.

The output of the consolidation model consists of real and transformed heights and profiles of density (which is directly related to porosity, given a uniform particle density), permeability and total, stress. The parameters for Biot acoustic transmission are then filled in as follows: The pore fluid is water and so standard values are taken for the density, viscosity, and bulk modulus. Sediment grain parameters of density and bulk modulus are taken from Ogushwitz (1985). Frame parameters are estimated using the self-consistent theory (SCT) of Berryman (1980). This theory was chosen because it converges to known exact results at both zero and unity concentration, and its results lie within rigorous theoretical limits for general ellipsoid particles (including the limiting cases of needle and disk). Most importantly, it successfully matches laboratory determinations of velocity and attenuation in conditions resembling those modeled here (Ogushwitz, 1985). (Laboratory porosity ranged from 60 to 97%, and acoustic frequency ranged from 50 kHz to 1.5 MHz. Here, porosity goes from 80 to 91% and the frequencies modeled are 300 kHz, 1 MHz, and 3 MHz.)

In self-consistent theory, density, bulk modulus, and shear modulus for a system with inclusions are given implicitly by (Berryman, 1980)

$$\rho = \sum_{i=1}^{N} c_i \rho_i \tag{12}$$

$$\sum_{i=1}^{N} C_i(K_i - K_b)P^{(i)} = 0 \tag{13}$$

and

$$\sum_{i=1}^{N} C_i(\mu_i - \mu_b)Q^{(i)} = 0 \tag{14}$$

Here, we are dealing with a two-parameter system, and so $N = 2$, the subscript b refers to the system as a whole, c is the concentration ($c_1 = 1 - \beta$ and $c_2 = \beta$), and P and Q are functions of particle shape. To obtain the frame moduli, the second constituent is taken to be air (or a vacuum), with ρ_2, K_2, and μ_2 all equal to zero.

Equation (12) is simply the composite density. The subsequent two equations have one finite solution and a degenerate solution of $\mu = 0$. In particular, for spherical inclusions, the finite solution becomes singular at a porosity of 60%, and above that *in the absence of forces due to flocculation*, the frame moduli become zero. Since modeling the effect of flocculation on acoustic transmission is beyond the scope of this work, K_b and μ_b are both set to zero. Under such conditions, H, C, and M in Equation (10) all become

$$K = (\beta/K_f + (1 - \beta)/K_r)^{-1} \tag{15}$$

and Equation (10) simplifies to

$$\ell = \sqrt{\frac{\rho\omega^2}{K} \frac{(a/\beta - \rho_f/\rho)\rho_f - i\eta F/\omega k}{(a/\beta - 2)\rho_f + \rho - i\eta F/\omega k}} \tag{16}$$

It is noted that in the limit as $\omega \rightarrow 0$, Equation (16) becomes Wood's (1941) equation for velocity of propagation through a suspension.

Although it would be appealing for the sake of the integrity of the modeling exercise to employ the same permeability in acoustics calculations and consolidation calculations, Hovem (1980) argues that permeability from soil mechanics is not applicable to acoustic transmission. The former is a static measure of steady flow whereas the latter is a dynamic phenomenon. He employs instead the Kozeny–Carman relation

$$k = \frac{d^2}{36k_0} \frac{\beta^3}{(1 - \beta)^2} \tag{17}$$

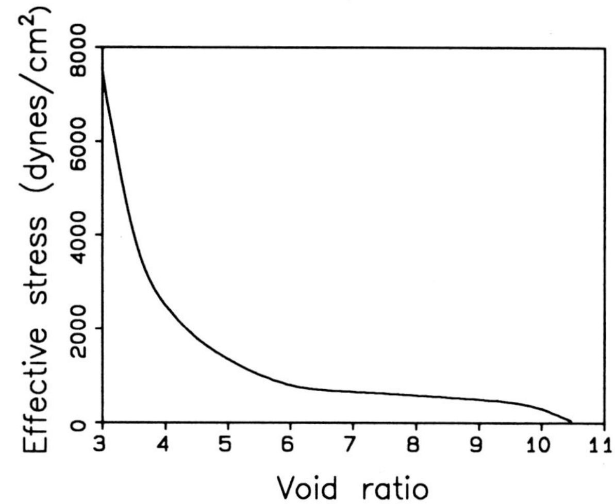

Figure 45.2. Spline fit to Been and Sills' effective stress vs. void ratio data.

where k_0 is an empirical parameter, assigned a value of 5.0, and d is the mean diameter of the sediment grains. The porosity comes directly from the results of the consolidation model. The pore-size parameter is calculated as (Hovem, 1980)

$$a = \frac{d}{3} \frac{\beta}{1 - \beta} \tag{18}$$

In Biot theory, the linkage between the motion of the frame and the pore fluid is given by the "tortuosity" parameter, m. The tortuosity, which is always greater the one, is a geometrical quantity that represents the ratio of the "effective" (in terms of inertial drag) pore length to true pore size. Biot (1956) suggested a value of $\sqrt{8}$; Stoll (1974) used 3 for clays. The parameter can be measured experimentally as the ratio of the conductivity of saltwater to conductivity in a saltwater-saturated sample of the sediment times the porosity. Various formulas also exist, including Berryman's

$$m = \frac{1}{2}(1 + 1/\beta) \tag{19}$$

and Kozeny–Carman's (Monte and Krizek, 1976)

$$m = CS^2 \frac{k\nu}{g} \frac{1 + e}{e^3} \tag{20}$$

in which C is a pore shape factor, S is surface area per unit volume of particles, and ν is the dynamic viscosity. The latter was initially preferred as it made greater use of the output of the consolidation model. It was rejected in favor of Berryman's relation, however, as it gave the wrong trend in tortuosity with depth.

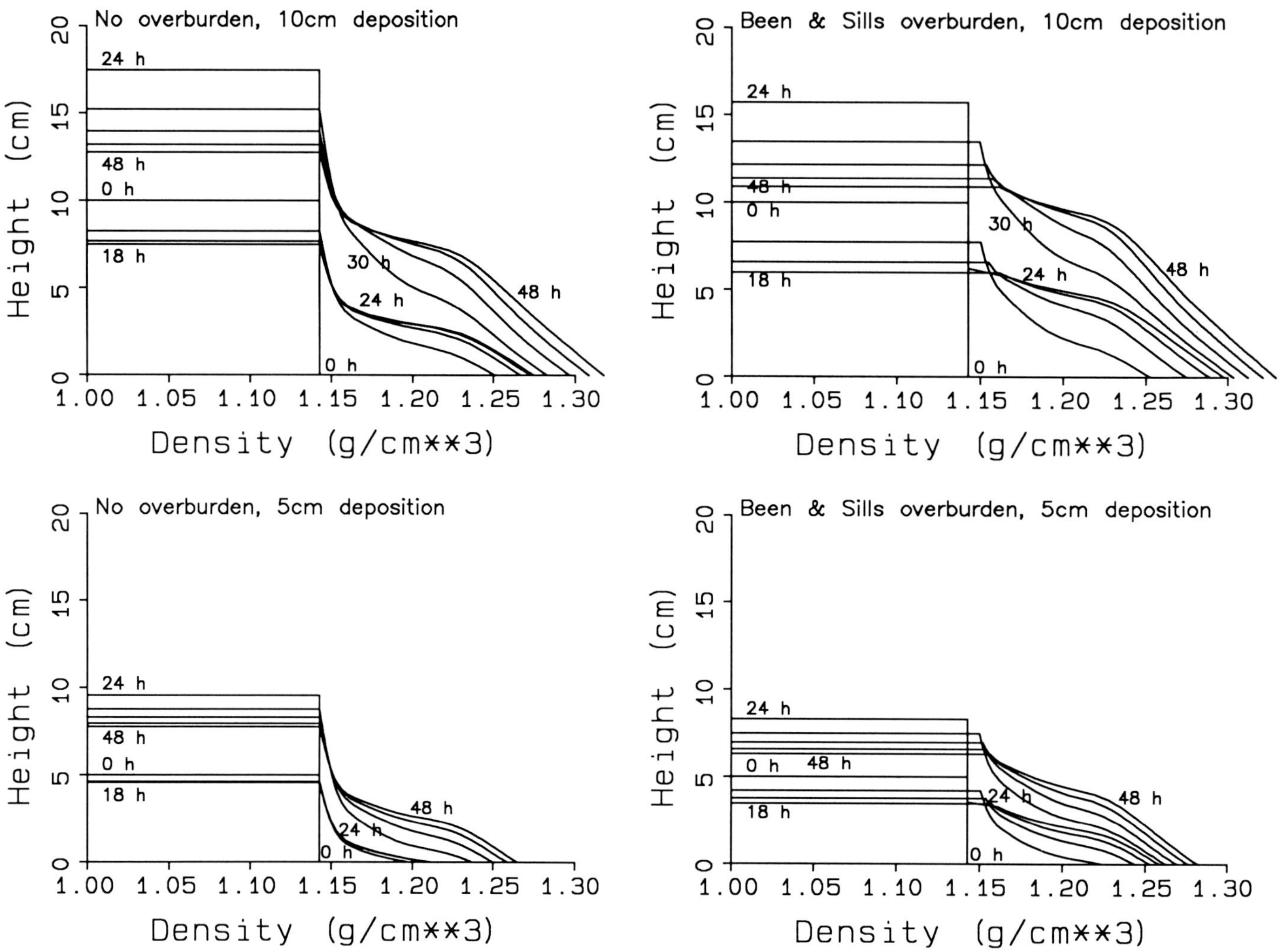

Figure 45.3. Density profiles.

To test the effect of the incident frequency on the reflectivity profile, three acoustic frequencies were simulated: 300 kHz, 1 MHz, and 3 MHz. At lower than 300 kHz resolution is lost and at greater than 3 MHz, scattering from individual particles becomes more important.

Results

The density profiles from the four simulations are depicted in Figure 45.3. The profiles are shown at 6-hr intervals, with each line corresponding to a specific time, and the flat horizontal lines representing the surface of the sediment. Taking as an example the first simulation, the horizontal line at 10 cm denotes the initial level of the first deposition. A family of horizontal lines falls just below that, corresponding to the receding of the surface under the self-weight consolidation. Then at 24 hr, the surface line jumps up to 17 cm with the second addition of material, giving rise in turn to a second family of surface recession curves.

In each of the four simulations the lower part of the profile is slightly concave downward, flattening out over time. Such a linear profile of void ratio vs. Lagrangian height is predicted by Gibson et al. (1967) in the case of an imposed weight upon a thin layer (that is, self-weight is unimportant compared to imposed load). Gibson defines a linearizing consolidation coefficient

$$C_F = -\frac{k(e)}{\gamma_w}\frac{1}{1+e}\frac{d\sigma'}{de} \tag{21}$$

which is assumed (with experimental support) to be constant, so that the consolidation becomes an analog of the heat equation

$$C_F\frac{\partial^2 e}{\partial z^2} = \frac{\partial e}{\partial t} \tag{22}$$

Similarly, Been and Sills achieve a linear form by assuming not only a constant consolidation coefficient, but also linear relation between effective stress and void ratio and proportionality between permeability and the void ratio plus one, but retaining the predominance of self-weight over any imposed load. Clearly, the linear assumption is invalid in the relations employed here, especially the large increase in void ratio for effective stresses below 0.15 kN/m². Been and Sills state that below that point, not only is the void ratio dramatically larger, but it is also subject to large variability, so much so that the authors state that there is no single well-defined void ratio for zero effective stress. In Monte and Krizek's (1976) derivation of consolidation theory, they appeal to the notion of the stress-free state as equivalent to the fluid limit. The fluid limit is then defined as that concentration at which clay particles begin to interact with each other. This then becomes an important concept because the motivation behind the field experiment described at the beginning is to examine the earliest phases of the soil formation, in which interfaces evolve and the erosion–resistance profile is laid down.

The most notable feature of the density profiles, however, is the concentration gradient that intensifies as it rises through the profile. Such an authigenic gradient, sometimes referred to as a "concentration wave," has been noted both in laboratory and field studies (Been and Sills, 1981; Parker and Kirby, 1982). This "front" also becomes a strong acoustic reflector, and is discussed in greater detail below.

The acoustic response profile is depicted in Figure 45.4a,b, and c, for incident acoustic frequencies of, respectively, 300 kHz, 1 MHz, and 3 MHz. Response is denoted in shades of gray, with black the highest values. The 300-kHz image is shaded so that each increment in gray scale represents a 1×10^{-4} difference in acoustic response, for a maximum displayable response of 1.6×10^{-3}. To partially compensate for the larger attenuation at the higher frequencies, the effective gain of the gray scale image is doubled for 1 MHz and again for 3 MHz. The profile is given vertically and the time history extends across the horizontal axis. The heavy asymptotic line is the acoustic reflection from the sediment surface. There are two of these descending curves in each plot representing, as before, the two deposition/consolidation cycles. The reflectivity evident in these plots is primarily the influence of density differences in Equation (11). Although Hampton and Urick (Hovem 1980) both noted in their laboratory studies that velocity decreased with decreasing porosity, reaching a minimum at $\approx$ 80%, here the larger mean particle diameter, and the consequent larger permeability, tends to flatten out the velocity variation.

The most obvious feature is the reflection caused by the concentration front, which forms near the bottom of the profile and rises asymptotically over time. In the case of 10 cm deposition, the front rises at an initial rate of 0.33 cm/hr to settle at a height of 3 cm after 18 hr. After the second addition, the reflectivity dissipates somewhat as it resumes its rise to settle out at 8 cm after 42 hr. The 10-cm deposition with artificial overburden exhibits a much greater rate of surface recession. Its front rises at a slightly higher rate and settles at 4.5 cm, just 1 cm below the surface, and virtually indistinguishable from the surface reflection when sensed with lower frequency devices. In addition, a weaker front is visible approximately 3–4 cm above the primary front. Not visible in this representation is the increase in surface reflection over time in the artificial overburden case as the sediment surface changes in character. By contrast, the 5-cm deposition with no overburden exhibits a reflection maximum that fails to rise significantly from the base over the first 24-hr period, and achieves a height of 2.5 cm in the latter. With the overburden, the front climbs steadily throughout the two depositional periods to a final height of 6.2 cm. Notably absent from all the results is an evident persistent interface between the two depositional events. In the two cases with no overburden, since the top of the sediment always has a void ratio equal to the initial void ratio, the interface becomes acoustically transparent immediately on the second deposition. In the case with an artificial overburden, the interface persists no more than an hour or two, being quickly eradicated by the diffusion-like second derivative in the governing equation. The short-lived interface is evidenced in the gray-scale plots by a small knob at the end of the primary recession curves.

The similarity of the climbing front to a progressive wave invites a characteristic analysis of the governing equation; that is, of the form

$$\frac{\partial e}{\partial t} + A\frac{\partial e}{\partial z} = R \tag{23}$$

Brought to this form,

$$A = \left(\frac{\gamma_s}{\gamma_w} - 1\right)\frac{d}{de}\left[\frac{k(e)}{1+e}\right] + \frac{d}{de}\left[\frac{k(e)}{\gamma_w\,(1+e)}\frac{d\sigma'}{de}\right]\frac{\partial e}{\partial z} \tag{24}$$

and

$$R = C_F\frac{\partial^2 e}{\partial z^2} \tag{25}$$

where A is the characteristic velocity, R is a diffusive term, and C_F is the consolidation coefficient defined in Equation (22). In this formulation it is noted that one effect of Gibson's and Been and Sills' linearizations is to set the characteristic velocity to zero everywhere. The general case, however, produces a family of characteristic lines whose velocity can vary both in space and time. For example, the characteristic velocity profile for the 10 cm, no overburden case at 8 hr (as the front is forming) is shown in Figure 45.5. It is seen that the upper part of the profile has characteristics proceeding downward and the lower part has characteristics proceeding upward with no movement at all at a level of 3 cm, just below the front. In fact, the negative derivative of velocity with height would cause the characteristics to cross, producing a double-valued solution (similar to the underside and crest of a breaking wave) were it not for the damping

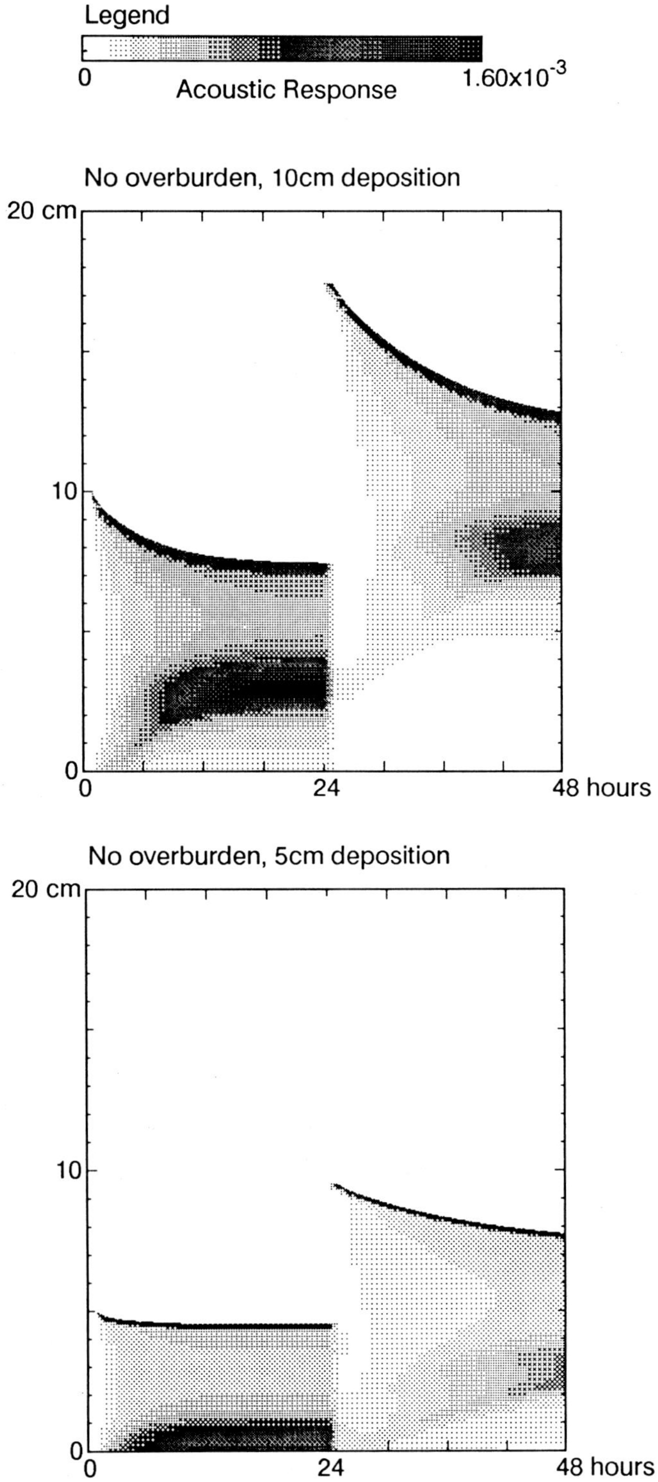

Figure 45.4. (a) Acoustic response history, 300 kHz.

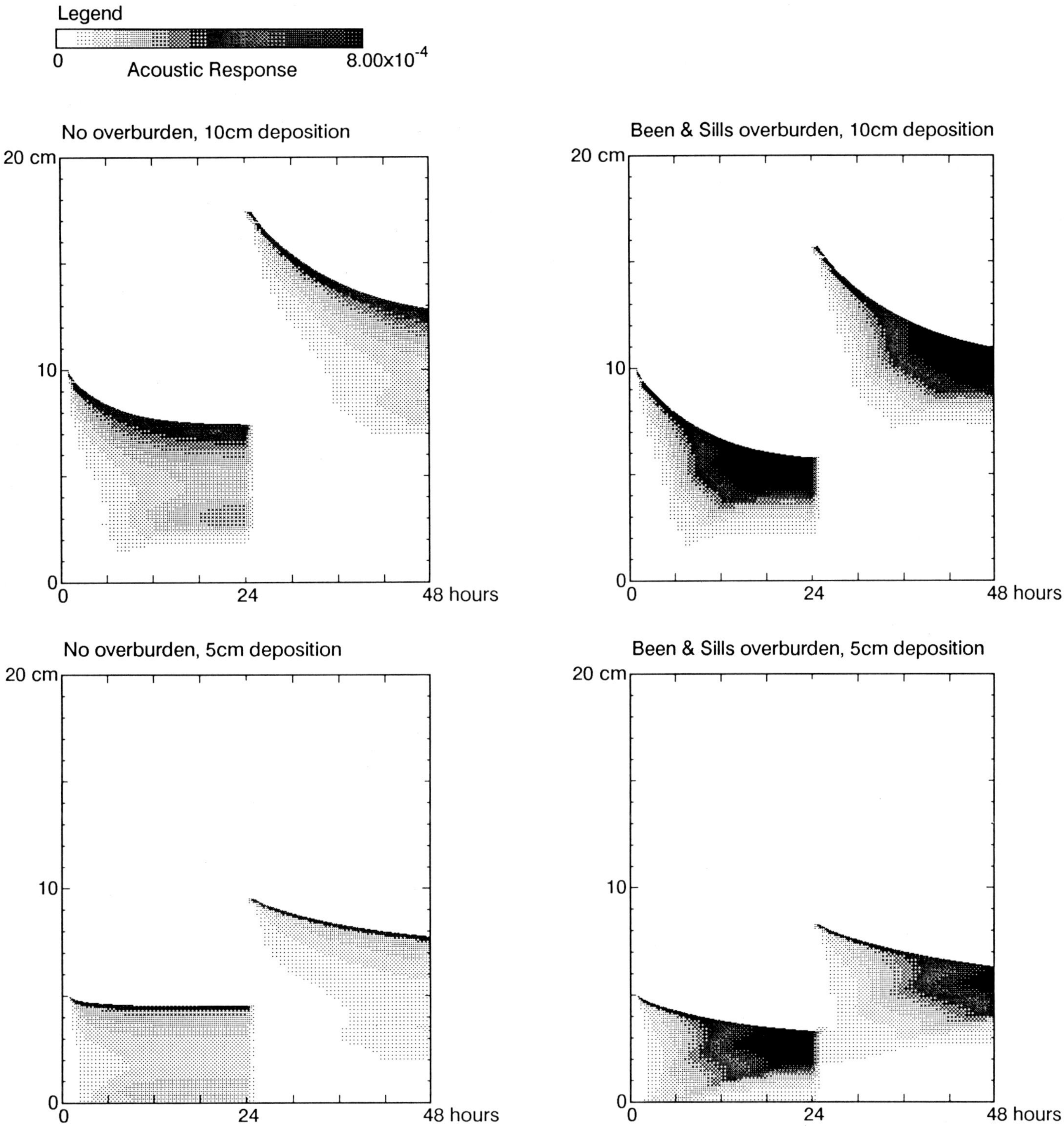

Figure 45.4. (b) Acoustic response history, 1 MHz.

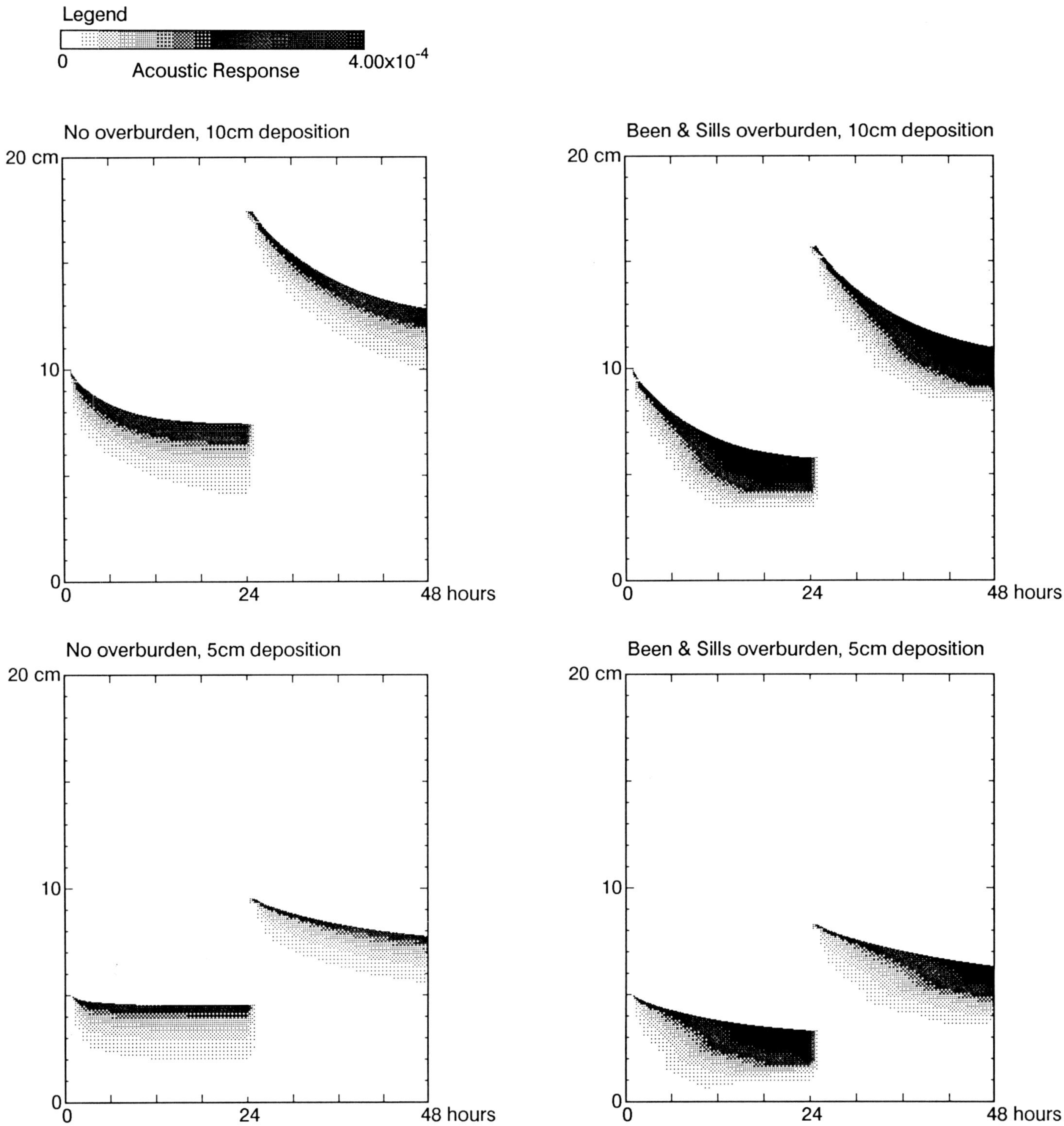

Figure 45.4. (c) Acoustic response history, 3 MHz.

effect of the second derivative (R in the characteristic form) (Lax, 1973). This is the mechanism behind the formation of the front, one that would be expected to be obtained except under linearizing restrictions in the form of the k vs. e and σ' vs. e relations.

The chemical engineering literature also contains reference to shocks that may arise spontaneously in the settling and consolidation of fine materials. Kynch (1952) presented a model of hindered settling in which settling velocity was a unique function of volumetric concentration. In this scheme, shocks are produced immediately, depending on the initial concentration distribution. Auzerais et al. (1988), expanded on the model to include interparticle contact stresses, compressibility of the sediment at the bottom of the vessel, and explicit methods for patching a solution within a shock region. In Auzerais' model, two parameters control when recognizable shock formation can occur; these are

$$v_t^2/gL \ll 1 \text{ and } \frac{d\sigma/d\phi}{\rho_s gL} \ll 1$$

where v_t is the terminal settling velocity of a lone particle in an infinite medium, g is acceleration due to gravity, L is the depth of the initial suspension, σ is the interparticle contact stess, ρ_s is the density of the particles, and ϕ is the volume fraction [$1 - \beta$ or $1/(1 + e)$]. The larger of these two parameters determines the thickness of the shock region and whether inertial effects or stress relations will control the solution. Analogy may be immediately drawn between, respectively, permeability at zero effective stress and v_t (there are interchangeable because it is the *relative* velocity of the particles and the medium that is physically important), and local effective stress and interparticle contact stress.

The analogy allows comparisons between the shock criteria and the present simulations. The value of the velocity parameter is 2.2×10^{-11} for 10-cm deposition and 4.4×10^{-11} for 5-cm deposition. The stress parameter, depending as it does on the local effective stress/void ratio relation, varies with void ratio and therefore within the profile. For the 10-cm deposition, it falls below 0.5 at a void ratio of 9.3 (density 1.16), reaches a minimum value of 0.15 at a void ratio of 6.9 (density 1.21), and climbs back above 0.5 at a void ratio of 6.1 (density 1.23). Indeed, referring to Figure 45.3, the domain of the density gradient lies between the densities of 1.16 and 1.23. For the 5-cm deposition, the stress criterion falls below 0.5 only between the void ratios of 7.9 and 6.5 (density 1.22 and 1.19) to reach a minimum of 0.3. This may serve to explain why in the 5-cm deposition simulation the high density gradient never rises above the sediment base.

In the 300-kHz image, the full range of this activity is visible. Acoustic penetration changes dramatically from one frequency to the next, however, so that in the 1-MHz image, returns are received from, at a maximum, 6 cm below the surface. At that frequency, the concentration gradient is visible only between hours 12 and 24 of the 10-cm depositions. Returns from the sediment base are also greatly reduced, with maximum response (occurring immediately prior to the second deposition) of the same order of magnitude as response from the internal concentration gradient. At 3 MHz, only the sediment surface, and the near-surface to a depth of 3 cm, are visible. Even the sediment base exhibits a response of only 10^{-12}. Reflection from the sediment surface, by contrast, is slightly enhanced at 3 MHz over the lower frequencies.

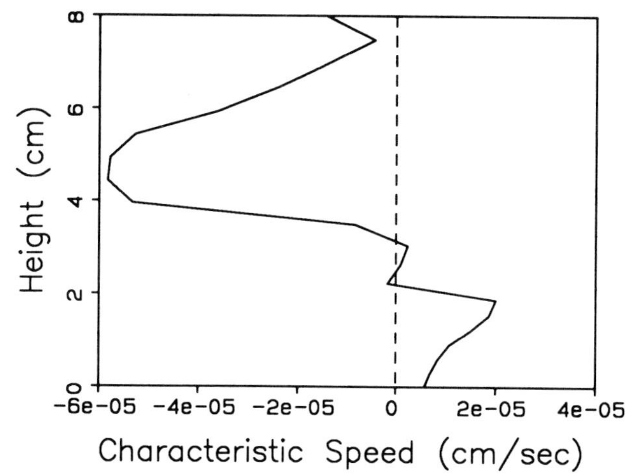

Figure 45.5. Characteristic velocity profile, 8 hr, 10-cm deposition, no overburden.

Although acoustic penetration at low frequencies is generally greater for silt and clay than for sand, in the range of frequencies employed here, the fall-off in signal with increased frequency is much greater with the finer grained materials. Due mostly to the low permeability, and the resultant linkage between fluid and sediment frame motion, losses in silts and clays are due primarily to the frame motion, and attenuation (in nepers per meter) increases as the first power of frequency. By contrast, dissipation in sands is to a large extent due to viscous losses in the pore fluid, for which attenuation rises as the one-half power of the frequency.

Conclusions

To recapitulate the questions posed in the introduction, based on the simulations performed here:

How does a newly formed unconsolidated layer appear acoustically? The strongest reflectors are the hard base beneath the sedimentation, if it exists, and the surface of the newly deposited layer. An internal gradient may form, which can be a strong acoustic reflector, in the case where the void ratio vs. effective stress or void ratio vs. permeability relations have strong nonlinearities. By contrast, under the assumption that the weight of the overburden is the only agent of consolidation, intervals of periodic deposition produce only transient acoustic signatures, if any at all. Other candidates for internal interfaces include

(1) fractionation within the deposited sediment (Been and Sills, 1981). That is, the initial deposition is so dilute that it resembles a suspension in character and the larger particles are able to fall through the finer particles to form their own layer at the surface of the more resistant sediment. In the experiments of Been and Sills, this tended to occur at initial void ratios of greater than 16, especially when the suspension was chemically deflocculated. Fractionation can also be aided by wave activity or turbulent pressure fluctuations. (2) Alternating sources of sediment. For instance, a storm at its peak intensity may supply coarser material with its slack supplying finer material. (3) Chemical or biological sources can, over time, change the properties of the sediment, in terms of effective stress or surface "armoring." Even in the absence of sediment volume change, ligatures between sediment particles can affect the stiffness of the frame and thereby alter a layer's acoustic properties (Johnson, 1986).

How thick must a layer be to be resolved? Resolution is proportional to acoustic wavelength. The Nyquist criterion suggests two samples per wavelength; however, a resolution distance of four full wavelengths is more commonly employed. This works out to approximately 2 cm at 300 kHz and 0.2 cm at 3 MHz. In the simulations performed here, however, only the 300 kHz signal was able to pick out the full range of activity, with the 1 MHz signal limited to approximately 6 cm of penetration, and the 3 MHz to 3 cm. Selection of an optimal frequency then becomes a very difficult problem that depends, in part, on the dominant mechanisms of attenuation in the sediment type being observed.

How does the acoustic signature change on consolidation? The most obvious effect is the subsiding of the sediment–water interface as the newly deposited material compacts. Under the imposition of an artificial overburden, the intensity of the surface reflection intensifies over time, otherwise it remains unchanged. The reflection from the base of the consolidating layer diminishes over time as the consolidating sediment becomes more acoustically opaque, and less differentiated from the base. Finally, nonlinearities in the consolidation process can cause a reflective front through the sediment column over time.

To form a sufficient model of the consolidation and scattering process, what field data are necessary, and what can be inferred? For calibration of the acoustic signal, particle size and initial porosity must be known from the field data. If initial porosity is unavailable, then at least some porosity must be known against which the acoustic signal and the consolidation model may be calibrated. To model the consolidation process, permeability vs. void ratio and effective stress vs. void ratio relations must be available, as well as sediment particle density. The remaining parameters of Biot transmission can either take generally accepted values for the sediment type at the site (for the bulk modulus of the sediment particles, for instance) or are inferred from other parameters (for instance bulk modulus of the frame is inferred from the total stress).

Given sufficient data against which to calibrate the models and the device, the simulations performed here indicate that those features that stand out in the consolidation process are also visible acoustically.

Acknowledgments

Support for this work comes in part from the Army Corps of Engineers Project No. DACW-39-88-K-0040 Dredge Materials Research Program and Office of Naval Research Contract No. N00014-88-K-0252. This support is very much appreciated.

References

Abramowitz, M., and I.A. Stegun, 1964. Handbook of Mathematical Functions with Formulas, Graphs and Mathematical Tables. U.S. Dept. of Commerce, National Bureau of Standards, Applied Mathematics Series 55, 1046 p.

Auzerais, F.M., R. Jackson, and W.B. Russel, 1988. The resolution of shocks and the effects of compressible sediments in transient settling. Journal of Fluid Mechanics, v. 195, p. 437–462.

Been, K., and G.C. Sills, 1981. Self-weight consolidation of soft soils: an experimental and theoretical study. Géotechnique, v. 31(4), p. 519–535.

Berryman, J.G., 1980. Long-wavelength propagation in composite elastic media, I and II. Journal of the Acoustical Society of America, v. 68, p. 1809–1819 and 1820–1831.

Biot, M.A., 1956. Theory of propagation of elastic waves in a fluid-saturated porous solid. I. Low-frequency range and II. Higher frequency range. Journal of the Acoustical Society of America, v. 28(2), p. 168–192.

Bruckstein, A.M., and T. Kailath, 1987. Inverse scattering for discrete transmission-line models. SIAM Review, v. 22(3), p. 359–389.

Cargill, K.W., 1985. Mathematical model of the consolidation/desiccation processes in dredged material. US Army Corps of Engineers, Waterways Experiment Station, Technical Report D-85-4, 68 p.

Gibson, R.E., G.L. England, and M.J.L. Hussey, 1967. The theory of one-dimensional consolidation of saturated clays. Géotechnique, v. 17, p. 261–273.

Hamilton, E.L., 1974. Prediction of deep-sea sediment properties: state-of-the-art. In: Inderbitzen, A.L. (ed.), Deep-Sea Sediments: Physical and Mechanical Properties. Plenum Press, New York, p. 1–43.

Hayter, E.J., 1986. Estuarial sediment bed model. In: Mehta, A.J. (ed.), Estuarine Cohesive Sediment Dynamics. Lecture Notes on Coastal and Estuarine Studies, v. 14. Springer-Verlag, Berlin, p. 326–359.

Hovem, J.M., 1980. Viscous attenuation of sound in suspensions and high-porosity marine sediments. Journal of the Acoustical Society of America, v. 65(5), p. 1559–1563 (errata: v. 68, p. 1531).

Johnson, D.L., 1986. Recent developments in the acoustic properties of porous media. In: Sette, D. (ed.), Frontiers in Physical Acoustics. North-Holland, Amsterdam, p. 255–290.

Krone, R.B., 1974. Engineering interest in the benthic boundary layer. In: McCave, I.N. (ed.), The Benthic Boundary Layer. Plenum Press, New York, p. 143–156.

Kynch, G.J. 1952. A theory of sedimentation. Transactions of the Faraday Society, v. 48, p. 166–176.

Lax, P.D., 1973. Hyperbolic systems of conservation laws and the mathematical theory of shock waves. Society for Industrial and Applied Mathematics, 48 p.

Lee K., and G.C. Sills, 1981. The consolidation of a soil stratum, including self-weight effects and large-strains. Numerical and Analytical Methods in Geomechanics, v. 5, p. 105–428.

Monte, J.L., and R.J. Krizek, 1976. One-dimensional mathematical model for large-strain consolidation. Géotechnique, v. 26(3), p. 495–510.

Ogushwitz, P.R., 1985. Applicability of the Biot theory. I. Low-porosity materials, II. Suspensions. Journal of the Acoustical Society of America, v. 77(2), p. 429–451.

Parker, W.R., and R. Kirby, 1982. Time dependent properties of cohesive sediment relevant to sedimentation management—European experience. In: Kennedy, V.S. (ed.), Estuarine Comparisons. Academic Press, New York, p. 573–589.

Stoll, R.D., 1974. Acoustic waves in saturated sediments. In: Hampton, L. (ed.), Physics of Sound in Marine Sediments. Plenum Press, New York, p. 19–39.

Stoll, R.D., and G.M. Bryan, 1970. Wave attenuation in saturated sediments. Journal of the Acoustical Society of America, v. 47(5), p. 1440–1447.

Wood, A.G., 1941. A Textbook of Sound. G. Bell and Sons, London, 578 p.

an aggregate bulk modulus of 35 dyn/cm² × 10¹⁰ (Holland and Brunson, 1988); this value agrees well with other reported moduli for fine-grained, "soft" sediments (e.g., Matthews, 1980; Stoll and Kan, 1981). There was no need to correct k_s for *in situ* conditions, because while there is an increase with increasing hydrostatic pressure and decreasing temperature, this change in the modulus is very small, and masked by a corresponding and much larger change in the pore water bulk modulus (Hamilton, 1971b).

A frame bulk modulus (k_f) is needed to realistically model grain-to-grain contacts, which are important in providing rigidity. Dynamic frame bulk moduli were computed using the regression equation of Hamilton (1971a), relating k_f to porosity (n), where

$$\log k_f = 3.73580 - 4.25075\ (n) \quad (9)$$

with k_f in dyn/cm² × 10⁸ for deep-sea silt clays. Again, no correction was necessary to simulate *in situ* conditions, because increases in pore water moduli overshadow any change in the frame moduli caused by differences in intergranular pressure, that is, changes in the buoyed weight of the mineral grains.

The bulk modulus of the pore water (k_w) was determined by the equation, $k = \rho V_p^2$. For this computation, the *in situ* density of the pore water was interpolated from Hamilton (1971b). The *in situ* velocity of the pore water was assumed to be the same as the bottom water at the sediment–water interface. This value was calculated using the *in situ* temperature, pressure, and salinity determined from CTD measurements and archived data. As Hamilton (1971a) noted, all of the input data used to compute elastic properties have margins of error, some are known and some are not, and, as such, all of the computed elastic properties should be considered approximations. The number of decimal places shown in the tables are for purposes of comparison between various computations and should not be taken as estimates of accuracy.

Results and Discussion

Table 46.2 lists the measured physical properties and calculated elastic parameters of the samples examined in this study. It is puzzling to note that only 15 of the 38 samples demonstrated rigidity, that is, only these samples were capable of sustaining shear waves. Three cores (8, 24, and 41) demonstrated rigidity in all samples from top to bottom within the 30-cm sampling interval, while six cores (7, 20, 22, 27, 45, and 46) contained no samples with dynamic rigidity.

No immediate pattern is discernible relating the geographic location of sediment cores with "rigid" samples to those with "nonrigid" samples. The cores with rigid sediments are found at different latitudes from 6°S to 17°S, with nonrigid cores located at latitudes between these cores. While four of these nonrigid cores (45, 46, 7, and 27) are all located near the center of the study area, these cores are immediately adjacent to Core 8, whose samples all demonstrated dynamic rigidity. Also, water depth does not appear to correlate directly with sediment rigidity, as both rigid and nonrigid cores are found throughout the total depth range of all cores examined here.

Table 46.3. Averaged physical and elastic properties calculated in this study.*

	Measured physical properties				
	Number of samples	V_p (m/sec)	Density (g/cm³)	Porosity (%)	Clay (%)
All samples	38	1494	1.40	76.7	87.9
With rigidity	15	1489	1.41	78.4	85.5
Without rigidity	23	1497	1.38	79.2	89.4
Abyssal hill†	–	1499	1.34	81.1	78.9
Abyssal plain†	–	1504	1.36	79.8	79.6

	Calculated elastic properties			
	Bulk modulus (dyn/cm² × 10¹⁰)	Shear modulus (dyn/cm² × 10¹⁰)	V_s (m/sec)	Poisson's ratio
This study	3.1167	0.0161	92	0.497
Abyssal plain†	3.0108	0.0421	173	0.493
Abyssal hill†	2.9474	0.0512	194	0.491

*Clay percentages for the abyssal hill and plain provinces were taken from Hamilton (1971b).
†Hamilton (1974b).

No apparent relationship exists between the depth of an individual sample within a core and sediment dynamic rigidity. In Core 6, the shallowest sample was nonrigid, while the two deeper samples showed rigidity, and in two gravity cores (5 and 23) only the deepest sediment sample could sustain shear waves, a pattern that might suggest the development of sediment rigidity with consolidation during burial. However, in Core 21, the shallowest two samples were rigid, while the deepest sample in this core was nonrigid. Furthermore, if dynamic rigidity was related to burial consolidation, one would expect to see significant differences in sediment porosity between rigid and nonrigid samples, but no such pattern is apparent in Table 46.2.

The reason for the lack of rigidity in over half of the analyzed samples does not appear to be directly related to sedimentology: There was no significant difference in density, porosity, or clay percentages between the rigid and the nonrigid samples. Thus, the reasons for the lack of dynamic rigidity in over half of the samples analyzed remain a mystery.

The "averaged" values of the elastic constants for these samples are shown in Table 46.3, along with individual averages for the two different classes (rigid and nonrigid) of samples. These values may be compared to the "abyssal" values computed by Hamilton (1974b), which are also listed in Table 46.3. (Note: all of Hamilton's computations are for sediments in the laboratory at one atmosphere pressure and velocities corrected to 23°C.)

Bulk Modulus

The strong inverse relationship between bulk modulus and porosity, as expected from Eq. (8), is demonstrated in Figure

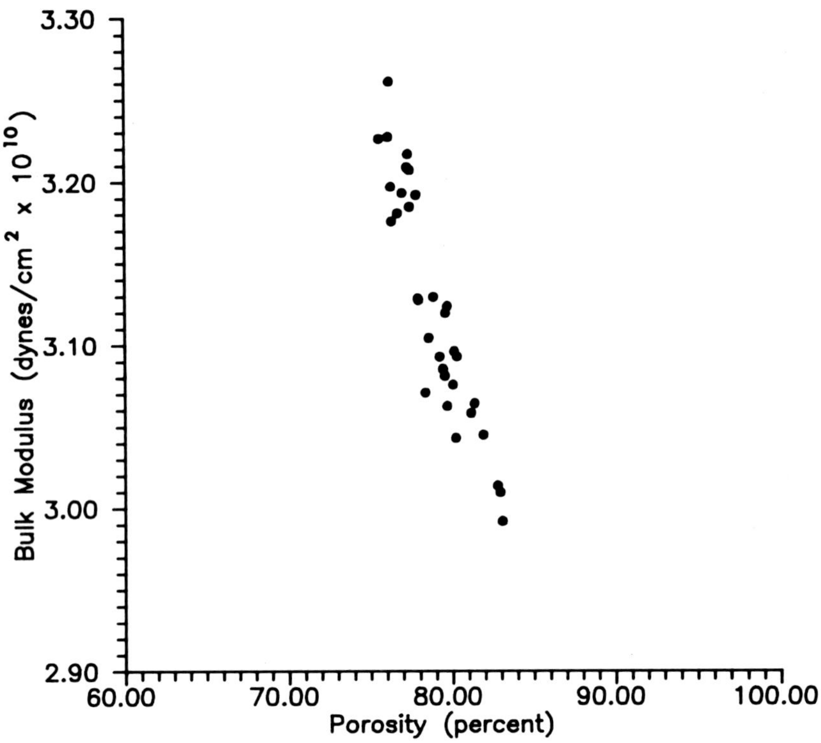

Figure 46.2. Porosity versus bulk modulus. The regression equation for this relationship is $k = -2.50993[\ln(n)] + 14.0853$.

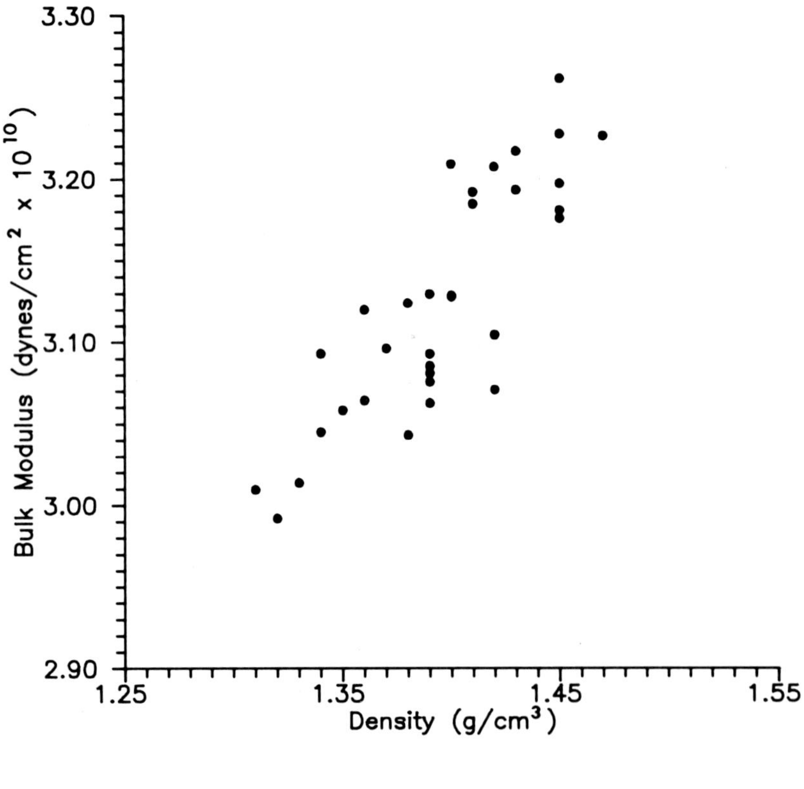

Figure 46.3. Density versus bulk modulus. The regression equation for this relationship is $k = 1.43928(\rho) + 1.11396$.

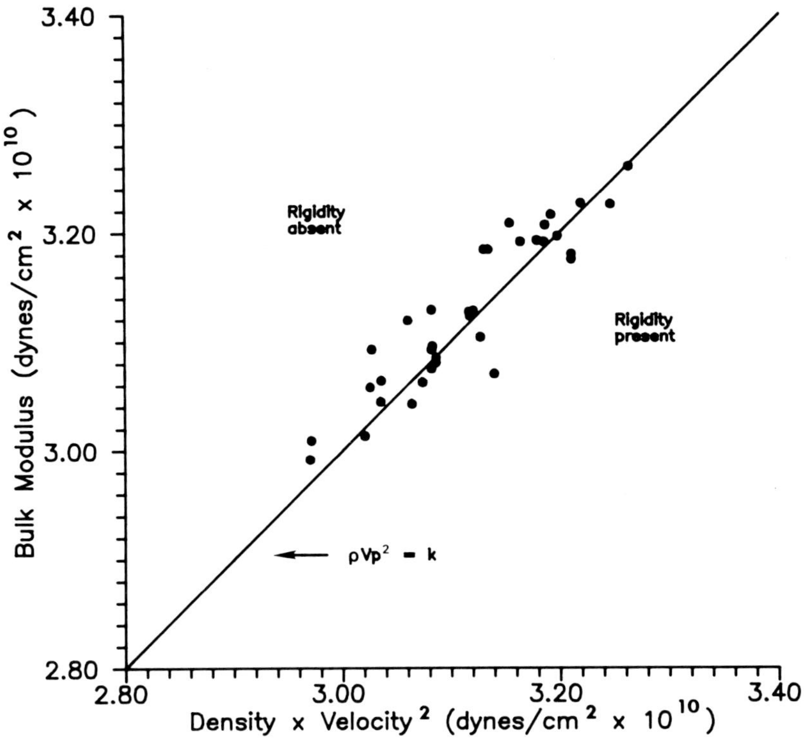

Figure 46.4. Density (compressional wave velocity)² versus bulk modulus. The relationship, $\rho(V_p)^2 = k$, expresses the situation where rigidity is absent. Regression equations relating these parameters are $k = 0.859776[\rho(V_p)^2] + 0.445273$ (for all samples) and $k = 0.957931[\rho(V_p)^2] + 0.110559$ (for rigid sediments only). Limiting values of $\rho(V_p)^2$ are 2.95 and 3.25 dyn/cm² × 10^{10}.

46.2. The scatter of the computed moduli is small with a mean of approximately 3.12 dyn/cm² × 10^{10} (Table 46.3). This relationship is best described by the regression equation:

$$k = -2.50993 [\ln(n)] + 14.0853 \quad (10)$$

where the units of k are dyn/cm² × 10^{10}, and the limiting values of (n) lie between 70 and 90%. Density also exhibits a strong correlation with bulk modulus, but with greater scatter (Fig. 46.3). The regression equation between 1.30 and 1.50 g/cm³ is

$$k = 1.43928 (\rho) + 1.11396 \quad (11)$$

where k is again expressed in dyn/cm² × 10^{10}. There is a strong dependence of aggregate and frame moduli on porosity (and hence, density). The relationship between density and porosity is

$$\rho_{sat} = n\, \rho_w + (1 - n)\, \rho_s \quad (12)$$

(Hamilton, 1971a). In this relationship, ρ_{sat} is the saturated bulk density, n is the fractional porosity, ρ_w is the pore water density, and ρ_s is the bulk density of the mineral grains.

The relationship between bulk modulus and $\rho(V_p)^2$, shown in Figure 46.4, is important in the determination of dynamic rigidity and shear wave probability. If $\rho(V_p)^2$ is less than the computed bulk modulus, rigidity is absent, no shear wave propagation is possible, and Poisson's ratio is at its maximum of 0.50 (for a fluid). Less than one-half of the samples of this study exhibited $\rho(V_p)^2$ greater than k (Fig. 46.4), in contrast to the conclusions of Hamilton (1971a).

Dynamic Rigidity/Shear Wave Velocity

A common question and important concern in underwater acoustics and marine geotechnology is "Do marine sediments exhibit sufficient rigidity for shear wave propagation?" Although most acousticians and geophysicists agree that these sediments do transmit shear waves to some extent, quantifying rigidity moduli and shear wave magnitudes have proved difficult. For a number of reasons, measurements or calculations of the elastic properties (i.e., dynamic rigidity and shear wave velocity) of fine-grained abyssal sediments are not common. Previous studies, however, have shown that rigidity decreases with decreasing grain size, increasing porosity, and decreasing density (Hamilton, 1971a).

Dynamic rigidity in our samples averaged about 0.016 dyn/cm² × 10^{10} (Table 46.3), and is shown versus mean grain diameter in Figure 46.5, and versus porosity in Figure 46.6. Also shown on these graphs are approximations of the variability of dynamic rigidity (i.e., 3 × standard error of the mean) for representative fine-grained sediment types (Hamilton, 1974b). In both cases shown, the computed rigidity is lower than that given by Hamilton (1974b). The mean grain diameters

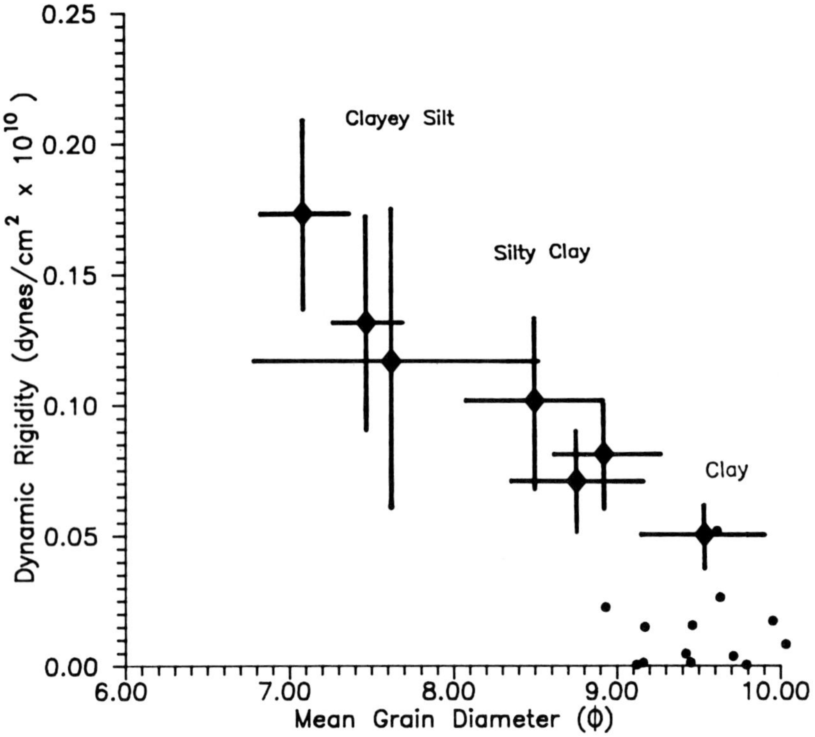

Figure 46.5. Dynamic rigidity versus mean grain diameter. Error bars (from Hamilton, 1974b) represent the variability of properties for the typical fine-grained sediment types (three times the standard error of the mean value for each sediment type), using the classification of Shepard (1954). Computed dynamic rigidity values for all but one sample in the present study are lower than the values calculated by Hamilton.

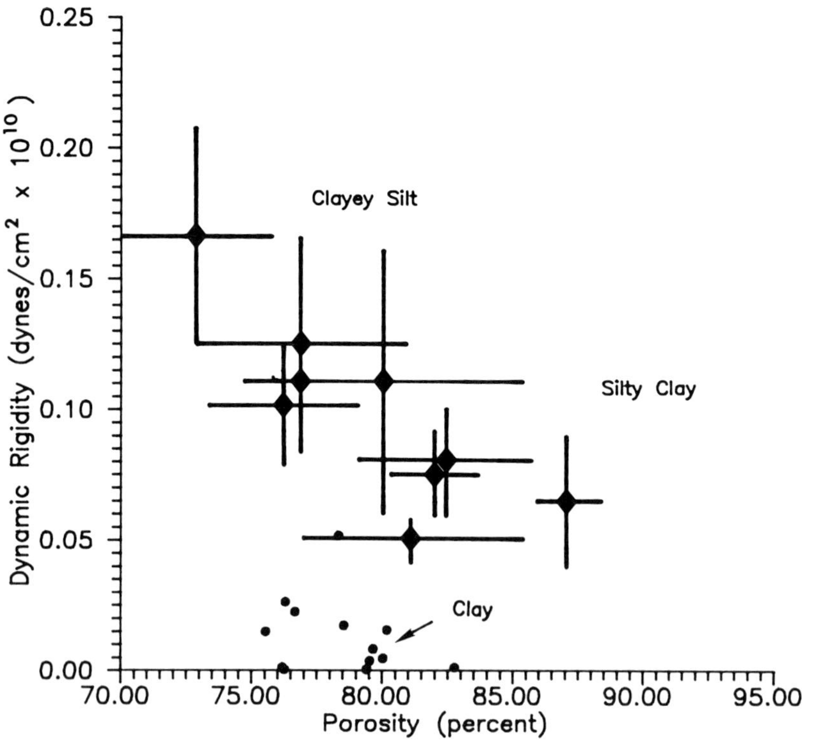

Figure 46.6. Dynamic rigidity versus porosity. Error bars are from Hamilton (1974b), and are defined in Figure 46.5. As in Figure 46.5, calculated dynamic rigidity values for sediments in the present study are lower than those calculated by Hamilton.

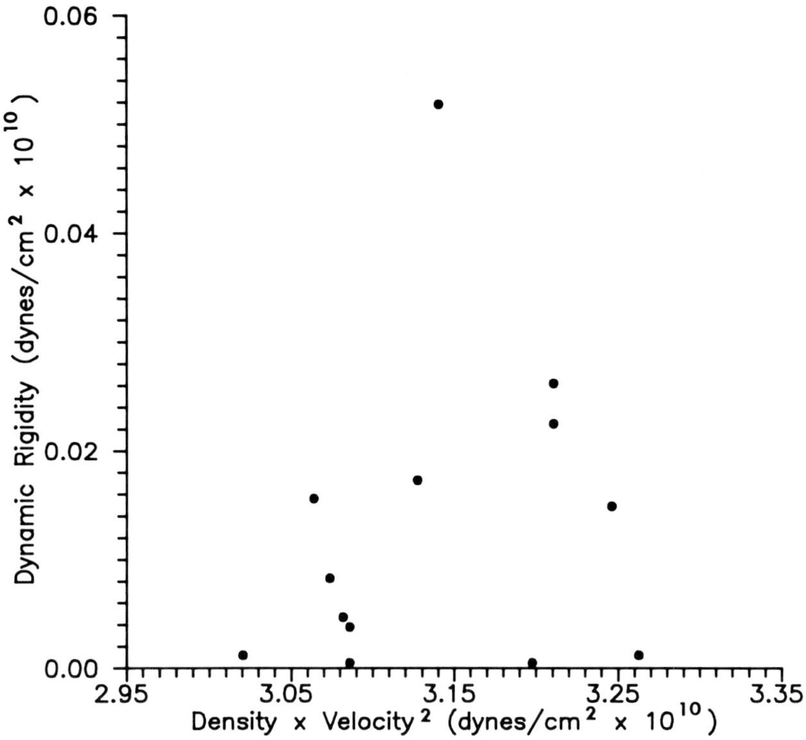

Figure 46.7. Dynamic rigidity versus $\rho(V_p)^2$.

of this study are in good agreement with predicted values; porosities, however, are somewhat lower.

Hamilton (1971a) considered $\rho(V_p)^2$ of abyssal-plain sediments, and percent clay (or percent sand-silt) of abyssal-hill sediments, to be the best indices to predicting dynamic rigidity. Detailed examination of the relationships between dynamic rigidity and $\rho(V_p)^2$ (Fig. 46.7) and clay percent (Fig. 46.8) shows little correlation within the limits of this presentation. However, this is still consistent with the observation of Hamilton (1971a) that large scatter is usually present in any plot of rigidity (or shear wave velocity) versus other physical properties.

Shear wave velocities for the samples of this study average 92 m/sec, and fall within the range of velocities commonly reported. Table 46.4, although not exhaustive, demonstrates the large variance of shear wave velocities that can be found in the literature.

Poisson's Ratio

Poisson's ratio is the ratio of the transverse to longitudinal strain of a medium under an applied stress, and related to V_p/V_s by

$$\sigma = \frac{[(V_p/V_s)^2 - 2\ ]}{2\ [(V_p/V_s)^2 - 1]} \qquad (13)$$

The ratio varies from 0.00 to 0.50 with its upper limit characteristic of a fluid (i.e., no shear wave transmission). Most rocks and sediments possess sufficient rigidity to sustain shear waves, and Poisson's ratio is usually less than 0.50.

As previously mentioned, we could calculate dynamic rigidity for less than half of the samples of this study (Fig. 46.4). Of those samples, Poisson's ratio averaged 0.497 (Tables 46.3 and 46.5). This is slightly higher than those reported by Hamilton (1971a), but compares very well with suggested values given in Hamilton (1979). These higher values are significant, because they imply that the sediments are more fluid-like and characterized by lower shear wave velocities. The differences result from the lower V_p used in this study, and the relationship between Poisson's and the V_p/V_s ratios in Eq. (13). A plot of Poisson's ratio versus density is shown in Figure 46.9, demonstrating the small variability that exists between the samples.

An interesting inverse relationship was noted between Poisson's ratio and dynamic rigidity (Fig. 46.10), in which the ratio decreases linearly as dynamic rigidity increases. This condition is explained by the V_p/V_s ratio [Eq. (13)]. An increase in dynamic rigidity causes a corresponding increase in V_p as predicted by the $(4\mu/3)/\rho$ component in Eq. (4), and an increase in V_s from Eq. (6). The overall effect is a decrease in the V_p/V_s ratio, and as the V_p/V_s ratio decreases, so does Poisson's ratio [Eq. (13)].

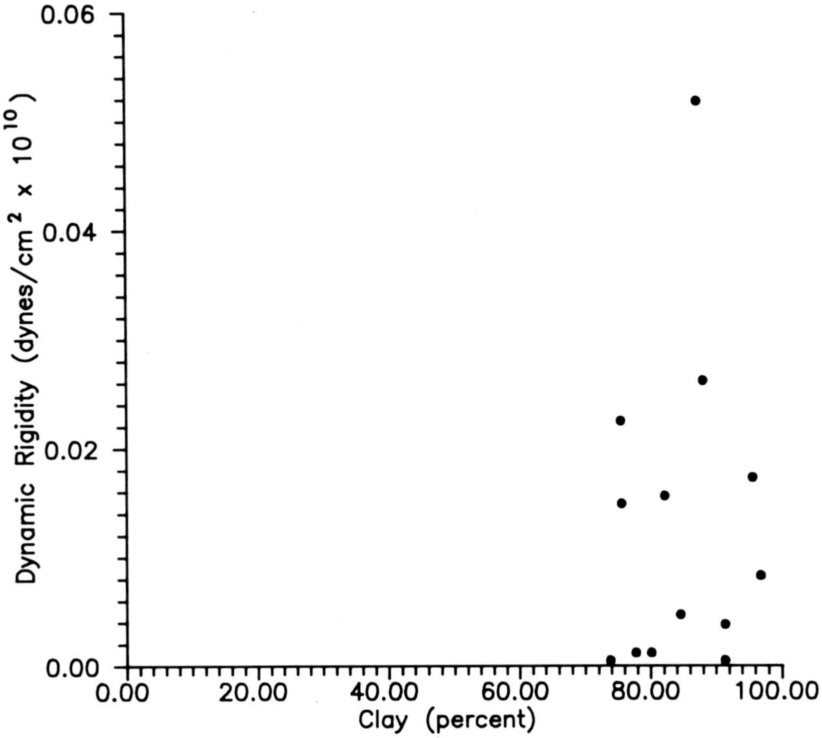

Figure 46.8. Dynamic rigidity versus percent clay.

Predictive Indices for Elastic Constants

As expected, the results of bulk modulus (k) computations show strong correlations with porosity and density, and it is suggested that these values be used in absence of V_p measurements. In contrast to Hamilton (1971a), $\rho(V_p)^2$ and percent clay show little usefulness as indices to dynamic rigidity. It is suggested that the equations of Hamilton (1971a, 1974b) and an estimate of bulk modulus, either through calculation or from the regression equations, be used instead. Because bulk modulus and the other elastic constants are critically dependent on measured laboratory values, we stress caution in applying regression equations and diagrams, particularly in light of the large variance in laboratory measurements of V_p.

Clay Fabric

The elastic properties of clay minerals, in contrast to those of seawater and many of the rock-forming minerals, are poorly documented (Matthews, 1980). Clay minerals are unique because of their affinity for water, an attraction attributable to hydrous layers that constitute the crystalline structure. Cohesion, an inherent property of the phyllosilicates, results from physiochemical forces that provide the clay sediment with structural strength by giving rise to particle-to-particle, as well as particle-to-water bonds. The structural fabric of clay sediment (e.g., Bennett et al., 1981) is also an important contributor to its strength, and, consequently, its rigidity. Hamilton (1970) concluded that the cardhouse-flocculated form is the strongest configuration, determined primarily by interparticle forces and the number of contacts.

Environmental factors, such as depositional rates, may also affect the strength of clay sediments (Hamilton, 1971a). In areas influenced by rapid sedimentation, such as deltas and harbors/bays, the bottom boundary layer may consist of little more than clay particles in suspension. These sediments exhibit negligible rigidity, and will not sustain shear waves. At the other extreme, in certain deep-sea sedimentary environments, such as abyssal hills, pelagic deposition occurs as slow particle-by-particle accumulation, and shear strength increases with increasing age of sediments. Sediment rigidity is supplied by increased interparticle bonding similar to cementation, where the age of the sediment simply implies increased exposure to "cementing" agents, such as iron oxides.

Unfortunately, clay mineralogical or fabric analyses were not conducted for this study. However, a tentative model, based on published clay mineral distribution patterns of the Brazil Basin and the physical properties presented in this study, is proposed. Bowles (1968) and Bowles et al. (1969) concluded that the clay fabric of most undisturbed marine sediments from the Gulf of Mexico was characterized by loose, open, random arrangements of particles. The clay fabric model of Bennett et al. (1981), derived from investigations of samples obtained from the equatorial

Table 46.4. Representative shear wave velocities of fine-grained sediments.* (modified from Hamilton, 1971a).

Sediment type	V_S (m/sec)	References†	Comments
			In situ (land)
Clay	230	1	Borehole; 3-m depth; explosive
Clay	120	2	0–20 m depth; (seismic logging); V_p/V_S
Clay	170	2	0–10 m depth; (seismic logging); V_p/V_S
			In situ seafloor [Stoneley waves (3, 4); probe (5)]
Silt clay	50–190	3	Deep Indian Ocean
Silt clay	101	4	San Diego Trough
Silt clay	107	4	San Diego Trough
Silt clay	87	4	San Diego Trough
Mud	20	5	La Spezia, Italy
			Laboratory measurements
Clay	2–40	6	Kaolinite clay (bimorph crystal transducers)
			Theoretical
Silt clay	116	7	Composite based on literature
			Computed average values
Silty clay	287	8	Continental Shelf
Silty clay	232	8	Abyssal Hills (Pacific Basin)
Clay	195	8	Abyssal Hills (Pacific Basin)
Silty clay	240	8	Deep Abyssal Plain (adjacent to Pacific Basin)
Clay	92	This study	Deep Abyssal Plain (Brazil Basin)

*From Hamilton, Journal of Geophysical Research, p. 591, 1971. Copyright by the American Geophysical Union.

† References:
1. Molotova (1963), in Hamilton (1971a).
2. Molotova and Vassil'ev (1960), in Hamilton (1971a).
3. Davies (1965).
4. Hamilton et al. (1970).
5. Richardson et al. (this volume).
6. Shirley and Hampton (1978).
7. Hamilton (1979).
8. Hamilton (1971a).

Pacific and Mississippi Delta, is used to predict the structure of abyssal sediments from the Brazil Basin. Although they stress that the model may not be applicable to all submarine clays, it is used with caution in this study to relate computed void ratio to clay microstructure, and may provide insight into the processes accounting for the lack of shear strength in many of the samples.

Illite is the predominant clay mineral in Atlantic deep-sea sediments in the <2 μm, and probably the 2–20 μm size fractions (Biscaye, 1965). The sedimentary regime of the Brazil Basin, particularly along the continental margin, is dominated by gravity-controlled mass flows (Summerhayes et al., 1976), and, to a much larger degree, is affected by sediment reworking by the deep thermohaline Antarctic Bottom Water (Damuth and Hayes, 1977). The surface sediments of the Brazil Basin are very low in

Table 46.5. Averaged values for Poisson's ratio in this study and for other fine-grained abyssal sediments (Hamilton, 1971a).

	Poisson's ratio		
Environment	Maximum	Minimum	Average
Abyssal Plain*	0.499	0.496	0.497
Abyssal Plain†	0.496	0.466	0.484
Abyssal Hill†	0.499	0.467	0.487

From Hamilton, Journal of Geophysical Research, p. 599, 1971, Copyright by the American Geophysical Union.
*This study.
† Hamilton (1971a).

calcium carbonate (Biscaye et al., 1976), and are composed predominantly of illite and smectite (Emery and Uchupi, 1985).

Void ratio (e), defined as the ratio of the volume of voids to the volume of solid constituents, was computed from porosity (Table 46.2) by the relationship, $e = n/(1 - n)$. In all instances, void ratios were very high (i.e., 3.68 for the rigid sediments and 3.84 for nonrigid sediments), and in either case, fall within the Bennett model comprised of "single plate-like particles and chains" (Fig. 46.11). Two explanations are possible to account for the lack of rigidity demonstrated in Figure 46.4, and low computed dynamic rigidity of the other samples (Table 46.3): Either the sediments are truly fluid-like, and do not possess significant structural moduli, or the sediment fabric was disturbed during the coring process. Although this model does not provide strong evidence for either condition, disturbances of the "weak" clay microstructure during the coring process (penetration and retrieval) can be easily envisioned.

In summary, lower than expected rigidity moduli, and subsequent elastic constants, probably result from a partial disturbance of the clay's structure during coring. Further, the lack of dynamic rigidity is believed to result from complete destruction of the sediment's frame. Unfortunately, it is not possible to determine if coring disturbances are present in these sediment cores, or, if present, to quantify their magnitudes.

Conclusions

The results of this study of sediment samples from the Brazil Basin permit a few preliminary conclusions regarding the elastic properties of fine-grained, abyssal sediments:

1. Computed moduli, shear wave velocities, and Poisson's ratio were slightly lower than those previously reported, and probably result from the lower compressional wave velocities used here. Average values include bulk modulus = 3.12 dyn/cm^2 × 10^{10}; shear (rigidity) modulus = 0.0161 dyn/cm^2 × 10^{10}; shear wave velocity = 92 m/sec; and Poisson's ratio = 0.497.
2. The best indices for estimating bulk modulus are porosity and density. Other previously reported correlations, such

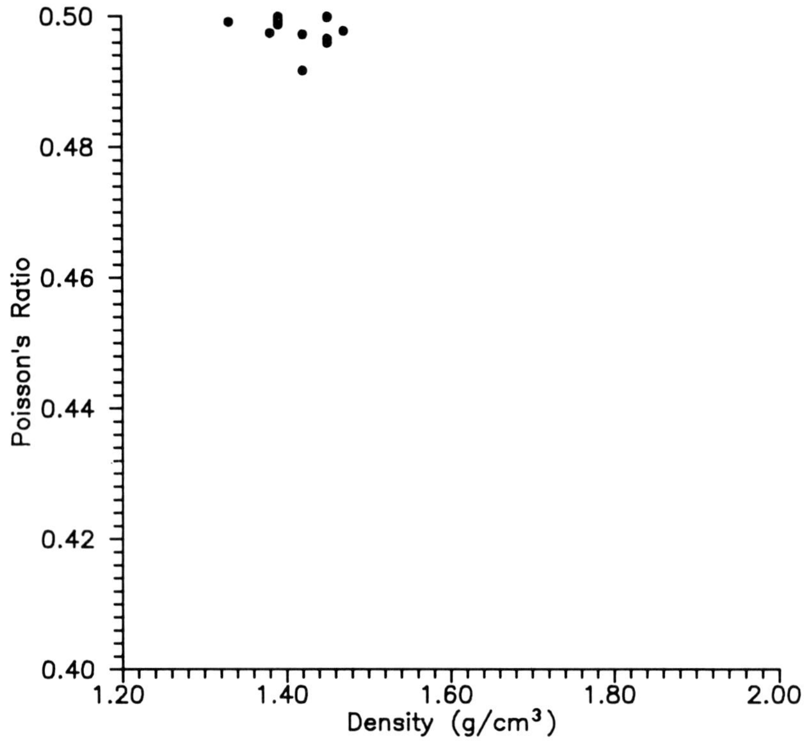

Figure 46.9. Poisson's ratio versus density.

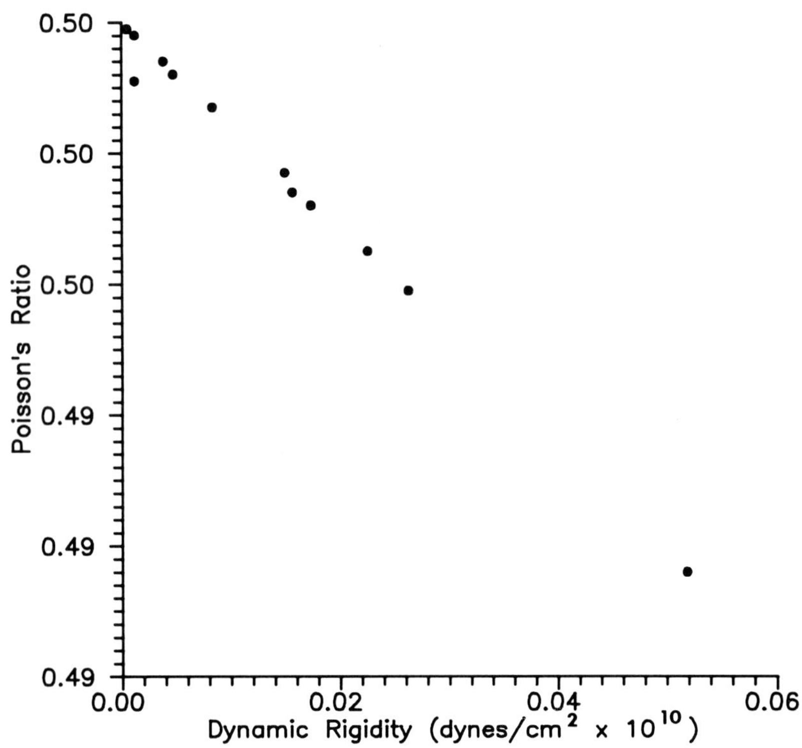

Figure 46.10. Dynamic rigidity versus Poisson's ratio.

Figure 46.11. Clay fabric models for smectite- and illite-rich submarine sediments (Bennett et al., 1981).

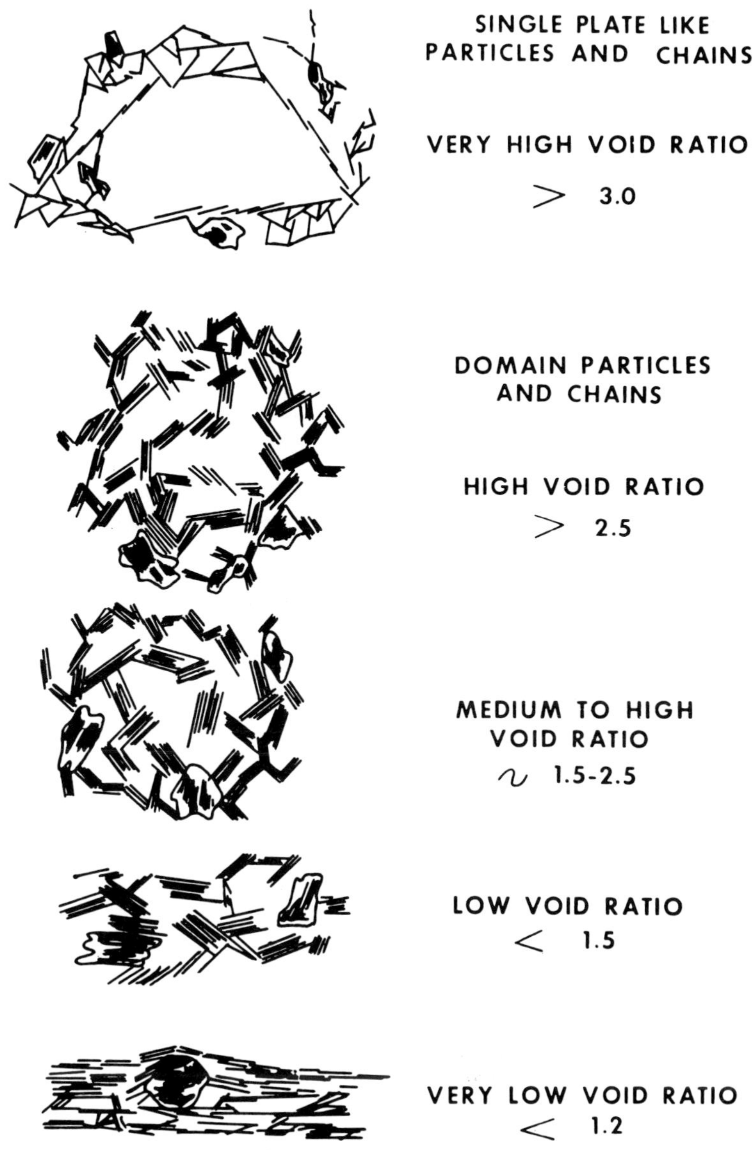

as $\rho(V_p)^2$ and percent clay versus dynamic rigidity, were not observed in this study. For estimates of rigidity, it is suggested that published regression equations be used.

3. A speculative clay fabric model, based on average void ratio, is presented (after Bennett et al., 1981) for the samples of this study. The model consists of single plate-like particles and chains, and characterized by very high void ratios.
4. Of the samples examined in this study, less than half exhibited rigidity. Two hypotheses are presented to explain this absence. Either the sediments are truly fluid-like, and do not possess significant structural moduli; or the sediment fabric was disturbed during the coring process. Based on the proposed clay fabric model, coring disturbances are envisioned as the probable mechanism for reducing or destroying the shear strength, rigidity, and ability to sustain shear wave propagation.

Acknowledgments

The U.S. Naval Oceanographic Office (NAVOCEANO) Marine Geological Laboratory is gratefully acknowledged for making this study possible. J. Bunce and C. Ingram supplied the analytical results, and G. Kelly, C. Ross, and J. Bowman performed the analyses and provided many helpful suggestions. R.H. Bennett of the Naval Ocean Research and Development Activity (NORDA) is thanked for extending the invitation to contribute to this book. The manuscript benefited significantly from critical reviews by E.L. Hamilton and R.T. Bachman of the Naval Ocean Systems Center (NOSC) in San Diego, California, J.I. Arvelo of the Naval Surface Warfare Center in Dahlgren, Virginia, and M. Badiey of the Port and Harbour Research Institute, Ministry of Transport, Nagase, Yokosuka, Japan.

References

American Society for Testing and Materials, 1964. Procedures for Testing Soils. ASTM, Philadelphia, Committee D-18, 4th edition, 540 p.

Bachman, R.T., 1985. Acoustic and physical property relationships in marine sediments. Journal of the Acoustical Society of America, v. 76, p. 616–621.

Bell, D.W., and D.J. Shirley, 1980. Temperature variations of the acoustical properties of laboratory sediments. Journal of the Acoustical Society of America, v. 68, p. 227–231.

Bennett, R.H., and D.N. Lambert, 1971. Rapid and reliable technique for determining unit weight and porosity of deep-sea sediments. Marine Geology, v. 11, p. 201–207.

Bennett, R.H., and T.A. Nelsen, 1983. Seafloor characteristics and dynamics affecting geotechnical properties at shelfbreaks. Society of Economic Paleontologists and Mineralogists, SEPM Special Publication No. 33, p. 333–355.

Bennett, R.H., W.R. Bryant, and G.H. Keller, 1981. Clay fabric of selected submarine sediments: fundamental properties and models. Journal of Sedimentary Petrology, v. 51, p. 217–232.

Biscaye, P.E., 1965. Mineralogy and sedimentation of Recent deep-sea clay in the Atlantic Ocean and adjacent seas and oceans. Geological Society of America Bulletin, v. 76, p. 803–832.

Biscaye, P.E., V. Kolla, and K.K. Turekian, 1976. Distribution of calcium carbonate in surface sediments of the Atlantic Ocean. Journal of Geophysical Research, v. 81, p. 2595–2603.

Bowles, F.A., 1968. Microstructure of sediments: investigations with ultrathin sections. Science, v. 159, p. 1236–1237.

Bowles, F.A., W.R. Bryant, and C. Wallin, 1969. Microstructure of unconsolidated and consolidated marine sediments. Journal of Sedimentary Petrology, v. 39, p. 1546–1551.

Briggs, K.B., M.D. Richardson, and D.K. Young, 1985. Variability in geoacoustic and related properties of surface sediments from the Venezuela Basin, Caribbean Sea. Marine Geology, v. 68, p. 73–106.

Clay, C.S., and H. Medwin, 1977. Acoustical Oceanography: Principles and Applications. Wiley, New York, p. 486–491.

Damuth, J.E., and D.E. Hayes, 1977. Echo character of the East Brazilian continental margin and its relationship to sedimentary processes. Marine Geology, v. 24, p. 73–95.

Davies, D., 1965. Dispersed Stoneley waves on the ocean bottom. Bulletin of the Seismological Society of America, v. 55, p. 903–918.

Emery, K.O., and E. Uchupi, 1985. The Geology of the Atlantic Ocean. Springer-Verlag, New York, 1020 p.

Gassmann, F., 1951. Über die elastizität poröser medien. Vierteljahrsschr. Naturforsch. Ges. Zürich, v. 96, p. 1–23.

Hamilton, E.L., 1970. Sound velocity and related properties of marine sediments, north Pacific. Journal of Geophysical Research, v. 75, p. 4423–4446.

Hamilton, E.L., 1971a. Elastic properties of marine sediments. Journal of Geophysical Research, v. 76, p. 579–604.

Hamilton, E.L., 1971b. Predictions of in-situ acoustic and elastic properties of marine sediments. Geophysics, v. 36, p. 266–284.

Hamilton, E.L., 1972. Compressional wave attenuation in marine sediments. Geophysics, v. 37, p. 620–646.

Hamilton, E.L., 1974a. Geoacoustic models of the sea floor. In: Hampton, L. (ed.), Physics of Sound in Marine Sediments. Plenum Press, New York, p. 181–221.

Hamilton, E.L., 1974b. Predictions of deep-sea sediment properties: state-of-the-art. In: Inderbitzen, A.L. (ed.), Deep-Sea Sediments, Physical and Mechanical Properties. Plenum Press, New York, p. 1–43.

Hamilton, E.L., 1979. V_p/V_s and Poisson's ratios in marine sediments and rocks. Journal of the Acoustical Society of America, v. 66, p. 1093–1101.

Hamilton, E.L., 1980. Geoacoustic modeling of the sea floor. Journal of the Acoustical Society of America, v. 68, p. 1313–1340.

Hamilton, E.L., and R.T. Bachman, 1982. Sound velocity and related properties of marine sediments. Journal of the Acoustical Society of America, v. 72, p. 1891–1904.

Hamilton, E.L., H.P. Bucker, D.L. Keir, and J.A. Whitney, 1970. Velocities of compressional and shear waves in marine sediments determined *in situ* from a research submersible. Journal of Geophysical Research, v. 75, p. 4039–4049.

Holland, C.W., and B.A. Brunson, 1988. The Biot-Stoll sediment model: an environmental assessment. Journal of the Acoustical Society of America, v. 84, p. 1437–1443.

Horn, D.R., B.M. Horn, and B.N. Delach, 1968. Correlation between acoustical and other physical properties of deep-sea cores. Journal of Geophysical Research, v. 73, p. 1939–1957.

Ingram, C., and R. Rogers, 1984. Field determination of approximate in situ sound velocity. Naval Oceanographic Office, Stennis Space Center, MS, NAVOCEANO Marine Geological Laboratory, unpublished manuscript, 2 p.

Ingram, C., C. Ross, K. Heffner, G. Meitzler, E. Kelly, R. Rogers, and D. Kekko, 1985a. Results of laboratory analyses performed on cores from the South Atlantic during USNS Wilkes cruise 330283-III, April 1983, 8 volumes. Naval Oceanographic Office, Stennis Space Center, MS, NAVOCEANO Marine Geological Laboratory Report 580.

Ingram, C., C. Ross, K. Heffner, G. Meitzler, E. Kelly, R. Rogers, and D. Kekko, 1985b. Results of laboratory analyses performed on cores from the South Atlantic during USNS Wilkes cruise 330483-I, June–July 1983, 8 volumes. Naval Oceanographic Office, Stennis Space Center, MS, NAVOCEANO Marine Geological Laboratory Report 583.

Krumbein, W.C., 1934. Size frequency distributions of sediments. Journal of Sedimentary Petrology, v. 4, p. 65–77.

Lavoie, D., and J.E. Matthews, 1983. South Atlantic geoacoustic models. Naval Ocean Research and Development Activity, Stennis Space Center, MS, NORDA unpublished manuscript, 153 p.

Matthews, J.E., 1980. Heuristic physical property model for marine sediments. Journal of the Acoustical Society of America, v. 68, p. 1361–1370.

McManus, D.A., 1982. Phi and sediment analysis: discussion. Journal of Sedimentary Petrology, v. 52, p. 1011–1026.

Molotova, L.V., 1963. Velocity ratio of longitudinal and transverse waves in terrigenous rocks. Bulletin (Izv.) Akad. Sci. USSR, Geophysical Series, English Translation No. 12, p. 1074–1080.

Molotova, L.V., and Y.I. Vassil'ev, 1960. Velocity ratio of longitudinal and transverse waves in rocks, 2. Bulletin (Izv.) Akad. Sci. USSR, Geophysical Series, English Translation No. 8, p. 731–743.

Naval Oceanographic Office, 1966. Handbook of Oceanographic Tables. NAVOCEANO Special Publication No. SP-68, p. 302–318.

Pierce, J.W., and R.R. Graus, 1981. Use and misuse of the ϕ-scale: a discussion. Journal of Sedimentary Petrology, v. 51, p. 1348–1350.

Richardson, M.D., 1983. The effects of bioturbation on sediment elastic properties. Bulletin of the Society of Geology of France, v. 25, p. 505–513.

Schultheiss, P.J., 1981. Simultaneous measurement of P and S wave velocities during conventional laboratory soil testing procedures. Marine Geotechnology, v. 4, p. 343-367.

Shepard, F.P., 1954. Nomenclature based on sand-silt-clay ratios. Journal of Sediment Petrology, v. 24, p. 151–158.

Shirley, D.J., and L.H. Hampton, 1978. Shear-wave measurements in laboratory sediments. Journal of the Acoustical Society of America, v. 63, p. 607–613.

Stoll, R.D., and T.K. Kan, 1981. Reflection of acoustic waves at a water-sediment interface. Journal of the Acoustical Society of America, v. 70, p. 149–156.

Summerhayes, C.P., R. Fainstein, and J.P. Ellis, 1976. Continental margin off Sergipe and Alagoas, northeastern Brazil: a reconnaissance geophysical study of morphology and structure. Marine Geology, v. 20, p. 345–361.

Tucholke, B.E., 1980. Acoustic environment of the Hatteras and Nares abyssal plains, western North Atlantic Ocean, determined from velocity and physical properties of sediment cores. Journal of the Acoustical Society of America, v. 68, p. 1376–1390.

Tucholke, B.E., and D.J. Shirley, 1979. Comparison of laboratory and in situ compressional-wave velocity measurements on sediment cores from the wester North Atlantic. Journal of Geophysical Research, v. 84, p. 687–695.

Urick, R.J., 1947. A sound velocity method for determining compressibility of finely divided substances. Journal of Applied Physics, v. 18, p. 983–987.

Wentworth, C.K., 1922. A scale of grade and class for clastic sediments. Journal of Geology, v. 30, p. 377–392.

II

Applied Clay Microstructure

C. Applications: Present Requirements

CHAPTER 47

Applications: Present Requirements, Waste Disposal, Containment, and Packing Material

Thomas F. Lomenick and J.D. Kasprowicz

The impact of microstructure on engineering and scientific problems is illustrated with problem solutions in a series of studies of fine-grained sediments and rocks. Special environmental problems involve the utilization of natural and artificially prepared geological materials for containment of waste materials. Disciplines of rock mechanics, geohydraulics, geotechnics, and remote sensing play critical roles in assessing the geological materials and formations selected for waste utilization and recovery of natural resources. Both of the activities impact the environment through time and space. It is essential that research advances made in understanding sediment microstructure, the underpinning of the mass physical and mechanical properties, be transitioned to engineering problems and applications in a timely manner.

The chapters in this section deal with the microstructure of clays and other marine soils, disposal of radioactive and hazardous wastes into shales and clay-rich rocks, and engineering and rheological properties of clays and oil well cements. Four of the chapters deal with the disposal of hazardous wastes. The authors of these have evaluated a variety of sedimentary rocks as potential host media for radioactive and hazardous wastes. Clay-rich rock (shale) was selected as the sedimentary rock with the greatest potential for containment based on technical aspects related to geology, geochemistry, hydrology, thermal performance, rock mechanics, natural resources, waste package material degradation, repository costs, and systems studies. Four chapters present the results of investigations pertaining to clay-rich rocks as a potential host for radioactive and hazardous wastes.

In Disposal of Radioactive and Hazardous Wastes into Clay-Rich Rocks, Lomenick and Kasprowicz consider unique mine stability problems that might occur in these rocks as a result of their use as repository hosts. Extensive laboratory creep tests on scale models of rock pillars and their surrounding rocks were conducted. Detailed description of tests, with initial pillar stresses ranging from 1000 to 10,000 psi and temperatures ranging from ambient to 100°C are presented. The simulation of mine pillars and model sealing is discussed, along with the development of a power-law transient creep equation. A comparison of deformational response of shale model pillars fabricated from clay-rich rock is presented along with that of other rocks. The scoping study on model pillar shortening suggests that the weaker, more deformable clay-rich rocks would probably be most useful as a potential host medium for non-heat-generating wastes, while the stronger rocks could accommodate a wider variety of wastes at deeper depths.

In two related chapters on nuclear and hazardous waste disposal, Nataraj reports on the investigations concerned with deep boreholes with waste emplacement from the surface. The influence of several geotechnical properties on strength and deformation of host material is examined, along with the stability of shales and clays. Based on preliminary work, the unconfined compression strength and cohesion of rock mass are estimated to be at least 2.5 and 2 MPa, respectively, for adequate stability of boreholes in indurated shales. For the weaker shales and clays, support pressure would be required to reduce deformation.

Quigley and Fernandez report laboratory studies of the relationship of the hydraulic conductivity of compacted clays to the fabric, microstructure, and porosimetry. Their results demonstrate and rationalize the marked increase in hydraulic conductivity that may occur if barrier clays are compacted in the presence of organic liquids. These results emphasize the importance of characterizing clay barriers under conditions representative of the fluid media that will be present, and worry that failure of containment may occur as the clay microstructure is altered when organic liquids permeate into the barrier.

Grabowski and Gillott conducted investigations to determine the effects of various types of silica and temperature on ambient and hydrothermal engineering properties of cement silica blends. The changes in strength, permeability, and other engineering properties are attributed to the differences in microstructure inferred from SEM micrographs.

Burkett, Bennett, and Bryant present the results of investigations on illitic-rich clays using transmission electron microscopy in their chapter entitled The Role of the Microstructure of Pacific Red Clays to Radioactive Waste Disposal. These studies were conducted on remolded samples consolidated to *in situ* porosities in the laboratory. The response of the fabric of red clay to thermal and mechanical influences resulting from the testing of simulated radioactive waste is described. It was stressed that no significant microfabric or grain size changes occurred in the material in response to mechanical or thermal disturbances that would be representative of radioactive waste emplacement. However, residual strength increase, localized decrease in water content, porosities, and permeabilities were observed. This chapter illustrates the significance of fabric studies in the feasibility evaluation of radioactive waste disposal within fine-grained sediments of ocean basins by carefully designed emplacement methods.

Some marine sediments are known to have high sensitivity, and, on disturbance, are transferred into a liquid phase and are usually known as quick clays. The resistance to flow of suspensions of soil is of great importance in submarine failures as shown in the chapter Influences of the Rheology of Marine Sediments Composed of Low-Activity Minerals. In this work, Torrance presents the results of investigations using a coaxial viscometer. The influence of physical, chemical, and mineralogical factors on rheological response of marine clays is described.

Clays and shales are used as permeability barriers in several geotechnical projects including waste disposal landfills. The integrity of clay barriers is of extreme importance in the prevention of leachate migration. In The Geotechnical Importance of Clay Flexibility, Aughenbaugh examines the importance of flexibility of clay that is deformed plastically without cracking when clay is used as a permeability barrier. A method to evaluate flexibility of compacted clays is presented. In the work, it is suggested that clay liners be compacted at optimum moisture content to achieve maximum flexibility.

Two chapters deal with clays and cements and are concerned primarily with microfabric, engineering, and rheological properties. In The Microfabric of Some Hong Kong Marine Soils, Tovey describes the mineralogy, chemistry, and microfabric of marine soils. Using scanning electron microscopy and image analysis techniques, the changes in fabric during the consolidation process are explained. The application of an intensity gradient technique to determine a quantitative measure of fabric (by defining the index of anisotropy) is explained. It is emphasized that the principle and application of the technique would be useful to researchers in microfabrics.

In the engineering analysis of nearshore/offshore marine sediments, conventional geotechnical investigations and laboratory tests are reported to be generally inadequate to provide the required information. To gain insight and correlate the engineering properties of these sediments, investigations related to their microstructure were initiated to provide the required data. Guorui presents one such approach in his chapter Application of Microstructure Classification of Marine Sediment to Engineering Geological Evaluation. Based on detailed study of SEM micrographs of marine soils in China, four types of microstructures are identified. Correlation curves for various types of microstructure with allowable bearing capacity and modulus of compressibility are presented.

In the study Microstructural and Mineralogical Characterization of Selected Shales in Support of Nuclear Waste Repository Studies, Lee, Hyder, and Alley describe the microstructural and mineralogical characterization of four shales. Petrographic microscopy, scanning electron microscopy, high-resolution transmission electron microscopy, including energy dispersive X-ray spectrometry, and X-ray diffraction analysis were used in the characterization. The results for samples at various temperature ranges are presented and discussed. A semiquantitative approach to estimate mineral composition of shale is suggested.

CHAPTER 48

Disposal of Radioactive and Hazardous Wastes into Clay-Rich Rocks

Thomas F. Lomenick and J.D. Kasprowicz

Introduction

Because of its unique containment properties, shale is believed to be an extremely desirable host rock for the disposal of high-level radioactive wastes. The principal advantage of utilizing shale as a disposal medium for high-level wastes is its very low permeability to the flow of groundwater. This property is due to its small sizes of matrix pore and its characteristic of deforming plastically at relatively shallow depths and/or low overburden loads (which promotes the self-healing of fractures). However, with emplacement of heat-generating wastes in shale strata unique mine stability problems may occur as a result of the elevated temperatures. To predict flow in the rocks, scale models of shale pillars and their surrounding rooms are being fabricated and tested from cores taken in four shales. These are believed to represent mineralogical extremes of shale types that may be suitable for repository utility.

The primary objective of this work is to provide data for the development and validation of creep laws for shales that are representative of strata that could be candidates for the disposal of high-level radioactive wastes. In the future, these tests will be designed to systematically study the influence of (1) the material, (2) stress state, (3) temperature, (4) bedding plane orientation, and (5) pillar shape (width/height ratio) on the time-dependent deformational response of the pillar. This report gives the results of a scoping study.

These results and future work on shales will provide data on the sensitivity of the mechanical response (pillar shortening) to increases in temperature and/or average pillar stress. That information will be useful during a national survey of shales when depth and lateral extent of various shale strata will be considered.

Although many shales may possess the geotechnical characteristics for containing and controlling emplaced wastes, our initial efforts are being directed toward four compositional end members, namely, the Devonian and Cambrian age deposits in the eastern United States and the Cretaceous and Cenozoic age sediments of the western United States. For this report, test data are reported for samples of the Huron shale of Devonian age, Conasauga Group (member unknown) of Cambrian age, Pierre Shale of Cretaceous age, and Green River Formation (oil shale) of Tertiary age. Composition and mechanical properties of three of the four shales considered here are given in Table 48.1, modified from Hansen and Vogt (1987).

Summary of Previous Work

Analytical and numerical methods for simulating the behavior of rock pillars (required for stability) were not well developed in the early 1960s when laboratory work began on the concept of disposal of high-level radioactive waste in conventionally mined cavities in rocks. Consequently, a series of tests on physical models of mine pillars was initiated at Oak Ridge National Laboratory (ORNL). Obert (1964) developed a laboratory-scale, model-pillar, constant-load creep test that accounted for the lateral confinement of the roof and floor rocks by steel bands. This test allowed for the study of the response of pillars with different width-to-height ratios and salt pillars with and without shale partings as described by Lomenick and Bradshaw (1969a). Furthermore, the test could be run at elevated temperatures to simulate the conditions in a repository containing heat-generating nuclear waste.

Lomenick and Bradshaw (1969b) reported on an extensive testing program that included the effects of load, temperature, temperature elevation after initial loading, shale partings, pillar shapes, salt samples from different localities, reproducibility of

Table 48.1. Average composition and mechanical properties of shales.*

Mineral	Shale type† (wt%)		
	Green River	Conasauga	Pierre
Illite	3.9 ± 0.6	53.4 ± 3.4	3.5 -13.6
Chlorite	0.0	20.4 ± 3.0	0.0 - 2.3
Kaolinite	0.0	0.0	1.9 - 8.7
Smectite	0.0	0:0	13.1 -25.1
Nonclay/detrital	31.3	24.0	14.6 -32.9
Total carbonates (wt%)	50.5	0.0	56.5 -19.1
Moisture content (wt%)	0.58	NA	18.4
Mechanical properties‡			
Ambient temperature			
C_0, MPa	94.8 ± 12.1	NA	7.2 ± 0.3
E (GPa)	10.4 ± 2.3	NA	0.6 ± 0.0
ν	0.31 ± 0.1	NA	0.12 ± 0.02
150°C			
C_0, MPa	24.7 ± 2.5	NA	NA
E, GPa	0.8 ± 0.3	NA	NA
ν	NA	NA	NA

*Upper Huron Formation shale data not available.
†Values ± standard deviations. Data are from Hansen and Vogt (1987).
‡C_0 = unconfined compressive strength, E = Young's modulus, ν = Poisson's ratio, NA = not available. 1 MPa = 145 psi, 1 GPa = 145,000 psi.

results, ultimate strength, mechanisms of deformation, and long-term creep. That study resulted in the well known power-law pillar creep formula:

$$\varepsilon = 1.30 \times 10^{-37}\, T^{9.5}\, \sigma^{3.0}\, t^{0.3} \qquad (1)$$

where ε is the pillar shortening of a model pillar initially 25 mm (1 in.) high, T is absolute temperature (K), σ is the initial average pillar stress (psi) (1 MPa = 145 psi), and t is time in hours. Equation (1) holds for cylindrical model pillars with a width-to-height ratio of 4 fabricated from salt core from Lyons, Kansas.

Most of the test data used in developing Eq. (1) were obtained during tests of less than 1000 hr duration. The longest testing time reported by Lomenick and Bradshaw (1969b) was slightly less than 30,000 hr at room temperature and indicated that the deformation rate was still declining, which was consistent with observations in mine openings in the Lyons, Kansas salt mine. Longer term model pillar results have been analyzed by Russell and Lomenick (1984).

Description of Tests

To simulate pillar, roof, and floor conditions that would exist in mined cavities in shales, sample specimens in an air-dry condi-

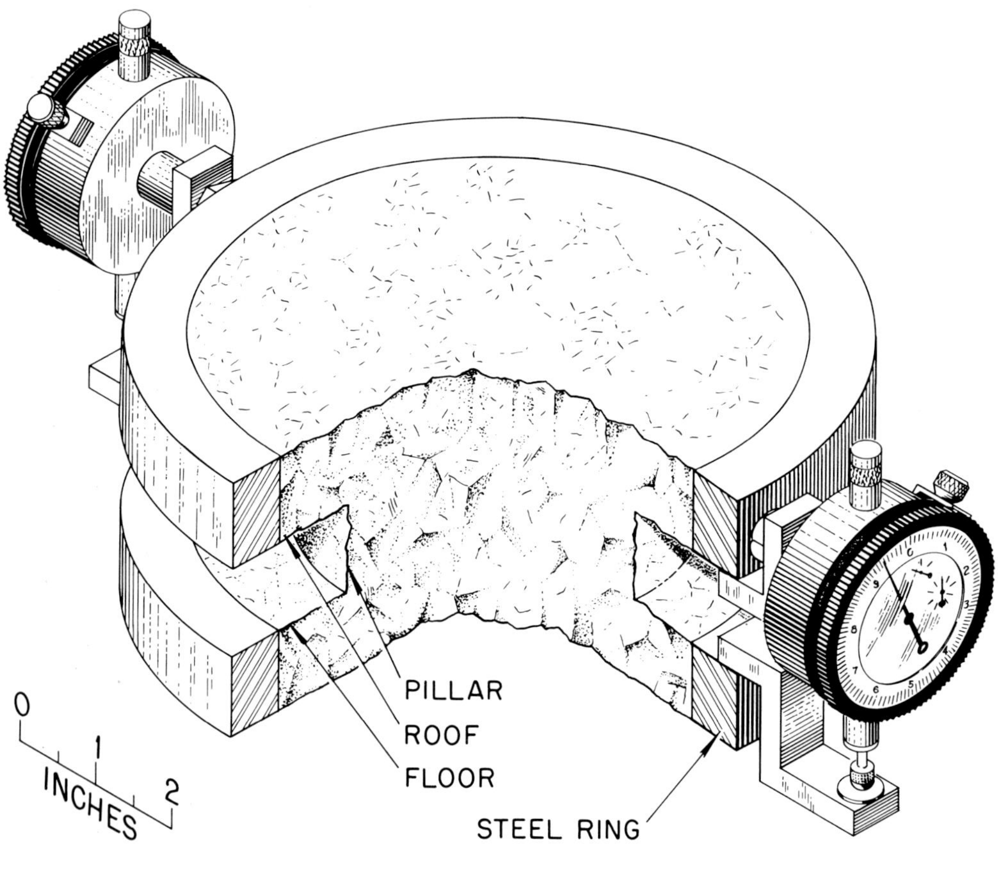

Figure 48.1. Pillar model illustrating geometry, placements of steel rings, and displacement measuring gauges.

tion are fabricated to represent scale models of shale pillars and their surrounding rooms. The test specimens used in this work are cylindrical, with a portion of the center ground out to form the pillar and surrounding rooms. By "epoxying" steel rings around the ends of the samples, effective confining pressure is applied to the roof and floor portions of the models when they are loaded (see Fig. 48.1). Constant uniaxial loads are applied to the models by hydraulic compression testers having capacities up to 1.3 MN (300,000 lb). Cavity closure is measured by mounting two dial gauges 180° apart on the rings. Elevated temperature tests are performed with the specimens inside a cylindrical heating jacket and with barrier heaters on top and bottom between the specimen and the platens. No barriers are present to prevent moisture loss during the test.

The short-term tests reported here are performed on model pillars fabricated from cores of shale with dimensions given in Table 48.2. Refer to Table 48.3 for sample identification, depth, and test conditions. For model preparation, all cores were first cut to their approximate desired length by band saw and then finished by machine sanding. Next, steel confining rings (ASTM A-108, Grade 1018) were affixed to the tops and bottoms of the cores using Ciba Araldite 502 epoxy and Ciba hardener 951. For samples heated up to 100°C, Ciba Araldite 6005 epoxy and Pyrometallitic Dianhydride (PMDA) hardener are used. The unconfined center portion of the sample is then ground out, using a sanding disc attached to a standard drill press, to form the size pillar desired. Sheet teflon greased with a mixture of silicon grease and graphite is inserted between the tops and bottoms of the samples and the platens of the compression machines to reduce the friction between the rock and the platens to an insignificant amount. Chromel-Alumel thermocouples are emplaced in 1.5-mm (0.06-in.)-diameter holes drilled into the center of the pillar, as well as at the outer edge to ensure a uniform temperature within the heated model specimens. A 12-point strip-chart recorder with a 0–350°C range is used to trace the temperature curves. A constant voltage transformer is used in supplying 118 V ac to adjustable autotransformers that are used as temperature controls for the cylindrical heating jackets, and the heaters in the platens at the top and bottom of the sample. In this manner, temperatures are regulated to within 1–2°C. In these scoping tests, no attempt has been made to determine the uniformity of temperature.

In general, the same test procedure is used for all specimens to ensure consistency. The procedure consists of rapidly applying a compressive load corresponding to 75% of the testing load and releasing it, then applying a load of 85% of the test load, which is also immediately released, and then applying 100% of the load. A base reading for the dial gauges is then taken. These initial loadings serve to seat the dial gauges, used to measure deformation, and to set the platens firmly on the sample. The deformation is recorded, first at intervals of seconds, then of minutes, and then at progressively longer intervals.

Table 48.2. Specimen and pillar dimensions.

	Specimen		Pillar		
Rock type	Diam. (in.) (mm)	Length (in.) (mm)	Width (in.) (mm)	Height (in.) (mm)	Width/height
Green River Formation	6 (152)	5 (127)	4 (102)	1 (25)	4
Conasauga Formation	2.4 (61)	4.4 (112)	1.6 (41)	0.4 (10)	4
Pierre Shale	3.5 (89)	4.6 (117)	2.4 (61)	0.6 (15)	4
Huron Shale	3.5 (89)	4.6 (117)	2.4 (61)	0.6 (15)	4

Table 48.3. Model pillar test matrix.

Rock type	Sample ID	Sample depth ft (m)	Av pillar stress MPa (psi)	Temperature (°C)	Start date	Duration (hr)
Green River Formation	CR/86/	470(143)	28(4000)	22.5	1-13-86	1029
			42(6000)	22.5	3-3-86	1005
Conasauga Group	C/86/	725(221)	28(4000)	22.5	4-21-86	1055
			41(6000)	22.5	4-10-86	1055
			69*(10,000)	22.5	5-28-86	1655
			28(4000)	100	5-12-86	2015
Pierre Shale	P/84/	216(66)	7(1000)	22.5	2-26-86	1010
			14(2000)	22.5	1-16-86	1010
			7(1000)	60	1-20-86	1012
Huron Shale	H/84/	515(157)	28(4000)	22.5	6-20-85	4059
		516(157)	41(6000)	22.5	7-8-85	4198
		523(159)	69(10,000)	22.5	8-13-85	15,952
		528(161)	42(6000)	100	7-10-85	2470
			28(4000)	100	3-18-86	1008

*Model pillar data not used in empirical model development because of apparent material failue. All model pillars have a width-to-height ratio of 4.

Table 48.4. Comparison of shortening data for two specimens.

Load		Pillar shortening (%)	
(MPa)	(psi)	Model shale pillar with roof and floor	Shale pillar without roof and floor
14	2,000	0.03	0.3
28	4,000	0.32	3.1
41	6,000	0.6	8.5
55	8,000	1.4	Catastrophic failure
69	10,000	2.7	
83	12,000	8.9	
97	14,000	12.1	
110	16,000	16.5	
124	18,000	21.3	
138	20,000	29.6	

Simulation of Mine Pillars and Model Scaling

The deformation characteristics of rock are extremely important considerations in the simulation of mine conditions in scale-model tests. To determine the stability of mine pillars composed of plastic materials like shale, it is absolutely necessary to simulate in the models, not only the mine pillar, but also the roof and floor conditions. The importance of the simulated roof and floor in the model pillars of shale can be illustrated by comparing test data from a model pillar with data obtained from testing a cylindrical specimen (without roof and floor constraint). In both cases, specimens were fabricated from 3.44-in.-diameter (87.5 mm) cores of shale. For the model pillar specimens the center portion of the specimen was ground out to form the pillar [0.6 in. (15 mm) high × 2.4 in. diameter (60 mm)] and the surrounding roof and floor. Steel rings were then "epoxied" to the top and bottom portion of the samples to restrain laterally the rock in the roof and floor (as would be the case in actual mine workings). On the other hand, for the specimens without roof and floor simulation, a cylindical sample of shale having a diameter of 3.44 in. (87.5 mm) and a height of 0.6 in. (15 mm) (pillar only) was used for the test.

Pillar shortening data for the two specimens are given in Table 48.4. In both cases the rate of loading was 2.3 MPa (333 psi)/min. For shale samples tested at room temperatures, it was found that the model pillar specimen deformed nonlinearly with increasing stress; however, even after loading to 138 MPa (20,000 psi) the model continued to creep without catastrophic failure (pillar shortening reached about 30%). In comparison, the cylindrical specimen without roof and floor failed suddenly at 55 MPa (8000 psi). These data indicate that actual mine conditions are simulated best when the roof and floor portions of the models are constrained laterally. This condition allows shear stress at the roof and floor to be transmitted into the pillar (at least for samples having a width-to-height ratio of 4), thereby increasing the mean stress in the pillar interior and decreasing the effective stress, which makes the pillar stronger.

Model pillar tests on rock salt were first presented by Obert (1964). The essential feature of these model pillars is the simulation of the role of the roof and floor material and stiffness in the axial creep of a pillar, which is not accomplished by the common compression test. In the usual unconfined uniaxial compression test, an attempt is made to make the stress state as uniform as possible so that strength and/or deformation properties may be inferred as a function of stress. Such uniaxial tests are not designed to simulate mine pillars because they ignore roof–pillar and floor–pillar interactions. Length-to-diameter ratios normally range from 2 to 2.5 to isolate the center portion of the sample from end effects. These uniaxial tests do not, by themselves, simulate pillar behavior; but these data may be used in conjunction with confined triaxial test results to develop a constitutive model that can be used, with a computer code employing continuum mechanics principles, to simulate pillar behavior numerically. Model pillar data may be used as a first step in validation of the constitutive model and computer code.

In reviewing the model pillar tests and results, it should be kept clearly in mind that these are tests on models of a major structural component of the ground support system in a repository and are *not* materials tests per se. However, it is possible to infer parameters for creep laws from model pillar data as was done by Thomas (1973) and later by Wahi et al. (1977).

Another point to keep in mind about the model pillar tests is that they are originally designed for test durations of the order of months, not years. Furthermore, because of the inherent specimen-to-specimen variability in rocks, in a scoping study it is desirable to test many samples with less accuracy of measurement, rather than a few samples with greater accuracy, to capture the essential (first-order) changes in response caused by geometry, load, and temperature.

As with other tyeps of structural modeling, the laws of similitude between the model and the prototype must be satisfied if the model is to exhibit the desired response. In reality, it is almost impossible to satisfy all the laws of similitude, and we are forced to attempt to simulate the desired response by maintaining similitude in the dominant variables. In this report, similitude is discussed as it relates to geometry, load, and temperature. For a more complex discussion of similitude as it relates to geology and structures in rock, the reader is referred to Hubber (1937) and Obert and Duvall (1967).

A model is geometrically similar to a prototype if the ratios of all length dimensions in the model to the same model-to-length dimensions in the prototype are identical. If this is the case, all dimensions relative to a characteristic length in the model will be the same as in the prototype. In mine pillars, it is well known that the pillar width-to-height ratio is of primary importance; consequently, pillar height is chosen as the characteristic length. Salt mine pillars with width-to-height ratios less than unity have been known to fail in a brittle mode, while those with a ratio greater than 2 tend to flow rather than fracture (Obert and Duvall, 1967), although pillar spalling is observed if pillar shortening is large. Some shale pillars may exhibit similar behavior.

Lomenick and Bradshaw (1969a) reported on model pillar tests with width-to-height ratios from 2 to 4. All results presented herein are for a ratio of 4:1 which is representative of preliminary repository designs. Future work may consider other ratios and/or geometries.

As previously noted, one of the primary features of the model-pillar test is that it incorporates the influence of the roof and floor material and stiffness. In this case, the choice of a dimensionless ratio similar to the prototype is not as clear because the roof rocks extend to the surface and the floor rocks at least through the earth's crust. Thus, maintaining a similar ratio of depth to pillar height, for example, is realistically impossible. Lomenick (1969a) reports on a study of this question and concludes that maintaining a roof thickness-to-pillar height ratio of 2:1 is adequate to give reproducible results. The same ratio is maintained for the floor thickness-to-pillar height. This ratio is consistent with elasticity theory where one would expect the influence of stress concentrations caused by pillar-to-roof and floor intersections to be negligible a few characteristic lengths away.

The ratio of the annular width of the roof and floor to the pillar height has been maintained at ~1:1 for the tests reported here and, in this range, should not be a particularly significant parameter as far as pillar response is concerned, i.e., one would not expect much different pillar shortening response if this ratio were 1.3:1, provided that steel rings of adequate stiffness were provided. Note that these tests do not simulate roof and floor response. Lomenick and Bradshaw (1969a) measured the tangential strain in the steel roof rings to estimate the amount of confinement provided by these rings to the salt roof and floor. Similar measurements have not been made for model pillars of shale but are being considered for future tests.

The radius of curvature between the pillar vertical surface and the horizontal roof and floor surfaces has a significant effect on the elastic stress concentration in the immediate neighborhood of the intersection. The importance of this parameter has not been studied. If the radius is smaller for a particular model, plastic flow and stress redistribution will occur at lower values of vertical stress. Of course, the average vertical pillar stress remains the same and is the parameter used in the empirical pillar creep law.

If actual mine pillars are considered, the pillar-to-roof or floor radius ratio depends on the excavation method and may be somewhat variable for mines using a drill-blast-load-haul system. This radius is much more uniform if a borer-type continuous miner is used. In this case, the radius may be approximately 0.3 m, and assuming two passes of the miner, the pillar height may be about 6 m leading to a radius-to-height ratio of 0.3:6 = 0.05. For a model pillar 25 mm high, this would correspond to a radius of 1.25 mm. In the scoping study reported herein, no attempt is made to machine the specimens to a particular radius.

With respect to grain size, American Society for Testing and Materials (ASTM) standards for compression tests recommend a ratio of sample diameter to average grain size of 10 to provide a statistically representative cross section to help minimize sample-to-sample variability. Because of the dominance of clay size particles in shale rocks, this is not a matter of great concern for shales.

The most obvious geometric deviation of the model pillars from actual mine pillars is the circular horizontal cross section of the model compared with the likely rectangular cross section of proposed repository pillars. The round models would probably simulate a square cross-sectional pillar better than a rectangular one with agreement becoming worse as the cross section becomes more oblong. Bradshaw and McClain (1971) note that the pillar shape and size can probably be accounted for in the leading coefficient in the power law creep equation with the exponents remaining approximately the same as in the model pillars. Table 48.2 lists overall and model pillar geometries for each of the four shales considered here.

If a pillar near the center of a repository with lateral repository dimensions greater than the depth (i.e., wider than critical width) is considered, the average pillar stress can be estimated from the following tributary area equation:

$$\sigma_{av} = \gamma h/(1 - R) \tag{2}$$

where σ_{av} is the average pillar stress, γ is the weighted average specific height of the overlying rocks, h is the depth from the surface, and R is the areal extraction ratio (area removed/total area). Equation (2) holds regardless of temperature (assuming that it is also widely distributed and no arching can take place. The average pillar stress given by Eq. (2) is assumed to be that applied on the model pillar because the model material is the same as the prototype material. The total load applied to the model pillar is σ_{av} times the cross-sectional area of the model pillar. The total load simulates the weight of the rocks overlying the pillar and the area midway to the surrounding pillars. This total load remains constant during the life of the prototype and consequently is held constant during the pillar test even though the model pillar increases in diameter as it deforms and the actual average pillar stress decreases.

In view of the above considerations and subject to the assumptions made, the model pillar test maintains similitude with respect to load to the degree that the load is held constant over the duration of the test.

Similitude with respect to temperature is maintained to the degree that the temperature can be maintained uniform in time and space. In the case of the repository, the temperatures are not constant in either time or space. Spatial gradients will exist throughout the repository because of the heat production of the waste in the canisters to be placed in holes in the floor of the rooms. Temperature gradients will depend on design parameters such as (1) the thermal power per canister, (2) spacing and pitch of canisters, (3) room size, (4) pillar dimensions, and (5) heat conduction properties of the shale and overlying and underlying rocks. Temporal changes in temperature in a repository are expected because of the decaying nature of the thermal sources.

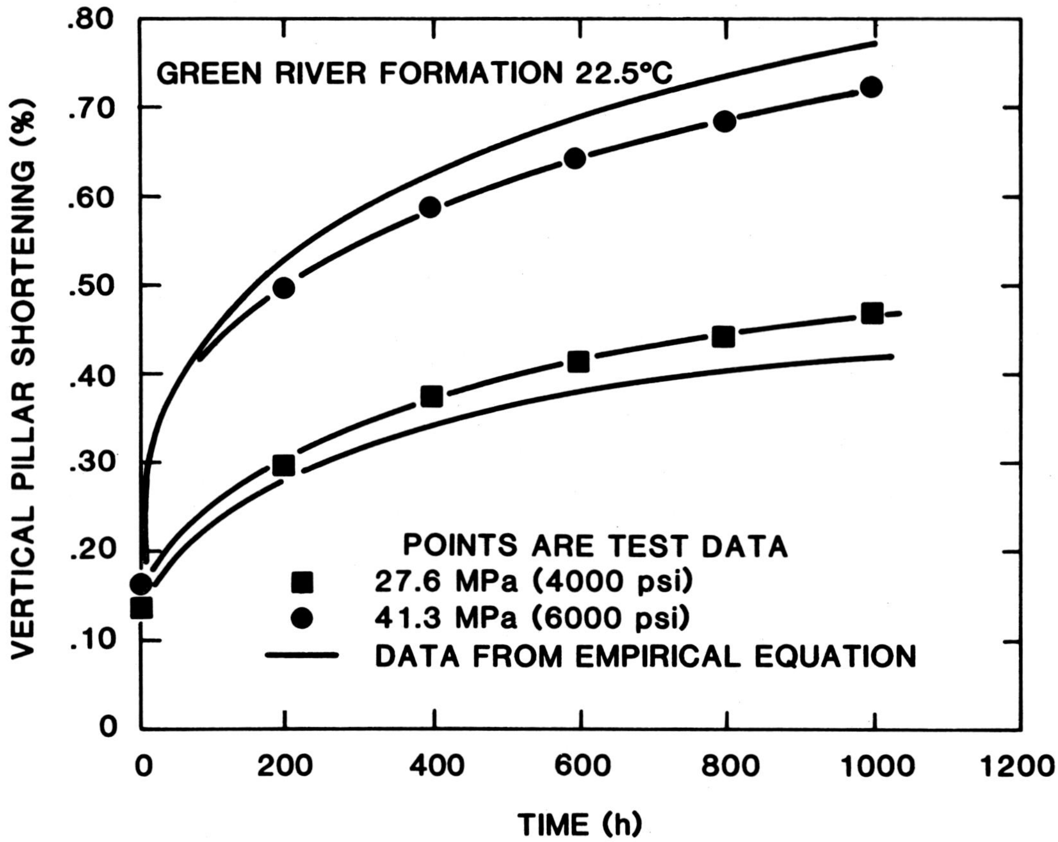

Figure 48.2. Pillar shortening for Green River Formation at 22.5°C.

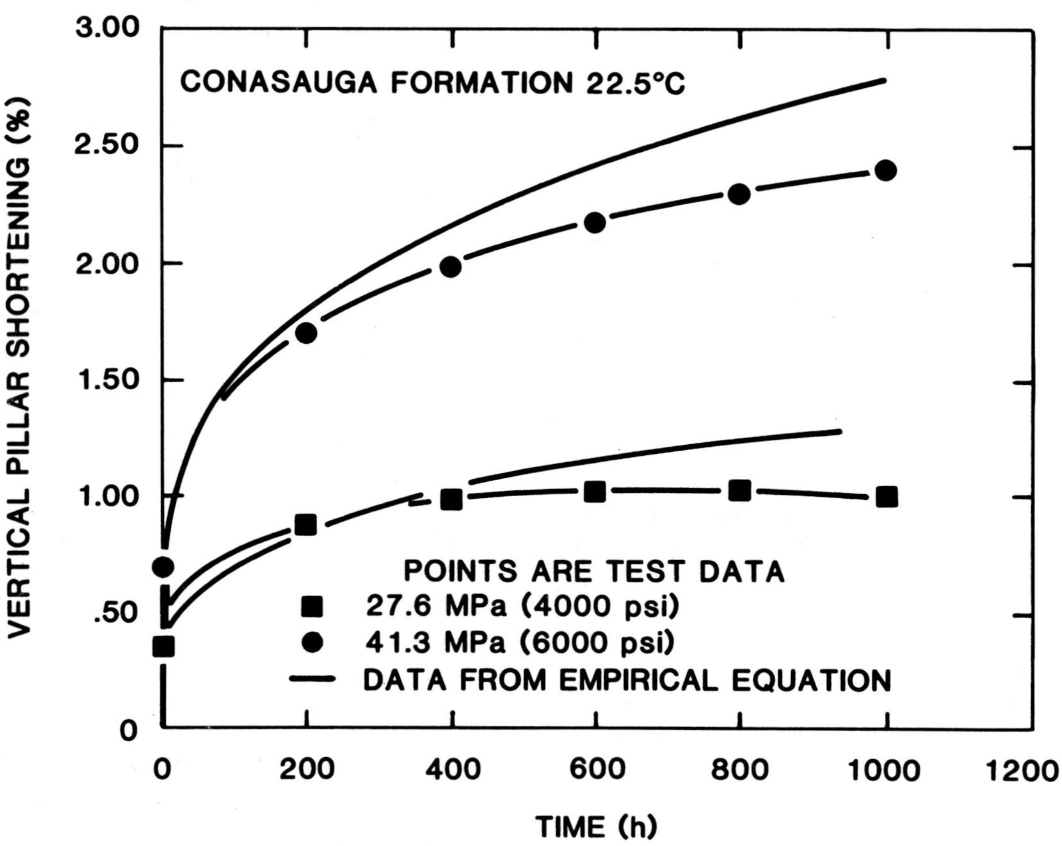

Figure 48.3. Pillar shortening for Conasauga Formation at 22.5°C.

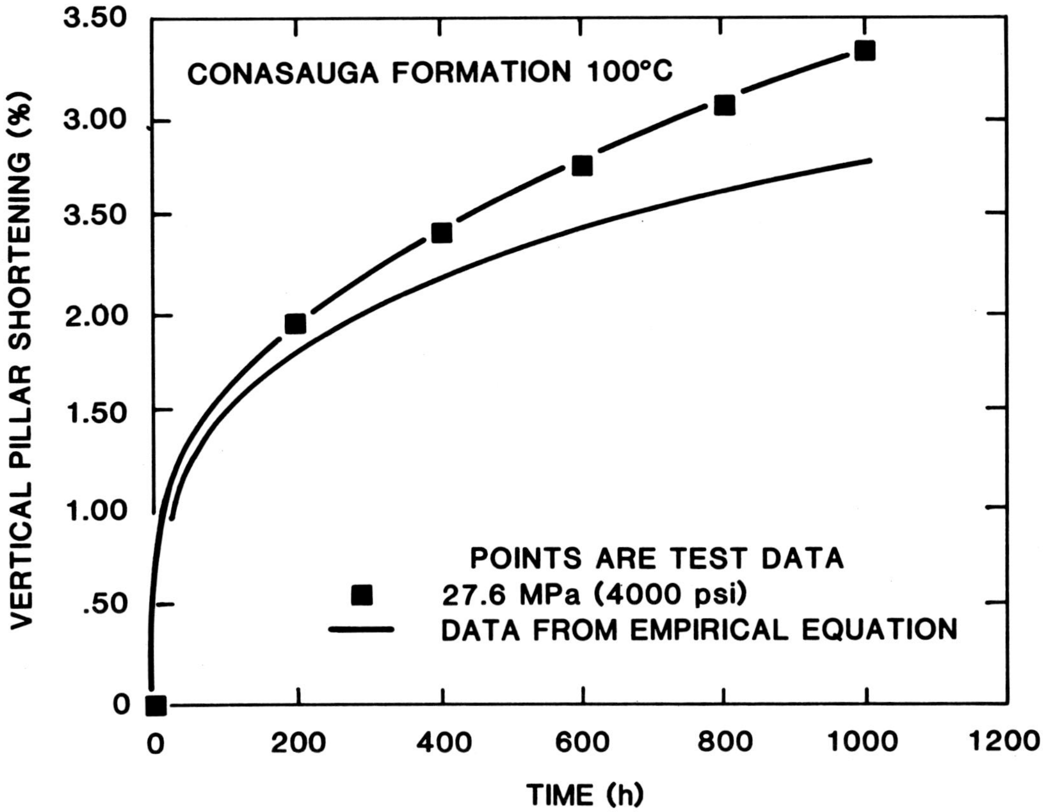

Figure 48.4. Pillar shortening for Conasauga Formation at 100°C.

Heated-model pillar tests are not designed to simulate the exact response of a prototype pillar, but rather to provide information, such as the increase or decrease in pillar-shortening rate to be expected if the average pillar temperature is increased or decreased. Drying produces changes in thermal properties of shale that changes the time-dependent thermal response. Drying-induced changes in the transient thermal response are not exactly simulated by model pillars. Nevertheless, model pillar data are expected to be useful in validating the constitutive models and computer codes.

Salt mine pillars in areas of high extraction exhibit pillar slabbing or spalling. In such cases, the pillar may take on an approximate hourglass shape formed by conjugate sets of shear fractures. Model pillars show similar tendencies, except that, in the case of salt, the material generally spalls off as individual grains. Lomenick and Bradshaw (1969a) presented photographs showing a salt model pillar and spalled material.

Another phenomenon sometimes observed in mine pillars is punching; i.e., a stiff and strong pillar penetrates into a soft and weak floor and/or roof. Punching into the roof normally causes roof control problems that can sometimes be corrected by reducing the pillar size, and therefore stiffness, even though the extraction ratio may be held the same. The extent to which model pillars of salt punch into the simulated floor and roof is discussed by Lomenick and Bradshaw (1969a) who presented a photograph of the vertical cross section of a previously tested model pillar showing this phenomena. Similar results are expected at least for the softer shales.

Test Results

Table 48.3 presents the shale model pillar test matrix for the model pillar geometries given in Table 48.2. Raw data from these tests were entered into a microcomputer for data reduction and analysis. Data reduction consisted of converting the dial gauge readings to incremental changes in pillar height and accumulating these changes. The cumulative displacements read from dial gauges 3 and 4 (gauges 1 and 2 are backup gauges) were averaged for each time and the average converted into nondimensional shortening by dividing by the initial pillar height, H, given in Table 48.2. Results for 13 of the 14 tests discussed are shown graphically in Figures 48.2 to 48.9. Figures 48.2 to 48.9 indicate that most of the pillar creep response curves appear to have the form referred to as primary creep, i.e., the creep rate continues to slow down with time. Consequently, a transient creep response function of the form used for rock salt, Eq. (1), has been fitted to the shale data.

The empirical model chosen for preliminary analysis has the form

$$\varepsilon = A\sigma^n T^u t^m \qquad (3)$$

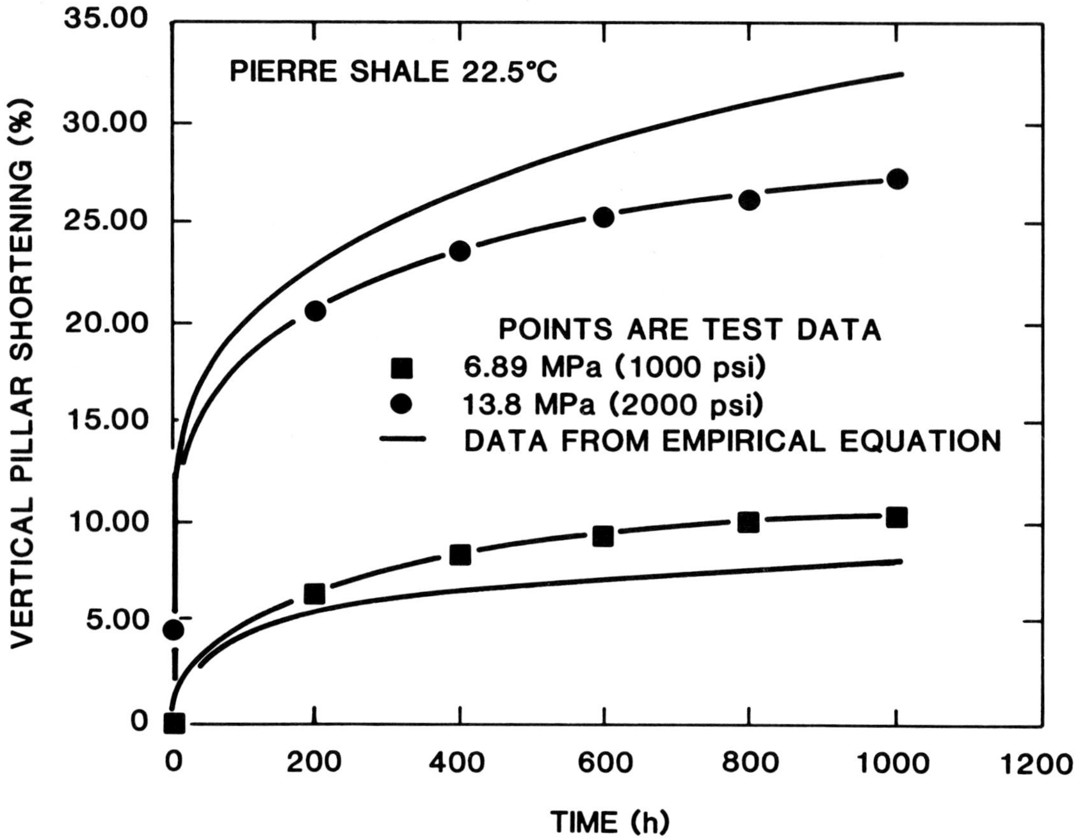

Figure 48.5. Pillar shortening for Pierre Shale at 22.5°C.

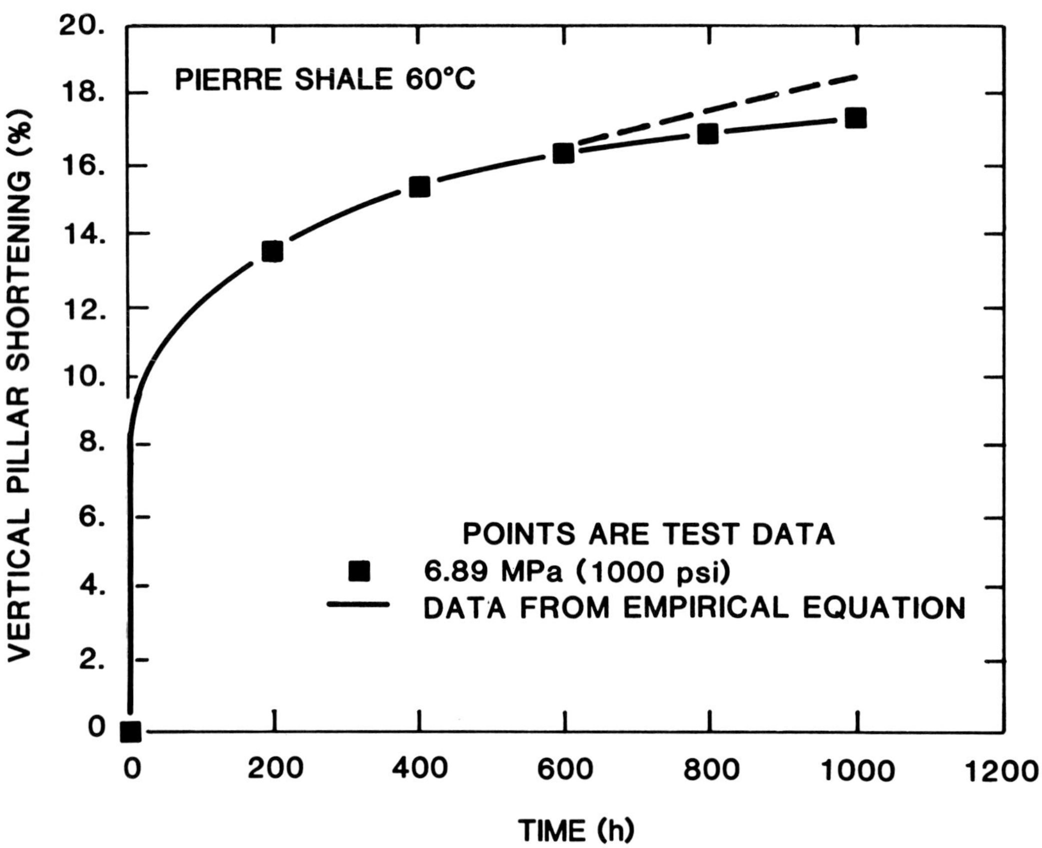

Figure 48.6. Pillar shortening for Pierre Shale at 60°C.

Figure 48.7. Pillar shortening for Huron Shale at 27.6 and 41.3 MPa.

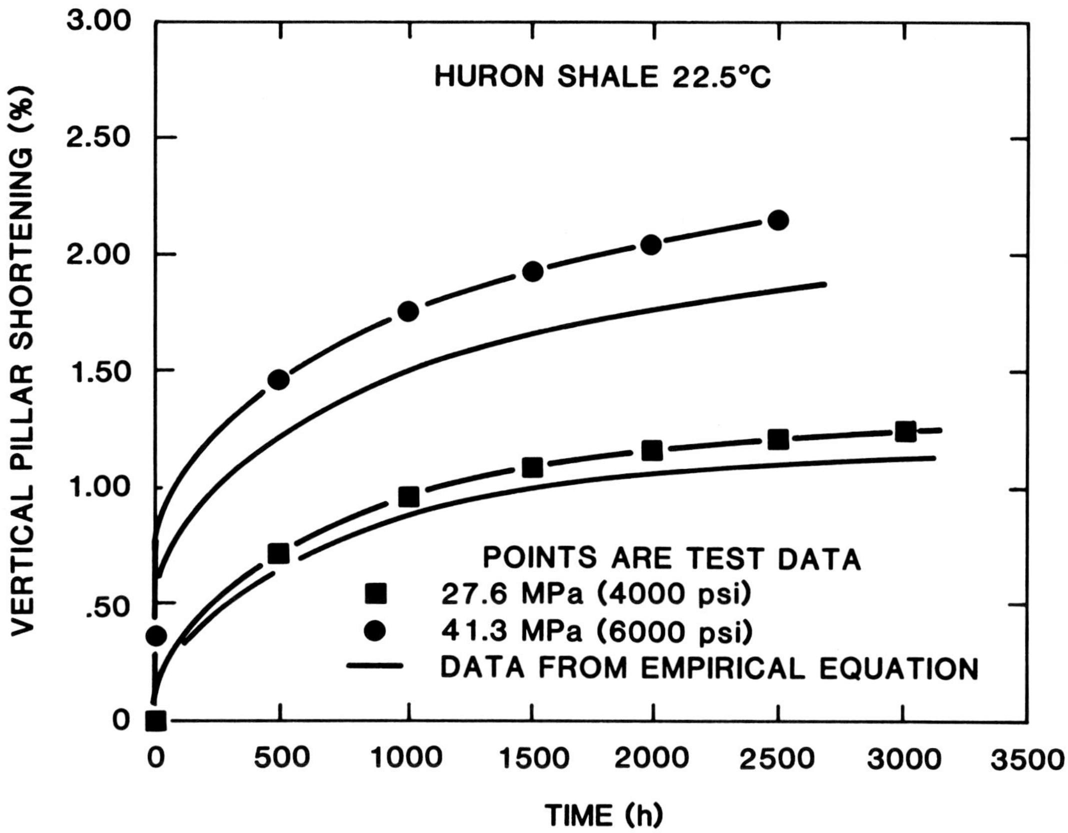

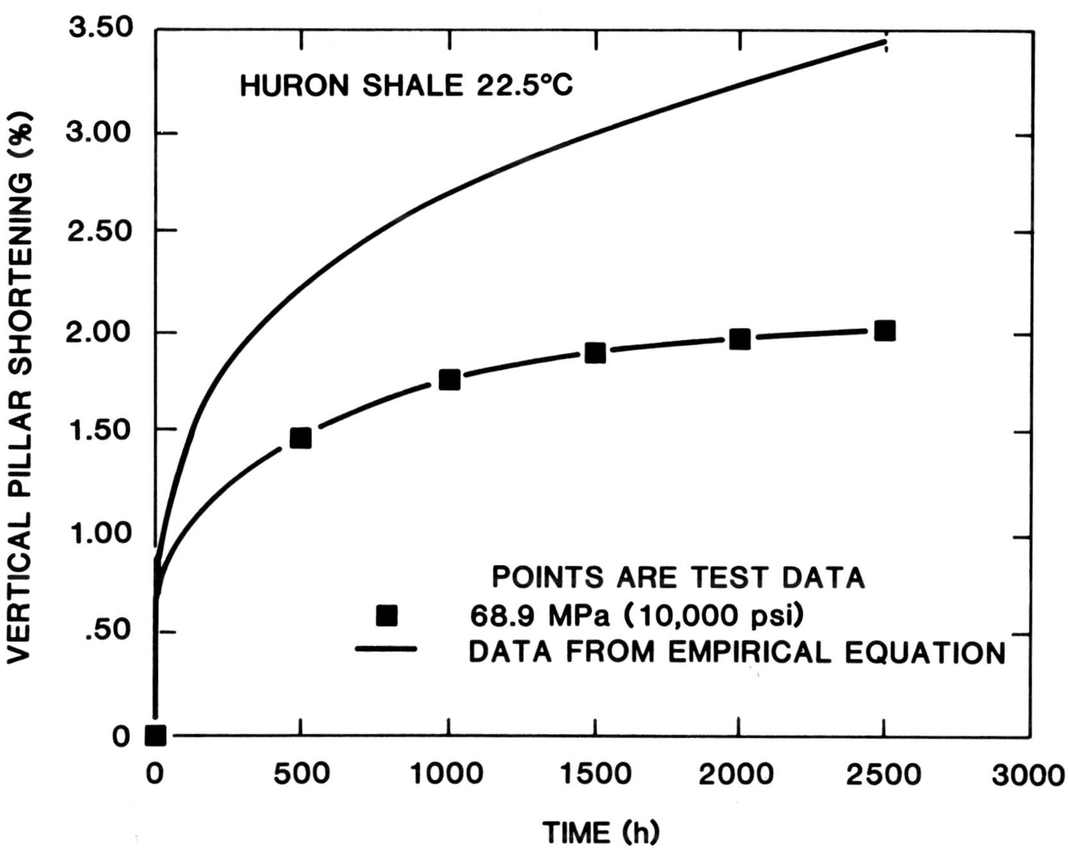

Figure 48.8. Pillar shortening for Huron Shale at 22.5°C.

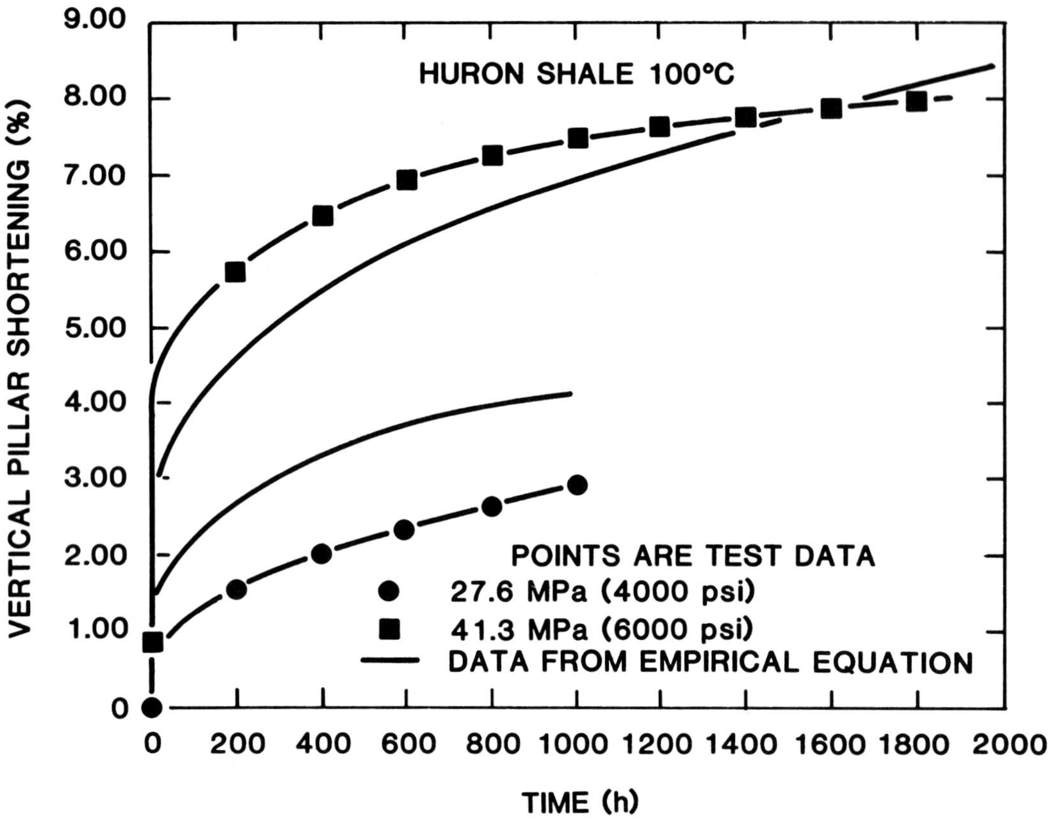

Figure 48.9. Pillar shortening for Huron Shale at 100°C.

where ε is the fractional pillar shortening (which is dimensionless), σ is the initial average pillar stress in psi (1 MPa = 145 psi), T is the absolute temperature in K, and t is the time in hours. The parameters (A, n, u, and m) used in Eq. (3) may be found by either (1) a nonlinear curve fitting method, such as the path of steepest descent, or (2) by first linearizing the equation (by taking logarithms of both sides) and determining the parameters by multiple linear regression. Because of the scoping nature of this study, the second approach was chosen to provide preliminary values for the parameters.

The logarithms of Eq. (3) becomes

$$\log(\varepsilon) = \log(A) + n \log(\sigma) + u \log(T) + m \log(t) \quad (4)$$

which is suitable for multiple linear regression.

The parameters for the model of Eq. (3) have been estimated for each of the four shales tested in this study and are shown in Table 48.5 along with the applicable ranges of initial stress, temperature, and time. Note that these parameters hold for the geometries given in Table 48.2, particularly for $W/H = 4$. Also, these parameters are limited to the shales described in Table 48.1. These parameters are likely to be sensitive to changes in shale composition (particularly the amounts of smectite and other clays, quartz, feldspar, and kerogen for the Green River Formation), moisture content, and porosity. In view of these limitations and the small size of the data base used, parameter values given in Table 48.5 must be considered very preliminary.

Note that the time range of applicability of some of the tests, as given in Table 48.5, is somewhat less than the duration of the corresponding tests shown in Table 48.3. This difference is due to irregularities in the data caused by power outages and pressure leaks.

The room temperature, 69 MPa (10,000 psi) test on the Conasauga Formation was not used in the data base because initial fractures developed on the pillar surface during loadup and because its behavior was considerably different from the remaining Conasauga tests. The room temperature test at 14 MPa (2000 psi) on the Pierre Shale also caused spalling on the pillar surface during loadup. We note that the unconfined compressive strength of the Pierre Shale from Table 48.1 is 7.2 MPa or about 1044 psi; therefore, it is not surprising that localized failures occur when the average pillar stress is 14 MPa (2000 psi). In this case, the 2000 psi test was left in the data base, and the results shown on Figures 48.5 and 48.6 indicate that the empirical model does a relatively good job of tracking the data.

The room temperature test at 69 MPa (10,000 psi) for the Huron Shale (Fig. 48.8) was left in that data base, but Figure 48.9 indicates that its behavior is not consistent with the other Huron Shale results. If these test data were eliminated from the data base, it is likely that the model parameters would change slightly, and the fits to the remaining data would improve. In this case, initial fractures were not reported during loadup, and we have no data on the unconfined compressive strength of the Huron Shale.

Table 48.5. Parameters for empircal model.*

Rock type	A (psi)$^{-n}$(K)$^{-u}$(h)$^{-m}$	n	u	m	Stress range (psi)†	Temperature range (°C)	Time range (hr)
Green River Formation	3.2669×10^{-9} ‡	1.5013	0	0.23338	4000–6000	22.5	0–1000
Conasauga Group	2.7739×10^{-18}	1.8796	3.2760	0.26878	4000–6000	22.5–100	0–1000
Pierre Shale	3.4719×10^{-25}	1.9902	6.7885	0.20995	1000–2000	22.5–60	0–1000
Huron Shale	2.2630×10^{-25}	1.2906	6.9335	0.26199	1000–4000	22.5–100	0–3145

*$\varepsilon = a\sigma^n T^u t^m$ where ε is dimensionless, σ is in psi, T is in K, and t is in hours. Stress, temperature, and time ranges of applicability are given.

†1 MPa = 145 psi.

‡Five digits are shown for the parameters in the empirical models to provide consistent results for shortening, because rounding of parameters may introduce inconsistencies in model results. Model values for shortening should be reported to three significant digits.

The temperature exponent given in Table 48.5 for the Green River Formation is zero because no data are available at elevated temperatures.

The only statistics calculated for the comparisons between the empirical model and the pillar shortening data (shown on Figs. 48.2 to 48.9) are the root mean square (RMS) differences which are given in Table 48.6. Note that these RMS values are relative to the magnitude of the shortening for each test and cannot be directly compared unless they are normalized by dividing by the maximum fractional shortening for each test. In general, small RMS differences imply good fits to the data.

Several general observations can be made about the response of the shale model pillars. In general, the transient response of shale model pillars is similar to that of model pillars of rock salt (Lomenick and Bradshaw, 1969, 1986). In both cases, higher stress and/or temperature lead to higher displacement rates and more cumulative displacement. However, as can be seen from Table 48.5, the stress exponents for the shales tested tend to be significantly lower than those for rock salt, as shown in Eq. (1). The Pierre and Conasauga samples show the highest stress sensitivity, with exponents of about 2 and 1.9, while the Huron, with an exponent of about 1.3, shows the least stress sensitivity to shortening.

The temperature sensitivity of the Pierre and the Huron Shales are about the same with exponents of about 6.8 and 6.9, compared with exponents of about 3.3 for the Conasauga Group Shale and 9.5 for rock salt. Consequently, the temperature sensitivity of transient creep response of the shales tested was less than rock salt but is still very significant. Table 48.1 lists both ambient and 150°C strength data for the shales from the Green River Formation (Hansen and Vogt, 1987). The higher temperature apparently weakens the Green River Formation Shales dramatically. Unfortunately, we have no model pillar data at elevated temperatures for the Green River Formation Shales to use for comparison.

Time exponents for the four shales show a remarkable consistency, ranging from a low of about 0.21 for the Pierre to a high of about 0.27 for the Conasauga. These values compare with a nominal value of 0.3 for rock salt. The lower value of the average time exponent for shale indicates that the creep rate for shale is likely to decrease faster than that for rock salt. In fact, for some cases, notably the Conasauga Group Shale at 22.5°C and 28 MPa (4000 psi), shown on Figure 48.3, the convergence rate is about zero for the last several hundred hours of the test. A similar situation is observed for the Huron Shale at 100°C and 41 MPa (6000 psi), shown on Figure 48.9, and the room temperature Huron Shale at 41 MPa (4000 psi), shown on Figure 48.7. This behavior indicates that the transient creep model used in this study and previously used in the study of closure of model pillars of salt may not be appropriate for the time dependent

Table 48.6. Root Mean Square (RMS) differences between empirical model and test data.

Rock type	Stress [MPa (psi)]	Temperature (°C)	RMS difference in shortening	Approx. maximum short, ε_{max}	$(RMS/\varepsilon_{max}) \times 100$ (%)
Green River Formation	28(4000)	22.5	2.06×10^{-4}	4.5×10^{-3}	4.6
	41(6000)	22.5	4.13×10^{-3}	7.2×10^{-3}	5.7
Conasauga Formation	28(4000)	22.5	1.17×10^{-3}	1.0×10^{-2}	11.7
	41(6000)	22.5	2.61×10^{-3}	2.4×10^{-2}	10.9
	28(4000)	100	2.20×10^{-3}	3.5×10^{-2}	6.5
Pierre Shale	7(1000)	22.5	1.14×10^{-2}	1.0×10^{-1}	11.4
	14(2000)	25.5	3.26×10^{-2}	2.7×10^{-1}	12.1
	7(1000)	60	4.21×10^{-3}	1.7×10^{-1}	2.5
Huron Shale	28(4000)	22.5	5.50×10^{-4}	1.2×10^{-2}	4.6
	41(6000)	22.5	2.63×10^{-3}	2.2×10^{-2}	12.0
	69(10,000)	22.5	6.37×10^{-3}	2.0×10^{-2}	31.9
	41(6000)	100	8.93×10^{-3}	8.0×10^{-2}	11.2
	28(4000)	100	1.01×10^{-2}	2.8×10^{-2}	36.1

Table 48.7. Comparison of approximate dimensionless pillar shortening at 22.5°C and 1000 hr for two stresses.

Rock type	29 MPa (4000 psi)	41 MPa (6000 psi)
Conasauga Group Shale	0.010	0.024
Huron Shale	0.008	0.017
Green River Formation Shale	0.0046	0.0072

response of model pillars of shale. A more appropriate time function for the empirical model may be

$$1 - \exp(t/t_c) \qquad (5)$$

where t_c is the time required for the pillar shortening, ε, to approach its final value. It is interesting to note that Stage I of the triaxial creep test reported by Hansen and Vogt (1987) (Fig. E-1, p. 79) apparently exhibits this same phenomenon, while stages II and III at higher temperature and stress difference, respectively, indicate continued straining with time. Hansen and Vogt's unconfined creep test (Fig. E-2, p. 80), however, again indicates the approach of a terminal strain at room temperature. At this time, we have no firm understanding of the microscopic creep mechanisms in these shales and, consequently, can come to no firm conclusions, especially in view of the small data base in the scoping study.

The shales tested during the scoping study may be ranked according to how much their model pillars shortened at a constant stress, temperature, and time. Pierre Shale model pillars were tested at lower stresses than the other shales, but they deformed more. Consequently, the Pierre Shale is the most easily deformed of the group tested. The remaining three shales were tested at 28 MPa (4000 psi) and 41 MPa (6000 psi), which allows direct comparison at 22.5°C and 1000 hr, as shown on Table 48.7. Note that for the above conditions, the Conasauga Group Shale pillars deformed the most, followed closely by the Huron Shale and the Green River Formation Shale, which is much less deformable for the stated conditions.

Conclusions

Based on model pillar tests of four different shales, we conclude that model pillars of these shales exhibit time-dependent closure and that the rate of closure is increased by both increases in the initial pillar stress and the test temperature. The power-law transient creep equation has been used as the empirical pillar creep model for these shales, and the comparisons between empirical model results and laboratory data are reasonably good for the range (stress, temperature, and time) of applicability of the empirical model. We have compared the deformational response of the shale model pillars with that of model pillars of rock salt and found both similarities and differences: (1) salt pillars appear to be more sensitive to both temperature and stress than shale pillars; and (2) some shale pillar data suggest that after an initial transient phase, the pillar shortening approaches a steady value. When more data become available, a quantitative comparison will be possible.

The scoping study on model pillar shortening provides preliminary information useful in investigations for a national survey of potential shales to host a nuclear waste repository. The weaker, more deformable Pierre Shale would probably be most useful where strata of sufficient thickness occur relatively close to the surface where overburden stresses and temperatures are likely to be lower. On the other hand, repository depths for the other shale types considered, could vary over a much wider range. The scoping data are too few to set any limits at this time. For example, at 22.5°C, the Green River Formation Shale appears to be the strongest and least deformable of the shales considered herein; however, Hansen and Vogt (1987) report a decrease in strength from 94.8 MPa (13,700 psi) at room temperature to 24.7 MPa (3580 psi) at 150°C.

Future work on shale should consider the role of (1) moisture content, (2) moisture loss through heating, (3) pore pressure increases caused by heating and rapid stress increases, and (4) the possible influence of pore fluid chemistry on strengths and deformation of the shales. The influence of each of these considerations is likely to vary with the clay mineral content of the shales. In particular, the smectite-rich Pierre Shale may shrink significantly when drying during a test at elevated temperature. Shrinking caused by drying would presumably add to the strain induced by the stress until most of the water has been forced out. To the author's knowledge, the above noted effects of moisture on deformation have not been systematically studied as they related to pillars in shale.

References

Bradshaw, R.L., and W.C. McClain, 1971. Project Salt Vault: A Demonstration of the Disposal of High Activity Solidified Wastes in Underground Salt Mines. ORNL-4555, Oak Ridge National Laboratory.

Gonzales, S., K.S. Johnson, and P.E. Potter, 1985. Geologic Characteristics of Shale, Appendix E.2. Evaluation of Five Sedimentary Rocks Other than Salt for High-Level Waste Repository Siting Purposes, ORNL/CF-85/2/V2.

Hansen, F.D., and T.J. Vogt, 1987. Thermomechanical Properties of Selected Shales. RSI-0305, Re/Spec, Inc.

Hubbert, M.K., 1937. Theory of scale models as applied to the study of geological structures. Geological Society of American Bulletin, v. 48, p. 1459–1520.

Lomenick, T.F., and R.L. Bradshaw, 1969a. Deformation of rock salt in openings mined for the disposal of radioactive wastes. Rock Mechanics, v. 1, p. 5–30.

Lomenick, T.F., and R.L. Bradshaw, 1969b. Model pillar tests for evaluating the structural stability of openings in rock salt utilized for the disposal of radioactive wastes. Nuclear Engineering and Design, v. 9, p. 269–278.

Obert, L.E., 1964. Deformational behavior of model pillars made from salt, trona, and potash ore. Proceedings of Sixth Symposium on Rock Mechanics, Rolla, MO, p. 539–560.

Obert, L., and W.I. Duvall, 1967. Rock Mechanics and the Design of Structures in Rock. Wiley, New York.

Table 49.1. Properties of clays.*

Property	Porters Creek clay	Yazoo clay
Bulk density	0.02 mn/m³	0.023 mn/m³
Plasticity index	40	60
Unconfined compressive strength	0.18–1 MPa	0.64–1 MPa
Cohesion	0.1–1.5 MPa	1.1 MPa
Friction angle	0–27°	0–25°
Modulus of elasticity	5–10 MPa	65–115 MPa
Failure strain	8–20%	3–10%

*Thickness of Porters Creek and Yazoo clay is about 305 and 105 m, respectively (Hansen and Brekken, 1988; Gonzales and Johnson, 1985). The Porters Creek clay, with a plasticity index of 40%, appears to be normally consolidated, and cohesion is about 1.5 MPa, with a friction angle of about 27° (Hunt, 1986; Bowles, 1988; Peck et al., 1974). The Yazoo clay, with a plasticity index of 60%, also appears to be normally consolidated, and cohesion is about 1.2 MPa, with a friction angle of about 25° (Hunt, 1986; Bowles, 1988; Peck et al., 1974).

Table 49.2. Stability of clays.

Clay*	Stability factor (N)	Squeeze ratio (R)	Remarks
Porters Creek clay (assuming UCS = 1 MPa and c = 1 MPa)	5.84	5.84	For $N \leq 3.0$, UCS = 2 MPa For $N \leq 4.5$, UCS = 1.3 MPa For $R \leq 6.0$, c = 1 MPa
Yazoo clay (assuming UCS = 1 MPa and c = 1 MPa)	3.5	3.5	For $N \leq 3.0$, UCS = 1.2 MPa For $N \leq 4.5$, UCS = 0.8 MPa For $R \leq 6.0$, c = 0.6 MPa

*UCS, unconfined compressive strength; c, cohesion.

literature, it appears Porters Creek clay has less stability compared to Yazoo clay. More investigation concerning the properties is needed before any definite conclusions can be made. The values of unconfined compressive strength necessary to give an adequate stability factor are also shown in Table 49.2 under the Remarks column. Based on values presented in Table 49.2, it can be seen that for clays the unconfined compressive strength be at least 1.5 MPa, for adequate borehole stability. This agrees with the European practice of very limited use of clays for repositories when the unconfined compressive strength of clay is less than about 1 MPa (Chapman et al., 1984).

In another approach (Bowles, 1988), the cohesion is used as a main property and uses the concept of squeeze ratio, R. If the squeeze ratio is greater than 6, squeezing takes place. If it exceeds 8, then squeezing takes place very rapidly, and use of supporting liners or slurry is suggested. The squeeze ratio for the two clays is also shown in Table 49.2. The value of cohesion necessary to give an adequate squeeze ratio, say less than 6, is also shown in Table 49.2 under the Remarks column. Based on values presented in Table 2, it can be seen that for clays, the cohesion be at least 1 MPa for adequate stability.

An estimate of bearing capacity of Porters Creek and Yazoo clays was made using Skempton's classical procedure (DeMello, 1969). The allowable bearing capacity (for a cohesion value of 1 MPa) is estimated to be 13 MPa for Porters Creek clay and 10 MPa for Yazoo clay. A simple way, although approximate, to look at the significance of this is to consider the estimated stress at the bottom of the borehole. Considering approximately 150 waste containers (each weighing about 600 kg) are emplaced in the borehole, backfilled, the intensity of stress at the bottom of the borehole will be about 10 MPa. That is, the allowable bearing pressure of clay is just sufficient to support the canisters. The concept of critical depth for deep excavations (Bowles, 1988) can be used to estimate the value of cohesion that is necessary to make a borehole in these two clays. This computation indicates that cohesion of about 1.4 and 1.1 MPa, respectively, for Porters Creek and Yazoo clay is necessary. This is in agreement with the values estimated based on the stability factor and squeezing ratio presented in Table 49.2.

Summary

The primary purpose of this study was to make a preliminary engineering evaluation of a borehole facility in selected clay deposits. Based on the limited properties reported in the available literature, it appears that Porters Creek clay is relatively less stable than Yazoo clay, though definite conclusions cannot be made without sufficient data on geotechnical properties. The unconfined compressive strength and cohesion of clays are estimated to be at least 1.5 and 1 MPa, respectively, to provide sufficient borehole stability. The allowable bearing pressure of Porters Creek and Yazoo clay is estimated to be 13 and 10 MPa, respectively. This allowable bearing pressure is found to be just adequate to support the waste canisters.

Acknowledgments

The writer wishes to express his sincere appreciation to Dr. T.F. Lomenick, Oak Ridge National Laboratory, for his guidance during the present study. This study was carried out under contract No. 19X-SC-206V.

References

Bonne, A., and P. Manfroy, 1986. Factors involved in updating the concept of a repository for HLW, ILW, and alpha-bearing wastes in a deep clay formation on land. Proceedings of Symposium on Siting, Design, and Construction of Underground Repositories for Radioactive Wastes, Hannover, p. 507–519.

Bowles, J.E., 1988. Foundation Analysis and Design, 4th ed., McGraw-Hill, New York, 1004 p.

Chapman, N.A., F. Gera, and Riccioni, 1984. Feasibility study for the disposal of high-level heat emitting, cladding hull, and other wastes in the Plio-Pleistocene blue clays, Rome, Practica, no. 2403, 210 p.

Croff, A.G., T.F. Lomenick, R.S. Lowrie, and S.H. Stow, 1986. Evaluation of five sedimentary rocks other than salt for geologic repository siting. ORNL-6241, 3 volumes, Oak Ridge National Laboratory, Oak Ridge, TN.

deBeer, E., R. Carpentier, P. Manfroy, and R. Heremans, 1977. Preliminary studies of an underground facility for nuclear waste burial in a tertiary clay formation. Proceedings of the First International Symposium on Storage in Excavated Rock, Stockholm, v. 3, p. 771–780.

deMello, V.F., 1969. Foundations of buildings in clay. Proceedings of the Seventh International Conference on Soil Mechanics and Foundation Engineering, Mexico, State of the art volume, p. 49–80.

Gonzales, S., and K.S. Johnson, 1985. Shales and other argillaceous strata in the United States. ORNL/Sub-64794/1, Oak Ridge National Laboratory, Oak Ridge, TN, 596 p.

Hansen, F.D., and G.A. Brekken, 1988. Baseline mechanical properties of Yazoo and Porters Creek clay, RSI-0337, Report to Oak Ridge National Laboratory, 40 p.

Horseman, S.T., M.G. Winer, and D.C. Entwistle, 1986. Geotechnical characterization of boom clay in relation to the disposal of radioactive waste. FLPU-86-12, British Geological Survey Report.

Hudson, J.A., and J.B. Boden, 1981. Geotechnical and tunnelling aspects of radioactive waste disposal. Proceedings of International Tunnelling Conference, Brighton, United Kindgom.

Hunt, R.E., 1986. Geotechnical Engineering Analysis and Evaluation. McGraw-Hill, New York, 730 p.

International Atomic Energy Agency, 1986. Siting, design, and construction of underground repositories for radioactive wastes. Proceedings of Int. Symposium, Hannover, 615 p.

Lomenick, T.F., and S. Gonzales, 1987. Clay deposits in the United States and Europe as potential hosts for high-level radioactive waste repositories. Draft II. Oak Ridge National Laboratory, Oak Ridge, TN, 26 p.

Organization for Economic Cooperation and Development, 1984. Geological Disposal of Radioactive Waste, 111 p.

Peck, R.B., W.F. Hanson, and T.H. Thornburn, 1974. Foundation Engineering, 2nd ed. Wiley, New York, 514 p.

Rousset, G., A. Bonne, and R.A. Jehan, 1986. Mechanical behavior of galleries in deep clay - study of a concrete example. Proceedings of International Symposium on High-Level Nuclear Waste Disposal, Washington, D.C.

CHAPTER 50

Hydrocarbon Liquids and Clay Microstructure

Robert M. Quigley and Federico Fernandez

Introduction

Extensive studies of the interactions between organic liquids and clay minerals have been carried out over the past 10 years. The work has dealt primarily with hydraulic conductivity, k, and the influence that organic liquids of low dielectric constant have on measured values of k (Michaels and Lin, 1954; Lambe, 1956; Mesri and Olson, 1971; Brown et al., 1984; Acar et al., 1985; Fernandez and Quigley, 1985, 1988; Foreman and Daniel, 1986; Quigley and Fernandez, 1989). In most cases, the above articles have dealt with double layer theory and resultant physicochemical effects, with little attention given to the role of soil fabric and microstructure.

The purpose of this chapter is to feature the importance of soil fabric in interpreting hydraulic conductivity test results where liquid hydrocarbons might be employed as molding fluids and/or permeants. The chapter presents previous work by the authors from a different perspective. The studies were carried out on a weathered silty clay from Sarnia, Ontario, Canada, which contains ~20% quartz and feldspar, ~10% carbonate, ~50% illite, ~8% chlorite, and ~12% smectite. For mineralogical details see Ogunbadejo and Quigley (1976) who discussed a similar but less clayey surface soil in the same area.

Microstructure of Dry Clay/Organic Mixtures

Photomicrographs showing the flocculation characteristics of a dried natural clay mixed with water (dielectric constant, $\varepsilon = 80$), commercial alcohol ($\varepsilon = 32$), and benzene ($\varepsilon = 2$) are presented in Figures 50.1 and 50.2. Under the microscope (Fig. 50.1) the clay in water is dispersed, the clay in alcohol is flocculated, and the clay in xylene is flocculated, aggregated, and agglomerated. The scanning electron photomicrographs in Figure 50.2 were obtained on k-test specimens molded with water, alcohol, and benzene and show even more clearly the fabric of the vertical surfaces of posttest, freeze-dried samples. The water-compacted sample displays a clay platelet parallelism characteristic of dispersion whereas both the alcohol and benzene-molded clays are highly flocculated and characterized by macropores that are especially well developed in the case of benzene ($\varepsilon = 2$).

According to the Gouy-Chapman theory of potential, liquids having a low dielectric constant cause very thin double layers compared to liquids of higher dielectric constant as illustrated in Figure 50.3. This in turn causes reductions in interparticle repulsive forces, resulting in flocculation.

Hydraulic Conductivity of Dry Clay/Organic Mixtures

The influence of the above discussed microstructural features on hydraulic conductivity, k, is illustrated in Figure 50.4 which presents six curves of k vs. void ratio (e) for the test soil molded and permeated with water, ethanol, and xylene. Two curves are presented for each liquid; one for k values measured using a constant flow rate permeameter (solid), and one for k values calculated from oedometer tests (dashed). The curves clearly show that at a void ratio of 0.8 (void ratio = volume of voids ÷ volume of solids), k increases from about 4×10^{-9} cm/sec for water to almost 1×10^{-4} cm/sec for xylene. These trends are in agreement with extensive studies by Mesri and Olson (1971) and Fernandez and Quigley (1985). Although normally correlated with a decreasing dielectric constant, the above trend is more

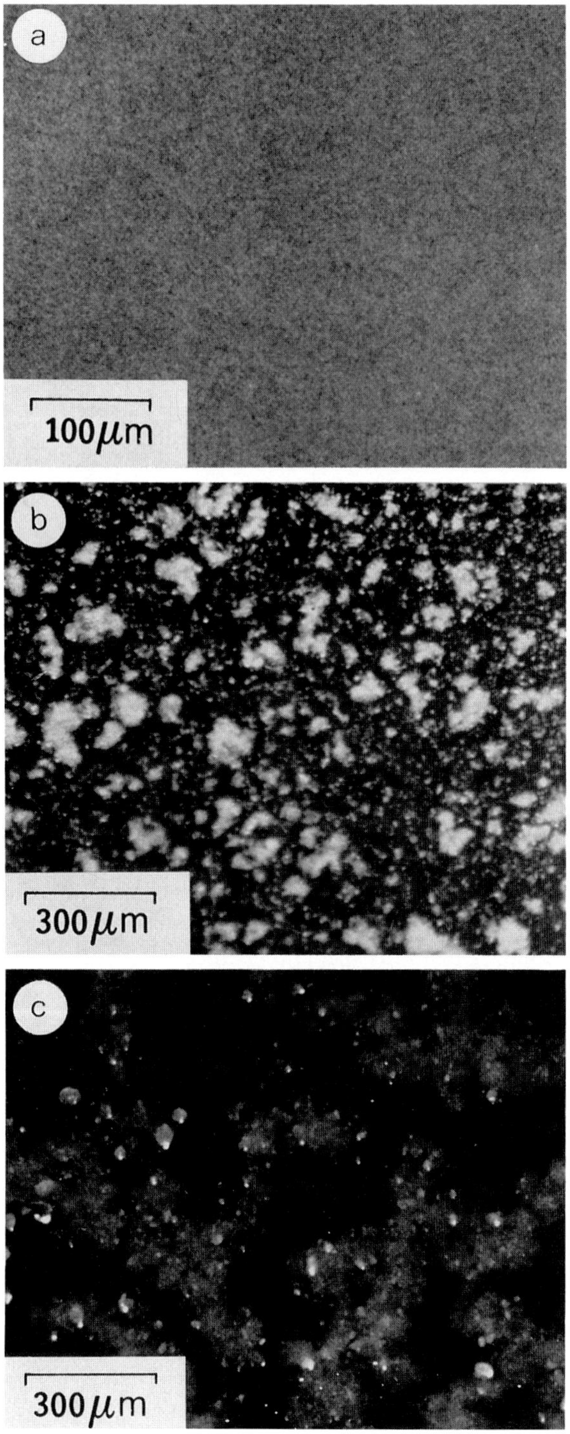

Figure 50.1. Flocculation characteristics of suspensions of 0.5 g clay in 30 ml of liquid as seen through a microscope: (a) water, (b) ethanol, (c) xylene. (Directly from Fernandez and Quigley, 1985.)

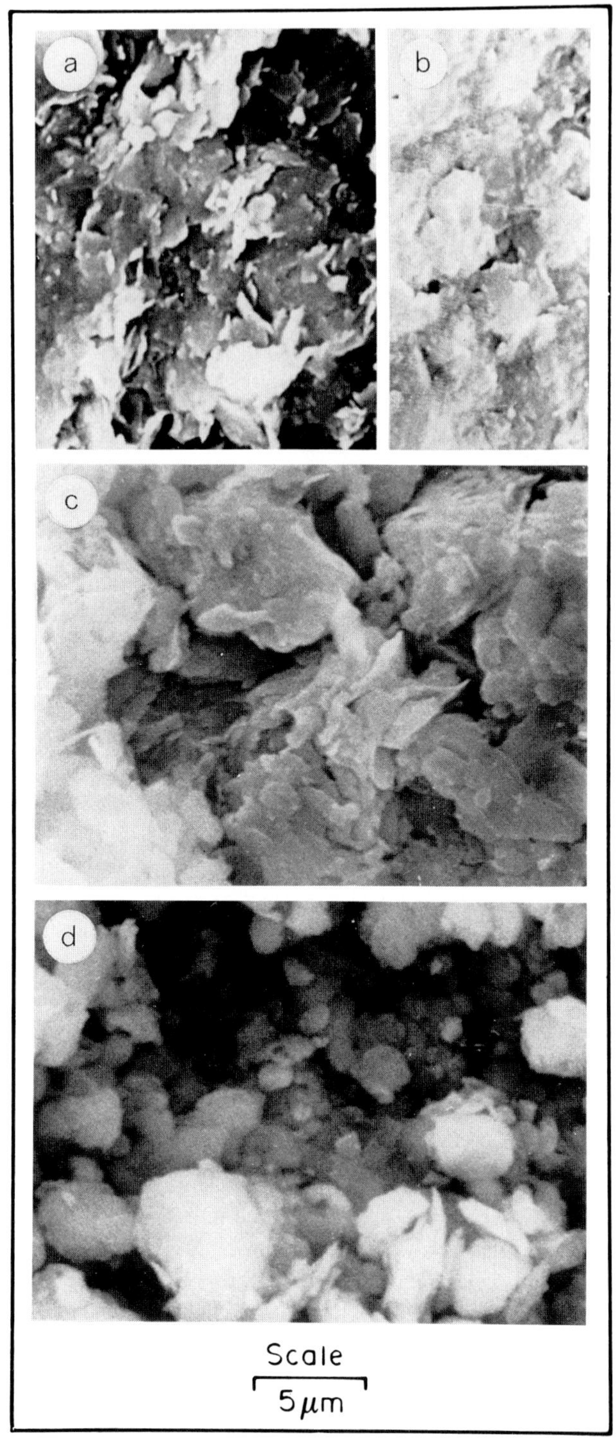

Figure 50.2. Scanning electron photomicrographs of vertical surfaces of *k*-test samples: (a) fractured, water-molded clay, (b) cut and smeared water-molded, (c) alcohol-molded, (d) benzene-molded. (Directly from Fernandez and Quigley, 1985.)

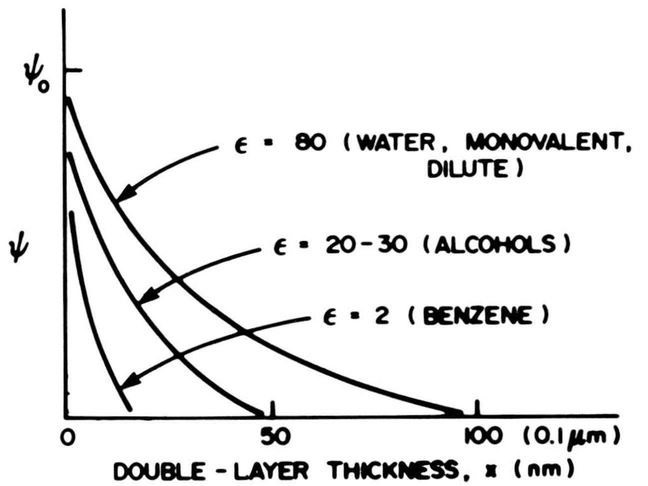

Figure 50.3. Illustration of double layer thickness as a function of dielectric constant of the pore fluid. (From Fernandez and Quigley, 1985.)

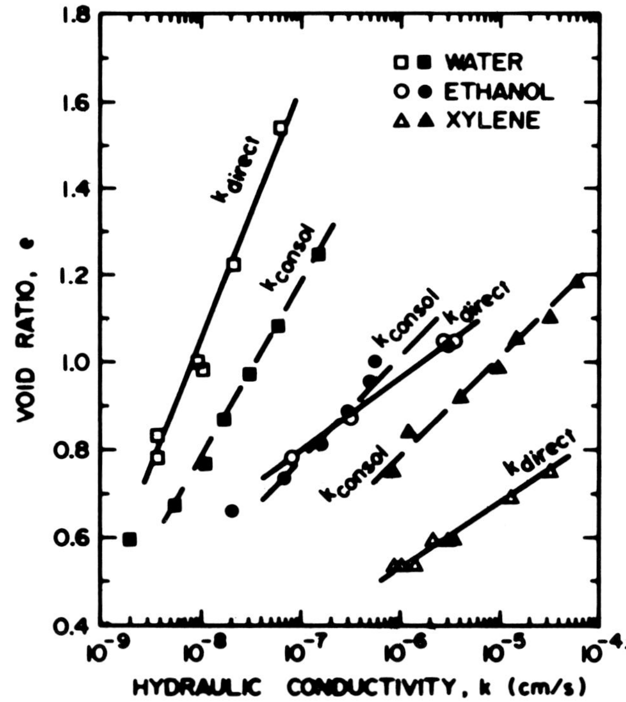

Figure 50.4. Hydraulic conductivity determined by direct measurement and by calculation from consolidation tests on permeant-molded soils at various void ratios. (Reprinted from Fernandez and Quigley, 1989, Proceedings of the Twelfth International Conference on Soil Mechanics and Foundation Engineering, Rio de Janeiro, 13–18 August 1989, volume 3).

directly related to soil microstructure, especially the large macropores that form in soils molded with low dielectric liquids (Fig. 50.2).

A flocculated soil structure is normally considered to be stronger than a more dispersed structure, and the three sets of curves in Figure 50.4 seem to support this. For example, the k_{consol} values for xylene ($\varepsilon = 2$) are almost an order of magnitude smaller than the directly measured values (k_{direct}) suggesting that a structural stiffness has interfered with the speed of oedometer consolidation. The k_{consol} values for the water-molded clay on the other hand are higher than the measured k_{direct} values. This is believed to reflect the greater compressibility of the more dispersed soil–water system. Another factor may be the tortuosity created by kneading compaction of the water-wet clay that produces the horizontal fabric illustrated in Figure 50.2a. Finally, the ethanol-molded clays ($\varepsilon = 32$) yield almost identical values of k whether calculated from oedometer consolidation tests or by direct measurement with the constant flow permeameter.

Porosimetry of Compacted Clays

Samples of air dry Sarnia clay powders were mixed with water and statically compressed to a void ratio of 0.80 (moisture content = 29%). The pore size distribution of freeze-dried specimens was then obtained by mercury intrusion porosimetry. A reproducible dominant pore diameter of ~0.11 μm was measured as illustrated by the solid curve in Figure 50.5. Corresponding clay powders were molded with aqueous solutions of 10–100% ethanol and pressed to the same void ratio. The two curves for 50% ethanol and 100% ethanol are also presented in Figure 50.5 and show a steady increase in the dominant size to 0.32 μm.

Another set of samples was water-molded and compacted by kneading, then permeated with various mixtures of ethanol and domestic waste leachate. The pore size distribution curves for the water samples (solid) show two broad modal sizes at ~3 and ~0.08 μm (Fig. 50.6). Permeation with the ethanol/leachate combinations invariably resulted in loss of the large pores and creation of an even stronger modal size of 0.08 μm. It is not known to what extent the changes represent artifacts of freeze drying and mercury intrusion.

A summary plot (hatched area) for 14 mixtures of clay compacted with ethanol/water or ethanol/leachate is presented in Figure 50.7. Also plotted are four open data points for the water-compacted clays subsequently permeated with ethanol-bearing leachate. The porosimetry for all specimens was run on post-k-test samples. The apparent porosimetry, therefore, corresponds to the soil when it yielded the final values of k. The plot clearly shows that ethanol/water and especially ethanol/leachate-molded samples have a significantly larger dominant pore size than the water-molded samples. The larger modal pore size of the clays molded with organic permeants is related to flocculation on mixing, and, all other factors constant, should result in higher k values than the water-molded samples.

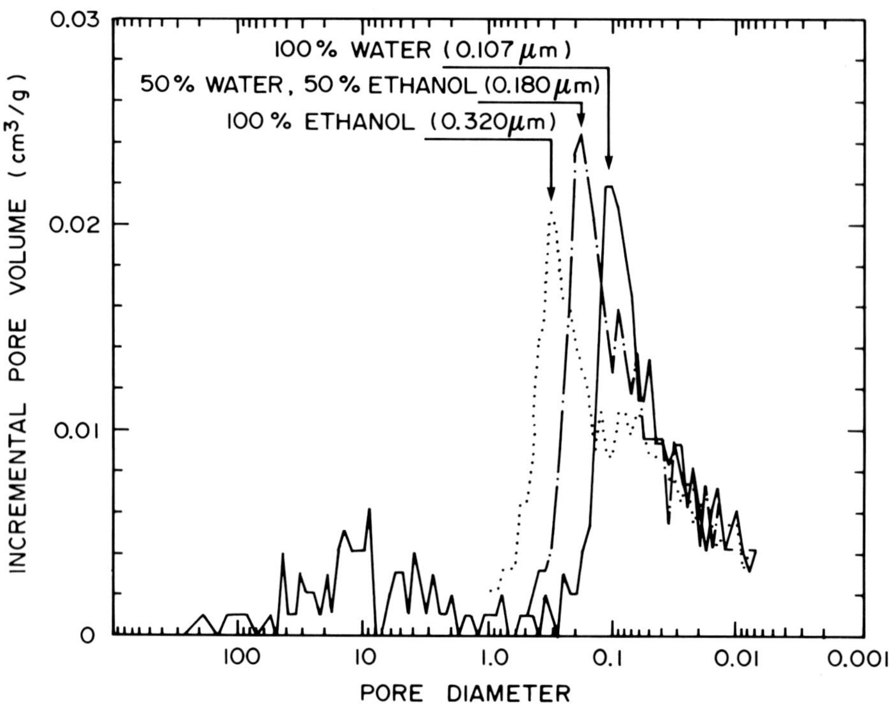

Figure 50.5. Pore size distribution of brown Sarnia soil mixed with 100% water, 50:50 water:ethanol and 100% ethanol at a void ratio of ~0.8.

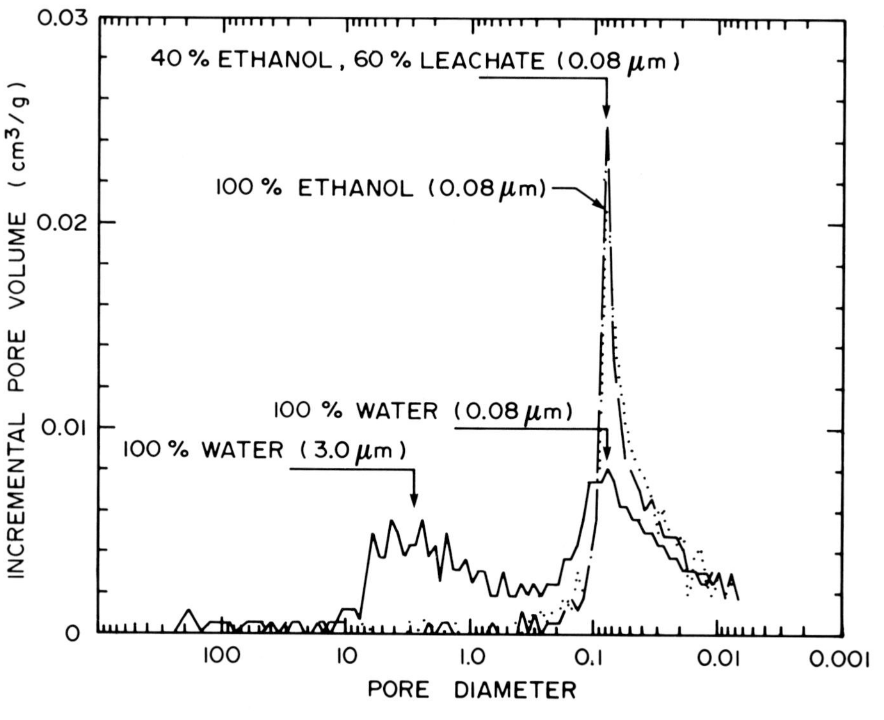

Figure 50.6. Comparison of pore size distribution of test soils molded and compacted with water, then permeated with water or mixtures of ethanol and leachate.

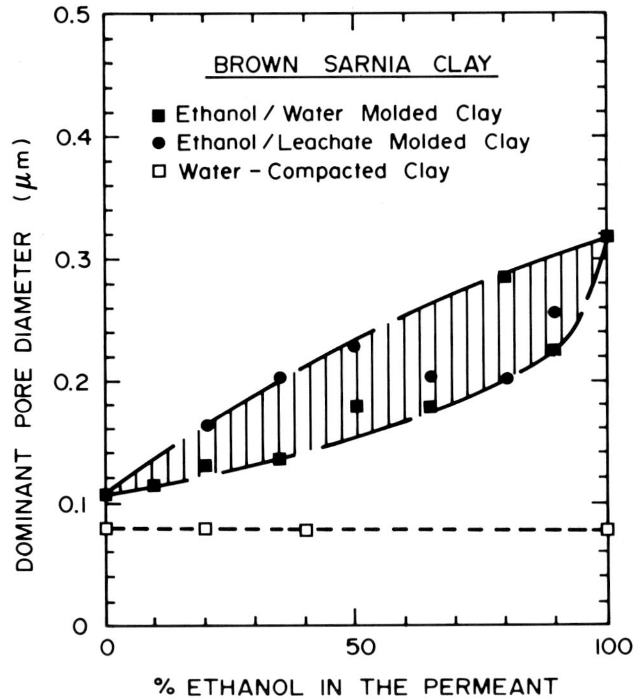

Figure 50.7. Dominant pore diameter for dry clays mixed and permeated with aqueous solutions of ethanol, and water-compacted clay permeated with the aqueous solutions of ethanol. (All porosimetry at end of permeation.)

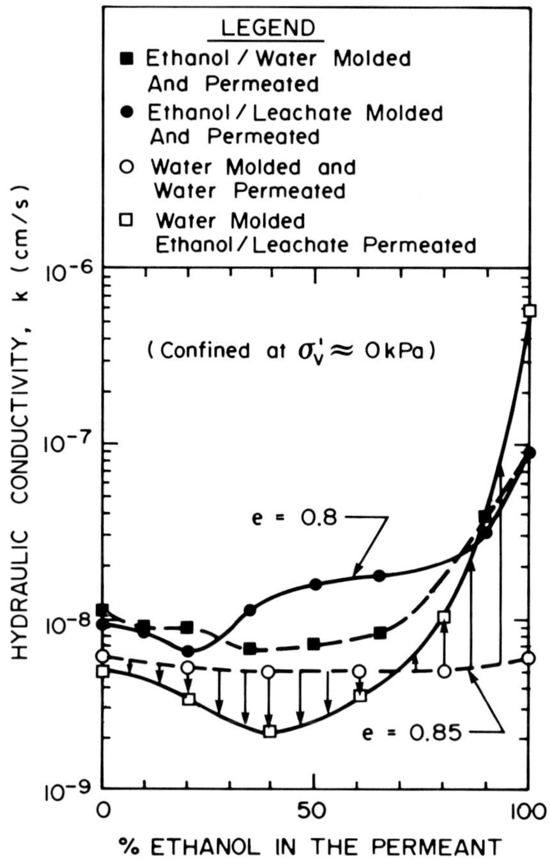

Figure 50.8. Hydraulic conductivity vs. percent ethanol in permeant for water-compacted and permeant-compacted Sarnia clay samples: (●,■) compacted by static compression; (○,□) compacted by kneading with a rod.

Hydraulic Conductivity of Permeant and Water-Compacted Clays

Summary plots of the hydraulic conductivity vs. percent ethanol in the permeants are presented in Figure 50.8. The open circles represent water-compacted clays permeated with water, and thus are reference values for the rest of the data. The reference k_{water} values for the dispersed soils with horizontal fabric oriented perpendicular to flow were normally $\sim 5 \times 10^{-9}$ cm/sec. The open squares represent water-compacted samples subsequently permeated with ethanol/leachate mixtures. The closed circles and squares represent k values obtained on the more flocculated samples molded and compacted with ethanol/leachate mixtures.

Passage of aqueous solutions of ethanol through the water-compacted samples caused a *decrease* in k at all concentrations up to ~70% (open squares). These reductions are caused by increases in viscosity which reach three times that of water at 50% ethanol concentration. Above 70% ethanol, the viscosity decreases again and the effects of a reduction in dielectric constant from 80 (water) to 32 (commercial ethanol) result in double layer contraction, an increase in free pore space at constant volume, and large increases in k. The interrelationship of viscosity to dielectric constant was discussed at length by Fernandez and Quigley (1988) and will not be repeated here. From a porosimetry point of view, however, it is important to note that the dominant pore size at the end of ethanol permeation seems to be very similar to that formed by water molding. This assumes that mercury intrusion porosimetry can be used on freeze-dried specimens of soils as complex as these.

Up to ~80% ethanol, the samples molded with ethanol-enriched permeant yielded k values consistently higher than those molded with water only. At 40–50% ethanol, where the lowest k values of $\sim 2 \times 10^{-9}$ cm/sec were obtained on the water-compacted clays, the permeant-compacted clays yielded k values of 7 to 20 $\times$ 10^{-9} cm/sec. These higher values can be logically attributed to macropores and the larger modal pore size related to the flocculated structure of those samples molded with ethanol-enriched water or leachate.

On the basis of the above observations, it is apparent that water contaminated with salts or soluble organics should not be used as molding fluid for compaction of clayey liners.

Summary

Fabric, soil microstructure, and porosimetry have been shown to be important factors in interpretation of the hydraulic conductivity of barrier clays. Concentrated organic liquids and even organically polluted water result in soil flocculation if used as molding liquids for compaction. The resultant soil structure may produce significantly higher values of hydraulic conductivity than will be obtained if potable water is used and a dispersed soil structure is obtained within the barrier clay.

Acknowledgments

The fundamental work presented in this chapter has been funded by grants from the Natural Sciences and Engineering Research Council of Canada. The practical section on ethanol/leachate mixtures was funded by the Ontario Ministry of the Environment. Both sources of funding are gratefully acknowledged.

References

Acar, Y.B., A Hamidon, S.D. Field, and L. Scott, 1985. The effect of organic fluids on hydraulic conductivity of kaolinite. In: Hydraulic Barriers in Soil and Rock. ASTM Special Technical Publication 874, p. 171–187.

Brown, K.W., J.C. Thomas, and J.W. Green, 1984. Permeability of compacted soils to solvent mixtures and petroleum products. In: Land Disposal of Hazardous Waste. Proceedings of the 10th Annual Research Symposium, Ft. Washington, PA, EPA-600/9-84-007, p. 124–137.

Fernandez, F., and R.M. Quigley, 1985. Hydraulic conductivity of natural clays permeated with simple liquid hydrocarbons. Canadian Geotechnical Journal, v. 22(2), p. 205–214.

Fernandez, F., and R.M. Quigley, 1988. Viscosity and dielectric constant controls on the hydraulic conductivity of clayey soils permeated with water-soluble organics. Canadian Geotechnical Journal, v. 25(3), p. 582–589.

Fernandez, F., and R.M. Quigley, 1989. Organic liquids and the hydraulic conductivity of barrier clays. Proceedings 12th International Conference on Soil Mechanics and Foundation Engineering, Rio de Janeiro, Brazil, Vol. 3, p. 1867–1870.

Foreman, D.E., and D.E. Daniel, 1986. Permeation of compacted clay with organic chemicals. Journal of Geotechnical Engineering, ASCE, v. 112(7), p. 669–681.

Lambe, T.W., 1956. The storage of oil in an earth reservoir. Journal of the Boston Society of Civil Engineers, v. 43, p. 179–241.

Mesri, G., and R.E. Olson, 1971. Mechanisms controlling the permeability of clays. Clays and Clay Minerals, v. 19(3), p. 151–158.

Michaels, A.S., and C.S. Lin, 1954. Permeability of kaolinite. Industrial and Engineering Chemistry, v. 46(6), p. 1239–1246.

Ogunbadejo, T.A., and R.M. Quigley, 1976. Till geology, mineralogy and geotechnical behavior, Sarnia, Ontario. In: Legget, R.F. (ed.), Glacial Till: An Interdisciplinary Study. Royal Society of Canada Special Publication 12, p. 336–345.

Quigley, R.M., and F. Fernandez, 1989. Clay/organic interactions and their effect on the hydraulic conductivity of barrier clays. Proceedings of the International Symposium on Contaminant Transport in Groundwater, Stuttgart, Germany, p. 117–124.

CHAPTER 51

Effects of Hydrothermal Treatment on the Engineering Properties, Microstructure, and Composition of Oilwell Cements

Eliza Grabowski and J.E. Gillott

Introduction

Under hydrothermal conditions normal portland cements show loss of strength and increase in permeability due to changes in phase equilibria and in the chemistry of hydration (Lea, 1971; Eilers et al., 1980). Cement blends with 40% silica flour or sand show improved hydrothermal characteristics; at ambient temperature, however, an inert filler degrades engineering properties. Improved ambient properties are achieved when silica fume, a highly reactive pozzolan (Buck and Burkes, 1980; Malhotra and Carette, 1982; Malhotra, 1984), replaces part of the cement. This is attributed to changes in composition and microstructure due to the pozzolanic reaction (Gallus and Pyle, 1978; Cheng-yi and Feldman, 1985a,b). In the case of oilwell cement–fume blends ambient properties are thus improved but hydrothermal behavior is more complex (Grabowski and Gillott, 1989).

Microstructural changes are amplified in cement–fume mortars, because the sand–paste interface plays an important role (Carette and Malhotra, 1982a; Cheng-yi and Feldman, 1985c,d). Engineering properties may depend therefore on the type of silica used. Other factors influencing engineering properties through changes in fabric and composition of thermally hydrated phases are period, temperature, and conditions of precuring prior to hydrothermal treatment. Evidently the degree of hydration, type of C–S–H, and availability of free silica and $Ca(OH)_2$ at the end of the precuring period influence the fabric and composition of hydrothermal products. The rate of both reactions, hydration and pozzolanic, is governed—among other variables—by curing temperature (Buck and Burkes, 1980; Carette and Malhotra, 1982b). Carbonation is reported to play some role in the increase of strength and decrease of permeability (Berger and Klemm, 1972; Young et al., 1974; Ho and Lewis, 1987), but on the other hand carbonic acid may also be responsible for corrosion of portland cement blends (Quan, 1984; Bruckdorfer, 1986). The objective of this work was to investigate the effects of types of silica, temperature, and period of precuring and effect of CO_2 on ambient and hydrothermal engineering properties of cement–silica blends. Changes in strength and permeability were related to changes in composition and fabric.

Materials and Methods

Oilwell cement Class G (American Petroleum Institute) was used as a basic cement from which slurries of thermal mixes containing three different types of silica were prepared. Chemical properties of the Class G cement used in this work are given in Table 51.1. Properties of typical oilwell cements are described by Smith (1976). This portland cement-based oilwell cement is intended for use from surface to a depth of 260 m (8000 ft). Between 1.5 and 3% of gypsum is added to the clinker to control the setting time of the cement on hydration. The three forms of silica used were condensed silica fume, flour, and sand. Silica fume, a by-product of the production of silicon metal or ferrosilicon alloys, is a highly reactive pozzolan composed of poorly crystallized or amorphous microspheres (~ 0.1 μm) with high specific surface area (~ 20 m^2/g). Thus, fume has a very high water demand (Mehta, 1983), so a solid superplasticizer was incorporated in the mixes to maintain adequate consistency at reasonable water–cement ratios. The superplasticizer was used in amounts of 3–6% by weight of silica fume, depending on the proportion of fume in the mix. Silica flour is pulverized quartz with a uniform minus 200 mesh particle size (< 75 μm). Ottawa sand (> 150 μm) was used as a coarse silica.

Table 51.1. A chemical analysis of the cement Class G used in this work.

Insoluble	0.24%	MgO	3.60
SiO_2	22.43	Na_2O	0.17
CaO	63.72	K_2O	0.40
Free CaO	0.83	SO_3	1.95
Al_2O_3	2.75	Ign. loss	1.50
Fe_2O_3	3.63		
		Total	100.15

Preparation of Slurry

Five mixes were designed each of which contained 40% silica by weight of cement. A constant water–solids ratio of 0.5 was used to facilitate comparison of engineering properties, composition, and microstructure of mixes made with different forms of silica. Proportions used in laboratory batches of 600 ml of slurry are shown in Table 51.2. Various proportions of three forms of silica have been used as a replacement for silica flour (fume:flour, 3:1, fume:sand, 1:1 and 3:1, and fume only). Total silica content was kept constant at 40% by weight of cement.

The mixing procedure followed was that specified by API for oilwell cements. The amount of superplasticizer for a given proportion of cement to silica fume was based on consistency measurements in an atmospheric consistometer. Slurries containing fume required longer mixing in the Waring blender (up to 2 min) than the 35 sec recommended in the API procedure. Specimens for strength or water permeability testing were cast in rectangular (2.5 × 2.5 × 27.5 cm) or truncated conical disc-shaped (6.2 × 5.6 × 1.8 cm) molds respectively. After 24 hr specimens were demoulded and cured at 22°C, 100% RH (in fogroom) for specified periods. Specimens were divided into two batches, one of which was tested after ambient curing and the second batch was tested after ambient precuring and hydrothermal treatment. Specimens treated hydrothermally were held at 230°C, 2.75 MPa (saturated steam pressure) for 6 days in specially designed pressure vessels; 1 extra day was allowed for cooling to ambient temperature and pressure.

Compressive strength tests were carried out on 2.5 cm cubes cut from the rectangular bars. Samples for permeability testing were presaturated under vacuum with deaired distilled water and subjected to a differential water pressure of 1.38 MPa (200 psi); permeability was calculated from measurements of the flow rate, as an average of not less than six tests.

Mineralogical composition was determined by X-ray diffraction and infrared spectroscopy (IR) using KBr discs. Microstructure was investigated by examination of freshly fractured, gold-coated surfaces on the scanning electron microscope (SEM) equipped with an EDAX energy dispersive analyzer.

Table 51.2. Mix proportions (g) used to prepare 600 ml of slurry.*

Mix	Cement	Fume	Flour	Sand	Water	Superplasticizer
SF	515	–	205	–	360	–
SFFB	515	173	50	–	369	8.5
F	515	216	–	–	365	12
FS	515	113	–	103	365	3.5
FSB	515	173	–	50	369	6

*W/(C + S) = 0.5, W/C = 0.7, C + S = cement + silica.
Total silica content = 40%.

Results and Discussion

Ambient Conditions

A cement blend with fume only (F) showed the highest compressive strength at all periods and conditions of precuring. It also showed the lowest permeability, except at early age curing in a CO_2 atmosphere. The second best values were recorded for the blend made with fume and sand at a 3 to 1 ratio (FSB). Mixes SFFB (fume:flour, 3:1) and FS (fume:sand, 1:1) showed relatively high strength and low permeability, while a commercial blend with flour only (SF) gave the lowest strength and highest permeability values at all periods and conditions of curing (Figs. 51.1, 51.2, and 51.5). Generally, strength increased and permeability decreased with period of curing; some differences, however, resulted from different proportions of mixes and conditions of curing. These differences are attributed to changes in mineralogy and fabric during hydration of the cement–silica blends. Increase in effective water–cement ratio due to replacement of 40% of cement with inert silica flour decreased strength and increased permeability. Fabric became more open (Fig. 51.7) and the amount of hydrated phases decreased. Replacement of part or all of the flour with pozzolanic fume changed the path of hydration. Fume is known to react rapidly with $Ca(OH)_2$ to form the low calcium type of calcium silicate hydrate, C–S–H I ($C/S \leq 1.5$); thus, larger pores are transformed into finer pores (20), and strength gain and permeability decrease are proportional to the amount of fume used. The needle-like fabric of high calcium C–S–H II ($C/S > 1.5$) was replaced by a finer, granular fabric of C–S–H I (Fig. 51.8).

A slow pozzolanic reaction between flour and $Ca(OH)_2$ led to slightly higher strength and much lower permeability at later ages for the cement blend with fume–flour than for the corresponding blend with fume–sand. Permeability values for mix FSB (fume:sand, 3:1) were of the same order of magnitude as those for mix SF (flour); thus pore structure and degree of interconnection in these two mixes may be similar. Comparison of the fabric (Fig. 51.8) confirms the above assumption.

The suggestion that a delayed pozzolanic reaction occurs between flour and $Ca(OH)_2$ is verified by the steady increase in strength at ages up to 270 days shown by mix SF; other mixes reached a strength plateau at about 56 days. Early strength of

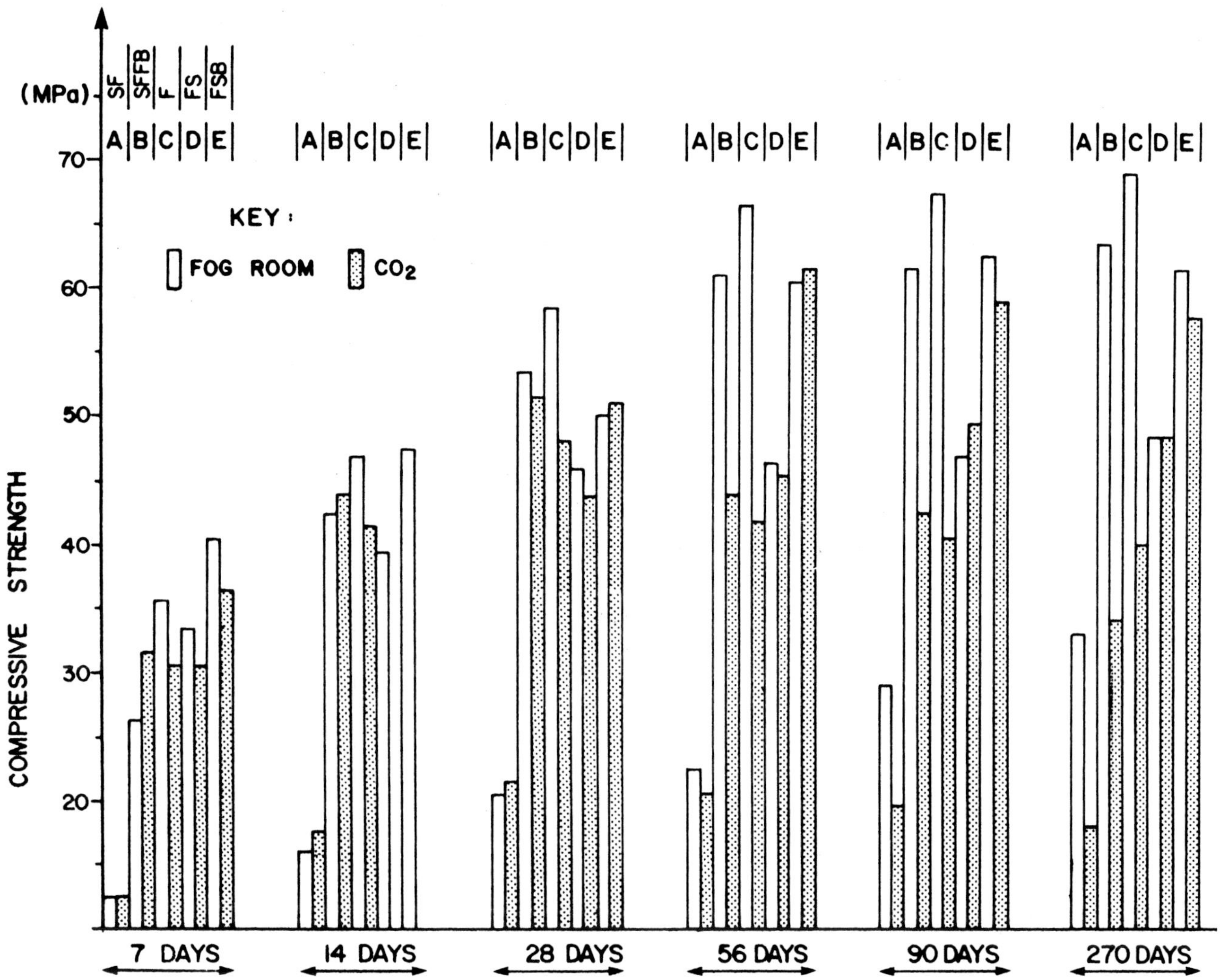

Figure 51.1. Effect of type of silica, curing period, and environment on compressive strength of cement–silica blends.

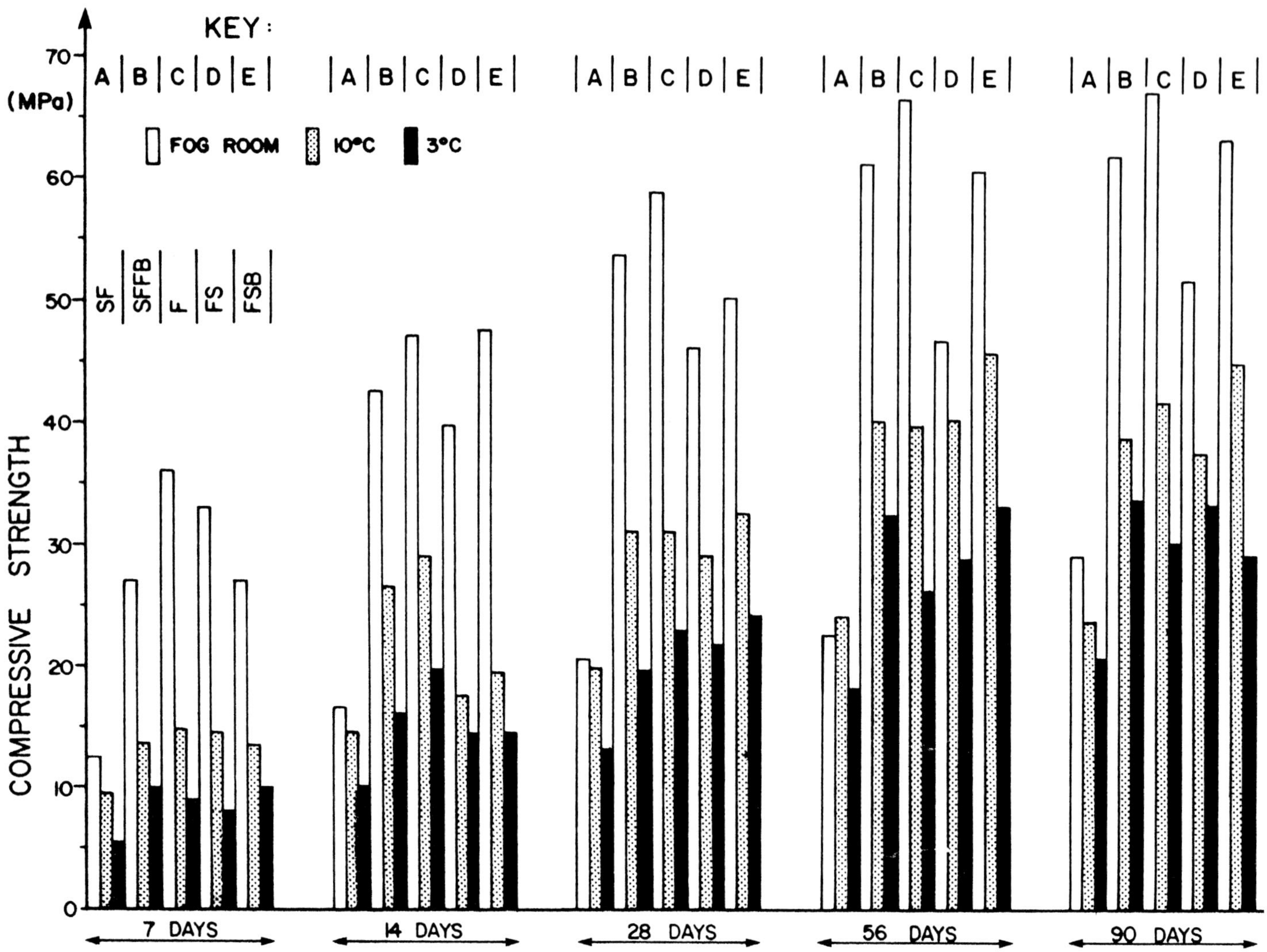

Figure 51.2. Effect of type of silica, curing period, and temperature on compressive strength of cement–silica blends.

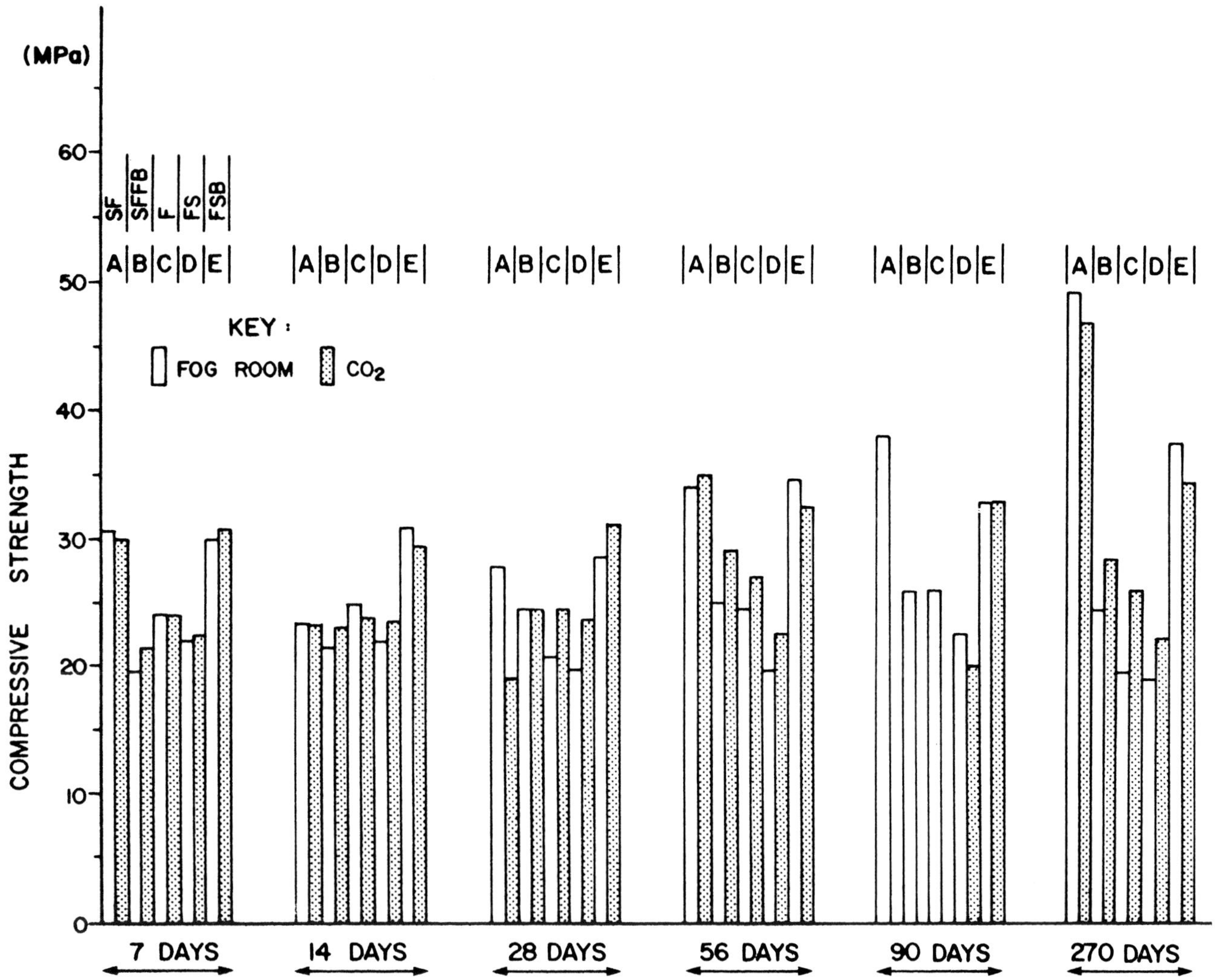

Figure 51.3. Effect of type of silica, precuring period, and atmosphere on compressive strength of cement–silica blends treated hydrothermally.

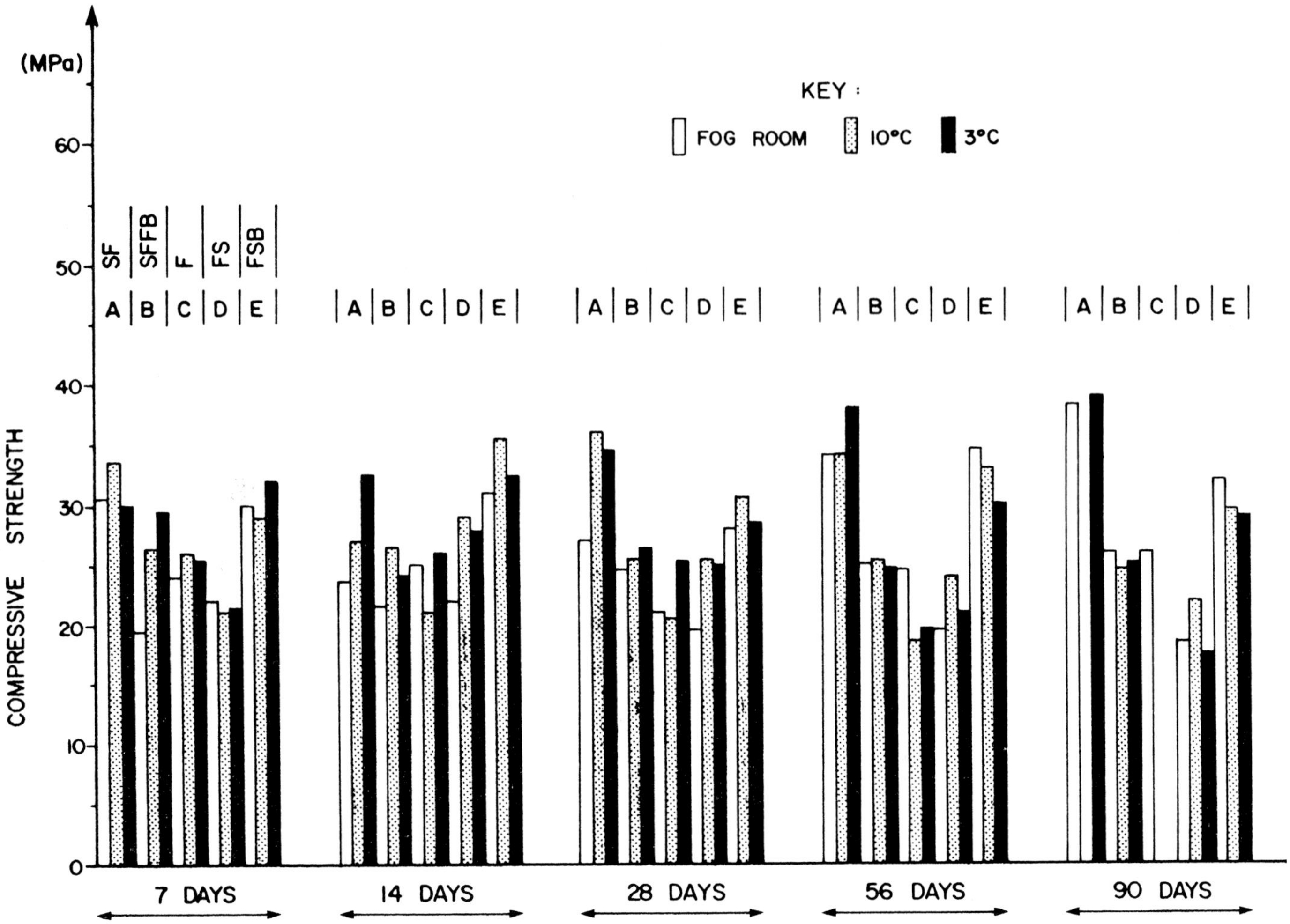

Figure 51.4. Effect of type of silica, precuring period, and temperature on compressive strength of cement–silica blends treated hydrothermally.

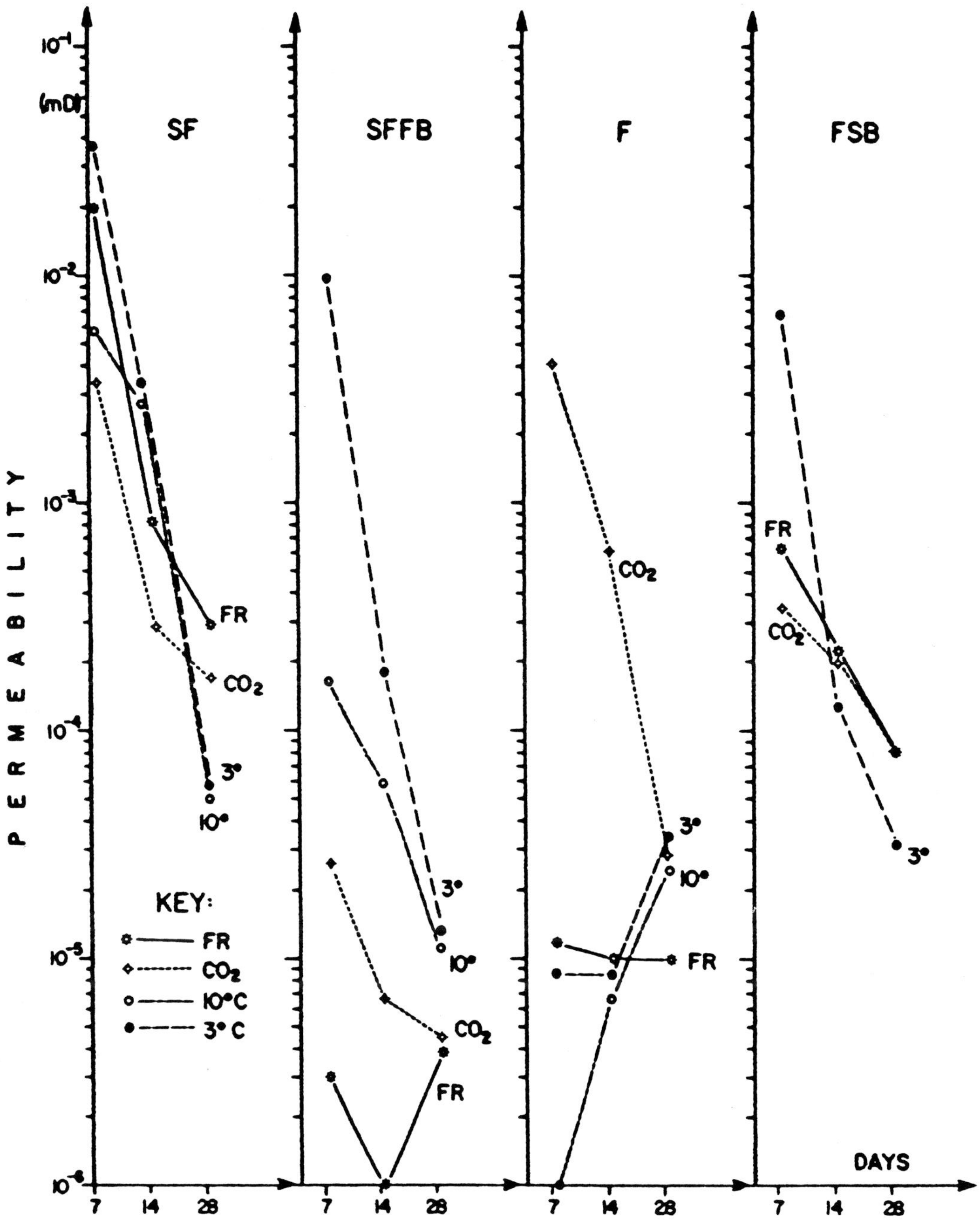

Figure 51.5. Effect of type of silica, curing period, and environment on permeability of cement-silica blends.

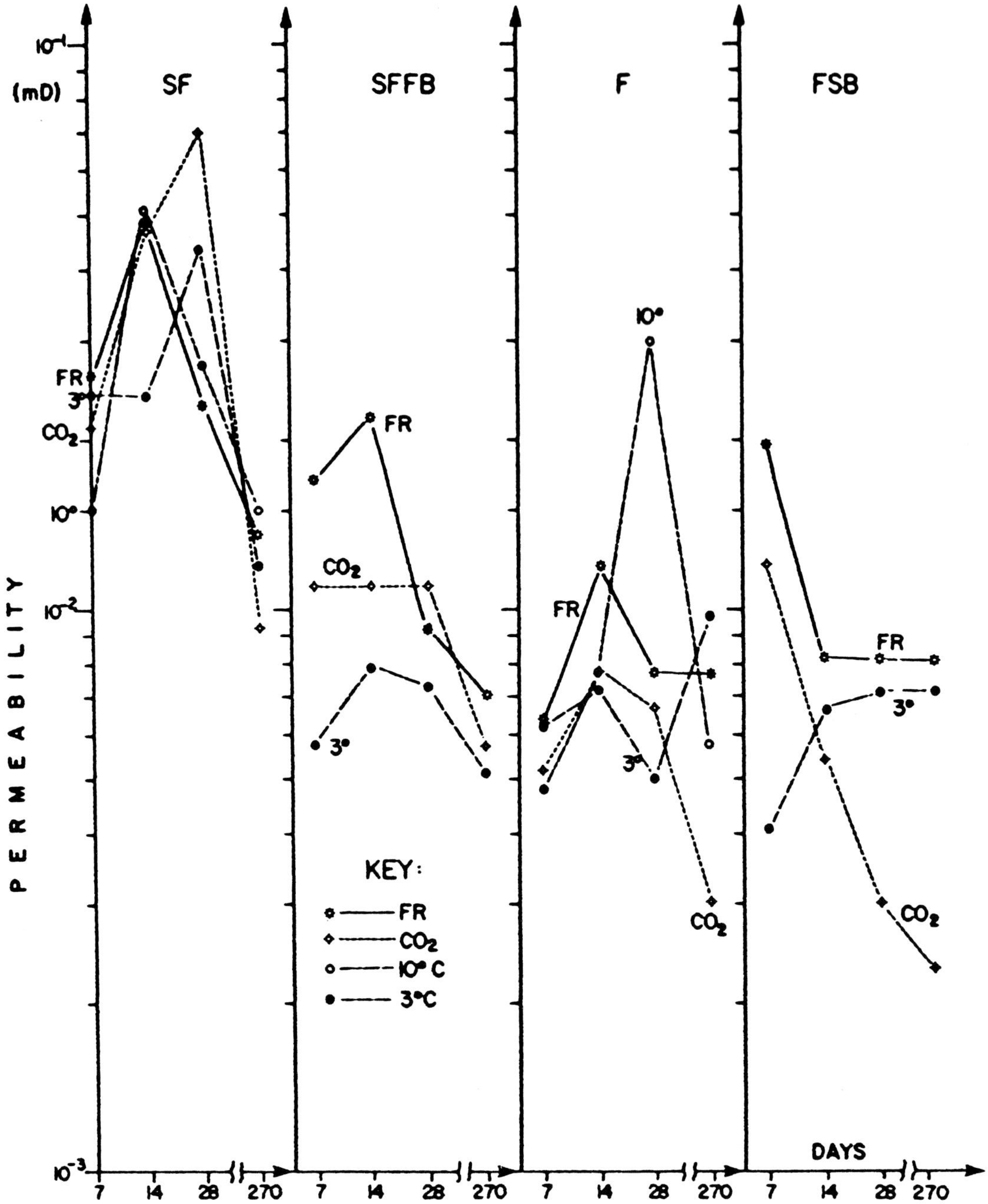

Figure 51.6. Effect of type of silica, curing period, and environment on permeability of cement–silica blends treated hydrothermally.

fume-containing mixes was high due to formation of C–S–H I during the rapid pozzolanic reaction between fume and $Ca(OH)_2$; $Ca(OH)_2$ was detectable by analytical methods only up to a curing age of 7 days. In the presence of CO_2 both the slow and rapid pozzolanic reactions are partially inhibited by competition between CO_2 and silica for $Ca(OH)_2$. In addition C–S–H itself is attacked by carbonic acid (Quan, 1984; Bruckdorfer, 1986). Since C–S–H makes a much greater contribution to strength increase (and permeability decrease) than calcium carbonates in a static CO_2 atmosphere, further (>28 days) strength development is hampered in mixes with flour and/or fume. Reaction between CO_2 and $Ca(OH)_2$ is slowed down in mixes with fume and sand by the reported tendency for better crystallized $Ca(OH)_2$ to form at sand–cement interfaces. The qualitative mineralogical composition was not changed significantly by curing in CO_2; most samples contained slightly more aragonite relative to calcite.

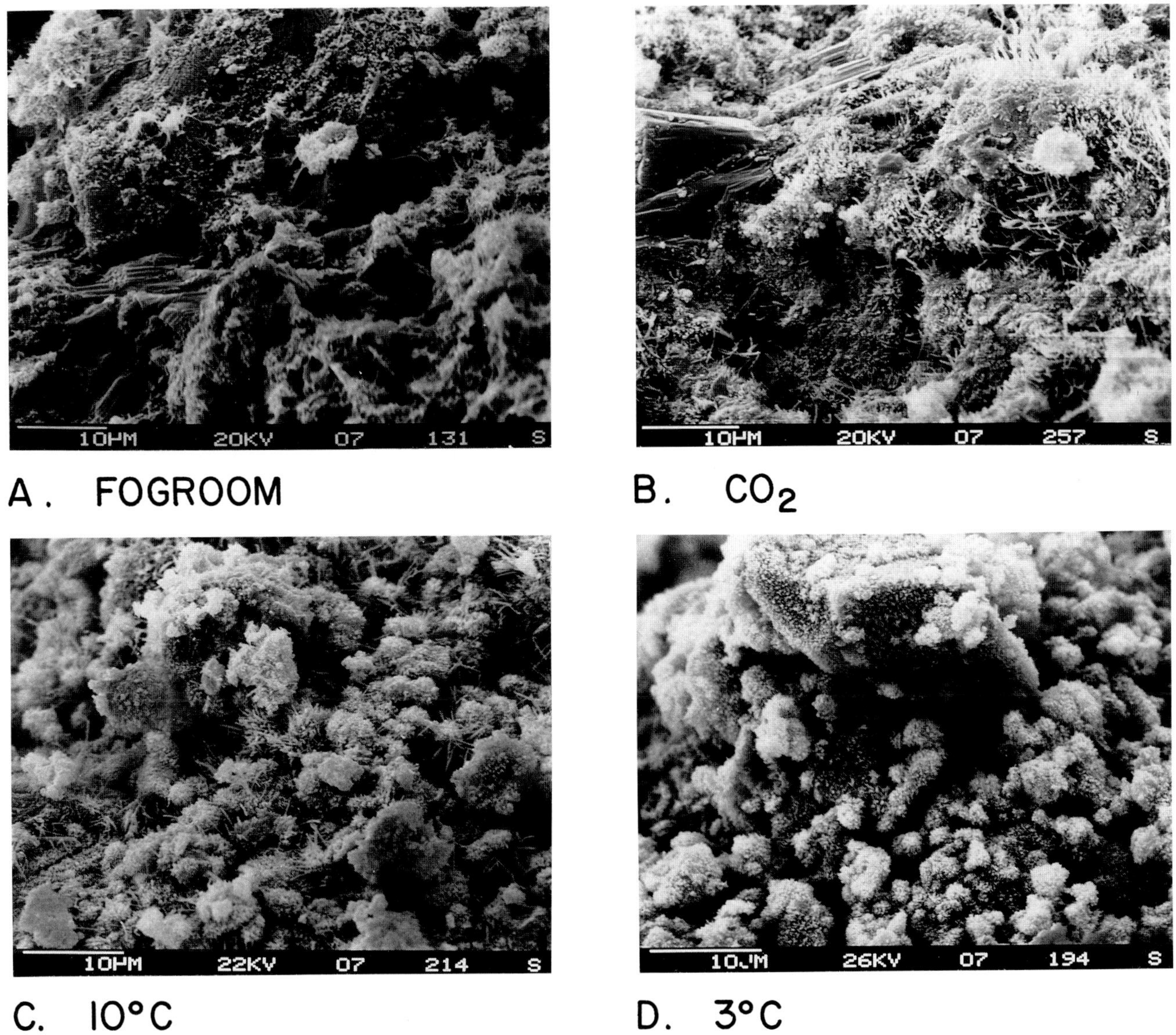

Figure 51.7. SEM micrographs of mix SF cured 1 day at different conditions.

Decrease in curing temperature was found to have the most pronounced effect on mixes containing the largest amounts of silica fume. Strength and permeability depend mainly on the degree of hydration, i.e., amount of C–S–H present. It is evident that temperature has more effect on the rate of the pozzolanic reaction than on the rate of cement hydration. In all mixes SEM observations confirmed the presence of $Ca(OH)_2$, unhydrated cement and silica particles after longer periods of curing at lower temperature than after normal temperature curing.

Hydrothermal Conditions (7 days at 230°C, 2.75 MPa, and 100% RH)

The cement blended with silica flour only (SF) and the mix blended with fume and sand in a ratio of 3 to 1 (FSB) showed the highest strength (Figs. 51.3 and 51.4) at all periods and conditions of precuring prior to 7 days of hydrothermal treatment. Permeability values (Fig. 51.6) were lower for all fume-containing mixes than for cement blended with silica flour; the highest values were recorded in most cases at total ages of between 14 and 28 days. The maximum permeability values were usually accompanied by minimum strength values; the strength "trough" (between 14 and 28 days) was deepest for mix SF and was absent for mix FSB. Precuring and hydrothermal treatment in a static CO_2 atmosphere did not significantly affect posthydrothermal properties (Fig. 51.3). Lower precuring temperatures resulted in slightly higher strength values and lower permeabilities, especially in the case of cement blended with silica flour. Comparison with values after corresponding periods of ambient curing shows that compressive strengths for all hydrothermally treated fume-containing mixes were lower,

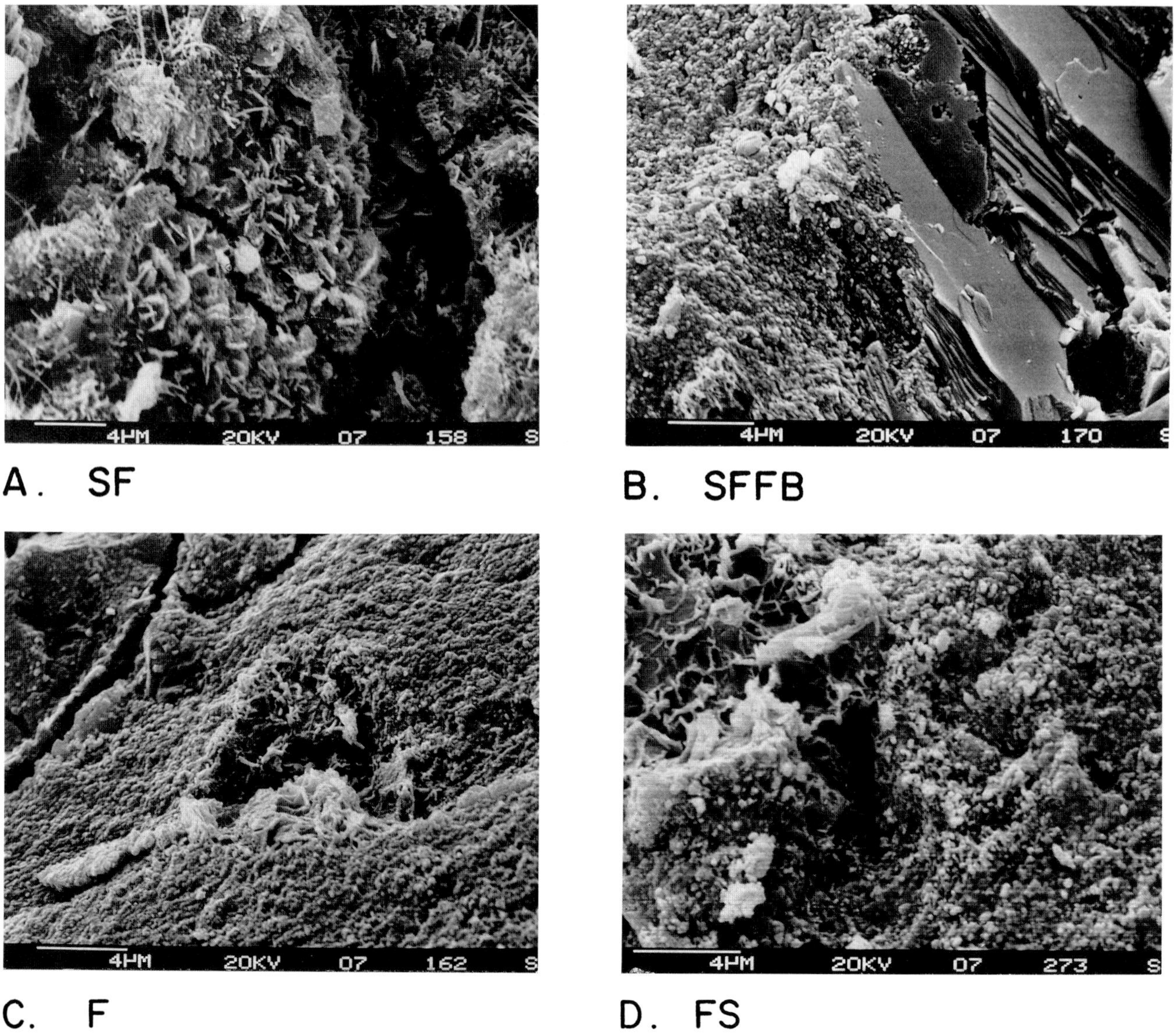

Figure 51.8. SEM micrographs of mixes cured 7 days in fogroom.

while blends with flour had higher strength. Permeability of the hydrothermally treated samples was one to three orders of magnitude higher for all mixes; values depended on the content of silica fume.

The trend toward increasing strength with increasing amounts of silica fume observed under ambient conditions was reversed after hydrothermal treatment (HT). Hydrothermal properties depended on the type of xonotlite (C_6S_6H) formed ($C = CaO$, $S = SiO_2$, $H = H_2O$). This is influenced by the initial mix composition (C/S ratio), by the type and amount of free silica, C–S–H, and availability of $Ca(OH)_2$ for hydrothermal reaction. The needle-like fabric of C–S–H II (Fig. 51.7) was retained in mix SF (silica flour) after HT (Fig. 51.9), probably due to epitaxial formation of xonotlite. The presence of fume at ambient conditions led to formation of finer, granular C–S–H I; this type of fabric was also retained after HT. The material was less porous and the associated lower strength is attributed to microcracking, which may have resulted from build-up of pressure on partial dehydration of C–S–H in the low permeability material. Single phase liquid conditions and associated high pressures may have developed within the micropores on conversion of more hydrous calcium silicates to xonotlite and other low hydrate phases.

A slight improvement in strength of mixes with flour and/or fume, precured and treated hydrothermally in CO_2, is most probably due to preferential formation of small amounts of scawtite ($C_7S_6\bar{C}H_2$; $\bar{C} = CO_2$).

Posthydrothermal properties of cement–silica blends were usually improved when samples were precured at 10 and 3°C;

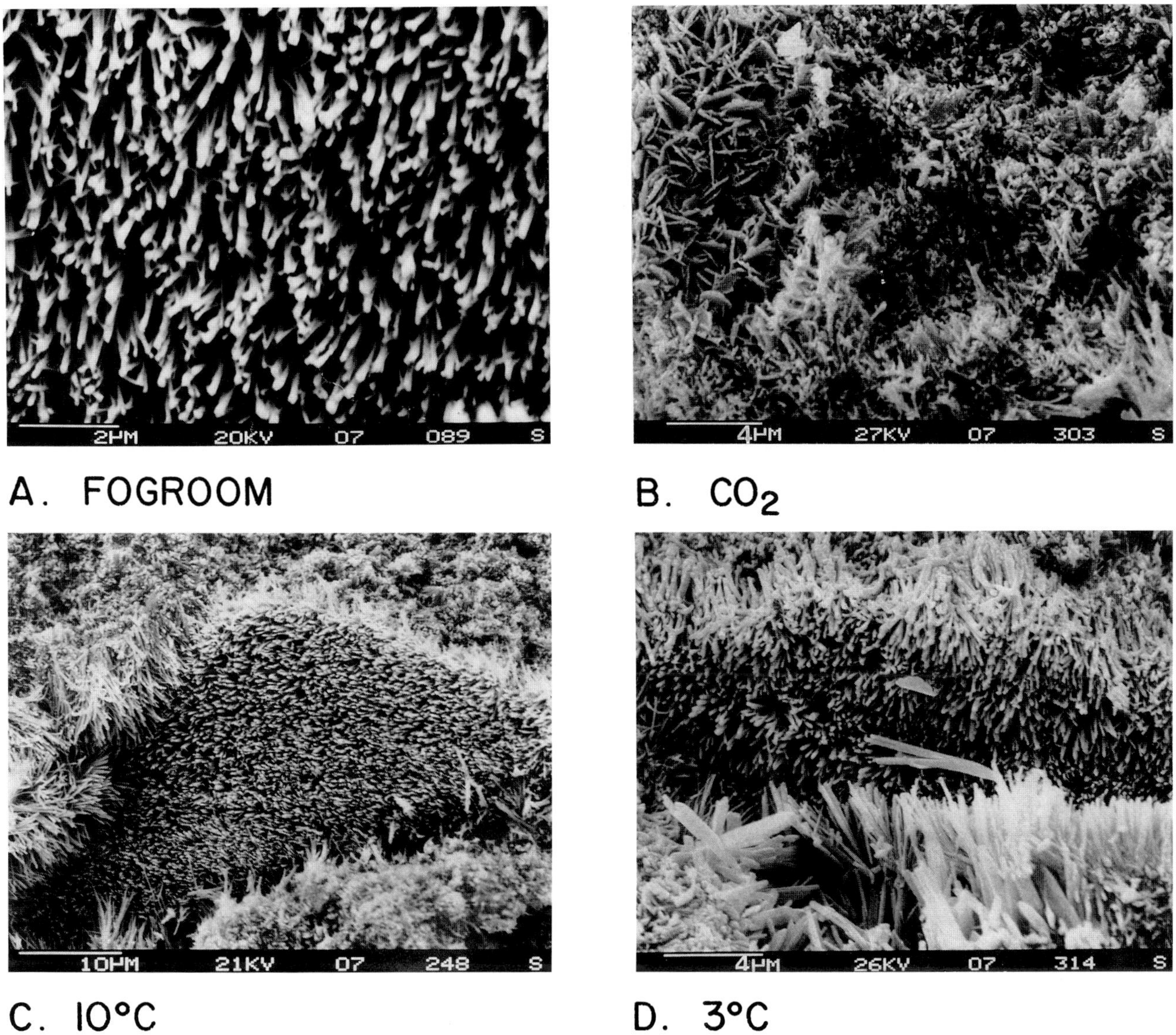

Figure 51.9. SEM micrographs of hydrothermally treated mix SF, precured at different conditions for 21 days.

the improvements were most significant for the cement blended with flour. Low temperatures slow down both hydration and pozzolanic reactions, and at a given period of precuring more unhydrated phases, free $Ca(OH)_2$ and silica are available to form the stronger form of needle-like xonotlite. The effect of precuring temperature on hydrothermal properties was most pronounced at ages of up to 28 days. Cements blended with fume and sand at a 3 to 1 ratio (FSB) showed superior hydrothermal properties to corresponding blends with fume and flour (SFFB) and flour only (SF). Strength was independent of precuring period but permeability decreased with time of precuring. Mineralogical investigation and SEM observations showed that all silica reacted with $Ca(OH)_2$ to form a needle-like form of xonotlite (Fig. 51.10). This is generally associated with high strength.

Conclusions

1. Control of mix proportions with respect to types of silica, period, and temperature of precuring prior to hydrothermal treatment makes it possible to achieve optimum strength and permeability under both ambient and hydrothermal conditions.
2. Under ambient conditions improvements in strength and permeability are governed by the amount of fume with respect to other forms of silica.
3. Normal hydration of calcium silicates in the presence of inert silica leads to formation of the needle-like C–S–H II; however, during the pozzolanic reaction granular C–S–H I is formed.
4. The difference in fabric between C–S–H I and C–S–H II is mainly responsible for differences in strength and permeabil-

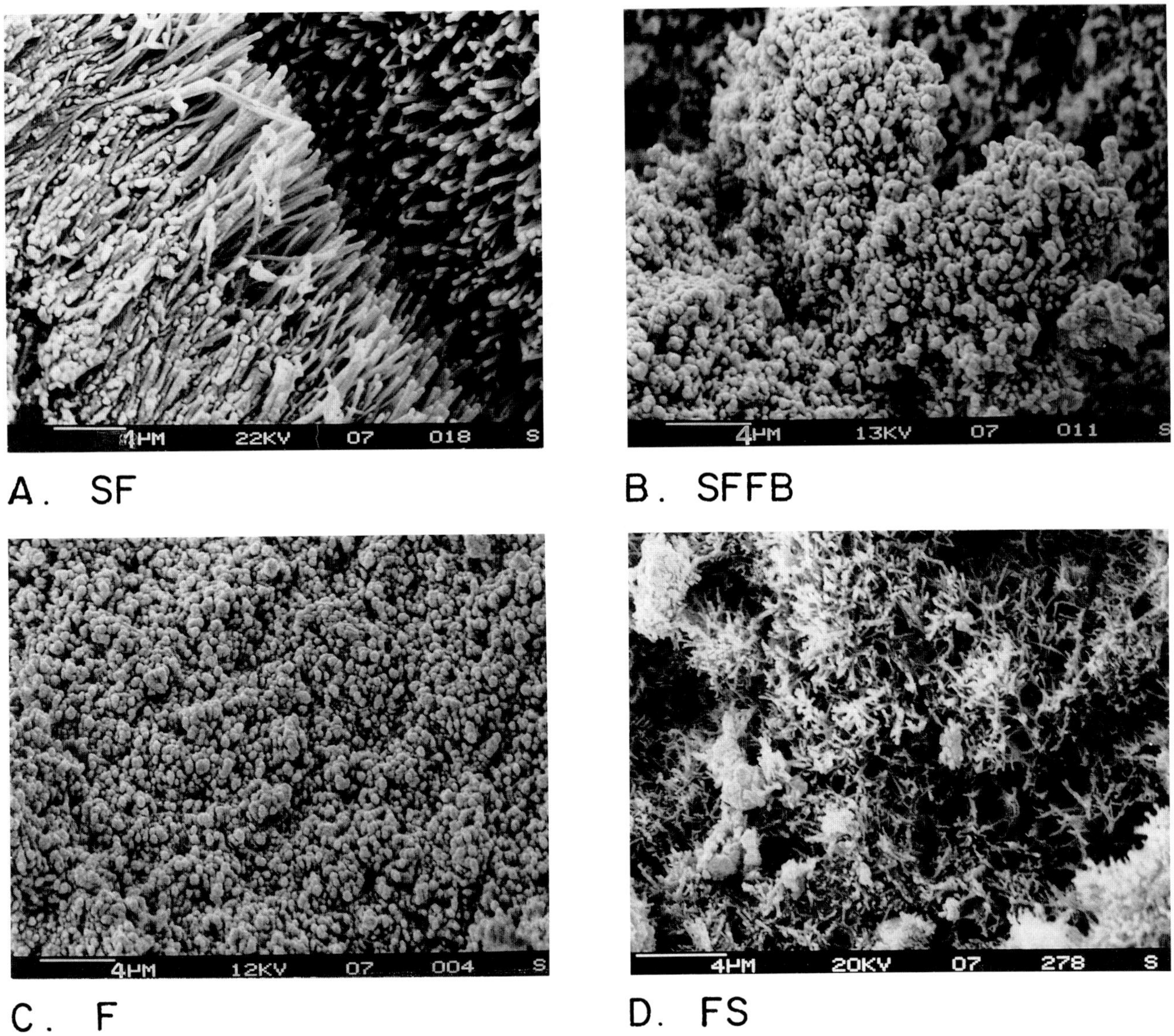

Figure 51.10. SEM micrographs of hydrothermally treated mixes, precured in fogroom for 1 day.

ity between cement blends containing various forms of silica and hydrated under normal or low temperature conditions.

5. The type of fabric formed under ambient conditions was retained after transformation of C–S–H I or II into xonotlite; the needle-like variety is stronger but more permeable than the granular type.
6. Precuring temperatures affect the rate of both the hydration and pozzolanic reaction and hence the proportions of hydrated to unhydrated phases; correspondingly properties are affected after hydrothermal treatment.
7. A static CO_2 atmosphere does not significantly influence the qualitative mineralogy or fabric of ambient or hydrothermally formed products.

Acknowledgments

The authors wish to express sincere thanks to AOSTRA and NSERC for financial support and NOWSCO, Calgary, for equipment loan. The technical assistance of Ken Velcic and secretarial help of Rene Wollin is acknowledged.

References

Berger, R.L., and W.A. Klemm, 1972. Accelerated curing of cementitious systems by carbon dioxide. Cement and Concrete Research, v. 2, p. 647–652.

Bruckdorfer, R.A., 1986. Carbon dioxide corrosion in oilwell cements. Society of Petroleum Engineers, SPE 15176.

Buck, A., and J.P. Burkes, 1980. Characterization and reactivity of silica fume. Proceedings, 3rd International Conference on Cement Microscopy (Houston), ICMA, Duncanville, p. 279–285.

Carette, G.G., and V.M. Malhotra, 1982a. Early-Age strength development of concrete incorporating fly ash and condensed silica fume. CANMET Report No. MRP/MSL 82-102, (OP & J) Draft, CANMET, Energy, Mines & Resources Canada, Ottawa, 25 p.

Carette, G.G., and V.M. Malhotra, 1982b. Use of silica fume in concrete - preliminary investigations. CANMET Report No. 82-1E, Energy, Mines & Resources Canada, Ottawa, 15 p.

Cheng-yi, H., and R.F. Feldman, 1985a. Properties of portland cement-silica fume pastes: I. Porosity and surface properties. Cement and Concrete Research, v. 15, p. 765–774.

Cheng-yi, H., and R.F. Feldman, 1985b. Properties of portland cement-silica fume pastes: II. Mechanical properties. Cement and Concrete Research, v. 15, p. 943–952.

Cheng-yi, H., and R.F. Feldman, 1985c. Dependence of frost resistance on the pore structure of mortar containing silica fume. Proceedings of American Concrete Institute, v. 82, p. 740–743.

Cheng-yi, H., and R.F. Feldman, 1985d. Influence of silica fume on the microstructure development of cement mortar. Cement and Concrete Research, v. 15(2), p. 285–294.

Eilers, L.H., E.B. Nelson, and L.K. Moran, 1980. High temperature cement composition: pectolite, scawtite, truscottite or xonotlite—which do you want. Society of Petroleum Engineers, SPE 9286.

Gallus, J.P., and D.E. Pyle, 1978. Performance of oilwell cementing compositions in geothermal wells. Society of Petroleum Engineers, SPE 7591.

Grabowski, E., and J.E. Gillott, 1989. Effect of replacement of silica flour with silica fume on engineering properties of oilwell cements at normal and elevated temperatures and pressures. Cement and Concrete Research, v. 19, p. 333–344.

Ho, D.W.S., and R.K. Lewis, 1987. Carbonation of concrete and its prediction. Cement and Concrete Research, v. 17, p. 489–504.

Lea, F.M., 1971. The Chemistry of Cement and Concrete, Chemical Publishing Co., Inc., New York 3rd ed. Chapter 9, p. 177–202.

Malhotra, V.M., 1984. Use of mineral admixture for specialized concretes. Concrete International, p. 19–24.

Malhotra, V.M., and G.G. Carette, 1982. Silica fume: A pozzolan of new interest for use in some concretes. Concrete Construction, v. 27(5), p. 443–446.

Mehta, P.K., 1983. Pozzolanic and cementitious by-products on mineral admixtures for concrete—a critical review. 1st International Conference on the Use of Fly Ash, Silica Fume, Slag and Other Mineral By-Products of Concrete, Montebello, Canada, v. 1, p. 1–46.

Quan, D.D., 1984. Effects of supercritical carbon dioxide corrosion on well cements. Society of Petroleum Engineers, SPE 12593.

Smith, D.K., 1976. Cementing. A monograph series of the Society of Petroleum Engineers of AIME, 184 p.

Young, J.F., R.L. Berger, and J. Breese, 1974. Accelerated curing of compacted calcium silicate mortars on exposure to CO_2. Journal of American Ceramic Society, v. 57(9), p. 396–397.

CHAPTER 52

The Role of the Microstructure of Pacific Red Clays in Radioactive Waste Disposal

Patti J. Burkett, Richard H. Bennett, and William R. Bryant

Introduction

The objective of this study was to examine the microfabric of red clay sediments from the Central Pacific deep-sea basin using transmission electron microscopy by assessing the response of the fabric to thermal and mechanical influences resulting from the testing of simulated radioactive waste. The affected samples analyzed were cored from remolded, laboratory consolidated, illitic-rich red clay used in the *In Situ* Heat Transfer Experiment Simulation test (ISIMU) (Percival, 1982). The fabric of these samples was compared to models, and to undisturbed samples designated RAMA, obtained from gravity cores at the MPG-1 site located in the Central Pacific Ocean. Specific objectives of this study included an analysis of the microfabric characteristics as a result of (1) mechanical disturbances from sediment remolding and heater probe insertion, (2) heating of the sediment causing induced thermal gradients, and (3) comparison of the "undisturbed" red clays to the remolded material. In addition we studied the changes in the physical properties, such as undrained shear strength, water content, void ratio, and mineralogy as they relate to the microfabric.

Background

The objectives of the Subseabed Disposal Program (SDP) were to evaluate the feasibility of disposing of high-level radioactive waste within fine-grained sediments of the world ocean basins by carefully designed emplacement methods, and to design and evaluate the technology required for the feasibility of subseabed disposal (Hollister et al., 1981; Percival, 1983; Percival et al., 1987). This concept is not to be confused with dumping of waste into the water column (prohibited by Article IV1a of the London Dumping Convention, 1978), or placement of waste on the sea floor. The SDP criterion for possible nuclear waste disposal considers four major factors. First, the site should be tectonically stable including no erosional evidence and minimal recent faulting. Second, the geologic processes and changes must be reasonably predictable within the same time scale as the rate of radionuclide decay, i.e., millions of years, and regionally uniform both horizontally and vertically. A third significant qualifier is the remoteness of available natural resources resulting in a marine "desert" with restricted biological activity and few mineral resources. Finally, man's activities and harmful climate effects must be distant from the site.

In accordance with established SDP criteria, the selected MPG-1 (mid-Pacific lithospheric plate, mid-gyre between the great currents) study area was located approximately 600 miles north of Hawaii in the Central North Pacific Ocean between the Mendocino and Murray Fracture Zones (Hollister, 1977). The sediment, collected in a water depth of 5705 m, is composed of cohesive, chocolate brown colored, predominantly clay-size particles referred to as abyssal red clays (Bryant and Bennett, 1988). They are composed of authigenic materials, windblown particles, and volcanic and insoluble biogenic debris.

The *In Situ* Heat Transfer Experiment (ISHTE) is a phase of SDP and the objectives are to provide data on the *in situ* thermal, fluid dynamic, thermochemical, and mechanical sediment responses to the emplacement of radioactive waste in the seabed. Measurements of the thermal response of the sediment were compared with numerical predictions to validate a thermomechanical model for deep-sea sediments (Olsen and Harrison, 1979; Percival et al., 1980). An integral part of ISHTE is the development and demonstration of the technology required to perform waste isolation experiments in the seabed in water depths of 6000 m (Percival et al., 1980, 1987). A scaled laboratory

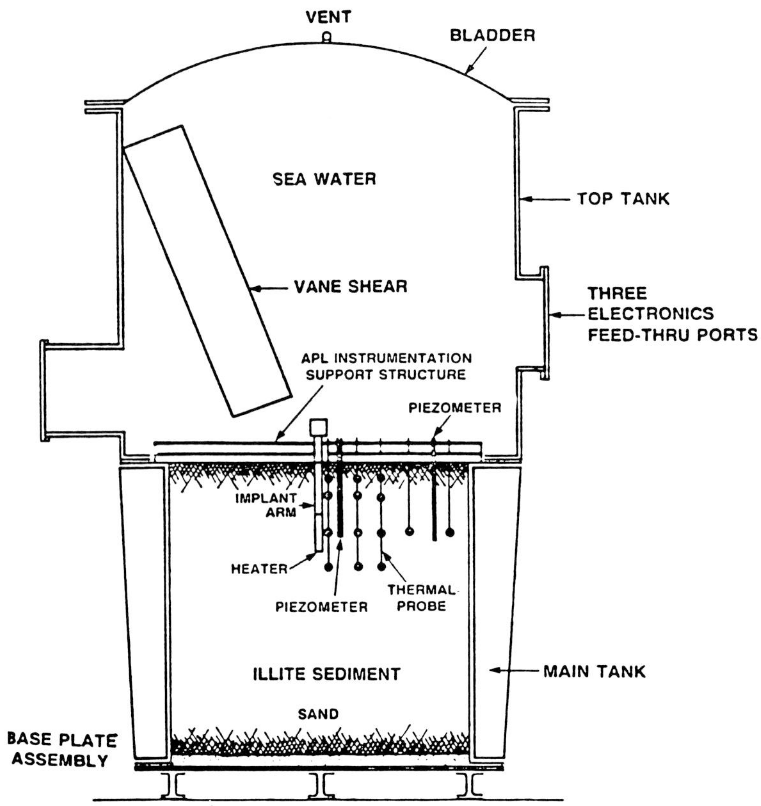

Figure 52.1. ISIMU test tank setup with reconstituted illitic sediment subjected to equivalent deep-sea ambient pressure in a hyperbaric chamber. (Redrafted from Silva and Jordan, 1983.)

ISHTE Simulation experiment (ISIMU) was conducted in 1981 by Sandia National Laboratories and participating institutions (Percival, 1982). ISIMU was designed to simulate predicted environmental field conditions and to test some of the instruments of a full-scale ISHTE experiment. The test was performed in a 1-m^3 cylindrical container filled with remolded, laboratory consolidated, illitic red clay recovered by dredged hauls from the proposed ISHTE test site (Fig. 52.1) (Silva and Jordan, 1984). ISIMU was conducted in a hyperbaric chamber at the David Taylor Naval Ship Research and Development Center, Annapolis, Maryland. A calculated scale of approximately 0.28:1.0 allowed adequate thermal equilibrium to be reached within 1 month. The dredged sediment was reconstituted in a rotary mixer with 34.2 ppt salinity seawater until a homogeneous slurry was formed. A series of thin (approximately 20 cm thick) layers of the slurry was added to the tank, and a vacuum was applied to each layer to remove excess air. Consolidation, occurring over a 10-week period, simulated *in situ* porosities of the deep ocean sediments (Silva and Jordan, 1983). The tank, sediments, and emplaced instrumentation were pressurized to 55 MPa at 4°C and the experiment was run for 1 month. After cool down, the heater probe was overcored and a second core adjacent was collected. The 12 fabric subsamples used in this analysis were collected normal to the longitudinal axis of the two cores at selected distances from the heater's edge and at measured depths below the mudline (Fig. 52.2). The RAMA control samples were subcored from undisturbed gravity core samples obtained from the MPG-1 site.

Methods

Grain-Size Determination

A remolded, laboratory-consolidated sample remote from the heater probe and a RAMA control sample were analyzed for grain size using Micromeritics Model 5000 particles size analyzer for the 4ϕ to 10ϕ (ϕ = negative log to the base 2 of the size in mm) silt and clay-size fractions. The calculations and resultant graph of cumulative percent versus phi distribution (ϕ) utilizes the principle of Stokes' law. The statistical data including the mean phi, standard deviation, skewness, kurtosis, and normalized kurtosis were calculated according to Folk and Ward (1957) for analysis and validation of the data in comparison to conventional data.

X-ray Diffraction

X-ray diffraction (XRD) was performed using techniques modified from Jackson (1969). Carbonates, organic matter, man-

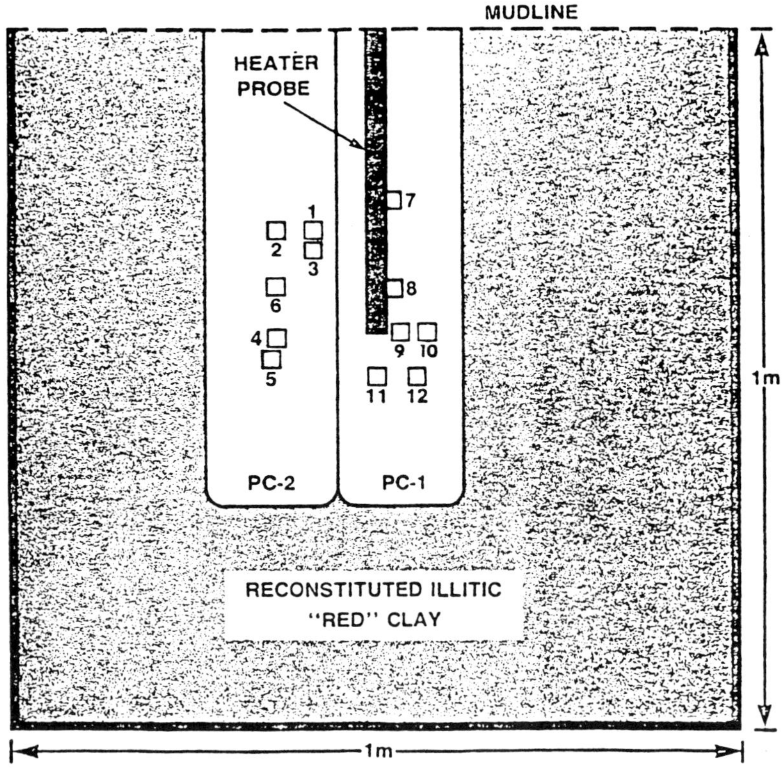

Figure 52.2. Diagram showing position of ISIMU heater probe in relationship to piston cores 1 and 2, and the subsamples used for clay microfabric analysis.

ganese dioxide, and free iron were removed. This was necessary since they act as binding agents, inhibit dispersion, and thus interfere with fractionation for mineral identification. Fractionation was accomplished by optimizing the pH, centrifuging with pH 10 Na^2CO^3 and flocculating in 1 *N* NaCl. The samples were saturated with magnesium and potassium. Oriented specimens were prepared by air drying a slurry of the clay on a glass slide. Vicor glass was used for the heated samples. The Mg^{2+} saturated samples were X-rayed at 25°C (room temperature), then glycerol solvated, and X-rayed again. The K^+ saturated sample slide was X-rayed following heat treatment of 25, 330, and 550°C.

Selected Area Diffraction

X-ray diffraction and selected area diffraction (SAD) analysis can provide information on lattice parameters and crystal symmetry of a mineral. SAD uses an aperture located in the column of the TEM to block all diffraction patterns except those obtained in a very small selected area of the specimen (Veblen, 1986). Different orientations of crystals can be recorded on SAD patterns, and varying intensities also can provide structural information. The mineral, illite, a phyllosilicate whose polycrystals yield a distinct ring pattern with varying diameters, was compared with the powder diffraction file d-spacings (distance between the crystal lattice comprised of periodic parallel planes of atoms) (ASTM Index, 1969), matching Miller indices and thus identifying the mineral species (Andrews et al., 1971).

Rose Diagram Orientation

Fabric orientation analysis was performed using a Rose diagram and frequency histogram. The direction of elongation of the particles and domains was measured from the micrographs in reference to an arbitrary straight line, in 10° increments from 0° to 180°. The particles measured were categorized by size. More than 1000 particles and small domains < 1 μm, and more than 300 elongated domains $\geq$ 1–5 μm, were examined. The resulting plot from the measured orientations, the Rose diagram, and frequency histogram provided a data base for comparison between samples. The frequency histogram was derived by calculating the percentage of particles in 10° increments from the total count and plotting frequency (%) versus degree (°).

Microfabric Sample Preparation

The sample preparation of fine-grain illitic-rich clay sediment used for the TEM microfabric analysis employed the methods first described by Bennett (1976). A detailed description of the

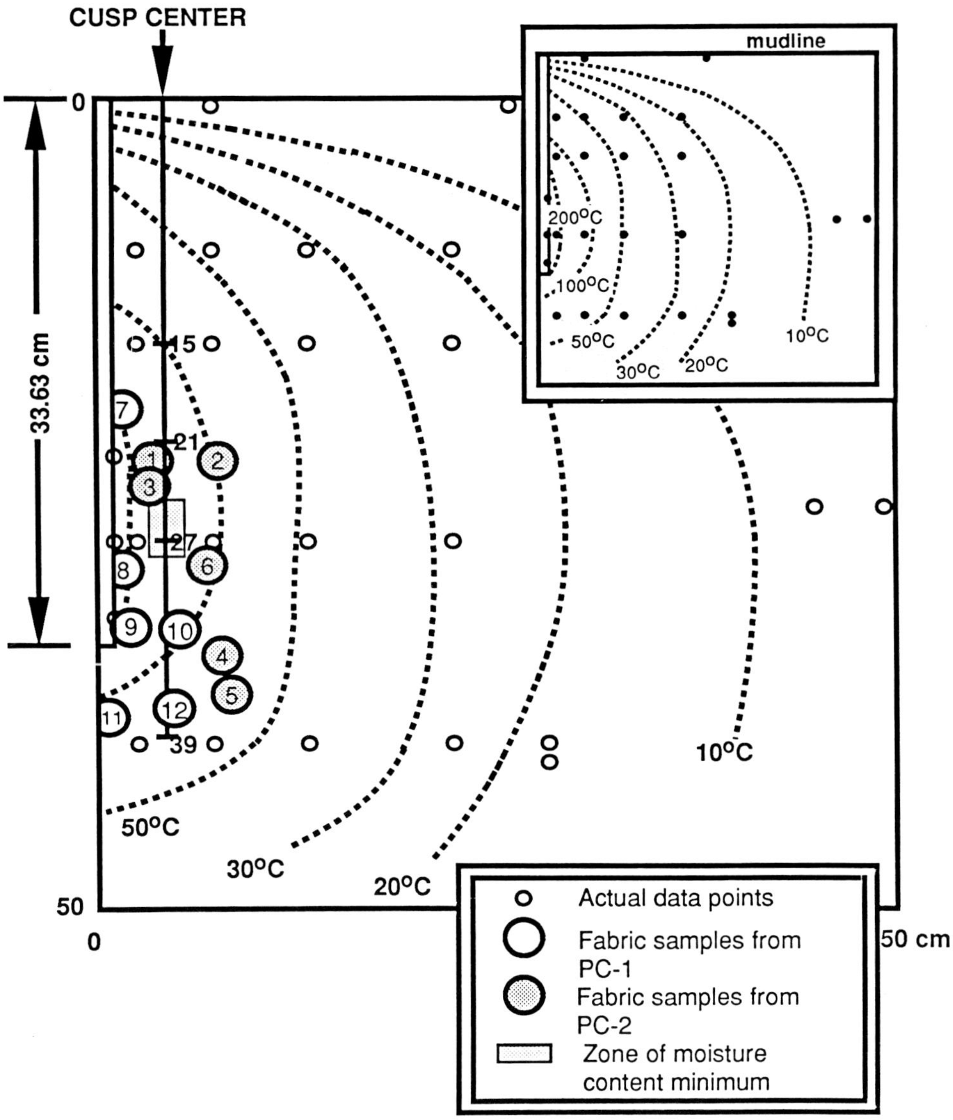

Figure 52.3. Steady state isotherm (drafted by H. Li, interpolated from experimental data, Miller et al., 1982) showing position of 12 microfabric samples, and the cusp samples used for X-ray diffraction analysis.

techniques is presented in this volume by Baerwald, Burkett, and Bennett. Micrographs using TEM were taken at magnifications of 8000× and 16,000×.

Results and Discussion

Thermal Effects

The significant effects resulting from this analysis as a result from thermal exposure include (1) mineral transformation, (2) physical properties alterations, and (3) diagnostic microfabric features observed in TEM.

The steady-state isotherm diagram (Fig. 52.3) shows the thermal fields generated by the heater probe for a total power dissipation of 160 W. The highest temperature generated, 296°C, occurred in a narrow region surrounding the lower 15 cm of the heater probe. Fabric sample B-8 originated from the highest intensity thermal zone, 200–296°C. Rapid decreases in temperatures were recorded radially away from the heater. Locations of the fabric samples relative to the distance from the heater edge and to the mudline depth are found in Table 52.1. Fabric samples B-1, B-3, B-7, and B-9 fell within the 100–200°C isotherm. Samples B-2, B-6, and B-10 were located proximal to the 100°C isotherm. Fabric samples B-4, B-5, B-11, and B-12 were exposed to temperatures within the 50–100°C zone.

The water content (expressed as percent dry weight) of the sediment from the cusp between cores PC-1, PC-2, and PC-3 varies with sediment depth. Figure 52.4 illustrates the 10% reduction (102% to 92%) in water content from 8 to 33 cm below

Table 52.1. ISIMU fabric samples by identification number, core number, distance from the edge of the heater, and depth below mudline.

Sample number	Core number	Depth below mudline (cm)	Distance from heater edge (cm)
B-1	PC-2	21.4–22.9	1.5
B-2	PC-2	21.4–22.9	5.5
B-3	PC-2	23.1–24.6	1.5
B-4	PC-2	33.4–34.9	5.5
B-5	PC-2	35.9–37.4	6.0
B-6	PC-2	27.5–28.0	5.5
B-7	PC-1	18.1–19.6	0.0
B-8	PC-1	27.8–29.3	0.0
B-9	PC-1	32.8–34.3	0.5
B-10	PC-1	32.8–34.3	3.5
B-11	PC-1	37.3–38.8	Directly under heater
B-12	PC-1	37.3–38.8	2.5

the sediment surface (mudline). Beyond the end of the probe at 33 cm, the water content increased slightly less than 10%, which is close to the value at 8 cm. Baking, in the ISIMU experiment (assuming 100% saturation of the sediment), is followed by consolidation of sediments, loss of pore fluid, and an accompanying decrease in void ratio. Radially, water contents are low between 2 and 4.7 cm, including the cusp zone created between cores 1, 2, and 3. Beyond approximately 5 cm, the water content fluctuates. The grided area of Figure 52.4 represents the zone of minimum water contents. No fabric samples were collected from this zone, but assuming axial symmetry of the thermal gradients (100–200°C), sediment of samples B-1, B-3, B-9, and B-10 probably experienced the same thermal effects from heating as the grided zone. The significant decrease in the water content near the heat source may have resulted from induced thermal effects resulting in pore water diffusion, consolidation, and chemical reaction.

The undrained shear strength profiles are depicted for three vertical test sequences: pretest unheated, heated, and post-test cooled conditions (Silva and Jordan, 1984 and Fig. 52.5). The large shear strength increase between 20 and 32 cm sediment depth during heating corresponds directly to the water content decrease (Fig. 52.4). The 38 kPa (375 g/cm²) shear strength maximum, occurring at 170°C corresponds to the 33 cm depth at the end of the heater probe and probably results from localized dewatering as a result of heating. The comparison of the maximum shear strength profile during heating and the profile obtained before heating indicates a 10-fold increase in shear strength. At the termination of the test (after cool-down), the sediment shear strength appears to have decreased, but not back to its original pretest strengths. This implies a slight thermally reversible or a residual characteristic strength following cool-down of this sediment type.

Samples located at various depths approximately 3.5 cm from the heater centerline in the cusp between cores PC-1, PC-2, and PC-3 were analyzed for clay mineralogy (Fig. 52.6). At a depth

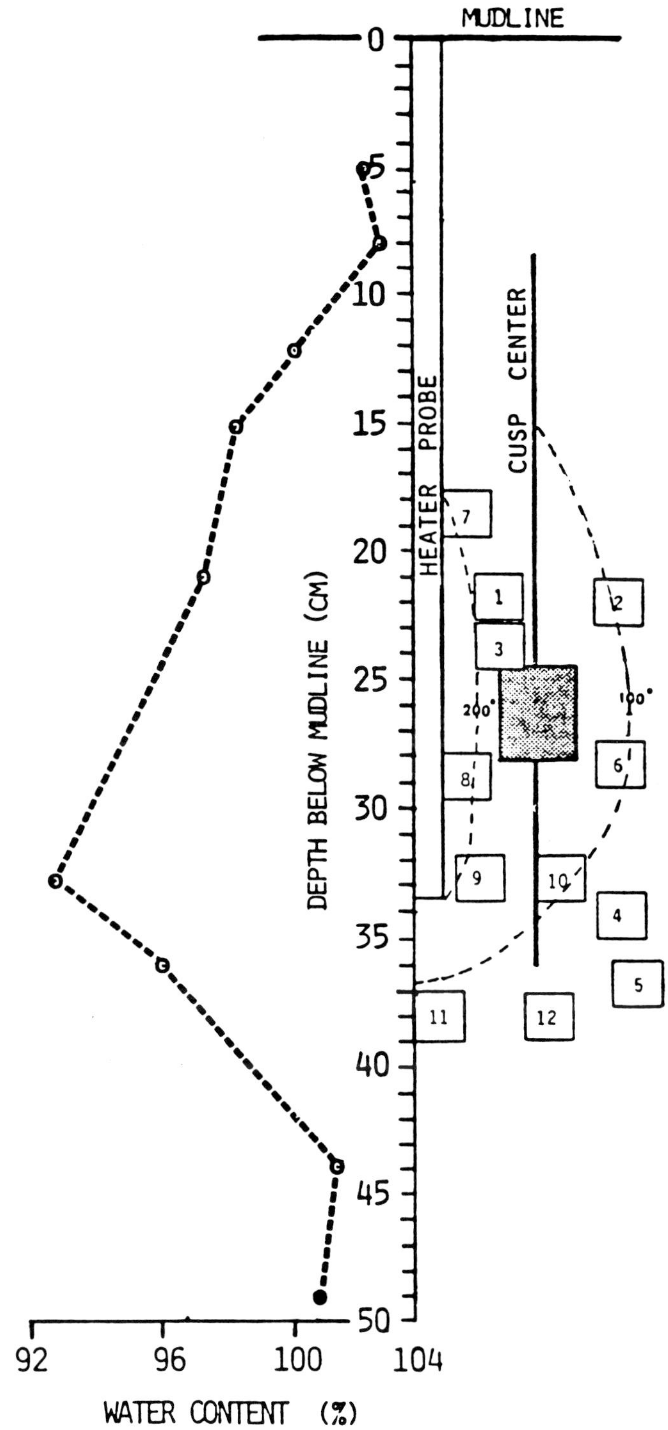

Figure 52.4. Water content versus cm depth profile for ISIMU post-test samples showing water content minimum corresponding to the end of the heater probe.

of 7 cm, the sediment was comprised of approximately 5% smectite, 34% chlorite, 48% illite, and 13% kaolinite. At 24 cm the mineral composition changed to 40% smectite, 20% chlorite, 40% illite, and trace amounts (less than 5%) of kaolinite. After

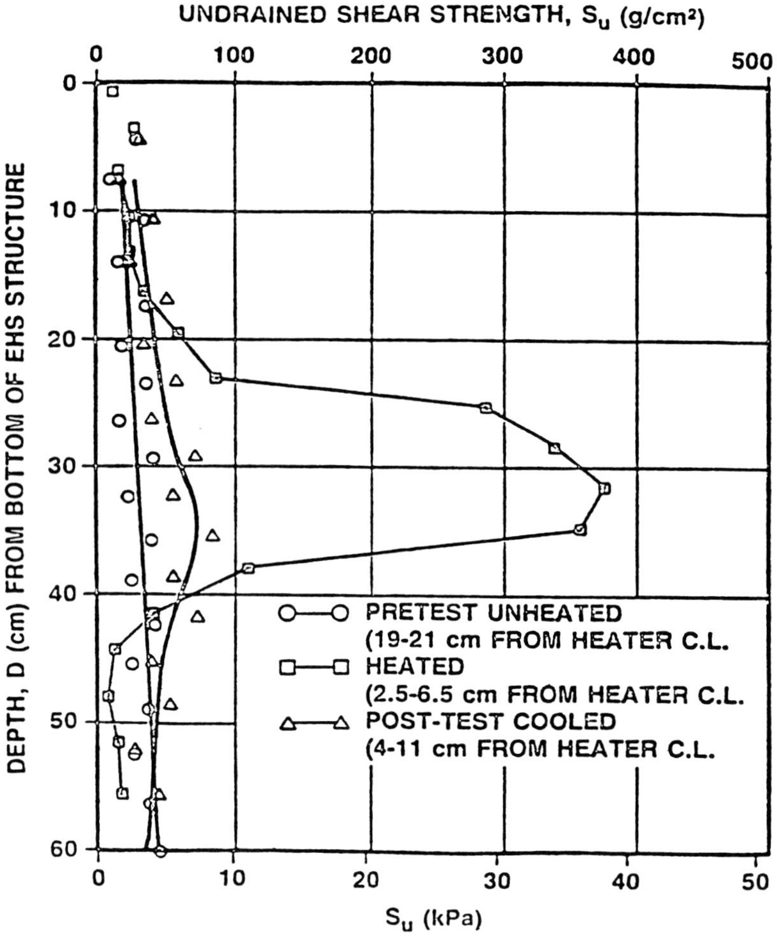

Figure 52.5. ISHTE simulation combined profile. Note the substantial increase in shear strength associated with the heated zone. (Redrafted from Silva and Jordan, 1983.)

consideration of the limits of instruments detection (at least 5% of a mineral in question is necessary for confirmation) and possible errors in calculation, a significant smectite increase accompanied by a change in color from brown to gray is evident. The 7 cm depth is associated with a thermal gradient between 50 and 100°C and water content of 103% (Fig. 52.3). At 24 cm the temperature recorded during the experiment increased to 170°C, with an associated post-test water content of 97% (Fig. 52.4). The water content is a measure of the intergranular pore water and free water or interlayer water not strongly attracted to the mineral skeleton. Routine water content determination procedure dries the sample in a 105°C oven (Lambe, 1951). Greater temperatures remove more of the loosely bonded water, although temperatures of approximately 500°C are required to remove the adsorbed water layer of smectite, therefore collapsing its structure. The detection of increased quantities of smectite is observed by an increase in the peak height on the X-ray patterns. This was evident in the cusp sample at 24 cm (Fig. 52.6). Thorton (1983) studied the transition of a sediment's mineralogy from mixed smectite–illite to an increase in smectite and decrease in mixed-layer clays, after being maintained in a 200–300°C hydrothermal apparatus. Other laboratory syntheses of smectite at elevated temperatures and pressures were carried out by Noll (1930), Sedletski (1937), Harward and Brindley (1964), and Seyfried and Bischoff (1981). Potassium is the cation associated with illite, which binds successive layers. From chemical analysis of the pore fluids, an increase in potassium level was detected in an area adjacent to the cusp. A thermally induced release of interstitial potassium may have resulted in the transformation to smectite, although other geochemical processes and other minerals may have played a role in smectite formation.

Orientation of particles was studied by examining the micrographs of samples B-4, B-5, and B-7 compared to the RAMA samples for possible thermal heating and/or mechanical remolding effects of the sediment (see Table 52.1 for exact location relative to the induced thermal fields generated). Because RAMA was not subjected to the heat source, its ambient *in situ* temperature would have been approximately 4°C.

Analysis of more than 1000 particle measurements on the RAMA mosaic, showing an area of approximately 2500 μm² (Fig. 52.7), revealed a random arrangement of particles (Fig.

52.8) as seen in the histogram and Rose diagram. From similar analysis of samples B-7 (nearfield), B-4 (far-field), B-5 (far-field) has the strongest degree of preferred orientation demonstrated by its Rose diagram and frequency histogram but was subjected to less heat (for further details see Burkett, 1987). Thus particle orientation appears to be independent of sample location relative to the generated thermal gradient determined from orientation analysis. Additional analysis of remolded, nonheated fabric is needed to evaluate the dominant parameter causing particle alignment, heating, or remolding or both. If the remolded samples do demonstrate preference of particle alignment, as indicated from the individual micrographs, perhaps the movement of pore fluids could be sufficient to move particles or small domains to a preferred orientation. Consolidation, the decrease in volume of the sediment resulting from the loss of pore fluid, is an expected outcome from exposure to intense heat. The flow of water originating in the "baked" zone could have affected the samples in the cooler zones. The processes of consolidation and fluid transport during ISIMU have been thoroughly documented analytically by McTigue et al. (1986).

Mechanical Disturbance

Mechanical disturbances of the sediment includes effects from probe insertion and remolding. (Detailed analysis and discussion of laboratory consolidation procedures are found in Bryant and Bennett, 1988.) Remolding can be considered in terms of primary remolding during preparation of the red clay sediment before emplacement in the ISIMU test tank (Fig. 52.1), and a secondary or localized remolding from heater probe penetration into the prepared consolidated sediment. Clay fabric disturbance, as a consequence of probe insertion, is difficult to assess using TEM micrographs. Extreme care was taken when selecting fabric subsamples to avoid any obvious preexisting physical disturbances such as fractures or microcracks in the sediment. The fabric samples were collected near the end of the probe at depths of ~21 to ~38 cm, therefore surface deformation effects of insertion are not considered. Little difference in traction stresses on probe insertion was found with varying textures (0.13–2.16 μm) of the Inconel probe's surface (Silva and Jordan, 1984). A 2 μm surface texture finish for the Inconel probe was chosen to provide good adhesion (forming a good seal) between the heater and remolded sediment, thus preventing water migration along the heater wall. Post-ISIMU, the heater was overcored, the core liner cut, then the sediment was sliced normal to the probe for sampling. If the probe had been extracted, suction forces may have been sufficient to result in large localized sediment deformation. Only fabric samples B-7 and B-8, adjacent to the probe, were expected to show any effect from traction force during insertion. Samples B-7 and B-8 are closest to the heater wall (Fig. 52.2) and generally would be expected to exhibit the most obvious signs of deformation from probe insertion. The fabric subsamples were extracted from the central portion of the larger samples, and those samples in the plastic zone would be expected to have characteristics such as smearing of the probe–sediment interface and possibly some lateral deformation from probe insertion.

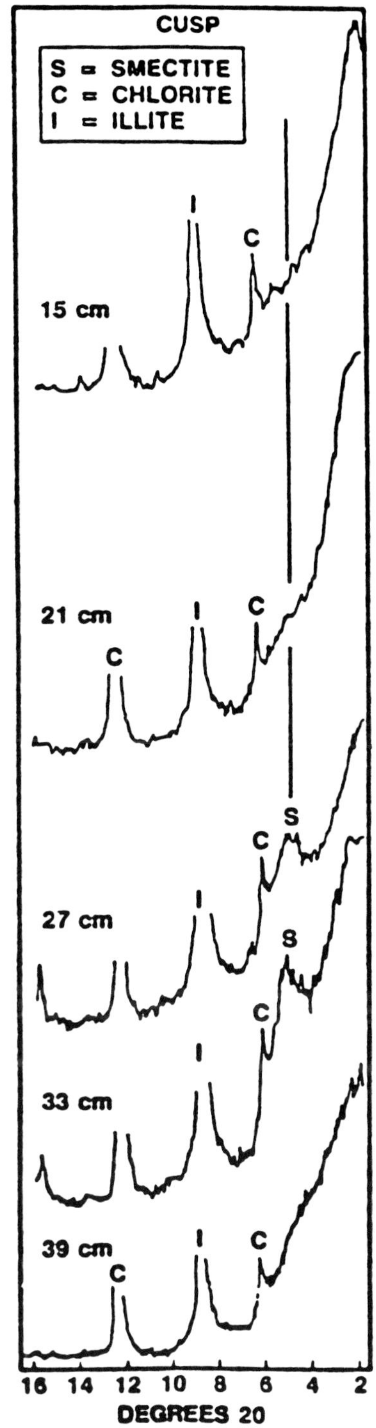

Figure 52.6. X-Ray diffraction pattern showing pronounced increase in the smectite peak occurring at the 27 cm and 33 cm depths.

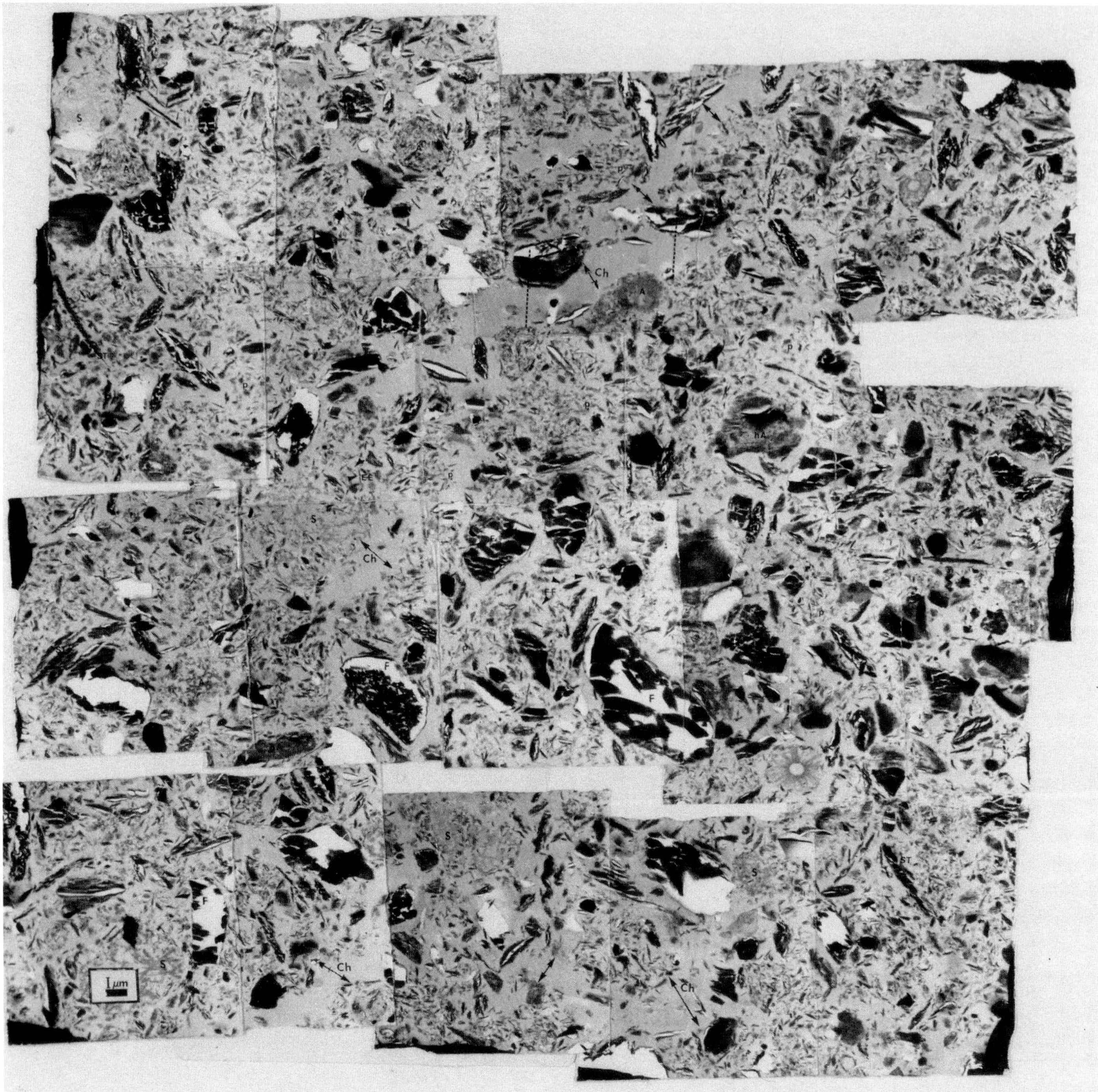

Figure 52.7. Mosaic of the RAMA undisturbed illitic-rich Pacific red clay sediment showing complex assemblage of particles comprising the *in situ* microfabric. Some features include a channel (CH) for transport of pore fluids, fractured illite argillites (F), high void ratio smectitic aggregates (S), denser aggregates (AA), and matrix of ≤ 1 μm domains and small particle (1 μm bar scale in upper left).

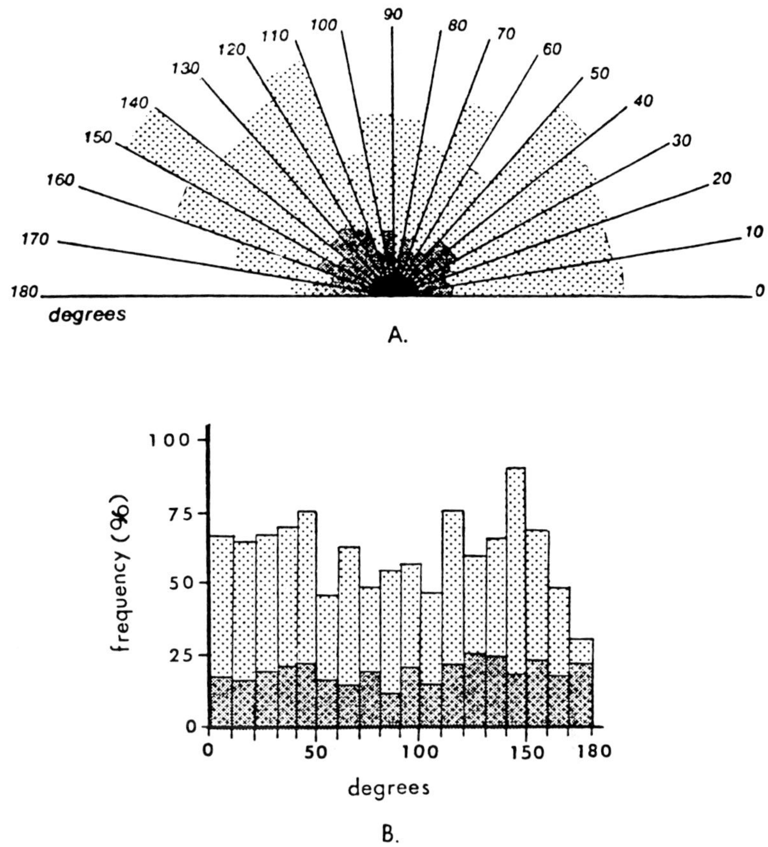

Figure 52.8. Rose diagram (A) and frequency histogram (B) for RAMA mosaic showing a completely random distribution of particle arrangements. The dark areas represent 2–5 μm domains, and the lighter areas represent ≤1 μm small particles and domains.

Primary remolding from mechanical disturbances occurs as the illitic-rich dredged sediment is reconstituted and then churned in a cement mixer before consolidation to *in situ* overburden pressures. The expected results from remolding would be breakage, swirling, or bending of long chains of particles or elongated domains and collapse and breakage of fragile flocs and aggregates. These effects would be more pronounced on longer particles than on individual particles or subrounded dense aggregates. If significant, primary remolding effects should be evident in all cores, without regard to the location in reference to the heat source.

Grain size analysis was performed on two samples. RAMA represented the control and core PC-5 was used to assess the effects of reconstituting and laboratory remolding of sediment for experimental purposes. Core PC-5 represented a far-field sample, and was located well beyond the effects from intense heat such as a decrease in water content or an increase in shear strength. The grain size determination resulted in a striking similarity between the two samples (Fig. 52.9). They were both homogeneous (well sorted) with only trace amounts of sand (greater than 4ϕ). The predominate particles were clay-size (≤3.9 μm) and comprised ~90% of the sample. The silt content of both samples was ~10%, and the remaining portions consisted of sand. These results imply negligible effects on the clay-size particles from reworking of the illitic-rich pelagic sediment. For all practical purposes the two samples analyzed are nearly identical.

Microfabric Analysis

The major mineral constituent was identified using selected area diffraction techniques. The analysis was performed by comparing polycrystal patterns to an ASTM standard and revealed the presence of the clay mineral illite. Figure 52.10 shows the measured, calculated parameters and percentage differences between the ASTM file values of *d*-spacings and the measured ring diameters of the unknown. The two values range from 0.8 to 6.0% deviation. The *hkl* values can then be assigned to the respective ring. Although polycrystals, suitable for good SAD patterns, located within the mixed assemblage of the *in situ* fabric are not common, many similar and fractured crystals are prevalent. This is significant because illite, the most abundant mineral as determined by X-ray diffraction analysis, now can be positively identified mineralogically and morphologically on the microscale. The identification of individual clay-size particles from mixed mineral assemblages can provide a better understanding of the microstructure and its physicochemical forces that control the sediment's behavior and response to static and dynamic loads.

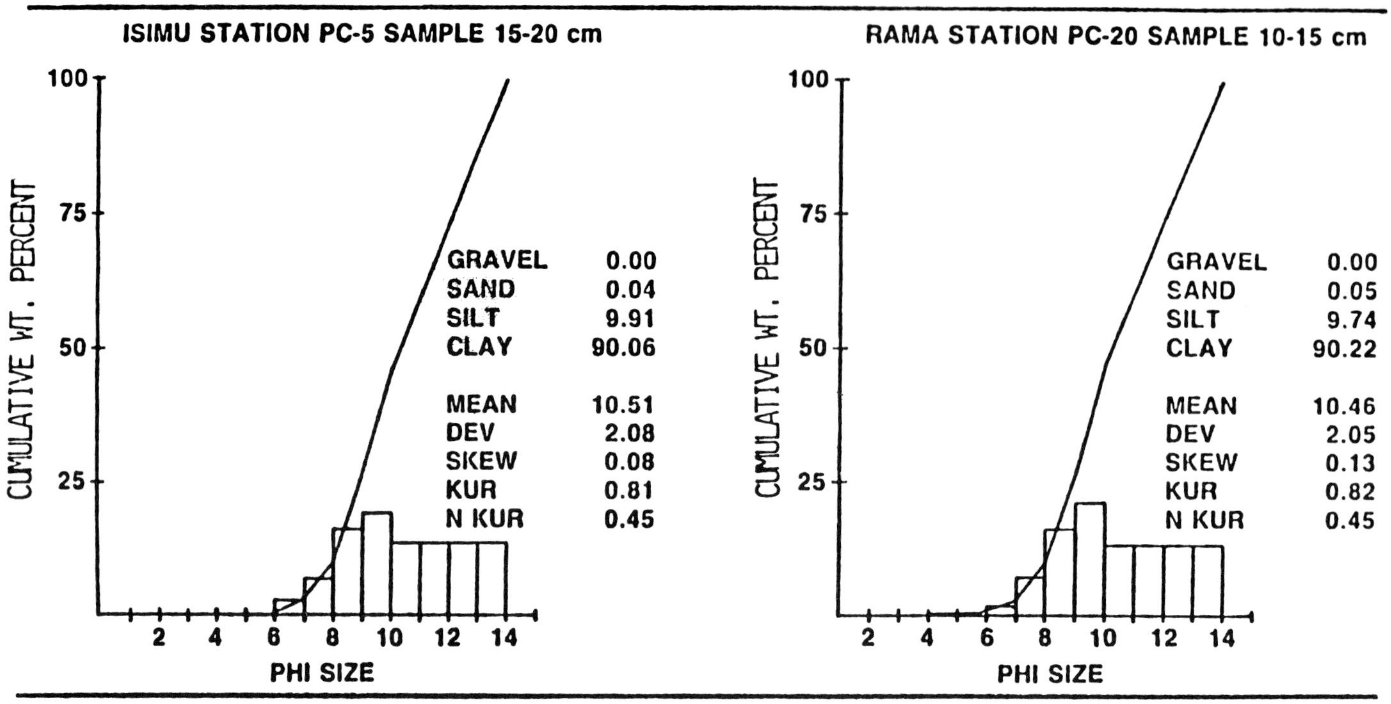

Figure 52.9. Grain size comparison of RAMA sample to sample PC-5, remolded and laboratory consolidated to equivalent *in situ* porosity. Notice the striking similarity of the grain sizes.

The sediment response from the ISIMU experiment was evaluated by examining TEM micrographs of the 12 fabric samples from ISIMU cores PC-1 and PC-2. Evidence of disturbance from probe insertion was seen in near-field fabric samples B-7 and B-8 adjacent to the heater wall. Sample B-8 shows a slightly denser arrangement of particles than far-field samples such as B-5 (Fig. 52.11A). The fabric has the appearance of broad bands between which are long zones of oriented small domains. These appear to be compression features resulting from probe insertion, and look like microfractures, slip planes, or shear planes. Figure 52.11B, of sample B-11 located under the probe, is an enlargement of the localized oriented chains, predominantly face-to-face, forming a band about 0.5 μm across. Figure 52.11C demonstrates how the shear planes become aligned with chains of oriented domains. The microfracture sample of Figure 52.12A (sample B-8) is subtle because of the lack of separation between the slip planes, but demonstrates the zone of locally oriented particles. Figure 52.12B shows evidence of disturbance as seen by the gap lined with oriented particles. Notice the bent domains, possibly from shearing with planes containing obstacles consisting of larger, subrounded argillites (aeolian shale domains fractured during ultrathin sectioning).

A possible diagnostic feature of primary remolding is evident as chains of swirled domains caused by initial mixing of the sediment before emplacement in the pressure tank. Figure 52.13A of sample B-8 shows how the chains are wrapped around the central group of particles. The voids formed by the swirls seem interconnected and might provide a pathway for movement of pore fluids. Figure 52.13B shows an enlarged swirled or bent domain, a primary remolding feature. These disturbance features have been observed in other studies (Bennett et al., 1977; Bohlke and Bennett, 1980). Evidence of mixing or stirring is readily apparent in long chains, but the dominant particle type in this fabric is short chains; thus this type of feature is not obvious or generally apparent in the severely remolded microfabric. Therefore primary remolding effects would be expected to be present in all the ISIMU samples, but subtle. Figure 52.13C of the RAMA sample depicts an *in situ* undisturbed fabric. Notice the straight chain of edge-to-edge domains not influenced by processes involving remolding or probe insertion.

Although the overall fabric of the near-field sample appears slightly denser than the RAMA fabric (Fig. 52.13C) as a result of probe insertion, frequent large void forms (Fig. 52.14A–F) exist that are lined with oriented particles and small domains (≤1 μm) arranged stepped face-to-face and end-to-end. These void forms can be thought of as "quasiexpansion" features and appear to be cross sections of microchannels formed by pore fluids moving through the sediment after expansion from ther-

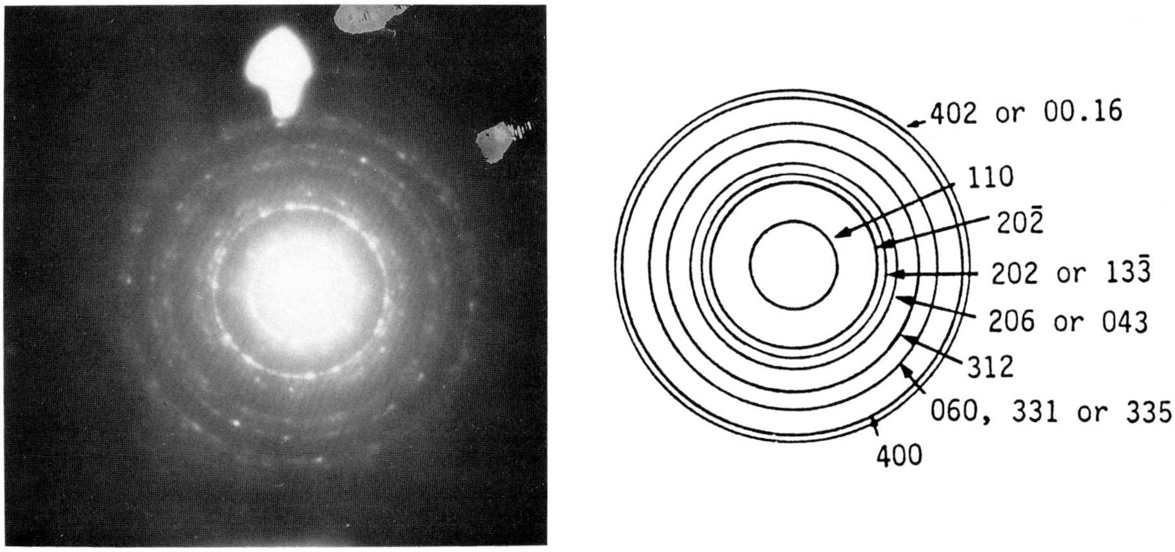

Measured Dia. mm	Camera Constant Divided by Dia.	ASTM File Dia.	% Deviation (ASTM-Calc/Calc)	hkl
12.5	4.736	4.46	6	110
22.3	2.653	2.57	3	202
24.5	2.416	2.45	1.4	202
		or		133
27.0	2.193	2.14	2.5	206
		or		043
34.2	1.731	1.65	2	312
38.6	1.534	1.50	2.3	060
		or		331
		1.345	2	335
45.0	1.316	1.297	1.5	400
47.0	1.259	1.269	0.8	402
		or		
		1.245	1.1	00.16

Figure 52.10. Illite identification using selected area diffraction (SAD). (A) TEM micrograph of SAD pattern from polycrystalline illite. (B) Drawing of the ring patterns with their Miller indices (*hkl*). (C) Table showing the $<6\%$ deviation between the measured diameters of the unknown sample and the ASTM file values.

mal forces. The pore fluids on heating may transport and push the smallest particles forming the lining of the channels. These conduits are relatively free of small particles, except those too large to be transported, as seen in Figure 52.14B–E. The resin-filled void forms are not to be confused with the white holes in the ultrathin sections (Fig. 52.15A, upper left-hand corner) caused by brittle domains being plucked during microtoming.

An interesting feature more prevalent in the high temperature zones is the appearance of "onion-skin" layering (first described by Brewer, 1964) of stepped face-to-face (Fig. 52.15) particles surrounding the low-density "smectitic" aggregate depicted in Figures 52.15 and 52.16. Cation exchange capacity (CEC) is the measure of a mineral's negative charge being balanced by exchangeable cations. Smectite has a high CEC exemplified by the "onion-skin" layering. Beyond the thin layer of particles lining the smectites are microchannels (Fig. 52.16), important in the transport of pore fluids and small particles. In Figures 52.15B and 52.16B, notice different types of aggregates having the "onion-skin" around them. These predominantly illitic aggregates have a lower CEC, and therefore do not attract small particles of opposite charge. The sediment distribution coefficient (K_d) is the ratio between the sorbed and dissolved metal cations located in sediment pore water (Heath, 1977). Thus increased smectite inhances the barrier effect of the sediment associated with an increased K_d. The aligned smectites exist in the RAMA samples as well, but not to as great an extent (Fig. 52.7).

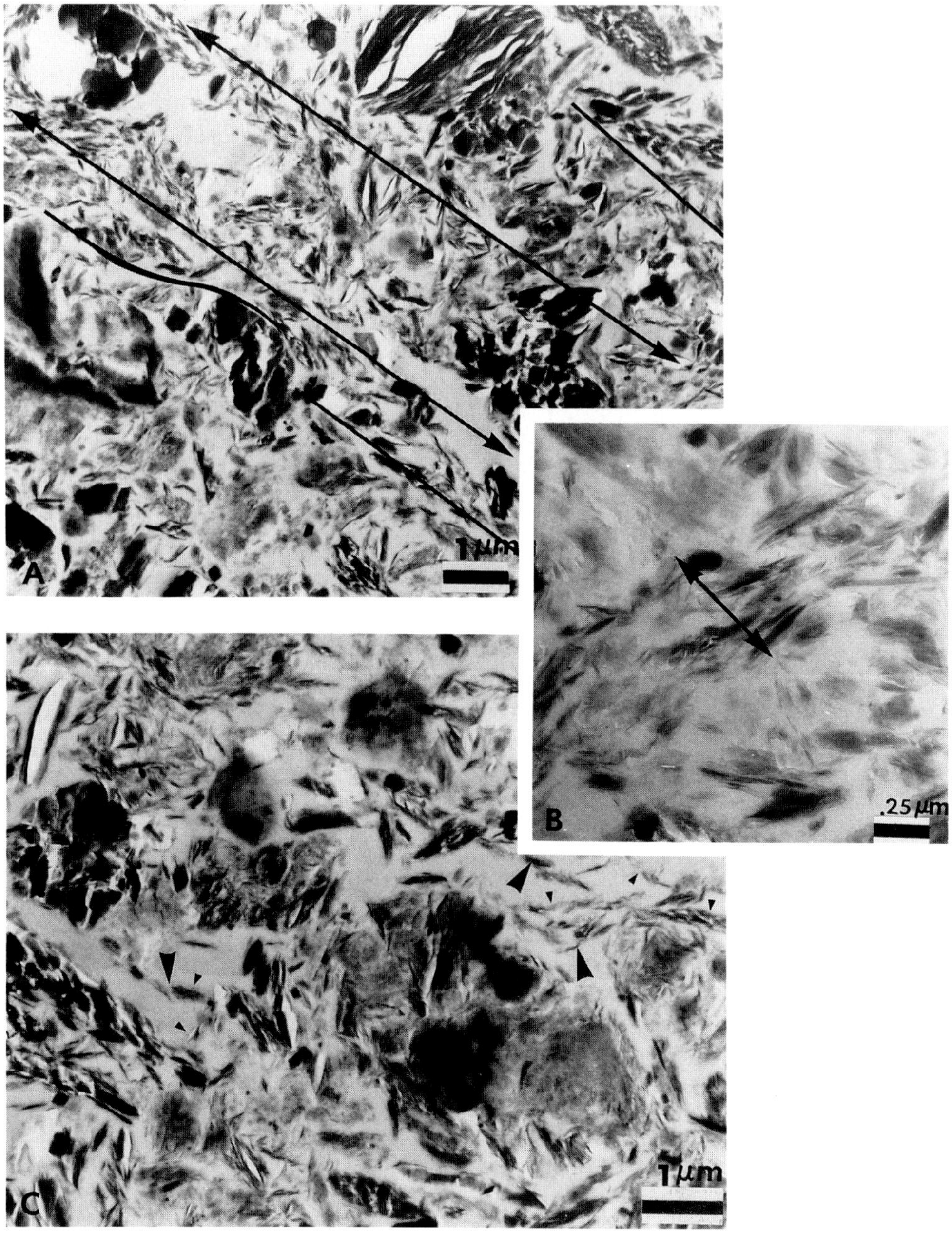

Figure 52.11. Probe insertion disturbance features. (A) Denser arrangement of particles forming broad zones defined by microshear planes in the near-field fabric sample B-8. (B) An enlargement of the microfracture showing ~0.5 μm band of oriented small F-F domains. (C) A microfracture lined by oriented domains (arrows).

Another important aspect of this study was to determine if this remolded, reconstituted, illitic-rich clay sediment is a reasonable material for laboratory study and testing of geotechnical properties. Does it adequately simulate the *in situ* properties and behavior of the red clay deposits? The RAMA mosaic (Fig. 52.7), composed of 20 micrographs representing an area of 2500 μm², demonstrates the complexity of the assemblage of clay minerals comprising the fabric. The RAMA mosaic shows the high void ratio fabric typical of Pacific red clays at MPG-1. The prevalent pore spaces of this high porosity fabric are represented by the light-colored background. Figure 52.17 (Bennett et al., 1977) demonstrates models of varying void ratios. RAMA compares to a void ratio of 23.5 (Fig. 52.18) (The heated fabric, described above, looks similar to a medium void ratio of 1.5–2.5). From features in the RAMA mosaic (Fig. 52.7), and Figure 52.18, the sedimentary matrix is dominated by individual particles and small domains behaving as individual particles approximately 1 μm in size. Larger domains (2–5 μm) are fewer in number. This is in contrast to Mississippi-type smectite-rich sediment comprised of domains as the basic framework of the fabric (Bennett et al., 1977). The red clay larger domains are, for the most part, illitic argillites fractured during sectioning with a diamond knife. Figure 52.19A is an enlargement of a subrounded "fractillite" (Bryant and Bennett, 1988) of aeolian origin. The fracture patterns are indicative of the direction of knife motion shown by the arrow. Flexure of these brittle grains during ultrathin sectioning leaves holes (white spaces), but does not affect the integrity of the fabric (Figs. 52.11A, 52.12B, 52.15A, and 52.16B). Elongated or "stringer" fractured illite domains are also common as seen in remolded sample B4 in Figure 52.19B and in the undisturbed RAMA sample. The different appearances of illite are a result of the random deposition orientation of the domains and matrix. Also characteristic of this fabric type is naturally occurring channels providing pathways for pore fluid flow evident in the RAMA mosaic. The channel in Figure 52.7 (upper-left to lower-middle) is 1–2 μm wide and 50 μm long; this is much larger than the cross sections of the microchannels induced from heating (Fig. 52.14).

An array of particle configurations typifies the red clays. Common arrangements include edge-to-edge and edge-to-face particle contacts clearly visible in the RAMA mosaic (Fig. 52.7). Many short chains, predominantly stepped face-to-face and edge-to-edge, are prevalent throughout the fabric, with a lesser amount of longer chains and edge-to-face particle arrangements. Several aggregate types also are common. Sample B4 (Fig. 52.15B) shows a large, subrounded, oriented, low intravoid aggregate. Adjacent is a high intravoid (low density), randomly oriented aggregate of a different mineralogy. This low-density mineral aggregate, probably smectite, has a characteristic small particle (≤ 0.1 μm) size. Smectite is recognized by its "fluffy" appearance (Fig. 52.12B) and is probably authigenic. In aggregate form, the smectitic structural integrity is demonstrated by the lack of tearing or damage common in the fragile ultrathin sections. This is due to the high adhesiveness of smectite and the numerous particle contacts. All the mineral aggregates are of a high strength because of the numerous particle contacts and attractive forces at these contacts. Tearing or breakage of the fabric tends to occur between aggregate types, not through them. Chains are also susceptible to breakage, as is evident in the remolded samples. Pusch (1970) observed aggregates moving as units, when subjected to unconfined compression test, and deformation occurring in the connecting links or chains.

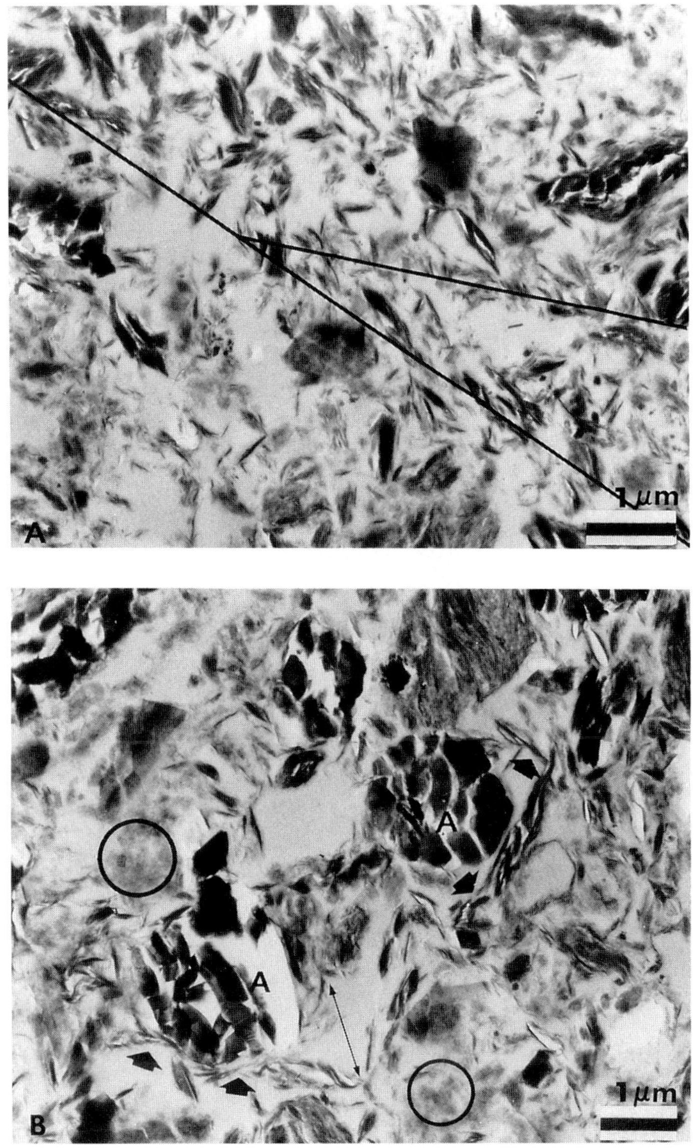

Figure 52.12. Disturbance features from probe insertion. (A) Less obvious shear or microfracture (dark lines) identified by chains of stepped F-F small domains. (B) A more easily recognizable microfault lined by oriented domains. The swirled domains (D) are possibly caused by remolding or shearing adjacent to larger argillites (A). The "fluffy," fine particles (inside circle) are perhaps authigenic smectite.

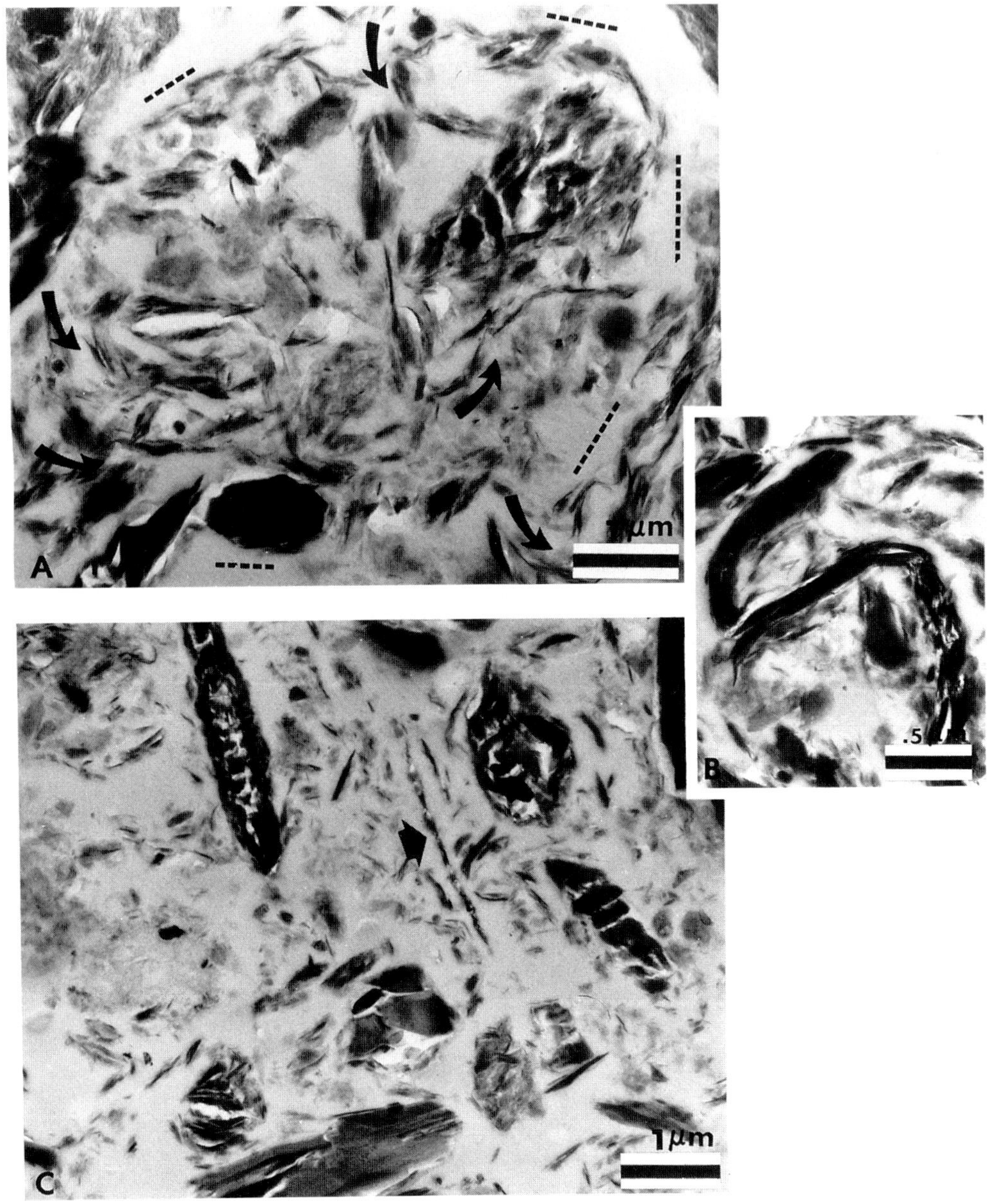

Figure 52.13. (A) Primary remolding causing swirling of domains (arrow) and forming voids performing as pathways (dotted lines) for pore fluids and small particle transport. (B) Enlargement of chain of domains bent, not subjected to remolding or probe insertion stress, demonstrating characteristic straight, long chains of domains (arrow).

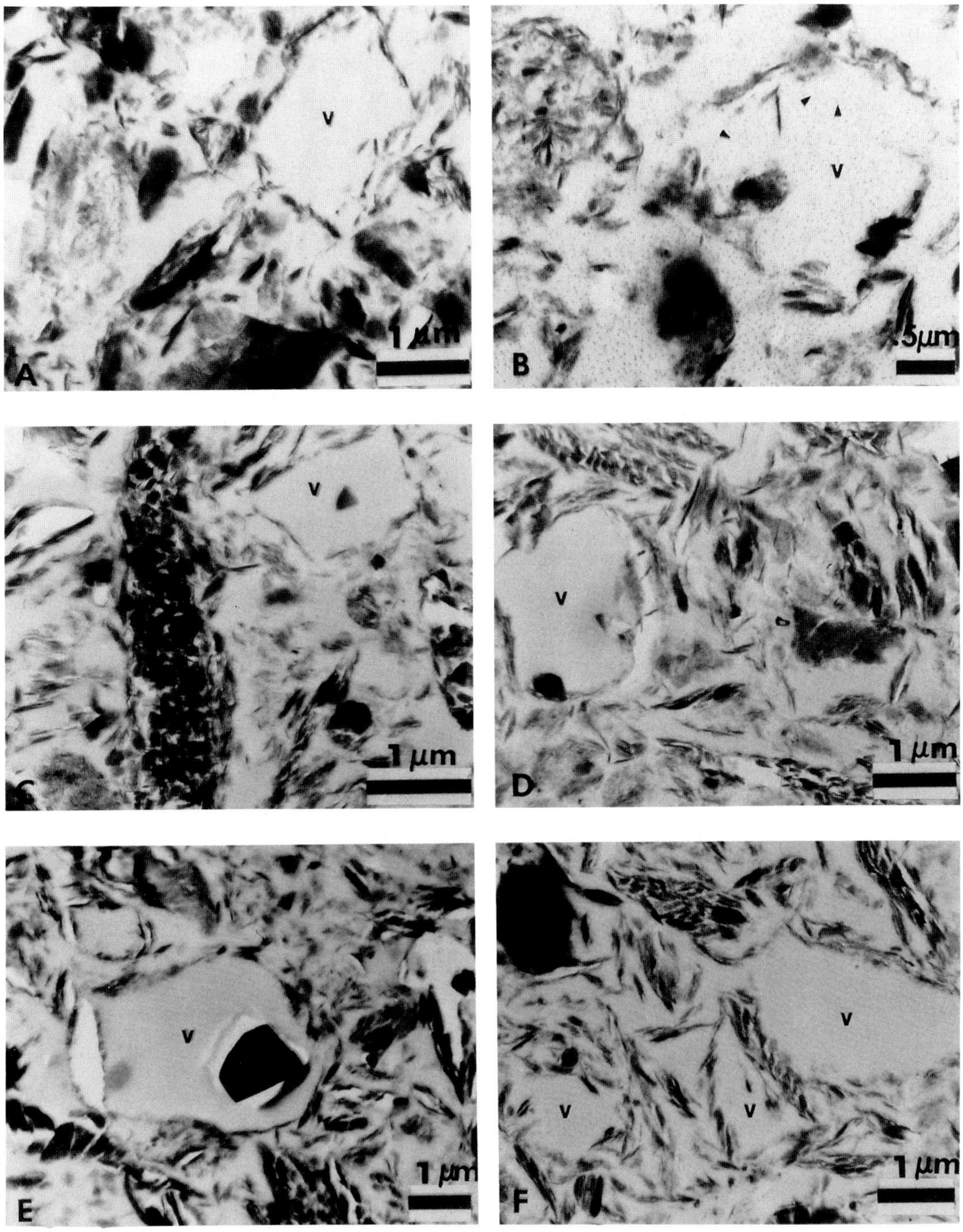

Figure 52.14. Examples of cross-sections of "quasiexpansion" features in near-field samples B-7, B-8, and B-10 which are voids acting as conduits for pore fluids. These voids (V) are lined with E-F and E-E particles and small domains. Micrographs B, C, D, and E show larger particulates which during transport process become trapped within the voids.

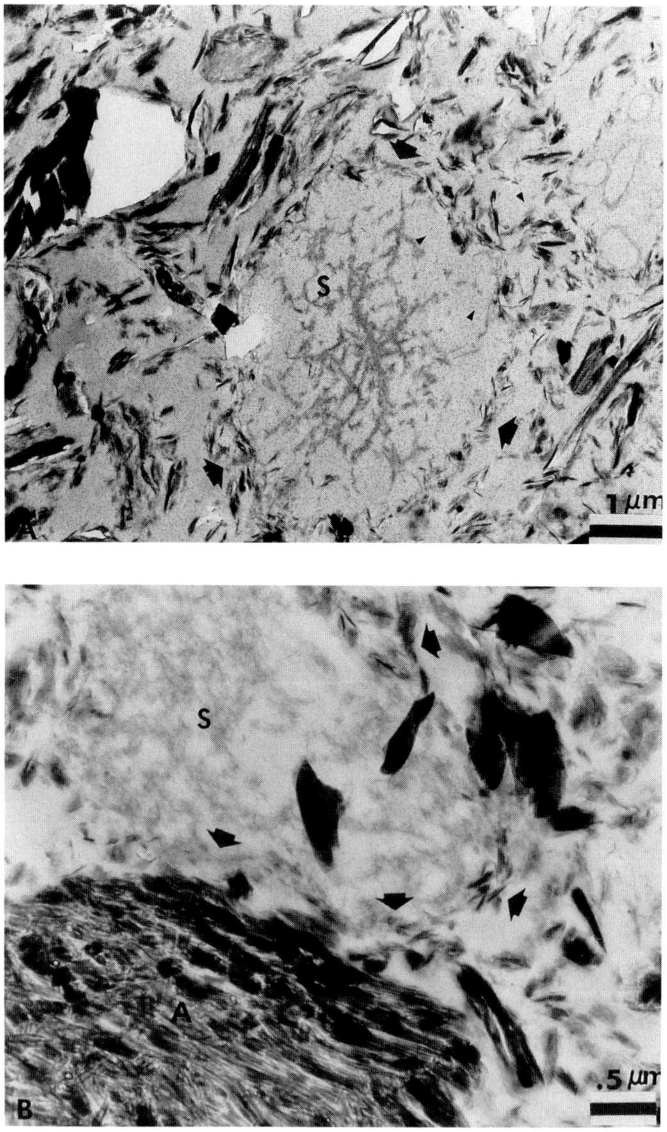

Figure 52.15. (A) Low-density smectite aggregate (S) of sample B-1 demonstrates the "onion-skin" layering of particles typical of heated fabric samples. The tiny dark dots are innocuous contaminates within the embedding resin. In the upper left portion, the micrographs show a hole in the resin as a result of a fractured argillite falling out during ultrathin sectioning. This frequent event does not affect the structural integrity of the fabric. (B) Two aggregate types of contrasting densities (A—high density, S—low density) show the affinity of small domains (arrows) for the high void smectitic aggregates (S).

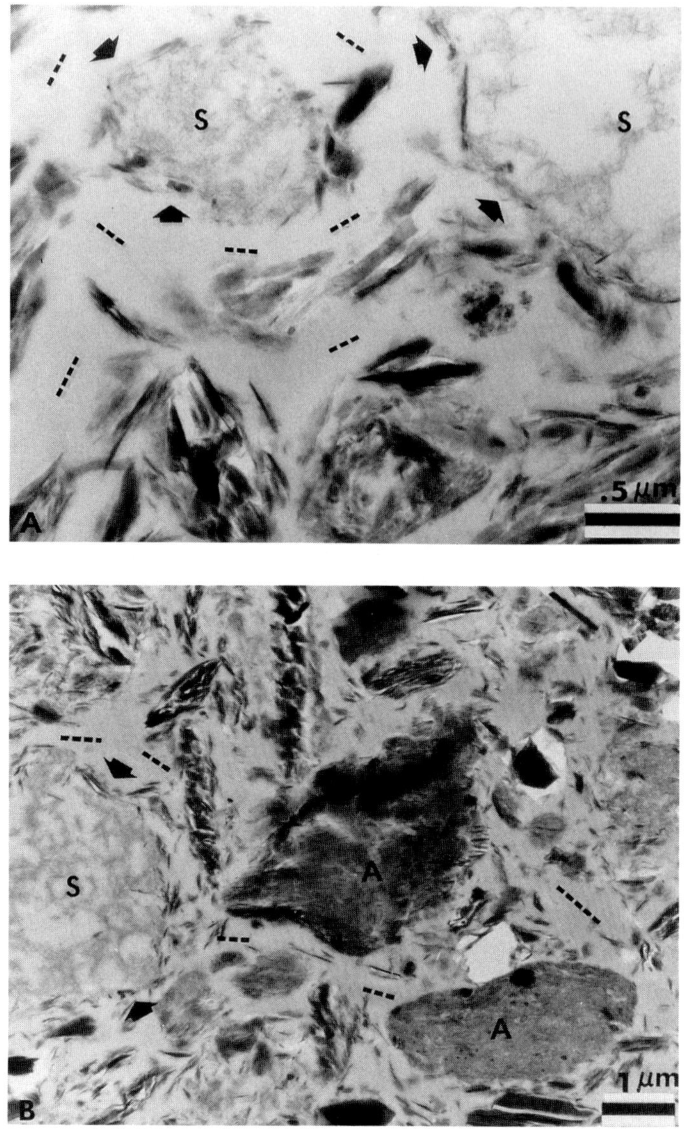

Figure 52.16. (A) Sample B-3 shows a low-density aggregate, probably smectite (S), surrounded by "onion-skin" layering of E-E and E-F particles (arrows), beyond which are channels (dotted line) for pore fluid transport. (B) Small particles and domains line the low density aggregate (S). Other aggregate types (A) have relatively few particles adhering to their surfaces.

Conclusions

Clay fabric was the foci for assessing the effects of thermal exposure and mechanical disturbances on this illitic sediment as a result of probe insertion, remolding, and varying temperature fields induced during ISIMU. In response to heating, several physical property parameters varied; shear strength increased and water content decreased. A localized transformation of the clay mineralogy from illite to smectite in the "baked zone," subjected to high temperatures, relates to an increase in the pore fluid potassium levels associated with its release from the interlayer of illite.

Fabric analysis did show some sediment response from primary remolding during reconstitution of the sediment before the ISIMU experiment. This was evidenced by the swirling, twisting, and breakage of longer chains of domains. Secondary remolding and traction stress caused by mechanical disturbance induced by probe insertion resulted in localized compression features in the near-field samples, demonstrated by an increase

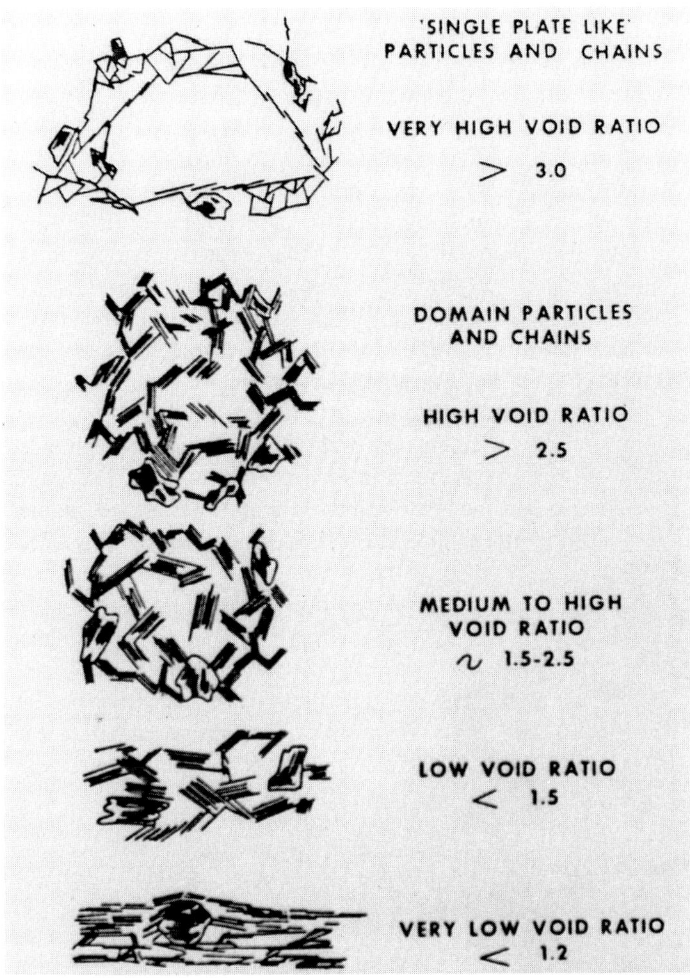

Figure 52.17. Clay fabric models for submarine sediments (Bennett et al., 1977).

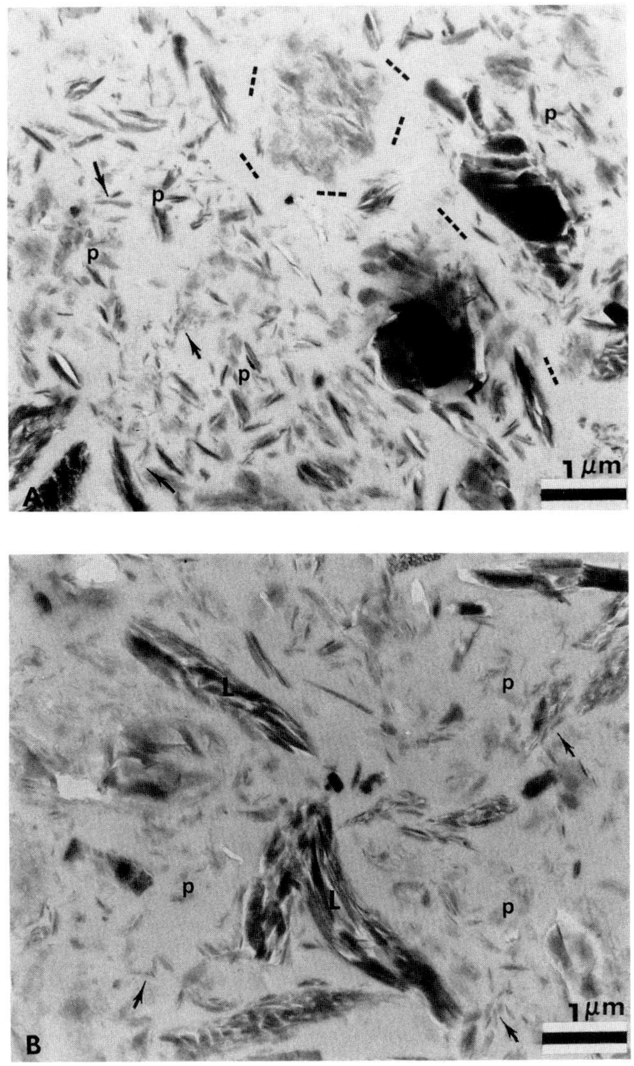

Figure 52.18. High void ratio "undisturbed" fabric of *in situ* RAMA sample showing microchannels (dotted line). The matrix is predominantly of particles and small domains (p) arranged stepped F-F and E-E, and forming short chains (arrows). (A) Notice the microchannels for fluid transport. (B) Numerous large argillite domains (L) exist which fracture during ultrathin sectioning.

in fabric density and alignment of broad bands of particles. The RAMA sample, the control not subjected to remolding or laboratory reconstituting effects, displayed longer, straighter chains of particles forming a less dense fabric.

Thermal forces resulted in "quasiexpansion" features defined as small particle-lined voids created by heated pore fluid moving through the sediment and pushing small particles aside forming the channel walls. The movement of fluids also brought small particles into contact with the smectitic-appearing aggregates bearing a large CEC, and this resulted in increased layers of stepped, face-to-face particles adhering to the aggregates in the near-field, remolded samples. Smectitic aggregates in the RAMA samples had decreased amounts of "onion-skin lining" of particles, as seen in the TEM micrographs.

The complexity of this Pacific red clay fabric type is striking, as seen by the many types of particles, such as large 2–5 μm argillite domains, aggregates of varying densities, and the channels for fluid movement approximately 2 μm in width. Various arrangements of particle contacts are observed, including edge-to-edge, stepped face-to-face, and lesser amounts of edge-to-face arrangements. The dominant particles comprising the matrix are small domains ~ 1 μm in length. The *in situ* orientation of particles comprising the fabric is random and forms a high void ratio sediment. The grain size of the far-field samples varies little from the nonremolded control.

Pacific illitic-rich red clay appears to be an excellent material for laboratory use and comparative geotechnical studies of *in situ* properties of sediment responses. This is evident by the remarkable similarity of the microfabric features between the Pacific illitic-rich red clay and the same sediment-type subjected to the stress of mechanical disturbance and heating. The

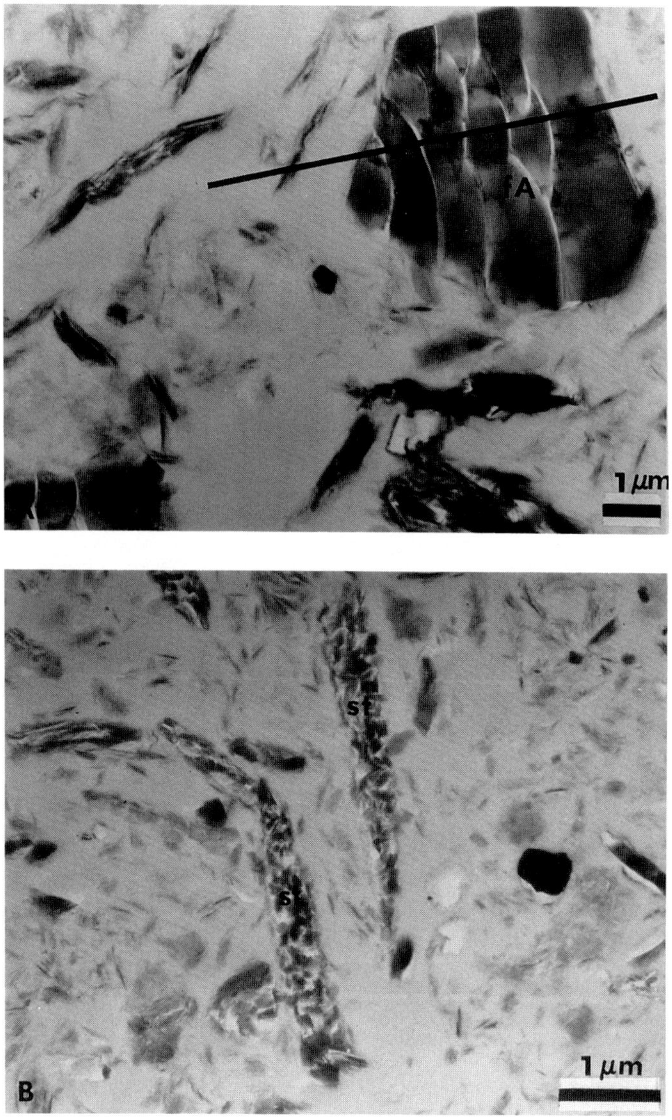

Figure 52.19. (A) A large subrounded argillite (FA) indicative of eolian origin. The fracture pattern is normal to the direction of sectioning (dark line) with diamond knife. (B) Characteristic fractured large "stringer" argillites (st) commonly found.

dominant particle type exemplified by the ≤ 1 μm matrix of domains and short chains is the single most important characteristic that makes this a resilient and durable material for both *in situ* and laboratory use. It is the microstructure, in addition to grain size and mineralogy, which controls the response and the physical properties including porosity, permeability, isotropy, and anisotropy of the sediment.

The role of microfabric of the Pacific illitic red clay sediment to radioactive waste disposal is summarized below. This provides an excellent example of how fabric studies can address practical problems and furnishes insight into engineering applications:

1. Enhanced geotechnical properties such as a residual shear strength, lower porosities, and permeabilities in the plastic zone positively impact the predictive capability of waste-containing canisters.
2. Although remolding features were observed such as swirled chains of particles and bent domains, no microfabric or grain size changes, which significantly affected the bulk physical properties, occurred in spite of the severe remolding in the laboratory.
3. Although small-scale microshears and localized orientation of particles in the direction of stress occurred, no significant fabric changes were observed in the zone of plastic deformation as predicted by theory.
4. Enhanced positive mineralogical changes ensue by enhanced K_d from the transformation of illite to smectite. The larger the coefficient, the better the radionuclide cation is contained in the sediment because of increased isolation capacity of the sediment.
5. No significant fabric changes due to thermal changes up to 296°C.
6. The gross nature and appearance of the undisturbed sediment compared to the remolded sediment are strikingly similar. The Pacific red clay is virtually an isotropic sediment, as evidenced by its microfabric.

Acknowledgments

Sandia National Laboratories provided funding under the Subseabed Disposal Program, contract DE-AC04-76DP00789. The Naval Ocean Research and Development Activity (NORDA) has designated this paper as NORDA Contribution JA360: 036:89, and supplemented the funding for this study under Program Element Number 61153N, administered by Dr. H. Eppert. This document has been reviewed and is approved for public release. We acknowledge the contributions of Dr. L. Shephard (Sandia National Lab) for his insight into interpretation and x-ray diffraction analysis, Dr. M. Percival (SNL) for collection of the ISIMU samples for electron microscopy analysis, Dr. H. Li (NOARL) for all his guidance, and L. Nastav (NOARL) for her graphics and editing expertise.

References

Andrews KW, D.J. Dyson, and S.R. Keown, 1971. Interpretation of Electron Diffraction Patterns. Plenum Press, New York, 239 p.

Bennett, R.H., 1976. Clay fabric and geotechnical properties in selected submarine sediment cores from the Mississippi Delta. Ph.D. dissertation, Texas A&M University, College Station, TX, 269 p.

Bennett, R.H., W.R. Bryant, and G.H. Keller, 1977. Clay fabric and geotechnical properties of selected submarine sediment cores from the Mississippi Delta. NOAA Professional Paper No. 9, 86 p.

Bohlke, B.M., and R.H. Bennett, 1980. Mississippi prodelta crusts: a clay fabric and geotechnical analysis, Marine Geotechnology, v. 4, p. 5–82.

Brewer, R., 1964. Fabric and Mineral Analysis of Soils. Wiley, New York, 470 p.

Bryant, W.R., and R.H. Bennett, 1988. Origin, physical, and mineralogical nature of red clays: the Pacific Ocean basin as a model. Geo-Marine Letters, v. 8(4), p. 189–249.

Burkett, P.J., 1987. Significance of the microstructure of Pacific red clays to nuclear waste disposal. M.S. thesis, Texas A&M University, College Station, TX, 79 p.

Folk, R.L., and W.C. Ward, 1957. Brazos River Bar: a study in the significance of grain size parameters. Journal of Sedimentary Petrology, v. 27, p. 3–27.

Harvard, M.E., and G.W. Brindley, 1964. Swelling properties of synthetic smectites in relation to lattice substitutions. Clays and Clay Minerals, v. 13, p. 209–222.

Heath, G.R., 1877. Barriers to radioactive waste migration. Oceanus, v. 20(1), p. 26–30.

Hollister, C.D., D.R. Anderson, and G.R. Heath, 1981. Subseabed disposal of nuclear waste. Science, v. 213, p. 1321–1326.

Index to Powder Diffraction File, 1969. American Society for Testing and Materials, Philadelphia, PA, p. 1385.

International Atomic Energy Agency, IAEA Inf. Circ. 205/Add.1/Rev. 1, 1978, 1972 Convention on the Prevention of Marine Pollution by Dumping of Wastes and Other Matter, 26 U.S.T. 2403, TIAS No. 8165, entered into force 1975, reprinted in 11 Int. Leg. Mat., 1292 (1972); D.A. Deese, Nuclear Power and Radioactive Waste (Health, Lexington, MA, 1978), p. 65–67.

Jackson, M.L., 1969. Soil Chemical Analysis Course, 2nd ed., 8th printing. Department of Soil Science, University of Wisconsin, Madison, WI, 895 p.

Lambe, T.W., 1951. Soil Testing for Engineers. Wiley, New York, 165 p.

McTigue, D.F., J. Lipkin, and R.H. Bennett, 1986. Consolidation under an isotropic total stress increase: part I, analysis for compressible constituents, Geotechnique, v. 36(1), p. 1–9.

Miller, J.B., V.W. Miller, and L.O. Olson, 1982. ISHTE Simulation APL-UW engineering report, University of Washington, Seattle, WA, 83 p.

Noll, W., 1930. Syntheses von Montmorilloniten. Chem. Erde., v. 10, p. 129–154.

Olson, L.O., and J.G. Harrison, 1979. Sea floor system for an *In Situ* Heat Transfer Experiment, Proceedings, OCEANS '79, IEEE Pub. 78 Ch 14 78-7 OEC, p. 421–423.

Percival, C.M., 1982. Laboratory simulation for a deep-ocean *In Situ* Heat Transfer Experiment, OCEANS '82, MTS/IEEE Conference Record, Sept. 20–22, Washington, DC, 6 p.

Percival, C.M., 1983. The Subseabed Disposal Program—*In Situ* Heat Transfer Experiment (ISHTE), SAND80-0202, Sandia National Laboratories, Albuquerque, NM, 47 p.

Percival, C.M., D.F. McVey, L.O. Olson, and A.J. Silva, 1980. *In Situ* Heat Transfer Experiment (ISHTE), OCEANS '80 NTS/IEEE Conference Record, Sept. 20–22, Washington, DC, p. 567–573.

Percival, C.M., L.O. Olson, R.H. Bennett, F.L. Sayles, and A.J. Silva, 1987. The Subseabed Disposal Program *In Situ* Heat Transfer Experiment (ISHTE) final design report. SAND86-2008-UC-70, Sandia National Laboratories, Albuquerque, NM., 104 p.

Pusch, R., 1970. Microstructural changes in soft quick clay at failure. Canadian Geotechnical Journal, v. 7, p. 1–7.

Sedleski, I.D., 1937. Genesis of minerals from soil colloids of the montmorillonite group. C.R. Academy of Science, U.S.S.R., v. 17, p. 375–377.

Seyfried, W.E., Jr., and J.L. Bischoff, 1981. Experimental seawater-basalt interaction at 300°C, 500 bars, chemical exchange, secondary mineral formation and implications for the transport of heavy metals: Geochemica Cosmochimica Acta, v. 45, p. 135–147.

Silva, A.J., and S.A. Jordan, 1984. Consolidation properties and stress history of some deep sea sediments. In: Denness, B. (ed.), Seabed Mechanics. Graham & Trotman, London, p. 25–39.

Thorton, E.C., 1983. Experimental and theoretical modeling of sediment-seawater hydrothermal interactions at constant temperature and in a thermal gradient: implication for the diagenesis and metamorphism of marine clay and the subseabed disposal of nuclear waste. Ph.D. dissertation, University of Minnesota, 193 p.

Veblen, D.R., 1986. Transmission electron microscopy: scattering processes, conventional TEM, and high-resolution imaging. In: Electron Microscopy and Microprobe Techniques in Clay Mineral Analysis Workshop, Clay Minerals Society Annual Meeting, Jackson, MS, 42 p.

CHAPTER 53

Influences on the Rheology of Marine Sediments Composed of Low-Activity Minerals

J. Kenneth Torrance

Introduction

Rheology is the science of the deformation and flow of matter. Extensive rheological investigations of soil materials have been conducted in the form of standard stress–strain tests and Atterberg limit tests. Relatively little rheological investigation of soils of geotechnical interest has been conducted at water contents above the liquid limit. This results partly because of the limited number of soils that naturally have such high water contents and partly because the strengths of these materials are below the limit of applicability of most standard geotechnical testing apparatus. The flow behavior of soils at water contents above the liquid limit is of interest with only a limited number of materials. These include the postglacial marine quick clays and soft unconsolidated sediments in marine or freshwater bodies. The current investigation of the influence of chemical, mineralogical, and physical factors on rheological response was conducted using a low-activity postglacial marine clay from the site of the South Nation River landslide which occurred in 1971, in Eastern Canada (Eden et al., 1971).

Materials and Methods

The South Nation landslide clay used in these experiments is typical of the Eastern Canadian quick clays. The clay content is 63% and its mineralogy is dominated by quartz, feldspars, illite, chlorite, and amphiboles (Fig. 53.1), all of which are low-activity minerals. The clay mineral content of these postglacial marine clays is generally less than 30%, and the clay size fraction generally has a greater abundance of primary minerals than clay minerals (Torrance, 1988). These soil materials accumulated approximately 10–12,000 years ago as sediments of glacially ground rock particles, predominantly of Canadian shield origin, in the marine waters of the Champlain Sea. The sediments have subsequently been uplifted by isostatic rebound, and freshwater has displaced much of the original saline pore water. These sediments have a flocculated structure, as a consequence of their marine origin. This structure has survived the salt removal process and is important to their exhibiting a substantial undisturbed strength and to their being stable under most circumstances. The higher sensitivities and liquid behavior on disturbance, required for designation as quick clays, develop as a consequence of the removal of salt from these coils but can be aggravated by other factors. Quigley (1980) reviewed the factors leading to high sensitivity. Torrance (1983) synthesized those that lead to quick clay development into a model expressed in terms of material properties, physical factors, and chemical factors, which are important at the time of deposition or exert their influence later.

The yield stress–water content investigations were conducted using a Haake Rotovisco RV12 coaxial viscometer and the MV series coaxial cylinder sensors. After appropriate pretreatments, the soil was sheared at rotation rates from 512 to 1 rpm, according to the pattern 512, 256, 512, 128, ... The yield stress in all cases was calculated by assuming that the soil behaved as a Bingham plastic material and that the yield stress was the intercept on the yield stress axis of the best fit straight line through the shear resistance values at the five fastest shear rates. The resultant yield stress–water content relationships provided the basis for comparison of the effects of various factors that affect the flow behavior.

The yield stress–water content responses of the Na-saturated, but otherwise unchanged, material at a range of salinities were accepted as the standards for comparison. The rheological response was investigated in the Na-saturated state after oxide

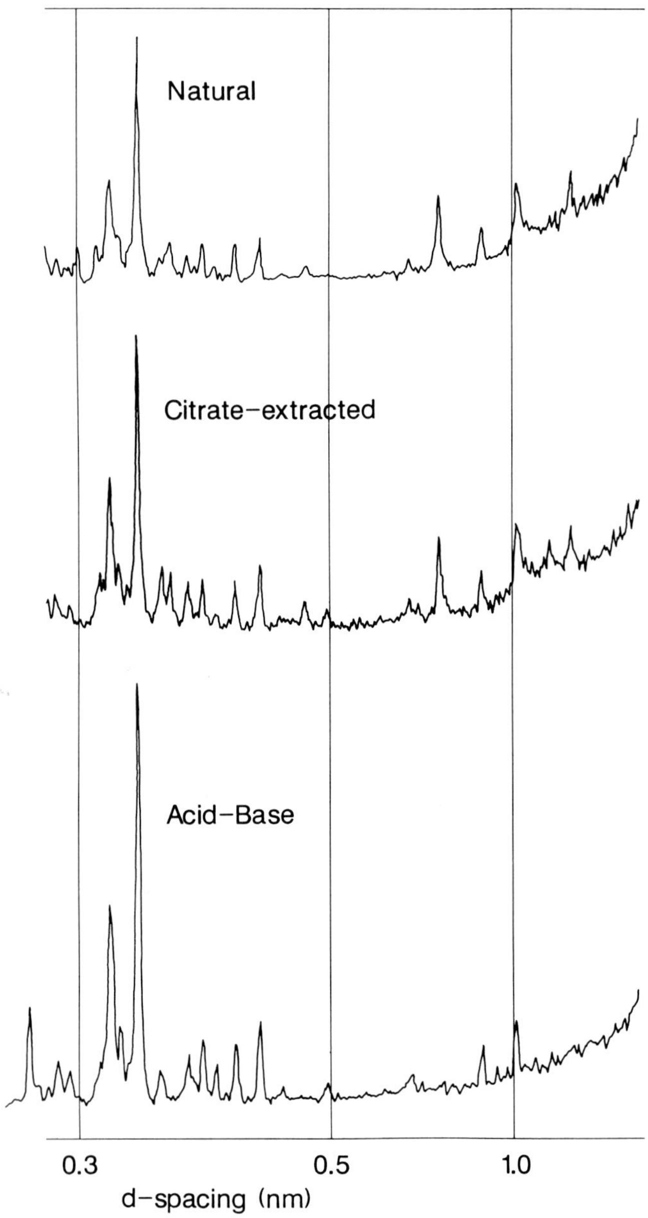

Figure 53.1. X-Ray diffraction patterns for natural, citrate-extracted (oxide-depleted), and acid-base extracted (phyllosilicate-depleted) soil materials.

depletion by dithionite–citrate–bicarbonate (DCB) extraction, phyllosilicate depletion by alternating strong acid–strong base extraction, and textural manipulation to create more silt-rich materials. The influence of Ca saturation was also investigated. All materials were tested salinities of 20, 5, and 2‰.

Results

Yield stress–water content relationships for the Na-saturated material are presented in Figure 53.2. The indistinguishable relationships at 20 and 60‰ indicate that the effect of salinity has reached the saturation point by 20‰. Below 20‰, the yield stress at any given water content decreased as the salinity decreased and the curves are sequentially shifted towards the origin. The *in situ* material, which had a salinity of 2.4‰, plotted at the position that would be expected if it is Na dominated. The strong influence of pore water salinity on the remolded behavior of this marine sediment, composed of low-activity minerals of silt and clay size, is readily apparent. In geotechnical terms, the remolded strength at constant water content is greatly reduced and the sensitivity greatly increased by salt removal.

The influences of Ca saturation and other imposed changes are presented in Figures 53.3, 53.4, and 53.5. At 20‰, the Ca-saturated curve is displaced to higher yield stress values than that of the Na-saturated material. Of practical significance is the observation that its position is unchanged by salinity reduction to 2‰. Further results, not shown, indicated that Ca-saturated material at 0.1‰ responded in the same manner as 10‰ Na-saturated material. Pore water calcium content must be very low for the natural material to behave as a fluid when disturbed. This is consistent with the Ca ion being a much stronger flocculating agent than the Na ion.

Textural manipulation to decrease the clay content shifted the curves toward the origin, with the extent of the shift being increasing as the clay content decreased. In all cases, the texturally manipulated materials remained susceptible to the effects of salinity, with a decrease in pore water salinity leading to a shift of the curves toward the origin. The effect of decreasing the clay content is not surprising. The only apparent inconsistency is that the extent of the change in response between 63 and 49% clay is greater than is consistent with that between lower clay contents. The most probable explanation of this inconsistency relates to the method of clay depletion. The clay depletion was accomplished by a sedimentation procedure whereby fine material was removed by decantation after an appropriate period of settling. The first decantation would remove a disproportionate share of the finest clay, which was the most phyllosilicate rich, compared to that removed in the subsequent decantations. Although the mineral suite would not change during textural manipulation, the proportion of phyllosilicates decreased more rapidly than the clay size fraction decreased. Textural manipulation thus imposed both a textural change and a change in the relative proportions of the minerals. With the phyllosilicates exhibiting higher plasticities than the other minerals present, the observed results are not surprising.

The DCB extraction process removes ferrihydrite (amorphous iron oxides), hematite, and geothite from soils, while also possibly having a small effect on iron-bearing silicates (Borggaard, 1988). In these marine clays, DCB would remove the hematite (Torrance et al., 1986) and less crystalline iron oxides. It may also remove a small amount of iron from illite and chlorite. The XRD trace (Fig. 53.1) indicates relatively little effect on silicate mineralogy although evidence for the production of a vermiculitic mineral is present. DCB extraction of the original 63% clay

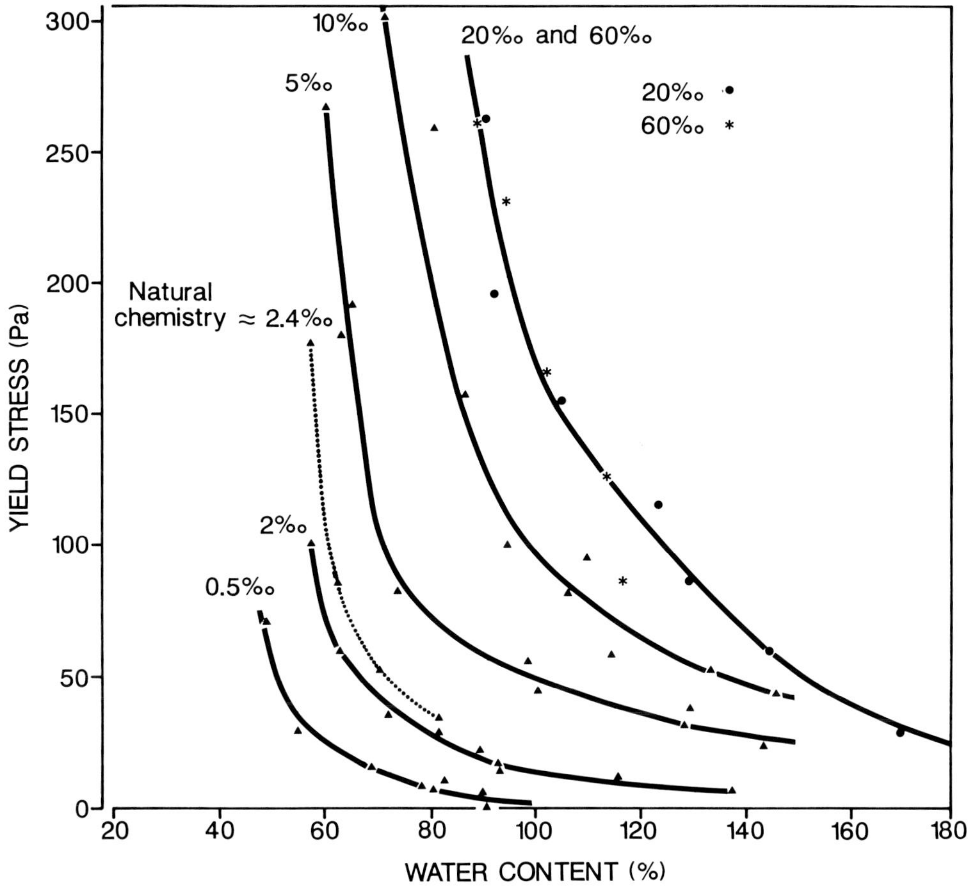

Figure 53.2. Yield stress–water content relationships for Na-saturated soil at various salinities and for the soil with its natural pore water chemistry.

material and of the 23% clay material caused a shift of the yield stress–water content curve toward the origin. The shift was relatively small but was consistently observed and was greater than the margin of error in the measurements. The shift toward the origin indicates that attractive forces between the particles are weaker. These results are consistent with the view of Quigley (1980) that the oxides act as "thickeners" and with the observations of Hendershot and Carson (1978) that their removal decreased the liquid limit. The oxides appear to be less important in the remolded materials than they are in the undisturbed soil. In the latter, good evidence exists that the oxides have a cementing effect that increases the undisturbed strength (Quigley, 1980).

Depletion of phyllosilicates was accomplished using a very harsh extraction of strong acid followed by strong base, and repeated four times. This extraction eliminated chlorite and greatly decreased the proportion of illite in the soil (Fig. 53.1). The extent to which the primary silicates were attacked cannot be accurately assessed, but the quartz and feldspar peaks retained about their same relative proportions while the amphibole peak was slightly decreased in proportion. The clay-size fraction decreased from 63 to 46%. This extraction thus made the material both siltier and more dominated by primary minerals. Phyllosilicate depletion had a dramatic effect on the response of the soil. At 20‰ (Fig. 53.3), the phyllosilicate-depleted soil behaved similarly to the 49° clay sample. This suggests that at high salinity the textural effect dominates and the difference in clay-size mineralogy (more complete depletion of phyllosilicates with the extraction) has a relatively minor influence. At 5‰ (Fig. 53.4), the effect of the phyllosilicate depletion greatly exceeded what can be accounted for by the change in texture alone. Only about one-half the change appears to be texture related. At 2‰ (Fig. 53.5), the phyllosilicate-depleted soil was strongly shear thickening and no yield stress could be determined. At water contents slightly below 30%, the soil registered no shear resistance at low shear rates and the shear resistance exceeded the capacity of the viscometer at high shear rates. At 2‰, less than one-half of the behavioral change was texture related. A comparison of the effects of phyllosilicate depletion at the different salinities indicates that the phyllosilicate are more important to the behavior of the soil at low salinity than at high salinity. In other terms, the clay-size primary minerals and the clay-size phyllosilicates have relatively similar behaviors under high salinity conditions, but at low salinities their behaviors differ with the phyllosilicates being the major contributors to the plasticity of the soil.

Factors other than those reported may also be important in these soils. Torrance and Pirnat (1984) investigated the influence of pH change and reported that decreasing the pH below

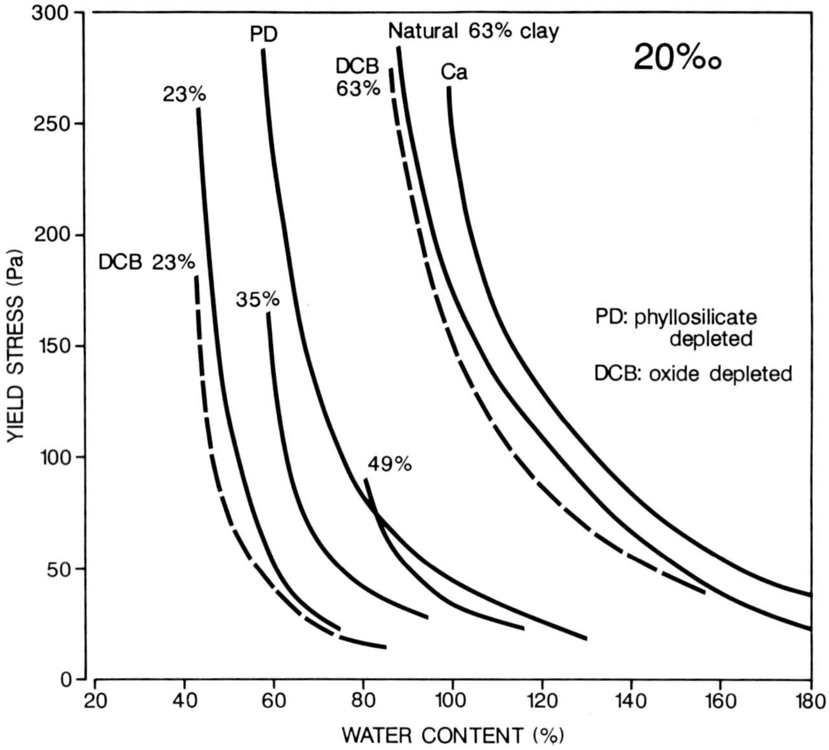

Figure 53.3. Yield stress–water content relationships at 20‰. (Ca = Ca saturated; DCB = oxide depleted; PD = phyllosilicate depleted; % = clay content in texture-manipulated experiments.)

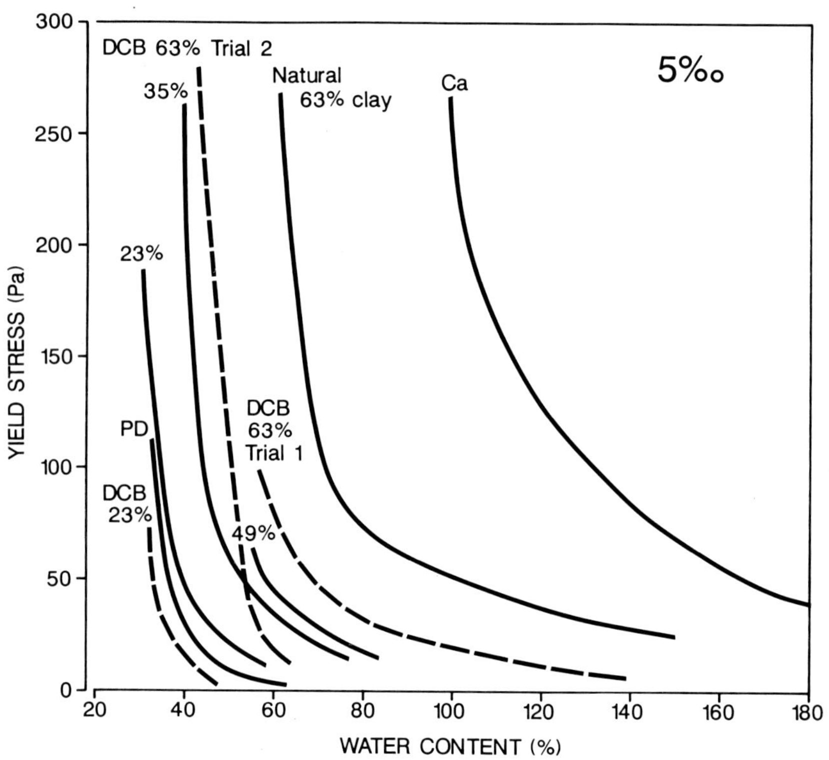

Figure 53.4. Yield stress–water content relationships at 5‰. (Ca = Ca saturated; DCB = oxide depleted; PD = phyllosilicate depleted; % = clay content in texture-manipulated experiments.)

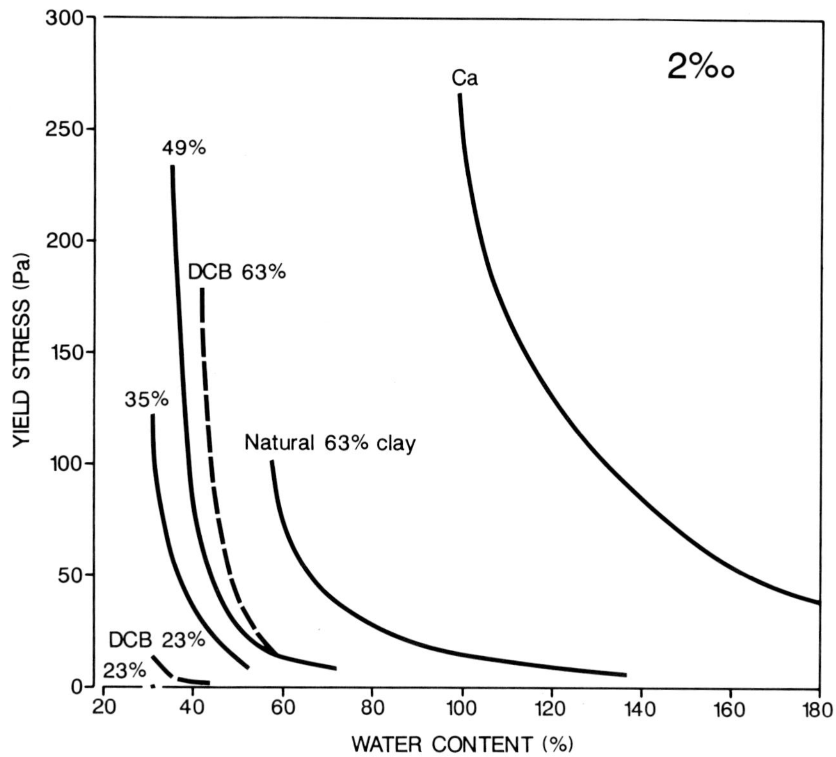

Figure 53.5. Yield stress–water content relationships at 2‰. (Ca = Ca saturated; DCB = oxide depleted; PD = phyllosilicate depleted; % = clay content in texture-manipulated experiments.)

approximately pH 7.5 increased the yield stress of the remolded soil material. This increase occurred regardless of whether the soil contained carbonates (in which case a change from Na saturation toward Ca saturation occurred) or was carbonate free. Small amounts of swelling clay minerals have been reported in the Eastern Canadian marine clays and their influence is currently being investigated by an examination of the rheological change caused by addition of a strongly swelling bentonite to these low activity sediments. Preliminary results indicate that addition of swelling clay shifts the curves away from the origin and that at some bentonite content, dependent on the clay content of the soil, the effect of decreasing salinity changes from decreasing the yield stress to increasing the yield stress. This is consistent with experiments done with pure bentonite that have demonstrated that it responds to salinity changes in the opposite manner to the low activity clay in that, it exhibits higher yield stresses at low salinity than at high salinity.

Discussion

The results demonstrate that the remolded behavior of soils, and hence the postfailure behavior of soft sediments, can be greatly affected by physical, chemical, and mineralogical factors. In these low-activity soil materials, the effects of salinity reduction, oxide depletion, phyllosilicate depletion, and decreasing clay contents have all been to decrease the shear resistance at constant water content. Calcium saturation, pH decrease, and addition of swelling clay minerals have the opposite effect. The yield stress in these experiments has been used as an index of the behavioral response of the soil material to imposed changes. It also should give an indication of the forces required to maintain a disturbed material in motion after failure has occurred. It is risky, in the lower portions of the water content ranges investigated, to use the shear resistance–shear rate curves to estimate the viscosity of the failed materials. Regardless the knowledge gained from experiments of this type should also be useful in the estimation of runout distances for quick clays or for failures in soft submarine materials. The coaxial viscometer offers considerable promise for providing information that will assist in the analysis of the postfailure behavior of soft sediments. Anyone attempting to follow this approach is cautioned to remember that the physical and chemical conditions of experimentation must be carefully controlled and that they should mimic the natural conditions as closely as possible.

Acknowledgments

The research was partially funded by the Natural Sciences and Engineering Research Council of Canada, Grant No. A8503.

References

Borggaard, O.K., 1988. Phase identification by selective dissolution techniques. In: Stucki, J.W., B.A. Goodman, and U. Schwertmann (eds.), Iron in Soils and Clay Minerals. NATO ASI Series C, v. 217, D. Reidel, p. 83–98.

Eden, W.J., E.B. Fletcher, and R.J. Mitchell, 1971. South Nation River Landslide, 16 May, 1971. Canadian Geotechnical Journal, v. 8, p. 446–451.

Hendershot, W.H., and M.A. Carson, 1978. Changes in the plasticity of a sample of Champlain clay after selective chemical dissolution of amorphous material. Canadian Geotechnical Journal, v. 15, p. 609–616.

Quigley, R.M., 1980. Geology, mineralogy and geochemistry of Canadian soft soils: a geotechnical perspective. Canadian Geotechnical Journal, v. 17, p. 261–285.

Torrance, J.K., 1983. Towards a general model of quick clay development. Sedimentology, v. 30, p. 547–555.

Torrance, J.K., 1988. Mineralogy, pore-water chemistry and geotechnical behavior of Champlain Sea and related sediments. In: Gadd, N.R. (ed.), The Late Quaternary Development of the Champlain Sea Basin. Geological Association of Canada, Special Paper 35, p. 259–275.

Torrance, J.K. and M. Pirnat, 1984. Effect of pH on the rheology of marine clay from the site of the South Nation River, Canada, landslide of 1971. Clays and Clay Minerals, v. 32, p. 384–390.

Torrance, J.K., S.W. Hedges, and L.H. Bowen, 1986. Mössbauer spectroscopic study of the iron mineralogy of post-glacial marine clays. Clays and Clay Minerals, v. 34, p. 314–322.

CHAPTER 54

The Geotechnical Importance of Clay Flexibility

N.B. Aughenbaugh

Introduction

Clays and their indurated equivalents, shales, mudstones, etc., are the earth materials most sought after and used as permeability barriers in geotechnical projects. These deposits are utilized both in their *in situ*, undisturbed state and as artificially recompacted masses. In some cases, *in situ* clays and shales are overlain by compacted clay lifts to form composite permeability "liners" to prevent liquid migration from a site.

Clays and shales serve as permeability barriers for a broad spectrum of geotechnical uses, which include lagoons, containment ponds, landfills, secure trenches, earth dams, levees, tailings impoundments, and heap leaching.

For this chapter clay will be used as the term that includes all argillaceous deposits regardless of their induration and clay content. The term also will be used to describe both the undisturbed and recompacted states. Most of the discussion will focus upon the remolded or recompacted state. The principles and material behavior that are outlined for compacted clays are valid for their *in situ* counterparts. The terms hydraulic conductivity and permeability will be used interchangeably throughout the article as the flow rate of a fluid through a material.

Clays as Permeability Barriers

Clays are the most ideal earth material for preventing fluids (and gases) from migrating rapidly from one point to another. In their undisturbed state, homogeneous clays (no joints or fractures) can have hydraulic conductivities of 10^{-7} to 10^{-10} cm/sec. When excavated and recompacted, laboratory hydraulic conductivities of 10^{-8} and 10^{-9} cm/sec are not uncommon.

Because of their low hydraulic conductivities clays are specified as the *in situ* containment material for landfills, secure trenches, lagoons, and containment ponds where the groundwater in the underlying aquifers must be protected from the contaminated fluids of the stored material. Therefore, the containment structure designs are based on the mass permeability of the underlying clay and its thickness, between a landfill and the first aquifer. Arbitrarily long periods of time (i.e., 500 years) often are set as minimum standards for the containment effluent to reach the aquifer. Thus, for example, if 500 years is the minimum time allowed, a clay deposit with a mass permeability of 10^{-8} cm/sec would have to be at least 5 ft thick. If the clay thickness beneath the excavation was less than 5 ft, a compacted clay liner of sufficient thickness and permeability would have to be placed over the *in situ* clay to achieve the designed retardant time. In recent years, geomembranes also have been used as a liner material to insure a greater factor of safety.

An alarming number of containment sites designed and constructed with good engineering care have failed in 5, 10, and 15 years. Contaminated fluids that were not supposed to seep from the sites or to reach underlying aquifers or adjacent surface water bodies for 300, 500, or more years were, in fact, polluting these water sources in a very few years. Back calculations of the containment clays hydraulic conductivities using the failed travel times revealed vast differences with the design permeabilities as established by laboratory tests. Actual *in situ* permeabilities have been found to be from 100 to 1000 times greater than the laboratory test data in many cases.

A number of reasons have been put forth for this wide discrepancy in laboratory hydraulic conductivity tests results and the actual documented rate of fluid flow in the field. In some cases the site geology was defined inaccurately and "thick,

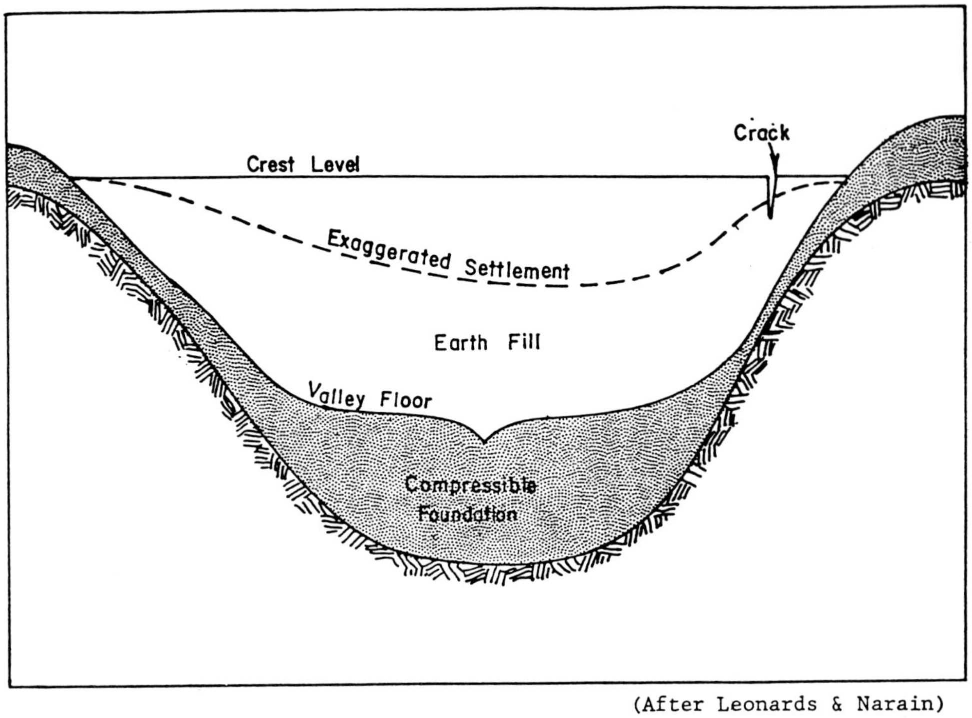

Figure 54.1. Flexural cracking of earth dams due to differential settlement. (After Leonards and Narain, 1963.)

homogeneous clay deposits" were in fact interstratified clays, silts, and sands. Where the geology was accurate, investigators have recognized weathering and jointing can increase the mass permeability (Williams and Farvolden, 1967; Stohr et al., 1988). Other factors geotechnical engineers have recognized or documented are desiccation cracks, freeze–thaw, poor construction control, and reaction of the leactate with the clay creating a more permeable fabric (Smith, 1976; Anderson, 1982; Daniel, 1985).

The one property factor that has not been recognized to date as an explanation for the increased permeability is flexural cracking of both the compacted liners and the *in situ* clay due to differential deformations. Relative to compression, clays are extremely weak in tension. As a result, a small amount of differential movement can cause fracturing transversely across the compacted liner and perpendicular to the sides and floor of the *in situ* clay. Such fractures breach the integrity of the clay as a hydraulic barrier by allowing fluids in the containment structure to escape rapidly through them. The term given to the flexural strength of fine-grained material is flexibility.

Flexibility

The term flexibility was first introduced by Leonards and Narain (1963), as a property parameter in their investigation on the cracking of earth dams. They defined flexibility as the tensile strain at cracking when compacted clay beams were subjected to simple bending.

The investigation was initiated because of excessive leakage due to internal cracks and the development of transverse cracks across the crown of many earthen dams. The author observed a large transverse crack across an earth dam in Utah in the 1960s that required immediate action to prevent erosion and failure of the dam.

The primary reason for the occurrence of internal and surface cracking in earth dams is that flexural strain exceeds the tensile (flexural) strength of clay because of differential settlements (Fig. 54.1).

Since the laboratory investigation by Leonards and Navain relatively little research has been made on the tensional and flexural properties and behavior of fine-grained soils. The few studies that have been made focused on earth dams and evaluating the tensile strength of compacted laboratory samples (Fang and Chen, 1971; Kezdi, 1973; Sherard, 1973; Ajaz and Parry, 1975). Also the research has been confined to the 1960s and 1970s. The author is not aware of any research or application of clay flexibility to landfills, secure trenches, lagoons, or containment ponds. Yet in the area of waste disposal and storage, the integrity of compacted and natural clay liners with respect to their hydraulic conductivities is the most essential design factor in protecting the adjacent surface and groundwaters from pollution.

Clay liners are subjected to many forms of flexural movement. Figure 54.2 is a schematic cross section of a municipal landfill. When excavating to establish the pit bottom and sides, the *in situ* clay will rebound on unloading. Dilation of the exposed clay into the excavation can result in flexural cracking perpendicular to

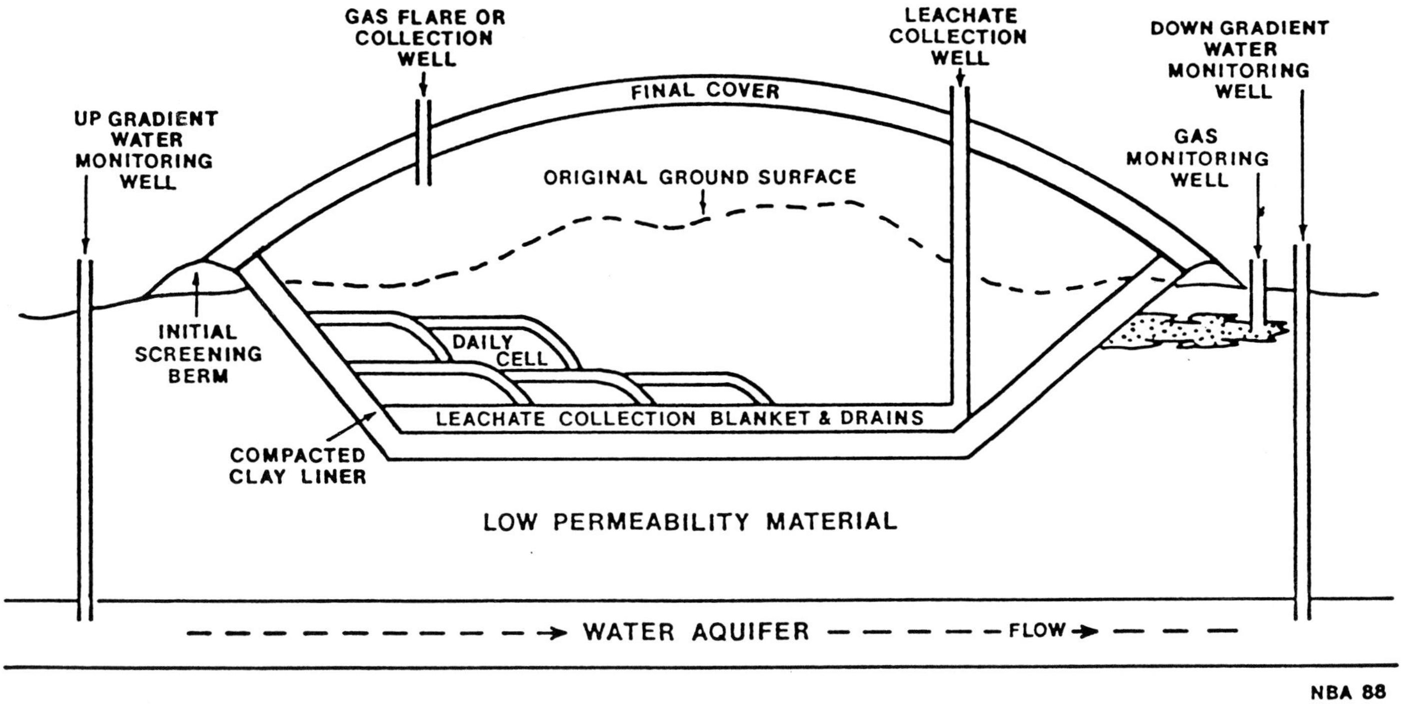

Figure 54.2. Exaggerated schematic of a modern landfill showing the *in situ* and compacted liners.

the surface. If exposed to rain, some clays will swell, which also causes differential movements. As waste is placed in the landfill the clay is unevenly reloaded and densified. The resulting flexural movements can cause cracking of the compacted clay liner especially at the junction of the pit walls with the floor (invert). Thus it is important to evaluate the rebound and swell potential of the *in situ* clay and to design the compacted liner to achieve the optimum permeability and flexibility.

Flexibility and Water Content

Typical specifications for compacted clay liners in waste disposal require a compaction level of 95% of optimum (the water content at maximum density) for Standard Proctor. For this compaction level most clays will have hydraulic conductivities of 10^{-7} cm/sec or less in laboratory permeability tests. However, this specification does not ensure the compacted clay can undergo moderate amounts of flexural movement without cracking.

The plasticity of a clay increases with compaction water content. Also, the amount and type of clay mineral influence both the extent of the plastic state (i.e., the plasticity index) and degree of plasticity a clay will exhibit. Flexibility is directly correlative to plasticity and thus is influenced by the compaction water content, the amount of clay minerals, and the clay mineral type.

Figure 54.3 is a standard compaction curve and a deflection curve at the initiation of cracking for a two-point flexural test of

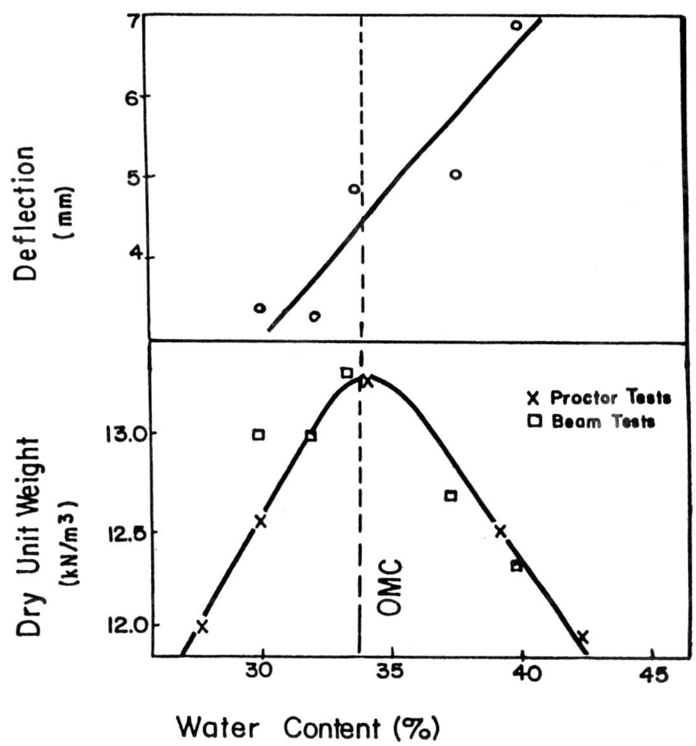

Figure 54.3. Relationship of a soil's flexibility to the standard compaction curve.

compacted soil beams. The deflection curve is representative of how a clay varies in flexibility with water content. On the dry side of optimum a clay is brittle and will deflect very little before cracking, whereas, on the wet side a clay may flexurally deform double of that of the dry side before cracking.

By specifying 95% of Standard Optimum, the flexibility property of the compacted liners can vary widely from one place to another depending on what side of optimum the contractor compacts the lifts. Also, depending on conditions, a contractor may compact most of the liner on the dry side of optimum, with the result that the clay liner will be brittle and susceptible to widespread cracking with only small differential movements. Clay liners should always be compacted at or wet of optimum to achieve maximum flexibility.

Flexibility Tests

The author has developed a laboratory flexibility apparatus to test compacted clays for flexibility. Because of patent considerations details and dimensions of the apparatus cannot be disclosed. Therefore, only a general description of the apparatus and test procedure will be given. Research on refining the test and analysis of various property parameters are continuing.

The flexibility test is patterned after the Standard Compaction (Proctor) test except the mold is beam-shaped rather than cylindrical. The clay soil is prepared by ASTM Standards for the Standard Compaction. The loose soil is placed in the mold in three layers and compacted to density using the standard compaction hammer. The number of blows per layer is such that the beam density correlates well with the standard compaction curve as illustrated in Figure 54.3.

After the compacted beam is extracted from the mold, a two point flexural test is made on each sample to failure by cracking. The deflection is measured at the midpoint of the beam length and recorded at the instant the crack initiates.

Summary

Flexibility is an important property parameter when using clays for permeability barriers. This property of plastically deforming without cracking when flexurally deformed was first recognized and investigated in the early 1960s with respect to earth dams. Since then very little research has been done on flexibility and no consideration has been given to its influence on clay liners used to contain polluted liquids in landfills, secure trenches, lagoons, and containment ponds. Since these structures are susceptible to differential movements, the integrity of the clay liner is breached if cracking due to flexural deformation takes place. The author has developed a laboratory apparatus and test to evaluate the flexibility of compacted clays.

References

Ajaz, A., and R.H.G. Parry, 1975. Stress–strain behavior of two compacted clays in tension and compression, Geotechnique, v. 25(3), p. 495.

Anderson, D., 1982. Does landfill leachate make clay liners more permeable? Civil Engineering, p. 66, September.

Daniel, D.E., 1985. Can clay liners work? Civil Engineering, v. 55(4), p. 48, April.

Fang, H.Y., and W.F. Chen, 1971. New method of determination of tensile strength of soils. Highway Research Record, n. 345, p. 62.

Kezi, A., 1973. Tensile and flexural strength of earth dam materials. 11th International Meeting on Earth Dams, Proceedings, v. III, p. 165.

Leonards, G.A., and J. Narain, 1963. Flexibility of clay and cracking of dams. Journal of Soil Mechanics and Foundations, v. 89(SM2), ASCE, p. 47.

Sherard, J.L., 1973. Embankment dam cracking. In: Hirschfeld, R.C. and Poules, F. (eds.), Embankment-Dam Engineering. Wiley, New York, p. 271.

Smith, R.H., 1976. Assessing clay as a landfill liner material. Waste Age, p. 59, September.

Stohr, C., W-I. Su, L. Follmer, P. DuMontelle, and R.A. Griffin, 1988. Engineering geology investigations of a hazardous-waste landfill in West Central Illinois, USA. Bulletin, International Association of Engineering Geology, n. 37, p. 77.

Williams, R.E., and R.N. Farvolden, 1967. The influence of joints on the movement of ground water through glacial till. Journal of Hydrology, v. 5, p. 163.

CHAPTER 55

The Microfabric of Some Hong Kong Marine Soils

N.K. Tovey

Introduction

Approximately 6 million people live and work in the 1050 sq km of Hong Kong. Much of the Territory is of rugged terrain with steep slopes rising rapidly from the sea and with 40% of the land area of Hong Kong Island having slopes exceeding 30°. The mean annual rainfall of 2225 mm is mostly concentrated in a 5-month wet season from May to September. During this period numerous landslides frequently occur (e.g., in 1982 when 3248 mm of rain was recorded over 1000 landslides occurred). Brand et al. (1984) described some of these events and, in particular, examined the relationship between rainfall and landslide occurrence.

The geology of Hong Kong has been described by Allen and Stephens (1971), and consists of two major rock types of granitic and volcanic origin in most of the territory with some sedimentary rocks in the northern part of the New Territories. The rocks are weathered to a considerable depth and in places reaches 50 m thick. Much of the coarse-grained granitic rock is covered with colluvium, alluvium, or recent marine deposits.

With the shortage of land, there has been extensive reclamation of land from the relatively shallow waters of Hong Kong Harbour over the last 100 years. In more recent years attention has also turned to those areas surrounding parts of the New Territories, particularly in the vicinity of the New Towns like Shatin, Tsuen Wan, and Tuen Mun. However, attention has also been given to the area adjacent to the north of Lantau Island near Chek Lap Kok as the site of a possible replacement airport, and also to the area near Kwai Chung. Extensive engineering evaluation of the former site, which is near the confluence of the Pearl River and the South China Sea, was undertaken including the construction of a heavily instrumented test embankment (Foott, 1982; Foott et al., 1987; Koutsoftas et al., 1987). In conjunction with these engineering studies, selected samples were examined for their chemistry, mineralogy, and microfabric. In this chapter, it is the last topic that is of most interest, and this extends the work previously reported in Tovey (1986a). In particular, some quantitative image analysis techniques have been employed in an attempt to relate particle orientation with stress level.

A total of nine separate samples from seven different boreholes, five from the Chek Lap Kok area and two from the Kwai Chung area, were studied. The locations of the boreholes are shown in Figure 55.1. Boreholes CLKB8 (two depths), CLKB9, CLKB9a, and CLKB12 were in the test embankment site at Chek Lab Kok, while borehole CLKM34 was approximately 1 km WSW of the embankment. Boreholes KC5 (two depths) and KC6 were approximately 250 m offshore in Lai Chi Kok Bay in a WSW direction from the Mei Foo Sun Chuen housing estate. Details of the vertical positions of the samples ares given in Table 55.1, while Figure 55.2 shows simplified logs from the boreholes.

Core samples were obtained by piston sampling and sections approximately 100 mm long and 95 mm diameter were used in this investigation. The disturbed samples from the ends of each tube were used for the mineralogical and chemical analyses, while for microfabric analyses, undisturbed subsamples approximately 12 mm in diameter and 4–5 mm thick were cut from the center of the cores for freeze-drying and critical point drying in the manner described in Smart and Tovey (1982) and Tovey (1986b). For each specimen, the major part of the freeze drying was done at temperatures below −50°C to minimize the effects of migratory recrystallization. Fresh fracture or peeled surfaces on approximately horizontal planes were then prepared and then coated with 10 nm of carbon followed by 20 nm of gold-palladium, for normal SEM observation, or 30 nm of carbon if microchemical analyses using the energy-dispersive X-ray system were needed.

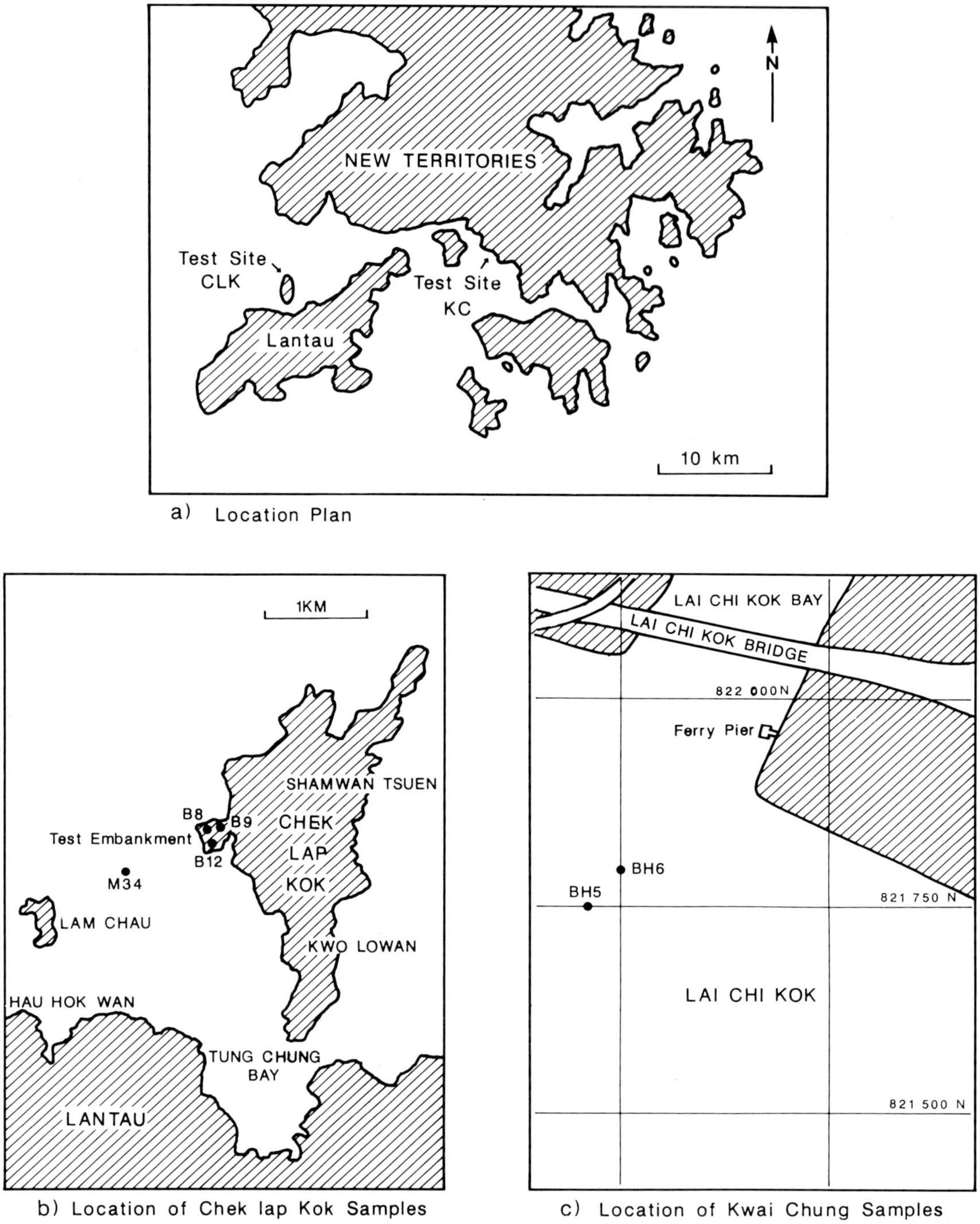

Figure 55.1. Maps showing locations of boreholes.

Mineralogy of the Samples

In all samples the most common clay mineral in the <2-μm fraction was illite, which was present in quantities ranging from 30 to 40% for the upper marine samples from 44 to 50% for the samples in the lower marine and alluvial crust deposits. These quantitative estimates were obtained by a technique described in Tovey (1986b) developed from Biscaye (1965). Smectites were present in concentrations usually exceeding 20% in the upper marine deposits, but only at about 5% in the deeper samples. Kaolinite was present in all samples: 24% in all Kwai Chung samples and 29–36% in the Chek Lap Kok samples. There were also

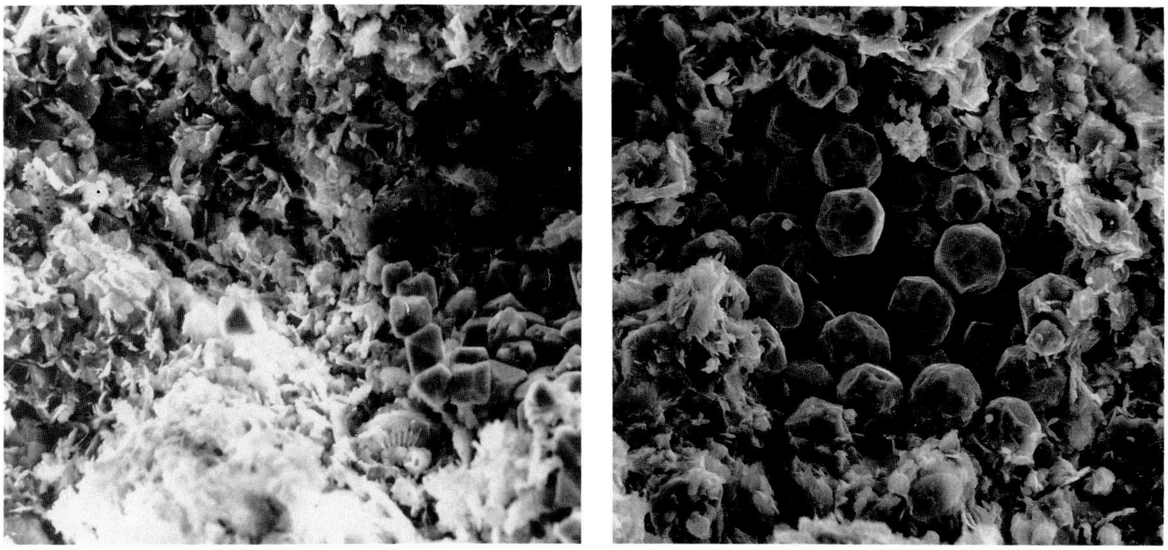

A B

Figure 55.6. Typical framboids of iron pyrite. In (B) no evidence of sulfur was detected suggesting oxidation. Picture widths (p): A = 69 µm; B = 28.5 µm.

of iron pyrite framboids (e.g., Fig. 55.6A and B). Interestingly, energy dispersive traces on the framboids in Figure 55.6B showed only iron and no sulfur indicating some oxidation here.

Quantification of the Soil Fabric

Although there appeared little qualitative evidence of preferred orientation in any of the samples, the opportunity was taken to determine a quantitative measure of fabric using the intensity gradient technique first suggested by Unitt (1975) who used a micrograph of soil fabric from the present author to demonstrate the technique. The technique was further developed by Tovey (1980), Tovey and Sokolov (1981), and Tovey and Smart (1986), and finally a fully general intensity gradient algorithm was developed by Smart and Tovey (1988). Some preliminary applications are reported in Tovey et al. (1989).

Many image analysis techniques that are currently available rely on the ability to adequately threshold an image into two or more separate gray levels. Subsequent analysis for particle size and shape and orientation is relatively easy. However, the thresholding will often create difficulties in images where the background intensity level varies across the image as it does in the case of secondary electron scanning electron micrographs. Some adaptive thresholding filters may be used to assist in the acquisition of stepwise images, but operator intervention is often needed to segregate touching particles or to deal with particles that have centers that are darker than the background. If suitable binary images can be obtained by thresholding, then one of several erosion and/or dilation kernels may be used to provide further information on features in the image. For instance, Ehrlich et al. (1984) used such a technique to analyze pore shapes and sizes.

Unlike many of the above image analysis techniques no thresholding is required in the intensity gradient techniques as it utilizes changes in intensity rather than absolute values and is thus not affected by general changes in a background intensity level.

The intensity gradient technique locates edges of features and in this respect is an alternative to the many other edge detection algorithms that have been proposed (e.g., Marr and Hildreth, 1980; Haralick, 1984; Zhou et al., 1989). However, unlike these other papers that concentrate on the detection of edges, the algorithms described in this chapter look specifically at the orientation of features.

Each image may be digitized, either directly from the microscope, or via a densitometer into a series of pixels on a 0–255 gray scale. The micrographs in question were digitized on a Joyce Loebl densitometer into an array 500 × 500 pixels. The choice of the number of pixels is a compromise between spatial resolution of the object, magnetic storage space, and computing time. Such an image occupies 256 kbytes, whereas a 1000 × 1000 array would occupy 1 Mbyte. The magnification of the micrograph must also be carefully chosen so as to be consistent with the resolution of the microscope at that magnification, the resolution required to separate individual features, and the pixel density on the densitometer. A further consideration is the total area covered by the micrograph. Thus, if only the individual particle alignments in a localized region are required, then a relatively high magnification should be used, whereas if the regional variation of fabric is the subject of study, the lowest magnification consistent with the above constraints should be adopted. In

22	15	10	14	21
16	6	2	5	13
11	3	0	1	9
17	7	4	8	20
23	18	12	19	24

Figure 55.7. Pixel numbering arrangement for intensity gradient technique.

the present study, a magnification of 500× is just compatible with the thickness of kaolinite particles.

The features in an image are differentiated from one another by changes in intensity, and this change will be most rapid in a direction orthogonal to the alignment of a feature. Thus by evaluating the direction of the intensity gradient vector at all pixels within an image, a rosette histogram may be constructed that will clearly show any preferred orientation that may be present. A further step from this is to transform this rosette diagram into the best fitting ellipse, which may then be used to quantitatively define an index of anisotropy such as the ratio of the principal axes of the ellipse. In this study the index of anisotropy I_a was expressed as

$$I_a = \frac{\text{length of major axis}}{\text{length of minor axis}} \tag{1}$$

There are clearly other possible definitions, including the reciprocal of the above, but this one has been adopted here, mainly from its simplicity. Thus a value of $I_a = 1.0$ represents a randomly orientated sample, while a high value is to be expected from anistropic samples.

Full details of the equations needed in the intensity gradient technique may be found in the papers listed above, but for the present purposes the following is sufficient. Figure 55.7 shows the arrangement of pixels in a 5 × 5 array around the picture point of interest. These are carefully numbered in the sequence shown for ease of manipulation in the fully general analysis (Smart and Tovey, 1988). If the intensity at the nth point is specified by I_n, the change in intensity in the y direction is given by

$$\frac{\delta I}{\delta y} = \frac{I_2 - I_0}{h} \tag{2}$$

where h is the spacing between pixels.

A similar expression may be written for the change in intensity in the x direction, while the orientation of the intensity gradient vector may be expressed as

$$\tan\theta = \frac{\delta I/\delta y}{\delta I/\delta x} \tag{3}$$

while the magnitude of the intensity gradient vector V_i may be defined as

$$V_i = \sqrt{\left(\frac{\delta I}{\delta x}\right)^2 + \left(\frac{\delta I}{\delta y}\right)^2} \tag{4}$$

In theory these formulae may be used in the anlysis, but there are several advantages of using additional points from the array in Figure 55.7. The above equations are correct to the first order, and a simple extension will also include points 3 and 4. For higher order solutions it is important that sufficient points are included to evaluate the key parameters $\delta I/\delta x$ and $\delta I/\delta y$ and to eliminate all other terms. Thus for a second-order solution, five points in addition to the central pixel are needed, for a third-order analysis, points 0–9 (or the equivalent) are needed as a minimum. If more points are actually used than are needed for a solution of a given order, then least-square techniques may be used. A nomenclature of the form **S,T** has been used by Smart and Tovey (1988) to denote a particular method of analysis from the many available. Here **S** refers to the number of pixels in the vicinity of the central pixel that are used in the analysis, and **T** refers to the minimum number of equations needed to solve the equations to a given order. Thus the **20,14** method would thus imply that 20 of the surrounding pixels are used to solve for the 14 variables needed in a fourth-order solution. Some methods using this notation cannot be solved (e.g., **9,9** method or **20,20** method) as the matrix generated is singular, but most other methods may be used. Clearly, with the general formulation of Smart and Tovey (1988), there are many different **S,T** methods that may be used, and the actual choice depends on several factors including how noisy the image is. In the current analysis, the **24,14** method was used throughout.

Tovey (1980) and Tovey and Smart (1986) noted that when the intensity varied little from one pixel to another such as in the center of a particle, problems could arise in that certain angles were preferentially selected because of the integer nature of the digitized image. They recommended that if class widths of 5° are used then vectors with magnitudes less than 2.0/h should be omitted from the aggregate analysis, particularly if a method of low order is used. In this work, this value of 2.0/h was also used as the threshold criterion. Although weighting of the vectors according to their magnitude is an alternative approach to this problem, this was not done here. One reason for this is that more contrasting sections of the image would then be weighted more strongly. A nonlinear weighting function would seem appropriate, but some experimentation would be necessary.

A total of 50 micrographs from the 9 samples were digitized. With representative micrographs from both the horizontal and vertical surfaces, this meant that there were normally at least three separate estimates of the degree of anisotropy of fabric in the vertical surfaces. The values determined were generally low,

Figure 55.8. Micrographs of vertical surfaces of lower marine clay from Chek Lap Kok (A and B), and upper marine clay from Kwai Chung (C). Picture widths: A and B = 137 μm; C = 28.5 μm.

confirming the general impression gained from direct observation. The highest individual value noted was 1.741 in one of the samples from the lower marine deposit at Chek Lap Kok.

Figure 55.8 shows typical micrographs of vertical surfaces selected for quantification. Of note are Figure 55.8A and B, which is from the same vertical surface of the same sample of lower marine clay. It indicates the extent of the variability of soil microfabric over just short distances. Figure 55.9 shows several of the rosette histograms from the analysis on Figure 55.8A. Several separate rosette diagrams from different magnitudes of the intensity gradient vectors are shown and all indicate the extent of the orientation present in the micrograph. The aggregate rosette diagram (Fig. 55.9f) should be compared with the diagram (Fig. 55.10a) corresponding to the micrograph in Figure 55.8B where the difference in the degree of orientation is clearly evident.

Despite the general appearance of aggregates of face-on kaolinite particles in Figure 55.8C, this micrograph shows a vertical surface. Of interest is the associated rosette histogram corresponding to a vector magnitude in the range 20/h–50/h in Figure 55.10b. This diagram shows that the edges of the approximately hexagonal shaped plate-like particles are aligned somewhat to one another as there is a general absence of vectors in directions close to the *y* direction. In Figure 55.10c corresponding with the intensity gradient vectors of relatively high magnitude, there is a pronounced preferred orientation perpendicular to the principal stress direction, and this corresponds to the several particles with edges nearly parallel to the *x* axis of the

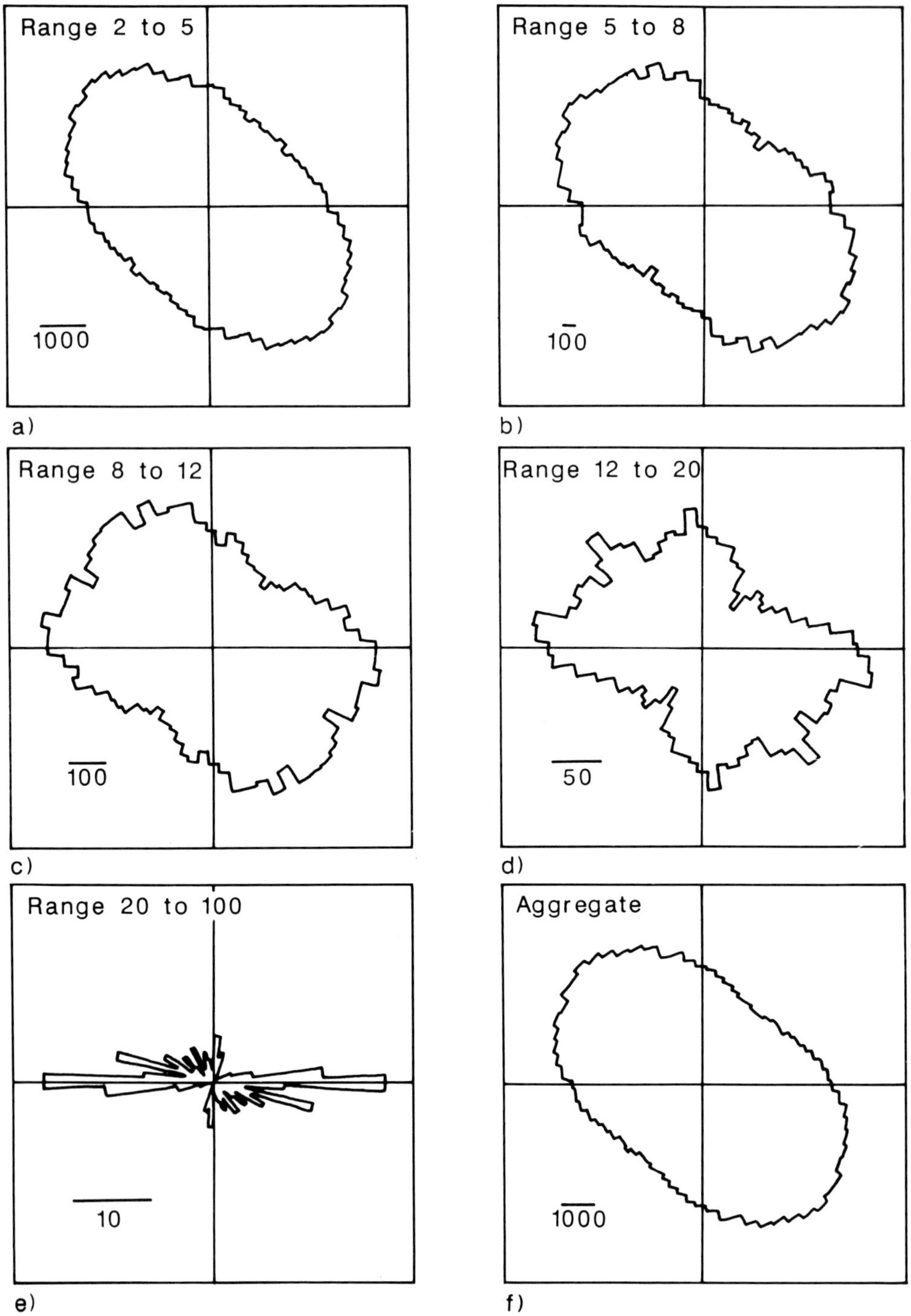

Figure 55.9. Rosette diagrams showing orientation present in Figure 55.8A. The various diagrams are for different ranges of the intensity gradient vectors.

micrograph. Neither of these two facts relating to the alignment of the face-on particles seen in the micrograph is readily apparent from a purely qualitative appraisal of the fabric. The general shape of the rosette diagram (Fig. 55.10b) is somewhat similar to that in Figure 55.9c and probably relates to the shape of the individual kaolinite particles.

The *in situ* vertical stresses were assessed by the Hong Kong Government during the site investigations at both general locations, and from the quantitative analyses undertaken here it is possible to examine the variation of the index of anisotropy with vertical stress for these soils. These results are shown in Figure 55.11 where each data point represents the arithmetic mean of

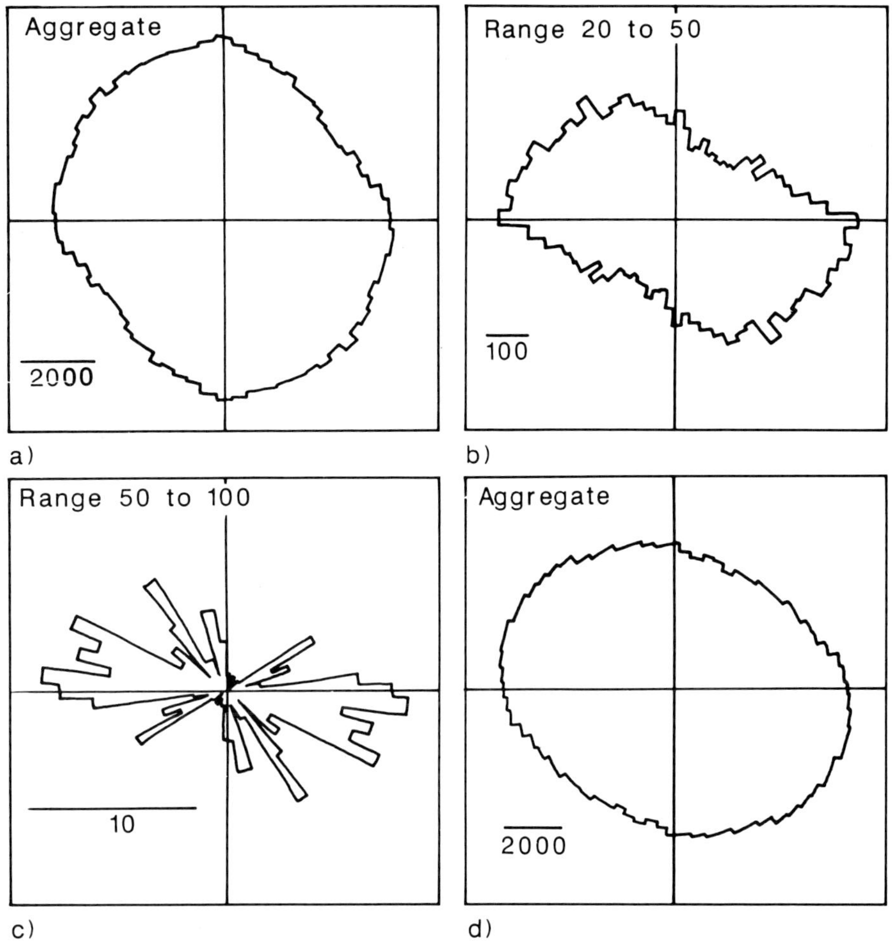

Figure 55.10. Rosette diagrams. (A) The aggregate diagram corresponding with b. (b–d) Associated with Figure 55.8C and show the diagrams for different intensity ranges.

the index of anisotropy from at least three separate micrographs. The indices of anisotropy at all stress levels are relatively low even at stresses as high as 300 kPa, and this confirms the general impression gained from qualitative interpretation of the micrographs. Such low values are not unexpected in these soils in view of the relatively high proportion of silt-sized particles present that tend to be more equidimensional in shape than the clay particles. This finding is similar to that of Curtis et al. (1980) who used X-ray texture goniometric methods to study the fabric of mudrocks.

Despite the general lack of preferred orientation there is a tendency for the degree of anisotropy to increase as the effective stress rises, although there is some scatter of the data points. In order to assess the likely variation in the value of I_a at any given stress level, a laboratory study of consolidated kaolin was made in which slurries of the clay were mixed at twice the liquid limit and then consolidated to a variety of stress levels. Over 50 micrographs were digitized at each stress level when it was found that a range of values in I_a of ± 0.2 is not uncommon. In the current observations the scatter seen in Figure 55.11 is thus not unreasonable when fabrics as diverse as those shown in Figure 55.8A and B are present on the same sample.

In determining the index of anisotropy, a straight arithmetic mean was derived. The intensity gradient technique may be readily modified to aggregate the results from several different areas in other ways such as combining individual histogram classes before the determination of an aggregated index of anisotropy. This alternative method may have merit in certain types of work, but it was not employed here.

Conclusions

The fabric of Hong Kong marine soils has been examined in the scanning electron microscope, and quantitative appraisal of the orientation patterns present indicates a weak trend of increasing anisotropy with increasing stress level. However, the indices of anisotropy derived from all samples were generally low and consistent with the high silt content in the soil. The nature of the orientation/stress level relationship does emphasize the

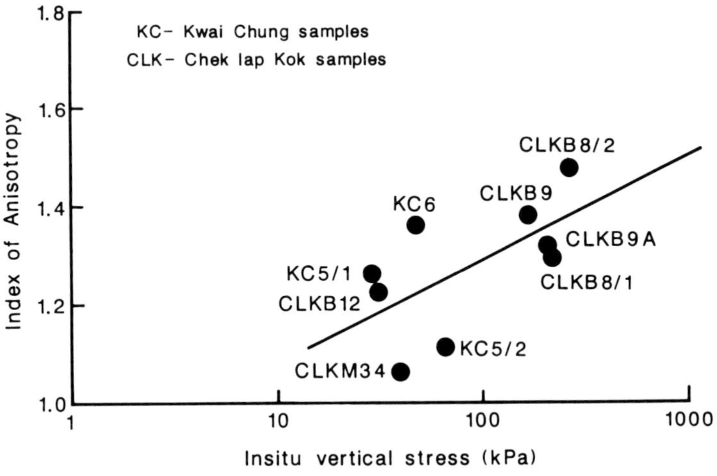

Figure 55.11. Variation of index of anisotropy with stress level composite plot incorporating all nine samples.

important role played in the fabric arrangements by the silt-sized particles.

Besides providing a direct quantitative measure of fabric, the techniques employed here have assisted in the purely qualitative interpretations of fabric as features that may otherwise be overlooked are sometimes clearly apparent in the rosette histogram diagrams.

Acknowledgments

The work described here was based on work undertaken for the Hong Kong Government. Support for the development of the quantitative work was provided by SERC Grant GRD/90574 and AFOSR Grant 87-0346.

References

Allen, P.M., and E.A. Stephens, 1971. Report on the Geological Survey of Hong Kong. Hong Kong Government Printer, 116 p.

Biscaye, P.E., 1965. Mineralogy and sedimentation of recent deep sea clay in the Atlantic Ocean, and adjacent seas and oceans. Geological Society of America Bulletin, v. 76, p. 803–832.

Brand, E.W., J. Premchitt, and H.B. Phillipson, 1984. Relationship between rainfall and landslides in Hong Kong. Proceedings of the 4th International Symposium on Landslides, Toronto, v. 1, p. 377–384.

Curtis, C.D., S.R. Lipshie, G. Oertell, and M.J. Pearson, 1980. Clay orientation in some Upper Carboniferous mudrocks, its relationship to quartz content and some inferences about fissility, porosity and compactional histsory. Sedimentology, v. 27, p. 333–339.

Ehrlich, R., S.J. Crabtree, S.K. Kennedy, and R.L. Cannon, 1984. Petrographic image analysis, I. Analysis of reservoir pore complexes. Journal of Sedimentary Petrology, v. 54, p. 1365–1378.

Foott, R., 1982. Moderator's report on settlement. Proceedings of the 7th South East Asian Geotechnical Conference, Hong Kong, v. 2, p. 353–363.

Foott, R., D.C. Koutsoftas, and L.D. Handfelt, 1987. Test fill at Chek Lap Kok, Hong Kong. Journal of Geotechnical Engineering, ASCE, v. 113, p. 106–126.

Haralick, R.M., 1984. Digital step edges from zero crossing of second directional derivatives. IEEE Transactions on Pattern Analysis and Machine Intelligence, v. 6, p. 58–68.

Koutsoftas, D.C., R. Foott, and L.D. Handfelt, 1987. Geotechnical investigations offshore Hong Kong. Journal of Geotechnical Engineering, ASCE, v. 113, p. 87–105.

Marr, D., and E. Hildreth, 1980. Theory of edge detection. Proceedings of the Royal Society of London Series, B, v. 207, p. 187–217.

Smart, P., and N.K. Tovey, 1982. Electron Microscopy of Soils and Sediments: Techniques. Oxford University Press, Oxford, 284 p.

Smart, P., and N.K. Tovey, 1988. Theoretical aspects of intensity gradient analysis. Scanning, v. 10, p. 115–121.

Tovey, N.K., 1980. A digital computer technique for orientation analysis of micrographs of soil fabric. Journal Microscopy, v. 120, p. 303–315.

Tovey, N.K., 1986a. Microanalyses of a Hong Kong marine clay. Geotechnical Engineering, v. 17, p. 167–210.

Tovey, N.K., 1986b. Microfabric, chemical and mineralogical studies of soils: techniques. Geotechnical Engineering, v. 17, p. 131–166.

Tovey, N.K., 1988. A mineralogical and microfabric study of Hong Kong marine soils. Report written for the Geotechnical Control Office, Hong Kong, v. 1, 112 p.

Tovey, N.K., and P. Smart, 1986. Intensity gradient techniques for orientation analysis of electron micrographs. Scanning, v. 8, p. 75–90.

Tovey, N.K., and V.N. Sokolov, 1981. Quantitative SEM methods for soil fabric analysis. Scanning Electron Microscopy 1981/1, SEM Inc, AMF O'Hare, p. 536–554.

Tovey, N.K., P. Smart, and M.W. Hounslow, 1989. Quantitative orientation analysis of soil microfabric. In: Douglas, L.A. (ed.), Soil Micromorphology: a basic and applied science, Elsevier, Amsterdam, p. 631–639.

Unitt, B.M., 1975. A digital computer technique for revealing directional information in images. Journal of Physics E., v. 8, p. 423–425.

Zhou, Y.T., V. Venkateswar, and R. Chellappa, 1989. Edge detection and linear feature extraction using a 2-D random field model. IEEE Transactions on Pattern Analysis and Machine Intelligence, v. 11, p. 84–94.

CHAPTER 56

Application of Microstructure Classification of Marine Sediment to Engineering Geological Evaluation

Gao Guorui

Introduction

As construction in the coastal area in China is rapidly developing, it is becoming more significant to evaluate the engineering geological properties of nearshore/offshore marine sediments. Conventional geotechnical investigations and laboratory tests, however, are not adequate to satisfy the engineering requirement along coastal areas or in sea areas and the special engineering properties of marine sediments.

A number of studies have revealed that the engineering properties of soils, including marine sediment, depend in a greater degree on their composition, and structure. Therefore, a number of the samples collected along the coastline in China were studied in terms of their composition, structure and the relation with their engineering properties (Gao, 1982, 1983, 1984, 1986). The result has shown that the microfabric features of offshore marine sediments are very complex, and are related to geological condition, burial depth, depositional environment, and material origin. The engineering properties were found to be closely related to microstructure. This presents the possibility that the engineering properties of marine sediment may be quantitatively evaluated and classified according to their microstructure.

From a large number of micrographs of marine sediments microfabric features have been classified based on mechanical strength and stability of the space structure system. The features of skeletal grains, cemented material, bond type, and arrangement mode in representative structure elements were analyzed as well as their effect on engineering properties. The classification system has been developed as one of the four classification systems that may show the properties of marine soils (Chaney and Fang, 1986).

To verify the feasibility of the application of this classification to engineering properties evaluation of marine sediments, 44 samples taken from each sea area in China have been studied and classified according to their microfabric features and compared with their engineering properties. Finally, correlation curves were established and further verified by case history. The result indicated that the correlation curves may provide the basis for quantitatively estimating certain engineering properties of marine sediments.

The classification of the microstructure of marine sediments, correlation curves of structure types, and engineering properties as well as a case history are introduced in this chapter.

The Classification of the Microstructure of Marine Sediment

From a large number of photo micrographs taken by scanning electronic microscopy, it has been observed that marine sediments consist of various kinds of skeletal grains in a space structure system. There are rigid microfragment grains and aggregate grains cemented by silicon and calcium, which possess high strength mechanical properties; while soft floccule and a clay matrix have poor mechanical properties; as well as containing a small amount of bioremains, diotomite, and pyrites. The bonding between the skeletal grains consists of the cementation bonding in which grains contact each other, and link bonding formed by far away grains. Cementation bonding possesses good structural strength and stability, whereas the structure of the link bonding is poor and may deform. The cementation and link bondings between grains are mainly composed of clays, salt crystals, and organic matter, with low levels of amorphous materials. Cementation bonding may be divided into salt crystal cementation and clay cementation; link bonding may be divided into long-link bonding and short-link bonding. In some soils, there are less skeletal grains and

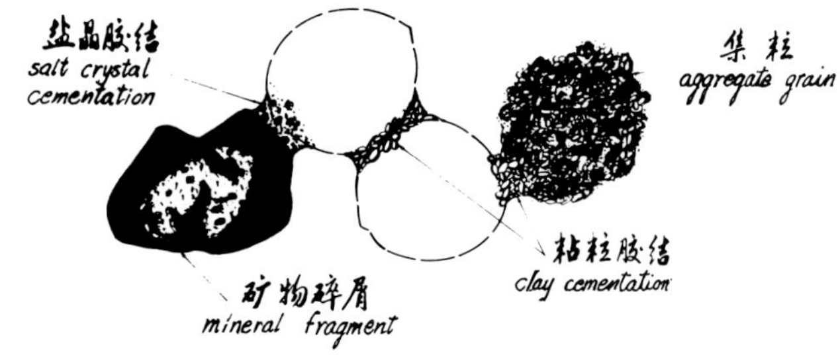

Figure 56.1. Schematic of granular cementation structure.

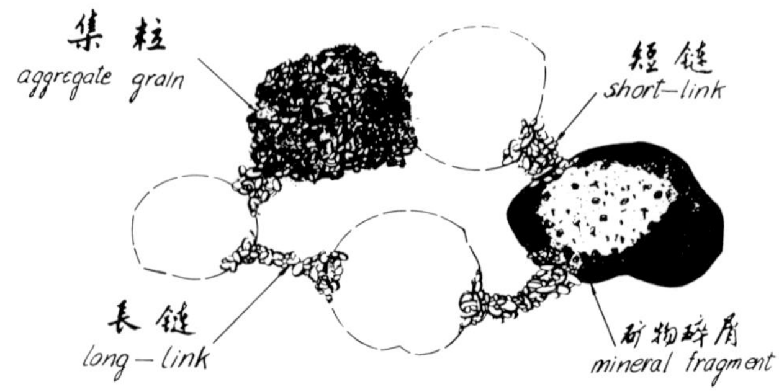

Figure 56.3. Schematic of granular link-bond structure.

most of the grains are buried in a clay matrix. The mechanical properties of this soil, therefore, depend mainly on the properties of the clay matrix.

Based on the above microfabric features and space arrangement modes, four types of microstructures have been classified as follows.

Granular Cementation Structure

The granular cementation structure consists of dominant sand grains, silt grains, and aggregate grains, which act as skeletal grains. Clay particles and amorphous material or organic matter act as cemented material, which exist either between grains or on the surface of grains. The cemented material adsorbed on the grain surfaces make the grains stable. The strength and stability of this structure depend not only on the strength of the grain and the cementation strength between grains, but also on the number of grains that contact each other; that is coordination number.

The higher the mean coordination number is, the stronger the structure. Therefore, the stability of this structure depends in a greater degree on the arrangement mode of the grains. The interlocked arrangement with a higher coordination number is far more stable than the space arrangement with a lower coordination number. The granular cementation structure may be found in the coastal soil layers of each sea area in China, generally existing in the Bohai Sea and the Yellow Sea more than in the South Sea and East Sea. It occurs as an interbed or layer in the delta area of river mouths. If the soil layer is thick enough, it may possess higher bearing capacity, and may sometimes be used for the supporting pile layer. This structure unit mode is shown in Figure 56.1 and its photo micrograph in Figure 56.2.

Granular Link-bond Structure

The granular link-bond structure contains less sand grains, silt grains, or aggregate grains. The center-to-center distance of the grains that are uniformly distributed in soils is larger than the mean diameter of these grains. As a result, the grains partly or incompletely contact each other and a link between grains is formed. The strength and stability between the grains are held by the link. These links have different lengths. Short links may

Figure 56.2. Photomicrograph of granular cementation structure.

Figure 56.4. Photomicrograph of granular link-bond structure.

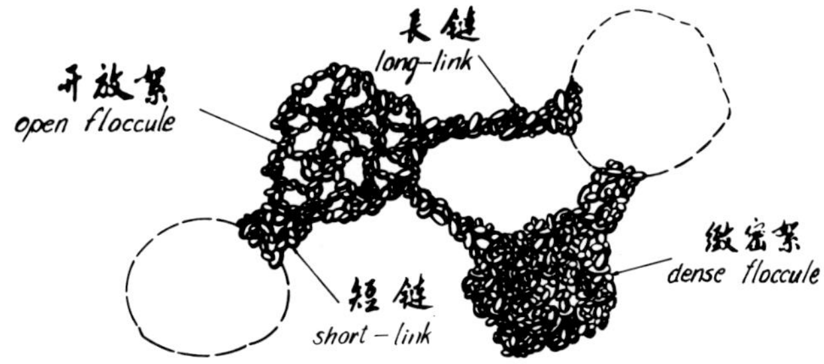

Figure 56.5. Schematic of flocculent link-bond structure.

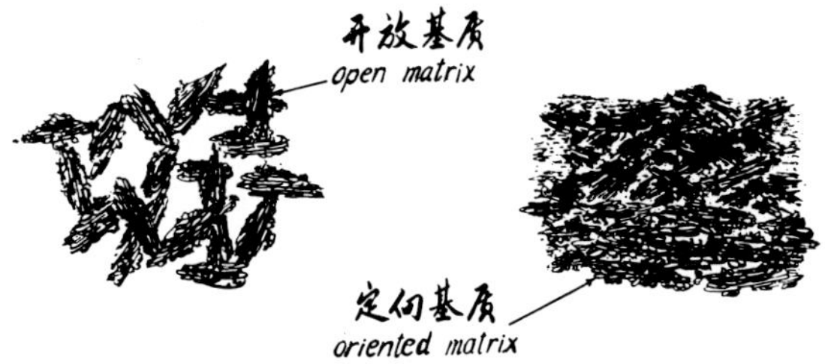

Figure 56.7. Schematic of clay matrix structure.

bear pressure and shear stress to a certain degree, whereas long links may bear only small tension and torsion. This structure exhibits lower strength and large compressive deformation as compared with the granular cementation structure. Deformation under pressure is gradually reduced with time and finally reaches a stable state. Marine sediments with the above structure may be used for the foundation of buildings that allow larger deformation and have small foundation pressure. This microfabric generally exists in shallow layered-marine-sediment. Its schematic is shown in Figure 56.3 and its typical photo micrograph in Figure 56.4.

Flocculent Link-bond Structure

For this type of structure, fewer sand grains, silt grains, and aggregate grains are distributed in the soil. Clay content and organic content are much higher. Floccules flocculated by clay and organic matter in the saltwater environment act as the skeletal frame. They are joined together by links of different lengths and thicknesses, forming the sponge-like flocculent link-bond structure for short links and the honeycomb-like flocculent link-bond structures for long links. This structure may be found only in marine mud soil. It has low strength and large deformation; especially under the effect of the long-period stress, there will be severe deformation. If no corrective measures are taken, it must not be used for the foundation of any buildings. The soil with this microfabric generally exists as the top layer of the marine sediment, and is subject to the effect of sea billow. This structure model is shown in Figure 56.5 and its typical photo micrograph in Figure 56.6.

Clay Matrix Structure

A clay matrix structure consists of a great many clay particles. These clay particles conglomerate to form clay "domains"; the clay domains then conglomerate to form regular or irregular aggregations, which in turn conglomerate to form clay clumps; finally the clay clumps further conglomerate to form a clay matrix structure. The few silt grains or sand grains that are entirely buried in the clay matrix do not act as skeleton grains. If the domains within an aggregation in the clay matrix are densely arranged and a parallel oriented face-to-face bond is

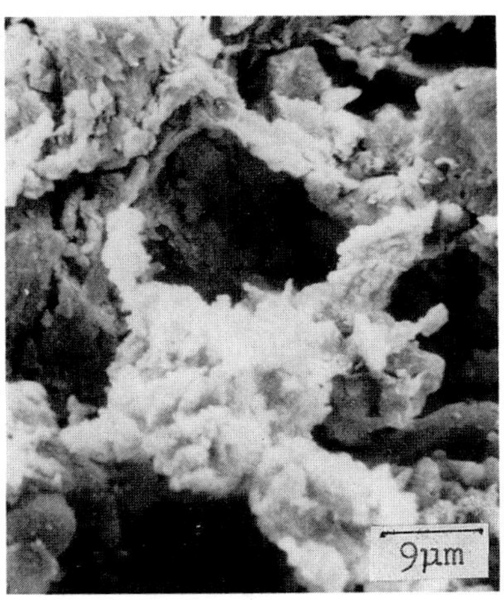

Figure 56.6. Photomicrograph of flocculent link-bond structure.

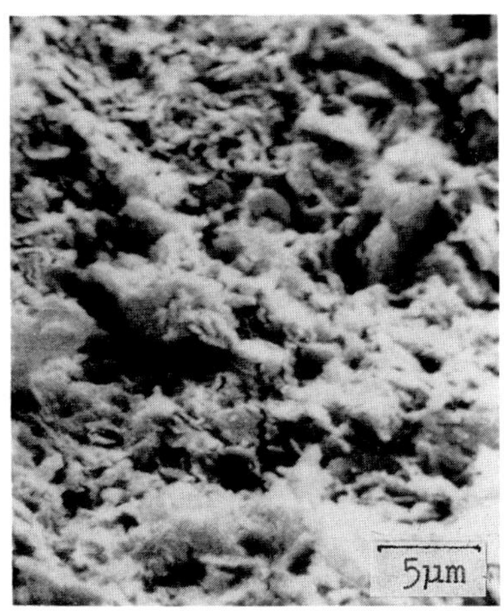

Figure 56.8. Photomicrograph of clay matrix structure.

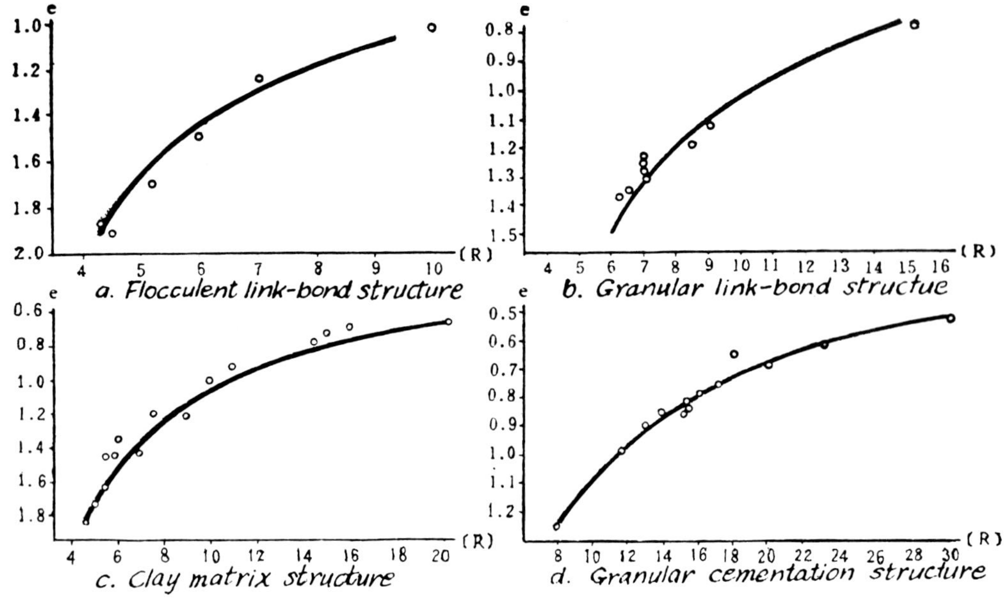

Figure 56.9. Void ratio versus allowable bearing capacity of four types of marine sediment structures.

formed, this oriented clay matrix structure may possess high bearing capacity. If the arrangement of the domains within the aggregation is loose and forms nonoriented edge-to-face bonds with more interlocking pores, this open clay matrix structure may possess very low bearing capacity, much greater plastic deformation characteristics, and higher sensitivity. The model of this structure is shown in Figure 56.7 and its typical photo micrograph in Figure 56.8.

The Correlation of the Microstructures with Engineering Properties

For the purpose of applying the classification of the microstructure to geological engineering evaluation, 44 samples of marine sediments that have been studied by scanning electronic microscopy were analyzed to relate their physical and mechanical properties to the various types of microstructure. It

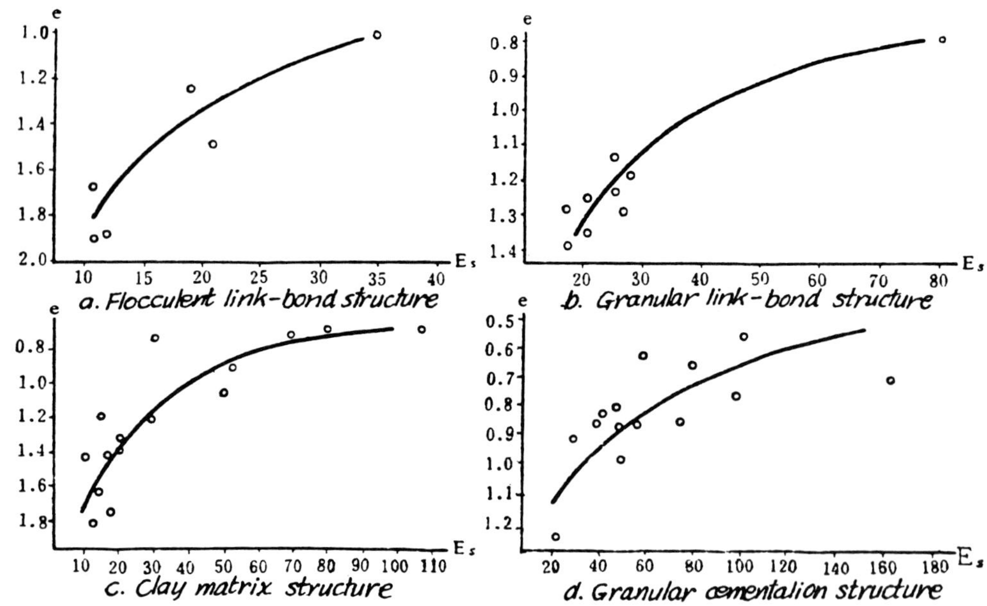

Figure 56.10. Void ratio versus modulus of compressibility of four types of marine sediment structures.

Figure 56.11. Photomicrograph of stratum 2 showing granular cement structure.

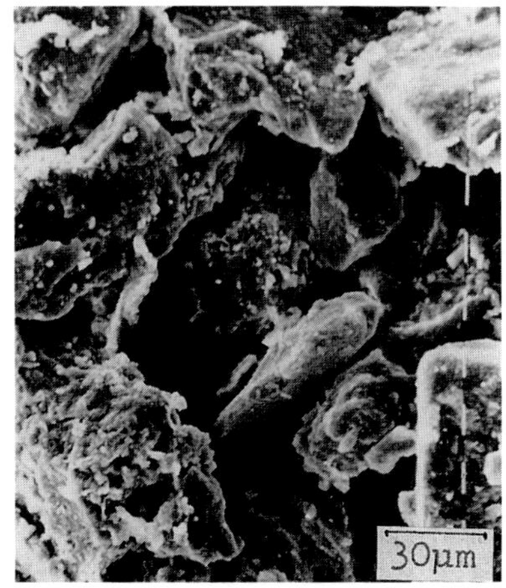

Figure 56.13. Photomicrograph of stratum 4 showing granular cement structure.

was found that the different microstructure types have various correlation curves of allowable bearing capacity [*R*], modulus of compressibility E_s, and void ratio (see Figs. 56.9 and 56.10). Though these curves are not ideal, because of the limited samples, they do show a certain regularity. An attempt to apply these curves to engineering analysis was made. The resulting estimated values of the bearing and deformation of marine sediments are basically consistent with the laboratory tests and *in situ* testing. This indicates that it is feasible to apply the classification of the microstructures to the evaluation of the engineering geological properties of marine sediments.

The Case History

The sediment profile in the coastal area in Tanggu district consists of six strata. The first stratum is mud scum, which is located in 0.5–1.5 m below the seafloor. The second stratum is a hard layer of sandy silt soil with a depth of 1.5–3.5 mbsf (meters below sea floor). The third stratum is gray-brown soft clay buried 3.5–10.5 mbsf. The fourth stratum is a sandy silty soil with stratified sand, which lies between 10.5 and 16.5 mbsf. The fifth stratum is gray-brown soft clay with a buried depth of 16.5–19.5 mbsf. The sixth stratum is yellow-brown sandy silty soil at a depth of 19.5 mbsf.

Figure 56.12. Photomicrograph of stratum 3 showing flocculent link-bond structure.

Figure 56.14. Photomicrograph of stratum 5 showing clay matrix structure.

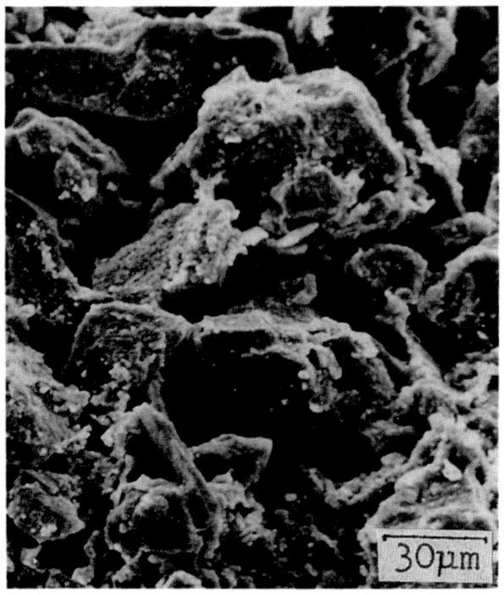

Figure 56.15. Photomicrograph of stratum 6 showing granular cement structure.

To have a quality method, combining *in situ* drilling with *in situ* testing has been used, including standard penetration test (SPT), static cone penetration test (CPT), and vane shear test. Because the mechanical properties of various strata are not the same, the drilling and *in situ* testing proved to be difficult. For example, in the mud or soft clay strata, such as the first or the third stratum, standard penetrometer (standard sample tube and drill rod) sank without being driven, so it failed to take data in these strata. The static cone penetrometer was not drilled through the sandy silt strata with stratified sand, that is, the fourth stratum, thus the data were taken from the first to the third strata. The vane shear test was not effective on the fourth stratum and below. Systematic data of all strata was obtained only from laboratory tests (LT). Good quality drilling and sampling at sea has proven difficult. In this case history, the engineering properties of the marine sediment must not depend only on the laboratory tests of the samples. To verify the effect of the above testing, microfabric analyses (MA) of the samples of each strata, except the first stratum, have been made by scanning electron microscopy. The result has shown that the hard layer of sandy silt soil, the second stratum, is a granular cementation structure (Fig. 56.11); the soft clay layer, the third stratum, is a flocculent link-bond structure (Fig. 56.12); the sandy silt soil layer, the fourth stratum, is a granular cementation structure (Fig. 56.13); the soft clay layer, the fifth stratum, is a clay matrix structure (Fig. 56.14); and the silty soil layer, the sixth stratum is a granular cementation structure (Fig. 56.15). The data of allowable bearing capacity and modulus of compressibility, according to the classification of the microstructure are shown in Figure 56.16.

Compared with various testing methods (Fig. 56.16) the microfabric analysis (MA) method not only reflects the engineering mechanical properties of each strata, but also makes up for the deficiency of some testing methods.

Conclusions

1. The microstructure of marine soils in China is very complex. The microstructure classification made from their microfabric features and the mechanical strength and stability of

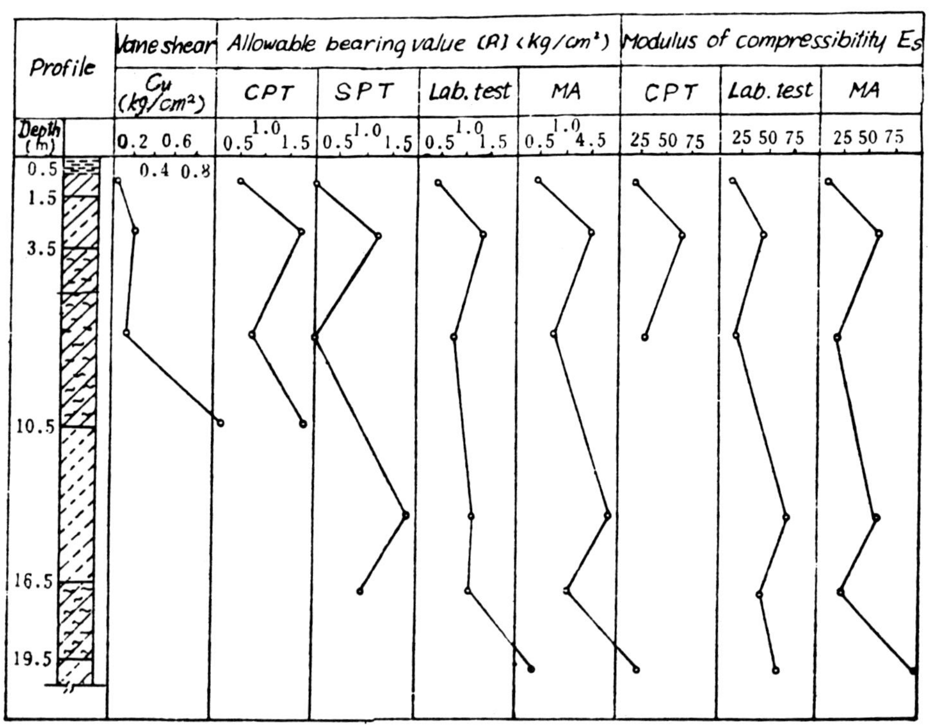

Figure 56.16. Data of vane shear, allowable bearing value, and modulus of compressibility.

their space structure system may well show the engineering properties of the marine sediments.

2. A certain correlation occurs between the various types of microfabric features of marine sediment and their engineering mechanical properties. According to the microstructure classification, it is possible to quantitatively evaluate the geological engineering properties. This provides a new way for geological engineering investigations of marine sediment.
3. It is proved from the case history that the microfabric analysis method not only may be used to quantitatively evaluate the engineering properties of marine sediment, but also may make up for the deficiency of some conventional testing methods.

References

Chaney, R.C., and H.Y. Fang, 1986. Summary. Marine Geotechnology and Nearshore/Offshore Structures. ASTM STP 923, p. 355–362.

Gao Guorui, 1982. Microstructure of the marine soils in China. Proceedings of the IVth Congress IAEG, New Delhi, India, v. 7, p. 263–273.

Gao Guorui, 1983. Microstructure of marine soils in China relative to engineering properties. Proceedings of the Shanghai Symposium on Marine Geotechnology and Nearshore/Offshore Structures, p. 38–53.

Gao Guorui, 1984. Microstructure of marine soils in China and their engineering properties. Scientia Sinica (Series B), n. 9, p. 849–860.

Gao Guorui, 1986. Microfabric features, composition, and microstructure classification of marine soils in China. Marine Geotechnology and Nearshore/Offshore Structures, ASTM STP 923, p. 178–195.

CHAPTER 57

Preliminary Geotechnical Evaluation of Deep Borehole Facilities for Nuclear Waste Disposal in Shales

Mysore S. Nataraj

Introduction

Recently Oak Ridge National Laboratory (ORNL) has completed a comparative evaluation of five sedimentary rocks (Croff et al., 1986), shales, sandstones, anhydrite, chalk, and carbonates, to determine their relative potential as host media for the disposal of radioactive wastes. It was determined that various rock mass characteristics of shale indicate many positive features with respect to emplacement of radioactive wastes within these formations. Shale possesses certain characteristics that are considered to be positive attributes in determining its potential as a host material for waste repositories: unfractured shale has (1) a very low permeability, (2) a high ion exchange capacity, and (3) is relatively plastic, a characteristic that can promote self-healing of fractures and ensure low permeability; in addition, shale strata in the United States are located at appropriate depths and have large lateral and vertical extent. However, it is to be noted that shales can have a low compressive strength and may creep readily. Further, the constitutive relationship for shale at great depths has not been completely investigated, and information on the mechanical properties of shale is very limited.

The present study is concerned with deep boreholes with waste emplacement from the surface. A schematic diagram of the concept of the borehole facility is shown in Figure 57.1. The basic geometry and configuration used was based on earlier studies (Chapman et al., 1984; Croff et al., 1986). The borehole is assumed to be of 0.52 m diameter and extending vertically to a depth of about 525 m. The depth from the surface to the disposal zone is about 300 m. The boreholes are laid out in hexagonal configuration, the reference spacing being 35 m. The concept of the deep borehole facility is a relatively inexpensive method for waste disposal. Its unique features are its complete flexibility and modular nature in operational conditions. A limited area is used at a time, and a completed borehole site is returned to normal use very quickly. This is a practical technique with efficient use of available expertise from civil and mining engineering.

Geotechnical Properties of Shale

A general evaluation of desirable characteristics of a suitable host material for waste disposal purposes includes study of hydrogeological, geochemical, mineralogical, thermomechanical, and geotechnical properties. In this study, the influence of a few geotechnical properties on strength and deformation of host material is briefly examined. The strength and stability are important characteristics because the physical integrity of the repository is not to be jeopardized under any circumstances.

A host material is considered to be stable when the load bearing capacity is sufficient to withstand external loads without undergoing inadmissable large deformations. In the case of a borehole facility, this principle may have many connotations. It may mean that the borehole will have to remain open for sufficient periods of time so as to enable the emplacement of waste material. The host material must be capable of supporting waste canisters over a long period of time and its load-bearing capacity must not be impaired as a result of stress changes either suddenly or slowly. The material properties of the host material are to be such that no inadmissible deformation occurs that may impair the usability of the facility.

Shale is an argillaceous rock with extreme variations in its geotechnical properties, and this variation makes the evaluation of its probable behavior difficult, although it is an important step in the engineering consideration of a waste repository. Large amounts of information on the properties of shale have been

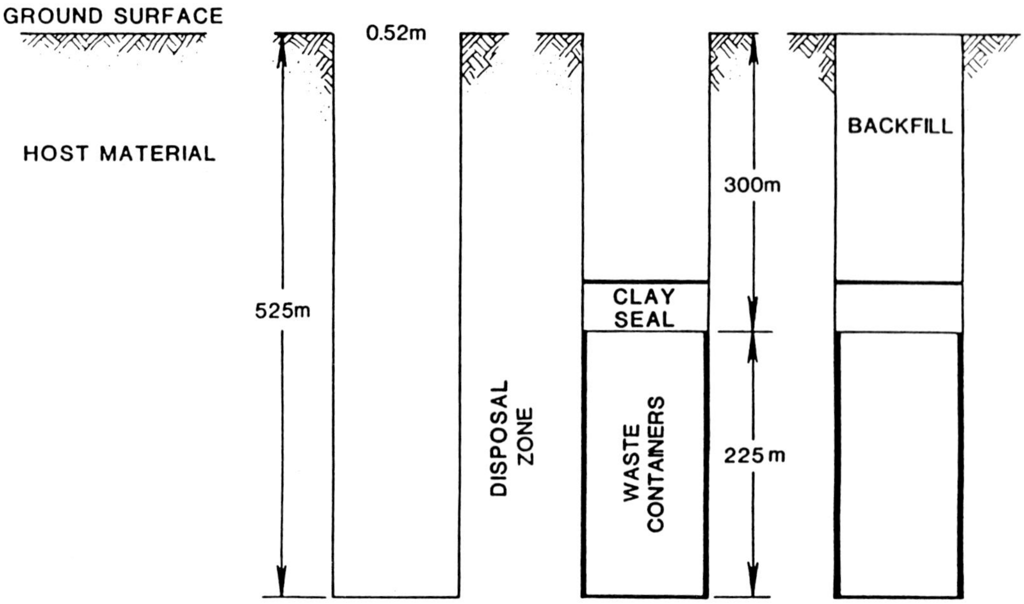

Figure 57.1. Schematic diagram of the concept of a borehole facility (not to scale).

collected in the past. Many reports and papers contain site-specific properties. This study is confined to Pierre shale, Rhinestreet shale, and a typical illite shale. In the present study, the properties used have been collected mainly from several reports using original references whenever possible (Abel and Gentry, 1975; Kibbe and Boch, 1978; Kopp, 1986; Gonzales and Johnson, 1985; Hansen and Vogt, 1987). In the selection of properties of rocks and rock mass, one has to consider the inherent limitations due to the nature of the data collected, organized, and presented by various researchers. Discussions of the scaling effect and relevance of laboratory and field tests to engineering practice are given in several publications (Goodman, 1980; Hoek and Brown, 1980; Bieniawski, 1984). In this study, it was necessary to assume values for some properties. These were estimated from values at a similar location, formation, or field condition. It should be emphasized that the estimated values are only approximations due to the nature of data available, from different test conditions and procedures, and scaling of values. In estimating the rock mass properties, suggestions made by various investigators (Kibbe and Boch, 1978; Goodman, 1980; Bieniawski, 1984; Hunt, 1986) have been used. A summary of the properties and their ranges used in this study for the three shales is presented in Table 57.1.

Preliminary Assessment of Support Capability

Rocks in general have adequate bearing capacity. However, it depends on rock type, nature of discontinuities and fracture, rock quality, and mineralogical composition. It is difficult to estimate the bearing capacity of rocks taking all the factors into consideration. Also, there is no rational procedure, except by conducting full-scale load tests, that can determine the allowable bearing pressure of rocks. However, various procedures have been suggested, some empirical and some theoretical, and used by many researchers (Stagg and Zienkiewicz, 1968; Goodman, 1980; Coates, 1981; Bowles, 1988). In this study, the approach suggeseted by Stagg and Zienkiewicz has been used.

Table 57.1. Properties of shale.

Shale property	Intact rock	Rock mass
Pierre		
Bulk density	0.02 MN/m^3	0.02 MN/m^3
RQD	28, 56, 84%	–
Unconfined compressive strength	10, 15 MPa	6.5 MPa
Cohesion	0.5, 1.5 MPa	1 MPa
Friction angle	5, 15, 30°	10°
Modulus of elasticity	1000 MPa	200 MPa
Poisson's ratio	0.2	0.2
Rhinestreet		
Bulk density	0.02 MN/m^3	0.02 MN/m^3
RQD	28, 56, 84%	–
Unconfined compressive strength	10, 25, 50 MPa	25 MPa
Cohesion	5, 10 MPa	10 MPa
Friction angle	5, 15, 30°	15°
Modulus of elasticity	17,500 MPa	3,500 MPa
Poisson's ratio	0.2	0.2
Typical illite		
Bulk density	0.02 MN/m^3	0.02 MN/m^3
RQD	28, 56, 84%	–
Unconfined compressive strength	10, 25, 70 MPa	30 MPa
Cohesion	20, 30, 40 MPa	20 MPa
Friction angle	15, 20, 30°	20°
Modulus of elasticity	11,000 MPa	2,000 MPa
Poisson's ratio	0.2	0.2

Figure 57.2. Equal allowable pressure curves for fair and good quality rock.

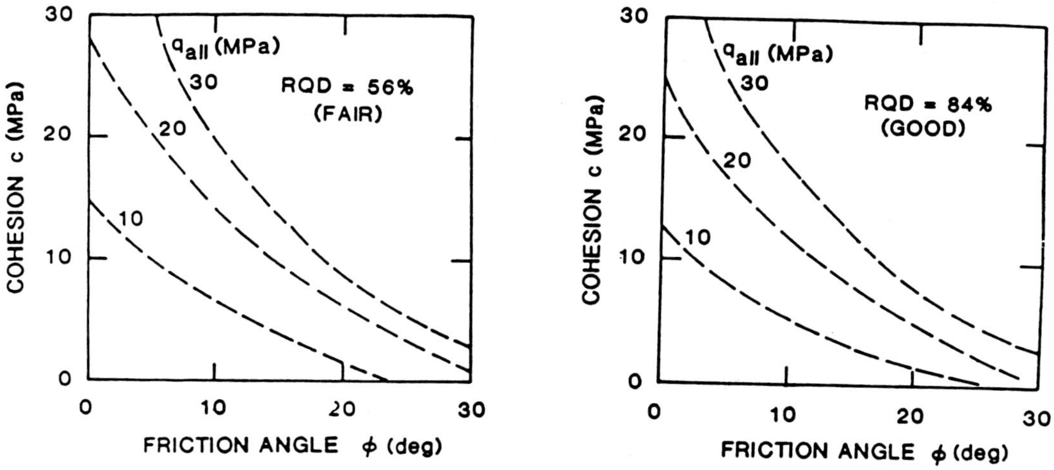

The ultimate bearing capacity was reduced based on rock quality designation (RQD) as

$$q_{\text{reduced}} = q_{\text{ultimate}} \times (\text{RQD})^2$$

A simple parametric study was carried out using the various values of cohesion and friction angle presented in Table 57.1

The bearing capacity depends on many factors, including the depth, friction angle, and cohesion. There are many suggestions for adopting an upper limit of allowable pressure on rock (Peck et al., 1974; Hunt, 1986; Bowles, 1988). To be simple, the maximum allowable pressure on intact rock is taken to be 1.5 times the unconfined compressive strength of samples determined in the laboratory. The maximum limit of the allowable bearing pressure be not more than 30 MPa for a rock with a RQD of 100%. Adapting this suggestion, the curves shown in Figure 57.2 were obtained from the results of parametric studies. These curves can be used in preliminary studies to estimate the allowable bearing pressure on a rock mass of specified cohesion and friction angle. However, it should be noted that as these curves have been obtained using approximate ranges for properties, the variations shown are only estimates. Another approximate procedure (Goodman, 1980) considering unconfined compressive strength and friction angle was also used. The salient results are presented in Figure 57.3, which shows the values of allowable bearing pressure for various values of unconfined compressive strength. The values of unconfined compressive strength assumed for rock mass are shown in parentheses. Since the

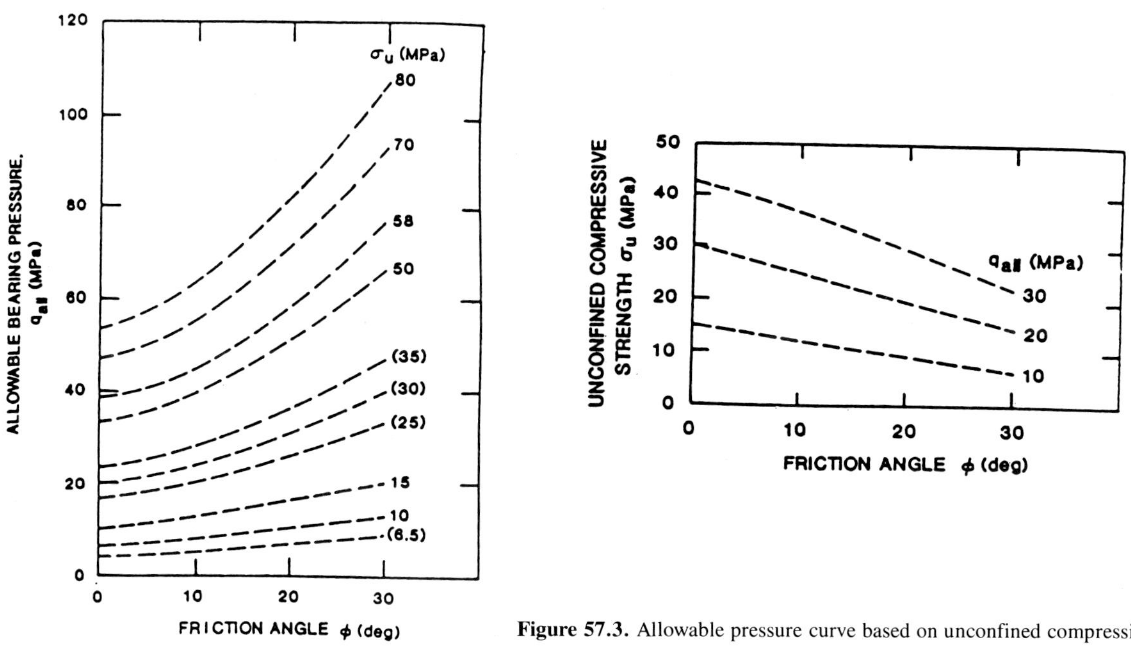

Figure 57.3. Allowable pressure curve based on unconfined compressive strength.

Table 57.2. Stability of shales.*

Shale	Stability factor (N)	Squeeze ratio (R)	Remarks
Pierre	1.62	10.5	For $N \leq 3.0$, UCS = 3.5 MPa
Rhinestreet	0.4	1.05	For $N \leq 4.5$, UCS = 2.33 MPa
Typical illite	0.35	0.53	For $R \leq 6$, c = 1.8 MPa

*UCS, unconfined compressive strength; c, cohesion.

strength of the rock mass controls the field behavior, it is desirable to know the allowable pressure of the rock mass. Equal allowable pressure curves are shown in Figure 57.3. These curves are useful for preliminary purposes to estimate the value of allowable pressure of rock mass, if friction angle, and unconfined compressive strength are known. Again, it is to be noted that these curves serve only as a guide for detailed analysis.

Another way of examining the stability of a borehole is to examine whether the artificial opening remains open or collapsed under the existing conditions. One method (Hudson and Boden, 1981) considers unconfined compressive strength and uses the concept of stability factor, N. Increasing values of stability factor, N, indicate decreasing stability. Hudson and Boden suggest a working value of N equal to three for radioactive waste repositories. However, values up to 4.5 may be quite appropriate (Chapman et al., 1984). The stability factor, N, for the shales under study are presented in Table 57.2. The values of unconfined compressive strength, necessary to give a stability factor of 3 and 4.5 for boreholes under consideration, are shown under the "Remarks" column. In another approach (Bowles, 1988), the cohesion is used as a main property and uses the concept of squeeze ratio, R. If the squeeze ratio, R, is greater than 6, squeezing takes place. If R exceeds 8, then squeezing takes place very rapidly and use of supporting liners or slurry is suggested. The squeeze ratio, R, for the three shales is shown in Table 57.2. The values of cohesion necessary to give a squeeze ratio of less than or equal to 6 for boreholes in the present study are shown

Table 57.3. Rock and rock mass data.

In situ stress	10.5 MPa
Unit weight of broken rock	0.02 MN/m³
Material constants for original rock mass of	$m = 1.0$
good quality	$s = 0.004$
Material constants for broken rock mass of	$m_r = 0.2$
fair quality	$s_r = 0.0001$
Poisson's ratio for rock mass	0.2
Modulus of elasticity of rock mass	
Typical illite shale	2,000 MPa
Rhinestreet shale	3,500 MPa
Pierre shale	200 MPa
Unconfined compressive strength of intact rock	
Typical illite shale	70 MPa
Rhinestreet shale	50 MPa
Pierre shale	15 MPa

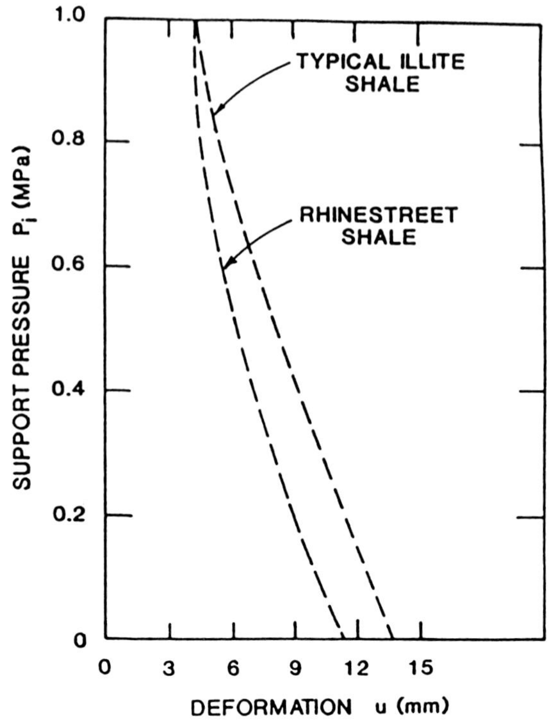

Figure 57.4. The variation of borehole deformation with support pressure.

in Table 57.2 under the "Remarks" column. Based on the values presented in Table 57.2, it can be seen that for shales, the unconfined compressive strength and cohesion of rock mass must be at least 2.5 and 2 MPa, respectively, for adequate stability of boreholes.

An approximate assessment of the likely response of the rock mass to a set of induced stresses was also attempted in this preliminary study. Many methods have been proposed and used by various researchers to obtain rock pressure curves (Goodman, 1980; Hoek and Brown, 1980; Coates, 1981; Bieniawski, 1984). Since the properties of rock mass for the shales are not described in the literature, it was necessary to assume some of them. In the analysis, it has been assumed for simplicity that original rock mass is of good quality and the broken rock mass is of fair quality. Hence, the analysis provides only a general idea about the response of the shales. The data used are presented in Table 57.3. The variation of borehole deformation with support pressure is shown in Figure 57.4 for Rhinestreet and typical illite shale. The effect of modulus of elasticity can be observed in that the typical illite shale undergoes relatively more deformation than Rhinestreet shale. However, the unconfined compressive strength of illite shale is slightly higher than that of Rhinestreet shale. Thus, the broken zone thickness for a given support pressure is less for illite shale than that for Rhinestreet shale, as shown in Figure 57.5. In contrast, the behavior of Pierre shale is catastrophic. The Pierre shale appears to collapse, as can be seen in Figure 57.5, and large support pressure is required to reduce

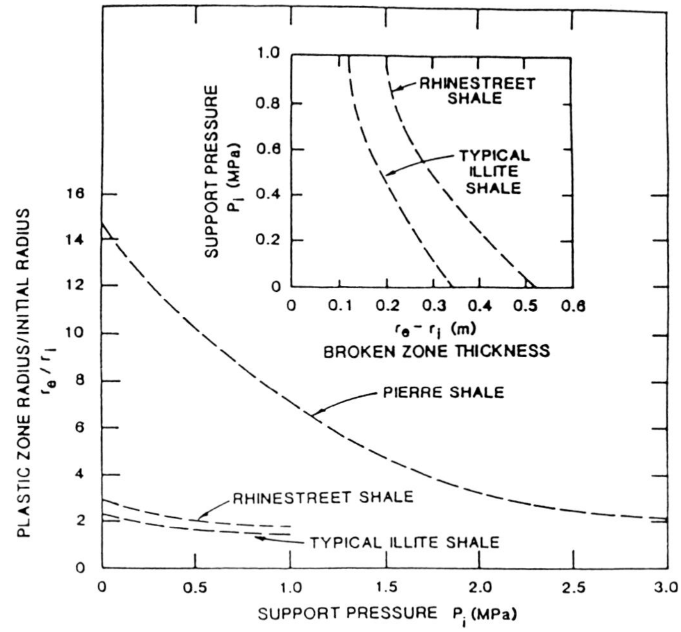

Figure 57.5. Comparison of behavior of three shales.

the deformation. The Pierre shale is a very low strength rock and can fail under relatively low stress values. Some form of support is necessary to prevent collapse of shale into drilled holes. This behavior of Pierre, Rhinestreet, and illite shale also was indicated earlier by the stability factor. In interpreting the above results, it should be kept in mind that these results are highly sensitive to the values of various parameters such as modulus of elasticity, unconfined compressive strength, and material constants for original and broken rock mass.

Summary

The primary purpose of this study was to make a preliminary engineering evaluation of a borehole facility in shales. Some of the geotechnical properties of Pierre, Rhinestreet, and typical illite shale have been collected from the limited existing literature. The properties were used to estimate the bearing pressure of shales, and equal pressure curves have been developed. It appears that Pierre shale is very unstable and requires support to prevent collapse. Typical illite shale is more stable than Rhinestreet shale, although it undergoes relatively more deformation than Rhinestreet shale. Based on preliminary computations, the unconfined compressive strength and cohesion of rock mass are to be at least 2.5 and 2 MPa, respectively, for adequate stability of boreholes in shales.

Acknowledgments

The author expresses his sincere gratitude to Dr. T.F. Lomenick, Oak Ridge National Laboratory, for his guidance during the present study. This study was carried out under contract No. 19X-SC-206V.

References

Abel, J.F., and D.W. Gentry, 1975. Evaluation of excavation experience: Pierre shale. ORNL/Sub-75/70347, Oak Ridge National Laboratory, Oak Ridge, TN, 101 p.

Bieniawski, Z.T., 1984. Rock Mechanics Design in Mining and Tunneling. A.A. Balkema Publishers, Rotterdam, 272 p.

Bowles, J.E., 1988. Foundation Analysis and Design, 4th ed. McGraw-Hill, New York, 1004 p.

Chapman, N.A., F. Gera, and Riccioni, 1984. Feasibility study for the disposal of high-level heat emitting, cladding hull, and other wastes in the Plio-Pleistocene blue clays, Rome, Practica, no. 2403, 210 p.

Coates, D.F., 1981. Rock Mechanics Principles. Monograph 874. Energy, Mines, and Resources, Canada, 412 p.

Croff, A.G., T.F. Lomenick, R.S. Lowrie, and S.H. Stow, 1986. Evaluation of five sedimentary rocks other than salt for geologic repository siting. ORNL-6241, 3 volumes, Oak Ridge National Laboratory, Oak Ridge, TN.

Gonzales, S. and K.S. Johnson, 1985. Shales and other argillaceous strata in the U.S. ORNL/Sub-64794/1, Oak Ridge National Laboratory, Oak Ridge, TN, 596 p.

Goodman, R.E., 1980. Introduction to Rock Mechanics. Wiley, New York, 478 p.

Hansen, F.D., and T.J. Vogt, 1987. Thermomechanical properties of selected shales. RSI-0305, Report to Oak Ridge National Laboratory, 91 p.

Hoek, E., and E.T. Brown, 1980. Underground Excavations in Rock. The Institution of Mining and Metallurgy, London, 527 p.

Hudson, J.A., and J.B. Boden, 1981. Geotechnical and tunneling aspects of radioactive waste disposal, Proceedings of International Tunneling Conference, Brighton, United Kingdom.

Hunt, R.E., 1986. Geotechnical Engineering Analysis and Evaluation. McGraw-Hill, New York, 730 p.

Kibbe, R.K., and A.L. Boch, 1978. Technical support for GEIS: radioactive waste isolation in geologic formations, v. 6. Y/OWI/TM-36/6, Oak Ridge National Laboratory, Oak Ridge, TN, 84 p.

Kopp, O.C., 1986. A preliminary assessment of mineralogical criteria on the utility of argillaceous rocks and minerals for high-level radioactive waste disposal. ORNL/TM-90979, Oak Ridge National Laboratory, Oak Ridge, TN, 271 p.

Peck, R.B., W.F. Hanson, and T.H. Thornburn, 1974. Foundation Engineering, 2nd ed. Wiley, New York, 514 p.

Stagg, K.G., and O.C. Zienkiewicz, 1968. Rock Mechanics in Engineering Practice. Wiley, New York, 442 p.

CHAPTER 58

Microstructural and Mineralogical Characterization of Selected Shales in Support of Nuclear Waste Respository Studies

S.Y. Lee, L.K. Hyder, and P.D. Alley

Introduction

The Sedimentary Rock Program at Oak Ridge National Laboratory (ORNL) has evaluated a variety of sedimentary rocks as potential host media for a U.S. civilian nuclear waste repository. Shale was selected as the sedimentary rock with the greatest potential, based on technical aspects related to geology, geochemistry, hydrology, thermal performance, rock mechanics, natural resources, waste package material degradation, repository costs, and systems studies (Croff et al., 1986). Other countries such as Belgium, Italy, Switzerland, and the United Kingdom are considering argillaceous formations as a potential host rock for high-level waste disposal (Gera, 1980).

Because shales are a diverse and highly variable type of rock, complete characterization of all types would be a formidable task. Therefore, four end-members that represent the spectrum of shales were identified based on differences in their chemical and mineralogical compositions (Stow and Croff, 1987). The end-members selected for characterization were organic-, carbonate-, smectite-, and illite-rich shales. The microstructure and mineralogy not only influence radionuclide migration but also influence the hydrologic and mechanical properties of the host rock (Weaver, 1979; Lomenick and Laughon, 1980; Lee and Tank, 1985; Milodowski et al., 1985; Von Damm, 1987; Burkett et al., 1987; Güven et al., 1988). Therefore, microstructure and mineralogical data should be acquired and compiled as a basis for further investigations. The objective of this study was to conduct the microstructural and mineralogical characterization of the end-member shales in support of nuclear waste repository studies.

Materials and Methods

Materials

Five core samples, representing the four end-member shale compositions, were selected for microstructural and mineralogical characterization studies. The shales selected were Chattanooga Shale (organic), Pierre Shale (smectitic), Green River Formation (carbonate), and Nolichucky Shale and Pumpkin Valley Shale (illitic). The Pierre Shale and Green River Formation core samples were obtained from RE/SPEC, Inc. (Rapid City, South Dakota) and samples of the other shales were collected by ORNL geologists.

The Chattanooga Shale sample was from the upper Dowelltown Member of the formation in Fentress County, Tennessee, at a depth of 141–142 m. This unit is generally described as interbedded, medium light gray claystone and dark gray shale beds, 3–12 cm thick (Conant and Swanson, 1961).

The Pierre Shale samples (PS/86/20U13-1T/2 and PS/86/20U13) were from the Mobridge Member of the Pierre Shale in Gregory County, South Dakota. They were retrieved from drill hole 84–20 at a depth of 88.2–88.9 m. The samples were described as claystone, thick bedded to massive, nonfissile, slightly to moderately calcareous, soft, moist, medium gray with a slight olive tinge, dense, solid, bedding at low angle, unweathered.

The Green River Formation samples (GR/86/V33-0/2-3/1 and GR/86/V22-0) were from Garfield County, Colorado. They were obtained from the roof of the Colony Mine from two separate but closely adjacent drill holes (about 1.8 and 0.9 m from the back of the mine). The samples were described as thinly bedded calcareous marl, very hard and compact.

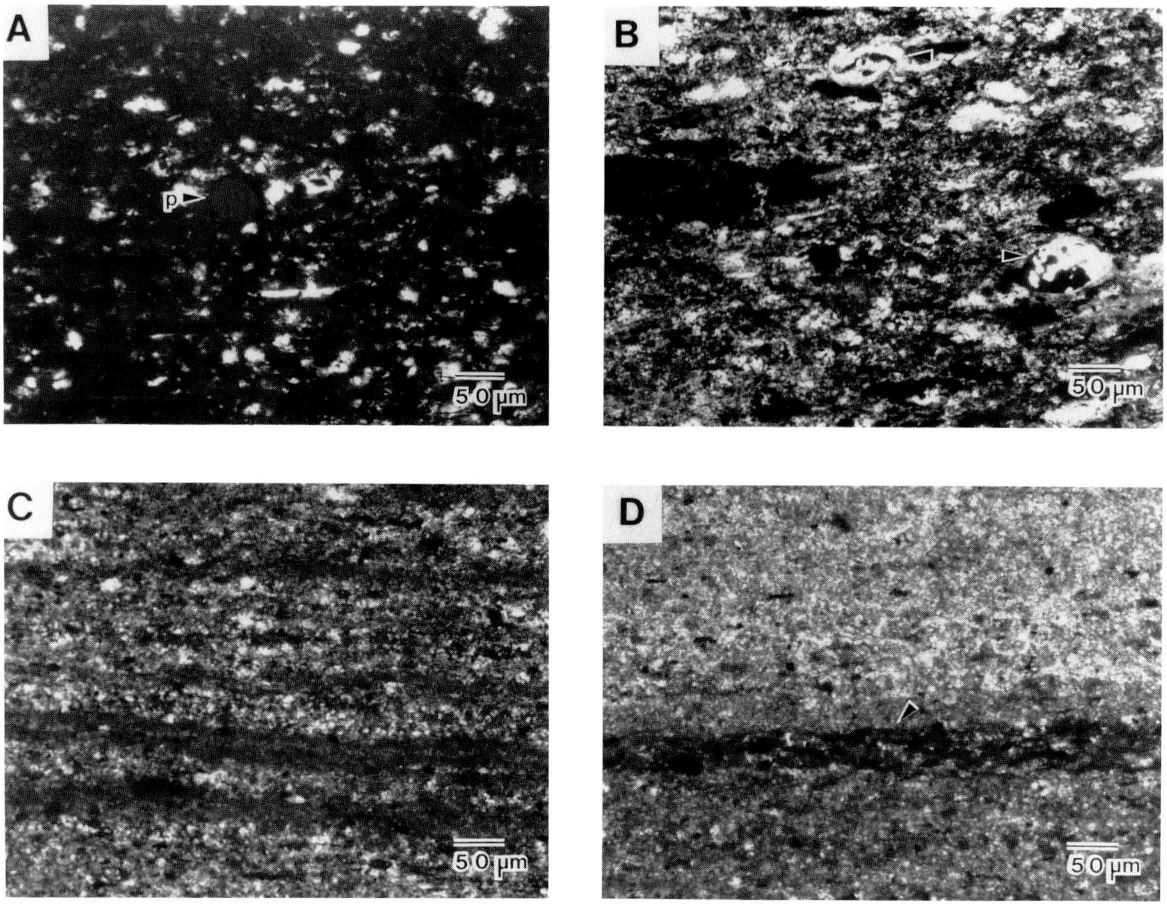

Figure 58.1. Photomicrographs of shales in transmitted light. (A) Chattanooga Shale with framboidal pyrite (p). (B) Pierre Shale with ostracods (arrows). (C) and (D) Green River Formation, showing heavy mineral concentrations (arrows).

The Nolichucky Shale and Pumpkin Valley Shale were from the Joy 2 well in Oak Ridge, Tennessee, at depths of 181–182 m and 604–605 m, respectively. Both shales belong to the Conasauga Group, a complex sequence of Middle to Upper Cambrian clastic and carbonate strata that outcrops throughout the Valley and Ridge Province. The Nolichucky section is described from the well log as a gray to brown shaley siltstone with discontinuous parallel bedding. The Pumpkin Valley section is a maroon-to-gray, glauconitic, laminated, silty mudstone.

Methods

Each core segment was vertically split into two half cores, and one of the half cores was broken with a rock hammer into small fragments (<3 cm). The fragments were crushed manually with mortar and pestle until they passed through a 2-mm-mesh sieve, and then they were pulverized for 10–15 min in a Siebtechnik mechanical "shatterbox" consisting of an agate liner, ring, and disk. The resultant rock powder was sieved through 0.18-mm mesh. Any material that did not pass through the sieve was repulverized by hand with an agate mortar and pestle and resieved. The remaining half of each core segment was used to make specimens for microstructural analyses.

The micromorphology and structure of the shales were examined by petrographic microscopy and scanning electron microscopy (JEOL JSM-840A). Semiquantitative chemical compositions of the surface and the polished cross-sectional surfaces of the shales were determined by a Tracor Northern energy-dispersive X-ray analyzer (EDX) attached to the scanning electron microscope. For high-resolution transmission electron microscopy, thin sections of the samples were cut perpendicular to the bedding plane to promote optimum orientation for lattice-fringe imaging of phyllosilicates. Following optical examination, selected areas (3 mm in diameter) were detached, ion-thinned on

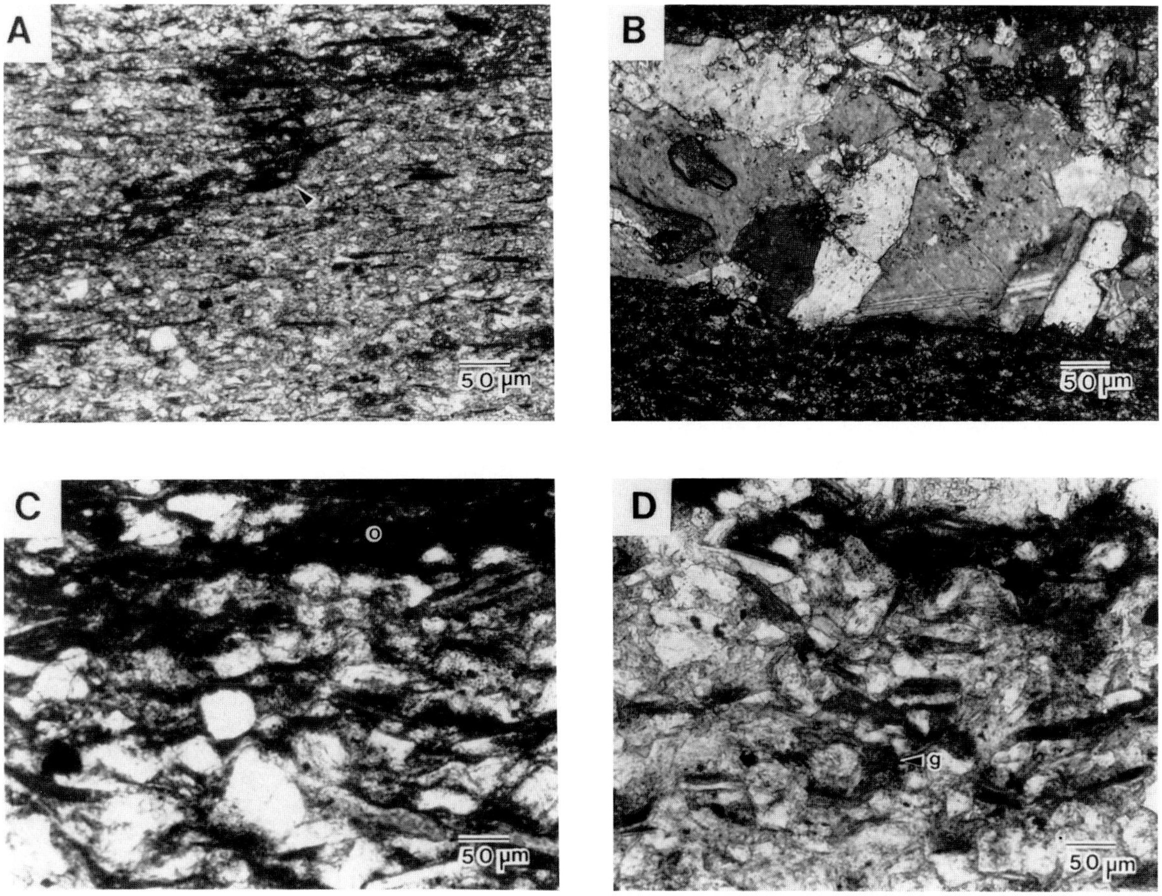

Figure 58.2. Photomicrographs of shales in transmitted light. (A) Nolichucky Shale with stair-step fracture. (B) Nolichucky Shale with calcite vein filling. (C) Pumpkin Valley Shale, showing organic matter, (o) and coarse detrital grains. (D) Pumpkin Valley Shale, showing fine detrital grains and glauconite (g).

a cold stage, and carbon coated. The ion-thinned samples were examined at 200 kV in a JEOL JEM-2000FX transmission electron microscope with attached Tracor Northern EDX system. One-dimensional lattice-fringe images were obtained by using (001) diffractions (Lee et al., 1975a).

Multielement neutron activation analyses of the pulverized shales were performed with the ORNL Research Reactor (Lee et al., 1987). After acid dissolution of the samples, selected elements were analyzed via atomic absorption by the ORNL Analytical Chemistry Division. Total sulfur and sulfide were determined with an automatic LECO titrator. The amount of inorganic carbonate in the shales was determined by measuring the volume of CO_2 that evolved from the reaction of carbonates with excess HC1 (Lee et al., 1987). Surface area was determined by measuring the amount of nitrogen adsorbed at liquid nitrogen temperature (standard BET method) on the surface of shale samples. Thermal gravimetric analyses and differential thermal analyses were performed on the samples over a temperature range of 30–1100°C with a Netzsch STA-429 thermobalance (data were shown in Lee et al., 1987).

Ten grams of pulverized shale (< 180-µm grain size) was treated with 100 ml of 1 *N* sodium acetate in a warm water bath for 24 hr to disperse the carbonates for geochemical fractionation. The shale residue was digested by the addition of 30% hydrogen peroxide while being heated in a warm water bath until the reaction ceased (to remove organics). The supernates from this and the previous treatment were collected for later analysis. Iron minerals and coatings were removed by washing the shale residue with sodium acetate, then treating it with dithionite in three 1-g increments while stirring in 50 ml of 1 *M* sodium bicarbonate and 0.3 *M* citrate solution (Jackson, 1975). The supernate from this treatment was held for analysis, and the residue was filtered through a 53-µm sieve. A 9-cm column of the < 53-µm fraction was suspended in a dilute sodium carbonate solution, centrifuged at 750 rpm for 2.9 min, and the < 2-µm fraction of supernate decanted. Similarly the < 2-µm

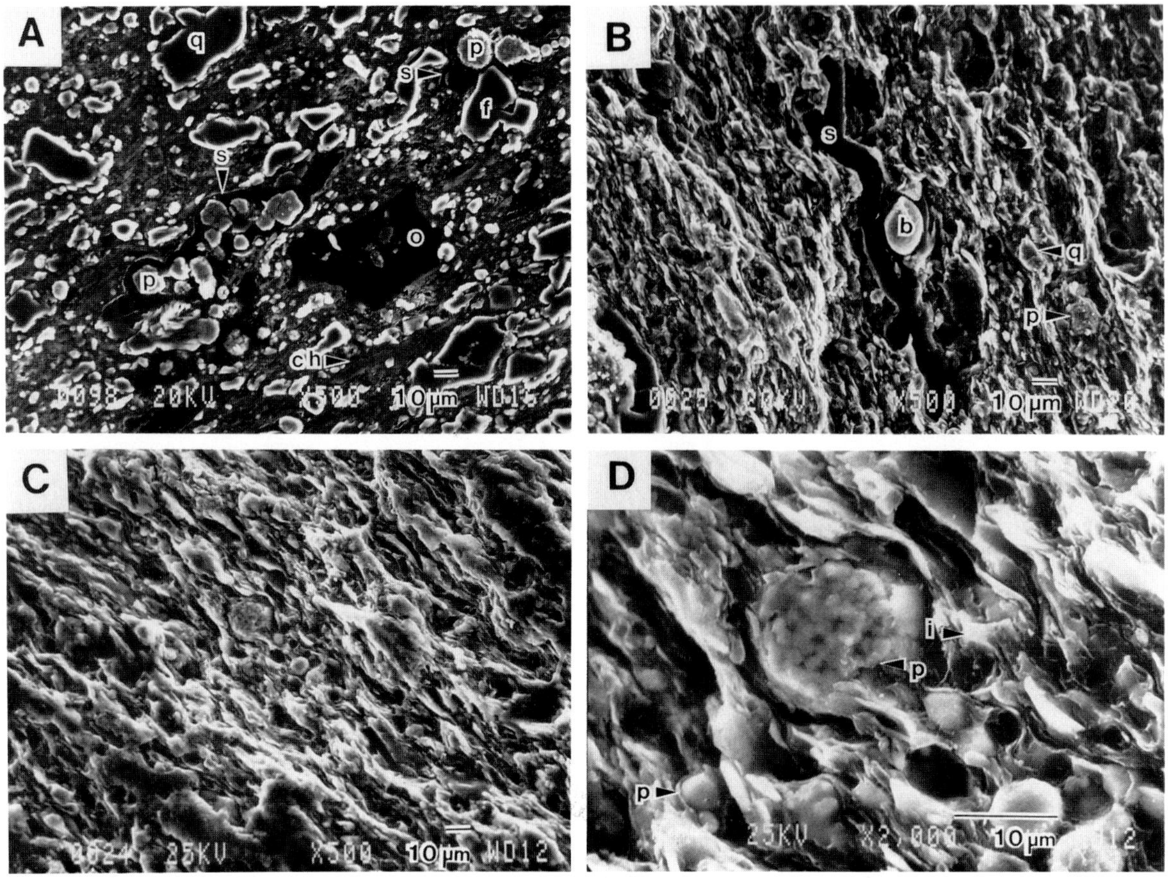

Figure 58.3. Scanning electron micrographs of Chattanooga Shale. (A) Backscattered electron image of polished section, showing quartz (q), feldspar (f), chlorite (ch), pyrite (p), organics (o), and spores (s). (B) Secondary electron image of fractured rock, showing spores (s), barite (b), quartz (q), and pyrite (p). (C) Secondary electron image, showing bedding. (D) Pyrite (p), surrounded by illite (i).

fraction was centrifuged at 2400 rpm for 31.4 min to separate the <0.2-μm fraction. The <0.2-μm fraction was divided into two aliquots for X-ray diffraction analysis; one was treated with 0.1 *N* potassium chloride and the other with 0.1 *N* magnesium chloride.

X-ray diffraction (XRD) analyses were performed on samples of the various heavy liquid and grain-size fractions with a Phillip's powder X-ray diffractometer. Samples were prepared for analysis by transferring potassium- or magnesium-saturated ethanol slurries of the sediments to glass slides, air drying them, then bombarding them with copper Kα radiation. The potassium-saturated samples were heated to 550°C after XRD and the magnesium-saturated samples were glycolated before being run in the diffractometer to aid in identification.

Results and Discussion

Petrographic Microscopy

Thin sections of the selected shale samples were examined to observe mineral constituents in their natural, undisturbed arrangements. Since shales consist almost exclusively of clay minerals and very fine particles, it was extremely difficult to identify all of their components petrographically.

The Chattanooga Shale was relatively uniform in texture and composition (Fig. 58.1A). The color in transmitted light was predominantly dark brown to black, representing mostly amorphous organic matter (kerogen). Compressed spores, frequently filled with framboids and individual crystals of pyrite, were common. Very thin beds, rich in silt-size (20- to 40-μm) detrital quartz grains alternated with a clay- and organic-rich matrix so as to create discontinuous laminae, 10–30 μm thick. Occasionally, a microfracture filled with calcite, secondary quartz, and/or pyrite was observed. Clay minerals were mixed uniformly with organic matter throughout the area examined, except for a burrow that was filled with randomly oriented clay and detrital minerals.

The Pierre Shale consisted of mottled brown, black, and yellow areas, with translucent grains in transmitted light (Fig. 58.1B). The black areas were concentrations of amorphous organic material and varying sizes (10–50 μm) of pyrite crystals.

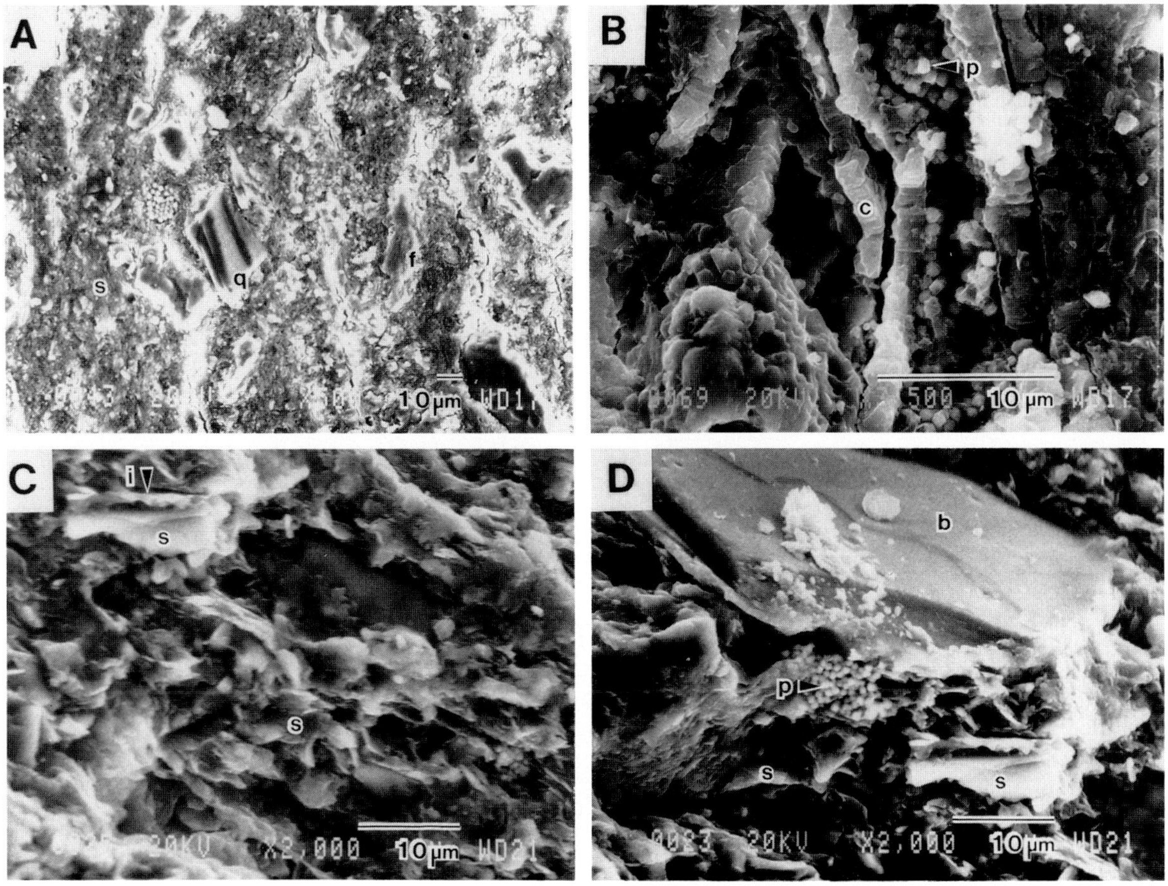

Figure 58.4. Scanning electron micrographs of Pierre Shale. (A) Smectite (s), feldspar (f), and quartz (q) grains. (b) Calcite (c) and pyrite (p) in microfractures subparallel to bedding. (C) Illite (i) in smectite (s) matrix. (D) Detrital biotite (b) associated with a fracture, plus pyrite (p) and smectite (s) grains.

The brown, black, and yellow areas were a mixture of clays (40–50%) and detrital grains. The translucent grains were quartz and calcite. Blocky calcite filled thin fractures and pores. Silt-size quartz particles were uniformly distributed. Gastropod and ostracod shells were present and associated with pyrite and amorphous kerogen. The shells were often filled with blocky calcite. The shale had a uniform texture with no apparent lamination, but the mineral grains displayed a preferred orientation.

The Green River Formation was extremely fine grained (< 10 µm) and composed mainly of dolomite, quartz, kerogen, and feldspars. The sample was uniform in texture but could be divided into two areas based on color in transmitted light. The first region was distinctly laminated (Fig. 58.1C). The laminations were 20–150 µm thick; ranged from red to orange, brown, and yellow; and were continuous, cross-bedded, and disrupted only by soft-sediment deformation structures. Quartz grains were uniformly distributed. The second region was uniform in grain size (about 10 µm), yellow in color, and devoid of observable lamination. Detrital quartz grains were evenly distributed throughout the area. Heavy minerals were uniformly distributed in both zones, but occasionally a higher concentration was found at the interface between laminated and homogeneous regions (Fig. 58.1D).

The Nolichucky Shale consisted of yellow to brown clay with stringers of black amorphous organic matter and scattered translucent detrital grains. The detrital grains of calcite, quartz, and feldspars were 10 µm or less in size. Clays were most abundant and preferentially oriented along bedding planes. Thin (< 10-µm) and short (50- to 300-µm) stringers, apparently filled with organic material, were observed. Stair-step fractures were another prominent feature of the Nolichucky Shale (Fig. 58.2A). Some linear fractures, generated during thin section preparation, occurred where the stringers were abundant. Several fractures, up to 300 µm in size, were filled with pseudospar calcite and secondary silica (Fig. 58.2B). A burrow, 120 µm wide and filled with calcite, was observed. Pyrite and other heavy minerals were present in minor amounts.

The Pumpkin Valley Shale contained two distinct types of laminations. The first type consisted of brown clay and black amorphous organic matter with few detrital grains (Fig. 58.2C).

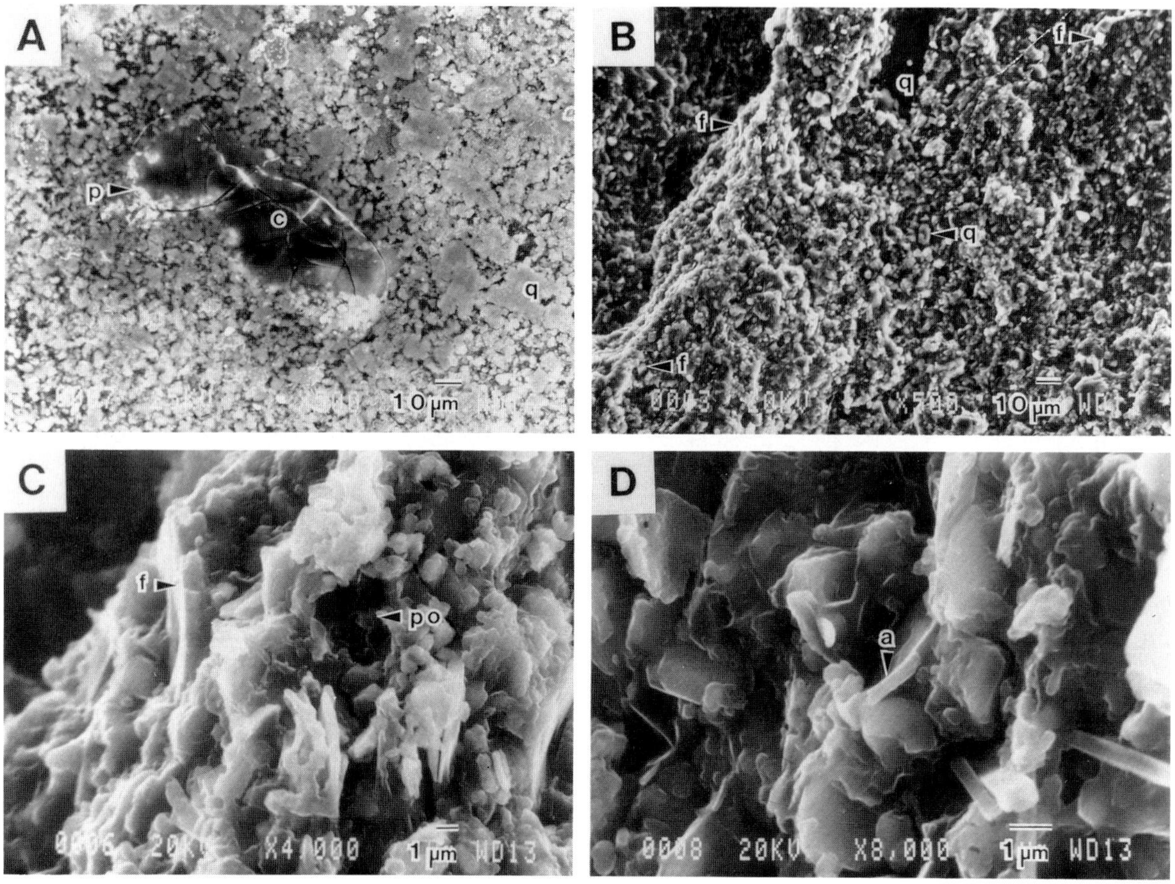

Figure 58.5. Scanning electron micrographs of Green River Formation. (A) Polished section, showing pyrite (p), calcite (c), and quartz (q) in dolomicrite matrix. (B) Fractured surface, showing quartz (q) and feldspar (f) in dolomicrite. (C) Enlargement of B, showing feldspar (f) and micropore (po). (D) Possible aragonite (a) in dolomicrite matrix.

The second type was translucent and was composed of detrital quartz grains, micas, and glauconite (Fig. 58.2D). The detrital grain laminations were further subdivided into finer (30- to 50-μm) and coarser (90- to 200-μm) grain sizes. These two subtypes were separated by the organic and clay-rich laminae. The fine-grained lamina was slightly darker in color than the coarse-grained lamina, due either to its fine-grained nature or to the presence of more organic material and iron oxide coatings. Fractures filled with secondary quartz or calcite were common in the shale. Some glauconite peloids (100–200 μm) were common in the detrital laminae; many were partially altered and had changed color from green to brown.

Scanning Electron Microscopy

In the Chattanooga Shale, relatively coarse, subangular grains (20–50 μm) of quartz and feldspars were observed under backscattered electron beams (Fig. 58.3A). The feldspars were mostly potassium rich and partially weathered along the edges. Fine (< 10-μm) flaky illite was the major component of the shale. The illite was densely compacted, with most of the micropores filled by amorphous organics and occasionally by calcium and barium sulfates. Compressed spore cases, 10–30 μm wide and 100–200 μm long, were aligned parallel to the general direction of bedding. Some spores were filled with secondary quartz, barite, or pyrite crystals. Amorphous organics occupied pores of various sizes and shapes (Fig. 58.3A and B). Cubic and framboidal pyrite was common and often associated with organics and conodonts (Fig. 58.3C and D).

The Pierre Shale was composed of loosely packed fine grains (<4 μm) of smectite and large grains (10–50 μm) of quartz (Fig. 58.4A). The sample had a large number of micropores, which might have been generated by partial shrinkage of the smectite due to the loss of sorbed water during sample preparation. Blocky calcite layers (20 μm thick) and fine (< 1 μm) pseudocubic pyrite grains in between the calcite layers occupied microfractures subparallel to bedding (Fig. 58.4B). Flaky smectite particles were oriented somewhat randomly (Fig. 58.4C) and showed fair amounts of calcium and iron in EDX analysis.

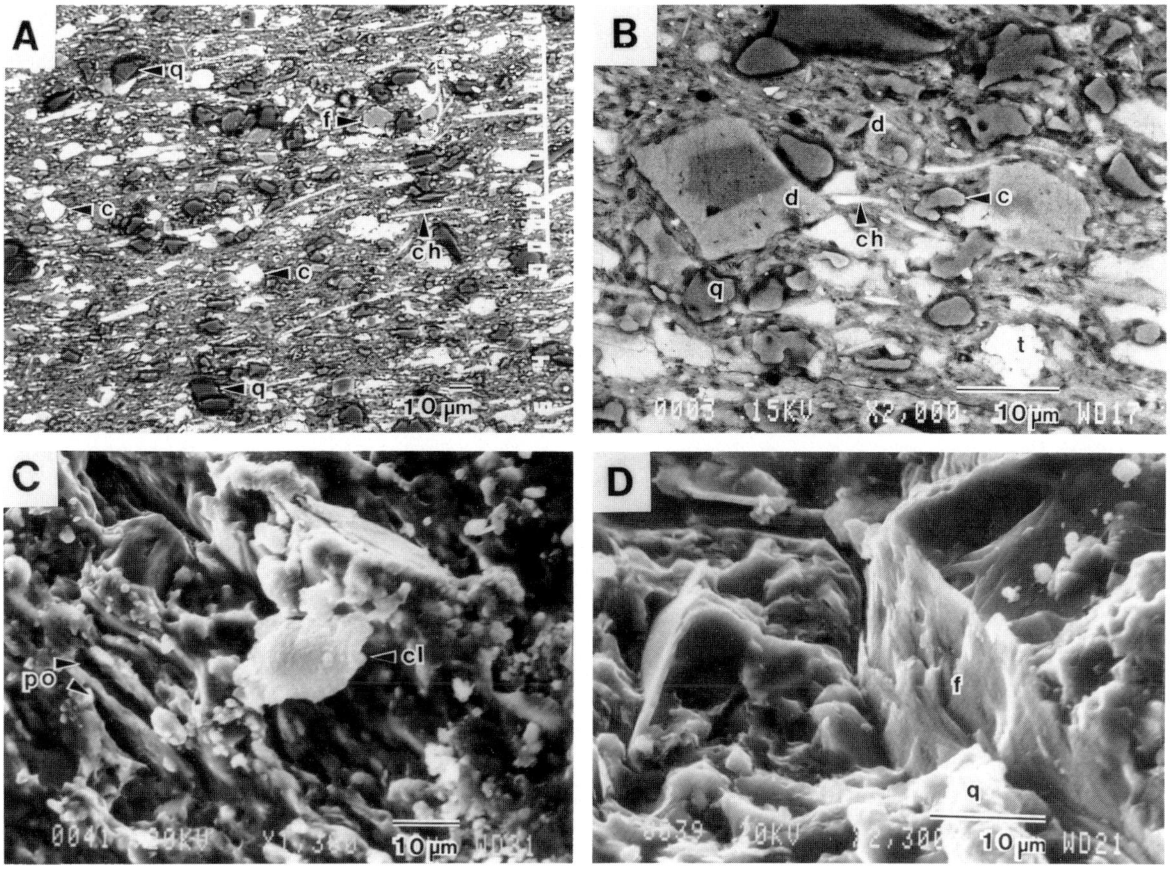

Figure 58.6. Scanning electron micrographs of Nolichucky Shale. (A) Polished section, showing detrital quartz (q), calcite (c), feldspar (f), and chlorite (ch). (B) Polished section, showing detrital calcite (c), chlorite (ch), quartz (q), dolomite (d), and a titanium mineral (t). (C) Fractured surface, showing clinochlorite grain (cl) and pores (po) in illite matrix. (D) Fractured surface, showing quartz (q) and feldspar (f).

The presence of calcium and absence of sodium suggested that the exchange sites of the smectite were occupied mainly by calcium. Detrital biotite grains, identified by the presence of K as well as Fe, Mg, Al, and Si, were observed in a fractured sample (Fig. 58.4D).

The Green River Formation was drastically different from the other shales in terms of morphology and mineralogy. The polished section of the formation showed densely packed aggregates of finely textured (<2-µm) carbonate grains and irregularly shaped and sized pores filled with kerogen. In the carbonate grains, calcium was the dominant cation, but there were considerable amounts of magnesium and occasionally iron and manganese. The matrix appeared to be a mixture of micrite and dolomicrite. A recrystallized iron-carbonate (ankerite) grain, with pyrite concentrated on the surface and in the fractures, was observed (Fig. 58.5A). Relatively fine grains (<20 µm) of quartz and feldspars were distributed throughout the matrix. SEM examination of a fractured surface demonstrated again the relative uniformity of texture in the formation (Fig. 58.5B). The grains were aggregated and cemented by calcium carbonate and siliceous material, based on detection of Si with Ca and Mg spectra in EDX analysis. In the close-up micrographs, randomly oriented mineral grains of variable sizes and shapes were observed, along with the micropores occupied by organics (Fig. 58.5C and D).

In the polished sections of the Nolichucky Shale, fine-silt-size detrital quartz, feldspars, dolomite, and calcite were observed in a clay mineral matrix (Fig. 58.6A and B). Illite was the dominant matrix mineral, but considerable amounts of chlorite and micas were present. In the fractured sample, microfractures and pores were observed between illite packets (Fig. 58.6C and D). Chlorite was much more prominent. It had a relatively high iron content and was tentatively identified as an iron-rich chlorite.

The Pumpkin Valley Shale was similar in mineralogical composition to the Nolichucky Shale. However, illite grains in the Pumpkin Valley Shale were much more randomly oriented (Fig. 58.7A). Relatively long packets (10–100 µm long and <10 µm thick) of iron-rich chlorite were common in the sample. The SEM examination showed thick chlorite packets with relatively clean surfaces and illites as aggregates with extensive coatings.

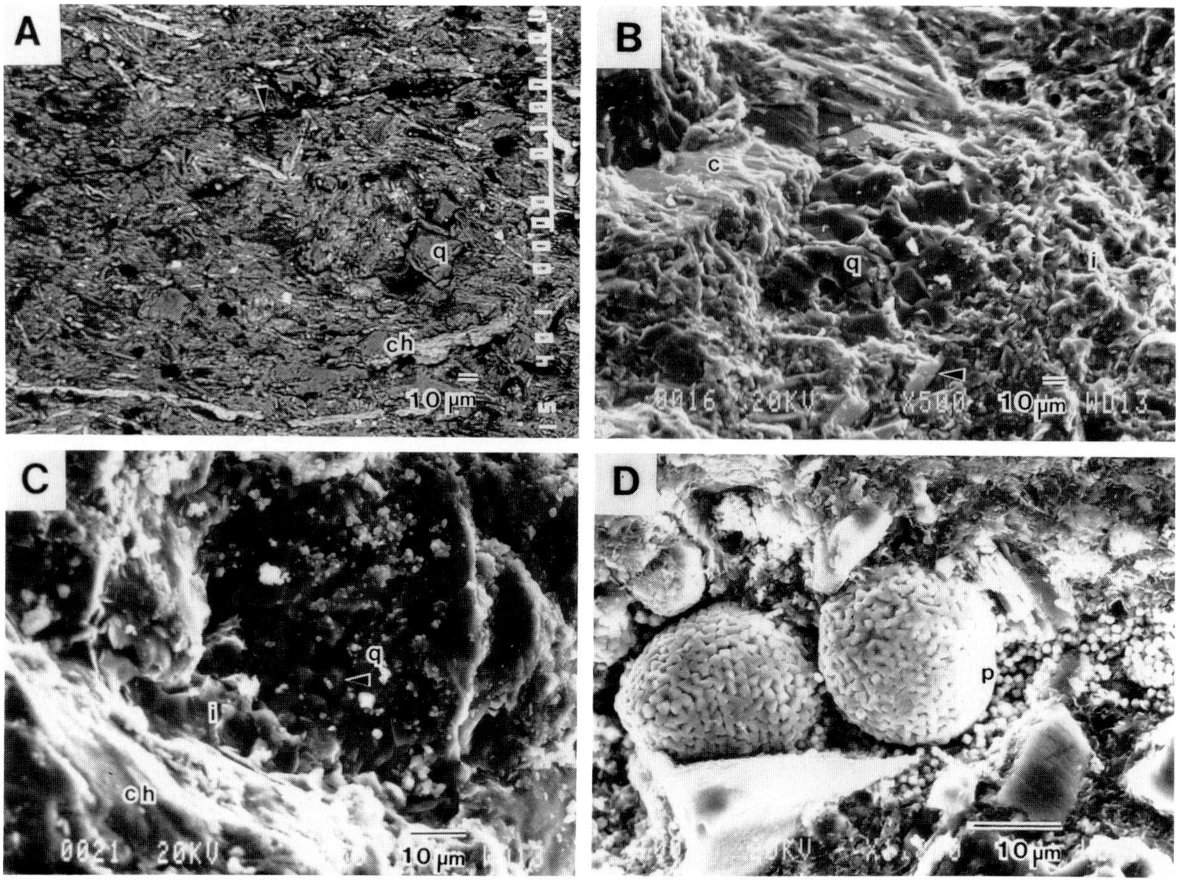

Figure 58.7. Scanning electron micrographs of Pumpkin Valley Shale. (A) Polished section, showing quartz (q), chlorite (ch) and a microfracture (arrow) in a matrix of randomly oriented illite. (B) Fracture-filling calcite (c) and quartz (q) with detrital calcite grain (arrow) in an illite matrix. (C) Authigenic quartz (q), chlorite (ch), and illite (i) in a micropore. (D) Framboidal pyrite (p).

There were a considerable number of pores that appeared to be filled with organic matter. Calcite was present in detrital form and as a fracture-filling material (Fig. 58.7B). Authigenic quartz and framboidal pyrite were common in micropores and microfractures (Fig. 58.7C and D).

High-Resolution Transmission Electron Microscope Analysis

High-resolution transmission electron microscopy (HRTEM), including energy dispersive X-ray spectrometry of undisturbed shale samples, provides crucial information on mineral composition and structure, matrix composition, and micromorphology, which cannot be determined by other techniques (Lee et al., 1975a,b; Bell, 1986; Klimentidis and Mackinnon, 1986). The observed lattice fringes in the micrographs correspond to the (001) *d*-spacings of the clay minerals. They were formed by combining the beams diffracted by Rutherford interactions with crystal planes and zero-order transmitted beams. Besides the fringe image measurements, layer silicate minerals were also identified by differences in their chemical composition, as analyzed by qualitative EDX. Besides Si, which was common in all the silicates, illite had moderate amounts of Al and Fe with K, muscovite had Al with K, biotite had high amounts of Fe with K, chlorite had high amounts of Mg and Fe with very low amounts of Al and K, and kaolinite had high amounts of Al and no Mg, Fe, or K.

Clays and detrital grains in the Chattanooga Shale were closely packed, and pores were filled with organic matter (Fig. 58.8A). Layer silicate minerals were preferentially aligned along bedding planes when they were not interrupted by larger detrital grains. Organic materials were unstable and progressively formed damaged spots under the electron beam. Many thin packets (5–20 nm) of illite having 0.5-nm (002) lattice fringes were observed. Occasionally, a thicker (>50-nm) detrital muscovite having 0.5-nm (002) lattice fringes was observed. Chlorite with 1.4-nm fringes was also observed in the sample (Fig. 58.8B).

Preparation of Pierre Shale specimens by ion milling was very difficult because the sample had very weak cementation between clay particles. In future studies, a low-viscosity bonding agent will be introduced prior to ion milling. HRTEM analysis of the

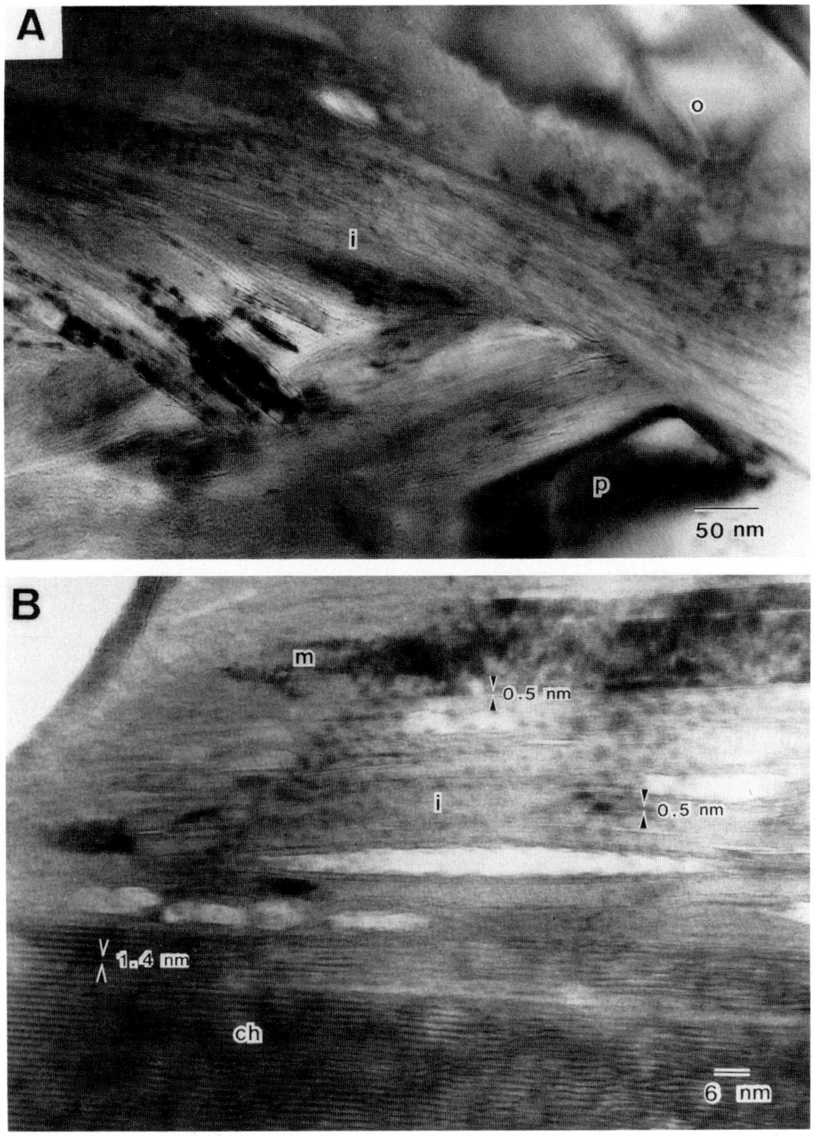

Figure 58.8. High-resolution transmission electron micrographs of Chattanooga Shale. (A) Illite packets (i) with organic matter (o) and pyrite (p). (B) Illite packets (i) between mica (m) and chlorite (ch).

sample indicated that smectite aggregates were randomly oriented, showing both (hkO) and (001) planes in the same area (Fig. 58.9A). Some of the clays showed 1.2-nm lattice fringes, but 1.0-nm fringes were more common in the sample (Fig. 58.9B). The clays were composed mainly of Si, Al, and Ca, with trace amounts of Fe and K, which indicated that they were predominantly dioctahedral calcium–montmorillonite. There was no evidence of interstratified illite–smectite mineral in the sample. Since the montmorillonite can be thermally altered, further detailed study is warranted for understanding waste-induced thermal effects on swelling clays.

The Green River Formation was hard and mechanically stable during specimen preparation. It had no preferred fracture orientation, as did the other samples. The minerals were apparently cemented by amorphous silica (Fig. 58.10A). The characteristics of the cemented phase should be reexamined by other methods because it is important for understanding the chemical stability of the formation. The dominant minerals were rhombohedral dolomite, quartz, calcite, and micas. A higher magnification of the mica particles showed well-defined 1.0-nm fringes (Fig. 58.10B). The chemical composition of the mica was close to muscovite, but it contained a small amount of iron. A thin particle with morphological features close to illite was observed next to the large mica particle shown in Figure 58.10B.

The Nolichucky Shale had many thin (10- to 100-nm) packets of illite that parallel or cross one another (Fig. 58.11A). The large spaces (1.2–1.4 nm) occurring between the packets could be a chlorite fringe, but such spaces could result from other causes. Thin illite packets (1.0-nm lattice fringes) occurred between thick (>100-nm) mica particles, which also had well-

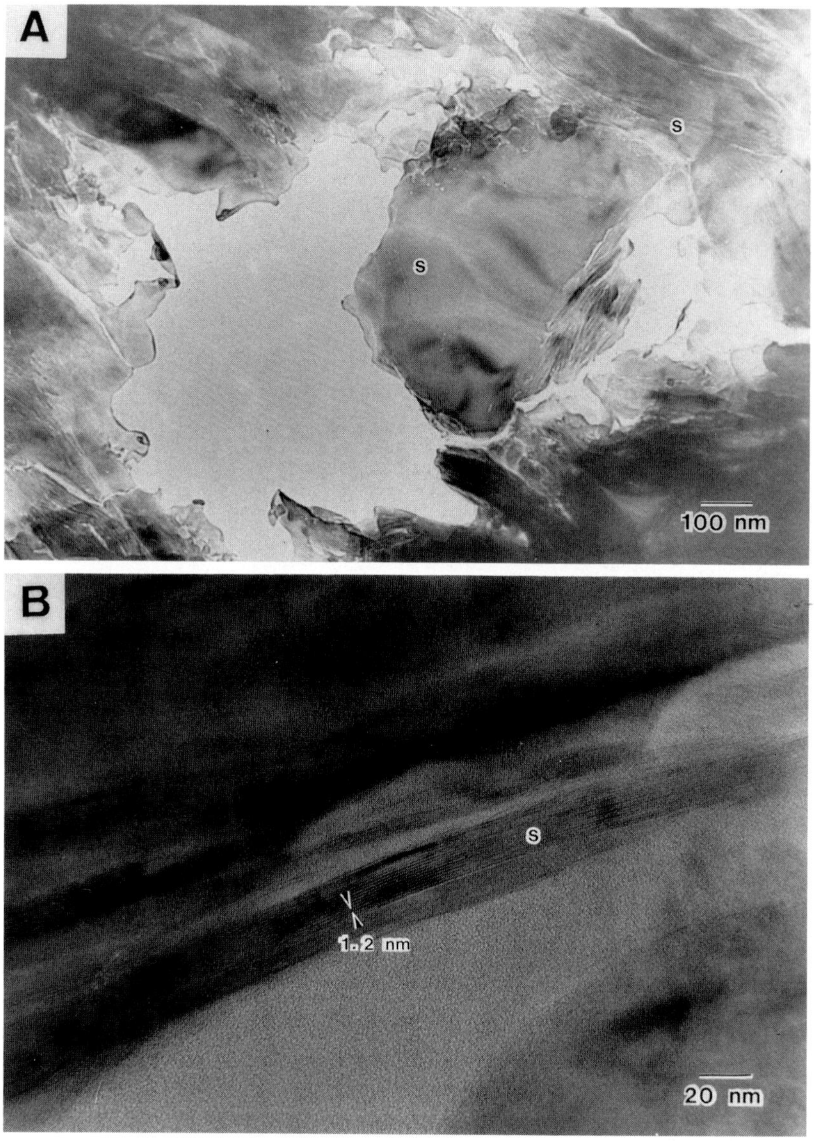

Figure 58.9. High-resolution transmission electron micrographs of Pierre Shale. (A) Random orientation of smectite aggregates (s). (B) 1.2-nm lattice fringes of smectite.

defined 1.0-nm lattice fringes (Fig. 58.11B). Illite was differentiated from mica only by morphological features, since both phases show similar chemical compositions in EDX analysis. Disappearance of fringes between illite packets was caused by the misalignment of the illite packet toward the electron beam in those areas. Carbonate cementation was observed but was not common in this specimen.

The Pumpkin Valley Shale was also a difficult sample for HRTEM analysis because of the lack of cementation between minerals and the presence of abundant micropores between randomly oriented illite packets (Fig. 58.12A). There were few areas in which illite particles were properly oriented toward the electron beam. These illites, like those in the Nolichucky Shale, had thin packets of 1.0-nm lattice fringes (Fig. 58.12B).

The noticeable differences in microstructure of the shales caused large differences in their BET surface areas. The surface areas of the samples ranged from 23.79 to 1.92 m^2/g, with the following order: Pierre > Nolichucky > Pumpkin Valley > Chattanooga > Green River (Table 58.1). The lower surface areas of the Chattanooga and Green River samples resulted from the extensive cementation observed between primary particles, which prevented penetration of nitrogen gas to the primary particle surfaces, thus reducing the measured surface areas.

The surface area measurements suggest that radionuclides and leaching solutions would have greater access to the surfaces of primary particles of the Pierre, Nolichucky, and Pumpkin Valley shales and lesser access to the surfaces of primary particles of the Chattanooga and Green River shales. Therefore, the

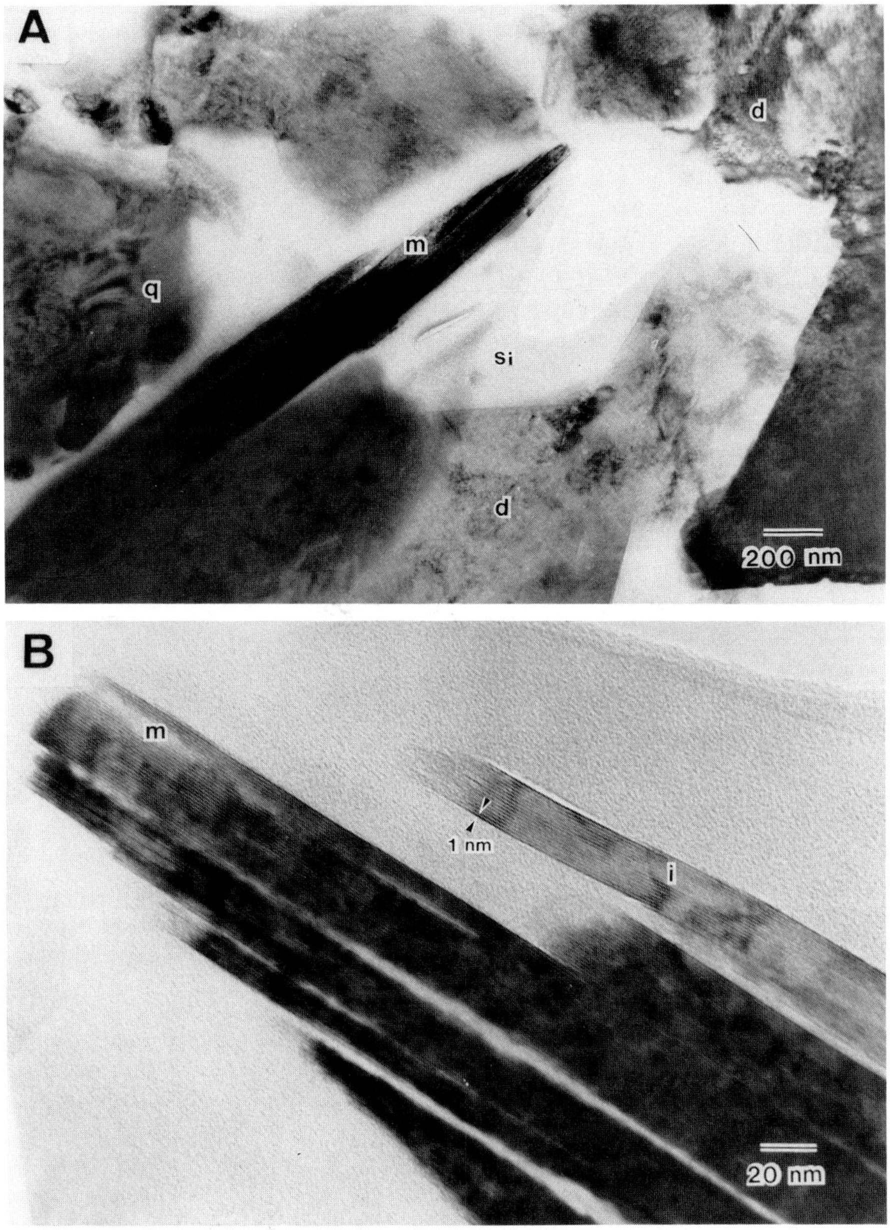

Figure 58.10. High-resolution transmission electron micrographs of Green River Formation. (A) Dolomite (d), quartz (q), and mica (m) grains and siliceous cement (Si). (B) 1-nm fringes on mica grain (m) and illite grain (i).

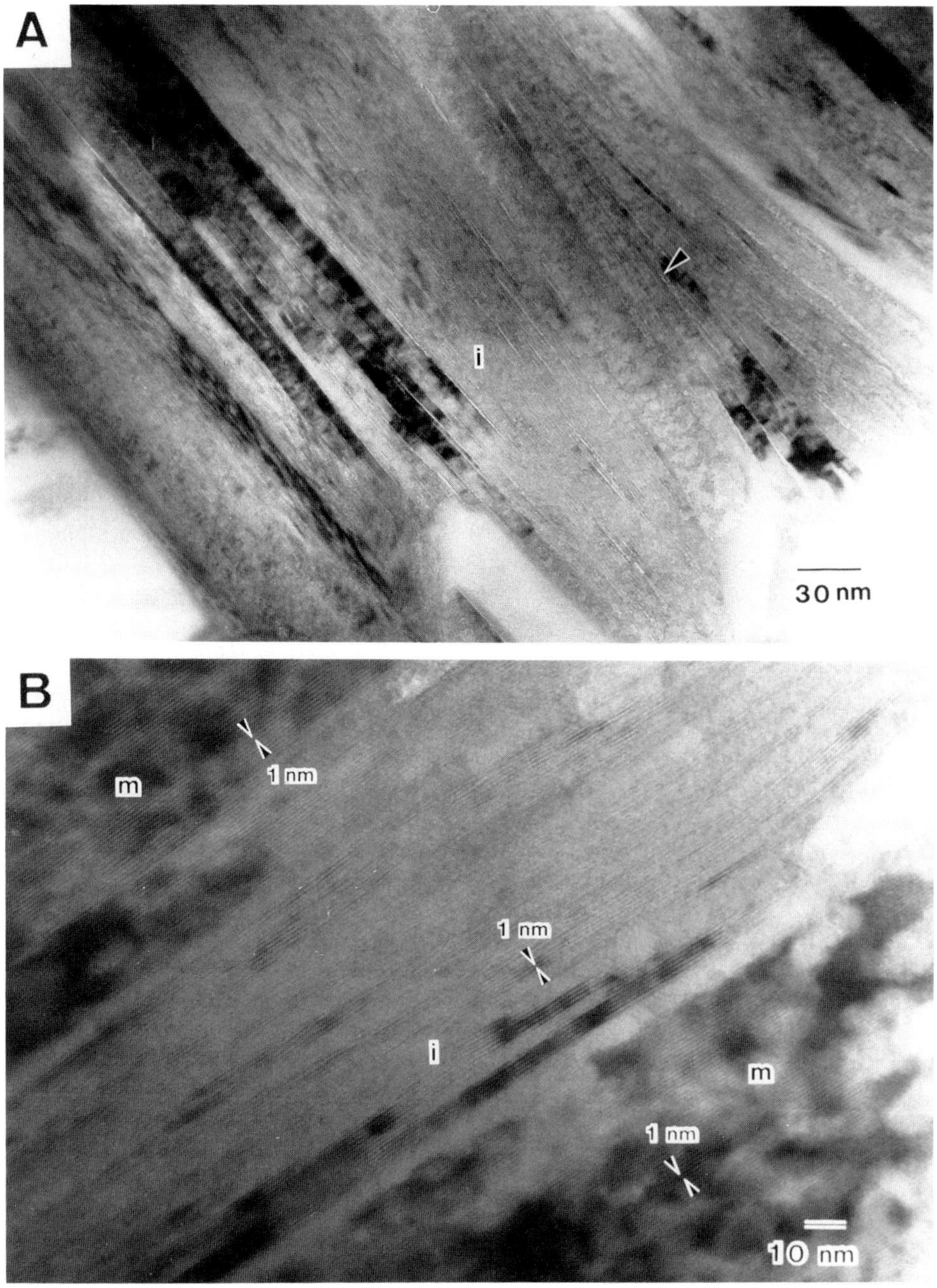

Figure 58.11. High-resolution transmission electron micrographs of Nolichucky Shale. (A) Illite packes (i) with discontinuities (arrow). (B) Illite packets (i) between mica particles (m).

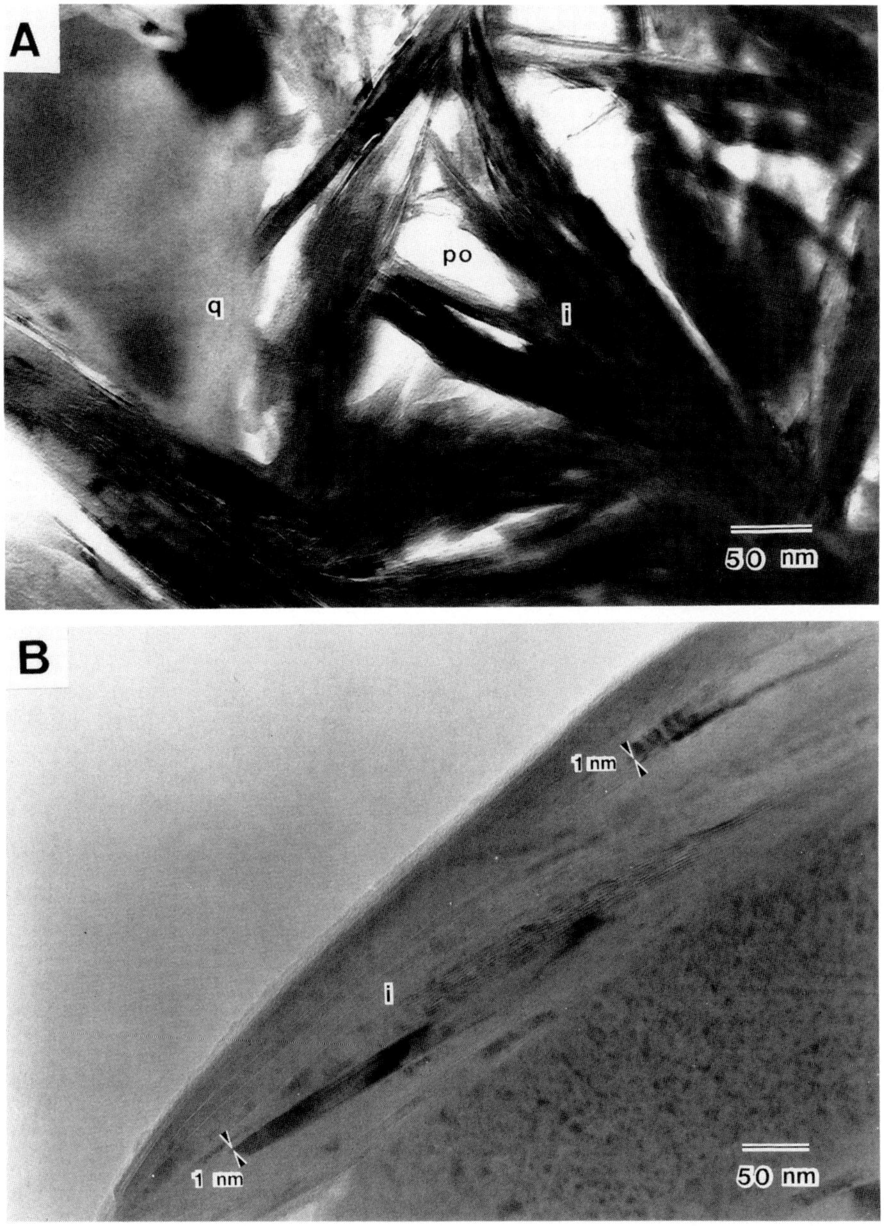

Figure 58.12. High-resolution transmission electron micrographs of Pumpkin Valley Shale. (A) Micropores (po), quartz (q), and randomly oriented illite packets (i). (B) 1-nm lattice fringes.

Table 58.1. Chemical composition, particle-size distribution, and surface area of whole-rock samples.

	Chattanooga Shale	Pierre Shale	Green River	Nolichucky Shale	Pumpkin Valley
Chemical composition (weight percent)					
SiO_2	57.78	49.22	34.24	40.66	62.06
Al_2O_3	13.61	13.99	7.56	13.04	20.79
K_2O	3.95	2.27	3.25	3.41	5.37
Na_2O	0.38	0.66	0.70	0.34	0.66
FeO	5.79	5.40	2.70	3.99	6.69
TiO_2	1.35	2.17	6.35	2.34	2.00
MgO	1.35	2.17	6.35	2.34	2.00
CaO	0.27	8.58	13.16	6.02	0.29
CO_3	<0.50	10.50	26.40	9.60	<0.50
SO_4	0.32	0.26	0.12	0.18	0.10
Sulfide	4.22	0.89	0.30	0.12	0.26
Particle-size distribution (weight percent)					
180–53 μm	1	1	1	1	7
53–2 μm	64	31	54	49	63
2–0.2 μm	26	19	14	21	18
<0.2 μm	9	50	32	30	12
Surface area (m^2/g)					
Surface area	4.6	22.8	2.0	15.2	12.6

Table 58.2. Estimated mineralogical composition (weight percent) of whole-rock samples of shale.*

	Chattanooga Shale	Pierre Shale	Green River	Nolichucky Shale	Pumpkin Valley
Organic matter	11	5	13	t	t
Chlorite + kaolinite	4	t	t	14	15
Illite	49	t	10	43	57
Micas	t	4	t	t	t
Smectite	nd	59	nd	nd	nd
Carbonates	t	15	42	11	t
Quartz-feldspars	25	11	28	29	22
Pyrite	6	2	t	t	t
Weight loss (105°C)	1	4	2	2	2

*t, trace (<2%), observed from thin section and electron micrographs; nd, not detectable.

diffusion rates through the shales and the kinetics of sorption of radionuclides onto the shales should be carefully evaluated. If some equilibrated radionuclides diffuse into the aggregates during sorption experiments, reversibility should be also determined under the same conditions. Sorption kinetics can be determined through standard sorption experiments, but microscopic analysis of radionuclide-equilibrated block samples would be more useful and realistic.

Semiquantitative Mineral Composition

Semiquantitative estimation of the mineralogical composition of shales is a very difficult task, but it is necessary for many reasons. In nuclear waste disposal studies, data on mineralogical composition would be helpful for predicting groundwater composition, interpreting sorption and desorption results and physical properties, and providing the data base for modeling radionuclide transport. There are many different approaches to quantitative mineralogical analysis (e.g., Schultz, 1964; Jackson, 1975), but there is no standard method established by the clay mineral research community. The major problem encountered, when an attempt is made to establish a standard method, is the extreme variability among sedimentary rocks in mineral composition, in mineral crystallinity, and in the effects of diagenesis and weathering on clay minerals, as illustrated by the shales selected for this investigation. Therefore, a combination of specifically applicable analytical techniques for the semiquantitative estimation of the mineral composition of the five shale samples was used in this study.

The results presented in Table 58.1 are semiquantitative in nature, and careful refinement of each analytical method, with cross-checking by independent methods, is needed. Organic matter and water contents of the shales were estimated from thermal analyses. Carbonate contents were calculated from CO_2 evolution data obtained through acid dissolution and thermal analysis. Pyrite contents were calculated from the sulfide contents. The chemical compositions of whole-rock samples of the shales analyzed by wet chemical methods and neutron activation are summarized in Table 58.2. Quartz plus feldspar contents were determined by SEM backscattered electron image analysis of thin section slides. Feldspars, a minor component in these samples, were included in the quartz contents. Phyllosilicates were estimated from chemical analysis data (potassium content), XRD peak area ratios, and both backscattered electron image and X-ray dot-mapping analyses. The estimates were recalculated with the particle-size distribution results (Lee et al., 1987).

Semiquantitative mineralogical analyses showed that the Chattanooga Shale was composed of 49% illite, 25% quartz plus feldspar, 11% organic matter, 6% pyrite, and 4% chlorite plus kaolinite. Microcline-orthoclase was dominant among the feldspars, and both chlorite and kaolinite were present in the sample. The Chattanooga Shale was an illite-rich shale but one that also was relatively rich in organic matter and pyrite in comparison with other illitic shales. Both constituents would play an important role in its geochemical behavior.

The Pierre Shale contained 59% smectite, which appeared to be calcium-saturated montmorillonite. Other minerals occurring in the shale were 15% carbonates, 11% quartz plus feldspars, and 5% organic matter among the carbonates. Feldspars were insignificant in the sample. The Pierre Shale is expected to have a high cation-exchange capacity and swelling potential compared with the other shales.

The Green River Formation was selected as a carbonate end-member shale, but it also contained a considerable amount of organic matter. The shale was composed of 42% carbonates, 28% quartz plus feldspars, 13% organic matter, and 10% illite. Dolomite appeared to be the dominant carbonate, but a considerable amount of calcite was present. The organic matter, in the form of kerogen, coated the mineral components and filled the micropores. The hydrophobic nature of organic-coated minerals would hinder their interaction with aqueous solutions. Therefore, the Green River Formation is not a typical carbonate-rich shale, and the program may need to identify another carbonate shale that is relatively free of kerogen for evaluation. The difference in the degree of maturity of the organic matter in the Green River Formation and that in the Chattanooga Shale makes for a useful comparison of alternative organic-bearing shales.

The Nolichucky Shale is an illitic shale containing carbonate but without organic matter. The sample was composed of 43% illite, 29% quartz plus feldspars, 14% chlorite plus kaolinite, and 11% carbonates. The carbonate in the sample was mostly fracture-filling secondary calcite. Iron-rich chlorite was the dominant phase in the chlorite plus kaolinite component. The feldspar phase was not fully analyzed. Since carbonate is the most soluble component in the sample, its possible effects on the geochemical behavior of this illitic shale should be determined.

The Pumpkin Valley Shale was composed of 57% illite, 22% quartz plus feldspars, and 15% chlorite plus kaolinite. Analyses indicated the occurrence of both chlorite and kaolinite, but a separate estimation of each individual mineral was not attempted. The chlorite in the shale behaved as iron-rich chamosite. Further analysis of feldspars is also needed. Nevertheless, the Pumpkin Valley Shale appears to be an excellent example of an illitic end-member shales which is relatively low in organic matter, carbonates, and pyrite.

The semiquantitative mineralogical analyses indicated that the shales selected as end-members of carbonaceous, illitic, smectitic, and carbonate shales appear to fulfill programmatic objectives. They are not true end-members by any means, but they are probably better than standard mineral samples for our purposes. If there are compositional parameters that should be of high priority in the ultimate selection of a particular type of shale for a repository site, information gained from this study could be invaluable in helping to selectively narrow down the number of potential shale units for further study during a national survey.

Summary

Shale is a complex rock and can have many different mineral, organic, and elemental components. Each component has a different role in radionuclide retardation and groundwater–shale interaction, and each has different thermal, mechanical, and hydrologic properties.

Four end-member shales were selected on the basis of their composition for comprehensive microstructural and mineralogical characterization: the Chattanooga Shale was selected as representing a carbonaceous shale, the Pierre Shale as a smectitic shale, the Green River Formation as a carbonate-rich shale, and the Nolichucky Shale and Pumpkin Valley Shale as illitic shales.

The Chattanooga Shale had a relatively low surface area (4.6 $m^2\ g^{-1}$) considering its moderately high clay content (35%), suggesting extensive compaction and cementation of the particles. It was composed of 49% illite, 25% quartz plus feldspar, 11% organic matter, 6% pyrite, 4% chlorite plus kaolinite, and other accessory minerals ($\sim$5%). Microscopic analyses indicated that the mineral components of the shale were highly compacted, and micropores were filled with organic matter. Carbonate and sulfate cements were present but were not common. The organic matter was apparently unstable under a prolonged low-temperature heat treatment (250°C for 6 months), as indicated by the disappearance of an exothermic peak representing organics in the differential thermal analysis after the heat treatment. The high organic and pyrite contents but near absence of carbonate in this shale are significant characteristics because they may at least partially control solution chemistry when they are equilibrated with radionuclide-containing groundwater.

The loosely packed Pierre Shale had a high surface area (23 $m^2\ g^{-1}$) as well as a high clay content (69%). It contained 59% smectite, 15% carbonate, 11% quartz plus feldspar, 5% organic matter, 3% mica, and 2% pyrite. Micropores and fractures were frequently observed in the sample. Carbonate minerals had grown in the fractures but did not cement the clay minerals. Calcite was the principal carbonate in the sample. The smectite, identified as montmorillonite, has a high cation-exchange capacity. The cation-exchange sites in the sample were saturated with calcium that could be replaced by radioactive cations released from waste packages. Montmorillonite also has a very high swelling capacity, an important engineering characteristic for seal and backfill materials. The micropores and fractures observed in a dehydrated state of this shale would be sealed by the swelling of montmorillonite clays when they are rehydrated by groundwater.

The shale sample from the Green River Formation was hard and structurally stable and had a very low surface area (2 $m^2\ g^{-1}$) for the rather large amount of clay-sized particles (46%). It was composed of 42% carbonates, 28% quartz plus feldspar, 13% organic matter, and 10% illite. The Green River Formation was selected primarily as a carbonate end-member shale, but the analyses indicated that it also contained a considerable amount of organic matter. The organic matter appeared to be a liquid hydrocarbon, rather than the solid matter found in the Chattanooga Shale. Hydrocarbon had coated the mineral components and filled the micropores in the shale, which hindered interaction of the minerals with aqueous solutions because of the hydrophobic nature of the coating. Dolomite was the dominant carbonate, but a considerable amount of calcite was present.

Dolomite and other crystalline mineral components were cemented by amorphous silica and carbonates. Such extensive cementation gave physical stability to the shale and locked the organic matter in isolated micropores.

The Nolichucky Shale had a moderate surface area (15 m^2 g^{-1}) for its clay content (51%). It was composed of 43% illite, 29% quartz plus feldspar, 14% chlorite plus kaolinite, and 11% carbonates, which is typical for illitic shales with carbonates and without organic matter. Thin illite packets with 1-nm lattice fringes, which were observed under high-resolution transmission electron microscopy, could provide sorption sites for radionuclides. Iron-rich chlorite was the dominant phase in the chlorite plus kaolinite component. A major portion of the carbonate was fracture-filling secondary calcite, but detrital dolomite and calcite were also observed.

The Pumpkin Valley Shale, a laminated mudstone and siltstone, had a moderate surface area (13 m^2 g^{-1}) and moderate clay content (30%) and contained 57% illite, 22% quartz plus feldspar, and 15% chlorite plus kaolinite. This was a more representative illitic shale, since it had only trace amounts of carbonates, organic matter, and pyrite. Clay particles in the shale were loosely packed and randomly oriented. Cementation was rarely observed in the mudstone but was common in the siltstone laminae. The presence of micropores was frequently observed among phyllosilicate packets. Both chlorite and kaolinite were present in the sample.

This investigation revealed that the shales selected as end-members of carbonaceous, smectitic, carbonate, and illitic shales have a wide range of microstructural and mineralogic characteristics that would be related to the performance of shale as a nuclear waste repository. Distribution of micropores and fractures, degree of cementation, mineral orientation, and mineralogical composition are a few of the features that could control the mobility of radionuclides released from wastes and geochemical conditions related to the stability of waste package materials. The results of the microstructural and mineralogical analyses are critical data for the interpretation of thermal properties, hydrologic behavior, and sorption properties of shale formations.

Acknowledgments

Research sponsored by the Geoscience Technology Support Program, U.S. Department of Energy, under contract DE-AC05-84OR21400 with Martin Marietta Energy Systems, Inc. Publication No. 3376, Environmental Sciences Division, Oak Ridge National Laboratory.

References

Bell, T.E., 1986. Microstructure in mixed-layer illite/smectite and its relationship to the reaction of smectite to illite. Clays and Clay Minerals, v. 34, p. 146–154.

Burkett, P.J., R.H. Bennett, H. Li, F.L. Nastav, W.R. Bryant, L.E. Shephard, and W.-A. Chiou, 1987. Microstructure of red clay from the central Pacific Deep-Sea Basin: Significance to subseabed nuclear waste disposal. SAND86-2492, Sandia National Laboratories, Albuquerque, New Mexico, 41 p.

Conant, L.C., and V.E. Swanson, 1961. Chattanooga Shale and related rocks of central Tennessee and nearby areas. U.S. Geological Survey Professional Paper 357. U.S. Government Printing Office, Washington, D.C.

Croff, A.G., T.F. Lomenick, R.S. Lowrie, and S.H. Stow, 1986. Evaluation of five sedimentary rocks other than salt for geologic repository siting. ORNL-6241, Oak Ridge National Laboratory, Oak Ridge, Tennessee, p. 1–2.

Gera, F., 1980. Summary of replies to the questionnaire on R&D activities relevant to the disposal of radioactive wastes in argillaceous formation. Proceedings, Workshop on the Use of Argillaceous Materials for Isolation of Radioactive Waste, NEA/OECD, Paris, 1979, p. 11–13.

Güven, N., C.R. Landis, and G.K. Jacobs, 1988. Characterization of clay minerals and organic matter in shales: applications to high-level nuclear waste isolation. ORNL/TM-10759, Oak Ridge National Laboratory, Oak Ridge, TN.

Jackson, M.L., 1975. Soil Chemical Analysis—Advanced Course, 2nd ed. Department of Soil Science, University of Wisconsin, Madison, p. 12–50.

Klimentidis, R.E., and I.D.R. Mackinnon, 1986. High-resolution imaging of ordered mixed-layer clays. Clays and Clay Minerals, v. 34, p. 155–164.

Lee, S.Y., and R.W. Tank, 1985. Role of clays in the disposal of nuclear waste: A review. Applied Clay Science, v. 1, p. 145–162.

Lee, S.Y., M.L. Jackson, and J.L. Brown, 1975a. Micaceous vermiculite occlusions in kaolinite observed by ultramicrotomy and high resolution electron microscopy. Clays and Clay Minerals, v. 23, p. 125–129.

Lee, S.Y., M.L. Jackson, and J.L. Brown, 1975b. Micaceous vermiculite, glauconite, and mixed-layered kaolinite-montmorillonite examination by ultramicrotomy and high resolution electron microscopy. Soil Science Society of America Journal, v. 39, p. 793–800.

Lee, S.Y., L.K. Hyder, and P.D. Alley, 1987. Mineralogical characterization of selected shales in support of nuclear waste repository studies. ORNL/TM-10567, Oak Ridge National Laboratory, Oak Ridge, TN, p. 2–56.

Lomenick, T.F., and R.B. Laughon, 1980. A regional characterization of the Devonian Shales in the Eastern U.S. for the storage/disposal of radioactive wastes. Proceedings, Workshop on the Use of Argillaceous Materials for the Isolation of Radioactive Waste, NEA/OECD, Paris, 1979, p. 121–128.

Milodowski, A.E., A.J. Bloodworth, and R.D. Wilmot, 1985. Long term effects on potential repository sites: The alteration of the Lower Oxford Clay during weathering. FLPU 85-13, British Geological Survey, London, p. 40–42.

Schultz, L.G., 1964. Quantitative interpretation of mineralogical composition of X-ray and chemical data for the Pierre Shale. Geological Survey Professional Paper 391-C, U.S. Government Printing Office, Washington, D.C.

Stow, S.H., and A.G. Croff, 1987. The potential for use of clay-rich strata as a host medium for a high-level nuclear waste repository. Abstract. 24th Clay Mineral Society Meeting, Socorro, New Mexico, October 20, 1987, p. 124.

Von Damm, K.L., 1987. Geochemistry of shale groundwaters: survey of available data and postulated mineralogic controls on composition. ORNL/TM-10488, Oak Ridge National Laboratory, Oak Ridge, TN, p. 1–5.

Weaver, C.E., 1979. Geothermal alteration of clay minerals and shales: diagenesis. ONWI-21, Office of Nuclear Waste Isolation, Battelle Memorial Institute, Columbus, Ohio.

III

Future Research Directions and Recommendations: Basic and Applied

CHAPTER 59

Research Recommendations of the Clay Microstructure Workshop

Matthew H. Hulbert

A central objective of the Clay Microstructure Workshop was the development of recommendations for future research, both basic and applied. The Workshop, held during October 1988 at the Stennis Space Center in Mississippi in conjunction with a conference titled "The Microstructure of Fine-Grained Terrigenous Marine Sediments—from Muds to Shale," brought together an international group of experts in various disciplines in which clay microstructure is important. The participants were well representative of these disciplines (with the exceptions of agriculture and ceramics) and of the ranges of interest from applied to theoretical and from field to laboratory orientation. Experts from academia were somewhat overrepresented. The research recommendations this expert group developed and prioritized will, it is hoped, serve to highlight important problems to be addressed and optimize research efforts.

In this chapter, the Workshop's research recommendations are presented; also included is a description of the methodology employed to arrive at the recommendations. A special approach to the task of developing research recommendations was required because of two special characteristics of clay microstructure: (1) it may be understood only by the application of a diversity of scientific disciplines; and (2) it is operationally important in a large number of technical fields that are ordinarily distinct.

Different and significant aspects of clay microstructure are illuminated by studies typically associated with diverse disciplines: geology, geotechnology, chemistry, sedimentology, physics, ceramics, soil mechanics, mineralogy, microbiology, oceanography, and a number of others. In the somewhat independent development of these disciplines, not only have very different techniques of investigation come to be preferred by the various disciplinary communities, but also radically different technical languages have evolved. Within a discipline, its own technical language, or jargon, makes communication precise and efficient. Between disciplines, the jargons make communication of ideas extremely difficult and subject to hidden risks—especially in the not uncommon instance of two disciplines defining a single term quite differently. A similar difficulty occurs in the application of knowledge of clay microstructure to practical problems: here an even more diverse group of technical fields are involved, including agriculture, petroleum recovery, civil engineering, and waste disposal.

The communication problem due to differences in the technical languages of the fields is compounded by the tendency of researchers and practitioners in the different fields to be isolated from one another—not participating in the same meetings, nor reading the same journals, nor working side by side. A special attribute of the Clay Microstructure Workshop is that it deliberately attempted, and by and large succeeded, to bring together outstanding workers from the whole range of appropriate disciplines and technical fields. This very success increased greatly both the importance of and the challenge of ensuring full participation of all attendees (while avoiding domination by one or a few of the eminent participants) in the development of research recommendations.

A novel four-phase attack was undertaken to derive, enunciate, refine, and rank the recommendations. The derivation process began well before the Conference: All participants were informed of this central task of the Workshop and charged to consider how the field of clay microstructure could best be advanced. This charge was reiterated when the Conference convened and formed a thread through all sessions in which papers were presented. Typically papers were presented from the point of view of a single discipline or a small cluster of disciplines and immediately discussed, in brief, by the entire group of Workshop participants. Technical presentations were divided, somewhat arbitrarily, into those with a principal focus on the fundamentals of clay microstructure and those dealing with its

Table 59.1. Outline of technical program.

I. Basic clay microstructure
  A. Microstructure: signatures
    Sedimentary sources and source materials
    Recent sedimentary environments
    The sedimentary column
    Ancient sedimentary environments
    The rock column
  B. Environmental processes: a continuum
    Atmospheric aerosols: fluvial particulates
    Coastal and open marine suspensates
    Sediment-water interface dynamics
    Postdepositional processes
II. Applied clay microstructure
  A. Modeling: past and present—new directions
    Physicochemistry
    Microfabric
    Porometry
    Isotropy/anisotropy
    Mass physical and mechanical properties
    Rheology
  B. Measurements: techniques and sampling strategy
    Scanning and transmission electron microscopy
    Organic-clay interactions
    Computers: hardware/software
    Field methods
    Laboratory methods
  C. Applications: present requirements
    Waste disposal: containment/packing materials
    Clay behavior
    Petroleum recovery: primary/secondary migration
    Stability of deposits
    Geophysical techniques

applications; this structure is shown in outline form in Table 59.1. The papers presented give an overview of the present state of development of studies of clay microstructure and of its applications.

A final session of the Conference was devoted to enunciation, refinement, and ranking of research recommendations. As a point of departure, participants were provided with 17 general questions (e.g., Where are we now? How can we improve our observational capabilities?) and asked to consider (1) leading questions in science and engineering, (2) required techniques and instrumentation, (3) new models—conceptual and numerical, and (4) deficiencies in data bases.

The Workshop session began with each participant choosing to work in one of several small (and somewhat fluid) groups. Five groups were organized according to the main categories of the technical program: microstructure signatures, environmental processes, modeling, measurement and sampling, and applications. Participants were admonished not to allow the structure to impede development of the best set of research recommendations the assembled experts could provide. Proof that this admonition was taken seriously came almost immediately: A sixth working group focusing on mechanisms of particle interactions arose spontaneously. Each working group discussed and developed recommendations in its focus area. The Workshop reconvened as a single body and these recommendations, with their rationales, were reported. All recommendations were reported and justified, even those with only a single champion. Brief versions of the recommendations were recorded on markerboards surrounding the room. Each participant was then encouraged to express his or her recommendations (including new ones in response to those presented by others), to argue the merits of each particular recommendation, and to respond to questions. The discourse was lively, intense, and inclusive. It proved highly effective in bringing out sometimes opposing views and especially in clarifying the recommenders' intentions across the variety of disciplinary lines represented in the Workshop.

After the process of developing and communicating the recommendations (Table 59.2) was completed, participants ranked the 40 recommendations by priority. In this stage of the process, each participant had an equal voice. Each was provided a "resource bank" of seven circles of adhesive-backed paper and instructed to spend these resources by putting them next to recommendations on the markerboard. All seven could be placed on a single recommendation, or the "resources" could be spread out (without subdividing the circles) among recommendations in any other way. Observation of participants at this task suggested that they indeed took allocation of the scarce "resources" quite seriously, but with excellent humor. There appeared to be consensus that nearly all of the 40 recommendations were worthy of adoption; it was quite evident that selection of seven or fewer required painful omissions. The substantial accumulation of bright green circles by some recommendations (and not by others) gave the participants immediate and graphic feedback.

Results

The results of the ranking process demonstrated the effects of difficult choices: two recommendations, both of which had been presented and defended with some vigor, did not receive any votes. One recommendation received overwhelming support: 47 of 301 votes affirmed the necessity to study the influence of microstructure on the macroscopic behavior of sedimentary materials. Macroscopic behavior to be included covered the range of geotechnical and geoacoustical properties such as permeability, strength, and acoustic response. This recommendation was put forward independently by participants focusing on applications of clay microstructure in one small group session and by those focusing on measurements in another. An additional 16 votes were cast for investigating the impact of diagenesis and of bioturbation on microstructure and in turn on macroscopic behavior. Together these four closely related recommendations received over 20% of the votes cast. The recommendation that state-of-the-art computational facilities be used to scale-up microstructure analysis to describe a functional sedi-

Table 59.2. Research recommendations of the Clay Microstructure Workshop.*

Recommendation	Votes	Recommendation	Votes
Microstructure influence on macroscopic behavior, e.g., permeability, strength, acoustics	32	Identification of the key structural properties needed for interpretation	6
Microstructure must be related to macrostructure	15	Computerization of image interpretation—to make sure that enough work is done	6
Role of diagenesis in influencing fabric and consequently geoacoustics and geotechnics	10	Papers on statistical analysis—with emphasis on how much work is needed	1
Role of bioturbation in geoacoustics and geotechnics; pattern recognition of fabrics produced by trace fossils	6	Convergence of evidence from all relevant techniques	6
Utilize state-of-the-art computing methods to scale up microstructural analysis for description of a functional volume	7	Coupled modeling: integration of consolidation and deformation; inclusion of parameters such as temperature, stress history, and viscosity	3
Modeling for remote sensing: acoustical, electrical properties through microstructure	2	Modeling the influence of the differences in measurements (e.g., strain dependence and permeability)	1
Identify and quantify mass transfer processes at the sediment-water interface	16	Study interactions in mixed systems (first-order physicochemistry of pure systems is known)	3
Microhydrodynamics of particle aggregation and sedimentation	13	Realistic modeling needs measurement of real systems (e.g., oxides, humics, clays)	1
Dynamic aspects of particle-particle interactions in fundamental details	10	Modeling to understand what is presently happening and to use as tool to predict future	5
Physicochemical interaction when particles move relative to each other (possibly using a new environmental chamber)	3	Use expert system to integrate existing knowledge	3
Structure of flocs and sediment as function of physical, chemical, and biological conditions including static and hydrodynamic stresses	15	Determine the structural effects of the interfaces on the interstitial fluid (the water)	7
Study fundamentals of structure development	11	Bridge the gap between molecular force and microstructure as in statistical mechanics	8
Fundamentals of structure development by modern physicochemical techniques of colloid and interface science	2	Applications of microstructure, microfabric relationship with mineralogy and pore fluid chemistry to geotechnics and waste disposal	12
Study physical, chemical, and biological effects on flocs in the water column	8	Improve sampling methods to eliminate damage and carry improvements through to testing, despite expense involved	6
Identifying sedimentary processes and environments from fabric	23	Standardized approaches for sampling and testing (especially for field applications); formulate reference samples	8
Fabrics and environment (e.g., fabric signatures of fresh and saltwater sediment/rock)	3	Environmental chamber should be used for all electron-sensitive materials	2
Modifications of structures during early burial	8	Improve communication: annotated bibliography; journal editors not recognizing relevance	8
Modification of structures during diagenesis and lithification	8	New preparation techniques for undisturbed samples for environmental chamber	0
Quantitative descriptions of microstructure	11	Measure grain size as well as possible	0
Need for fabric classification scheme and need to quantify fabric	8		
Do we have fundamental information required to develop a useful model?	4		

*This table contains the abbreviated form of recommendations followed by the number of votes cast for each.

ment volume received seven votes. In this context, a functional sediment volume refers to a quantity of sediment observed at a macroscopic scale in which many microfabric elements respond in concert to environmental stresses such as consolidation loading. Prediction of macroscopic behavior as detected by remote sensing received another two votes.

Sixteen votes were cast for the recommendation that processes of mass transfer at the sediment-water interface be identified and quantified. This was one of several recommendations concerned with developing better understanding of what happens during the period in which a relatively few particles (or even one particle) become associated with enough other particles to form a significant mass. Dynamic aspects of this interaction received particular attention: microhydrodynamic studies of particle aggregation and sedimentation received 13 votes; 10 additional votes were cast for studies of fundamental details of the dynamic aspects of particle–particle interactions; 3 votes were cast for attempting this using an environmental cell with an electron microscope. Another recommendation with a stress on modeling also pointed to the general problem of development of structure received 15 votes. It suggested investigation of the structure of flocs and sediment as a function of physical, chemical, and biological conditions and specifically called for the inclusion of static and hydrodynamic stresses. Eleven votes were cast for general fundamental studies of particle–particle interactions and an additional two for doing this by applying techniques used in modern colloid and interface science. Eight votes went for studies of the impact of physical, chemical, and biological driving forces on flocs suspended in the water column.

A third area of work was strongly supported with 26 votes: determining how fabric can be used to identify sedimentary environments (e.g., fresh versus saline water) and processes. An additional eight votes each went to recommendations that modifications of structures be studied during early burial and during diagenesis and lithification.

The next most strongly supported set of recommendations arises from the perceived complexity of the systems under investigation and the need for reduction to more easily manipulated mathematical representations: 11 votes for quantitative descriptions of microstructure and 8 votes for a quantitative scheme to classify fabric. Clearly, however, there is doubt that at present we are sufficiently knowledgeable to formulate quantitative approaches: four votes were cast for the proposition that we do not yet have the fundamental information needed to develop models useful for engineering applications and six votes recommending identification of the key structural properties that are needed for interpretation of sediment behavior. The recommendation that image interpretation be computerized to ensure that a sufficiently large volume of sample is examined received six votes and an appeal for papers on statistical treatment of measurements to emphasize the large amount of work needed received one vote.

Another recognition of the complexity of the task was reflected by six votes for the recommendation that all relevant techniques be used in investigations and that the interpretation represent the convergence of the results. For understanding of real systems and for realistic modeling, it was recommended that coupling of energy sources be considered as well as the typical deviation from chemical or mineralogical purity (presence of organic material, surface oxides, and so on). By implication this level of complexity has not been integrated into present treatments. Five votes were cast for developing models aimed at understanding what is happening in the sedimentary material and at predicting future behavior. A suggestion of the difficulty of constructing such models with even the admittedly inadequate knowledge existing at present is reflected in three votes recommending that expert systems be used to integrate existing knowledge. The effects of interfaces on the structure of the pore fluid was recommended for study by seven votes. Eight votes were cast for an effort to bridge the gap between molecular force considerations and the microstructure by using techniques analogous to those used in statistical mechanics. An example of the type of result hoped for from this unorthodox approach is prediction of the elastic modulus by a statistical mechanical treatment derived from liquid-state theory.

Participants were not entirely committed to attempting to deal with the complexity of the systems by construction of models from fundamentals, however: for application of clay microstructure to geotechnical and waste disposal studies, 12 votes were cast for developing working relationships between the microstructure and the sediment's mineralogy and pore-fluid chemistry.

An important need for better methods of sampling and preservation of intact samples was reflected in the six votes recommending that these be developed despite their anticipated additional expense. Eight votes called for the standardization of approaches to sampling and testing (especially for field applications) and for the formulation of reference samples. Two additional votes were cast for use of environmental cells for study of all samples subject to damage in preparation/examination by electron microscopy.

Finally, eight votes were cast for a recommendation that communications be improved for those applying clay microstructure to practical problems (e.g., waste disposal, sediment stability) by publication of annotated bibliographies and by persuading journal editors of the relevance of clay microstructure to their readership.

Conclusion

Development of recommendations by intensive interaction of a sizable and representative group of experts in various aspects of clay microstructure was facilitated by the process of following technical presentations with small-group discussion and then structured advocacy of individual's suggestions. A powerful way of obtaining and communicating a consensus from highly divergent starting positions was ranking the resulting recommendations by allowing all participants to assign limited (symbolic) resources. This consensus stresses the importance of physical scale: employing microstructure to develop an understanding of the macroscale behavior of sedimentary materials. The consensus recognizes the incompleteness of our fundamental understanding of microstructure. It urges attacking this with fundamental studies and with incorporation of realistic levels of complexity in description of components and energy sources. It stresses the requirement for better modeling and sampling methodologies and for better communication among the scientists and engineers in this rapidly evolving field.

Index